Principles of Health and Safety in Agriculture

Editors

James A. Dosman, M.D., F.R.C.P.(C)
Director and Professor of Medicine
Centre for Agricultural Medicine
University of Saskatchewan
Saskatoon, Saskatchewan, Canada

and

Donald W. Cockcroft, M.D., F.R.C.P.(C)
Professor of Medicine
Department of Respiratory Medicine
University of Saskatchewan
Saskatoon, Saskatchewan, Canada

CRC Press, Inc.
Boca Raton, Florida

Library of Congress Cataloging-in-Publication Data

Principles of health and safety in agriculture / editors, James A.
Dosman and Donald W. Cockcroft.
p. cm.
Bibliography: p.
Includes index.
ISBN 0-8493-0160-2
1. Agricultural laborers--Health and hygiene, 2. Agriculture-
-Health aspects. 3. Lungs--Dust diseases. I. Dosman, James A.
II. Cockcroft, Donald W., 1946-
RC965.A5P75 1989 89-9914
363.11'963--dc20 CIP

Direct all inquiries to CRC Press, Inc., 2000 Corporate Blvd., N.W., Boca Raton, Florida, 33431.

Cover designed by Gary Bennett, George Fotiadis, and Chris Pearl

International Standard Book Number 0-8493-0160-2

Library of Congress Card Number 89-9914
Printed in the United States

PREFACE

The papers in this volume have been prepared for health-care workers interested in various aspects of health risks in agriculture and its associated industries. In planning the Symposium that generated the majority of these papers, and in organizing the material into this volume, it was not possible to be all inclusive. However, we have attempted to use examples of the principal methodologies and to review health and safety risks in agriculture in their various phases.

To accomplish these goals, we have collected material on the principal respiratory risks in agriculture, the expanding knowledge on grain dust and health, the important area of occupational asthma, particularly as it might apply to workers in these industries, and we use several examples of environmental assessment and health promotion approaches in these areas. As respiratory dysfunction is so important in these workers, we have attempted to place in perspective, epidemiologic principles appropriate to assessing the health effects of these kinds of exposures. It was therefore important to evaluate the spirogram as a basis for lung function assessment, as well as to deal with tests that assess airways responsiveness and gas diffusion in the lung. Other methodologies, including the assessment of structural alterations, the emerging area of bronchoalveolar lavage, and the use of biological assays have the potential of enhancing our ability to estimate risk. In the area of grain dust and health, much new knowledge has taken place since the publication of *Occupational Pulmonary Disease: Focus on Grain Dust and Health,* edited by J. A. Dosman and D. J. Cotton (Academic Press, New York, 1980). In the same years, advancement of understanding of the pathogenic mechanisms of occupational asthma, and new methods of assessment and treatment have developed in an important and exciting manner. Assessment and diagnosis have as their goal prevention of disability through improvement in working conditions. Such improvements depend upon accurate and dependable assessments of dusts and other substances in the working environment. Finally, health surveillance and health promotion programs for workers remain the most obvious social and medical means of maintaining good health.

As the volume deals with a variety of health and safety risks, it has not been possible to tie the material into a completely smooth framework. However, this does not detract from the importance of each of the identified risk areas. In each case, including respiratory, chemicals, cancer, skin problems, safety, injuries, and stress, we have attempted to collect material that best outlines the principles involved. For example, in the section of chemicals, we have avoided dealing with specific toxicologies of each farm chemical or fertilizer, as that information, to the extent that it exists, is available elsewhere. Rather, we have dealt with the most important principles of toxicology regulation and assessment. We placed considerable emphasis on injuries and safety because we believe that this is an extremely important area that has been largely untouched by research and which represents an important area for advancement of knowledge and prevention. It is our feeling that the section on stress and psychiatric problems is woefully inadequate, considering the importance of the problem. Yet, so little methodical research has been carried out in this area that a comprehensive treatment of the subject was not possible. However, what information we have is much appreciated and indicates the need for a systematic approach to research, case identification, and treatment.

Some of the material presented at the Symposium is not presented here, as it did not meet the editorial criteria or pass the review process, but the editors are grateful to those who presented this information, as it added to the proceedings, and assisted in stimulating thought and discussion.

It is our hope that the information in this volume will assist those who wish to define new risks, assess, treat, and counsel patients from a clinical point of view, structure health maintenance programs, and to prepare research protocols designed to answer the may questions that remain.

INTRODUCTION

The International Symposium on Health and Safety in Agriculture took place because of an increasing awareness that farmers and their families, and workers in related industries and rural areas make up a very large population exposed to occupational health and safety risks peculiar to the workplace in which those populations function. The collection of papers in this volume can be roughly grouped into two categories. The first category includes information with regard to health and safety risks to the respiratory system as a result of airborne particulate and gaseous contamination in the workplace. The second category deals with health and safety risks as a result of chemicals, cancer, skin problems, accidents, and injuries, as well as stress and psychiatric problems in rural areas.

The text of this volume commences with an overview of methodologies that represent the tools of respiratory occupational surveillance and research. Such tools involve epidemiologic principles, estimations of longevity and mortality, the use of spirometric measurements for lung function testing in exposed workers, as well as the use of newer methodologies including airways responsiveness, measurements of diffusing capacity, assessment of structural changes, assessment of airways inflammation by bronchoalveolar lavage, and the use of biological assays in the assessment of toxicity of inhaled substances. Subsequently, specific risks to the respiratory system as a result of a variety of exposures in various types of farming operations are outlined. These papers, collectively, confirm the idea that farmers, as a group, are exposed to respiratory dysfunction, and give credibility to results of epidemiologic investigations indicating that farmers are at risk of developing lung dysfunction. As the effects of grain dust and lung function have received considerable attention in recent years, the material in this volume adds to previous knowledge by concentrating on mechanisms of dysfunction and ends with recommendations for a threshold limit value for grain dust. As farmers and others exposed are considered to be at risk of developing occupational asthma, a systematic overview of that condition on farmers and grain workers is presented. Methods of environmental assessment and discussions of health surveillance programs round out this volume.

In this introduction, we summarize our interpretation of the material in key areas. These are methodologies in respiratory occupational surveillance, respiratory risks in agriculture, grain dust and health, occupational asthma, and environmental assessment and health promotion. In addition, we present information on health effects of chemicals used in agriculture, suspected relationships between cancer and agricultural activities, dermatological and other skin problems peculiar to agricultural workers, the very important area of safety and injuries, and the recent recognition of stress and psychiatric problems peculiar to farmers, their families, and rural dwellers.

Methodologies

Methodologies in respiratory occupational surveillance have become more standardized during the last decade. Principles of epidemiology permit an orderly and systematic approach to the identification and measurement of disease and dysfunction. The purpose of such identification is to determine causes of dysfunctions that result in setting of hygiene standards. The understanding of mechanisms is important in the identification of risk factors and all of these are important in the final common pathway which is the minimization of known risks and prevention The process of identifying new or unsuspected risks classically begins with the clinician and this was the case in identifying obstructive lung disease in grain workers. Epidemiological inquiry provides the framework for the identification of disease or dysfunction in populations and is usually the stimulant for remedial action in the workplace. Types of epidemiology include descriptive epidemiology, cross-sectional epidemiology, analytical epidemiology, and experimental epidemiology. In this framework, multiple factors such as host susceptibility and other environmental exposures such as cigarette smoke may contribute to dysfunction in exposed workers.

While epidemiologic principles are important in structuring observations aimed at determining dysfunction, specific tests of lung function are necessary to identify the presence of dysfunction or disease, and to predict mortality. Menkes and Beckett discuss pulmonary dysfunction as an adequate predictor of mortality, not only because individuals may die of their lung disease but because reductions in forced expired volumes may reflect a general debility not centered in the lungs. Menkes and Beckett present the interesting hypothesis that normal lungs are a necessary protection of the body as a whole from exogenous or endogenous toxic substances, and that when the lungs fail, this protection is lost. Thus, if tests of lung function measure the health effects of airborne environmental substances, such tests, when used in epidemiological studies, could indicate the safety or danger of exposure to specific environmental contaminants, and provide early indications of successful intervention. Of the tools available for the assessment of lung function in exposed workers, the spirogram remains the most basic, effective, and inexpensive tool for the assessment of dose-exposure relationships. In addition, the forced vital capacity has been shown to correlate with mortality. However, the slope of phase III of the single-breath nitrogen test may be a more sensitive indicator of mortality than is the spirogram. The forced expired volume in 1 sec (FEV_1), should be considered as the basis for respiratory occupational assessment. Recent efforts to standardize spirometric measurements in order to meet NIOSH/HLBI criteria should enhance reproducibility of tests. However, variation within and between subjects must be taken into consideration in the interpretation of results, and factors influencing the annual change in FEV_1 are such that multiple measurements over time are necessary in order to establish a decrement line for pulmonary function values in individual cases.

Recent advances in the understanding of pulmonary physiology have resulted in a number of new tests that are useful in the occupational setting. Of these, the assessment of airway responsiveness by bronchoprovocation tests utilizing either inhaled aerosolized methacholine or inhaled histamine are important in identifying occupational asthma in exposed workers and are useful in advising individual patients and in establishing cause and effect in the laboratory setting. The use of the diffusing capacity test, particularly in interstitial lung disease, has been made more useful by efforts to standardize methodology. Such tests have been marked by significant intra- and inter-individual variability that has detracted from their utility in utilizing this test in a dose-exposure manner. Standardization efforts should serve to reduce inter-laboratory variability and to emphasize and define the magnitude of variation introduced by various factors such as physiologic variability, variability in technique, variability in the measurements of gas concentrations, variations due to errors in the measurement of volumes and times during the test, and variations created by calculation algorithms.

Recent interest in structural changes that take place as a result of occupational exposures to the lung has resulted in a better understanding of disease processes. For example, exposure to toxic gases can result in bronchiolitis obliterans, while other exposures, such as the well-known exposure to NO_2 in silo-filling operations, can result in acute pulmonary edema with subsequent interstitial fibrosis. Other exposures are less well known. For example, paraquat-exposure damages the lung by increasing the number of free oxygen radicals during its cyclic oxidation and reduction causing diffuse alveolar damage with a subsequent intense fibroblastic proliferation and subsequent cyst formation. Exposures, such as silica, produce a more predictable pattern of interstitial fibrosis. Pathological lesions as a result of grain dust have not been described for humans, but in mice grain dust exposure produces abnormalities of alveolar macrophages and in rabbits, interstitial granulomata. Classical "Farmer's Lung" is perhaps the most common type of extrinsic allergic alveolitis. Many of the exposures are not well worked out.

Recent developments in cell biology have provided the opportunity to estimate lung inflammation by bronchoalveolar lavage. While the science of these assessments, as they apply to the clinical situation and to epidemiology is quite new, it may now be possible to identify occupationally induced inflammation of the lungs in a new and more definitive manner. In the future, it is possible that biological assays may be useful in the short-term assessment of inhaled substances, but considerable work remains to be carried out before such methodologies can be accurately applied to human exposure.

Respiratory Risks in Agriculture

As farmers, their family members, and various workers are exposed to a variety of substances, the response of the lung to such exposures may vary. In this volume, Warren illustrates that dust levels may range from 12.6 mg/m^3 while feeding hogs, to 22.0 mg/m^3 while drying grain, and 31.4 mg/m^3 while dressing seeds; all levels substantially higher than the recommended maximum dust levels of 10 mg/m^3 for the grain industry in Canada. In addition, exposure to endotoxins may be considerable, particularly when exposed to such substances as moldy hay and organisms such as *thermophilic* actinomycetes. As three-quarters of farmers studied had farming operations that included both grain and animals, a variety of responses in the lung might be expected to include chronic obstructive pulmonary disease, occupational asthma, and interstitial lung disease. Our own observation in 1824 farmers indicate that there are reductions in both forced vital capacity (FVC) and in forced expiratory volume in 1 sec (FEV_1) but that a similar change in the ratio of FEV_1 to FVC indicated the possibility of a mixed restrictive and obstructive component in these groups. This is a possibility in view of the exposures involved. For example, toxic gases from manure pits, located within or under animal-confinement building, may include up to 150 types of gases, including methane (CH_4), ammonia (NH_3), carbon dioxide (CO_2), hydrogen sulfide (H_2S), and carbon monoxide (CO). Merchant and Donham have described a number of pulmonary manifestations in workers exposed to dusts and gases in animal confinement units including pulmonary edema, asthma, bronchitis, bronchiolitis, airways obstruction, and organic dust toxic syndrome. The etiologies of these manifestations are not well understood. Similar exposures may occur in poultry farmers. Swedish studies have demonstrated the presence of dusts, endotoxins, and multiple microbial forms in swine production buildings. Although swine producers and other workers within confinement buildings would appear to be at higher risk, our own work, and that of Holness and Nethercott, indicate that swine producers have similar lung function values as compared to other farmers, although in our own work, both groups had lower values than rural-dwelling, nonexposed control subjects. Organic dust toxic syndrome (pulmonary mycotoxicosis) in men exposed to the thermophilic group and *Aspergillus* group of fungi may result in pulmonary edema, and lavage fluid that consists primarily of neutrophils and histiocytes. Further investigation to study the significance of these exposures is required. As chronic and acute symptoms have been described in farmers in various geographic areas, cooperative studies to ascertain mechanisms, as well as dose-exposure relationships will be necessary.

Grain Dust and Health

Over the last 10 years, there has been a considerable amount of research both as to the epidemiology of respiratory dysfunction as a result of exposure to grain dust, and of structural changes and pathogenic mechanisms. Several studies have demonstrated reductions in FEV_1 in grain workers relative to control subjects. Longitudinal studies have demonstrated a direct relationship between longitudinal decrease in pulmonary function and both decrease in pulmonary function over the work shift and methacholine responsiveness determined at the 6-year follow-up examination. Thus, the presence of an acute work-related decrease in pulmonary function may prove to be a useful approach for predicting longitudinal loss of pulmonary function on grain handlers. However, exposure response relationships have still not been well documented. In several studies, years of employment as a grain handler were found to correlate positively with the prevalence of pulmonary symptoms and negatively with pulmonary function.

There has been some progress in studies of mechanisms. Stepner et al. exposed rabbits to grain dust particles and demonstrated a granulomatous interstitial pneumonitis which centered on the membranous and respiratory bronchioles and that extended to the alveolar regions. Exposure of mice to grain dust by Armanious et al. resulted in quantitative and qualitative changes in alveolar macrophages, but there were airway changes that were not demonstrated consistently. It has been demonstrated that extracts of grain dust contain substances that can stimulate the cells of the immune system in an immunologically nonspecific manner, and that grain dust extracts, as well as silage materials, can activate complement in normal human serum *in vitro*. There are similarities between tannic acid and grain dust extract, consistent with the possibility that the complement-fixing activity of grain dust is due to a tannin or tannin-like material. It is also possible that contaminating material such as fungal organisms, bacteria, and endotoxins could contribute to complement-fixing activity. Certainly, mycotoxins are present in grain dust as a result of preharvest infection by toxigenic fungi. Such infestations are more likely to occur in southern climates. Ochratoxin A, sterigmatocystin, and citrinin appear to be the main mycotoxin contaminants. Certain host factors may also influence the progression of lung dysfunction in exposed workers. Cotton et al. suggest that the effect of smoking, and the effect of grain working are about equal in pathogenesis of symptoms and changes in pulmonary function test values in exposed workers.

The considerable new information on grain dust and lung function has resulted in health surveillance programs and suggestions for regulation of the maximum permissible dust levels in the workplace. Recently, the American Conference of Governmental Industrial Hygienists (ACGIH) has recommended a threshold limit value of 4 mg/m^3 for grain handling facilities in the U.S. The ACGIH cites the work of doPico et al. as indicating that workers are unlikely to experience severe lung decrements in lung function at exposures below 5 mg/m^3. The problem in the setting of maximum permissible dust levels, and in assessing work-related dysfunction, is that grain is not a single entity and consists of many types of cereal fragments and contaminants. Further research to assess dose response relationships in the grain industry is a priority.

Occupational Asthma

The nature of farming exposures, with a multiplicity of organic, microbial, gaseous, and chemical exposures, puts farmers at risk of developing occupational asthma. Such developments are particularly troublesome for farmers as relocation into other occupations is often difficult. These circumstances make it incumbent upon the physician to carry out accurate clinical and environment assessments, careful treatment, and adequate counseling. Asthma can be defined as airway hyperresponsiveness to specific immunological, i.e., allergic or nonspecific, nonimmunologic agents. The diagnosis of occupational asthma requires definitive criteria for the diagnosis of asthma, evidence for an occupational cause or relationship, identification of specific immunological causal agent, and distinction from nonspecific effects. The nature of the asthmatic response will depend upon the immunological reactivity of the subject, the nature and source of causal agents, and circumstances of exposure. Pathogenic mechanisms involve the increased responsiveness of the airways to physical, chemical, and pharmacological stimuli. Possible mechanisms of airway hyperresponsiveness include decrease in baseline airway caliber, increased responsiveness of the smooth muscle itself, changes induced by components of inflammation, abnormality of the autonomic nervous control of the smooth muscle, and increases in the accessibility of the stimuli to target cells. Implicating cause and effect in occupational asthma is often difficult because there may be both early and late responses. Recently, a variant of occupational asthma called "reactive airways dysfunction syndrome" (RADS) had been described in which a single exposure to a very high concentration of irritant results in a syndrome simulating asthma, which is characterized by cough, dyspnea, wheezing, and hyperresponsive airways, which may lead to persistent asthma and airway hyperresponsiveness.

In assessing occupational asthma, it is important to recognize that many cases of occupational asthma may involve nonallergic airway hyperresponsiveness as a result of exposure to gas, dusts, fumes, or vapors, which are encountered in the workplace and result in sensitization. In these and other cases, the use of methacholine or histamine challenge studies in order to quantitate airway responsiveness has been a recent and important advance. Serial measurements of airway responsiveness in relation to work and work-free periods are particularly useful. Often, these assessments with and without exposure may be used to infer dose-effect relationships, which are particularly important when assisting in the assessment for Worker's Compensation, or in job relocation. Specific substance challenge remains the most definitive way of documenting occupational asthma. Patient assessment may also be assisted by careful history, serial measurements of ambulatory peak-flow rates, and serial measurements of nonallergic airway responsiveness. These data make it incumbent upon the physician to assist the patient in making successful choices. Once a diagnosis of occupational asthma has been confirmed, the patient must decide whether to continue work or to find alternate employment. Job relocation may be particularly important because increased responsiveness of the airways does not necessarily revert to normal following cessation of exposure.

Farmers and grain workers may present with occupational asthma soon after commencement of their employment, or occasionally after many years of exposure. There is virtually no information on the prevalence of occupational asthma as a result of exposure to grains and other dusts and substances in the workplace. However, several studies show that some grain workers demonstrate an acute shift change in lung function that appears to be an obstructive defect directly related to the level of dust exposure and unrelated to age, smoking, duration of employment, atopic status, or skin test reactivity to grain dust antigens, and associated with an increase in leukocyte count. It is our view that while the incidence of grain dust asthma is relatively high in grain workers, prevalence is low because these workers select themselves out of the industry. However, among farmers, where mobility is limited, grain dust asthma appears to be more common and constitutes an important area for research.

Environmental Assessment and Health Promotion

The purpose of epidemiologic inquiry is to identify dysfunction or disease, identify dose-effect relationships, and by environmental alteration, to effect preventative measures. Because of the high cost of obtaining dust level measurements, adequate dose-effect relationships are not present in the grain industry at the present time. However, a considerable proportion of dust particles are below a mean aerodynamic size of 3.5 μm at 50% collection efficiency of a size-selective sampler, and are therefore considered "respirable dust", capable of passing through the terminal bronchioles and into the alveolae. Grain dust is a pleomorphic mix of particles of grains, plants, starch, soil, and cones of the tips of trichome particles. In addition, grain dust may include chemical residues and other substances, as well as various trace elements. In order to adequately delineate exposures, dust sampling instruments that are capable of measuring particles over a wide spectrum are needed to precisely determine worker exposure to respirable dust. In addition, the microflora of grain dust that evolves to harvesting, storage, and handling, requires delineation and assessment. Opportunities for reduction of dust levels are taking place with increasing automation of facilities and of consolidation of facilities. However, dust and microbial exposure during silo unloading, resulting in "organic dust toxic syndrome" are described as a result of inhalation of bioaerosols in silage dust. Similar principles of environmental assessment are being used to assess the working environment in the poultry industry and in other industries such as the nontextile cotton industry where airborne cotton dust levels may be high. The use of devices such as the Lumsden-Lynch vertical elutriator are of assistance in assessing the environment.

Assessments of worker environment and worker protection are necessary. Respirators and other protective devices against organic dusts require improvement in design and additional assessment. However, it is our opinion that protective devices cannot be used effectively throughout an 8-hr work day and that protective devices should not be considered a substitute for adequate environmental control. However, development of these protective devices is essential to protect the worker against acute short-term exposures which are difficult to control under any circumstances. Improvements in recent bioassay techniques to detect mycotoxins and other substances are of potential importance in protecting the worker.

Health promotion programs aimed at primary prevention are potentially productive means of achieving goals of health maintenance among workers, for an informed work force is the most reliable means of making gains. Our experience with such programs in the grain industry in Saskatchewan has been so positive that we feel it could be a model for other industries. For example, a health program conducted by the Saskatchewan Wheat Pool has included health management seminars conducted by an employee health service nurse, with the objectives of enhancing employee education about the effects of smoking in general and combined with the grain dust, and employee education with regard to measurements of lung function studies carried out during health surveillance programs. Counseling of individual workers by an occupational health nurse is perhaps the most effective means of education.

Industry, workers, and government should share the task of health maintenance programs, and a tri-part approach to health protection should be undertaken. Industry has an obligation to provide safe working environments and good employee health services, and such costs should be accepted. Government can lead by regulations concerning maximum dust levels, health surveillance programs, and providing information and other intra-structures. Workers appear ready to equip themselves with more information and understanding of ways and means of achieving safe working environments, better working practices, and lifestyle habits aimed at health protection.

Toxicology and Health Effects of Pesticides

In this section, no attempt was made to cover the specific toxicology of all agricultural chemicals and fertilizers, but rather to discuss a range of points from public health policy, to toxicology, to assessments. In Canada, pesticides are registered under the Pest Control Products Act which is under federal legislation. Products may be admitted and utilized in Canada under one of three categories including that of research permit, temporary registration, and full registration. Although the Canadian regulatory process is handled by the Pesticides Directorate of Agriculture Canada, consultations are obtained by the Pesticides Directorate from the federal Departments of Agriculture, Health, Environment and Fisheries. The assessment of pesticide toxicity, as outlined by Ritter, classifies pesticides into "domestic class" products for which regulation is more stringent, and into "commercial class" products for use in commercial operations such as agriculture, forestry, and industry, and for which there is provision for inclusion of more toxic products than in the domestic class. In order for the safety of the product to be assessed, toxicity testing is carried out and includes three kinds of studies: acute studies (single exposure); short-term studies (multiple exposure — 10% lifespan); long-term studies (multiple exposure —lifespan); and special studies. Protective equipment reduces such exposures. Theoretical maximum daily exposure limits are assessed from animal studies. The public has become much more interested in pesticides usage and regulation, and several recent controversies have heightened awareness. Governmental agencies sometimes appear to respond to community demands for regulation on the basis of the nature of the controversy as opposed to the toxicology of the chemical involved.

The widespread use of individual chemicals, such as the phenoxy herbicides, raises certain environmental questions. Such usage of herbicides has increased from approximately 2 million kg in the 1950s to over 17 million kg annually by the 1980s, over the Canadian prairie grain-growing areas. Airborne monitoring studies have established the presence of airborne residues of herbicides, such as 2,4-D esters, triallate, and trifluralin, both during and immediately following the spraying season. Some of these residues are present throughout the growing season. Several herbicides, such as 2,4-D, MCPA, Dicamba, and others are applied as postemergence treatments which means that large deposits occur on the vegetation canopy of both the crop and the competing weeds. Herbicides are also applied directly to the soil as pre-emergence treatments. Knowledge as to the ultimate fate of such substances in the atmosphere is incomplete. It is possible that some substances, such as 2,4-D, may induce hypersensitivity reactions and may therefore be associated with contact dermatitis. The herbicide, 2,4-D, is known to be readily degraded by soil microorganisms. However, the accumulation of dissociated 2,4-D in soils is enhanced by the acidification of soils, raising the possibility of longer term accumulation.

It is known that certain insecticides, such as organophosphate compounds, can act by irreversible inhibition of cholinesterase as a result of accumulation of acetyl choline at both muscarinic and nicotinic receptors, resulting in cholinergic crisis and neuromuscular blockade, causing death via asphyxia as a result of bronchospasm, upper airway obstruction, bronchorhea, respiratory muscle paralysis, and central apnea. Other compounds, such as fenitrothion (O, O-dimethy, O-[4-nitro-M-tolyl] phosphorothioate) have been used to control spruce budworms in the maritime provinces of Canada and have been the subject of controversy there. Unfortunately, adequate risk-benefit analyses have not been conducted. In New Brunswick, the annual aerial forest spraying was suspected by the general population as being causal in Reye's Syndrome (a serious and frequently fatal encephalopathy in children), cancer, and birth anomalies. The cost of demonstrating that fenitrothion was efficacious and a relatively safe insecticide has been high. Other forms of pesticide exposure, such as the use of ethylene dibromide (EDB) to control pest infestation in flour used in human consumption, and a spot fumigant in the grain industry, resulted in deregistering the compound by Agricultural Canada in 1984. Residual testing of products is carried out, and while traces of the compound are present in flour in Canada, it is considered that the level should not contribute to "significant" levels of EDB in the diet. Grain fumigants are used in the grain storage and handling industry in North America. This is of concern from the point of view of exposure to grain workers, and from residue present in cereal grains used in food components. These pesticides carry certain risks, as each step of the process involves risk of spill and fire. Other forms of food production, such and mushroom growing, may be a source of exposure to workers and consumers.

Toxicological methodology to adequately assess risk-benefit analyses remains inadequate and constitutes an important area of further research and develpoment.

Cancer Risks in Agriculture

The chapter on cancer and agriculture commences with a paper by Dr. Brewer on the possibility that pharmacogenetics and the role of inheritance may be important in producing variation in response to exposure to certain substances. Brewer points to the possibility that Aryl Hydrocarbon Hydroxylase (AHH) inducibility may be important in the induction of cancer of the lung and points to the use of paraoxon, the active metabolite in the organophosphorus insecticide pation as resulting in deaths in agricultural workers as a result of inhibition of essential acetyl cholinesterase, and indicates that such deaths may occur primarily in that portion of the population with genetically determined low activity of paraoxonase. McDuffie et al., using a population-based registry in Saskatchewan, suggest an environmental influence in the etiology of lung cancer in males. But although some pesticides utilized by farmers are suspected carcinogens, and the use of agricultural chemicals is extensive, the relative risk due to occupational exposure to these substances is small in relation to the risk from cigarette smoking. The data support the theory that lung cancer develops more often in damaged tissues as evidenced by the significantly higher frequencies of lung cancer in patients with bronchitis, emphysema, chest injury, and Farmer's Lung disease. In fact, in Dr. McDuffie's work, the lung cancer patients had significantly greater exposures to grains and fungicides than did control subjects, raising the possibility of an environmental component to the development of lung cancer in farmers. Proportional cancer mortality analysis show that farmers have significantly elevated risks for several cancers, including lip, stomach, pancreas, nasal sinus, prostate, and leukemia. The lowered risk of death from cancers of larnyx, lung, and esophagus, in Gallagher's studies, were explained by the lower levels of smoking and alcohol consumption in the farmers observed. Utilizing death certificates in Iowa, Bermeister observed that the odds ratios for mortality in farmers were significantly elevated for leukemia, multiple myeloma, non-Hodgkins, lymphoma, prostate, and stomach. In addition, Gallagher et al. noted an elevated risk for myeloma in farmers, that interestingly, appeared to be associated with prior allergies. Blair et al. report an interview study that may indicate an elevated risk for chronic lymphatic leukemia in non-Hodgkins lymphoma. Morrison and coworkers, using a census-based record linkage study on Canadian farmers, report a higher than expected mortality from stomach and prostatic cancer, as well as from multiple myeloma and aplastic anemia.

These data, albeit with some inconsistencies and contradictions, collectively point to the need for several broadly based, multiprovince, prospective case-control studies in Canada to establish risks here for various kinds of cancer in farmers.

Dermatoses and Zoonoses

The full nature and extent of dermatological problems in agriculture and related industries is not fully understood. However, agricultural workers make up a major proportion of workers suffering dermatological problems. For example, although skin disease accounted for 38.7% of all occupational illnesses in 1979 in California, the number of skin disease cases as a percentage of all illnesses within agriculture was disproportionally high, at 64.9%. The principle exposures relating to occupational dermatitis in these workers were inedible plant and animal products, agricultural chemicals, and edible food products. In addition, farmers may be at risk of developing zoonotic skin infections, particularly dermatophytosis (tanea or ringworm). Other infections transmittable to humans include *Microsporum canus* and *Thrichophyton verucosum*. Viral zoonotic skin infections occasionally occur in farm families, veterinary students, and clinicians. Bacterial zoonotic skin infections are considered to be relatively rare. Dermatophyte infections as a result of *Microsporum, Trichophyton,* and *Epidermophyton* of the fungi imperfecti (ringworms) are relatively common infections among farmers and agricultural workers. The most common infections occur in workers between the ages of 21 and 30 years, but children are also susceptible.

Cutaneous exposures to pesticides as spray applications constitute a major risk of skin problems in agricultural workers. Exposures are usually multiple and much more significant than is commonly recognized. Exposures may be reduced by the use of protective, disposable clothing, barrier creams, and the utilization of improved application systems. Priorities include efforts to improve surveillance data for pesticide-associated illness, as well as field work aimed at control of exposure and illness in high risk groups of agricultural workers.

An additional skin problem in agricultural workers is that of cancer of the skin because of the chronic exposure to the carcinogenic effects of ultra-violet light, and Saskatchewan has a high incidence of squamous cell carcinoma. The combined exposure to sunlight and various agricultural exposures requires further elucidation. Protective measures such as the use of sun blocks, and early diagnosis of skin lesions, particularly on the lips, can lead to improved survival.

Principles of Safety in Agriculture and Related Industries

As farmers and agricultural workers typically perform a wide range of tasks during regular hours and long work weeks during busy periods of the year, and as farm laborers and farmers are often poorly trained in safety procedures, performing tasks that are dangerous, these workers are at risk of being injured. National Safety Council (U.S.) data for 1982 indicate that 22.3% of agricultural injuries were incurred by hired help while the remaining 73.4% involved members of the farm family. Injury incidence rates for hired agricultural workers are higher than for the general private sector, most injuries taking place in men between the age of 15 and 30 years who have been on the job less than 6 months at the time of the injury and are often not well trained. Typically, these workers suffer from sprains/strains or cuts/lacerations involving the extremities with the source of injury most likely to be the work surface, livestock, or hand tools. In the Province of British Columbia, statistically significant elevated proportional mortality rates for farmers were found for motor vehicle accidents, and environmental accidents which include deaths by animals, exposure, machinery, gun-related accidents with increasing trends toward poisoning from gases and machinery accidents. In Ontario, a task force on health and safety in agriculture, established in 1983 in response to a specific request made by the farming industry to the Ministers of Agriculture and Food and of Labor, determined that farm tractors were a factor in 50% of all farm work fatalities and that children were involved in many cases. A second major area of safety risks is unsafe facilities in yards and fields. Job design can be utilized to reduce various injuries in this area, including the large number of back injuries.

Hearing loss is an important, widespread, and generally unrecognized cause of disability in farmers and their families. At the frequency of 4000 hz, hearing loss for male and female farmers is 54 and 32 db, respectively. At 4000 hz, the mean percent hearing loss in farmers over various age groups was in excess of 50% while that of nonfarmers was approximately 30%. Hearing loss is amenable to simple preventative measures.

The most effective treatment for farm or other related injuries lies in prevention. Such prevention is important whether it relates to mechanical devices, or to substances such as fertilizers or other chemicals. Protection standards for farm workers are being developed in some jurisdiction. For example, the Texas Department of Agriculture has adopted 50 standards to protect individuals from exposure to pesticides. An example of such a measure is that neighbors living adjacent to aerial field spraying must be notified prior to such spraying being initiated. In addition, regulations concerning re-entry intervals and protective devices reduce exposure for workers. Of course, the ultimate responsibility for farm safety and prevention of accidents lies in an educated and well-informed farmer and worker group. In Saskatchewan, the Saskatchewan Safety Council is initiating farm safety programs, similar to those offered by the Ontario Farm Safety Organization. The medical practitioner is often poorly equipped by virtue of knowledge and time to assist in preventative measures.

Compensation programs are available for farmers in some jurisdictions. In Canada, Workers Compensation is available to farmers in 9 of 11 provinces and territories. However, very few farmers will insure themselves or their children, and, in Saskatchewan, where approximately 3% of eligible farmers have enrolled with the Workers Compensation Board, the enrollment is often for employed farm laborers.

Injuries in Agriculture and Related Industries

The U.S. National Safety Council estimates indicate that agriculture is the single most dangerous major occupation, followed by mining and construction. In a prospective evaluation on 220 farmers, there were 140 injuries on 93 farms, in which two-thirds of the injuries required medical attention. Of the injuries, a majority involve the upper extremity, and many result in amputation. A major proportion of those injured are children, particularly with the widespread use of all-terrain vehicles. However, injuries from farm animals are particularly common among dairy workers and veterinarians. A major, largely unquantified problem refers to back injuries to which the farmer is particularly prone by virtue of the diversity of tasks and the solitary nature of his occupation. Disabling spinal cord injuries resulting in paraplegia and quadraplegia are frequent results of farm injuries and other specialty procedures. High-pressure injection equipment and electrical apparatus result in a wide diversity of injuries.

Injuries may well account for the largest single cause of work lost and human suffering in agriculture and related industries. A major program of basic research into epidemiology, causation factors, and methods of prevention is a high priority.

Stress and Psychiatric Problems in Rural Areas

The recent economic downturn in western agriculture has highlighted stress and other psychiatric problems leading to alcoholism, family break-up, and suicide. The lack of psychiatric and psychology skills in rural areas further compounds the difficulties of diagnosis and treatment. An understanding of the epidemiology of mental illness in rural areas is hampered by the rapid economic and social changes taking place in the farming communities. Urban/rural differences exist for a variety of psychiatric disorders. In the Province of Quebec, with higher unemployment in the rural areas, different patterns of depression have been demonstrated. The role of the rural woman has not been maximized. Mental-health services planning should take into account the problems in rural areas from the point of view of diagnosis and delivery, and should take into account the large number of persons who are overstressed and depressed. In some areas, such as in North Carolina, specific programs aimed at rural areas have been initiated. An additional, largely unexplored area of potential importance to rural areas, is the question of neurotoxicity of agricultural workers exposed to pesticides, and the possible mental effects of such exposure.

The entire area of stress and psychiatric problems in rural areas requires an intensive program of investigation in order to outline the principal problems, and to point the way to resource allocations for prevention and treatment.

James A. Dosman
Donald W. Cockcroft

EDITORS

James A. Dosman, M.D., F.R.C.P.(C) is the Director of the Centre for Agricultural Medicine, and is a Professor, Department of Medicine, and an Associate Member of the Department of Community Health and Epidemiology and the University of Saskatchewan.

Dr. Dosman received his M.D. degree from the University of Saskatchewan in 1963, and his Fellowship in the Royal College of Physicians of Canada in 1972 after completing his postgraduate training in respiratory medicine and respiratory research at the Royal Victoria Hospital in Montreal. He was the founding head of the Section of Respiratory Medicine, Department of Medicine, and was named Founding Director of the Centre for Agricultural Medicine, Department of Medicine at the University of Saskatchewan in 1986.

Dr. Dosman was co-chairman of the International Symposium on Grain Dust and Health in 1977 and co-author of an edited volume from that symposium published under the designation: Dosman, J. A. and Cotton, D. J., *Occupational Pulmonary Disease: Focus on Grain Dust and Health,* Academic Press, New York, 1980. In 1985, he was chairman of the International Symposium on Health and Safety in Agriculture which has led to this volume. Dr. Dosman is a Life Member of the Saskatchewan Anti-Tuberculosis League, a past president of the Canadian Thoracic Society, a past Scholar of the Medical Research Council of Canada, and a current Scholar of Health and Welfare, Canada.

His ongoing research interests include population-based studies of the pathophysiology and epidemiology of environmental and occupational lung problems.

Donald W. Cockcroft, M.D., F.R.C.P.(C) is a Professor in the Division of Respiratory Medicine, Department of Medicine, at the University of Saskatchewan.

Dr. Cockcroft graduated with a B.Sc. in Chemistry from the University of British Columbia in 1967, and received an M.D. from the same University in 1970. Specialty training included 2 years of General Internal Medicine and 1 year of Allergy and Clinical Immunology at the Royal Victoria Hospital (McGill University) in Montreal, 1 year of Clinical Respiratory Medicine at St. Paul's Hospital (UBC) in Vancouver, and 2 years of research experience in Chest Allergy (asthma) at St. Joseph's Hospital (McMaster University) in Hamilton, Ontario. A Specialty (fellowship) in General Internal Medicine was obtained from the Royal College of Physicians and Surgeons of Canada in 1975.

Dr. Cockcroft is a member of the Canadian Society of Clinical Investigation, Canadian Thoracic Society, Canadian Society of Allergy and Clinical Immunology, the American Thoracic Society, the American Academy of Allergy and Clinical Immunology, the American College of Chest Physicians, the Canadian Medical Association, the Saskatchewan Medical Association, the Saskatchewan College of Physicians and Surgeons, and the Canadian College of Physicians and Surgeons.

Dr. Cockcroft has presented approximately 25 invited lectures at international and national meetings. He has published in excess of 75 research papers, as well as in excess of 25 review articles or book chapters. His current interests include asthma and research into both allergic and nonallergic airway responsiveness.

EDITORIAL COMMITTEE

CONTRIBUTORS

David Abramson, Ph.D.
Research Scientist
Research Station
Agriculture Canada
Winnipeg, Manitoba, Canada

Craig L. Anderson
Department of Health and Human Services
Robert A. Taft Laboratories
Cincinnati, Ohio

Antonio Anzueto, M.D.
Research Fellow
Department of Pulmonary Medicine
University of Texas Health Science Center
San Antonio, Texas

M. Armanious
Department of Pathology and Medicine
College of Medicine
University of Saskatchewan
Saskatoon, Saskatchewan, Canada

Kenneth R. Armstrong, D.V.M.
Professor
Department of Herd Medicine and Theriogenology
University of Saskatchewan
Saskatoon, Saskatchewan, Canada

Barbara J. Ault
Marshfield Medical Research Foundation
Marshfield, Wisconsin

Donald W. Badger, Ph.D.
Research Physiologist
Applied Psychology and Ergonomics Branch
National Institute for Occupational Safety and Health
Cincinnati, Ohio

D. Baldwin
Department of Medicine
Division of Pulmonary Diseases
University of Alberta
Edmonton, Alberta, Canada

Pierre R. Band, M.D.
Department Head
Department of Epidemiology, Biometry and Occupational Oncology
Cancer Control Agency of British Columbia
Vancouver, British Columbia, Canada

Eric David Barnes
Research Associate
Department of Educational Resources
University of South Dakota Medical School
Vermillion, South Dakota

G. D. Barnett, M..D.
Retired
Saskatchewan Antituberculosis League
Fort San, Saskatchewan, Canada

R. Baumal
Gage Research Institute
Hospital for Sick Children
University of Toronto
Toronto, Ontario, Canada

W. Beckett
Division of Environmental Physiology
School of Public Health
Johns Hopkins University
Baltimore, Maryland

Glenn A. Bennett
Department of Mycotoxin Research
Agricultural Research Service
United States Department of Agriculture
Peoria, Illinois

Gilbert Berdine, M.D.
Associate Professor
Department of Pulmonary Medicine
University of Texas Health Science Center
San Antonio, Texas

Puham Bhasin, M.Sc.
Centre for Agricultural Medicine
University of Saskatchewan
Saskatoon,Saskatchewan, Canada

Leo M. Blade, Ch.E.
Industrial Hygienist
National Institute for Occupational Safety and Occupational Health
Cincinnati, Ohio

Aaron Blair, Ph.D.
Chief
Occupational Studies Section
National Cancer Institute
National Institutes of Health
Bethesda, Maryland

Barry R. Blakley, Ph.D.
Professor
Department of Veterinary Physiological Sciences
University of Saskatchewan
Saskatoon, Saskatchewan, Canada

William A. Blattner, M.D.
Chief
Viral Epidemiology Section
Environmental Epidemiology Branch
Environmental and Biostatistics Program
Division of Cancer Etiology
National Cancer Institute
National Institutes of Health
Bethesda, Maryland

Dan Blazer, Ph.D.
Professor Department of Psychiatry
Duke University Medical Center
Durham, North Carolina

C. Vaughan Bowen, F.R.C.S.C.
Associate Professor
Department of Orthopaedic Surgery
University of Saskatchewan
Saskatoon, Saskatchewan, Canada

Joseph D. Brain, M.D.
Cecil K. and Philip Drinker Professor of Environmental Physiology
Director of Respiratory Biology Program
Department of Environmental Science and Physiology
Harvard School of Public Health
Boston, Massachusetts

George Brewer
Department of Human Genetics
University of Michigan
Ann Arbor, Michigan

Irvin Broder, M.D.
Director
The Gage Institute
Toronto, Ontario, Canada

Stuart M. Brooks, M.D.
Professor and Chairman
Department of Environmental and Occupational Health
College of Public Health
Tampa, Florida

Leon F. Burmeister, Ph. D.
Professor and Vice-Chairman
Department of Preventative Medicine and Environmental Health
University of Iowa
Iowa City, Iowa

K. Cantor
Department of Environmental Epidemiology
National Cancer Institute
National Institutes of Health
Bethesda, Maryland

J. Carr
Department of Pathology and Medicine
University of Saskatchewan
Saskatoon, Saskatchewan, Canada

Andre Cartier, M.D.
Associate Professor
Department of Chest Medicine
University of Montreal
Montreal, Quebec, Canada

Moira Chan-Yeung, F.A.C.P.
Professor
Department of Medicine
Vancouver General Hospital
University of British Columbia
Vancouver, British Columbia, Canada

Leslie R. Chasmar, F.R.C.S.
Clinical Professor of Surgery
Head of the Division of Plastic Surgery
University Hospital
Saskatoon, Saskatchewan, Canada

M. Cheang, B.S.
Statistical Consulting
Biostatistical Consulting Unit
Faculty of Medicine
Winnipeg, Manatoba, Canada

Jerry L. Clere
National Institute for Occupational Safety and Health
Morgantown, West Virginia

Donald W. Cockcroft, M.D.
Professor
Department of Respiratory Medicine
University of Saskatchewan
Saskatoon, Saskatchewan, Canada

Murray L. Cohen, Ph.D.
Deputy Director
Division of Safety Research
National Institute for Occupational Safety and Health (USPHS)
Morgantown, West Virginia

Huynh Congly, M.Sc.
Laboratory Scientist
Mycology Section
Laboratories and Disease Control Service
Regina, Saskatchewan, Canada

David J. Cotton, M.D.
Professor
Department of Medicine
College of Medicine
University of Saskatchewan
Saskatoon, Saskatchewan, Canada

Robert O. Crapo, M.D.
Associate Professor
Department of Medicine
University of Utah
Salt Lake City, Utah

Diane Darrow
Herkimer, New York

Mark Darrow
Herkimer, New York

Ira Deep, Ph.D.
Professor
Department of Plant Pathology
Ohio State University
Columbus, Ohio

John E. Diem
Pulmonary Diseases Section
Tulane Medical Center
New Orleans, Louisiana

Kelley J. Donham D.V.M.
Professor
Department of Preventive Medicine and Environmental Health
Institute of Agricultural Medicine and Occupational Health
Oakdale, Iowa

Guillermo A. doPico, M.D.
Professor
Department of Medicine
Pulmonary Medicine Station
University of Wisconsin
Madison, Wisconsin

James A. Dosman, M.D.
Director and Professor of Medicine
Centre for Agricultural Medicine
University of Saskatchewan
Saskatoon, Saskatchewan, Canada

Pierre R. Dupuis, M.D.
Associate Professor of Surgery Department of Orthopaedics
University of Saskatchewan
Saskatoon, Saskatchewan, Canada

Donald J. Ecobichon, Ph.D.
Professor
Department of Pharmacology and Therapeutics
Faculty of Medicine
McGill University
Montreal, Quebec, Canada

Victor S. Ejercito, M.D.
Chairman
Department of Otolaryngology, Head and Neck Surgery
Marshfield Clinic
Marshfield, Wisconsin

George F. Elliott, M.D.
Director of Medical Services
Medical Department
Worker's Compensation Board
Regina, Saskatchewan, Canada

J. Mark Elwood, B.A.O.
Professor and Head
Department of Community Medicine and Epidemiology
University of Nottingham
Nottingham, England

Dean A. Emanuel, M.D.
National Farm Medicine Center
Marshfield Medical Research Foundation
Marshfield, Wisconsin

M. A. Emanuel
National Farm Medicine Center
Marshfield Medical Research Foundation
Marshfield, Wisconsin

Donald A. Enarson, M.D.
Professor
Department of Pulmonary Medicine
University of Alberta
Edmonton, Alberta, Canada

G. Everett
Department of Preventative Medicine and Environmental Health
University of Iowa
Iowa City, Iowa

Jean-Pierre Farant, Ph.D.
Associate Professor
School of Occupational Health
McGill University
Montreal, Quebec, Canada

Peter Fretz
Department of Veterinary Anesthesiology
Western College of Veterinary Medicine
University of Saskatchewan
Saskatoon, Saskatchewan, Canada

F. Froh (Deceased)

Richard P. Gallagher, M.A.
Head
Epidemiology Section
Division of Epidemiology, Biometry and Occupational Oncology
Cancer Control Agency of British Columbia
Vancouver, British Columbia, Canada

Linda K. George, Ph.D.
Professor
Department of Psychiatry
Duke University Medical Center
Durham, North Carolina

Mohammad A Ghafouri, Ph.D.
Medical Research Associate
Clinical Research Department
Boehringer Ingelheim Pharmaceuticals, Inc.
Ridgefield, Connecticut

R. Gibson
Department of Epidemiology
University of Minnesota
Minneapolis, Minnesota

Henry Glindmeyer, III, D.Engr.
Associate Professor
Department of Medicine
Tulane Medical School
New Orleans, Louisiana

Brian L. Graham, Ph.D.
Biomedical Engineer
Section of Respiratory Medicine
Department of Medicine
University of Saskatchewan
Saskatoon, Saskatchewan, Canada

Raj Grover, Ph.D.
Section Head
Section of Environmental Chemistry of Herbicide
Research Station
Agricultural Canada
Regina, Saskatchewan, Canada

Maria I. C. Gyongyossy-Issa, Ph.D.
Instructor
Douglas College
New Westminster, British Columbia, Canada

Walter M. Haag, Jr., M.S.
National Institute for Occupational Safety and Health
Cincinnati, Ohio

P. Haglind
Department of Environmental Hygiene
University of Gothenburg
Gothenburg, Sweden

Debbie Hall
Research Technician
Kelsey Institute
Saskatoon, Saskatchewan, Canada

Donald A. Hansen, M.Sc.
Audiologist
Department of Audiology
National Farm Medicine Center
Marshfield, Wisconsin

Lynda M. Haverstock, Ph.D.
Registered Psychologist
Department of Psychology
Saskatoon Mental Health Clinic
Saskatoon, Saskatchewan, Canada

Frank A. Herbert, M.Sc.
Professor
Department of Pulmonary Medicine
University of Alberta
Edmonton, Alberta, Canada

Daniel J. Hogan, M.D.
Associate Professor of Dermatology
Department of Medicine
University of Saskatchewan
Saskatoon, Saskatchewan, Canada

V. Holford-Strevens, Ph.D.
Research Associate
Department of Immunology
University of Manitoba
Winnipeg, Manitoba, Canada

William G. Hollister
Professor Emeritus
Department of Psychiatry
University of North Carolina School of Medicine
Chapel Hill, North Carolina

Dorothy Linn Holness, M.H.Sc.
Assistant Professor
Department of Occupational and Environmental Health
University of Toronto
Toronto, Ontario, Canada

Sandra L. Horne, Ph. D.
Assistant Professor
Department of Medicine
University of Saskatchewan
Saskatoon, Saskatchewan, Canada

Edward P. Horvath, Jr., M.P.H.
Medical Director
Division of Occupational Health
BP America, Inc.
Cleveland, Ohio

Evan Howlett, M.D.
Family Physician
Department of Orthopaedic Surgery
University Hospital
Saskatoon, Saskatchewan, Canada

P. M. Huang, Ph.D.
Professor
Department of Soil Science
University of Saskatchewan
Saskatoon, Saskatchewan, Canada

Don Jasper
Manager
Department of Science and Technology
Monenco Consultants, Ltd.
Calgary, Alberta, Canada

Clark D. Jeffries, Ph.D.
Visiting Associate Professor
Department of Mathematical Sciences
Clemson University
Clemson, North Carolina

Geoffrey H. F. Johnston, M. D.
Assistant Professor of Surgery
Department of Orthopaedic Surgery
University Hospital
Saskatoon, Saskatchewan, Canada

William G. Jones, Ph. D.
Certified Industrial Hygienist
Division of Respiratory Disease Studies
United States Public Health Service
Morgantown, West Virginia

C. William Kauffman, Ph.D.
Associate Professor
Department of Aerospace Engineering
University of Michigan
Ann Arbor, Michigan

George G. Khachatourians, Ph.D.
Professor
Department of Applied Microbiology and Food Science
University of Saskatchewan
Saskatoon, Saskatchewan, Canada

Mohamed Khan, M.D.
Professor of Neurosurgery
Department of Clinical Neurobiological Sciences
University of Saskatchewan
Saskatoon, Saskatchewan, Canada

John R. Kinley
Department of Health and Safety in Agriculture
Ontario Task Force
Toronto, Ontario, Canada

D. J. Klaassen, M.D.
Director
Cancer Control Agency of British Columbia
Vancouver, British Columbia, Canada

Norman E. Klein, Ph.D.
Professor
Center for Environmental Health
Department of Animal Science
University of Connecticut
Storr, Connecticut

Heather Koshinsky, B.Sc.
Graduate Student
Department of Applied Microbiology and Food Science
University of Saskatchewan
Saskatoon, Saskatchewan, Canada

John P. Kostuik, M.D.
Professor
Department of Orthopaedics
University of Toronto
Toronto, Ontario, Canada

Vivianne Kovess, M.D.
Assistant Professor
Department of Psychiatry and Epidemiology
McGill University
Montreal, Quebec, Canada

Van P. Kozak, M.Sc.
Chief
Pesticide Section
United States Environmental Protection Agency
Dallas, Texas

Michael J. Kryda, M.D.
Pulmonary Disease Section
Marshfield Clinic
Marshfield, Wisconsin

Samuel Kuperman, M.D.
Assistant Professor
Department of Psychiatry
University of Iowa College of Medicine
Iowa City, Iowa

Chris N. Larson, M.D.
Assistant Professor of Child Psychiatry
Western Psychiatric Institute and Clinic
University of Pittsburgh School of Medicine
Pittsburgh, Pennsylvania

Susan M. Leier, R.N.
Occupational Health Nurse
Medical Department
Consumers' Co-Operative Refineries, Ltd.
Regina, Saskatchewan, Canada

Alexander H. Leighton, M.D.
Professor of Psychiatry and Epidemiology
Department of Community Health and Epidemiology
Faculty of Medicine
Dalhousie University
Halifax, Nova Scotia, Canada

Bruce N. Leistikow, M.D.
Department of Occupational Medicine
University of Iowa
Iowa City, Iowa

Gary R. Leonardson, Ph.D.
Coordinator of Research
University of South Dakota School of Medicine
Rapid City, South Dakota

Merv Letts, M.D.
Head of Orthopaedics
Department of Surgery
University of Manitoba
Winnipeg, Manitoba, Canada

Daniel M. Lewis, Ph.D.
Immunologist
Immunology Section
National Institute for Occupational Health and Safety
Morgantown, West Virginia

K. Y. Robert Li, M.S.
Consulting Statistician
Health and Welfare of Canada
Bureau of Drug Research
Ottawa, Ontario, Canada

Herbert I. Linn, M.S.
Information Specialist
Division of Safety Research
National Institute for Occupational Health and Safety
Morgantown, West Virginia

Peter W. Maley, B.A.
Makakilo, Hawaii

J. Manfreda
Respiratory Hospital
University of Manitoba
Winnipeg, Manitoba, Canada

Yang Mao, Ph.D.
Head of Cancer Section
Bureau of Chronic Disease
Health and Welfare of Canada
Ottawa, Ontario, Canada

Rosemary Marchant
Occupational Environmental Health
Saskatoon, Saskatchewan, Canada

Tom Martin, M.A.
Associate Professor
Department of Microbiology
University of Saskatchewan
Saskatoon, Saskatchewan, Canada

James J. Marx, Jr., Ph.D.
Senior Research Immunologist
Marshfield Medical Research Foundation
Marshfield, Wisconsin

C. G. Toby Mathias, M.D.
Chief of Occupational Dermatology
Industry-Wide Studies Branch
National Institute for Occupational Safety and Health
Cincinnati, Ohio

John J. May, M.D.
Attending Physician in Pulmonary Medicine
Department of Medicine
Mary Imogene Bassett Hospital
Cooperstown, New York

Craig McClain, M.P.A.
Business Manager
Administrative Services
Vermillion School District 13-1
Vermillion, South Dakota

Helen H. McDuffie, Ph.D.
Assistant Professor
Center for Agricultural Medicine
Department of Medicine
University of Saskatchewan
Saskatoon, Saskatchewan, Canada

Gordon McKay, Ph.D.
Research Scientist
College of Pharmacy
University of Saskatchewan
Saskatoon, Saskatchewan, Canada

Harold Menkes (Deceased)

J. A. Merchant
Director
Institute of Agricultural Medicine and Occupational Health
University of Iowa
Oakdale, Iowa

Michael B. Moll, Ph.D.
Senior Mathematical Statistician
Division of Safety Research
National Institute for Occupational Safety and Health
Morgantown, West Virginia

Joanne Moody
Research Coordinator
Department of Epidemiology, Biometry and Occupational Oncology
Cancer Control Agency of British Columbia
Vancouver, British Columbia, Canada

Paula Morelli-Schroth, B.S.
Industrial Hygienist
PEI Associates, Inc.
Cincinnati, Ohio

Philip R. Morey, Ph.D.
Research Industrial Hygienist
National Institute for Occupational Safety and Health
Morgantown, West Virginia

Donald P. Morgan
College of Medicine
University of Iowa
Iowa City, Iowa

William Keith C. Morgan, M.D.
Professor
Department of Medicine
University of Western Ontario
London, Ontario, Canada

Howard I. Morrison, M.Sc.
Research Analyst
Laboratory for Disease Control
National Health and Welfare
Ottawa, Ontario, Canada

Kathy L. Morring, M.P.H.
Certified Industrial Hygienist
Office of Biosafety
Centers for Disease Control
Atlanta, Georgia

Roger Mulder, M.A.
Information Specialist
Pest Management Department
T.D.A.
Austin, Texas

Judith C. Mull, M.S.
Medical Technologist
Department of Immunology
National Institute for Occupational Safety and Health
Morgantown, West Virginia

James R. Nethercott, M.D.
Associate Professor
Department of Occupational and Environmental Health
University of Toronto
Toronto, Ontario, Canada

Stephen A. Olenchock, Ph.D.
Chief
Immunology Section
Division of Respiratory Disease Studies
National Institute for Occupational Safety and Health
Morgantown, West Virginia

Michael A. O'Malley
Epidemiology Section
Department of Health and Human Services
Cincinnati, Ohio

S. Wayne Ormrod
Pesticides Division
Agriculture Canada
Ottawa, Ontario, Canada

David A. Otto, Ph.D.
Research Psychologist
Neurotoxicology Division
United States Environmental Protection Agency
Research Triangle Park, North Carolina

John Owen, M.B.
Professor
Department of Social and Preventive Medicine
University of Saskatchewan
Saskatoon, Saskatchewan, Canada

Rodney R. Parry, M.D.
Professor
Department of Internal Medicine
University of South Dakota
Sioux Falls, South Dakota

Jack Pepys
Department of Immunological Medicine
Royal Postgraduate Medical School
Hammersmith Hospital
London, England

Yvonne Peterson
Research Assistant
Department of Environmental Hygiene
University of Gothenburg
Gothenburg, Sweden

William Robert Pettit, D.V.M.
Clive, Iowa

Greg M. Piacitelli, B.S.
Certified Industrial Hygienist
National Institute of Occupational Safety and Health
Morgantown, West Virginia

Willard E. Pierce
Biostatistician
Department of Epidemiology and Biostatistics
Marshfield Medical Research Foundation
Marshfield, Wisconsin

William Popendorf
Department of Preventative Medicine and Environmental Health
Institute of Agricultural Medicine
University of Iowa
Iowa City, Iowa

K. Post, D.V.M.
Associate Professor
Department of Veterinary Internal Medicine
Western College of Veterinary Medicine
University of Saskatchewan
Saskatoon, Saskatchewan, Canada

David S. Pratt, M.D.
Associate Professor of Clinical Medicine
Columbia University College of Physicians and Surgeons
New York, New York

Anne E. Prince, B.A.
Pulmonary Therapist
Department of Internal Medicine
Section of Pulmonary and Critical Care Medicine
University of Nebraska Medical Center
Omaha, Nebraska

Ronald S. Ratney, Ph.D.
Assistant Regional Administrator
Occupational Safety and Health Administrator
Boston, Massachusetts

Edward B. Reczek, M.Sc.
Corporate Industrial Hygienist
Human Resources Department
Texaco Canada Inc.
Don Mills, Ontario, Canada

Marilyn Reddy, R.N.
Supervisor
Employees Health Service
Saskatchewan Wheat Pool
Regina, Saskatchewan, Canada

L. D. Reed
Division of Safety Research
National Institute for Occupational Safety and Health
Morgantown, West Virginia

Stephen I. Rennard, M.D.
Associate Professor of Medicine
Department of Internal Medicine
Section of Pulmonary Critical Care
University of Nebraska Medical Center
Omaha, Nebraska

Leonard Ritter, Ph.D.
Chief of Pesticides Division
Health and Welfare of Canada
Ottawa, Ontario, Canada

Richard A. Robbins, M.D.
Assistant Professor
Department of Internal Medicine
Section of Pulmonary and Critical Care Medicine
University of Nebraska Medical Center
Omaha, Nebraska

Ronald C. Roberts, Ph.D.
Senior Research Biochemist
Department of Research
Marshfield Medical Research Foundation
Marshfield, Wisconsin

Ragnar Rylander, M.D.
Professor
Department of Environmental Hygiene
University of Gothenburg
Gothenburg, Sweden

Liora Salter, M.A.
Associate Professor and Chairman
Department of Communication
Simon Fraser University
Burnaby, British Columbia, Canada

Keith Schappert, Ph.D.
Post Doctoral Fellow
Department of Medical Genetics
University of Toronto
Toronto, Ontario, Canada

Evelyn Schlenker, Ph.D.
Assistant Professor
Department of Physiology and Pharmacology
University of South Dakota School of Medicine
Vermillion, South Dakota

E. Schram
Department of Medicine
Division of Pulmonary Diseases
University of Alberta
Edmonton, Alberta, Canada

Leonard M. Schuman, M.D.
Mayo Professor and Director
Department of Epidemiology
School of Public Health
University of Minnesota
Minneapolis, Minnesota

Danna Skea, Ph.D.
Graduate Student
Department of Medical Sciences
McMaster University
Hamilton, Ontario, Canada

J. Drew Smith, M.Sc.
Research Scientist
Department of Horticulture Science
University of Saskatchewan
Saskatoon, Saskatchewan, Canada

Robert Smith
Des Moines, Iowa

W. G. Sorenson
Honeywell, Inc.
Golden Valley, New York

John J. Spinelli, M.Sc.
Statistician
Department of Epidemiology, Biometry and Occupational Oncology
Cancer Control Agency of British Columbia
Vancouver, British Columbia, Canada

Marlin G. Stahl, M.D.
Assistant Professor
Department of Pulmonary Medicine
Internal Medicine Associates
Omaha, Nebraska

Lorann Stallones, Ph.D.
Assistant Professor
Department of Preventive Medicine and Environmental Health
University of Kentucky College of Medicine
Lexington, Kentucky

Nate M. Stepner, M.Sc.
Research Associate
Gage Research Institute
Toronto, Ontario, Canada

Austin B. Thompson, M.D.
Assistant Professor of Medicine
Department of Internal Medicine
Section of Pulmonary Critical Care Medicine
University of Nebraska Medical Center
Omaha, Nebraska

William J. Threlfall, B.Sc.
Programmer and Statistician
Department of Epidemiology, Biometry and Occupational Oncology
Cancer Control Agency of British Columbia
Vancouver, British Columbia, Canada

Teresa To, Ph.D.
Department of Social and Preventive Medicine
Center for Agricultural Medicine
University of Saskatchewan
Saskatoon, Saskatchewan, Canada

Mary W. Treuhaft, Ph.D.
Section Leader
Department of Biotechnology Assays
Schering Corp.
Bloomfield, New Jersey

Stephanie F. Van Lier, M.S.
Environmental Specialist III
Environmental Protection Division
Department of Natural Resources
Des Moines, Iowa

Paul D. Van Loon, M.Sc.
Health Educator
Saskatoon Lung Association
Saskatoon, Saskatchewan, Canada

Sverre Vedal, M.D.
Assistant Professor
Department of Medicine
University of British Columbia
Vancouver, British Columbia, Canada

C. Peter W. Warren, M.B.
Professor
Department of Medicine
University of Manitoba
Winnipeg, Manitoba, Canada

Hans Weill, M.D.
Schlieder Foundation Professor
Pulmonary Diseases Section
Tulane University School of Medicine
New Orleans, Louisiana

R. White
Denver, Colorado

Donald T. Wigle, M.P.H.
Chief
Surveillance and Risk Assessment Division
Bureau of Chronic Disease
Epidemiology Laboratory Center for Disease Control
Ottawa, Ontario, Canada

V. Woytowich
Department of Medicine
Division of Pulmonary Diseases
University of Alberta
Edmonton, Alberta, Canada

Joanne L. Wright, M.S.
Associate Professor
Department of Pathology
University of British Columbia
Vancouver, British Columbia, Canada

Ken Yong-Hing, M.B.
Department Head
Department of Orthopaedic Surgery
University Hospital
Saskatoon, Saskatchewan, Canada

Ken Yoshida, Ph.D.
Adjunct Associate Professor
Department of Community Health Sciences
Faculty of Medicine
University of Calgary
Calgary, Alberta, Canada

Dennish D. Zaebst, M.S.
Supervisory Industrial Hygienist
Division of Surveillance, Hazard Evaluations, and Field Studies
National Institute for Occupational Safety and Health
Cincinnati, Ohio

John N. Zey, M.S.
Industrial Hygienist
Central Missouri State University
Warrensburg, Missouri

ACKNOWLEDGEMENTS

We would like to express our sincere appreciation to the large number of friends, colleagues, and supporters whose generous assistance made possible both the International Symposium on Health and Safety in Agriculture, where much of this material was first presented, and the subsequent publication of this volume. Firstly, we would like to thank the many colleagues whose contributions are published in this volume. The editors would like to thank the International Symposium co-chairs, C. Vaughan Bowen, Brian L. Graham, Daniel J. Hogan, David L. Keegan, David J. Klaassen, and H. Bruno Schiefer. The financial support received from many agencies and companies was essential and appreciated. The Saskatchewan Lung Association made the initiation of the Symposium possible by the provision of an initial and major grant. Additional financial support was provided by Health and Welfare Canada Research Programs Directorate, the National Institute for Occupational Safety and Health (USPHS), Saskatchewan Workers' Compensation Board, Western Grain Elevator Association, Labour Canada, Saskatchewan Cancer Foundation, Massey Ferguson Ltd., Saskatchewan Health Research Board, Ciba-Geigy, Eli Lilly Canada Inc., Toxicology Research Centre - University of Saskatchewan, Society of Toxicology of Canada, Saskatchewan Medical Association, Canadian Lung Association, Monsanto Chemical Company, Canadian Centre for Occupational Health and Safety, and University Hospital - Saskatoon. In particular, we would like to thank members of the organizing and scientific committees, as well as the Symposium faculty. Several individuals stand out as having made unique contributions. Mr. Frank Froh, now deceased, former Executive Secretary of the Saskatchewan Lung Association, and Mr. J. W. Churchman, President of the Saskatchewan Lung Association were instrumental in the initiation of the Symposium. Dr. Pervis Major, Deputy Director, Division of Respiratory Disease Studies, National Institute for Occupational Safety and Health at Morgantown, West Virginia was instrumental in organizing a very significant financial and scientific input from that organization. Dr. O. E. Laxdal, Violet Ehman, and Miriam Hills of the Department of Continuing Medical Education, University of Saskatchewan assisted in planning and carrying out the Symposium. We would like to recognize in a special manner the contributions of Mrs. Ruth Day and Maura Gillis-Cipywnyk whose dedicated approach to co-ordinating the Symposium and obtaining the documentation was essential and appreciated. Many individual farmers and their families have provided us with valuable information and support. Mrs. J. Bramley prepared manuscripts. Finally, we wish to thank the staff of CRC Press for their assistance.

James A. Dosman
Donald W. Cockcroft

TABLE OF CONTENTS

DEDICATION

This work is dedicated with great respect to Frank Froh, J. Wilfred Churchman, and Dr. Leo Kristjanson, and with love to our families: Susan McKay Dosman and Sue Cockroft, Audrey Susan Dosman, Jill Marie Dosman, Andrea Jane Dosman, Cara Fairclough Dosman, John Harold Dosman, and Sean William Cockcroft.

Frank Froh (deseased, 1984) was Secretary and Executive Director of the Saskatchewan Lung Association, 1952 to 1984; J. Wilfred Churchman was President of the Saskatchewan Lung Association, 1978 to 1988; and Dr. Leo Kristjanson was President of the University of Saskatchewan, 1980 to 1989. They believed in the International Symposium on Health and Safety in Agriculture and the subsequent development of the Centre for Agricultural Medicine at the University of Saskatchewan.

We also dedicate this work to Harold Menkes, M.D. and his wife, Marilyn Menkes, Ph.D., both formerly of the Faculty of Johns Hopkins University, and who both died tragically on August 27, 1987. We dedicate this volume to them because Harold, in particular, believed in us. In his uniquely enthusiastic manner, Harold believed that the International Symposium on Health and Safety in Agriculture, and the subsequent development of the Centre for Agricultural Medicine at the University of Saskatchewan were necessary and important.

Methodologies in Respiratory Occupational Surveillance

Epidemiological Principles in the Identification of an Occupational Health Risk: Airways Obstruction in Grain Workers

J. A. Dosman and T. To

INTRODUCTION

Epidemiology may be defined as "the study of the distribution of a disease or physiological condition in human population and factors that influence this distribution".[1] The epidemiologist is concerned with the occurrence of disease by time and place, the incidence and prevalence of disease or dysfunction, and personal or other factors that influence disease or dysfunction.[2] The suspicion that farmers and agricultural workers may be at increased risk of respiratory dysfunction[3–6] stemmed from the utilization of epidemiological principles.

Principles of epidemiology permit an orderly and systematic approach to the identification and measurement of disease and dysfunction, the ascertainment of causes, the setting of hygiene standards, the interaction between risk factors and disease outcome, and minimization of known risks.[7] The process of identifying new or unsuspected risks often begins by observations in the clinical setting which in turn may lead to epidemiological inquiry. Types of epidemiology include descriptive, cross–sectional, analytical, and experimental.[7] Within this framework, multiple factors may be present that contribute to dysfunction in exposed workers. Workers, industry, government agencies, and the scientific community have an opportunity to work together to identify risk factors, control the environment, and promote good health in the workplace.

IDENTIFYING NEW OR UNSUSPECTED RISKS

New or unsuspected risks may be identified by either clinical observation or epidemiological inquiry.

Clinical Observation

Clinical observation remains the oldest and most important method of suspecting or identifying an occupational health risk. For example, in agriculture and related industries, a pathophysiological response that the lung may make as a result of a biological or other exposure is the development of chronic airways obstruction.

Case — Mr. O. Y., D.O.B. 1908 — This 68–year–old retired grain elevator agent presented at the Chest Clinic, October 4, 1976 complaining of severe shortness of breath and wheezing. He had been a farm worker until age 21 when he began employment in a grain elevator, which he continued to age 65. He began smoking cigarettes at age 30 and used 1/2 package of cigarettes daily until the age of 54 for a 12–pack–year history. Respiratory symptoms began at age 46 with cough and sputum, progressing to shortness of breath at age 54, and disabling shortness of breath by age 65. When seen, he was tachypneic at rest with a barrel chest and wheezing. Pulmonary function studies demonstrated a forced expired volume in 1 s (FEV_1) of 1.57 l (55% of the predicted value of 2.87 l), with a low ratio of FEV_1 to forced vital capacity (FVC) and a brisk increase of 17% in FEV_1 to 1.83l following 200 μg of inhaled Salbutamol, an increase considered unusual in cases of chronic obstructive lung disease. Chest X–ray demonstrated marked hyperinflation and loss of vasculature consistent with airways obstruction and emphysema.

This and other similar cases demonstrated more severe airways obstruction than is common with the smoking history in question, and alerted the physician that exposure to grain dust might result in airways obstruction. This suspicion led to a survey of workmen exposed to grain dust (epidemiological inquiry). In addition, the greater–than–expected increase in FEV_1 with an inhaled bronchodilator aerosol in these cases led to an evaluation of airways reactivity in a future study (analytical epidemiology).

Epidemiological Inquiry

The purpose of epidemiological inquiry is to identify disease or dysfunction in populations, to establish causes to assist in minimizing known risks, and to set hygiene standards. These steps have been identified by Schilling and are as follows: (1) to identify disease or dysfunction, (2) to identify the prevalence of disease or dysfunction, (3) to identify causes, (4) to minimize known risks, and (5) to set hygiene standards.[7]

These steps took place in the identification of chronic airway obstruction in grain workers. Following the clinical impression that disease might be present, surveys and cross–sectional studies to estimate the prevalence and severity of disease were carried out[8–12] and investigations into causes and mechanisms were begun.[13–16] In 1978 the regulatory agency set about minimizing risks by the introduction of a mandatory lung function testing program and the establishment of hygiene standards in the form of a maximum permissible dust level.[17]

TYPES OF EPIDEMIOLOGY

The four principle types of epidemiology include descriptive, cross–sectional, analytical (i.e., case control and cohort studies), and experimental.[7] All of these types of epidemiology have been used in delineating lung dysfunction in grain workers.

Descriptive Epidemiology

Descriptive epidemiology is derived from data that have usually been collected for other purposes such as vital statistics and information from surveys. These kinds of data will often provide useful background information for the formulation of hypotheses and the structuring of cross–sectional or prospective studies. As these data have not been collected to satisfy a specific hypothesis, they are often "inefficient" for the purpose at hand because a large body of data may have been collected for other reasons. The 1976 Canada Census revealed an increased risk of death from diseases of the respiratory system in males in census divisions in the western Canadian cereal–grain growing areas.[18] Although there might be other reasons to explain an increased risk, vital statistics can provide clues that a specific risk, such as exposure to grain dust, is present, particularly if other evidence, such as clinical observation, has alerted observers to the possible risk.

Survey data may also be useful in formulating hypotheses or planning future studies. For example, survey data were obtained on grain workers by the Saskatchewan Lung Association in 1972. No effort was made to establish a controlled epidemiological study. However, when matched with an existing population of nonworkers, this information was useful in understanding the potential prevalence and nature of dysfunction of workers in the grain industry.[8,19] Figure 1 shows information from the 1972 Saskatchewan survey and demonstrates maximum midexpiratory flow rate (MMFR), as a percent of that predicted, plotted against exposure years or equivalent age.[19] These survey data suggested an increased risk of lung dysfunction with years of employment in the grain industry,[8] and suggested that the effects of grain working and the effects of smoking might be more than additive. This survey was useful in planning a well–controlled cross–sectional study of grain workers, in which smoking and its effects were specifically accounted for in the study design.

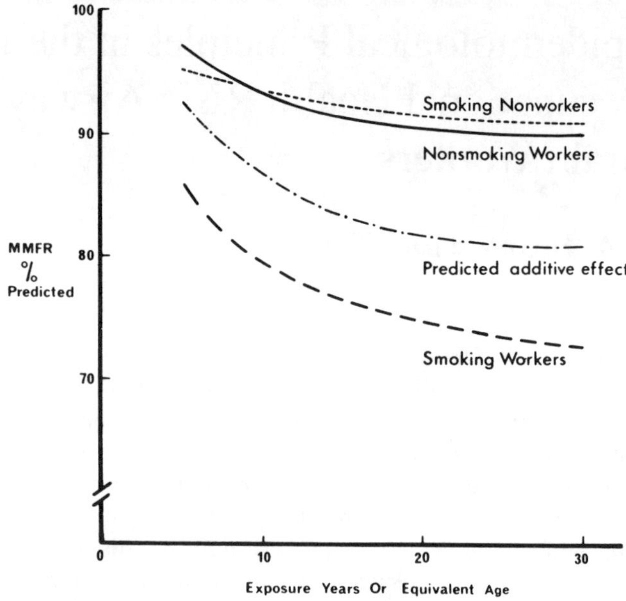

FIGURE 1. Mean MMFR for four groups plotted against years of exposure to grain dust or equivalent age. The MMFR is expressed as a percent of that predicted based on the data of Morris et al.[20] Solid line (——) represents grain workers who are lifetime nonsmokers; the finely dashed line (- - - -), smokers who do not work in the grain industry; the coarsely dashed line (— — —), the smoking workers; and the dashed line with dots (—·—·—), the predicted additive effect of both smoking and working in the grain industry. (Reproduced from Dosman, J.A., *Ann. Intern. Med.*, 87, 7784, 1977. With permission.)

Cross–Sectional Epidemiology

Cross–sectional studies usually refer to studies on diseases within an area during a well–defined period of time. If a higher frequency of cases of the abnormality or dysfunction is present in the population suspected to be at risk, compared to the control population, or if differences in mean values for variables (i.e., lung function test values) or frequency of respiratory symptoms that are deemed to be clinically significant are demonstrated, then an association between disease or dysfunction and the characteristic in question may be inferred. When the characteristic in question, such as exposure to grain dust, is suspected of being associated with a respiratory abnormality, it is common to use a standardized respiratory questionnaire, such as the American Thoracic Society Standardization Committee Questionnaire,[21] or pulmonary function studies that are performed according to certain criteria, such as those established by the National Institutes of Health.[22] Using such techniques, Cotton et al. structured carefully chosen groups of grain workers and control subjects in a cross–sectional study.[23] They matched a group of lifetime nonsmoking grain workers with a group of lifetime nonsmoking age– and sex–matched nonexposed community control subjects and they extended the match to include smoking grain workers and smoking nonexposed control subjects, all of whom were matched for age, sex, and — where appropriate — exposure years and smoking years.[23] Not all cross–sectional studies can be matched so extensively, and statistical analysis may be used to assess the effects of unequal matching. However, in the case of the grain workers' cross–sectional study reported by Cotton et al., inferences could be made about both the effects of grain dust exposure and smoking. Cotton et al. concluded that grain dust and smoking had nearly additive, or equal, effects on lung function, but that the effects of smoking were somewhat more pronounced.[23]

Cross–sectional data are frequently useful for describing relative risk for multiple categories. Predictions about the association between a disease and a risk factor are strengthened if information is available to support a dose–gradient relationship between degree of exposure ("dose") to a characteristic or disease or dysfunction. For example, Corey et al. have utilized such data to demonstrate dose–effect relationships between dust exposure and pulmonary function changes in grain workers.[24]

Analytical Epidemiology

Analytical epidemiology may consist of case control studies or cohort studies. When cases and controls are matched in pairs to make the two groups comparable with regard to one or more factors, a fourfold (2 × 2) table is derived. Status of cases with regard to the presence or absence of the characteristic is compared with its presence or absence in their respective controls.

Cohort or longitudinal studies are frequently the best way to assess the effect of an environmental exposure on human health, as changes in a given characteristic may be observed over time in an exposed as compared to a nonexposed population. Cohort studies suffer from the disadvantages of taking a long time to accomplish, being expensive, and suffering from different rates of subject loss from exposed and nonexposed groups because of illness induced by the exposure (i.e., the healthy–worker effect). However, in some circumstances, inferences can be drawn from longitudinal studies that are less clear in cross–sectional or retrospective studies. This point is exemplified by a longitudinal study[25] in which we made initial and 3–year measurements of the lung function variables, FVC and FEV_1, on four groups of workers and control subjects who had not changed work or smoking status: (1) 200 lifetime nonsmoking workers (mean age ±1 SD; 33.1 ± 12.0 years; exposure to grain dust 11.8 ± 7.7 years); (2) 353 smoking workers (age 36.8 ± 10.5 years; exposure years 13.0 ± 9.0; smoking pack years 18.0 ± 12.7); (3) 113 lifetime nonsmoking community control subjects (age 38.8 ± 12.7 years); and (4) 133 smoking control subjects (age 40.6 ± 11.8 years; smoking pack years 22.2 ± 12.9). We carried out an analysis of variance for FVC and FEV_1 to assess the importance of grain working and smoking over a 3–year period (Table 1).[25]

TABLE 1
3–Year Longitudinal Study of Grain Workers: Significance of Pulmonary Function Test Value Changes between Observation 1 and Observation 2

	Significance
FVC[a]	
Grain	$p < 0.000$
Smoking	—
g — s	$p < 0.054$
FEV_1[a]	
Grain	$p < -.000$
Smoking	—
g — s	$p < 0.060$

Note: Grain = effect of grain working on workers; smoking = effect of smoking on workers; g — s = effect of interaction of grain working and smoking on workers.

[a] Calculations based on percent expected values.

Reproduced from Dosman, J. A., *J. Public Health Policy,* 6, 185, 1985. With permission.

The interpretation of the longitudinal data was different than on a cross–sectional study carried out on essentially the same workers. The cross–sectional study reported by Cotton et al. suggested that the effects of smoking and the effects of grain working were nearly equal, with smoking being somewhat more important,[23] whereas the longitudinal study indicated that over a 3–year period, the effects of grain working were more important.[25] The public health regulatory implications of these differing interpretations of data from the same populations, obtained by longitudinal as compared to cross–sectional observations, are considerable.

Experimental Epidemiology

Experimental epidemiology may be utilized to test specific hypotheses. For example, clinical cases such as Mr. O. Y. described above and the work of Gerrard et al.[13] suggested that grain workers might have greater bronchial reactivity than control subjects. In order to test this hypothesis, 29 lifetime nonsmoking grain workers were matched with 29 nonexposed control subjects by age, sex, and baseline lung function.[14] Other tests such as allergy skin tests and indices of allergy indicated that there was no increased prevalence of allergy or atopy amongst the grain workers (in fact, there were two cases of asthma amongst the control subjects). Despite this weighting of allergy among the control subjects, there was a decrease in the histamine provocation concentration that reduced specific conductance by 35% (PC_{35}) among the workers. The PC_{35} was 5.9 ± 3.0 mg/ml for the control group and 4.5 ± 2.0 mg/ml for the grain workers ($p < 0.05$). Figure 2 demonstrates

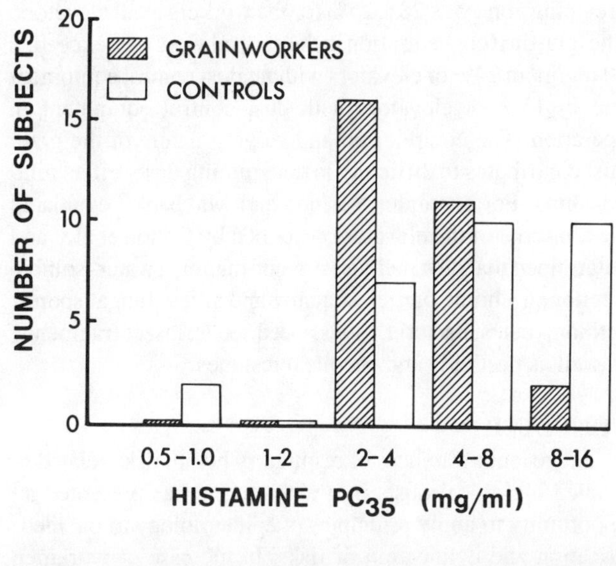

FIGURE 2. Distribution of histamine PC_{35} in 29 nonsmoking grain workers (hatched bars) and 29 nonsmoking control subjects (open bars). Horizontal axis is the histamine PC_{35} (mg/ml) and the vertical axis the number of subjects. (Reproduced from Mink, J.T., Gerrard, J.W., Cockcroft, D.W., Cotton, D.J., and Dosman, J.A., *Chest,* 77, 28, 1980. With permission.)

the frequency distribution of PC_{35} for the two groups. Twenty–seven control subjects had normal nonspecific bronchial reactivity and two control subjects had increased nonspecific bronchial reactivity (PC_{35} less than 2 mg/ml). On the other hand, although none of the grain workers had an increase in nonspecific bronchial reactivity as significant as that of the two asthmatic patients, the frequency distribution of PC_{35} for the grain workers was weighted towards the left, with 16 of the 29 workers having values between 2 and 4 mg/ml. This type of experimental epidemiology led Mink et al.[14] to suggest that chronic exposure to grain dust in nonallergic individuals appears to be associated with both increased prevalence of respiratory symptoms and increased nonspecific bronchial reactivity.

UNDERSTANDING THE ENVIRONMENT

Knowledge of the nature and duration of exposure is essential in understanding the pathogenesis of disease in exposed workers and in recommending hygiene standards. In the cereal grain industry, the maximum permissible exposure level of 10 mg/m^3 was set when grain dust was considered a nuisance dust.[17] In the study described by Cotton et al.,[23] the concentration of general work area dust in grain elevators depended on the operation of dust control equipment[26] in the elevators. In 54% of the elevators cited by Cotton et al.,[23] control equipment was in operation; in these elevators the mean (±SD) concentration of total dust in the general work area was 6.1 ± 9.2 mg/m^3, and in the breathing zone (8 h average) 1.9 ± 3.2 mg/m^3. In 46% of elevators, no dust equipment was in operation; in these elevators the dust concentration in the general work area was 13.6 ± 22.1 mg/m^3 and the 8–h average concentration in the breathing zone was 6.2 ± 14.3 mg/m^3. The average respirable fraction of dust in the breathing zone was 28 ± 25% for the workers in all elevators. The total dust concentration in the general work area exceeded 10 mg/m^3 in 24% of elevators without dust control equipment and in 15% of elevators with dust control equipment in operation. The complicated and varying nature of the grain dust contributes to difficulty in determining dose–effect relationships. For example, Yoshida and Maybank[27] evaluated dust in some of the elevators described by Cotton et al.[23] and determined that dust includes organic matter, a water–soluble fraction, trichome particles, quartz and silica, fungal spores, actinomycetes, bacteria, mites, weed seeds, insect fragments, as well as pesticide and chemical residues.[27]

CONCLUSION

The recent awareness of respiratory health risks related to inhaled biological dusts in grain workers has presented an opportunity to apply principles of epidemiology to the identification and delineation of risks. In the case of workmen exposed to cereal grain dust, these advances in understanding and enhancement of risk management took place over a relatively short time framework. Other environmentally associated occupational lung diseases, such as asbestosis and pneumoconiosis, were initially assessed in a day when the science of epidemiology was less well developed and knowledge, understanding, and preventive measures were slow to develop. With current understanding of epidemiology and biostatistics, and with modern ability to collect and analyze data and to assess the environment, workers, industry, investigators, and health care professionals have an opportunity to work together to accomplish the goals of efficient research, effective hygiene measures, health promotion, and disease prevention.

REFERENCES

1. **Lilienfeld, D.E.,** Definitions of epidemiology, *Am. J. Epidemiol.,* 107, 87, 1978.
2. **Lilienfeld, A.M. and Lilienfeld, D.E.,** *Foundations of Epidemiology,* Oxford University Press, New York, 1980.
3. **Madsen, D., Klock, L.E., Wenzel, F.J., Robbins, J.L., and Schmidt, C.D.,** The prevalence of Farmer's lung in an agricultural population, *Am. Rev. Respir. Dis.,* 113, 171, 1976.
4. **Cockcroft, D.W. and Dosman, J.A.,** Respiratory health risks in agriculture, *Ann. Intern. Med.,* 95, 380, 1981.
5. **Dosman, J.A., Cockcroft, D.W., and Hoeppner, V.H.,** Airway obstruction in occupational pulmonary disease, *Med. Clin. North Am.,* 65, 691, 1981.
6. **Popendorf, W., Donham, K.J., Easton, D.N., and Silk, J.,** A synopsis of agricultural respiratory hazards, *Am. Ind. Hyg. Assoc. J.,* 46, 154, 1985.
7. **Schilling, R.,** Problems in the identification of occupational disease, in *Occupational Pulmonary Disease: Focus on Grain Dust and Health,* Dosman, J.A. and Cotton, D.J., Eds., Academic Press, New York, 1980, 3.
8. **Sheridan, D., Deutscher, C., Tan, L., Maybank, J., Gerrard, J., Horne, S., Yoshida, K., Barnett, D.G., Cotton, D.J., and Dosman, J.A.,** The relationship between exposure to cereal grain dust and pulmonary function in grain workers, in *Occupational Pulmonary Disease: Focus on Grain Dust and Health,* Dosman, J.A. and Cotton, D.J., Eds., Academic Press, New York, 1980, 229.
9. **Becklake, M.R., Jedoin, G., Lefort, L., Rose, B., Mandl, M., and Frazer, R.G.,** A respiratory health study of grain handlers in St. Lawrence river ports, in *Occupational Pulmonary Disease: Focus on Grain Dust and Health,* Dosman, J.A. and Cotton, D.J., Eds., Academic Press, New York, 1980, 239.
10. **doPico, G.A., Reddan, W., Flaherty, D., Tsiatis, A., Peters, M., Ras, P., and Rankin, J.,** Respiratory abnormalities among grain handlers: a clinical physiologic and immunologic study, *Am. Rev. Respir. Dis.,* 115, 915, 1977.
11. **Broder, I., Mintz, S., Hutcheon, M., Corey, P., Silverman, F., Davies, G., Leznoff, A., Peress, L., and Thomas, P.,** Comparison of respiratory variables in grain elevator workers and civic outside workers of Thunder Ray, Canada, *Am. Rev. Respir. Dis.,* 119, 193, 1979.
12. **Chan–Yeung, M., Schulzer, M., MacLean, L., Dorker, E., and Grzybowski, S.,** Epidemiologic health survey of grain elevator workers in British Columbia, *Am. Rev. Respir. Dis.,* 121, 329, 1980.
13. **Gerrard, J.W., Mink, J., Cheung, S.–S.C., Tan, L.K.–T., and Dosman, J.A.,** Nonsmoking grain handlers in Saskatchewan: airways reactivity and allergic status, *J. Occup. Med.,* 21, 392, 1979.
14. **Mink, J.T., Gerrard, J.W., Cockcroft, D.W., Cotton, D.J., and Dosman, J.A.,** Increased bronchial reactivity to inhaled histamine in nonsmoking grain workers with normal lung function, *Chest,* 77, 28, 1980.
15. **Olenchock, S.A., Mull, J.C., Gladich, M.E., Peach, M.J., and Major, P.C.,** Airborne grain dust activation of complement, in *Occupational Pulmonary Disease: Focus on Grain Dust and Health,* Dosman, J.A. and Cotton, D.J., Eds., Academic Press, New York, 1980, 263.

16. **Horne, S.L., Clarke, M.A., and Barnett, G.D.,** Alpha–1–antitrypsin deficiency: its effect on pulmonary function in Saskatchewan grain buyers, in *Occupational Pulmonary Disease: Focus on Grain Dust and Health,* Dosman, J.A. and Cotton, D.J., Eds., Academic Press, New York, 1980, 349.
17. Labor Canada 1978, Guidelines for Environmental and Medical Surveillance Program in the Grain Industry, Ottawa Occupational Safety and Health Branch File No. 895–7–11, January 1978.
18. Census of Canada, 1976.
19. **Dosman, J. A.,** Chronic bronchitis and smoking in grain workers, *Ann. Intern. Med.,* 87, 7784, 1977.
20. **Morris, J., Koskie, A., and Johnson, L.C.,** Spirometric standards for healthy nonsmoking adults, *Am. Rev. Respir. Dis.,* 103, 57, 1971.
21. **Ferris, B.G.,** Epidemiology Standardization Project, *Am. Rev. Respir. Dis.,* 118, 1, 1978.
22. National Heart and Lung Institute, Recommended Standardized Procedures — NHLI Division of Lung Diseases, Epidemiology Studies, Bethesda, 1971 (revised 1973).
23. **Cotton, D.J., Graham, B.L., Li, K.Y.R., Froh, F., Barnett, G.D., and Dosman, J.A.,** Effects of grain dust exposure and smoking on respiratory symptoms and lung function, *J. Occup. Med.,* 25, 131, 1983.
24. **Corey, P., Hutcheon, M., Broder, I., and Mintz, S.,** Grain elevator workers show work–related pulmonary function changes and dose–effect relationships with dust exposure, *Br. J. Ind. Med.,* 39, 330, 1982.
25. **Dosman, J.A.,** Occupational exposure and cigarette smoking: respiratory health protection in the workplace, *J. Public Health Policy,* 6, 185, 1985.
26. **Halter, G.S.,** Dust control in grain elevators, in *Occupational Pulmonary Disease: Focus on Grain Dust and Health,* Dosman, J.A., and Cotton, D.J., Eds., Academic Press, New York, 1980, 545.
27. **Yoshida, K. and Maybank, J.,** Physical and environmental characteristics of grain dust, in *Occupational Pulmonary Disease: Focus on Grain Dust and Health,* Dosman, J.A. and Cotton, D.J., Eds., Academic Press, New York, 1980, 441.

Airborne Environmental Exposure, Lung Function, and Mortality

H. Menkes and W. Beckett

INTRODUCTION

Chronic exposure to airborne environmental agents such as cigarette smoke, cotton dust, and grain dust can lead to pulmonary dysfunction. On the one hand, pulmonary dysfunction may reflect lung disease resulting in obstructive or restrictive ventilatory defects. On the other hand, pulmonary dysfunction may be produced by disease of other systems. This paper proposes a hypothesis stating that chronic airborne exposures that lead to pulmonary dysfunction cause an increase in mortality from nonpulmonary as well as pulmonary disease. The rationale for this hypothesis is based on observations indicating that pulmonary dysfunction is associated with increased mortality from nonpulmonary disease. If so, tests of lung function may be helpful in predicting nonpulmonary as well as pulmonary effects of airborne environmental agents, and such prediction could lead to preventive strategies.

RELATIONSHIP BETWEEN PULMONARY FUNCTION AND MORTALITY

In 1971 an interdisciplinary epidemiological study of airways obstruction was initiated by Cohen at Johns Hopkins.[1] Between 1971 and 1976 more than 4500 subjects including index groups (pulmonary disease patients, lung cancer patients, neighborhood controls, nonpulmonary patient controls, and healthy teachers) and their relatives were evaluated. Questionnaires and tests of lung function including pulmonary mechanics and gas exchange were administered. Of 2539 subjects not selected on the basis of their own health status, 115 died during a follow-up period that ended in 1981. When initial evaluation characteristics of subjects who died were compared with subjects who survived, a striking relationship between baseline lung function and mortality was observed.[2] Figure 1 shows that even after adjustments were made for smoking, mortality was higher in subjects whose initial forced expiratory volume as a percentage of forced vital capacity ($FEV_1\%$) was below 68% than in those whose initial $FEV_1\%$ exceeded 68%. Further strong evidence linking low forced expiratory volume with mortality has been provided by the Framingham study of coronary artery disease[4] and the Tecumseh study of lung disease.[5] Thus, there is ample evidence from multiple studies indicating that pulmonary dysfunction is an important risk factor for mortality.

WHY IS PULMONARY DYSFUNCTION LINKED TO MORTALITY?

The first, most obvious possible reason for a link between pulmonary dysfunction and mortality is that individuals with lung disease die because of their lung disease. However, this is not the case. Figure 2 indicates that although the 3 (of 115) who died of lung disease in the Hopkins study had pulmonary dysfunction when first evaluated, the vast majority of deaths occurred from nonpulmonary disease. In fact, no difference was detected in the relative rates of death from cancer, cardiovascular-renal disease, or other causes between those with initial pulmonary dysfunction and those without pulmonary dysfunction. This is consistent with reports from Framingham and Tecumseh supporting the notion that the increased mortality is not due to deaths from lung disease. It should be pointed out that this figure shows percentages of death caused by different causes and, thus, conceals the increase in each group associated with pulmonary dysfunction.

A second possible reason for the link between pulmonary dysfunction and mortality is that reductions in forced expiratory volumes reflect a general debility not centered in the lungs. In fact, this could explain results obtained in the Framingham and Tecumseh studies that were based on absolute measurements of lung volumes. However, Figure 2, taken from the Hopkins study, reflects differences in $FEV_1\%$. Insofar as flow limitation is effort independent, respiratory muscle

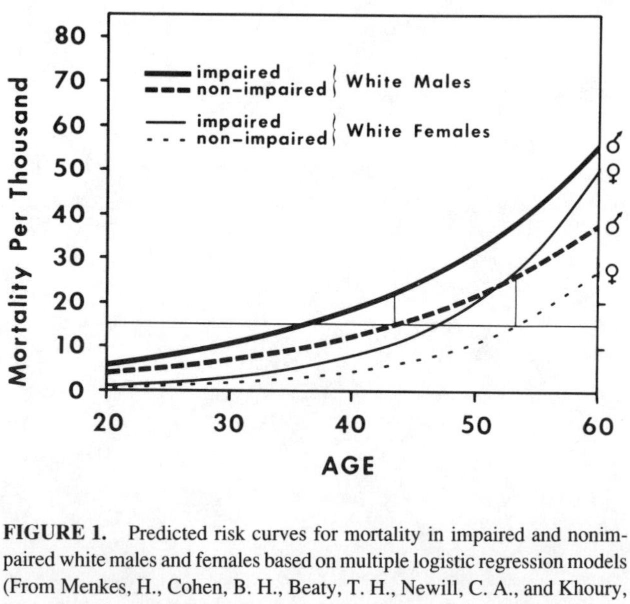

FIGURE 1. Predicted risk curves for mortality in impaired and nonimpaired white males and females based on multiple logistic regression models (From Menkes, H., Cohen, B. H., Beaty, T. H., Newill, C. A., and Khoury, M. J., *Genetics and Epidemiology of Coronary Heart Disease, Past, Present and Future,* Alan R. Liss, New York, 1984, 501. With permission.)

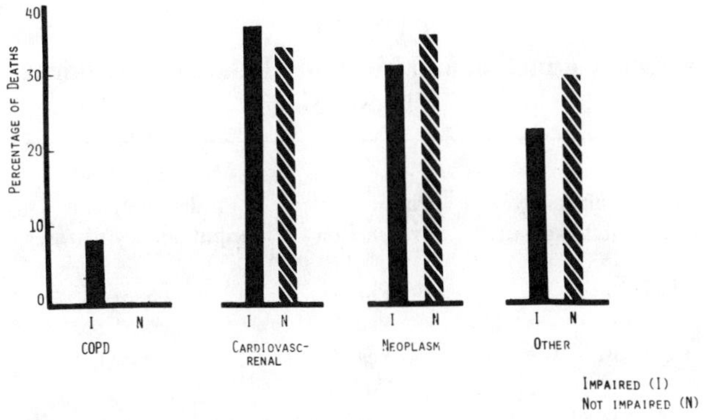

FIGURE 2. Percentage distribution of cause of death for deceased study participants by their initial-visit lung function. (From Menkes, H., Cohen, B. H., Beaty, T. H., Newill, C. A., and Khoury, M. J., *Genetics and Epidemiology of Coronary Heart Disease, Past, Present and Future,* Alan R. Liss, New York, 1984, 501. With permission.)

strength or vigor does not contribute to the result.[6] Further support for this conclusion is shown in Figure 3 where mortality vs. slope of phase III ($\Delta N_2/l$) from the single-breath nitrogen test is shown.[7] $\Delta N_2/l$ was measured during a maneuver where expiratory flows were maintained below 0.5 l/s. This test reflects the evenness of the distribution of inspiration to volume and the rate of exhalation from the peripheral lung.

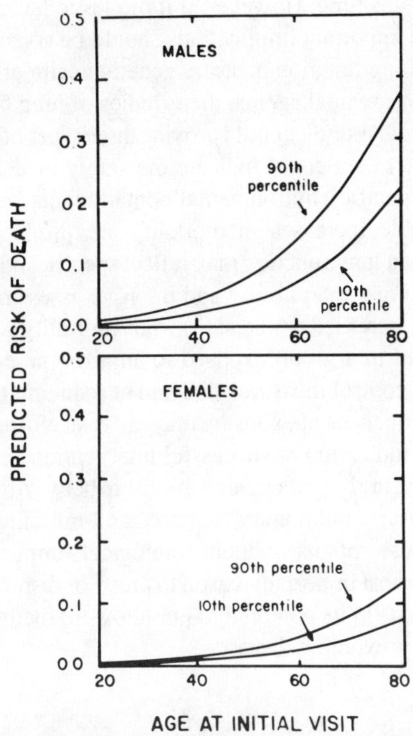

FIGURE 3. Predicted risk for dying during follow-up based on logistic regression, including age at initial visit, smoking, and N_2/l among males and females. Mean smoking levels were used here. (From Menkes, H. A., Beaty, T. H., Cohen, B. H., and Weinmann, G., *Am. Rev. Respir. Dis.,* 132, 115, 1985. With permission.)

Thus, there is good reason to believe that the relationship between lung dysfunction and mortality is related to a pulmonary abnormality and not to repiratory muscle strength.

A third possible reason for the link between pulmonary dysfunction and mortality is that normal lungs are necessary for the protection of the body as a whole from exogenous or endogenous toxic substances. When the lungs fail, this protection is lost. This attractive hypothesis forms the basis of a schema presented by Cohen in 1981[8] (Figure 4). However, this hypothesis remains unproven.

A fourth possibility is that abnormalities in lung function reflect extrapulmonary disease that, upon improvement, is associated with an improvement in lung function. In 1917 Peabody and Wentworth published a classical paper indicating that congestive heart failure impairs lung function.[9] When the lungs are normal, they provide a passageway for airborne environmental agents into the rest of the body and provide a window for the investigator to study secondary effects produced by disease in other organ systems in the body. Thus, measurements of lung function provide a noninvasive glimpse of the function of those organ systems. When the other organ systems begin to fail, abnormal lung function may be a sensitive, early indicator of disease. This possibility, like the one hypothesized by Cohen, is an intriguing one, but nonetheless an untested explanation for the consistently strong association between lung dysfunction and mortality.

HYPOTHESIS

Lung dysfunction is the necessary link between chronic airborne environmental exposures and mortality. The corollary hypothesis is that chronic airborne environmental expo-

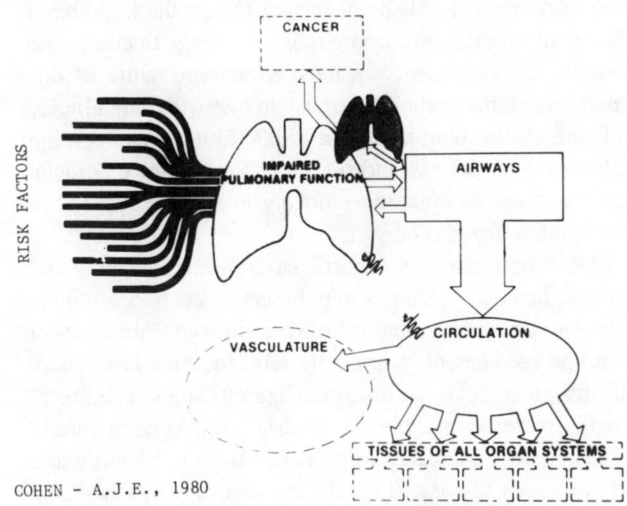

FIGURE 4. The lungs as a focus for clearing endogenous and exogenous "toxins" from the body. Normally, the lungs protect the body unless risk factors lead to impaired pulmonary function. In that case, abnormalities of the airways may contribute to progressive neoplastic and nonneoplastic disease of the lungs and to disease of other systems of the body which are linked to the lungs by the cardiovascular system. (From Cohen, B. H., *Am. J. Epidemiol.,* 112, 274, 1980. With permission.)

sures that produce lung dysfunction cause an increase in mortality from all causes.

This is not a new hypothesis. Indeed, every investigator who assesses lung function in an environmental or occupational setting does so with the hope that he defines relationships between exposures, morbidity, and mortality. If there is anything new in this paper, it is an attempt to look at mortality from nonpulmonary disease as well as from lung disease. In mortality studies, it is frequently difficult to distinguish between death caused by pulmonary and extrapulmonary disease. Furthermore, in some countries such as France, it is not possible to obtain information on the cause of death for individual subjects. If the hypothesis is true, the distinction is not important.

Why should the relationship between environmental exposures, pulmonary function, and mortality from nonpulmonary causes be emphasized when dealing with health and safety in agriculture?[2] Our experience with cigarette smoke has identified effects on lung function. Furthermore, according to the U.S. Surgeon General's Report — 1979, "the overall mortality ratio for all male current cigarette smokers irrespective of quantity, is about 1.7 (70 percent excess) compared to nonsmokers."[10] Indeed, there is no question that the vast majority of deaths caused by lung cancer, chronic bronchitis, and emphysema can be related to cigarette smoking. However, the Surgeon General's Report goes on to state that "Coronary heart disease is the chief contributor to the excess mortality among cigarette smokers." Thus, if investigators restricted their studies of cigarette smoking to lung function abnormalities and mortality related to lung disease, they would miss the major health effects of smoking. Whether this applies to airborne exposure, in general, will have to be determined. In order to do this, it will be necessary to test the hypothesis relating to chronic airborne exposures, lung function, and mortality. At first glance, a most attractive feature of this hypothesis is that it should be easy to test with data already available in the literature. But is it? Cautions concerning healthy-worker effects and questionable control of crucial variables such as cigarette smoking are frequently raised when studies are evaluated.

Table1 lists specific airborne environmental agents and relates exposure to changes in pulmonary function and mortality. Unfortunately, there is little certainty concerning most of the entries. The only entry in the table that should generate little disagreement is the first one. Cigarette smoking unquestionably increases pulmonary disease and has been clearly associated with increases in mortality from pulmonary and nonpulmonary disease. Other discussions shown in this table may be challenged; thus, definitive testing of the hypothesis awaits the acquisition of further data.

TABLE 1
Pulmonary Function and Mortality Related to Airborne Exposure

Chronic airborne environmental exposure	Pulmonary dysfunction	Mortality from pulmonary and nonpulmonary diseases
Cigarettes	Yes	Increased
Asbestos	Yes	Increased
Coal dust	Yes	Increased
CO	No ?	?
Cotton bract	Yes	?
Grain dust	Yes	?
Formaldehyde	?	?
Red cedar dust	Yes	?
TDI	Yes	?
Radon daughters	?	Increased
Arsenic	?	Increased
Benzene	?	Increased

IMPLICATIONS OF HYPOTHESIS — IF TRUE

The simple statement that airborne environmental exposure that causes pulmonary dysfunction increases mortality from lung disease is probably true. However, the hypothesis stating that pulmonary function is the focal link between chronic airborne toxic substances and mortality from pulmonary and nonpulmonary causes deserves consideration. It is certainly possible that some airborne toxins (e.g., CO, benzene) may increase mortality and have no measurable effects on lung function. Furthermore, it is possible for an airborne agent to produce pulmonary dysfunction, even moderately severe pulmonary dysfunction, and not lead to increased mortality. Thus, it is certainly reasonable to challenge the hypothesis at this time. However, if it proves to be "almost" correct, some important implications should be considered.

If tests of lung function measure general health effects of airborne environmental agents, then studies of lung function in epidemiological studies could provide three types of critical information: (l) they could indicate the safety or danger of exposure to specific environmental contaminants years before measurable increases in mortality are produced; (2) abnormalities in lung function may reflect specific individual exposures and/or susceptibility and improve assessments of individuals at increased risk; and (3) when mortality is shown to be increased in a group exposed to an airborne environmental agent, control measures directed at reducing the pulmonary consequences of exposure may prove useful and may provide early indication of successful intervention.

In summary, the hypothetical relationship between chronic airborne exposure, pulmonary function, and mortality is attractive because of its pathophysiological implications. However, the most important reason to prove or disprove this relationship rests in its potential to point toward methods for detecting and preventing disease.

REFERENCES

1. **Cohen, B.H., Ball, W.C., Jr., Brashears, S., Diamond, E., Kreiss, P., Levy, D., Menkes, H., Permutt, S., and Tockman, M.,** Risk factors in chronic obstructive pulmonary disease (COPD), *Am. J. Epidemiol.*, 105, 223, 1977.
2. **Beaty, T.H., Cohen, B.H., Newill, C.A., Menkes, H.A., Diamond, E.L., and Chen, C.H.,** Impaired pulmonary function as a risk factor for mortality, *Am. J. Epidemiol.*, 116, 102, 1982.

3. **Menkes, H., Cohen, B.H., Beaty, T.H., Newill, C.A., and Khoury, M.J.,** Risk factors, pulmonary function, and mortality, in *Genetics and Epidemiology of Coronary Heart Disease, Past, Present and Future,* Rao, Elston, Kuller, Feinlab, Carter, Havlik, Eds., Alan R. Liss, New York, 1984, 501.
4. **Ashley, F., Kannel, W.B., Sorlie, P.D., and Masson, R.,** Pulmonary function: relation to aging, cigarette habit, and mortality, *Ann. Intern. Med.,* 82, 739, 1975.
5. **Higgins, M.W. and Keller, J.B.,** Predictors of mortality in the adult population of Tecumseh, *Arch. Environ. Health,* 21, 418, 1970.
6. **Pride, N.B., Permutt, S., Riley, R.L., and Bromberger-Barnea, B.,** Determinants of maximum expiratory flow from the lungs, *J. Appl. Physiol.,* 23, 646, 1967.
7. **Menkes, H.A., Beaty, T.H., Cohen, B.H., and Weinmann, G.,** Nitrogen washout and mortality, *Am. Rev. Respir. Dis.,* 132, 115, 1985.
8. **Cohen, B.H.,** Chronic obstructive pulmonary disease: a challenge in genetic epidemiology, *Am. J. Epidemiol.,* 112, 274, 1980.
9. **Peabody, F.W. and Wentworth, J.,** Clinical studies of respiration. IV. The vital capacity of the lungs and its relationship to dyspnea, *Arch. Intern. Med.,* 20, 443, 1917.
10. Report of the Surgeon General — Smoking and Health, DHEW Publ. No. (PHS) 79-500066, U.S. Department of Health, Education and Welfare, U.S. Government Printing Office, Washington, D.C., 1979.

The Spirogram as the Basis for Lung Function Assessment in Epidemiological Studies

W. K. C. Morgan

INTRODUCTION

The forced expiratory volume maneuver may be recorded as time vs. volume or, alternatively, as volume vs. flow. Despite the enthusiasm of the manufacturers of medical equipment for the complexities and cost of the apparatus necessary for measuring flow-volume curves, in general, the simpler time vs. volume curve is to be preferred. There are a number of reasons for this, in particular, the relative inexpensiveness of the standard water-filled spirometer, its simplicity, and its robustness.

A large number of indices of ventilatory capacity can be derived from the forced expiratory volume maneuver whether measured as volume vs. time or flow vs. volume. These include the forced vital capacity (FVC), the forced expiratory volume in 1 s (FEV_1), the forced expiratory flow between 25 and 75% of the FVC (FEF_{25-75}), peak expiratory flow (PF) and expiratory flows at 25, 50, 75, and 90% of the FVC viz., FEF_{25}, FEF_{50}, FEF_{75}, and FEF_{90}.[1] Before applying or deciding to use any of these tests in an epidemiological study, the investigator needs to assess them according to criteria shown in Table 1. When this is done, the forced flows come out third best, being in the main far from reproducible,[2] overly sensitive, lacking in specificity, and requiring complex and often fickle equipment which may be difficult to calibrate.[3] Moreover, few generally accepted predicted values exist for forced expiratory flows, especially in older persons. Significant reductions in forced flows are frequently unrelated to the presence of respiratory symptoms, in particular, shortness of breath. Like closing volume and frequency dependence of dynamic compliance, forced flows do not accurately predict the development of disabling airways obstruction, and, although sensitive, frequently yield falsely positive results.[2,3] In contrast, the FVC and, in particular, the FEV_1 are acceptable to the subject being tested, and over 98% of subjects can, with suitable coaching, perform the tests in reproducible fashion, that is to say, with the two largest within 5% of the highest value. They are also good predictors of disability and if it comes to that, death.

Numerous long-term studies have been carried out using these indices and the mean annual decrement for the FEV_1 has been shown to range from 22 to 74 ml, depending on age and smoking habits.[4–9] The usual decline for the FEV_1, however, is between 22 and 32 ml/year, while for the FVC the decline is usually around 15 to 22 ml/year. Both of these indices decline in a linear fashion, at least over most of the adult life, although there is probably some speeding up after the age of 60.[9] In contrast, the FEF_{25-75} and the forced flows decline in an exponential fashion, and indeed, adequate data for the rate of decline of these indices are, for the most part, unavailable. For the above reasons, the FVC and FEV_1 are the preferred tests to be used in epidemiological surveys, although they may be supplemented by additional tests such as forced flows and by lung volumes where appropriate. Even so, the use of FVC and FEV_1 in long- term epidemiological studies is associated with a number of problems, the more important of which are described below.

EQUIPMENT AND PERSONNEL

A standard and acceptable spirometer that meets the NIOSH/HLBI criteria should be used.[10,11] A permanent record must be available. No change in equipment should occur once the study has started, especially in studies that rely on serial testing. Calibration must be carried out regularly and should be done not only for volume, but also for flows. The latter often presents a major problem. All personnel should be fully and adequately trained prior to the start of the study. Inadequate coaching and undue diffidence in encouraging the subjects to cooperate can lead to significant and persistent bias that is difficult, if not impossible, to detect and which leads to spurious results.

SUBJECT VARIATION

Although the decline in the FEV_1 of normal and diseased subjects is, for the most part, linear, this is not invariably the case, especially when the observations are made over a relatively short period, i.e., 2 to 3 years.[8,9] Clearly, obstruction and other diseases such as pneumonia, allergic alveolitis, and lung abscess can lead to a temporary excess decline in lung function. Often these correct themselves within 6 months to a year with an increase in lung function, but if the subject is

TABLE 1
Criteria for Tests

Acceptability — safety, discomfort
Simplicity — equipment and procedure
Objectivity — not influenced by subject cooperation or observer bias
Precision (reproducibility) measurement error
Equipment and observer
True biological variability
Accuracy or validity
Relation of quantity measured to what one wishes to know
Sensitivity
Specificity
Predictive value

tested shortly after an acute illness and then again later when he has recovered, spurious increments in lung function may be found. Even in the absence of acute disease, the rate of decline in subjects may vary greatly. Figure 1 shows the serial measurements of FEV_1 over a 4–year period in a relatively healthy heavy cigarette smoker with no overt disease.[1] Although such variation is not uncommon in any individual, it is rare to see such variation present in a large group of subjects. Exceptionally, significant increments or decrements may occur in the majority of the cohorts under study.[8,12] These may be related to the time of year in which the tests are performed, to a variety of environmental factors including severe air pollution, to influenza epidemics, and to other such episodes. Short-term changes in the FEV_1, viz., over 2 to 3 years, therefore tend to be unreliable whether in the individual or in a group of subjects. Occasionally, increments are observed to occur in the population and these are, for the most part, due to a learning effect or due to a change in the technician who may be more experienced or enthusiastic and has encouraged the subjects to greater efforts.

THE INTERPRETATION OF RESULTS

It is customary in longitudinal studies to try and assess the contributions of various factors to the observed decline in the FEV_1 and FVC. Thus, by a series of statistical manipulations often far removed from common- sense biology, the effects of cigarette smoking, dust, age, height, air pollution, and other factors are quantitated, and pronouncements later appear stating, for example, that cigarette smoking has three times the effect of coal dust.[13,14] While true for the whole population, in any specific individual such a statement is often misleading and usually completely incorrect.

Thus, studies of Fletcher et al. have shown that only 12 to 13% of cigarette smokers suffer from airways obstruction.[8] The vast majority are unaffected despite relatively long-term smoking. While it is true that the heavier the smoker, the greater the risk of the subject developing airways obstruction,

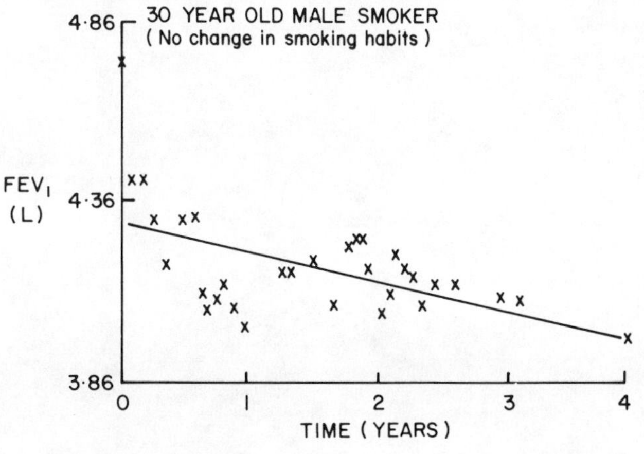

FIGURE 1. Serial estimations of the FEV_1 over a 4–year period in a heavy cigarette smoker.

there is no doubt that many heavy smokers go through their lives without serious effects on their lungs. This situation would suggest that the mean effects of smoking on the rate of decrement in the FEV_1 are so slight as to suggest that the effects of smoking would be of little consequence over the lifetime. This is erroneous in that while the mean effects of smoking on lung function are small because most smokers are unaffected and do not develop significant obstruction, a minority progress to severe obstruction which eventually disables them and, in some instances, leads to premature death. Thus, it is clear that there is a susceptible minority of subjects who are prone to develop impairment and disability related to cigarette smoking–induced airways obstruction. Nevertheless, it needs to be emphasized that susceptibility exists. Thus, were the whole population to smoke 20 cigarettes a day for 30 years, the resultant loss in ventilatory capacity would not have a normal distribution nor would it be linear, but the distribution would be markedly skewed with many showing no or relatively minor losses, but with a small group showing a clinically significant decline. Fletcher et al. have pointed out that such a situation is confirmed by the skewed distribution of the amount of emphysema measured at autopsy among 106 smokers studied by Ferris and colleagues.[5] Thus, spurious comparisons can easily be made. Where exposure to high concentrations of coal dust to induce a small decrement in ventilatory capacity in the majority of persons exposed, i.e., around 60 to 70% as the evidence suggests,[16] it would be possible to compare the mean decrement of the dust-exposed group to the mean decrement of a comparable group of nondust-exposed smokers of whom only 12 to 15% are affected, and to come up with a roughly comparable average decline in the FEV_1. A far more accurate gauge of the likelihood of respiration disability arising from dust exposure and cigarette smoking can come about by comparing the prevalence of severe and disabling impairment in nonsmoking men exposed to dust with the prevalence of similar type impairments in cigarette smokers. Moreover, other problems exist in regard to longitudinal studies carried out in such populations in that it is quite clear that those who leave the working population, in general, are smokers. An analysis of the smoking habits of most study populations shows that somewhere around 60% of those below the age of 30 are cigarette smokers, but in the 50 to 60 age group, the percentage of smokers has declined to around 30 to 35%.[16,17] This would indicate that the smokers drop out because of ill health — in short, because of the development of emphysema, lung cancer, or coronary artery disease — and that those who are not smokers are less likely to leave the work force. Furthermore, those smokers who remain in the work force are almost certainly resistant to the effects of cigarette smoking, and represent a survivor population.

Despite these drawbacks, many of which are unrecognized, there is little doubt that serial measurements of the FVC and FEV_1 can tell us much about the natural history of the disease, thereby enabling suitable measures to be taken that will permit early detection of disease, and, hence, allow the subject to be removed from further exposure. Even so, it is clear that in a subject who is a cigarette smoker, the confounding effects of this habit seriously limit the usefulness of serial measurements of these indices of ventilatory capacity.

REFERENCES

1. **Morgan, W.K.C.,** Epidemiology and occupation lung disease, in: *Occupational Lung Diseases,* 2nd ed., Morgan, W.K.C. and Seaton, A., Eds., W.B. Saunders, Philadelphia, 1984, chap. 5.
2. **McCarthy, D.S., Craig, D.B., and Cherniack, R.M.,** Intraindividual variablility in the maximal expiratory flow volume and closing volume in asymptomatic subjects, *Am. Rev. Respir. Dis.,* 112, 407, 1975.
3. **Morgan, W.K.C.,** Pulmonary disability and impairment: can't work? Won't work?, *Basics Respir. Dis,* 10, 1, 1982.
4. **Ashley, F., Kannel, W.B., Sorlie, P.D., et al.,** Pulmonary function: relation to aging, cigarette habit and mortality, *Ann. Intern. Med.,* 82, 739, 1975.
5. **Ferris, B.G., Higgins, I.T.T., Higgins, M.W., et al.,** Chronic nonspecific respiratory disease in Berlin, New Hampshire, 1961—1967. A follow-up study, *Am. Rev. Respir. Dis.,* 107, 110, 1973.
6. **Howard, P.,** A long-term follow-up of respiratory symptoms and ventilatory function in a group of working men, *Br. J. Ind. Med.,* 27, 326, 1970.
7. **Higgins, I.T.T. and Oldham, P.D.,** Ventilatory capacity in miners, A five-year follow-up study, *Br. J. Ind. Med.,* 19, 65, 1962.
8. **Fletcher, C., Peto, R., Tinker, C., et al.,** *The Natural History of Chronic Bronchitis and Emphysema,* Oxford University Press, New York, 1976.
9. **Bosse, R., Sparrow, D., Garvey, A.J., et al.,** Cigarette smoking, aging and decline in pulmonary function: a longitudinal study, *Arch. Environ. Health,* 35, 247, 1980.
10. **Gardner, R.M., Hankinson, J.L., and West, B.J.,** Evaluating commercially available spirometers, *Am. Rev. Respir. Dis.,* 121, 73, 1980.
11. **A.T.S. Statement,** Snowbird workshop in Standardization of Spirometry, *Am. Rev. Respir. Dis.,* 119, 831, 1979.
12. **Diem, J.E., Jones, R.V., Hendrick, D.J., et al.,** Five year longitudinal study of workers employed in a new toluene disocyante manufacturing plant, *Am. Rev. Respir. Dis.,* 126, 420, 1982.
13. **Love, R.G. and Miller, B.G.,** Longitudinal study of lung function in coal-miners, *Thorax,* 37, 193, 1982.
14. **Attfield, M.D.,** Longitudinal decline in FEV_1 in United States coalminers, *Thorax,* 40, 132, 1985.
15. **Ryder, R.C., Dunnill, M.S., and Anderson, J.A.,** A quantitative study of bronchial mucous gland volume, emphysema and smoking in a necropsy population, *J. Pathol. Bacteriol.,* 104, 59, 1971.
16. **Kibelstis, J.A., Morgan, E.J., Reger, R.B., et al.,** Prevalence of bronchitis and airway obstruction in American bituminous coal miners, *Am. Rev. Respir. Dis.,* 108, 886, 1973.
17. **Lainhart, W.S., Doyle, H.N., Enterline, P.E. et al.,** Pneumoconiosis in Appalachian bituminous coal miners, U.S. Department of Health, Education and Welfare, U.S. Government Printing Office, Washington, D.C., 1969.

Factors Influencing Estimates of Annual Change in FEV_1

H. W. Glindmeyer, J. E. Diem, and H. Weill

INTRODUCTION

Annual change in lung function can be estimated by cross-sectional or longitudinal analyses. Since these two methods of estimating annual change are influenced by a different set of factors, which include cohort effects and the precision with which the linear regressions fit the data, the investigator planning to evaluate a population must optimize the study design by independently evaluating, for each factor, their overall effect on the estimate. In order to provide a guide to optimizing a study design, the influence of cohort effects on the accuracy of cross-sectional estimates of annual change and the factors which influence the precision of longitudinal estimates of annual change will be quantitatively evaluated.

COHORT EFFECTS

Since cross-sectional and longitudinal analyses are sensitive to past and current factors which influence lung function, respectively, they can, when applied to the same data base, produce estimates of annual change which may not be consistent with each other, but, in both cases, are consistent with the temporal factors that influence the data. A population whose respiratory experience over time has not changed can produce cross-sectional and longitudinal slopes which agree, but if the current experience is different from the past, either estimate can be greater or less than the other.

The apparent inconsistency of slopes obtained when longitudinal or cross-sectional analyses are applied to the same data base can be understood by examining the results of a hypothetical population with data collected in 1975 and 1985 (Figure 1). This population consists of subjects A, B, and C who are 35, 40, and 45 years of age in 1975, respectively. All three subjects are of the same sex, race, and height, and all began smoking at age 25 and quit smoking in 1975. In 1975 the older subjects present a more depressed forced expired volume in 1 s (FEV_1) due to an increased length of smoking.

An analysis of these data indicates that cross-sectionally the age regression determined from both the 1975 and 1985 data produces the same abnormal slope (–50 ml/year); however, longitudinally, the data collected in 1975 and 1985 provide a more normal slope (-10 ml/year) due to the removal of the smoking effect. Therefore, the 1975 and 1985 cross-sectional data, influenced by the past effects of smoking, provide identical estimates of annual change that overestimate the observed annual change determined longitudinally.

Figure 2 presents the results of another hypothetical population, of the same age range presented in Figure 1, with tests conducted in 1975 and 1985. This population consisted of nonsmokers who began smoking after the 1975 test session and continued to smoke until the 1985 test session. The 1975 and 1985 cross-sectional analyses provide identical estimates of annual change (–10 ml/year) that underestimate the observed annual change determined longitudinally (–50 ml/year).

These two examples indicate that cross-sectional and longitudinal analyses can provide a substantially different view of estimated annual change. Longitudinal analyses are sensitive to current effects and can be insensitive to past effects, whereas cross-sectional analyses are sensitive to past effects and can be insensitive to current effects.

FACTORS INFLUENCING SLOPE PRECISION

Since cross-sectional analyses are sensitive to cohort effects, only longitudinal analyses can provide an estimate of current annual change in lung function. This longitudinal

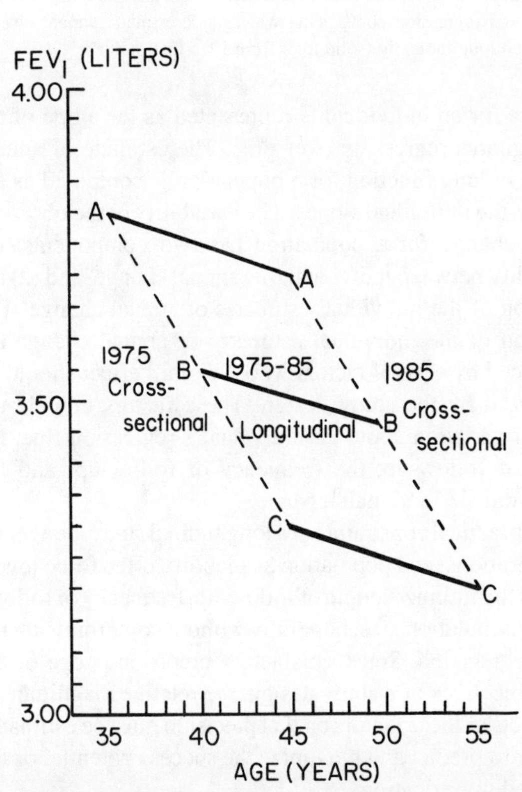

FIGURE 1. Cross-sectional regressions (dashed lines) of 1975 and 1985 FEV_1 data from a hypothetical population that quit smoking in 1975 produce age regression coefficients which overestimate annual change determined longitudinally (solid line) from 1975 to 1985.

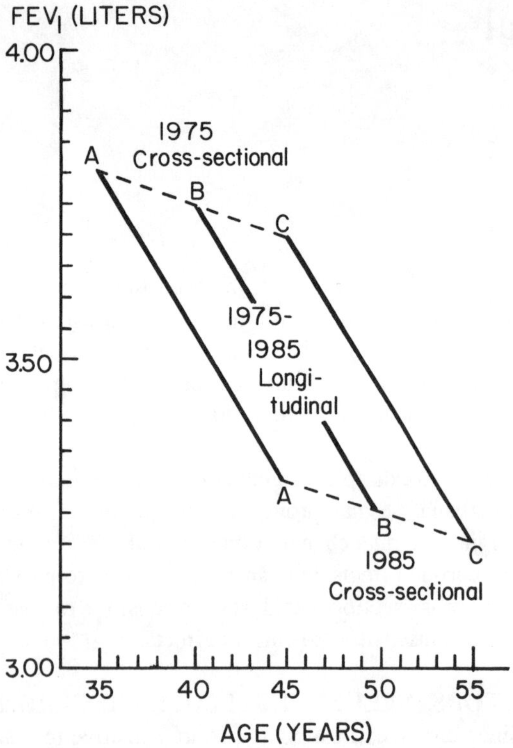

FIGURE 2. Cross-sectional regressions (dashed lines) of 1975 and 1985 FEV_1 data from a hypothetical population that began smoking in 1975 produce age regression coefficients which underestimate annual change determined longitudinally (solid line) from 1975 to 1985.

estimate for an individual is represented as the slope of the least–squares regression over time. The estimate of annual change in lung function for a population is computed as the mean of the individual slopes. The variability of the observed annual change for a population has two components: (1) variability between individual true annual change, and (2) the precision of the individual estimates of annual change. The precision of the individual estimates of annual change are influenced by several factors whose overall effect should be minimized by the study design. These factors include the residual variation about an individual's regression line, the length of follow-up, the frequency of follow-up, and the population size and maintenance.

The practical constraints of a longitudinal study, budgetary considerations, and population availability often force investigators to minimize length of follow-up, frequency of follow-up, and population size, hopefully without compromising the study's precision. Since satisfactory precision is one of the basic objectives of a study design, the relative magnitude of the effects of these factors on that precision must be estimated in order to predict and maximize the success potential of the proposed investigation.

To evaluate the magnitude of these contributing factors and provide an ordering of their importance, we evaluated their relative effect on a hypothetical population by applying a statistical model which assumes that the observed individual slope is equal to the true slope plus an error term. Therefore, the observed slope variance is equal to the true slope variance, expressed as $BISSD^2$, where BISSD equals the between-individual true slope standard deviation, plus the variance of the error term, expressed as $RVSD^2$/–Sum (ti – tbar)^2, where RVSD equals the standard deviation of the residual variation about an individual's regression line, ti equals the observation times, and tbar equals the mean time of observations. For a population, the observed mean slope variance (MSV) equals the observed individual slope variance divided by n, the population size. To evaluate the relative magnitude of these factors, each is independently varied throughout an expected range, and the effect on the MSV is computed.

The effect of RVSD on MSV was evaluated for a popuation of 100 subjects with seven annual measurements collected over a 6–year study and BISSD = 25 ml/year. The results (see Figure 3) indicate an approximate doubling of MSV with a doubling of RVSD. Values of RVSD in milliliters and the corresponding MSV in milliliters per year squared were 100 and 9.82, 125 and 11.83, 150 and 14.30, 175 and 17.19, 200 and 20.50, 225 and 24.33, and 250 and 28.57, respectively.

Since the true variance of the estimated mean slope is equal to 6.25 (ml/year)^2, $BISSD^2$/n, the percentage of variability in estimated mean slope due to variation between true slopes is 63.6, 52.8, 48.7, 36.3, 30.5, 25.7, and 21.9% for RVSD values of 100, 125, 150, 175, 200, 225, and 250 ml, respectively. Therefore, even for a 6–year study with annual measurements and BISSD = 25 ml/year, more than half the MSV is due to noise once the RVSD exceeds 150 ml.

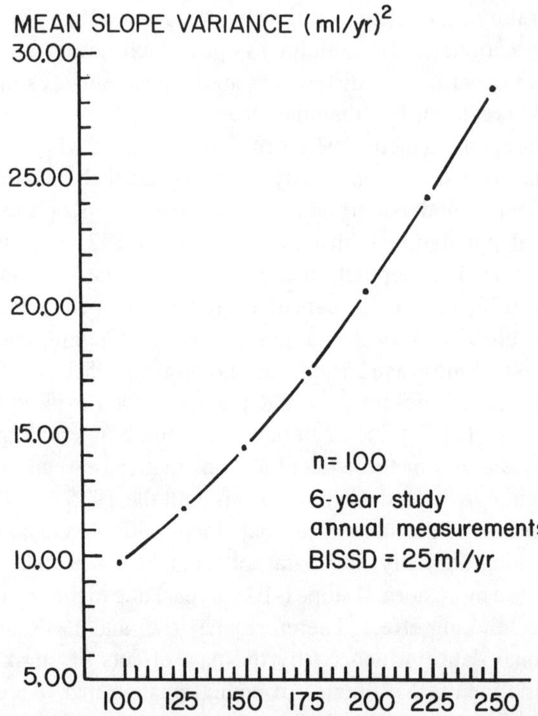

FIGURE 3. MSV more than doubles with increasing values of residual variation about individual regression lines.

The effect of follow-up in years on MSV is presented in Figure 4 for a population of 100 subjects with annual measurements, a BISSD of 25 ml/year, and RVSD values of 100 and 200 ml. Even for a relatively low value of RVSD (100 ml), the MSV for a 2–year follow–up is 56.25 (ml/year)^2; therefore, for a BISSD of 25 ml/year, the true variance is 6.25 (ml/year)^2, indicating that only 11% of total variance is true variance, with the remaining 89% due to noise. A dramatic reduction in variance occurs with increasing length of follow-up, reaching, for RVSD = 100 ml, 9.82 (ml/year)^2 or 64% true variance after 6 years. Continuing the follow-up past 6 years further reduces the variance; however, under actual circumstances, this improvement would probably be offset by population attrition.

Since MSV more than doubles with RVSD (see Figure 4), the length of follow-up must be proportionally increased if precision of mean slope is to be maintained with increasing RVSD. The length of follow-up required to assure that MSV includes at least 50% true variance for RVSD values of 100, 150, and 200 ml is 5, 7, and 9 years, respectively. However, due to population attrition, these estimates of length of follow-up underestimate those which would be required under actual conditions.

The effect of frequency of tests on MSV was evaluated for a population of 100 subjects with BISSD equal to 25 ml/year and RVSD equal to 100 ml. A 12–year study period was considered, since it allowed for evaluating testing at 1, 2, 3, 4, and 6 years. The results (Figure 5) indicate that variance is not very dependent on frequency of follow-up, since increasing the frequency from testing every 6 years to annual testing only reduces the variance from 7.64 to 6.80 (ml/year)^2. Therefore, for 3 or 13 tests conducted over 12 years, the percent true variance is equally high at 81.8 and 91.9, respectively. Though test frequency does not significantly influence variance, frequent tests are necessary to detect nonlinear trends, to maximize participation in light of population attrition, and reduce survivor bias.

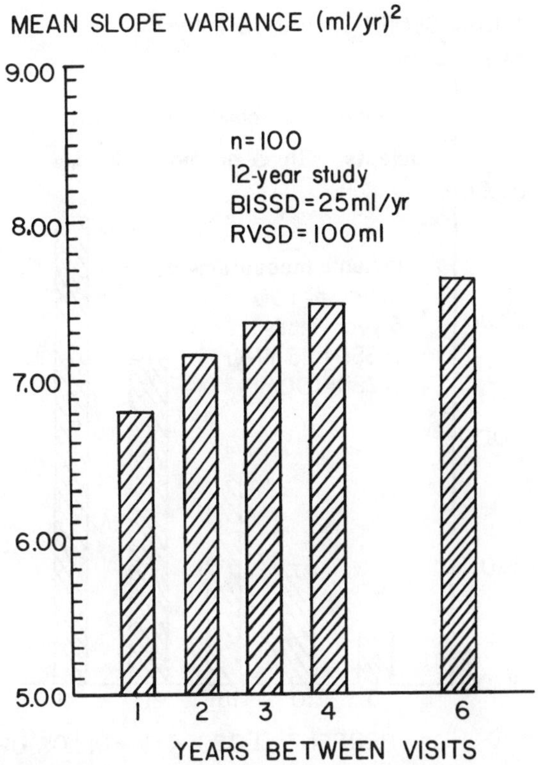

FIGURE 5. MSV is only slightly affected by test frequency, but frequent tests are necessary to detect nonlinear trends, maximize participation, and reduce survivor bias.

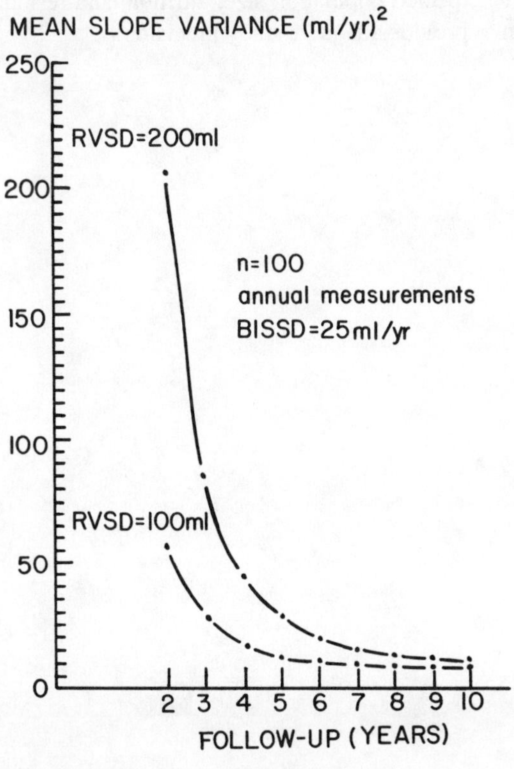

FIGURE 4. Significant reductions in MSV occur with increasing length of follow-up. Doubling RVSD from 100 to 200 ml requires an equal doubling in length of follow-up if precision of MSV is to be maintained.

To evaluate the effect of population attrition on MSV, MSV was calculated as the variance of the precision weighted average of slopes for subjects with three or more tests, and is presented (see Figure 6) as a function of the percent of the initial population lost to follow-up each year. Given a 6–year study with annual tests and an initial population of 100 (RVSD = 100 ml and BISSD = 25 ml/year), the number of subjects with at least three tests at the end of the study would be 90, 81, 72, 64, and 56, for annual attrition rates of 5, 10, 15, 20, and 25%, respectively. The corresponding MSVs are 9.82, 14.85, 18.52, 23.07, and 28.93 (ml/year)^2. Since the true MSV for 100% participation is equal to 6.25 (ml/year)^2, the precision drops below 50% when loss to follow-up exceeds 5% per year.

DISCUSSION

Annual change in FEV_1 can be estimated by cross-sectional or longitudinal analyses. Cohort effects can cause estimates of annual change from cross-sectional analyses (age

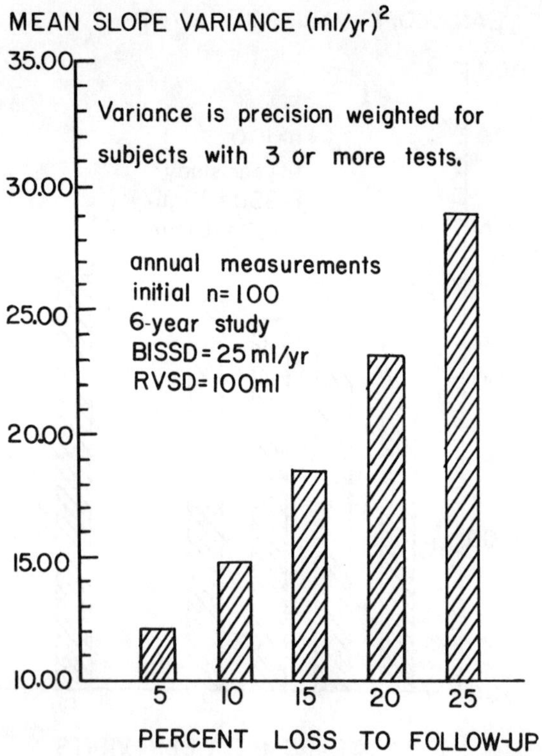

FIGURE 6. Population attrition significantly affects the precision of MSV due to overall reduction in length of follow-up.

regression coefficients) to be smaller or larger than actual longitudinal change, since cross-sectional studies are sensitive to past influencing factors, but are insensitive to current ones. Only when the past and current respiratory experience of a population has not changed would cross-sectional analyses properly estimate longitudinal course.

Longitudinal studies are necessary to determine annual change if the respiratory experience has varied. To maximize the precision with which longitudinal studies estimate annual change, the factors affecting the precision must be individually quantified in order to reduce the overall adverse effect on the estimate. The preceding analyses indicate that the length of follow-up is the most important determinant of precision. Population attrition affects precision by reducing average length of follow-up for the population. Frequency of follow-up does not signficantly affect precision, but frequent tests are necessary to detect nonlinear trends, to maximize participation, and to reduce survivor bias. Residual variation about individual slopes becomes less critical with increasing length of follow-up, but reductions in this variation are important if significant results are to be obtained with a 5– or 6–year study design.

Therefore, accurate estimates of annual change in lung function require a longitudinal follow-up with (1) a commitment to measurement standardization in order to minimize residual variation, (2) a commitment to population maintenance to obtain full follow-up on the maximum number of participants, (3) sufficiently frequent measurements to validate the application of linear statistical models and reduce survivor bias, and (4) sufficient length of follow-up in respect to the expected population size, attrition, and residual variation to provide adequate study precision.

Assessment of Airway Responsiveness in Working Populations

D. A. Enarson, S. Vedal, and M. Chan-Yeung

INTRODUCTION

The utility of bronchoprovocation testing with methacholine or histamine for the diagnosis of asthma has been well demonstrated.[1,2] Indeed, the presence of airway hyperresponsiveness to these substances is one of the hallmarks of asthma.[3] The identification of asthma in groups of workers exposed to sensitizing agents in their workplace is not straightforward. The standard techniques of respiratory questionnaire and spirometry have not been entirely satisfactory. The addition of the methacholine or histamine test to these standard methods would seem to be logical where workers are exposed to sensitizing agents.

In order to determine whether bronchoprovocation with methacholine identifies morbidity not already measured by the standard techniques of questionnaire and spirometry in prevalence studies of respiratory health in the occupational setting, we have added this test in our study of several working groups, some of which are exposed to known sensitizing agents.

STUDIES IN THREE GROUPS OF WORKERS

Three groups of workers were well characterized with standardized respiratory questionnaire,[4] prick skin tests with *D. farinae,* cat, histamine, and saline control, standardized spirometry[5] compared to normal predicted values,[6] and a standardized methacholine inhalation test.[7]

We studied 218 office workers, 646 grain elevator workers, and 511 sawmill workers comprising 82, 95, and 93% of eligible subjects, respectively. A sizable minority (390, 26%) were excluded from the test because of history of angina, hypertension, heart disease, epilepsy, or forced expired volume in 1 s (FEV_1) <1.5 l. They were older, were more likely to be ex-smokers, and had a lower level of FEV_1 (Table 1).

The two groups with exposure to known respiratory hazards were significantly more likely to report respiratory symptoms (Table 2). The excess was greatest among nonsmokers. Excess of productive cough was noted among grainhandlers and breathlessness was significantly more common in both exposed groups as compared with the reference group. The exposed groups also demonstrated significantly lower spirometric values compared to the reference group (Table 3), and the deficit was of a larger magnitude than that attributed to cigarette smoking.

In each of the studied groups the prevalence of airway hyperresponsiveness both at 8 and 2 mg/ml was higher in those with respiratory symptoms (Table 4). It varied inversely with the level of FEV_1 in each of the groups (Table 5). Prevalence of airway hyperresponsiveness was also highest in those with positive allergy skin tests in each of the three groups. Among those with negative allergy skin tests, the exposed groups had a higher prevalence of airway hyperresponsiveness (Table 6). Among those with positive allergy skin tests, there was no difference in prevalence of airway hyperresponsiveness among the three occupational groups.

Figure 1 demonstrates the prevalence of airway hyperresponsiveness in all the subjects of the populations studied, taking into account those factors demonstrated to be associated with the prevalence of airway hyperresponsiveness. Statistical analysis was undertaken using multiple logistic regression analysis, employing various models. Significant associations in all models were found with FEV_1 level and presence of respiratory symptoms. Positive allergy skin tests were significantly associated with the airway hyperresponsiveness among office and grain elevator workers, but not among sawmill workers. Neither smoking habit nor age was associated with airway hyperresponsiveness in any of the models employed.

In comparing the prevalence of airway hyperresponsiveness among the three working groups, we found no difference in the prevalence of airway hyperresponsiveness at 2 mg/ml, regardless of the model employed. In comparing the prevalence of airway hyperresponsiveness at 8 mg/ml, we found no difference among the groups when we included in the model only those variables known to be significant, including FEV_1, presence of symptoms, and allergy skin test results. However, when we introduced an interaction factor to account for the varying relationships between allergy skin tests and airway hyperresponsiveness, we found a significant interaction in the sawmill group (but not in grain elevator workers) with the allergy skin test results in the association between this group and the prevalence of airway hyperresponsiveness at 8 mg/ml. Moreover, we found a significant positive association between airway hyperresponsiveness at 8 mg/ml and duration of employment among grain elevator workers only.

RELEVANCE

Bronchoprovocation testing with methacholine or histamine is time-consuming and technically demanding, although it appears to be safe in epidemiologic surveys. To justify its use in such surveys, the test must be seen to provide information not supplied by the more routine tests such as a questionnaire and spirometry.

One such piece of information which might be usefully provided by the test would be a significant difference in the

TABLE 1
Characteristics of Participants Having or Being Excluded From the Methacholine Test in Three Working Groups

Group:	Office		Grain elevator		Sawmill	
Methacholine test:	No	Yes	No	Yes	No	Yes
Number	84	134	142	504	135	376
Age (years)	46 ± 11	41 ± 11[a]	44 ± 13	37 ± 13[a]	52 ± 11	42 ± 14[a]
Smoking (%)						
Ever	68	59	83	72[a]	73	67[a]
Now	32	26	49	44[a]	35	39[a]
Immediate skin reactivity (%)	21	37[a]	21	16	13	20
FEV_1 (ml)	3995 ± 746	4178 ± 710	3773 ± 975	3896 ± 744[a]	3649 ± 814	3798 ± 796[a]

Note: Values are mean ± SD; FEV_1 measurements are standardized for age, height, and smoking.

[a] $p \leq 0.05$.

TABLE 2
Risk (Rate Ratio) of Respiratory Symptoms in Exposed Groups as Compared with a Reference Group (Office Workers)

	Nonsmokers		Ex–smokers		Current smokers	
Symptom	Grain elevator n = 143	Sawmill n = 126	Grain elevator n = 141	Sawmill n = 102	Grain elevator n = 220	Sawmill n = 148
Any	2.08[a]	2.04[a]	1.94[a]	1.72[a]	1.33[a]	1.33[a]
Cough	1.47	1.67	1.94	3.52	1.33	0.95
Phlegm	3.05[a]	1.47	2.84[a]	1.17	1.20	1.10
Wheeze	1.86	1.56	1.33	1.09	1.09	1.16
Dyspnea	1.62	3.17[a]	3.25[a]	2.37[a]	1.40	1.65

[a] $p < 0.05$ from chi–square.

TABLE 3
Residual of Lung Function Values Associated with Statistically Significant Related Variables from Multiple Regression Analysis

Independent variable	FEV_1 (ml)	FVC (ml)	$FEF_{25-75\%}$ (ml/s)
Age	−33 ± 1	−26 ± 2	−56 ± 3
Height	43 ± 2	64 ± 3	16 ± 5
Current smoking	−168 ± 37	−118 ± 45	−342 ± 81
Grainhandlers	−300 ± 43	− 274 ± 52	−384 ± 94
Sawmill workers	−213 ± 44	−205 ± 53	−257 ± 96

prevalence of airway hyperresponsiveness associated with specific exposures, unexplained by differences in spirometry or prevalence of respiratory symptoms. It is well known, for example, that respiratory morbidity may be present which is not identified by a respiratory symptoms questionnaire, but can be demonstrated by spirometry. It is conceivable, then, that some workers without symptoms and with normal spirometry may have airway hyperresponsiveness. This is obviously true, as even a small proportion (4%) of those without symptoms and with "supernormal" (110+%) FEV_1 had bronchial hyperreactivity at 8 mg/ml (although none at 2 mg/ml). Nevertheless, this did not differ among the groups, so cannot be thought to represent occupational lung "disease".

A second possible piece of information which would support the use of the test would be the demonstration of an association between airway hyperresponsiveness and respiratory "illness", measured by the conventional tools, associated with specific exposure. Such an association would indicate the "type" of illness being measured and, if distributed differently among the groups, would suggest a specific type of illness associated with a specific exposure. In the case of the methacholine test, it would not be unreasonable to expect illness in groups exposed to sensitizing agents such as plicatic acid (the active agent in Western Red Cedar dust) to display airway hyperresponsiveness. These types of useful information could be obtained in a prevalence survey.

A third piece of information, which cannot be provided by a prevalence survey, but which would indicate the usefulness

TABLE 4
Prevalence of Airway Hyperresponsiveness at 8 and 2 mg/ml of Methacholine According to the Presence or Absence of Respiratory Symptons

	Office workers		Grain workers		Sawmill workers	
Symptoms	**8 mg/ml**	**2 mg/ml**	**8 mg/ml**	**2 mg/ml**	**8 mg/ml**	**2 mg/ml**
None	15 (9.3)	0 (0.0)	25 (10.7)	8 (3.4)	21 (10.6)	3 (1.5)
Any	13 (22.4)	5 (8.6)	52 (19.3)	18 (6.7)	53 (29.8)	16 (9.0)

Note: Percent is in parentheses.

TABLE 5
Prevalence of Airway Hyperresponsiveness at 8 and 2 mg/ml of Methacholine According to Level of FEV_1

	Office workers		Grain workers		Sawmill workers	
Level of FEV_1 (% predicted)	**8 mg/ml**	**2 mg/ml**	**8 mg/ml**	**2 mg/ml**	**8 mg/ml**	**2 mg/ml**
110+	5 (4.3)	0 (0.0)	5 (4.5)	0 (0.0)	8 (7.6)	0 (0.0)
90—109	14 (12.7)	4 (3.6)	34 (12.0)	5 (1.8)	42 (20.3)	5 (2.4)
<90	7 (63.6)	6 (9.1)	40 (36.7)	21 (19.3)	27 (42.2)	14 (21.9)

Note: Percent is in parentheses.

TABLE 6
Prevalence of Airway Hyperresponsiveness at 8 and 2 mg/ml of Methacholine According to Results of Allergy Skin Tests

	Office workers		Grain workers		Sawmill workers	
Allergy skin tests	**8 mg/ml**	**2 mg/ml**	**8 mg/ml**	**2 mg/ml**	**8 mg/ml**	**2 mg/ml**
Negative	9 (6.3)	1 (0.7)	63 (14.2)	19 (4.3)	48 (16.0)	14 (4.7)
Positive	16 (22.9)	4 (5.7)	17 (21.5)	7 (8.9)	15 (19.7)	5 (6.6)

Note: Percent is in parentheses.

of the methacholine test, would be the demonstration of airway hyperresponsiveness as a risk factor in the development of occupational lung disease. Evidence is available indicating that airway hyperresponsiveness may be associated with the development of chronic airflow obstruction.[8] No information is yet available to implicate airway hyperresponsiveness as a risk factor for occupatinal lung disease over and above its role in the development of airflow obstruction in a group of patients. Such information can only be provided by a prospective study.

What evidence is available from the current study regarding the utility of the measurement of airway responsiveness in prevalence studies of occupational lung disease? We have demonstrated an excess prevalence of airway hyperresponsiveness in sawmill workers exposed to Western Red Cedar dust which is over and above the excess of respiratory morbidity identified by spirometry and respiratory symptoms questionnaire. This excess, however, is present only in those with negative allergy skin tests and is only in the intermediate range of airway hyperresponsiveness (8 mg/ml) often associated with bronchitis or hay fever,[7] and not at the lower level (2 mg/ml) commonly seen in asthma patients. The clinical significance of this observation is not clear. The same is true for the association of airway hyperresponsiveness associated with duration of employment in grain elevator workers. The two observations, nevertheless, would indicate a specific occupational aspect of airway hyperresponsiveness, in addition to the general association of airway hyperresponsiveness with respiratory morbidity.

On the other hand, we did not demonstate an association between airway hyperresponsiveness and respiratory morbidity over and above its association in the unexposed reference group. Thus, the increment in prevalence of airway hyperresponsiveness associated with respiratory symptoms and/or airflow obstruction was not different among the groups of workers. This would suggest that the type of respiratory illness identified by questionnaire and spirometry is the same with respect to airway responsiveness in all groups.

It must be pointed out, however, that a significant minority of participants were excluded from the test and our inability to

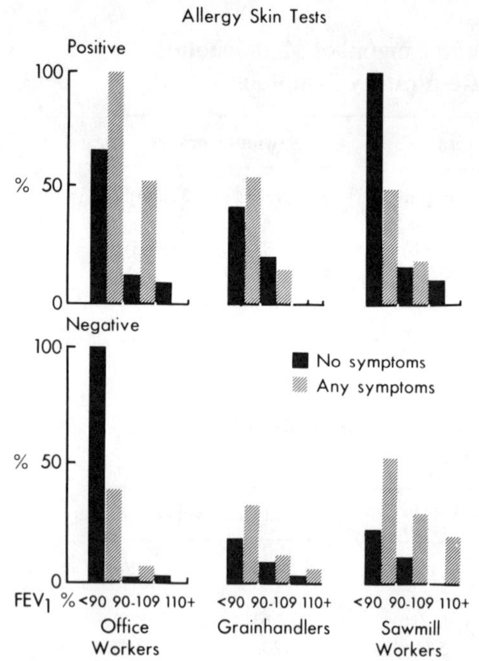

FIGURE 1. Prevalence of airway hyperresponsiveness at 8 mg/ml of methacholine according to level of FEV_1, presence of respiratory symptoms, and results of allergy skin tests in working populations.

demonstrate clear differences between the groups might have been affected by these exclusions. Nevertheless, the proportion and characteristics of excluded individuals were very similar among the groups, suggesting that comparisons between groups would be likely unaffected.

The determination of the role of airway hyperresponsiveness as a predictor of occupational lung disease is probably the most important in defining the role of this test in the occupational setting. This determination awaits the results of prospective, controlled studies of occupational groups.

REFERENCES

1. **Boushey, H., Holtzman, M., Sheller, J., and Nadel, J.,** Bronchial hyperreactivity, *Am. Rev. Respir. Dis.*, 121, 389, 1980.
2. **Hargreave, F.E., Juniper, E.F., Ryan, G., Dolovich, M., Cartier, A., Frith, P.A., Newhouse, M.T., and Hargreave, F.E., Eds.,** Clinical significance of nonspecific airway hyperreactivity, in *Airway Reactivity*, Astra Pharmaceuticals Canada Ltd., Mississauga, 1980, 216.
3. American Thoracic Society, Definition and clarification of chronic bronchitis, asthma, and pulmonary emphysema, *Am. Rev. Respir. Dis.*, 85, 762, 1962.
4. American Thoracic Society, Recommended respiratory disease questionnaire for use with adults and children in epidemiological research, *Am. Rev. Respir. Dis.*, 118 (Suppl.), 7, 1978.
5. American Thoracic Society, ATS Statement, Snowbird workshop on standardization of spirometry, *Am. Rev. Respir. Dis.*, 119, 831, 1979.
6. **Morris, J.F.,** Spirometry in the evaluation of pulmonary function, *West. J. Med.*, 125, 110, 1976.
7. **Lam, S., Wong, R., and Yeung, M.,** Nonspecific bronchial reactivity in occupational asthma, *J. Allergy Clin. Immunol.*, 63, 28, 1979.
8. **Minette, A., Marcg, M., and Gepts, L.,** Prognostic value of a positive acetylcholine test regarding VC and FEV_1 in coal-miners with a history of chronic bronchitis, *Bull. Eur. Physiopathol. Respir.*, 14, 167, 1978.

Carbon Monoxide Diffusing Capacity Standardization

R. O. Crapo

INTRODUCTION

Assessment of pulmonary gas exchange using the single-breath carbon monoxide diffusing capacity (DL_{CO} or DL_{COSB}) or transfer factor (TL_{CO}) is becoming more widely accepted in both clinical and industrial settings. Make et al. have outlined areas where DL_{CO} measurements may augment the information gained from spirometry and a few settings where they may be superior.[1] The DL_{CO} may be most useful in the setting where patients have, or are at risk of developing, diffuse infiltrative disease. In a retrospective study, Epler et al.[2] have shown that the DL_{CO} may be more sensitive than forced vital capacity (FVC) and total lung capacity (TLC) in identifying individuals with dyspnea, a normal chest radiograph, and biopsy-documented interstitial lung disease. Prospective information will be required to show the predictive value of a low DL_{CO} in the setting of a dyspneic patient with a normal chest radiograph. Epler et al. have also shown[3,4] that DL_{CO} correlated better than other spirometric and lung volume parameters with histologic abnormalities of asbestosis and provided one of the better indices of impairment and disability in individuals with interstitial lung diseases. While still controversial, there is at least a suggestion that DL_{CO} may also identify individuals with chronic obstructive lung diseases who will develop arterial desaturation during exercise.[5] The DL_{CO} may have some advantages in monitoring occupational exposures where workers are at risk of developing pulmonary fibrosis. Britton et al.[6] have demonstrated that a decline in DL_{CO} frequently preceded a decline in FVC in asbestos workers. The potential value of DL_{CO} in assessing impairment and disability in individuals with parenchymal fibrosis is recognized by proposals to include it in some disability evaluation.[7,8]

In spite of the potential, progress in establishing the utility of DL_{CO} has been slow. One of the primary reasons for the delay is the high variability in DL_{CO} results from lab to lab. This makes it very difficult, if not impossible, to compare results from one laboratory with those from another. The wide variability[1] among prediction equations also suggests that variation in the commonly reported "percent of predicted" may be an additional problem.

The magnitude of the variability has been documented by the report of Clausen et al.[9] on interlaboratory comparisons on three individuals. Two individuals were tested in 16 and 28 university or university-affiliated pulmonary laboratories. The intralaboratory coefficient of variation (CV) was about 5%; the interlaboratory CV was 13% with measured values ranging from 28 to 43 ml/min/mmHg for one individual. Predicted normal values for one individual ranged from 28 to 40 ml/min/mmHg and the reported percent of predicted values ranged from 84 to 131%, demonstrating that the laboratories had not selected prediction equations matching the values measured in their laboratories. It is important to emphasize that these were good laboratories with excellent reputations and that this experience is likely to underestimate the magnitude of the problem. The problem of a large interlaboratory variability has also been documented by Saunders,[10] who reported on his experience in sending two patients to six prominent pulmonary function laboratories in London. The DL_{COSB} was measured in five of six laboratories and varied from 10.5 to 20.4 ml CO per minute per millimeter of mercury. The DL_{CO} measurement was a notable exception to the usual comparability among these laboratories.

One individual was tested in four university-affiliated laboratories in the Salt Lake City area with a range of measured values of 30 to 63 ml/min/mmHg with a CV of 18%.[9] After the laboratory technique in all four laboratories was rigidly standardized to that proposed by the Intermountain Thoracic Society,[11] the interlaboratory CV fell to 5%. One year later, the CV had increased again to about 18% and significant variations in technique had crept in. This is a very limited experience, but does suggest that rigid standardization of technique can reduce interlaboratory variability for DL_{CO}. It also emphasizes that, without close monitoring, laboratories tend to alter their techniques in ways that can introduce large changes in the results and may also suggest that a year is too long a monitoring interval.

In recognition of the importance of standardization in reducing interlaboratory variation, there have been several attempts to provide standard methods. The Intermountain Thoracic Society published a standardized technique in 1975 (revised in 1984).[11,12] The Epidemiology Standardization Project (ESP) proposed a standard DL_{CO} method in 1978.[13] In 1983 the Working Party of the European Community for Coal and Steel (ECCS) proposed a standard method for the European community.[14] Morris and Crapo[15] summarized the standardization proposals in the literature and showed that not all of the items that affect variability have been addressed by all proposals. The problem of nonuniform reference values has not yet been adequately addressed. In 1984 the American Thoracic Society convened a conference of representatives from the U.S., Canada, and Europe at Alta, UT to develop a recommendation for a standard technique. The final report was published in November 1987.[16]

The selection of a standard recommended technique should not be confused with establishing a "correct" method of performing the test. The standardization will serve to reduce interlaboratory variability and to emphasize and define the magnitude of variation introduced by the various factors involved. It should serve as a marker for further research and, as a result, help define a better test in the future. Refinements of the single- breath DL_{CO} are already being tested that may offer advantages over current tests.[17,18]

SOURCES OF VARIABILITY

The sources of variability to be considered in the establishment of a standard technique are (1) physiologic variability (2) variations in the technique (3) variability or errors in the measurements of the gas concentrations (4) variations due to errors in the measurement of volumes and times in the test, and (5) variations created by the calculation algorithms.

Physiologic variability is introduced by factors that may change individual DL_{CO} values. Hemodynamic factors, including exercise or recent exercise, patient position (sitting, standing, supine), Mueller or Valsalva maneuvers, temperature (patient and/or ambient), and recent meals, are the most common.[14,19,20] DL_{CO} is increased by factors that increase the average pulmonary capillary blood volume (Vc) and decreased by factors that decrease Vc. Diffusing capacity is reported to decrease about 1.2% per hour from 9:30 a.m. to 5:30 p.m. and by 2.2% per hour from 5:30 p.m. to 9:30 p.m.[21]

Ingestion of alcohol may acutely decrease DL_{CO} by as much as 15%.[22] The DL_{CO} increases with growth, hypoxemia, and polycythemia; it decreases with anemia, and with age after maturity.[20] Current recommendations vary but generally suggest that the patient be quiet and relaxed in an upright seated posture with an ambient temperature of 16 to 23°C.[11–14,20] In longitudinal studies, the time of day should be standardized for each individual' s repetitive measurements.

Test technique factors may also cause increased variability in DL_{CO}. The most important factor appears to be the size of the alveolar volume (V_A).[13,19] DL_{CO} increases as alveolar volume increases but D_LCO does not increase proportionally with the increase in V_A and DL_{CO}/V_A does not, therefore, remain constant.[19] Most descriptions recommend that the inhaled volume of test gas be 90 to 100% of the largest previously measured vital capacity.[12–14] The method of measuring V_A will also affect the calculated results especially in patients with obstructive disease.[12,13] When V_A is measured by the single-breath inert gas technique, it is underestimated when significant airway obstruction and, hence, trapped gas are present. The ESP study suggested that underestimation of V_A was small until the FEV_1/FVC was below 0.4,[13] but the ECCS recommends measuring V_A with a technique that is not dependent on rapid gas mixing. It has been suggested that maximum and minimum values of DL_{CO} may be determined by using both the single breath V_A technique and a technique independent of rapid gas mixing.[20]

Other factors that must be controlled include the rate of inspiration, the duration of breath-hold, the volume of the dead space washout prior to sample collection, and the volume and duration of the sample collection.[12–15,23] The measured DL_{CO} will be higher if the patient tries to inhale during the breath-hold (Mueller maneuver) than if he/she relaxes against a closed glottis or valve during the breath-hold. The concentration of oxygen in the test gas will also alter DL_{CO}. The DL_{CO} decreases as the oxygen concentration increases, even over the narrow ranges of oxygen concentration commonly used for the test.[24] The ECCS recommends an oxygen concentration of 0.17,[12] while 0.21 is the most common concentration used in North America. For laboratories at altitudes other than sea level, the ITS recommends increasing the oxygen concentration in the test gas to create an inspired oxygen pressure of 150 mmHg.[12] Using current recommended oxygen concentrations, Kanner and Crapo's data[24] suggest that DL_{CO} could vary by 15%, depending on the concentration of oxygen in the test gas and the altitude of the laboratory. Corrections for CO backpressure must be considered. They are probably not necessary if the patient does not smoke for at least 12 h prior to testing. In active smokers, CO backpressure could account for a 5 to 15% reduction in DL_{CO}.

Cotes[25] has demonstrated a striking variability in DL_{CO} due only to variability in gas analysis. He distributed cylinders of gases simulating a typical DL_{CO} test to 12 different laboratories. The gases were analyzed in the various DL_{CO} systems. The interlaboratory CV for helium was 3.25 to 4% and for CO it was 8.9 to 12.6%.[25] Calculating DL_{CO} from the various reports holding all nongas factors constant resulted in variations in DL_{CO} from 6.7 to 24.8 ml/min/mmHg.[25] Since this study was reported in 1963, Cotes has repeated the study and found that the gas analysis problem has lessened significantly, but has not been eliminated.[27]

Variation due to errors in measuring volume and time are obvious, but the sources of these errors are not so clear. Measurement errors in tests done by individual technicians are common[26] and computers should be more accurate. Even this is not guaranteed, however. We recently encountered a commercial DL_{CO} system with a programming error that introduced a 10% timing error. There is also no justification for the all-too-common assumption that flow and volume sensors will remain calibrated. Even commercial systems must be frequently validated.

Major variations may also be introduced simply by selecting different calculation algorithms. Excluding all measurement errors, Morris and Crapo have shown that the choice of calculation algorithms alone could account for variations of more than 40%.[15] The magnitude and direction of changes depend upon whether corrections are made, whether the conditions are appropriate for the corrections made, and the method of making the calculations. The variables and the potential changes they may cause are summarized below:[15]

Calculation or correction factor	Maximum variation introduced (%)
Deadspace in the inspired volume	±8
Deadspace in the sample collection bag	±6
Inspired volume conditions (ATPD or ATPS)	±4
Hemoglobin concentration	±7

Calculation or correction factor	Maximum variation introduced (%)
Carboxyhemoglobin (COHb) concentration (backpressure)	±%COHB
CO_2 absorption from the sample gas prior to gas analysis	± 5
Method of calculating breath-hold time	±7
V_A conditions in DL/V_A ratio (BTPS or STPD) (variation in DL/V_A, not DL_{CO})	±17

These percentages are individually small and may be either canceling or additive errors depending upon how they are selected. If the algorithm was selected so that they are additive, the variation in DL_{CO} could be ±41%.[15] A review of major papers on reference values for DL_{CO} reveals that few authors give all the details of their calculation algorithm, making it difficult to compare one study with another.[15]

Two factors with an impact in occupational studies are the correction for hemoglobin concentration and the correction for carboxyhemoglobin. Measuring hemoglobin and carboxyhemoglobin makes any study more complex but is, perhaps, most inconvenient in "field" studies. At the same time, these may be the very situations in which hemoglobin and carboxyhemoglobin are most important. Smoking histories and anemia may be factors that separate groups being studied and an apparent difference in DL_{CO} may reflect the hemoglobin and carboxyhemoglobin and not a difference in gas exchange across the lung. Corrections for hemoglobin and carboxyhemoglobin have not been standardized and are a cause of controversy about whether and how such corrections will be calculated.

CONCLUSION

Test methods and calculation algorithms must be carefully outlined for each study so that readers can assure themselves that the differences reported do not represent methodologic factors. A copy of the ATS standard can be obtained from the American Thoracic Society, 1740 Broadway, New York, NY, 10019.

REFERENCES

1. **Make, B., Miller, A., Epler, G., and Gee, J.B.L.,** Single breath diffusing capacity in the industrial setting, *Chest,* 82, 351, 1982.
2. **Epler, G.R., McLoud, T.C., Gaensler, E.A., Mikus, J.P., and Carrington, C.B.,** Normal chest roentgenograms in chronic diffuse infiltrative lung disease, *N. Engl. J. Med.,* 298, 934, 1978.
3. **Epler, G.R., Samet, J.M., and Gaensler, E.A.,** Diffusing capacity in the diagnosis of asbestosis, *Chest,* 74 (Abstr.), 336, 1978.
4. **Epler, G.R., Saber, F.A., and Gaensler, E.A.,** Determination of severe impairment (disability) in interstitial lung disease, *Am. Rev. Respir. Dis.,* 121, 647, 1980.
5. **Owens, G.R., Rogers, R.M., Pennock, B.E., and Levin, D.,** The diffusing capacity as a predictor of arterial oxygen desaturation during exercise in patients with chronic obstructive pulmonary disease, *N. Engl. J. Med.,* 310, 1218, 1984.
6. **Britton, M.G., Hughes, D.T.D., and Wever, A.M.J.,** Serial pulmonary function tests in patients with asbestosis, *Thorax,* 32, 45, 1977.
7. Component Committee on Disability Criteria, American Lung Association Pediatric and Adult Lung Disease Committee, Evaluation of impairment/disability secondary to respiratory disease, *ATS News,* p. 20, Summer 1981.
8. **Morgan, W.K.C.,** Pulmonary disability and impairment: can't work? Won't work?, *ATS NEWS,* p. 30, Summer 1982.
9. **Clausen, J., Crapo, R., and Gardner, R.,** Interlaboratory comparisons of pulmonary function testing, *Am. Rev. Respir. Dis.,* 129 (Abstr.), A37, 1984.
10. **Saunders, K.B.,** Reporting lung function tests: criteria and methods, including the use of computers, *Proc. R. Soc. Med.,* 70, 162, 1977.
11. **Kanner, R.E. and Morris, A.H., Eds.,** *Clinical Pulmonary Function Testing: A Manual of Uniform Laboratory Procedures for the Intermountain Area,* Intermountain Thoracic Society, Salt Lake City, UT, 1975.
12. **Morris, A.H., Kanner, R.E., Crapo, R.O., and Gardner, R.M.,** *Clinical Pulmonary Function Testing: A Manual of Uniform Laboratory Procedures,* 2nd ed., Intermountain Thoracic Society, Salt Lake City, UT, 1984.
13. **Ferris, B.G.,** Principal investigator, epidemiology standardization project (American Thoracic Society), *Am. Rev. Respir. Dis.,* 118 (Suppl.), 62, 1978.
14. **Quanjer, Ph.H., Ed.,** Standardized lung function testing, *Bull. Eur. Physiopathol. Respir.,* 19 (Suppl. 5), 39, 1983.
15. **Morris, A.H. and Crapo, R.O.,** Standardization of computation of single-breath transfer factor, *Bull. Eur. Physiopathol. Respir.,* 21, 183, 1985.
16. **Crapo, R.O., Gardner, R.M., et al.,** Single breath carbon monoxide diffusing capacity (transfer factor), recommendations for a standard technique, *Am. Rev. Respir. Dis.,* 136, 1299, 1987.
17. **Graham, B.L., Dosman, J.A., and Cotton, D.J.,** A theoretical analysis of the single breath diffusing capacity for carbon monoxide, *IEEE Trans. Biomed. Eng.,* 27, 221, 1980.
18. **Graham, B.L., Mink, J.T., and Cotton, D.J.,** Improved accuracy and precision of single-breath CO diffusing capacity measurements, *J. Appl. Physiol.,* 51, 1306, 1981.
19. **Forster, R.E.,** Editorial, The single-breath carbon monoxide transfer test 25 years on: a reappraisal, *Thorax,* 38, 1, 1983.
20. **Cotes, J.E.,** *Lung Function: Assessment and Application in Medicine,* 4th ed., Blackwell Scientific, Oxford, 1979, 230.
21. **Cinkotai, F.F. and Thomson, M.L.,** Diurnal variation in pulmonary diffusing capacity for carbon monoxide, *J. Appl. Physiol.,* 21, 539, 1966.
22. **Peavy, H.H., Summer, W.R., and Gurtner, G.,** The effects of acute ethanol ingestion on pulmonary diffusing capacity, *Chest,* 77, 488, 1980.
23. **Ogilvie, C.M., Forster, R.E., Blakemore, W.S., and Morton, J.W.,** A standardized breath holding technique for the clinical measurement of the diffusing capacity of the lung for carbon monoxide, *J. Clin. Invest.,* 36, 1, 1957.
24. **Kanner, R.E. and Crapo R.O.,** Diffusing capacity: correction factor for changes due to altitude, *Am. Rev. Respir. Dis.,* 131 (Suppl., Abstr.), A89, 1985.
25. **Cotes, J.E.,** Effect of variability in gas analysis on the reproducibility of the pulmonary diffusing capacity by the single breath method, *Thorax,* 18, 151, 1963.
26. **Gardner, R.M., Crapo, R.O., Billings, R.G., Shigeoka, J.W., and Hankinson, J.A.,** Spirometry: what paper speed?, *Chest,* 84, 161, 1983.
27. **Cotes, J.E.,** personal communication.

Lung Disease in Farm Workers: Pathological Reactions

J. L. Wright

INTRODUCTION

Farmers, by the very nature of their occupation, are exposed to a variety of agents through inhalation or contact exposure which may cause pulmonary disease. This article addresses many of these diseases and is slanted toward their pathological features. Pathology by itself is not enough, and therefore some clinical and radiological associations are discussed, but, in general, this has not been performed in a detailed fashion. Although the descriptions have been based on data regarding the farming population, when necessary, lesions produced by the same entities, but in another occupation, have been discussed. Occupational asthma has not been considered, nor have the number of infective conditions to which the farmer is prone.

This discussion has been broken into sections based on the etiological agents. Table 1 shows the source of the exposure to the various etiological agents, as well as the agents themselves. Table 2 indicates the different types of pathological reactions, as well as the agents causing these reactions; note that many agents cause similar pathological processes. If the pathological process is nonspecific, such as interstitial fibrosis, bronchiolar fibrosis, or adult respiratory distress syndrome (ARDS), only knowledge of the agents to which the subject has been exposed will allow the correct diagnosis to be made. It is therefore extremely important for the pulmonary physician and the pathologist to talk to each other before the biopsy is performed and the pathological report is issued.

TOXIC CHEMICALS AND GASES

Farmers can be exposed to a number of toxic gases and chemicals. Liquid manures may contain high concentrations of CH_4, H_2S, NH_3, and CO_2. In addition, NH_3 (anhydrous ammonia) is a common fertilizer. Case reports have documented deaths due to asphyxiation or ARDS after accidents involving manure pits and sewage tanks.[1,2] Accidental exposures to some of these gases in other industries, with sequelae of ARDS as acute lesions, and bronchiolitis obliterans and bronchiectasis as chronic lesions, have also been reported.[3–6]

TABLE 1
Possible Causes of Lung Disease in Farmers

Fertilizers	Anydrous ammonia, silica
Herbicides and pesticides	Paraquat, arsenicals
Sewage	Methane, H_2S, ammonia, CO_2
Welding and diesel fumes	NO_2
Grain, hay, silage, and contaminants	Fungi, proteins NO_2, N_2O_4 Endotoxins (?)

TABLE 2
Causes of Various Types of Pathological Reactions

ARDS (diffuse alveolar damage)	Oxides of nitrogen, H_2S, ammonia Paraquat
Bronchiolitis obliterans	Oxides of nitrogen, ammonia Immunological reaction to proteins (extrinsic allergic alveolitis)
Bronchiolar fibrosis	Oxides of nitrogen Grain dust (?) Silicates, silica
Extrinsic allergic alveolitis (hypersensitivity pneumonitis)	Animal, vegetable, insect products
Pulmonary fibrosis (honeycombing)	Oxides of nitrogen, H_2S, ammonia Paraquat Animal, vegetable, insect products
Carcinoma	Pesticides–herbicides

Probably the most common toxic gas exposures in the farming community are the oxides of nitrogen. Exposure to oxides of nitrogen can be acute, of high dose, and overwhelming such as in silo filler's disease, or chronic, of low dose, and insidious such as occurs with exposure to diesel exhaust and welding fumes.

Silo Filler's Disease

Silo filler's disease is a common health hazard with a high fatality rate.[7] Data from the U.S.[8] suggest that 4.2% of the farm operators developed symptoms while working in or adjacent to silos. In the silage tower, enzymatic action on the nitrate content of plants results in the production of nitrites. There are variable amounts of nitrates present, depending on the amount of nitrates in the soil and the maturity of plants; immature plants have more nitrates. The nitrites react with acetic and lactic acids, produced by bacterial degradation of carbohydrates, to produce nitrous acid (HNO_2) which decomposes to H_2O, NO, NO_2, and N_2O_4 as the temperature rises. NO_2 and N_2O_4 are the gases responsible for pulmonary damage, and are generally present in a ratio of 30:70. Gas begins to form within a few hours after filling the silo, and reaches its peak concentration in 1 to 5 d.[9] Concentrations generally range from 150 to 1900 parts per million (ppm), although up to 4000 ppm have been found.[8] Since the gases are heavier than air, they concentrate in the depressions and at the surface of the silage. The farmer is exposed after entering the silo through the chute and jumping down on the silage surface.

Since NO_2 is poorly soluble, it can reach the small airways and alveoli before undergoing hydrolysis to nitrous and nitric acids. The degree of disease is proportional to both the concentration of NO_2 inhaled and the duration of exposure.

The mildest form of disease consists only of central airway irritation, and manifests as cough, dyspnea, and fatigue; this form is called reversible.[8] At the other end of the scale, there may be asphyxiation, intense bronchospasm, or reflex respiratory arrest. Lesser concentrations may produce what has been described as pulmonary edema,[10] but is most likely ARDS.[8,11] In this process, the acid hydrolysis products cause epithelial damage. There is also probable endothelial damage, since methemoglobulinemia occurs indicating a reaction between nitrite ions and hemoglobin. The early pathological features of ARDS consist of alveolar collapse with only the small airways and alveolar ducts remaining open[12] (Plate 1*). These are often lined by a thick proteinaceous layer which is termed a hyaline membrane (Plate 2*). As in most cases of ARDS, there can be either complete resolution or various degrees of organization and fibrosis (Plate 3*). It is important to note that organizing ARDS is not synonymous with the usual form of interstitial fibrosis (UIP, fibrosing alveolitis). Organizing ARDS is a repair reaction and has a relatively short time course. UIP is an idiopathic fibrotic reaction which occurs over a long time period.

Patients who recover from mild episodes of ARDS may develop bronchiolitis obliterans in 1 to 4 weeks, probably reflecting the concentration gradient of NO_2 down the respiratory tract.[10] Bronchiolitis obliterans is characterized by tufts of granulation tissue filling the respiratory and occasionally the membranous bronchioles and extending down into the alveolar ducts (Plate 4*).

In acute disease, patients show varying combinations of restriction and obstruction combined with decreased diffusing capacity and hypoxemia.[10] In the subacute and chronic phase, depending on the extent of the disease, a survivor of silo filler's disease may show no abnormalities of pulmonary function, a restrictive pattern indicating parenchymal fibrosis, an obstructive pattern indicating bronchiolitis obliterans, or a combination of restriction and obstruction.

Low Levels of NO_2

Low levels of NO_2 may be produced by metal arc welding or by exposure to diesel fumes. Experimental data[13] show that human subjects exposed to levels of approximately 5 ppm develop mild airflow obstruction and increased airway responsiveness to carbacol or histamine. These changes disappear in 24 to 36 h and no residual effects are noted. Studies on British coal miners exposed to diesel fumes with concentrations of NO_2 which ranged up to 2.3 ppm (mean shift concentrations varied from 0.03 to 0.84 ppm) did not show evidence of airflow obstruction,[14] although long-term animal exposures to diesel exhaust (NO_2 approximately 4 ppm) show definite bronchiolar fibrosis.[15] Cross-sectional studies of welders have shown controversial results, but, in general, if there is any evidence of airflow obstruction in welders, it is of minimal degree.[16-18] Casual pathological examination of lung tissue from welders may show mild focal bronchiolar thickening, but, to my knowledge, no formal pathological study of welders' lungs has been carried out.

Herbicides and Pesticides

Data relating herbicidal or pesticidal exposure to pulmonary disease have been largely concerned with the lesions caused by paraquat (dimethylbipyridylium dichloride). Paraquat acts by increasing the number of free oxygen radicals during its cyclic oxidation and reduction.[19] Human[20-22] and experimental[19,21] data show that the initial pathological lesion is diffuse alveolar damage (ARDS). This progresses with intense fibroblastic proliferation and subsequent cyst formation (Plate 3*). Although, in general, the etiology of ARDS cannot be determined from the histology, there is a form of fibrosis thought to be distinctive for paraquat.[20] The lung architecture remains intact, but the airspaces are filled by loose fibrous tissue (Plate 5*).

Although there has not been an increased tumor mortality in the survivors of paraquat poisoning, exposure to the herbicides amitrol or phenoxy acids may be associated with an increased risk of tumors, including lung cancer.[23] Carcinoma has been diagnosed in pesticide applicators[24] and in agricultural workers exposed to a variety of pesticides.[25,26] Arsenicals were implicated as the carcinogenic agent in a group of pesticide manufacture workers,[27] and similar suggestions were raised in the studies of agricultural workers.[25,26]

Silica and Silicates

A further environmental hazard is related to the presence of silica or silicates, components of soil additives, pesticides, or the soil itself. Early studies from the Soviet Union[28] identified "dust induced pulmonary affections" in subjects who worked as grain cleaners, in elevators, or in flour mills. This initial observation was followed by experimental data: intratracheal instillation of crude grain dust into rats was said to have incited an inflammatory airways reaction followed by sclerosis.[29] The grain dust was found to contain 8 to 18% free silica. A similar "silicate pneumoconiosis" was found in seven patients from California, five of whom were vineyard workers.[30] All displayed abundant crystalline particles in the lung interstitium associated with pulmonary fibrosis; no silicotic nodules were present. Heatley described a single case of a grain handler who developed advanced fibrosis after an 8-year history of moving grain in which the silica content was 9.9%.[31] Other epidemiological studies have shown radiological features of fibrosis in grain handlers or dock laborers handling grain.[32] Again, no silicotic nodules were described. Williams et al.[32] found that, in Saskatchewan, the total airborne grain dust contains 8% free silica by weight in oats, and 7.4% in wheat. Although the data are extremely confusing, they do suggest that farm workers, particularly those exposed to pesticides using clay silicates, may develop either airway inflammation and fibrosis, or interstitial pulmonary fibrosis, or both as a result of silica or silicate inhalation. Certainly, further data need to be

Plates 1 to 5 appear on pages 30 to 32.

collected before definite disease is ascribed to the silica content of grain dusts. In an experimental model in which mice were exposed to crude grain dust from elevators, Armanious et al. identified a peculiar needle-shaped crystal present in pulmonary alveolar macrophages;[33] the significance of these inclusions is unclear. Zettergrent described interstitial granulomata in his rabbit model, but appeared to be describing extrinsic allergic alveolitis rather than silicosis.[32]

Silo Unloader's Syndrome, Grain Fever, Pulmonary Mycotoxicosis

Silo unloader's syndrome,[11] pulmonary mycotoxicosis,[34] and grain fever[32,35] may all be similar pathophysiological reactions. Grain fever has been reported in many epidemiological surveys of grain workers, with incidences of up to 28%.[36] Silo unloader's syndrome and pulmonary mycotoxicosis appear to be synonymous since all reports involve workers with moldy silage. Symptoms of all three conditions are similar, with cough, breathlessness, and fever occurring during or within a few hours of exposure. Much in the pattern of byssinosis,[37] the syndrome may be due to an endotoxic reaction with complement activation.[11] A lung biopsy from one case in the acute stage showed an interstitial lymphoplasmacytic infiltrate and numerous fungal organisms resulting in the diagnostic term "mycotoxicosis".[34] It appears doubtful that these patients develop chronic disease.

EXTRINSIC ALLERGIC ALVEOLITIS

Extrinsic allergic alveolitis (hypersensitivity pneumonitis) is a generic term describing a pathophysiological reaction to a number of vegetable, animal, and insect products. Table 3, modified from Lopez and Salvaggio,[38] indicates the disease, antigen source, and probable antigen for agents to which a farmer is likely to be exposed. Extrinsic allergic alveolitis is the result of an immunologic reaction, probably a combination of types 3 and 4, to the offending antigen.[39,40] In general, it is thought that a heavy exposure to the antigen is needed; one study calculated that a farmer working with moldy hay could inhale and retain 7.5×10^5 spores per minute, 98% of which were actinomycetes and less than 1.0 μm in diameter.[38] Evidence for type 3 reaction includes the short time (6 to 8 h) between exposure and symptoms, and the presence of a precipitating antibody with increasing titers at the time of active disease, and decreasing titers on remission. There is, however, little consistent evidence for an immune vasculitis, although Seal[41] found intimal proliferation and capillary obliteration and Kawanami et al.[42] identified rare areas of vessel thrombosis and acute vasculitis. The latter workers also found ultrastructural changes with fenestrated, multilayered, and edematous endothelial cells. Type 4 immunologic reaction is acknowledged by the presence of delayed hypersensitivity, and cell-mediated immunity with *in vitro* lymphocyte transformation and MIF production. Granulomata, although not always present (see below), also suggest type 4 reaction.

TABLE 3
Antigens, Antigen Sources, and Popular Names of Causes of Extrinsic Allergic Alveolitis

Name	Source	Antigen
Farmer' s lung	Moldy hay	Thermophilic actinomycetes *Micropolyspora faeni* *Thermoactinomyces vulgaris*
Mushroom worker's lung	Moldy compost	Thermophilic actinomycetes *M. faeni* *T. vulgaris*
Malt worker's lung	Contaminated barley	Fungi *Aspergillus clavatus*
Grain measurer's lung	Cereal grain	Unknown
Humidifier lung	Contaminated humidifier	Thermophilic actinomycetes *T. candidus* *T. vulgaris*
Pigeon breeder's lung	Droppings	Pigeon serum protein
Duck fever	Feathers	Proteins
Turkey handler's lung	Turkey products	Proteins
Miller's lung	Wheat weavils	*Sitophilus granarius*

Farmer's lung is probably the most common type of extrinsic allergic alveolitis with a prevalence range from 4%[43] up to 12.6%[44] in various areas. Since the organisms grow in moldy hay, it is not surprising that there is a seasonal variation, which is different in different parts of the world.[10,45,46] Interestingly, the farmer is not the only farm dweller at risk, since the disease has also been reported in a cow; characteristic pathology was present, and the animal had antibodies to micropolyspora faeni.[47]

The acute-stage pathological reaction in extrinsic allergic alveolitis is similar regardless of the inciting agent (Plates 6 to 8*). An interstitial lymphoplasmacytic infiltrate is present in 100% of cases[42,48] and is also occasionally intraalveolar.[42] The lymphocytes may show a "hand mirror" configuration, suggestive of activated T cells.[42] Granulomata are present in approximately $^2/_3$ of cases; they are often described as "sarcoidal", but in reality the resemblance is only superficial, as they are generally small and rather loosely constructed. Bronchiolitis obliterans is present in $^1/_2$, and organizing pneumonia is noted in approximately 65% of the cases. Interstitial fibrosis is seen in 65 to 100% of cases; there is no predominate site in the lobule, and it is mild in the majority of cases.

In the chronic stages, interstitial inflammation is present in all cases to a variable degree. There may be fine interstitial fibrosis (Plate 9*), or frank honeycombing, which is said to have an upper lobe predominance.[41]

It is tempting to suggest relationships between the pathological response and the physiological measurement and chest radiograph. In the acute stage, loss of lung volumes and decreased compliance[49] could reflect the interstitial inflammatory response, seen on radiograph as a ground-glass appearance.[50] Airways obstruction, seen in a large number of acute cases,[51] as well as nodular or striated patchy opacities in

* Plates 6 to 9 appear on pages 33 and 34 .

the radiograph could represent areas of bronchiolitis obliterans and organizing pneumonia. Since this latter radiographic pattern is associated with chronic disease in 24% of cases,[50] interstitial fibrosis may also be an important component. Chronic disease, manifested by increased upstream resistance and small airways obstruction,[43,49,51] may represent a preponderance of bronchiolitis obliterans, while decreased compliance and decreased lung volumes suggest interstitial fibrosis.

GRAIN WORKERS

Although many exacting epidemiological surveys of grain workers and farmers[32,36,52-55] have found abnormal pulmonary function with airways obstruction, a detailed pathophysiological correlation has not been performed. Animal experimental data are also lacking; clearly, this is an area of importance and needs further investigation.

SUMMARY

In summary, farmers are at risk for developing pulmonary disease due to the inhalation of, or contact exposure to, a variety of substances in their environment. Although the pathology and, indeed, pathophysiological relationships of some diseases are well worked out, this is not true for all. It is unclear what the pathological reaction is after exposure to silica and silicates in grain dusts, grain dusts themselves, or large amounts of fungal spores. It is important to clearly define these reactions in order to determine adequate preventative measures before lung disease occurs, and to shed light on therapy and prognosis after pathology is discovered.

REFERENCES

1. **Cockcroft, D.W. and Dosman, J.A.,** Respiratory health risks in farmers, *Ann. Intern. Med.,* 95, 380, 1981.
2. **Osbern, L.N. and Crapo, R.O.,** Dung lung: a report of toxic exposure to liquid manure, *Ann. Intern. Med.,* 95, 312, 1981.
3. **Woodford, D.M., Coutu, R.E., and Gaensler, E.A.,** Obstructive lung disease from acute sulfur dioxide exposure, *Respiration,* 30, 238, 1979.
4. **Sobonya, R.,** Fatal anhydrous ammonia inhalation, *Hum. Pathol.,* 8, 293, 1977.
5. **Hoeffler, H.B., Schweppe, H.I., and Greenberg, S.D.,** Bronchiectasis following pulmonary ammonia burn, *Arch. Pathol. Lab. Med.,* 106, 686, 1982.
6. **Close, L.G., Catlin, F.I., and Cohn, A.W.,** Acute and chronic effects of ammonia burns of the respiratory tract, *Arch. Otolaryngol.,* 106, 151, 1980.
7. **Morrisey, W.L., Gould, I.A., Carrington, C.B., and Gaensler, E.A.,** Silo-Filler's disease, *Respiration,* 32, 81, 1975.
8. **Horvath, E.P., doPico, G.A., Barbee, R.A., and Dickie, H.A.,** Nitrogen dioxide-induced pulmonary disease: five new cases and a review of the literature, *J. Occup. Med.,* 20, 103, 1978.
9. **Commins, B.T., Raveney, F.J., and Jesson, M.W.,** Toxic gases in tower silos, *Ann. Occup. Hyg.,* 14, 275, 1971.
10. **Ramirez, R.J. and Dowell, A.R.,** Silo-Filler's disease: nitrogen dioxide-induced lung injury, *Ann. Intern. Med.,* 74, 569, 1971.
11. **Pratt, D.S. and May, J.J.,** Feed-associated respiratory illness in farmers, *Arch. Environ. Health,* 39, 43, 1984.
12. **Pratt, P.C.,** Pathology of adult respiratory distress syndrome, in *The Lung. Structure, Function and Disease,* Thurlbeck, W.M. and Abell, M.R., Eds., Williams & Wilkins, Baltimore, 43, 1978.
13. **Horvath, S.M.,** Nitrogen dioxide, pulmonary function, and respiratory disease, *Bull. N.Y. Acad. Med.,* 56, 835, 1980.
14. **Robertson, A., Dodgson, J., Collings, P., and Seaton, A.,** Exposure to oxides of nitrogen: respiratory symptoms and lung function in British coalminers, *Br. J. Ind. Med.,* 41, 214, 1984.
15. **Hyde, D.M., Plopper, C.G., Weir, A.J., Murnane, R.D., Warren, D.L., Last, J.A., and Pepelko, W.E.,** Peribronchiolar fibrosis in lungs of cats chronically exposed to diesel exhaust, *Lab. Invest.,* 52, 195, 1985.
16. **Akbarkhanzadeh, F.,** Long-term effects of welding fumes upon respiratory symptoms and pulmonary function, *J. Occup. Med.,* 22, 337, 1980.
17. **Keimig, D.G., Pomrehn, P.R., and Burmeister, L.F.,** Respiratory symptoms and pulmonary function in welders of mild steel: a cross-sectional study, *Am. J. Ind. Med.,* 4, 489, 1983.
18. **Kalliomaki, P.L., Kalliomaki, K., Korhonen, O., Nordman, H., Rahkonen, E., and Vaaranen, V.,** Respiratory status of stainless steel and mild steel welders, *Scand. J. Work. Environ. Health,* 1 (Suppl.), 117, 1982.
19. **Bus, J.S. and Gibson, J.E.,** Paraquat: model for oxidant-initiated toxicity, *Environ. Health Perspect.,* 55, 37, 1984.
20. **Thurlbeck, W.M. and Thurlbeck, S.M.,** Pulmonary effects of paraquat poisoning, *Chest,* 69, 276, 1976.
21. **Levin, P.J., Klaff, L.J., Rose, A.G., and Ferguson, A.D.,** Pulmonary effects of contact exposure to paraquat: a clinical and experimental study, *Thorax,* 34, 150, 1979.
22. **Rebello, G. and Mason, J.K.,** Pulmonary histological appearances in fatal paraquat poisoning, *Histopathology,* 2, 53, 1978.
23. **Axelson, O., Sundell, L., Anderson, K., Edling, C., Hogstedt, C., and Kling, H.,** Herbicide exposure and tumor mortality. An updated epidemiologic investigation on Swedish railroad workers, *Scand. J. Work. Envrion. Health,* 6, 73, 1980.
24. **Blair, A., Grauman, D.J., Lubin, J.H., and Fraumeni, J.F.,** Lung cancer and other causes of death among licensed pesticide applicators, *Natl. Cancer Inst.,* 71, 31, 1983.
25. **Barthel, E.,** Increased risk of lung cancer in pesticide-exposed male agricultural workers, *J. Toxicol. Environ. Health,* 8, 1027, 1981.
26. **Barthel, E.,** Cancer risk in pesticide exposed agricultural workers, *Arch. Geschwulstforsch.,* 51, 579, 1981.
27. **Mabuchi, K., Lilienfield, A.M., and Snell, L.M.,** Lung cancer among pesticide workers exposed to inorganic arsenicals, *Arch. Environ. Health,* 34, 312, 1979.
28. **Stepanov, S.A. and Karpova, M.I.,** Changes in the respiratory organs under the effect of grain dust, *Sov. Med.,* 29, 65, 1966.
29. **Stepanov, S.A.,** Characteristics in the development of pneumoconiosis during the exposure to grain dust, *Arkh. Patol.,* 33, 74, 1971.
30. **Sherwin, R.P., Barman, M.L., and Abraham, J.L.,** Silicate pneumoconiosis of farm workers, *Lab. Invest.,* 40, 576, 1979.
31. **Heatley, T.F., Dalton, K., and Rex, C.R.,** A case of silicosis caused by wheat dust, *J.A.M.A.,* 124, 980, 1944.
32. **Williams, N., Skoulas, A., and Merriman, J.E.,** Exposure to grain dust, *J. Occup. Med.,* 6, 319, 1964.
33. **Armanious, M., Cotton, D., Wright, J.L., Dosman, J., and Carr, I.,** Grain dust and alveolar macrophages: an experimental study of the effects on grain dust on the mouse lung, *J. Pathol.,* 136, 265, 1982.
34. **Emanuel, D.A., Wenzel, F.J., and Lawton, B.R.,** Pulmonary mycotoxicosis, *Chest,* 67, 293, 1975.
35. **Becklake, M.R.,** Grain dust and health: state of the art, in *Occupational Pulmonary Disease: Focus on Grain Dust and Health,* Dosman, J.A. and Cotton, D.J., Eds., Academic Press, New York, 1980.
36. **Tse, K.S., Warren, P., Janusz, J., McCarthy, D.S., and Cherniack, R.M.,** Respiratory abnormalities in workers exposed to grain dust, *Arch. Environ. Health,* 27, 74, 1973.
37. **Castellan, R.M., Olenchock, S.A., Hankinson, J.L., et al.,** Acute bronchoconstriction induced by cotton dust: dose-related responses

to endotoxin and other dust factors, *Ann. Intern. Med.*, 101, 157, 1984.

38. **Lopez, M. and Salvaggio, J.**, Hypersensitivity pneumonitis: current concepts of etiology and pathogenesis, *Annu. Rev. Med.*, 27, 453, 1976.
39. **Slavin, R.G. and Avioli, L.V.**, Hypersensitivity pneumonitis, *Arch. Intern. Med.*, 136, 352, 1976.
40. **Salvaggio, J.E. and Karr, R.M.**, Hypersensitivity pneumonitis; state of the art, *Chest*, 75S, 270, 1979.
41. **Seal, R.M.E., Hapke, E.J., Thomas, G.O., Meek, J.C., and Hayes, M.**, The pathology of the acute and chronic stages of farmer's lung, *Thorax*, 23, 469, 1968.
42. **Kawanami, O., Basset, F., Barrios, R., Lacronique, J.G., Ferrans, V.J., and Crystal, R.G.**, Hypersensitivity pneumonitis in man, *Am. J. Pathol.*, 110, 275, 1983.
43. **Madsen, D., Klock, L.E., Wenzel, F.J., Robbins, J.L., and Schmidt, C.D.**, The prevalence of farmer's lung in an agricultural population, *Am. Rev. Respir. Dis.*, 113, 171, 1976.
44. **Grant, I.W.B., Blyth, W., Wardrop, V.E., Gordon, R.M., Pearson, J.C.G., and Mair, A.**, Prevalence of farmer's lung in Scotland: a pilot survey, *Br. Med. J.*, 1, 530, 1972.
45. **Terho, E.O., Lammi, S., and Heinonen, O.P.**, Seasonal variation in the incidence of farmer's lung, *Int. J. Epidemiol.*, 9, 219, 1980.
46. **Hapke, E.J., Seal, M.E., Thomas, G.O., Hayes, M., and Meek, J.C.**, Farmer's lung, *Thorax*, 23, 451, 1968.
47. **Pirie, H.M., Dawson, C.O., Breeze, R.G., Wiseman, A., and Hamilton, J.**, A bovine disease similar to farmer's lung: extrinsic allergic alveolitis, *Vet. Rec.*, 88, 346, 1971.
48. **Reyes, C.N., Wenzel, F.J., Lawton, B.R., and Emanuel, D.A.**, The pulmonary pathology of farmer's lung disease, *Chest*, 81, 142, 1982.
49. **Warren, C.P.W., Tse, K.S., and Cherniack, R.M.**, Mechanical properties of the lung in extrinsic allergic alveolitis, *Thorax*, 33, 315, 1978.
50. **Monkare, S., Ikonen, M., and Haahtela, T.**, Radiologic findings in farmer's lung, *Chest*, 87, 460, 1985.
51. **Allen, D.H., Williams, G.V., and Woolcock, A.J.**, Bird breeder's hypersensitivity pneumonitis: progress studies of lung function after cessation of exposure to the provoking antigen, *Am. Rev. Respir. Dis.*, 114, 555, 1976.
52. **DoPico, G.A., Reddan, W., Flaherty, D., Tsiatis, A., Peters, M.E., Rao, P., and Rankin, J.**, Respiratory abnormalities among grain handlers, *Am. Rev. Respir. Dis.*, 115, 915, 1977.
53. **Corey, P., Hutcheon, M., Broder, I., and Mintz, S.**, Grain elevator workers show work-related pulmonary function changes and dose-effect relationships with dust exposure, *Br. J. Ind. Med.*, 39, 330, 1982.
54. **Cotton, D.J., Graham, B.L., Li, K.Y., Froh, F., Barnett, G.D., and Dosman, J.A.**, Effects of grain dust exposure and smoking on respiratory symptoms and lung function, *J. Occup. Med.*, 25, 131, 1983.
55. **Chan-Yeung, M., Schulzer, M., MacLean, L., Dorken, E., and Grzybowski, S.**, Epidemiologic health survey of grain elevator workers in British Columbia, *Am. Rev. Respir. Dis.*, 121, 329, 1980.

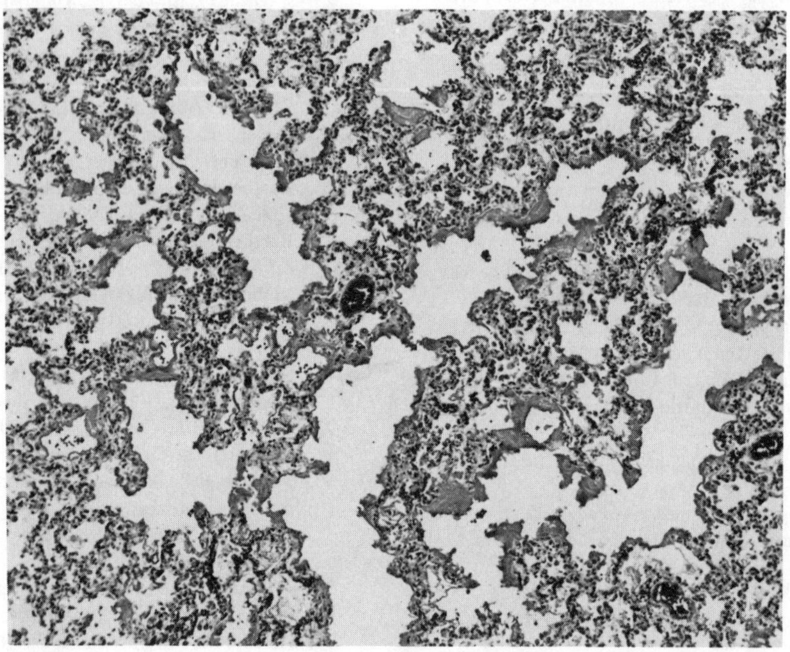

PLATE 1. The early stages of ARDS (diffuse alveolar damage), showing prominent alveolar collapse around the respiratory bronchioles and alveolar ducts which remain open. (H + E; magnification × 40.)

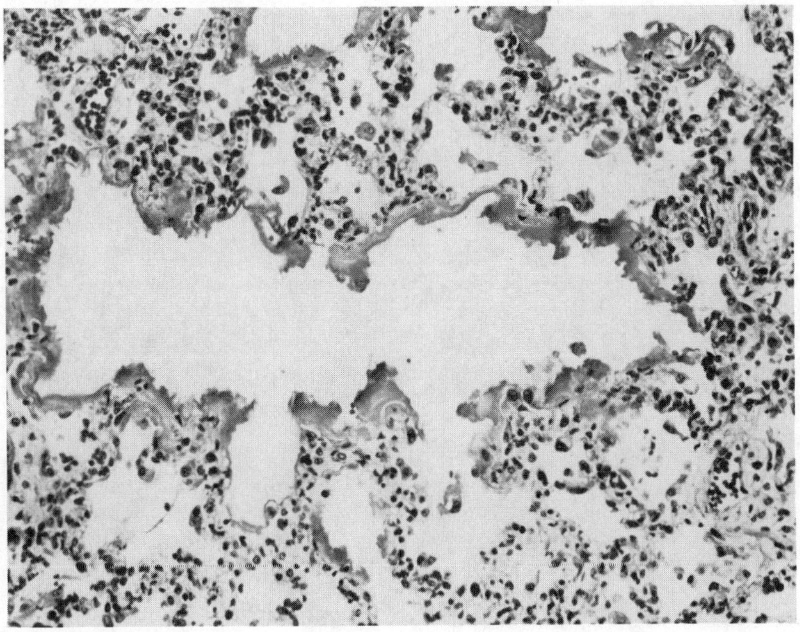

PLATE 2. The early stages of ARDS. Note that the hyaline membrane is lining the alveolar ducts. It is thick, amorphous, and brightly eosinophilic. (H + E; magnification × 64.)

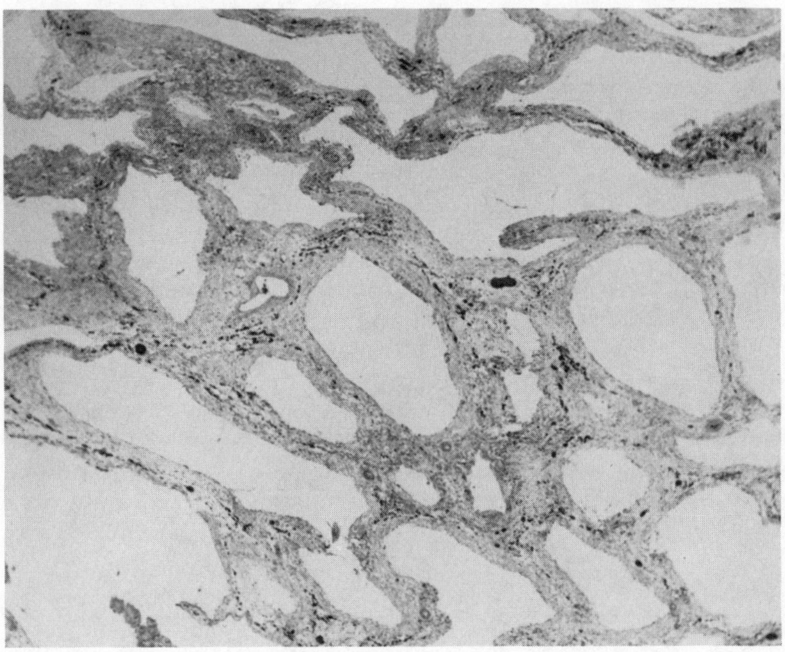

PLATE 3. Honeycombing, a nonspecific endstage repair reaction which can be produced by many different etiologies; this micrograph was taken from a case of paraquat poisoning. Note the thick walls of the cysts and the rearrangement of the architecture. (H + E; magnification × 20.)

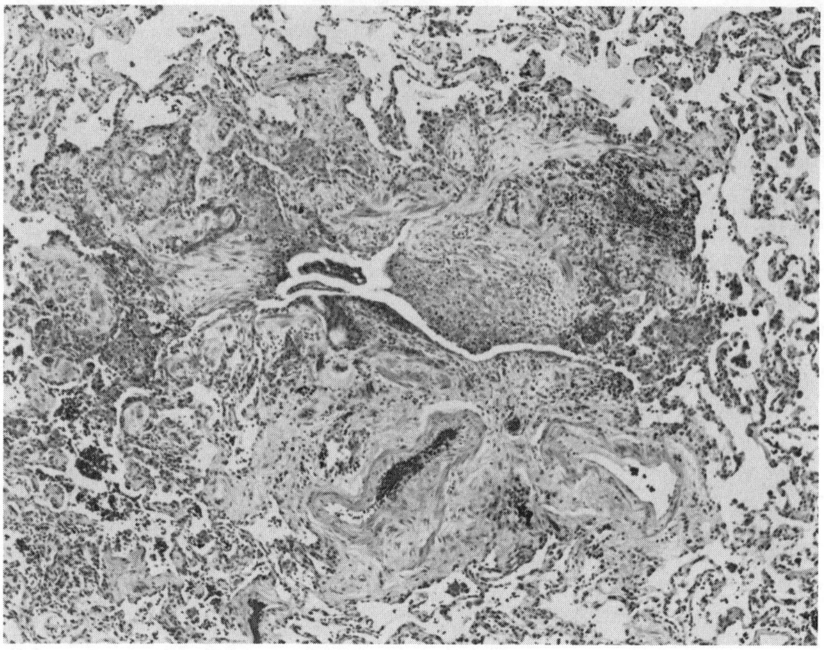

PLATE 4. Bronchiolitis obliterans which was caused by inhalation of noxious fumes. Note the tufts of granulation tissue which partially obliterate the lumen of the respiratory bronchiole and extend down the alveolar ducts. (H + E; magnification × 64.)

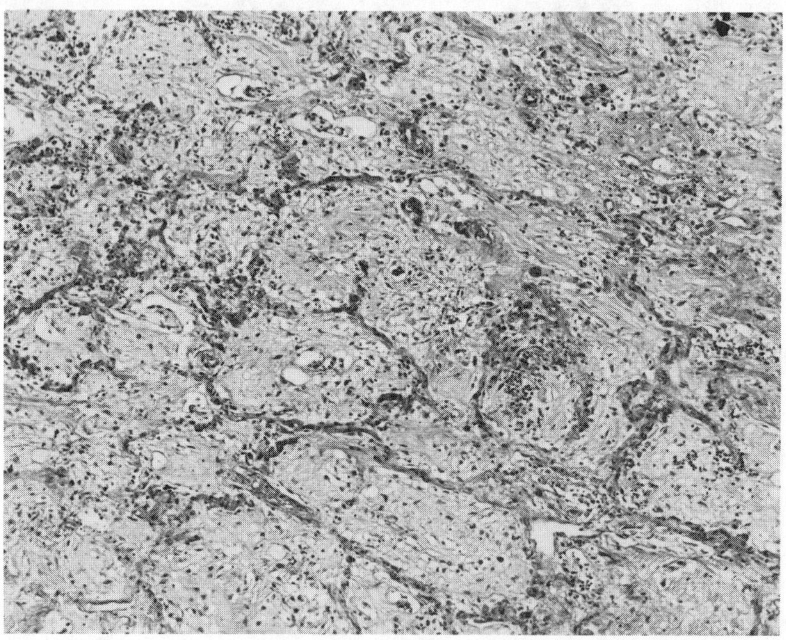

PLATE 5. Histological appearance which has been suggested to be diagnostic of paraquat poisoning. Although the architecture remains intact, as seen by the capillary pattern, the alveolar spaces are filled by loose granulation tissue. (H + E; magnification × 64.)

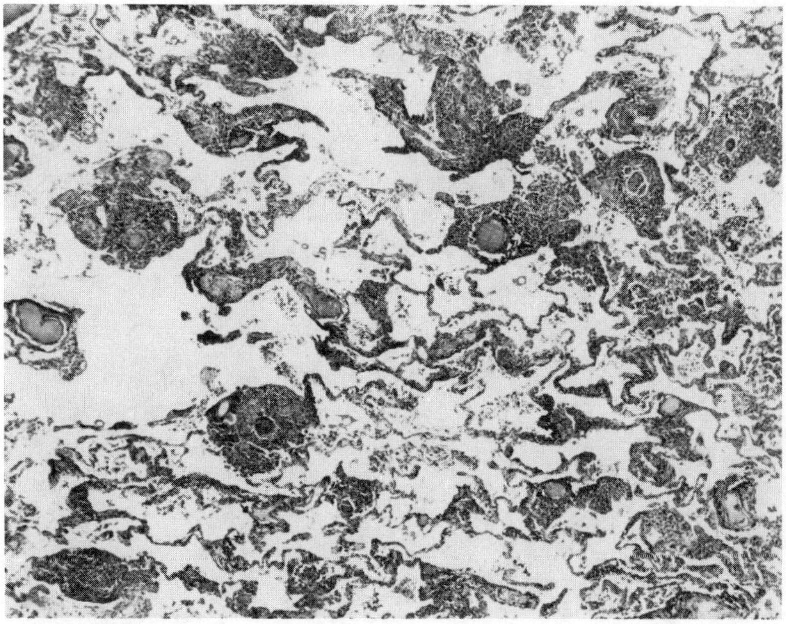

PLATE 6. A low-power view of extrinsic allergic alveolitis, showing the marked infiltrate in the alveolar septa, accompanied by loosely formed granulomata. (H + E; magnification × 40.)

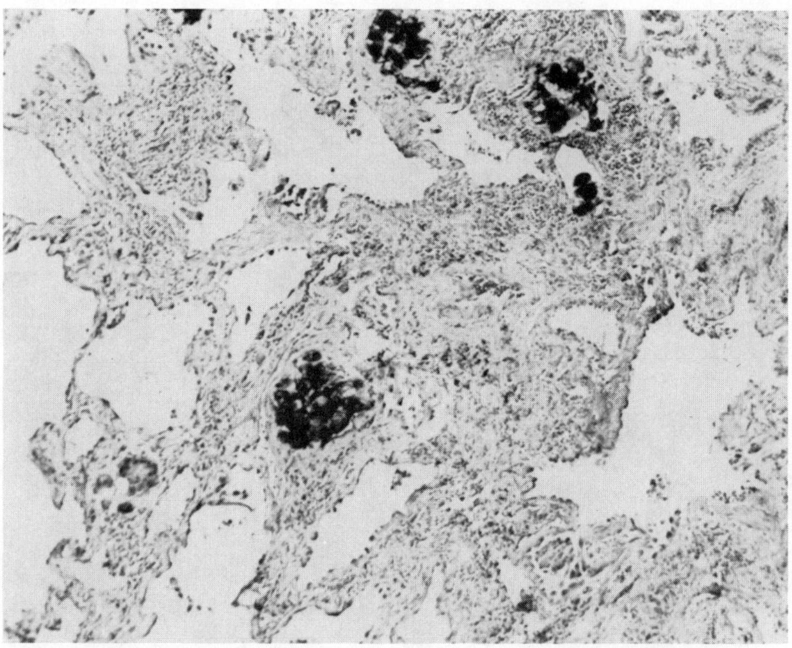

PLATE 7. Higher-power view which shows the loosely formed granulomata. Note the foreign body material in the multinucleated cells (arrows). The interstitial infiltrate is composed of lymphocytes and plasma cells. There is a mild intraalveolar infiltrate present. (H + E; magnification × 64.)

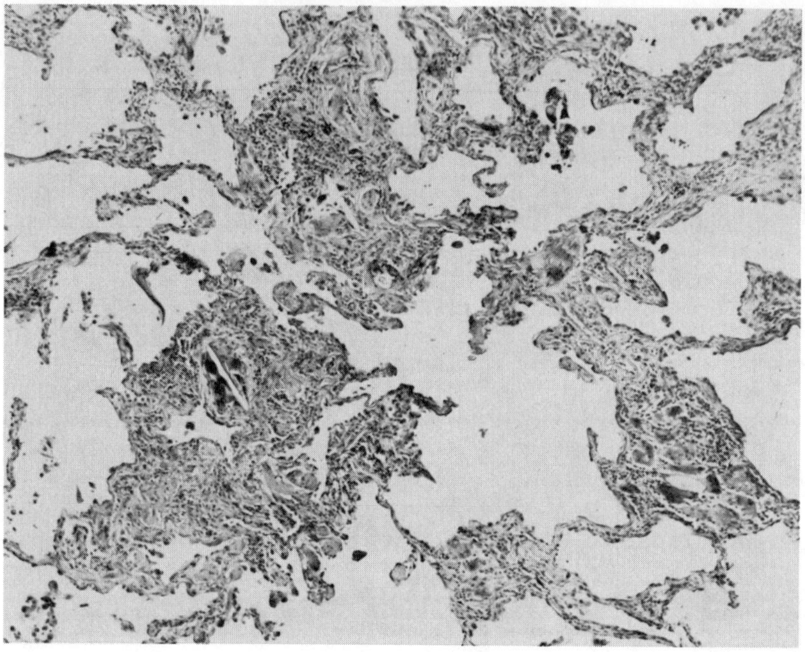

PLATE 8. High-power view which shows an area of interstitial fibrosis. There are small granulomata present as well as a mild interstitial inflammatory infiltrate. Foreign body material is present in some of the multinucleated giant cells. (H + E; magnification × 64.)

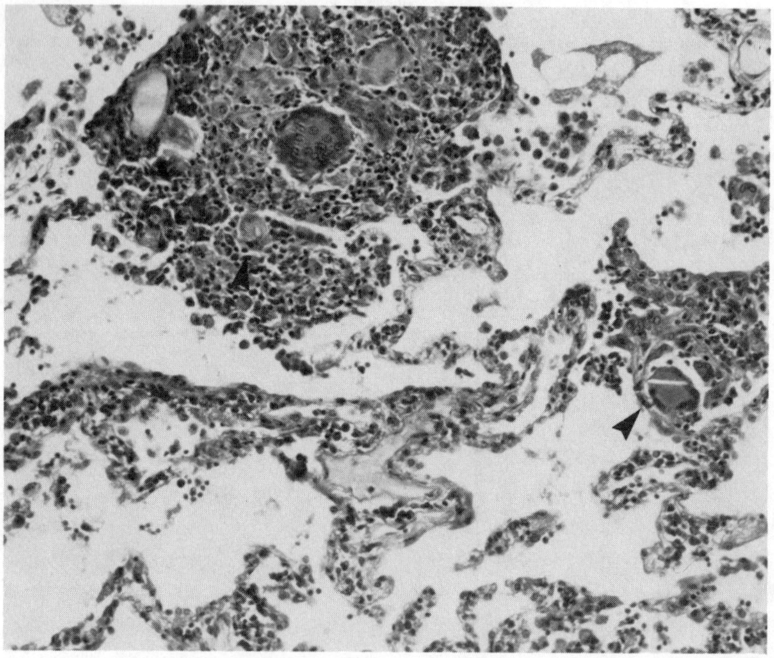

PLATE 9. Chronic stage of extrinsic allergic alveolitis with interstitial fibrosis. A chronic inflammatory cell infiltrate is present, but the granulomata are replaced by clusters of calcospherules. (H + E; magnification × 64.)

Assessment of Airways Inflammation by Bronchoalveolar Lavage

A. B. Thompson, M. A., Ghafouri, R. A. Robbins, M. G. Stahl, A. E. Prince, and S. I. Rennard

INTRODUCTION

The exposure of workers in grain elevators to grain dust frequently causes respiratory symptoms and may lead to progressive pulmonary impairment.[1–4] Farmers, like elevator workers, are often exposed to atmospheres contaminated by grain dust, and, as do grain handlers, farmers may complain of dyspnea, cough, sputum production, or wheezing.[5,6] However, farmers also have other common, significant exposures including pesticides and microbial antigens such as those arising in moldy hay or corn. Because of the variety and intermittency of exposures, characterization of the pathophysiology of farmers' respiratory complaints is a complex problem. Epidemiology and pulmonary function testing are invaluable for documenting associations between exposures, respiratory complaints, and impairment, but cannot be used to assess inflammatory responses. A method for characterizing the local pulmonary inflammation due to exposure to inhaled substances might allow the investigator to sort out mechanisms of the inflammation and thereby clarify the etiology of farmers' pulmonary complaints. Bronchoscopy with bronchoalveolar lavage promises to be such a method.

Bronchoalveolar lavage (BAL) is safe and easily performed on alert, locally anesthetized patients. The bronchoscope is wedged into a subsegmental bronchus and boluses of saline are instilled and recovered from the distal airways and alveoli. BAL has been used to evaluate lower respiratory tract inflammation and the immune responsiveness of cells recovered from the lower respiratory tract.[7–9] Characteristic changes in the number and types of cells recovered from lavage fluid have been described in idiopathic pulmonary fibrosis, eosinophilic granulomatosis, sarcoidosis, hypersensitivity pneumonitis, and collagen-vascular disease associated with interstitial fibrosis.[8,9] Quantification and characterization of cells recovered from the alveoli of patients, especially with sarcoidosis, are useful in confirming the diagnosis, staging the activity of the alveolitis, predicting response to therapy, and monitoring therapeutic effectiveness.[10] Thus, BAL has demonstrable utility in studying alveolar inflammation.

However, while farmers may suffer from hypersensitivity pneumonitis which causes parenchymal inflammation, the symptoms and pulmonary function changes due to grain dust exposure suggest primarily airways disease. Current evidence suggests that BAL can be used to study airways inflammation as well as alveolar inflammation. This evidence is summarized in Table 1 and indicates that lavage with small amounts of saline (5 to 20 ml) seems to preferentially recover cells and proteins which are derived from the airways. Merrill et al.,[12] using large aliquots (50 ml), found no statistically significant differences in sequential lavage protein content that could not be explained by progressive dilution of the bronchoalveolar contents by progressive lavages, but did note that IgA and free secretory component, when normalized to albumin, increased as the amount of fluid returned in the first aliquot decreased. Davis et al.,[11] also using large lavage aliquots (60 ml), found higher percentages of neutrophils and lymphocytes in the first aliquot and a higher percentage of macrophages in subsequent aliquots. The other three studies,[13–15], in which small, 5– to 20–ml aliquots were used, all found statistically significant differences between the cell differentials in small, first aliquots as compared to large or pooled subsequent aliquots. Epithelial cells, which arise in the airways, were more prevalent in the smaller aliquots, while lymphocytes, macrophages, and IgG, all thought to arise distally, were more prevalent in larger or pooled subsequent aliquots.

TABLE 1
Bronchoalveolar Lavage Fluid Processing Affects Quantitation of Cells and Proteins

Method (ml)	Results[a]	Ref.
4 × 60	Lymphocytes and neutrophils highest in the first aliquot, macrophages lowest in the first aliquot, unchanged second to fourth	11
6 × 50	Progressive fall in IgA and IgG IgA/alb and IgG/alb constant IgA/alb and FSC/alb tended to increase as size of first aliquot decreased	12
5 × 20	First aliquot had significantly higher neutrophils and epithelial cells, pooled second to fourth aliquots had significantly higher lymphocytes and macrophages	13
1 × 5, 10, 20, 50	Epithelial cells significantly higher in aliquots 10 ml or less and significantly lower in aliquots 50 ml or higher Macrophages were significantly higher in aliquots 20 ml or higher-IgG was higher in aliquots 100 ml or higher	14
5 × 50	Neutrophils significantly higher in the first aliquot	15

[a] Abbreviations: IgA, immunoglobulin A; IgG, immunoglobulin G; alb, albumin; FSC, free secretory component.

In summary, performing BAL with small sequential aliquots of saline and analyzing the first aliquot separately from the rest yield significant differences in both cellular and immunoglobulin content. The first small aliquot contains a predominance of cells and proteins from the airways and therefore seems to be enriched for bronchial epithelial lining fluid and its contents, while the later aliquots seem to be enriched for alveolar lining fluid and its contents.

A recent report of BAL findings in four adult patients with bronchiolitis[16] confirmed that airways diseases may alter the cellular composition of BAL fluid. BAL was performed by sequentially instilling and retrieving 5– to 20 ml aliquots of saline and pooling all of the samples before their analysis. The mean percentage of neutrophils in the BAL fluid from these patients was 53%, which contrasted to the 0.3% neutrophils found in normals.

Chronic bronchitis is a syndrome characterized pathologically by neutrophilic infiltration of airway walls and lumens.[17] It was hypothesized that since BAL can sample bronchial epithelial lining fluid, it can be utilized to assess the neutrophilic inflammation associated with chronic bronchitis.

STUDIES IN FOUR GROUPS OF PATIENTS

To test the hypothesis that BAL can be used to assess airways inflammation, data for four groups of subjects were retrospectively analyzed using the method of Ghafouri et al. (see below) to enrich for airways material. The four groups were smoking patients with a history of chronic, productive cough suggestive of chronic bronchitis (n = 18), smoking patients without a history of chronic bronchitis (n = 12), nonsmoking patients without a history of chronic bronchitis (n = 33), and, for comparison, normal, asymptomatic volunteers (n = 12). Patients were excluded if they had clinical or roentgenographic evidence for active lower respiratory tract inflammation other than chronic bronchitis. Patients underwent bronchoscopy and BAL for evaluation of suspected intrathoracic malignancy or as surveillance for occult infection prior to bone marrow ablative chemotherapy. The normal controls had volunteered under a protocol approved by the Human Subjects Institutional Review Board. The characteristics of the subject groups are summarized in Table 2. All of the patient groups were older than the controls. The smoking patients with chronic bronchitis had more packyears of smoking history than did the asymptomatic smokers. Spirometry showed that while all of the patient groups had smaller lung volumes and lower flows than the control group, on average, only the smoking patients with cough and sputum had obstruction to flow as reflected by a low forced expired volume in 1 s – to – forced vital capacity (FEV_1/FVC) ratio. While the mean FEV_1/FVC ratio of the smoking, but nonbronchitic patient group did not reflect airflow obstruction, some members of the group were obstructed. However, their low FEV_1/FVC ratios were masked by restrictive changes in others. Restriction in these patients was due to either underlying malignancy or prior chemotherapy.

Bronchoscopy and BAL were performed by the method of Ghafouri et al.[13] The bronchoscope was wedged into a subsegmental bronchus and the distal airways lavaged with five 20–ml aliquots of normal saline, each recovered by gentle suctioning. This procedure was repeated for a total of three lavage sites. The first aliquots from each of the subsegments, the bronchial sample, were pooled and processed separately from the final four aliquots, the alveolar sample, which were likewise pooled. Suctioning recovered 33% of the instilled saline for the bronchial samples from the normals and 20% of the bronchial samples from the patient groups. For the alveolar samples, 71% was recovered from the normals and 47% from the patient groups.

After the bronchial and alveolar samples were pooled they were strained through a single layer of nylon gauze and, when necessary to remove excess mucus, the bronchial samples were centrifuged (400 × *g*, 8 min), the mucus decanted, and the cells resuspended in sterile Hank's balanced salt solution. The cells in the resulting suspensions were then counted with a hematocytometer and fluid containing 100,000 cells was allocated for cytocentrifugation (165 × *g*, 10 min, Cytospin 2,

TABLE 2
Patient Characteristics

	Age (years)	Smoking history (pack years)	FVC (% pred)	FEV_1 (% pred)	FEV_1/FVC (%)	FEF_{25-75} (% pred)
Normal controls	27 ± 1.2[a]	0	112 ± 2.5	113 ± 2.5	84 ± 1.6	116 ± 5.9
Nonsmoking patients	42.4 ± 2.9[b]	0	93 ± 3.4[c]	92 ± 3.6[c]	82 ± 1.8	90 ± 6[c]
Smoking patients without bronchitis	45.9 ± 3.4	35.4 ± 8.4	82 ± 6.1	76 ± 5.2[c]	78 ± 2.3	63 ± 7.1[c]
Smoking patients with bronchitis	54.4 ± 3.7[d]	60.1 ± 9.8[b]	77 ± 4.4[d]	66 ± 4.7[e]	69 ± 3.4[c]	49 ± 7.6[e]

[a] Mean ± SE.
[b] Significantly higher than preceding value ($p < 0.05$, comparison by ANOVA and Fishers Least Significant Difference Test, except smoking history, *t* test).
[c] Significantly lower than the preceding value ($p < 0.05$).
[d] Significantly higher than the value for nonsmoking patients ($p < 0.05$).
[e] Significantly lower than the value for nonsmoking patients ($p < 0.05$).

Shandon). The resulting cell button was stained with a modified Wright Giemsa stain (Diff Quik, American Scientific Products) and differentials (including neutrophils, eosinophils, lymphocytes, macrophages, and epithelial cells) were performed on 200 cells. Total cell counts were obtained by multiplying the hematocytometer count by the volume of lavage fluid recovered for the sample.

Bronchial and alveolar lavage sample differentials were analyzed for correlation between lavage fluid neutrophils and the presence of chronic bronchitis. Separation of the bronchial and alveolar samples was crucial in this regard, because in all cases the number of cells in the bronchial sample was less than the number in the alveolar samples. For all subjects the mean number of cells in the bronchial sample ($3.4 \pm 0.4 \times 10^6$, mean ± SE) was much less than the number of cells in the alveolar sample ($33.3 \pm 3.3.2 \times 10^6$), accounting for only 9% of the total number of cells recovered (Figure 1). Thus, if the bronchial samples were not separated, small changes in the cellular content of the bronchial epithelial lining fluid could be masked by the overwhelming contribution to the alveolar epithelial lining fluid to the total number of BAL fluid cells.

The neutrophil content of the bronchial samples for each group was significantly higher than the neutrophil content of the alveolar samples (see Figure 2, $p < 0.05$ by Wilcoxon matched pairs test), confirming earlier observations (see Table 1).

Furthermore, the neutrophil content of the bronchial, but not of the alveolar, samples was associated with the presence of chronic bronchitis. The bronchial lavage samples from the smoking patient group with chronic bronchitis had a significantly higher ($p < 0.05$ for all comparisons by ANOVA) percentage of neutrophils (25.4 ± 4.1%) than the normals (9.1 ± 2.1%), the nonsmoking patients (8.2 ± 1.5%), and the smoking, but nonbronchitic patients (10.08 ± 3.8%). The percentages of neutrophils in the bronchial samples from the normals, nonsmoking patients, and smoking patients without chronic bronchitis did not differ significantly.

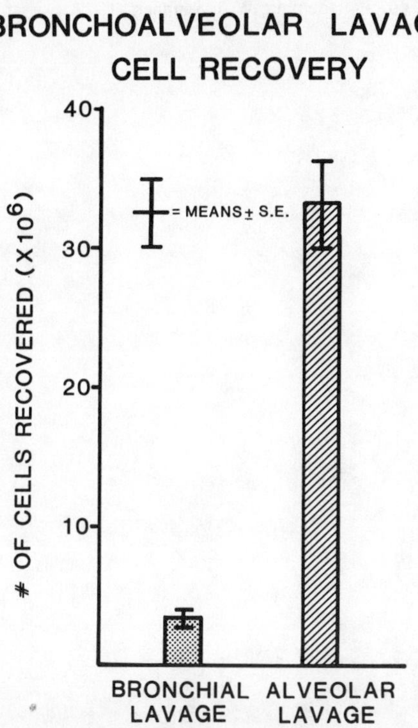

FIGURE 1. Total number of cells, in millions, recovered in the bronchial sample and in the alveolar sample. The number of cells in the bronchial sample only accounts for 9% of the total number of cells recovered by the lavage. The difference is statistically significant ($p < 0.05$, paired t test).

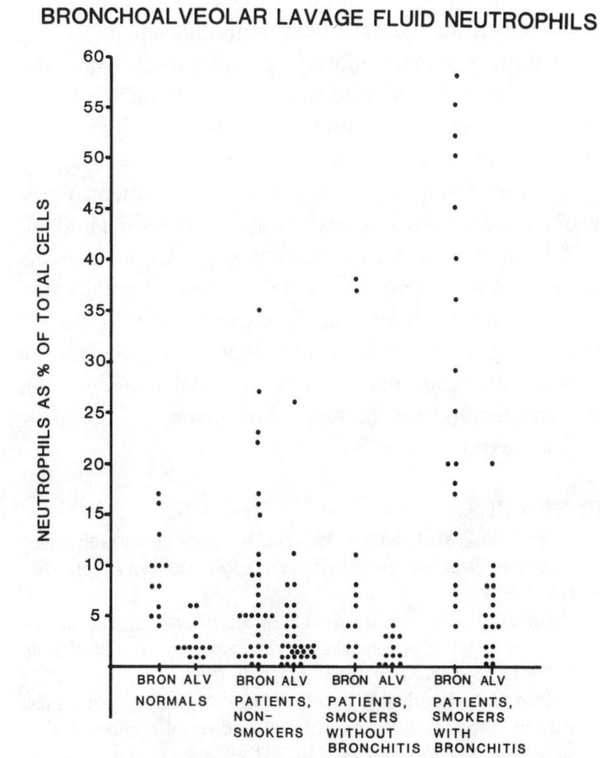

FIGURE 2. The number of neutrophils, as percent of total cells, recovered in the bronchial (bron) and the alveolar (alv) samples for normal volunteers, nonsmoking–nonbronchitic patients, smoking–nonbronchitic patients, and smoking patients with chronic bronchitis.

RELEVANCE

The results suggest that BAL can be used to sample bronchial epithelial lining fluid in such a way as to permit characterization of airways inflammation. This was supported by the increased percentages of neutrophils in the bronchial, but not in the alveolar, samples of patients with chronic bronchitis compared to nonsmoking normals and patients. Furthermore, the elevation of neutrophils was associated with the specific presence of chronic bronchitis and not just cigarette smoking. The group of smoking, but nonbronchitic patients had fewer neutrophils than the patients with bronchitis and did not differ in this regard from the nonsmoking patients.

While some of the nonsmoking patients and smoking, but nonbronchitic patients had percentages of neutrophils in their bronchial samples which were clearly higher than the range of values obtained for normals, those individuals had received chemotherapy for their underlying malignancies and this might have been the cause of their inflammation. Why they did not have cough or sputum production is not clear.

Farmers are exposed to a large number of respirable agents that may cause pulmonary symptoms and impairment. While some of these exposures cause diseases involving the interstitium of the lung, such as hypersensitivity pneumonitis, many are associated with symptoms and spirometric findings suggesting airways involvement. Application of BAL to the understanding of the cellular and humoral mechanisms of airways inflammation may provide insights enabling better control and therapy for the respiratory disorders caused by agricultural exposures.

REFERENCES

1. **Chan-Yeung, M., Schulzer, M., et al.,** Epidemiologic health survey of grain elevator workers in British Columbia, *Am. Rev. Respir. Dis.,* 121, 329, 1980.
2. **Cotton, D.J., Graham, B.L., et al.,** Effects of smoking and occupational exposure on peripheral airway function in young cereal grain workers, *Am. Rev. Respir. Dis.,* 126, 660, 1982.
3. **DoPico, G.A., Reddan, W., et al.,** Epidemiological study of clinical and physiologic parameters in grain handlers of northern United States, *Am. Rev. Respir. Dis.,* 130, 759, 1984.
4. **Chan-Yeung, M., Wong, R., et al.,** Respiratory abnormalities among grain elevator workers, *Chest,* 75, 461, 1979.
5. **Warren, C.P.W. and Manfreda, J.,** Respiratory symptoms in Manitoba farmers: association with grain and hay handling, *Can. Med. J.,* 122, 1259, 1980.
6. **Warren, C.P.W., Holford-Stevens, V., et al.,** Respiratory disease in grain farmers, in Occupational Lung Disease, Bernard, J., Raven Press, New York, 1984.
7. **Reynolds, H.Y., Fulmer, J.D., et al.,** Analysis of cellular and protein content of broncho-alveolar lavage fluid from patients with idiopathic pulmonary fibrosis and chronic hypersensitivity pneumonitis, *J. Clin. Invest.,* 59, 165, 1977.
8. **Weinberger, S.E., Kelman, J.A., et al.,** Bronchoalveolar lavage in interstitial lung disease, *Ann. Intern. Med.,* 89, 459, 1978.
9. **Hunninghake, G.W., Gadek, J.E., et al.,** Inflammatory and immune processes in the human lung in health and disease: evaluation by bronchoalveolar lavage, *Am. J. Pathol.,* 97, 149, 1979.
10. **Line, B.R., Hunninghake, G.W., et al.,** Gallium-67 scanning to stage the alveolitis of sarcoidosis: correlation with clinical studies, pulmonary function studies, and bronchoalveolar lavage, *Am. Rev. Respir. Dis.,* 123, 440, 1981.
11. **Davis, G.S., Giancola, M.S., Costanza, M.C., and Low, R.B.,** Analyses of sequential bronchoalveolar lavage samples from healthy human volunteers, *Am. Rev. Respir. Dis.,* 126, 611, 1982.
12. **Merrill, W., O'Hearn, E., et al.,** Kinetic analysis of respiratory tract proteins recovered during a sequential lavage protocol, *Am. Rev. Respir. Dis.,* 126, 617, 1982.
13. **Ghafouri, M.A., Rasmussen, J.K., et al.,** Use of sequential bronchoalveolar lavage to enrich for "bronchial" and "alveolar" material, *Clin. Res.,* 33, 464A, 1985.
14. **Lam, S., Leriche, J.C., Kijek, K., and Phillips, D.,** Effects of bronchial lavage volume on cellular and protein recovery, *Chest,* 88, 856, 1985.
15. **Martin, T.R., Raghu, G., Maunder, R.J., and Springmeyer, S.C.,** The effects of chronic bronchitis and chronic airflow obstruction on lung cell populations recovered by bronchoalveolar lavage, *Am. Rev. Respir. Dis.,* 132, 254, 1985.
16. **Dorinsky, P.M., Davis, W.B., et al.,** Adult bronchiolitis: evaluation by bronchoalveolar lavage and response to prednisone therapy, *Chest,* 88, 58, 1985.
17. **Spencer, H.,** Pathology of the Lung Excluding Pulmonary Tuberculosis, 3rd ed., Philadelphia, W.B. Saunders, 1977.

Use of Biological Assays in Short–Term Assessment of Inhaled Substances

Joseph D. Brain

INTRODUCTION

Workers in the agricultural industry are exposed to an exceptionally wide variety of inhaled particles. These include fertilizers, pesticides, and herbicides as well as resuspended soil. Moreover, the composition of the soil (for example, the fraction which is free silica) varies from place to place. Other workers are exposed to complex grain dusts, such as that coming from various cereal grains (wheat, barley, rye, oats, corn), as well as various contaminants such as insects, mites, rodent debris, and fungi. This wide array of complex dusts presents problems in assessing the potential risk of various occupational exposures in agriculture.

In order to understand such exposures, it is possible to measure responses at the molecular, cell, organ, or organismic level. All approaches reflect the need to evaluate the toxicology of materials to which agricultural workers are exposed so that we can take appropriate preventive action. Government, unions, and industry now face the difficult task of assessing the toxicology of a wide variety of new chemicals and especially complex mixtures. The creativity of chemists who synthesize new compounds, the availability of new technologies, and finally the competitiveness of agriculture ensure that there will be a continuing stream of new aerosol exposures whose potential for damage must be assessed. Since they are new, epidemiology fails to provide information about health effects. Nevertheless, a guide to potential toxicity is needed to help design both appropriate control strategies and medical surveillance studies for humans employed in agriculture.

How can the risk of human pulmonary disease caused by exposure to complex and often poorly characterized dusts in the agricultural industry be predicted? Risk assessment may include: (1) air monitoring and physical and chemical characterization of collected dusts; (2) epidemiologic studies of humans; (3) controlled experimental exposures of humans in the laboratory; (4) chronic lifetime animal studies; (5) short-term animal bioassays; and (6) *in vitro* tests of mammalian cells. This paper emphasizes the fifth method of analysis and discusses the use of short-term animal bioassay systems to determine the health effects of inhaled particulates.

Animal studies have numerous advantages since ethical problems are minimized. The possibility of more serious disease can be assessed, and there are few limits to the invasiveness of the diagnostic procedures used. For example, long-term inhalation exposures of animals, followed by functional or histopathological studies of their lungs, have been used to study asbestos,[1] crystalline silica,[2] and coal dust.[3] A problem is that such studies are costly and time consuming. A typical lifetime study in rodents costs between 0.5 and 3 million dollars and may take 3 to 5 years to plan and complete. It is also difficult to obtain quantitative estimates of toxicity using standard pathological analyses. Morphometric measures based on extensive sampling of lung tissue as well as physiological or biochemical assessment may be required.

Clearly, there is a need for short-term tests. If large numbers of materials are to be analyzed, it is essential to have assays that are relatively inexpensive and that yield results in weeks or months, not years. Many investigators have promoted the use of *in vitro* assays to assess the potential toxicity of inhaled aerosols.[4-7] *In vitro* systems have advantages of reproducibility, cost, and specificity. Several tissue culture systems have been developed.[8,9] However, because the human pulmonary response to inhaled particles is the result of complex interactions involving many different cell types within the lung, the results obtained may be spurious. For example, inflammation involves recruitment of neutrophils, platelets, and serum proteins to the injured lung. Fibrogenesis involves the action of fibrogenesis–stimulating factors secreted by one cell (e.g., a macrophage) on another cell (a fibroblast). These essential interactions are rarely reproduced in any *in vitro* system.

Short-term *in vivo* assays can be considered as an alternative to short-term *in vitro* tests, because the short-term response of small animals to dusts is sufficiently similar to the human response to have predictive values when properly calibrated and interpreted. The major mechanisms of lung injury[10] are common to most mammals.

THE HAMSTER BIOASSAY

The hamster bioassay features the use of bronchoalveolar lavage (BAL). During the last decade, BAL has been used increasingly to assess lung injury in animals and man. BAL has been employed to discriminate among toxic agents such as metal salts or mineral dusts.[11,12] Key issues in the application of BAL to inhalation toxicology are the specificity and sensitivity of the procedure. What is the smallest amount of dust which causes a measurable response? More important, what is the ability of BAL to discriminate among dusts of varying toxicities and those producing different resulting lesions? To what extent does BAL have predictive value? Can one examine acute events and describe long-term irreversible chronic changes?

We have developed a short–term (1 to 30 d postexposure) animal bioassay system in which the toxicity of a particular dust may be estimated by comparison to known dusts with a demonstrated range of toxicities for human pulmonary disease. We employ hamsters exposed to dusts by either inhalation or by intratracheal instillation[13] and quantify the response by measuring biochemical and cellular indicators in BAL fluid. The parameters measured represent a wide spectrum of possible responses to inhaled particles, including inflammation, pulmonary edema, cellular damage, cellular secretion, and endocytic capacity of pulmonary macrophages. We have calibrated the system with dusts for which there is considerable human experience. Cellular and biochemical changes were measured in BAL of hamsters after exposure to α-quartz, iron oxide, and aluminum oxide.[12] α–Quartz is a highly toxic, fibrogenic mineral dust, whereas aluminum oxide and iron oxide are both of low toxicity.

One day after exposure, the levels of β–*N*–acetylglucosaminidase were significantly elevated by exposure to the 0.75– and 3.75–mg doses of all three dusts (see Figure 1). However, the response to α–quartz was greater than the response to the other two dusts, especially at the highest dose. β–*N*-acetylglucosaminidase is an example of a lysosomal enzyme that is released from cells during phagocytosis, cell injury, or cell death.[14] Polymorphonuclear neutrophils (PMNs),[14] macrophages, and type II cells[15] all contain acid hydrolases. Excessive release of lysosomal enzymes may elicit unwanted proteolysis from cathepsins or membrane destruction by phospholipases.

α–Quartz also elevated albumin levels in lavage fluid at both 0.75– and 3.75–mg doses as shown in Figure 2. The highest dose caused a more than 40-fold increase above control levels. Aluminum oxide and iron oxide were also associated with an increase at 3.75 mg, but albumin levels clearly distinguished between these relatively nontoxic dusts and the highly fibrogenic α–quartz. Albumin is primarily a

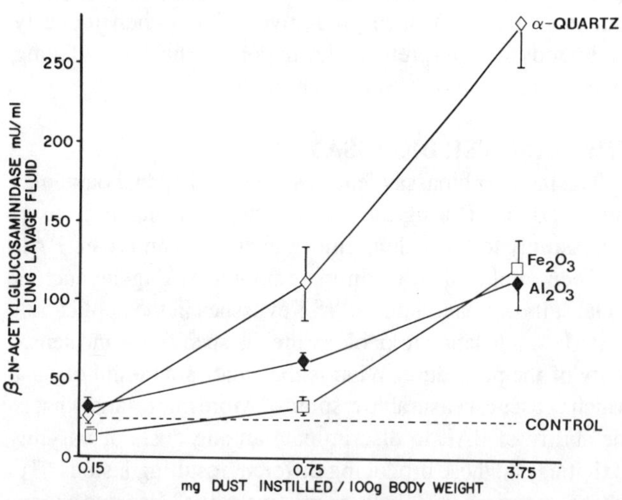

FIGURE 1. Dose-response curve for β–*N*-acetylglucosaminidase 1 d after instillation of particles. $p < 0.01$ for all points except 0.75 mg iron oxide and 0.15 mg aluminum oxide ($p < 0.05$), and 0.15 mg α–quartz (not significant). Values are mean ± standard errors. (Adapted from Beck, B. D., Brain, J. D., and Bohannon, D. E., *Exp. Lung Res.*, 2, 289, 1981. With permission.)

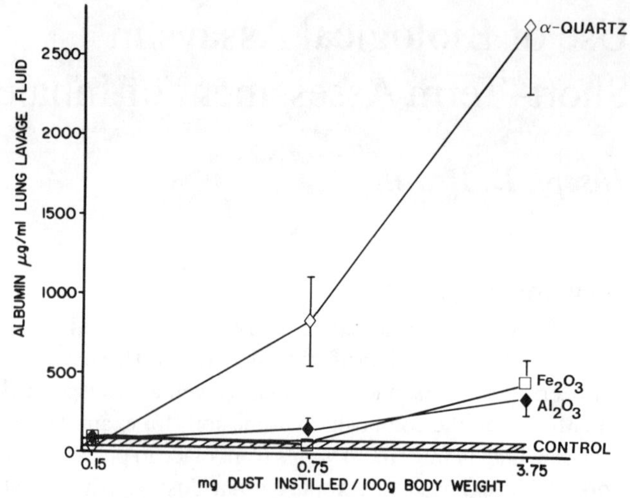

FIGURE 2. Dose-response curve for albumin in extracellular supernatant of lung lavage fluid 1 d after exposure to iron oxide, aluminum oxide, or α–quartz. The Wilcoxon rank-sum test was used to compare experimentals and saline–only controls. $p < 0.01$ for all points except 0.75 mg aluminum oxide and all 0.15–mg samples (not significant). Values represent mean ± standard errors. (Adapted from Beck, B. D., Brain, J. D., and Bohannon, D. E., *Exp. Lung. Res.*, 2, 289, 1981. With permission.)

serum protein whose presence in BAL is due to passage across damaged endothelial and epithelial barriers. Albumin is usually the most abundant protein in BAL.[16,17] Elevated albumin levels indicate pulmonary edema, a common manifestation of acute pulmonary injury.[12,18]

Figure 3 illutratres that α–quartz also causes depressed macrophage function. The lambda values shown are the fraction of radioactive gold colloid which was ingested 90 min after it had been instilled through the trachea. Brain and Corkery[19] provide details of this assay which estimate the endocytic activity of macrophages *in situ*. At a dose of 3.75 mg of α–quartz, less than 30% of the gold was ingested; iron oxide and aluminum oxide have no significant effect on lambda.

The full bioassay includes a number of other parameters such as peroxidase, elastase, hemoglobin, as well as the numbers of erythrocytes, neutrophils, and macrophages. An essential aspect of bioassays like this is to compare the responses of unknown dusts with other well–characterized standards. Both positive and negative controls should be used. The best calibrating materials would be those for which there is a considerable experience in humans such as the dusts shown in Figures 1 to 3. Then the type and intensity of response for a new unknown dust could be compared to these standards.

A key feature of assays utilizing lung lavage is the time course of the response. Some agents will yield similar responses when examined soon after exposure. However, the more toxic material may frequently exhibit a more persistent change in the cellular and enzymatic parameters than nontoxic controls. For example, there was a prolonged elevation in the numbers of macrophages and PMNs with quartz, but not with iron oxide. PMN numbers in the lung lavage fluid were

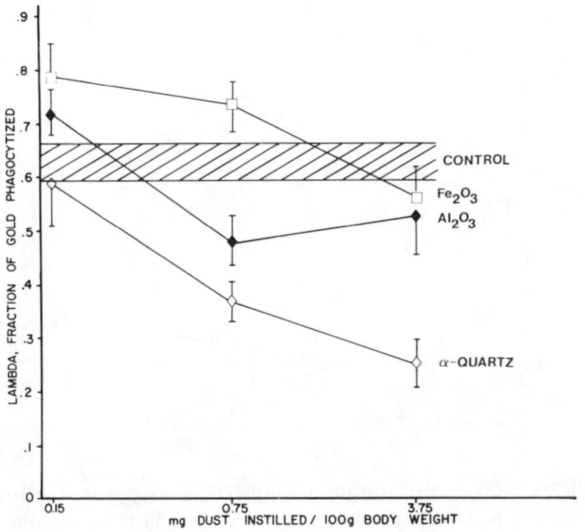

FIGURE 3. Dose-response curve for lambda assay 1 d after exposure to iron oxide, aluminum oxide, or α–quartz. The Wilcoxon rank-sum test was used to compare experimentals and saline only controls. $p < 0.01$ for 0.75 and 3.75 mg α–quartz, 0.75 mg aluminum oxide; $p < 0.05$ for 0.15 mg iron oxide. Values represent mean ± standard errors. (Adapted from Beck, B. D., Brain, J. D., and Bohannon, D. E., *Exp. Lung. Res.*, 2, 289, 1981. With permission.)

highest 4 d after exposure to α–quartz, although after 2 weeks they still had not approached control levels.[12]

A somewhat different pattern was observed for lactate dehydrogenase (LDH) in lavage fluid. This is a cytoplasmic enzyme involved in energy metabolism; its extracellular release is associated with cell injury or death. LDH levels in lung lavage fluid were highest 1 d after exposure to both iron oxide and α–quartz. In time, LDH levels declined signficantly in the quartz-exposed animals and only slightly in the iron oxide-exposed animals. Nevertheless, the levels in the quartz animals remained higher than those in the iron oxide-exposed animals at all times. These effects were observed at relatively low levels of quartz compared to levels used in animal models of chronic silicosis.

Application of this system to dusts produced by the eruption of Mt. St. Helens volcanic ash suggested that volcanic ash has low to moderate toxicity.[20] We concluded that adverse health effects in human populations are unlikely except with high or prolonged exposure. Surfactant levels in BAL in rats after quartz and Mt. St. Helens volcanic ash exposure have been studied by Martin and co–workers.[21] Quartz causes a prolonged elevation in PMN numbers and surfactant levels. The effects were much less marked with volcanic ash than with quartz. These observations are consistent with histopathological studies of lungs of exposed animals which demonstrated much greater fibrogenicity of α–quartz than of volcanic ash. These studies show the usefulness of BAL in providing a rapid evaluation of the toxicity of poorly characterized samples. Useful results can be obtained even when chemical analyses of epidemiological studies are not available for toxicity estimates.

IDENTIFYING SOURCES OF DAMAGE INDICATORS: LDH ISOENZYMES

We are searching for other ways of making the assay more interpretable. As discussed earlier, LDH is released from cells in response to toxic particles. However, if LDH is recovered in the cell–free supernatant of lung lavage fluid, where does it come from? Is the source inflammatory cells (macrophages or PMNs), serum, epithelial cells, or endothelial cells? Beck et al.[22] have used isoenzyme analysis to infer the sources of LDH.

To differentiate among types of injury, we monitored changes in LDH isoenzyme patterns in BAL after a range of injuries: α–quartz, hyperoxia, the detergent Triton X-100, and SO_2. The LDH isoenzyme patterns in BAL were evaluated and compared with patterns from hamster lung homogenates, red blood cells, macrophages, PMNs, type II cells, and serum. The isoenzyme pattern in BAL from quartz–exposed animals resembled that of the PMNs and macrophages, suggesting phagocytic cell death. In contrast, BAL from Triton X-100–treated animals had an isoenzyme pattern similar to that of the lung homogenate and red blood cells. Exposure to 100% O_2 for 4 d produced an isoenzyme pattern similar to serum, an observation consistent with the demonstrated effects of O_2 on the capillary endothelium.

Figure 4 presents graphically the percentage of each LD isoenzyme from serum or from lung lavage fluid of Syrian golden hamsters exposed to 100% O_2 for 96 h. The distribution of the five LD isoenzymes is similar and consistent with the hypothesis that oxygen toxicity caused damage to the air-blood barrier. Serum LD and other serum proteins leaked into alveolar spaces and were subsequently recovered by lavage.

In Figure 5 the LD pattern is shown for: (1) supernatant from BAL recovered from hamsters exposed to iron oxide aerosol and (2) hamster peritoneal PMNs. The LD patterns shown in Figure 5 are markedly different from those seen in Figure 4. For example, there is little LD1 (<3%), but a great deal of LD5 (~60%). The similarity in pattern suggests that the

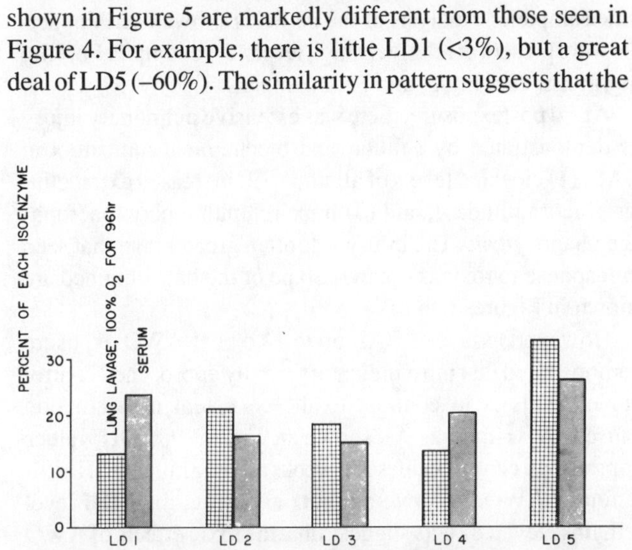

FIGURE 4. Comparison of LD isoenzyme patterns from hamster serum and from lung lavage fluid of hamsters exposed to 100% O_2 for 96 h. (Adapted from Beck, B. D., Gerson, B., Feldman, H. A., and Brain, J. D., *Toxicol. Appl. Pharmacol.*, 71, 59, 1983. With permission.)

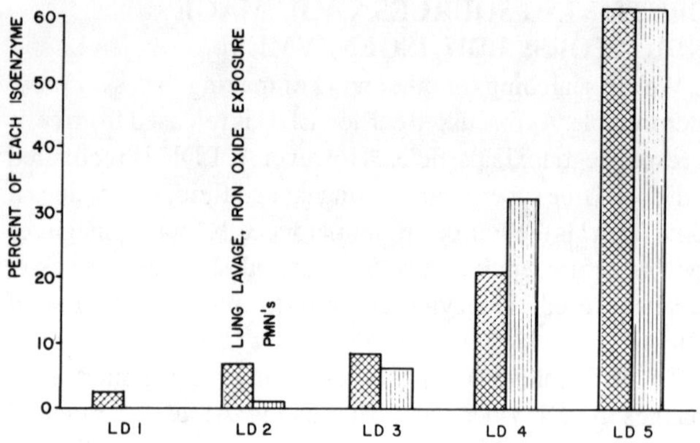

FIGURE 5. Comparison of LD isoenzyme patterns from hamster peritoneal PMNs and from lung lavage fluid of hamsters exposed to 3.75 mg iron oxide per 100 g body weight. (Adapted from Beck, B. D., Gerson, B., Feldman, H. A., and Brain, J. D., *Toxicol. Appl. Pharmacol.*, 71, 59, 1983. With permission.)

LD could be coming from PMNs. Macrophages have a similar LD composition, so they also may be a source.

AUTOMOBILE WASTE OIL COMBUSTION PRODUCTS

This assay is particularly suited for analyzing new complex agents which are just being introduced into the environment. We have recently investigated the pulmonary toxicity of respirable particulates from an air–atomizing oil space heater using automobile waste crankcase oil (AWO).[23] A combustion sample was prepared from AWO from a service station by Dr. R. E. Hall of the U. S. Environmental Protection Agency, using an air–atomizing oil burner rated at 250,000 BTU/h heat input. Respirable particulates were collected from a dilution tunnel by electrostatic precipitation using a massive air volume sampler.[24] Analysis of the particles showed certain metals were present at relatively high levels, for example: Pb, 75.6 mg/g; Zn, 23.0 mg/g; and Fe, 5.3 mg/g.

At 1 d postexposure, there was extensive pulmonary injury as demonstrated by cellular and biochemical indicators in BAL: (1) elevated levels of albumin, (2) increased extracellular glucosaminidase, and (3) impaired pulmonary macrophage phagocytosis. The injury was often greater than that seen in response to toxic α–quartz. Some of the data obtained are shown in Figures 6 to 8.

However, assays of BAL up to 14 d post-AWO exposure demonstrated that most indicators rapidly approached control values. This is in contrast to the persistent inflammation caused by α–quartz. As shown in Figure 9, LDH values approached control values at 2 weeks after intratracheal instillation of AWO. Following quartz exposure, the LDH level remains elevated. This suggests that the toxic effects of AWO stem from soluble components which are rapidly cleared. AWO may be less likely to cause chronic pulmonary disease than α–quartz unless exposure persists. Acute injury as manifested by bronchitis or increased susceptibility to infection may be a more likely outcome than fibrosis.

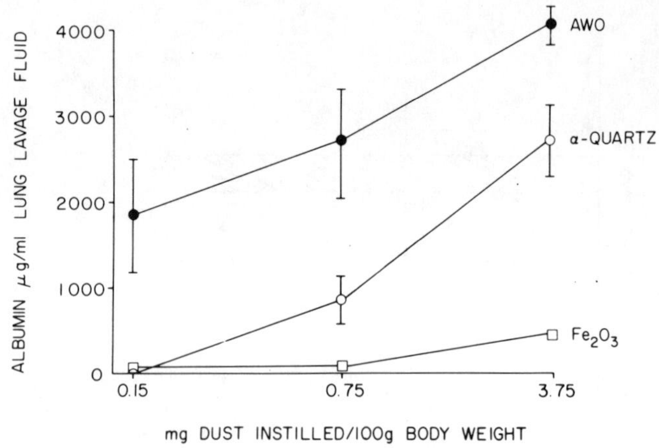

FIGURE 6. Concentration of albumin in the cell-free supernatant of BAL fluid. The effects of iron oxide, α–quartz, and combustion products of AWO are shown 1 d after intratracheal instillation. Values are mean ± standard errors. (Adapted from Beck, B. D., Brain, J. D., and Wolfthal, S. F., *Inhaled Particles IV*, Dodgson, J., Ed., British Occupational Hygiene Society, Edinburgh, Scotland.)

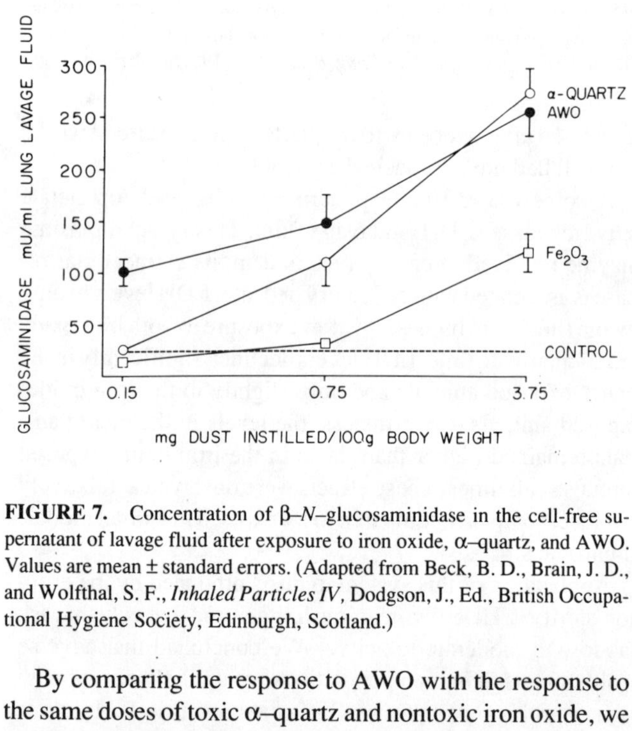

FIGURE 7. Concentration of β–*N*–glucosaminidase in the cell-free supernatant of lavage fluid after exposure to iron oxide, α–quartz, and AWO. Values are mean ± standard errors. (Adapted from Beck, B. D., Brain, J. D., and Wolfthal, S. F., *Inhaled Particles IV*, Dodgson, J., Ed., British Occupational Hygiene Society, Edinburgh, Scotland.)

By comparing the response to AWO with the response to the same doses of toxic α–quartz and nontoxic iron oxide, we conclude that the AWO combustion products have a high potential to cause acute lung injury. Both soluble and insoluble components of AWO can produce lung injury. Some, but not all, of these effects are due to acidity and divalent cations, such as lead, which are present at high levels.

CONCLUSION

Experimental pathology has frequently advanced because of the addition of new diagnostic tools. During the last decade, BAL has emerged as a very useful tool in the assessment of lung injury. It is applicable to both animal models exposed to inhaled particles and gases in a laboratory and to humans

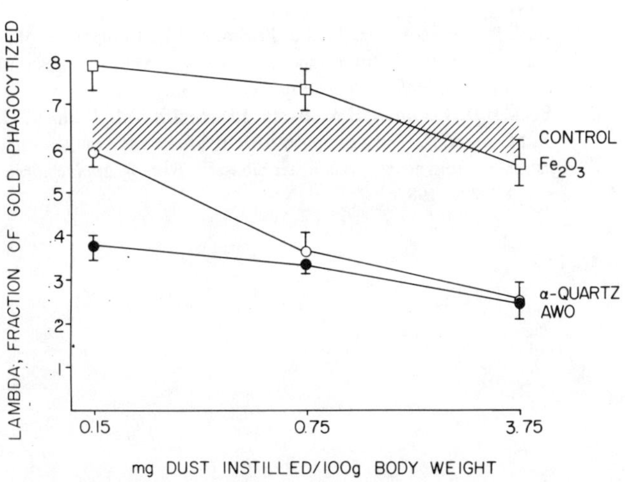

FIGURE 8. The fraction of gold particles, lambda ingested by macrophages *in situ*, is shown. Measurements were made 1 d after exposure to iron oxide, α–quartz, and AWO. Values are mean ± standard errors. (Adapted from Beck, B. D., Brain, J. D., and Wolfthal, S. F., *Inhaled Particles IV,* Dodgson, J., Ed., British Occupational Hygiene Society, Edinburgh, Scotland.)

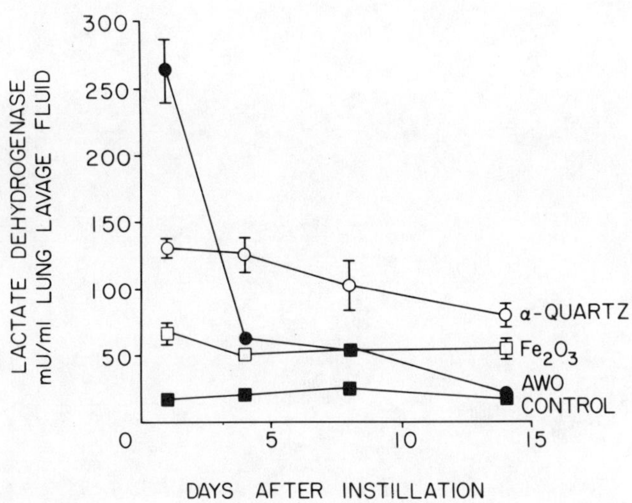

FIGURE 9. Time course for LDH in the extracellular supernatant fraction of lung lavage fluid after exposure to 3.75 mg iron oxide, α–quartz, or AWO per 100 g body weight. Values are mean ± standard errors. (Adapted from Beck, B. D., Brain, J. D., and Wolfthal, S. F., *Inhaled Particles IV*, Dodgson, J., Ed., British Occupational Hygiene Society, Edinburgh, Scotland.)

encountering exposures to the same agents in occupational and urban environments. Information can be gathered from BAL relating to the extent and type of lung injury and the mechanisms involved. Needed are more extensive comparisons of injury as judged by other approaches with the results of BAL. For example, short-term bioassay results can be integrated with industrial hygiene and epidemiology results as was done in a recent study of talc and granite dusts.[25] It is also likely that other constituents of BAL can be quantified which will help make bioassays utilizing BAL more specific and sensitive. The use of BAL in short-term animal assays can be an important source of information regarding the toxicity of new and poorly characterized inhaled particles.

REFERENCES

1. **Brody, A.R. and DeNee, P.B.,** Biological activity of inorganic particles in the lung, *CRC Crit. Rev. Toxicol.,* 7, 277, 1981.
2. **Gross, P., DeVilliers, A.J., and deTrevelle, R.T.P.,** Experimental silicosis, *Arch. Pathol.,* 84, 87, 1967.
3. **Busch, R.H., Filipy, R.E., Karagianes, M.T., and Palmer, R.F.,** Pathologic changes associated with experimenal exposure of rats to coal dust, *Environ. Res.,* 24, 53, 1981.
4. **Dean, J.H., Boorman, G.A., Luster, M.I., Adkins, B., Jr., Lauer, L.D., and Adams, D.O.,** Effect of agents of environmental concern on macrophage functions, in *Mononuclear Phagocyte Biology,* Volkman, A., Ed., Marcel Dekker, New York, 1984, 473.
5. **Liu, W.K., Tsao, S.W., and Wong, J.W.C.,** In vitro effects of fly ash on alveolar macrophages, *Conservation Recycling,* 7, 361, 1984.
6. **Snella, M.-C.,** Manganese dioxide induces alveolar macrophage chemotaxis for neturophils in vitro, *Toxicology,* 34, 153, 1985.
7. **Hatch, G.E., Boykin, E., Graham, J.A., Kewtas, J., Pott, F., Loud, K., and Mumford, J.L.,** Inhalable particles and pulmonary host defense: in vivo and in vitro effects of ambient air and combustion particles, *Environ. Res.,* 36, 67, 1985.
8. **Kaw, J.L.,** Tissue culture in pneumoconiosis, *CRC Crit. Rev. Toxicol.,* 5, 103, 1977.
9. **Miller, K.,** The effects of asbestos on macrophages, *CRC Crit. Rev. Toxicol.,* 5, 319, 1978.
10. **Fantone, J.C. and Ward, P.A.,** Mechanisms of lung parenchymal injury, *Am. Rev. Respir. Dis.,* 130, 484, 1984.
11. **Henderson, R.F., Rebar, A.H., Pickrell, J.A., and Neulton, G.J.,** Early damage indicators in the lung. III. Biochemical and cytological response of the lung to inhaled metal salts, *Toxicol. Appl. Pharmacol.,* 51, 123, 1979.
12. **Beck, B.D., Brain, J.D., and Bohannon, D.E.,** An in vivo hamster bioassay to assess the toxicity of particulates for the lungs, *Toxicol. Appl. Pharmacol.,* 66, 9, 1982.
13. **Brain, J.D., Knudson, D.E., Sorokin, S.P., and Davis, M.A.,** Pulmonary distribution of particles given by intratracheal instillation or by aerosol inhalation, *Environ. Res.,* 11, 13, 1976.
14. **Weissman, G., Smolin, J.E., and Korchak, H.M.,** Release of inflammatory mediators from stimulated neutrophils, *N. Engl. J. Med.,* 303, 27, 1980.
15. **Hook, G.E.R.,** Extracellular hydrolases of the lung, *Biochemistry,* 17, 520, 1978.
16. **Bell, D.Y., Haseman, J.A., Spock, A., McLennan, G., and Hook, G.E.R.,** Plasma proteins of the bronchoalveolar surface of the lungs of smokers and nonsmokers, *Am. Rev. Respir. Dis.,* 124, 72, 1981.
17. **Merrill, W., O'Hearn, E., Rankin, J., Naegel, G., Matthay, R.A., and Reynolds, H.Y.,** Kinetic analysis of respiratory tract proteins recovered during a sequential lavage protocol, *Am. Rev. Respir. Dis.,* 126, 617, 1982.
18. **Chichester, C.O., Palmer, K.C., Hayes, J.A., and Kagen, H.M.,** Lung lysyl oxidase and prolyl hydroxylase: increases induced by cadmium chloride inhalation and the effect of beta-aminopropionitrile in rats, *Am. Rev. Respir. Dis.,* 124, 709, 1981.
19. **Brain, J.D. and Corkery, G.C.,** The effect of increased particles on the endocytosis of radiocolloids by pulmonary macrophages *in vitro:* competitive and cytotoxic effects, in *Inhaled Particles IV,* Walton, W.H., Ed., Perfamon, New York, 1977, 551.
20. **Beck, B.D., Brain, J.D., and Bohannon, D.E.,** The pulmonary toxicity of an ash sample from Mt. St. Helens volcano, *Exp. Lung Res.,* 2, 289, 1981.
21. **Martin, T.R., Chi, E.Y., Covert, D.S., Hodson, W.A., Kessler, D.E., Moore, W.E., Altman, L.C., and Butler, J.,** Comparative effects of inhaled volcanic ash and quartz in rats, *Am. Rev. Respir.*

Dis., 128, 144, 1983.

22. **Beck, B.D., Gerson, B., Feldman, H.A., and Brain, J.D.,** Lactic dehydrogenase isoenzymes in hamster lung lavage fluid after lung injury, *Toxicol. Appl. Pharmacol.*, 71, 59, 1983.
23. **Beck, B.D., Brain, J.D., and Wolfthal, S.F.,** Assessment of lung injury produced by particulate emissions of space heaters burning automobile waste oil, in *Inhaled Particles VI,* Dodgson, J., Ed., British Occupational Hygiene Society, Edinburgh, Scotland, in press.
24. **Hall, R.E., Crooke, M.W., and Barbour, R.L.,** Comparison of air pollutant emissions from vaporizing and air atomizing waste oil heaters, *J. Air Pollut. Control Assoc.*, 33, 683, 1983.
25. **Beck, B.D., Feldman, H.A., Brain, J.D., Smith, T.J., Hallock, M., and Gerson, B.,** The pulmonary toxicity of talc and granite dust as estimated from an *in vivo* hamster bioassay, *Toxicol. Appl. Pharmacol.*, 87, 222, 1987.

Respiratory Risks in Agriculture

Overview of Respiratory Health Risks in Agriculture

C. P. W. Warren

From antiquity there are indications that the harmful effects of the dust of farm products on the lungs have been recognized.[1] The modern era began when Cadham of Winnipeg described asthma in farmers due to allergy to grain dust.[2] Eight years later Campbell described an outbreak of pneumonia in farmers, later called farmer's lung, which resulted from inhalation of the dust of moldy hay from the wet harvest of the previous year.[3] Thirty years later cooperative work by a team of immunologists and agricultural mycologists headed by Pepys and Gregory identified that farmer's lung is a hypersensitivity reaction to the microflora of the moldy hay.[4-7] The wet hay would ferment, heat, and so produce conditions which would favor the growth of the thermophilic actinomyces, especially *Micropolyspora faeni.* This filamentous bacterium produces spores of 3 to 4 μm which will penetrate the alveoli when inhaled. Prodigious numbers of them may be released when hay is shaken out for feed. The reaction in the lungs is called allergic alveolitis or hypersensitivity pneumonitis, a complex allergic reaction to the antigens. The possibility that a rural patient may be suffering from farmer's lung is often overlooked by physicians.

I will illustrate this overview with case histories of farmers seen by me.

C.P., aged 17 years, stopped eating and lost 9 kg. Anorexia nervosa was diagnosed but she had a cough. Her chest radiograph was normal. Her dyspnea on exertion had been attributed to the fatigue of her anorexia. Her doctor had neglected to ask what her job was — a helper on her brother's dairy farm. She had fine inspiratory crackles; lung function revealed a restrictive disorder. A transbronchial biopsy found granulomata. Her serum contained precipitating antibodies against farmer's lung antigens.

J.M., aged 19 years, developed pneumonia and was treated by his doctor in hospital with antibiotics. He recovered and returned to work only for his illness to relapse. He looked after 16,000 chickens. He had allergic alveolitis to chicken feathers.[8]

B.B., aged 14 years, developed a cough, fevers, weight loss, and dyspnea and miliary nodular infiltrates on chest radiograph. Despite negative TB skin test and cultures, TB was diagnosed and he was treated. He got better in hospital. He helped his father on his dairy farm. He had farmer's lung. Precipitins against *Thermactinomyces vulgaris* were positive.

The prevalence of farmer's lung is uncertain. Warren estimated that 8% of Manitoba dairy farmers had experienced it at some time.[9] In Finland a prevalence of farmer's lung of 1.7 and an incidence of 540 per 100,000 cattle farmers were found.[10] It is a serious disease; one third of sufferers are left with permanent abnormalities.[11] Farmer's lung is associated with hay because hay has a high water content and so molds relatively easily.

The introduction of silage as a feed for cattle has reduced the use of hay, and so of farmer's lung, but has introduced new hazards. The worst is silo-fillers lung. This is due to inhalation of oxides of nitrogen which form in silage. A farmer entering a silo may die of asphyxia. If rescued he may die of pulmonary edema in the next 24 h, or if this is survived, he may succumb to a progressive bronchiolitis obliterans in the next few weeks.[12] A newly recognized respiratory disorder of silos is the "silo unloaders syndrome" (SUS) which is part of the "organic dust toxic syndrome" (ODTS). This is a febrile flulike disorder, often associated with respiratory symptoms, that occurs when the moldy top layer of forage used to seal the top of the silo is removed; it is not farmer's lung.[13]

Grain rarely produces farmer's lung because it has a low water content and the appropriate organisms do not grow in it. Although allergic alveolitis has been reported from grain,[14] it is rare.

E.E., aged 53 years, cleaned out a bin of rotting flaxseed and some hours later developed a high fever, chills, cough, dyspnea, and myalgias. His chest radiograph showed lower lobe infiltrates. He recovered without treatment. It is assumed he had farmer's lung. The causative organism was not identified.

Similar stories have been recounted to the author by several farmers on clearing bins of rotten grain. This is grain fever, a variant of ODTS.

M.J., aged 42 years, cleaned out a bin of 1–year–old rotting rapeseed. Some hours later he developed chills, headache, fever, cough, phlegm, and some dyspnea. No abnormalities were found by his doctor and his chest radiograph was normal. He recovered in 36 h. Lung function done 4 d later was normal. Farmer's lung precipitins were negative.

The latter can also be chronic.

A.V., aged 25 years, had a winter job at a seed-cleaning plant. After a month he developed a nightly fever and extreme fatigue. He would go to bed every night at 7 p.m. His wife thought he was dying. He was referred after some 6 weeks of this illness. His physical examination, chest radiograph, and lung function were normal. All immunological tests were normal. He stopped work and all symptoms disappeared.

Although the molds of farmer's lung are rare in grain, other molds are very common and many are allergens that produce allergic asthma in farmers. *Cladosporium, Alternaria, Verticillium,* etc. are known as field fungi and produce asthma.[15] After harvest the flora change to storage fungi, mainly, *Aspergillus* sp. and *Penicillium* sp. These can produce allergic disease.

C.F., aged 53 years, would develop asthma and febrile infiltrates with eosinophilia often after shoveling stored grain. His chest radiographs revealed migratory infiltrates and a bronchogram proximal bronchiectasis. Investigations demonstrated he was sensitized to *A. fumigatus*.

Another source of potent allergens on farms is the arthropods. Mites, such as *Lepidoglyphus destructor,* have been shown to produce asthma in grain workers and in farmers.[16] Sensitization to common allergens, such as grasses, weeds, etc., is associated with acute symptoms such as wheezing which develop in farmers when exposed to grain, suggesting that atopy and perhaps sensitization to grain-related allergens are important in their pathogenesis.[17]

This paper has so far concentrated on the respiratory problems which occur on exposure to what we feed to animals. Modern animal husbandry is practiced in closed confinement buildings. As a result there may be little air exchange in it. Moreover, the manure falls through the slatted floor and is stored in a pit where it collects for up to 6 months before emptying. To empty it, the manure is agitated and huge quantities of H_2S may be released.[18]

R.W., aged 47 years, agitated his manure pit prior to pumping it out. He was in the barn and was overcome by the gas. Luckily he staggered to the door before losing consciousness and then recovered in the fresh air. Two of his cows died before he could turn off the agitator and get them out. Two days later when the provincial safety officer took measurements during agitation, H_2S levels of 600 ppm were found.

Such mistakes can be fatal and multiple, for would-be rescuers can also be overcome. The risk for H_2S is that, although at low levels it smells of rotten eggs, at high concentrations it paralyzes the olfactory system and the farmer is not aware of the danger.

Farming is clearly a hazardous occupation and the image it has of health and good fresh air may be inappropriate. Yet in many countries farmers have a greater longevity than other occupations. Part of this may be that farmers live healthy lifestyles. For instance, they tend not to smoke as much as the general population. This may not be wholly beneficial, for allergic alveolitis appears to be more common in nonsmokers. Warren reported that there was a higher proportion of nonsmokers in patients with allergic alveolitis than in comparable patients with sarcoidosis and fibrosing alveolitis.[19] Subsequent studies have confimed that the observation was correct, for in Finland[10] farmer's lung had twice the prevalence in nonsmokers than in smokers. Moreover, precipitins against farmer's lung antigens are detected twice as often in the serum of nonsmokers as they are in that of smokers in several studies.[20] The mechanism for this phenomenon is unclear.

The effect of these disorders on the general health of the rural community is uncertain. To clarify this we have examined the respiratory disorders of a rural population and related them to farming, smoking, and allergies. We performed a survey of all adults, aged 20 to 65 years in a grain-growing area. There were 1902 people who took part, 83% of the eligible population. The material and methods are reported elsewhere[21] in this book. There was a high frequency of acute disorders related to graindust exposure. Chronic respiratory symptoms of chronic cough and wheezing appeared to be more frequent in the farmers, but once corrected for age and other factors, this excess disappeared. Lung function was only worse in retired farmers who smoked more than did non- or current farmers. Thus, in this population of farmers, although acute disorders were experienced from exposure to grain, there was no evidence of accumulative effects in current farmers. Whether the retired farmers had been affected by grain dust is uncertain. An example of this may be the following.

G.A., aged 69 years, grain farmed from age 20 to 55 and smoked $^1/_2$ pack per day from age 20 to 44. He had always been short of breath when exposed to grain dust. He became dyspneic and was found to have severe bullous emphysema and severely abnormal lung function. α_1-Antitrypsin was normal.

There have been a few studies of the environment on farms, and yet it must be by improving the environment for the farmer that we will prevent their disorders. Total dust levels, colony-forming organism counts, endotoxin levels, toxic gas levels, and spore counts have been reported from various farm operations. There is one common theme to them all. Environmental conditions on farms would be unacceptable in any other industry. Recommended threshold- limit values for gases are often exceeded.[21] A summary of some observations on dust and endotoxin levels is shown in Table 1.[22,23]

Let us consider the following case histories: are they due to these hazards?

R.C., a lifetime nonsmoking dairy farmer, aged 47 years, had 3 years of dyspnea and no history suggestive of farmer's lung or asthma. Chest radiograph and lung function showed severe emphysema. His PO_2 was 65 falling to 57 on exercise. His α_1–antitrypsin phenotype was M.

A.F., a nonsmoking poultry farmer, aged 21 years, worked

TABLE 1
Total Dust[22] and Endotoxin[23] Levels in Some Farming Operations

	Mean dust (mgm/m³)	Endotoxin (µgm/m³)
Feeding		
Cows	5.6	—
Hogs	12.6	0.3 — 0.9
Chickens	6 — 7.2	0.6 — 1.1
Seed dressing	31.4	—
Ploughing	2.2	—
Sowing	7.0	—
Hay bale opening	9.9	6.7 (Moldy hay)
Threshing	4.0	
Graindrier	22.0	0.37 (Grain handling)

in a chicken plant handling feed and collecting eggs. He had recurrent evening cough and dyspnea without wheeze. Examination, chest radiograph, and lung function were normal. Allergy and precipitin tests were negative and he had no airway hyperresponsiveness.

E.W., a nonsmoking hog farmer, aged 57 years, got new barley and premix for his hogs and began to develop cough, chest tightness, and dyspnea. Examination and chest radiographs were normal, but he had mild airflow obstruction relieved by a bronchodilator. Allergy and precipitin tests were negative.

The first reaction of the farmer on learning of an occupational illness is often that it must have been caused by the "chemicals". Herbicides, pesticides, and fertilizer are known to be a great hazard. The farmer cannot believe that natural farm products could be more harmful to the lungs. One, of course, cannot deny that the chemicals may be harmful, but ingestion or absorption through the skin, not inhalation, is often their route of entry. Even paraquat, which produces severe lung fibrosis, enters through accidental swallowing.[25]

W.B., aged 43 years, was attempting to fix a broken valve on a tank of anhydrous NH_3, when there was an escape of the gas and he was overcome. He lay in the area for 30 min before rescue. He developed pulmonary edema and was ventilated. He was never weaned from the machine and died 3 months later of progressive pulmonary fibrosis and bronchiolitis obliterans.

There are many unique features to the occupational disorders of farmers. First, they may affect the whole family, for wives, children, or other relatives who help on the farm can develop the diseases. Moreover, when the farmer is sick he may have no help to run the farm. He is not usually covered by Worker's Compensation and rarely has disability insurance.. It is most important that research is continued by physicians, agricultural microbiologists, entymologists, and engineers to ensure that the conditions that produce these disorders on the farm do not occur and dust is kept to a minimum.

REFERENCES

1. **Hunter, D.,** *The Diseases of Occupations,* 6th ed., Hodder and Stoughton, London, 1978, 926.
2. **Cadham, F.T.,** Asthma due to grain rusts, *J.A.M.A.,* 83, 27, 1924.
3. **Campbell, J.M.,** Acute symptoms following work with hay, *Br. Med. J.,* 24, 1143, 1932.
4. **Gregory, P.H. and Lacey, M.E.,** Mycological examination of dust from mouldy hay associated with farmer's lung disease, *J. Gen. Microbiol.,* 30, 75, 1963.
5. **Pepys, J., Jenkins, P.A., Festenstein, G.N., Gregory, P.H., Lacey, M.E., and Skinner, F.A.,** Farmer's lung: thermophilic actinomyces as a source of farmer's lung antigen, *Lancet,* 2, 607, 1963.
6. **Williams, J.V.,** Inhalation and skin tests with extracts of hay and fungi in patients with farmer's lung, *Thorax,* 18, 182, 1963.
7. **Pepys, J. and Jenkins, J.A.,** Precipitin (FLH) test in farmer's lung, *Thorax,* 20, 21, 1965.
8. **Warren, C.P.W., Tse, K.S., and Cherniack, R.M.,** Extrinsic allergic alveolitis owing to hypersensitivity to chickens, *Am. Rev. Respir. Dis.,* 109, 672, 1974.
9. **Warren, C.P.W.,** Respiratory disorders in Manitoba cattle farmers, *C.M.A.J.,* 125, 41, 1981.
10. **Husman, K., Terho, E.O., and Vohlonen, I.,** Respiratory disorders among Finnish farmers, Int. Symp. on work-related respiratory disorders among farmers, Kuopio, Finland, 1985, *Eur. J. Respir. Dis.,* to be published.
11. **Braun, S.R., do Pico, G.A., Tsiatis, A., Horvath, E., Dickie, H.A., and Rankin, J.,** Farmer's lung disease: long term clinical and physiological outcome, *Am. Rev. Respir. Dis.,* 117, 185, 1979.
12. **Fleetham, J.A., Munt, P.W., and Tunnicliffe, B.W.,** Silo-fillers disease, *C.M.A.J.,* 119, 482, 1978.
13. **Pratt, D.S. and May, J.J.,** Feed-associated respiratory illness in farmers, *Arch. Environ. Health,* 39, 43, 1984.
14. **Dutkiewicz, J., Kus, E., Dutkiewicz, E., and Warren, C.P.W.,** Hypersensitivity pneumonitis in grain farmers due to sensitization to *Erwinia herbicola, Ann. Allergy,* 54, 65, 1985.
15. **Darke, C.S., Knowleden, J., Lacey, J., and Milford Ward, A.,** Respiratory disease of workers harvesting grain, *Thorax,* 31, 294, 1976.
16. **Warren, C.P.W., Sinha, R.N., and Holford-Strevens, V.,** Sensitization in a grainhandler to the stored grain mite, *Lepidoglyphus destructor, Ann. Allergy,* 50, 30, 1983.
17. **Manfreda, J., Holford-Strevens, V., Cheang, M., and Warren, C.P.W.,** Acute symptoms following exposure to grain dust in farming, *Environ. Health Perspect.,* 66, 73, 1986.
18. **Osbern, L.N. and Crapo, R.O.,** Dung lung: a report of toxic exposure to liquid manure, *Ann. Intern. Med.,* 95, 312, 1981.
19. **Warren, C.P.W.,** Allergic alveolitis: a disease commoner in nonsmokers, *Thorax,* 32, 567, 1977.
20. **Cormier, Y., Belanger, J., and Durand, P.,** Factors influencing the development of serum precipitins to farmer's lung antigen in Quebec dairy farmers, *Thorax,* 40, 138, 1985.
21. **Donham, K.J., Rubino, M., Thedell, T.D., and Kammermeyer, J.,** Potential health hazards to agricultural workers in swine confinement buildings, *J. Occup. Med.,* 19, 383, 1977.
22. **Louhelainen, K., Kangas, J., Husman, K., and Terho, T.E.,** Total dust exposure of farmers in different work environments, Int. Symp. on work-related respiratory disorders among farmers, Kuopio, Finland, 1985, *Eur. J. Respir. Dis.,* to be published.
23. **Rylander, R. and Morey, N.,** Airborne endotoxin in industries processing vegetable fibres, *Am. Ind. Hyg. Assoc. J.,* 43, 811, 1982.

Respiratory Health in Farmers: Symptoms and Pulmonary Function*

J. A. Dosman, B. L. Graham, H. H. McDuffie, D. Hall, P. Van Loon, P. Bhasin, and F. Froh

Previously, the occupation of farming has not been associated with respiratory dysfunction. Individual cases of farmer's lung and silo-filler's disease[1,2] and the causal agent in restrictive allergic alveolitis have been described.[3] An epidemiological study describing the prevalence of farmer's lung in the population has been reported.[4] The understanding that respiratory symptoms and obstructive airways dysfunction in agricultural workers could be related to inhaled biological dusts is relatively new.[5] Early studies implicated a relationship between respiratory dysfunction and exposure to cereal grain dusts[6–11] and more recent epidemiological studies have confirmed these relationships and extended understanding.[12–16] It has been suggested that the combined effects of occupational exposure to grain dust and inhalation of cigarette smoke may have a synergistic effect leading to small airways dysfunction in young cereal grain workers.[17] Alterations in pulmonary function in grain workers could be related to increases in nonspecific bronchial reactivity that are separate from allergy.[18] Animal exposure studies to grain dust have demonstrated an alveolitis with macrophage proliferation in mice[19] with both airways obstruction and alveolitis in rabbits.[20] Changes in lung function in humans exposed to biological dusts have been related to environmental factors.[21–23] Subsequently, it has been suggested that farmers, in general, may be prone to respiratory disease,[24,25] and that farmers might be exposed to a variety of respiratory health risks.[26]

We carried out a respiratory health survey on 1824 farmers in 16 municipalities in Saskatchewan. These data were compared with control data from a group of nonfarming community control subjects in a large town in the central area of the province, with the objective of determining whether farmers had more respiratory symptoms and/or poorer pulmonary function than nonexposed control subjects, controlling for age and smoking exposure.

Each male farmer and male control subject participating in the study completed a technician-administered questionnaire covering history of employment, respiratory health, and tobacco smoking and an assessment of working conditions. The respiratory questionnaire and smoking history were modified from the American Thoracic Society Epidemiology Standardization Questionnaire.[27] A trained technician assisted the farmers and control subjects in performing three acceptable maximum exhalation maneuvers using a computerized data acquisition and analysis system for testing as previously described.[16,17,28] Each variable was expressed as a percentage of expected value based on the farmer's or control subject's sex, height, weight, and age, using the regression equations of Dosman et al. for forced vital capacity (FVC), forced expired volume in 1 s (FEV_1), FEV_1/FVC, and maximum midexpiratory flow rate.[29] Data regarding presence or absence of farming, exposure, type of farming, and symptoms were obtained from the questionnaire. Chronic bronchitis was defined as sputum production on most days for 3 or more consecutive months each year for 2 or more years. The questionnaire also determined smoking habit and categorized individuals into nonsmokers (smoking history of less than 20 packs in a lifetime), current smokers, and ex–smokers (subjects who were previously regular smokers but with no smoking at all in the previous month). Number of pack years for smoking was calculated by multiplying the average number of packages of 20 cigarettes per day smoked by the number of years of smoking history. Analysis of variance was used to compare pulmonary function tests of farmers and control subjects and both continuous and discontinuous variables were corrected for both age and smoking.[30] A Mantel-Haenzel analysis was used to determine relative risk for discontinuous variables.[31]

Table 1 demonstrates mean values of anthropometric variables in the farmers and control subjects, listed according to three age categories. The farmers were significantly taller, heavier, and had a somewhat greater pack–year history of smoking than did the control subjects. Table 2 demonstrates the principle types of farming. Over 90% of farmers grew grain or other field crops, while 65.4% had beef cattle, and fewer numbers other types of animals. Symptom prevalence in farmers and control subjects is shown in Table 3. Morning phlegm was significantly more prevalent in the younger and middle group of farmers as well as in the group as a whole, being present in 14.5% of farmers and 9.2% of control subjects. Wheeze was more frequent in each category of farmers, relative to control subjects, and was present in 27.4% of all farmers as compared to 9.9% of control subjects. Similarly, shortness of breath was present more frequently in all age categories, as well as in 33.3% of the total group of farmers and 18.2% of the control subjects. The condition of chronic bronchitis was present in 11.1% of all farmers and 7.7% of control subjects. Mean values of pulmonary function

* Previously published in expanded form in *J. Occup. Med.*, 29, 38, 1987, and in part in the *Respiratory Times*.

TABLE 1
Mean Values of Anthropometric Variables in Farmers and Control Subjects

Age categories	≥20 years, <35 years		≥35 years, <50 years		≥50 years, <70 years		Total	
	No.	Mean ± SD	No.	Mean ± SD	No.	Mean ± SD	No.	Mean ± SD
Age (years)								
Farmers	497	28.4 ± 4.0	558	42.1 ± 4.4	769	58.0 ± 5.3	1824	45.1 ± 13.1
Control subjects	264	27.2 ± 4.2[a]	156	41.7 ± 4.3	136	56.4 ± 4.6[b]	556	38.4 ± 12.7[a]
Height (cm)								
Farmers	497	177.5 ± 6.3	558	176.9 ± 6.5	769	174.2 ± 6.5	1824	175.9 ± 6.6
Control subjects	264	174.7 ± 6.1[a]	156	173.4 ± 6.7[a]	136	172.1 ± 7.0[a]	556	173.7 ± 6.6[a]
Weight (kg)								
Farmers	497	83.5 ± 13.6	558	86.1 ± 13.1	769	83.7 ± 12.8	1824	84.4 ± 13.2
Control subjects	264	80.6 ± 12.6[b]	156	84.7 ± 12.6	136	84.5 ± 11.6	556	82.7 ± 12.5[b]
Pack years								
Farmers	494	5.4 ± 7.5	555	13.6 ± 14.6	768	20.9 ± 21.3	1817	14.5 ± 17.6
Control subjects	264	5.1 ± 6.4	156	15.0 ± 14.9	134	22.3 ± 19.7	554	12.0 ± 15.0[b]

[a] $p < 0.001$.
[b] $p < 0.01$.

TABLE 2
Commodities Handled or Grown in Last 5 Years

Commodity	≥20, <35 (n = 495)		≥35, <50 (n = 554)		≥50, <70 (n = 761)		Total (n = 1810)	
	No.	%	No.	%	No.	%	No.	%
Wheat	485	98.0	541	97.7	747	98.2	1773	98.0
Other field crops	486	98.2	528	95.3	704	92.5	1718	94.9
Beef	334	67.5	376	67.9	473	62.2	1183	65.4
Poultry	247	49.9	268	48.4	338	44.4	853	47.1
Horses	183	37.0	205	37.0	203	26.7	591	32.7
Hogs	183	37.0	157	28.3	164	21.6	504	27.8
Dairy	67	13.5	88	15.9	86	11.3	241	13.3
Sheep	30	6.1	37	6.7	21	2.8	88	4.9
Other livestock	22	4.4	20	3.6	14	1.8	56	3.1
Grain only	101	20.4	122	22.0	224	29.4	447	24.7
Livestock only	1	0.2	4	0.7	1	0.1	6	0.3
Grain and livestock	393	79.4	428	77.3	536	70.4	1357	75.0

test variables in farmers and control subjects are demonstrated in Table 4. The data are expressed as percent of expected values (adjusting for age) and are adjusted for pack years using an analysis of covariance. FVC and FEV_1 were both significantly lower in all categories of farmers and control subjects. Maximum midexpiratory flow rate was different between farmers and control subjects when the group as a whole was considered. The ratio of FEV_1 to FVC was significantly greater in the farmers than the control subjects in the age ≥20, <35 group, and this difference largely accounted for a difference in the group as a whole. There was no significant difference in $FEV_1/FVC \times 100$ for the middle and oldest age categories.

The data from this survey indicate that farmers in Saskatchewan have more respiratory symptoms and lower values for FVC and FEV_1 than do a group of nonfarming town-dwelling men from the same geographic area. There is evidence that farmers are at risk of developing acute lung trouble as a result of inhaled substances.[1,2] Such substances include organic antigens such as pollen, fungal spores, animal products, grain dust and grain dust mites, chemical fertilizers and herbicides, and gases associated with plant decomposition including nitrous oxide, hydrogen sulfide, methane, and ammonia. The respiratory manifestations of inhaling these substances may include interstitial lung diseases such as hypersensitivity pneumonitis; airways obstruction such as asthma, chronic airways obstruction, and bronchitis; acute laryngotracheobronchitis; and even acute noncardiogenic pulmonary edema.

Hypersensitivity pneumonitis (extrinsic allergic alveoli-

TABLE 3
Symptom Prevalence in Farmers and Control Subjects

	Age categories															
	≥20, <35				≥35, <50				≥50, <70				Total			
	Farmers (n = 497)		Control subjects (n = 264)		Farmers (n = 558)		Control subjects (n = 156)		Farmers (n = 769)		Control subjects (n = 136)		Farmers (n = 1,824)		Control subjects (n = 556)	
Symptoms	No.	%	No.	%	No.	%	No.	%	No.	%	No.	%	No.	%	No.	%
Morning phlegm	68	13.7	20	7.6[a]	78	14.0	12	7.7[a]	119	15.5	19	14.0	265	14.5	51	9.2[b]
Wheeze	131	26.4	26	9.8[c]	149	26.7	9	5.8[c]	220	28.6	20	14.7[c]	500	27.4	55	9.9
Shortness of breath	103	20.7	24	9.1[c]	182	32.6	33	21.2[b]	322	41.9	44	32.4[a]	607	33.3	101	18.2[c]
Chronic bronchitis	50	10.1	17	6.4	59	10.6	10	6.4	93	12.1	16	11.8	202	11.1	43	7.7[a]

[a] $p < 0.05$.
[b] $p < 0.01$.
[c] $p < 0.001$.

TABLE 4
Pulmonary Function Test Variables in Farmers and Control Subjects (Adjusted for Pack Years)[a]

Variables (% predicted)	Age categories ≥20, <35 (mean ± SE)	≥35, <50 (mean ± SE)	≥50, <70 (mean ± SE)	Total (mean ± SE)
FVC				
Farmers	94.5 ± 0.5	98.1 ± 0.5	99.5 ± 0.5	97.7 ± 0.3
Control subjects	103.5 ± 0.7[b]	107.3 ± 0.9[b]	109.8 ± 1.3[b]	106.1 ± 0.5[b]
FEV_1				
Farmers	94.9 ± 0.5	97.2 ± 0.6	95.8 ± 0.6	96.1 ± 0.3
Control subjects	102.3 ± 0.7[b]	104.5 ± 1.0[b]	103.3 ± 1.4[b]	102.8 ± 0.6[b]
MMFR				
Farmers	94.3 ± 1.1	91.7 ± 1.1	78.3 ± 1.1	87.0 ± 0.6
Control subjects	96.4 ± 1.4	92.8 ± 2.2	77.8 ± 2.5	89.8 ± 1.2[c]
$FEV_1/FVC \times 100$				
Farmers	100.0 ± 0.3	98.9 ± 0.3	95.5 ± 0.4	97.8 ± 0.2
Control subjects	98.3 ± 0.4[b]	97.6 ± 0.6	93.9 ± 0.9	96.7 ± 0.4[d]

[a] Abbreviations used are FVC, forced vital capacity; FEV_1, forced expired volume in 1 s; MMFR, maximum midexpiratory flow rate; $FEV_1/FVC \times 100$, ratio of FEV_1 to FVC.
[b] $p < 0.001$.
[c] $p < 0.5$.
[d] $p < 0.1$.

tis)[1] may present as an acute illness, which is diagnosed at the time. However, it is possible that many cases go undiagnosed and lead to long–term respiratory impairment of the restrictive type.[4] Hypersensitivity pneumonitis may present as an acute febrile illness with onset several hours after exposure or as a slowly progressive interstitial pulmonary process. There is usually inhalation of large amounts of antigen if particle size is within the respirable range and the immunopathogenesis involves formation of immune complexes, cell–mediated immunity, or both. Fungal spores of *Thermophilic actinomycetes* released from moldy hay or grain have been implicated as a major cause of "farmer's lung".[3] Farmers may be at risk of developing hypersensitivity pneumonitis from *Aspergillus* fungal sores, *Penicillium* fungal spores, or proteins contained in bird droppings.[32]

Asthma may be a relatively common condition in farmers as a result of exposure to pollen from cereal grain danders, livestock, fungal antigens in grain dusts and on live crops, and dust mites in organic dust. The reactions may include early asthmatic responses, late asthmatic responses, and recurrent nocturnal asthma for several days after exposure.[33] These conditions may be caused by or associated with allergen-induced increases in nonspecific bronchial reactivity or, as may be the case in some exposures to grain dust, increases in bronchial reactivity without increased atopy.[18] Some of the dusts, for example, grain dust, may cause asthma by pharmacological release of mediators from mast cells, and it is known that grain dust activates the alternative pathway of complement.[34] Organic phosphates used as insecticides may induce asthma by inhibition of cholinesterase that is thought to potentiate the effect of acetylcholine released from the vagal fibers inervating bronchial smooth muscle.[35] Also, farmers with asthma or with increases of bronchial reactivity may develop bronchospasm from the nonspecific irritant effects of various dusts, fumes, vegetable matters, and gases to which they are exposed.

The increased prevalence of respiratory symptoms in farmers compared with control nonfarmers suggests that exposure to various environmental dusts, chemicals, and other substances can cause effects on the tracheobronchial tree that are separate from the effects expected by exposure to cigarettes. It has been suggested by Fletcher and Peto[36] that young smokers suffer from both airflow obstruction and mucus hypersecretion and that these may represent unrelated disease processes. We have shown that the effects of cigarette smoking and the effects of being exposed to grain dust result in approximately equal amounts of dysfunction in grain workers.[28] The significance of the increase in respiratory symptoms in farmers as opposed to grain workers remains uncertain. It is possible that increased symptoms may relate more to the ongoing acute exposures to dusts or other environmental agencies, whereas the development of airflow obstruction may relate more to chronic accumulative effects.[28]

Recently, health risks from animal confinement units have been described.[37] It is possible that pulmonary edema, asthma, bronchitis, airways obstruction and hypersensitivity pneumonitis may all be more common in workers who spend time in animal confinement units.[37]

The findings that we have presented demonstrate differences in respiratory symptoms and pulmonary function test variables between farmers and nonfarming community control subjects. If substantiated, these findings could indicate the presence of a respiratory problem among the large number of farmers in North America and other areas. Further epidemiological studies to confirm our findings, and delineation of possible mechanisms, are indicated.

REFERENCES

1. **Rankin, J., Kobayashi, M., Barbee, R. A., and Dickie, H. A.,** Pulmonary granulomatoses due to inhaled organic antigens, *Med. Clin. North Am.*, 51, 459, 1967.
2. **Lowry, T. and Schulman, L.M.,** "Silo-filler's disease": a syndrome caused by nitrogen dioxide, *J.A.M.A.*, 162, 153, 1956.
3. **Pepys, J., Jenkins, P.A., Festenstein, G.N., Gregeory, P.H., Lacey, M.E., and Skinner, F.A.,** Farmer's lung: thermophilic actinomycetes as a source of "farmer's lung" hay antigen, *Lancet*, 2, 607, 1963.
4. **Madsen, D., Klock, L.E., Wenzel, F.J., Robbins, J.L., and Schmidt, C.D.,** The prevalence of farmer's lung in an agricultural population, *Am. Rev. Respir. Dis.*, 113, 171, 1976.
5. **Dosman, J.A. and Cotton, D.J., Eds.,** *Occupational Pulmonary Disease: Focus on Grain Dust and Health,* Academic Press, New York, 1980.
6. **Williams, N., Skoulas, A., and Merriman, J.E.,** Exposure to grain dust. I. A survey of the effects, *J. Occup. Med.*, 6, 319, 1964.
7. **Skoulas, A., Williams, N., and Merriman, J.E.,** Exposure to grain dust. II. A clinical study of the effects, *J. Occup. Med.*, 6, 359, 1964.
8. **Tse, K., Warren, P., Janusz, M., McCarthy, D.S., and Cherniack, R.M.,** Respiratory abnormalities in workers exposed to grain dust, *Arch. Environ. Health,* 27, 74, 1973.
9. **Kleinfeld, M., Messite, J., Swencicki, R.E., and Shapiro, J.,** A clinical and physiologic study of grain handlers, *Arch. Environ. Health,* 16, 380, 1968.
10. **Becklake, M.R., Jodoin, G., McDonald, J.C., Rose, B., and Fraser, R.G.,** Cardiorespiratory function in men exposed to grain dust, in *Occupational Pulmonary Disease: Focus on Grain Dust and Health*, Dosman, J.A. and Cotton, D.J., Eds., Academic Press, New York, 1980.
11. **Sheridan, D., Deutscher, C., Tan, L., Maybank, J., Gerrard, J., Horne, S., Yoshida, K., Barnett, G.D., Cotton, D., and Dosman, J.A.,** The relationship between exposure to cereal grain dust and pulmonary function in grain workers, in *Occupational Pulmonary Disease: Focus on Grain Dust and Health,* Dosman, J.A. and Cotton, D.J., Eds., Academic Press, New York, 1980, 229.
12. **doPico, G.A., Reddan, W., Flaherty, D., et al.,** Respiratory abnormalities among grain handlers: a clinical, physiologic, and immunologic study, *Am. Rev. Respir. Dis.,* 115, 915, 1977.
13. **Broder, I., Mintz, S., Hutcheon, M., et al.,** Comparison of respiratory variables in grain elevator workers and civic outside workers of Thunder Bay, Canada, *Am. Rev. Respir. Dis.,* 119, 192, 1979.
14. **Broder, I., Mintz, S., Hutcheon, M.A., et al.,** Effect of layoff and rehire on respiratory variables in grain elevator workers, *Am. Rev. Respir. Dis.,* 122, 601, 1980.
15. **Chan-Yeung, M., Schulzer, M., MacLean, L., et al.,** Epidemiological health survey of grain elevator workers in British Columbia, *Am. Rev. Respir. Dis.,* 121, 329, 1980.
16. **Dosman, J.A., Cotton, D.J., Graham, B.L., et al.,** Chronic bronchitis and decreased forced expiratory flow rates in lifetime nonsmoking grain workers, *Am. Rev. Respir. Dis.,* 121, 11, 1980.
17. **Cotton, D.J., Graham, B.L., Li, K.-Y.R., Froh, F., Barnett, G.D., and Dosman, J.A.,** Effects of smoking and occupational exposure on peripheral airways function in young cereal grain workers, *Am. Rev. Respir. Dis.,* 126, 660, 1982.
18. **Mink, J.T., Gerrard, J.W., Cockcroft, D.W., Cotton, D.J., and Dosman, J.A.,** Increased bronchial reactivity to inhaled histamine in nonsmoking grain workers with normal lung function, *Chest,* 77, 28, 1980.
19. **Armanious, M., Cotton, D.J., Wright, J., Dosman, J., and Carr, I.,** Grain dust and alveolar macrophages: an experimental study of the effects of grain dust on the mouse lung, *J. Pathol.,* 136, 265, 1982.
20. **Stepner, N., Broder, I., and Baumal, R.,** A microscopic study of the pulmonary effects of chronic grain dust exposure in rabbits, chap. 25, this volume.
21. **Farant, J.-P. and Moore, C.F.,** Dust exposures in the Canadian grain industry, *Am. Ind. Hyg. Assoc. J.,* 39, 177, 1978.
22. **Yoshida, K. and Maybank, J.,** Physical and environmental characterisics of grain dust, in *Occupational Pulmonary Diseases: Focus on Grain Dust and Health,* Dosman, J.A. and Cotton, D.J., Eds., Academic Press, New York, 1980, 441.
23. **Cotton, D.J. and Dosman, J.A.,** Grain dust and health. III. Environmental factors, *Ann. Intern. Med.,* 89, 420, 1978.
24. **Warren, C.P.W. and Manfreda, J.,** Respiratory symptoms in grain

farmers: a brief report, in *Occupational Pulmonary Disease: Focus on Grain Dust and Health,* Dosman, J.A. and Cotton, D.J., Eds., Academic Press, New York, 1980.

25. **Warren, C.P.W. and Manfreda, J.,** Respiratory symptoms in Manitoba farmers: association with grain and hay handling, *C.M.A.J.,* 122, 1259, 1980.
26. **Cockcroft, D.W. and Dosman, J.A.,** Respiratory health risks in farmers, *Ann. Intern. Med.,* 95, 380, 1981.
27. **Ferris, B.G.,** Epidemiology standardization project, *Am. Rev. Respir. Dis.,* 118, 1, 1978.
28. **Cotton, D.J., Graham, B.L., Li, K.-Y.R., Froh, F., Barnett, G.D., and Dosman, J.A.,** Effects of grain dust exposure and smoking on respiratory symptoms and lung function, *J. Occup. Med.,* 25, 131, 1983.
29. **Dosman, J.A., Cotton, D.J., Graham, B.L., Hall, D.L., Li, K.-Y.R., Froh, F., and Barnett, G.D.,** Sensitivity and specificity of early diagnostic tests of lung function in smokers, *Chest,* 79, 6, 1981.
30. **Steel, R.G.D. and Torrie, J.H.,** *Principles and Procedures in Statistics,* McGraw-Hill, New York, 1960.
31. **Mantel, N. and Haenzel, W.,** Statistical aspects of the analysis of data from retrospective studies of disease, *J. Natl. Cancer Inst.,* 22, 719, 1959.
32. **Hargreave, F.E., Pepys, J., Longbottom, J.L., et al.,** Bird breeder's (fancier's) lung, *Lancet,* 1, 445, 1966.
33. **Davies, R.J., Green, M., and Schofield, N.M.,** Recurrent nocturnal asthma after exposure to grain dust, *Am. Rev. Respir. Dis.,* 114, 1011, 1976.
34. **Olenchock, S.A., Mull, J.C., Major, P.C., et al.,** Activation of the alternative pathway of complement by grain, *Clin. Allergy,* 8, 125, 1978.
35. **Weiner, A.,** Bronchial asthma due to the organic phosphate insecticides, *Ann. Allergy,* 15, 211, 1961.
36. **Fletcher, C. and Peto, R.,** The natural history of chronic airflow obstruction, *Br. Med. J.,* 1, 1645, 1977.
37. **Merchant, J.A. and Donham, K.J.,** Health risks from animal confinement units, chap. 13, this volume.

Toxic Gases from Manure Pits

R. O. Crapo

Efforts to improve the productivity of livestock operations over the last 15 to 20 years have resulted in the proliferation of systems in which large numbers of animals are raised in small areas[1,2] or "confinement buildings". This creates the problem of dealing with animal wastes, usually stored in pits or tanks under or outside the confinement building. Manure pits vary in size, but Morse et al.[3] described a representative structure 2.4 m high with a capacity of 378,500 l. Outside slurry ponds are less frequently used because of transportation and odor problems. The manure pit is usually under the confinement building so waste material can be easily moved into the pit and stored as a liquid or slurry until it can be applied as fertilizer.

Confinement buildings and management of animal waste products create real and potential health hazards for confined animals and for humans working with and around them.[1,2,4,5] Stevens[4] summarized the hazards to include: (1) toxic gases from digestion of organic material by anaerobic bacteria; (2) pathogenic microbes on contaminated surfaces, in airborne particulates, or in contaminated livestock/livestock products; (3) indirect microbial problems including drug resistance that may result from the use of antibiotics; (4) airborne particulates which increase the risk of both irritant and hypersensitivity reactions;[1,5] (5) chemical residues in food additives which may accumulate and be transferred when waste products are used as feed or fertilizer; (6) methane gas which may reach explosive levels; and (7) large liquid pools which always present a potential for drowning.

This paper will focus on the problem of toxic gas exposures, which are the most dramatic of these hazards and may rapidly result in multiple deaths.

Injuries and deaths caused by sewage environment exposures were reported as early as the 18th century, and since then have been reported from many types of exposures to decaying organic materials.[6] Risk increases when exposure occurs in an enclosed or a poorly ventilated area. Sewage systems, abandoned cesspools, poorly ventilated barns, latrines, and the holds of fishing ships have all been implicated.[3,5,7,8] The potential problems have increased with the proliferation of the livestock confinement systems; in 1982 Donham et al.[2] estimated that 500,000 people work in or around such systems. The U.S. Center for Disease Control reported eight deaths and two near deaths associated with liquid manure system exposure between August 1979 and April 1981.[9] In the absence of a formal reporting system, the numbers are likely to be underestimated.

Approximately 150 gases may occur in such systems.[2] The "fixed" gases of primary importance, include methane (CH_4), ammonia (NH_3), carbon dioxide (CO_2), and hydrogen sulfide (H_2S). Carbon monoxide (CO) is also present. All are found in manure pits and confinement systems, at times in concentrations that exceed current threshold values.[2] The levels of toxic gases in manure pits may vary; agitation of the slurry in a pit may cause a rapid increase in the concentration of gases to potentially lethal levels.[1,2] Increased concentrations with agitation are likely due to increased surface area created by the agitation, the release of trapped bubbles containing gases, and the disturbance of the surface scum.[2] Gas levels may drop rapidly after the agitation stops. Hagley and South reported a drop in H_2S concentration in a manure pit from more than 150 to 70 ppm in 10 min.[10] Concentrations may also vary as a function of the ventilation in the confinement building. Anoxia and H_2S are the most likely causes of serious injury.

CO_2 produced by animals in a tightly enclosed confinement area could cause a reduction of oxygen levels to 10% after 6 to 8 h.[1] In spite of this, anoxia is generally not thought to be a common cause of injury. The potential for injury is greater for livestock because humans are able to leave the environment. On the other hand, acute anoxia is known to cause rapid impairment of judgement. In an enclosed manure pit, the potential for an anoxic injury exists, and current descriptions of injuries do not allow it to be excluded as a cause of injury.[9] Anoxia may act synergistically to potentiate other toxic injuries[5] or reduce the potential for escape.

The most serious toxic exposure is H_2S, thought to be the cause of most human toxicity in confinement buildings and manure pits; documentation that H_2S was the actual source of injury is often lacking.

Our experience with an exposure illustrates many of the features of such events and of H_2S toxicity.[11] A farmer entered a recently partially drained manure pit to retrieve a lid kicked into the pit by a cow, and rapidly became unconscious. Two sons entered the pit to attempt a rescue; one succumbed and the other escaped to call for help. The county sheriff responded accompanied by the barber who had been cutting his hair; they sequentially entered the pit and rapidly collapsed. An ambulance attendant also entered and collapsed, but was retrieved by a rope tied around his waist and revived rapidly. The barber was fished out with a rope around his arm and was revived by cardiopulmonary resuscitation. The others were retrieved by firemen using self-contained breathing apparatuses, and were pronounced dead at the scene; autopsies showed heart blood sulfide ion concentrations of 0.8, 3.6, and 5.0 mg/l, consistent with acute H_2S toxicity. Later analysis suggested the H_2S concentration at the time of the incident was over 500 ppm.[9] Two of the fatalities had massive manure aspiration and the other had severe pulmonary edema without aspiration. The barber survived a stormy course with adult respiratory distress syndrome (ARDS) complicated by multiple infections of his lungs and sinuses.[11] His lung function returned to normal, but he had a persistent recent memory defect.

This incident had characteristics similar to previous reports including recent manipulation or agitation of the slurry, rapid loss of consciousness, multiple deaths, sequential entry of rescuers without respiratory equipment and their sequential collapse, successful retrieval of victims by individuals with respiratory equipment, rapid revival of victims quickly removed from the exposure, and severe pulmonary edema in surviving and nonsurviving victims.[9] Aspiration of manure suggests that drowning played a role in the fatalities.

H_2S gas occurs in confinement areas and manure pits in concentrations varying from 10 ppm (common) to >500 ppm, a highly toxic level.[2] The dangers, recommended maximum exposure levels, and recommended safety requirements, detailed elsewhere,[6] are summarized briefly here.

The extreme toxicity of H_2S has been known since the late 1700s.[6] A colorless, flammable gas, H_2S is heavier than air and commonly occurs in many industrial settings including manure pits and confinement buildings.[6] H_2S has a characteristic odor of rotten eggs in low concentrations (10 to 20 ppm);[12,13] however, olfactory nerve endings are quickly paralyzed by high concentrations (>150 ppm) of the gas, and, thus, potential victims will not be warned by the odor when the danger is greatest.[14] Human toxicity usually results from acute inhalation exposure; there is uncertainty about whether a chronic injury entity exists.[4] The major toxic effect of H_2S appears to be the inhibition of cytochrome oxidase in a manner similar to cyanide.[13,14] Concentrations as low as 0.005% (50 ppm) may cause toxic symptoms and concentrations of 0.05 to 0.2% (500 to 2000 ppm) may be rapidly fatal.[12,13] The National Institute of Occupational Safety and Health (NIOSH) has recommended that areas with H_2S levels >47 ppm be immediately evacuated.[6,9] Toxicity appears to relate to concentration rather than duration of exposure because the body has good capacity to detoxify sulfides.[12] Low concentrations (50 to 200 ppm) result primarily in mucous membrane irritation with conjunctivitis, sore throat, and cough being common. Increased exposure leads to nausea, vomiting, diarrhea, abdominal cramps, dyspnea, choking, chest pain, headache, insomnia, and anorexia.[5,12] Exacerbations of asthma and chronic obstructive airways disease may also occur.

High concentrations rapidly cause unconsciousness, seizures, and death. Paralysis of the central nervous system with respiratory arrest may occur after one or two breaths.[12,13,15] Rapid loss of consciousness is, in part, responsible for multiple deaths as one would-be rescuer after another enters the toxic environment without adequate protection. Drowning is an additional risk as unconscious victims fall into the manure slurry. Recovery may also be rapid if resuscitation measures are instituted quickly. Pulmonary edema is common, but not universal, in victims of severe exposures; some show only evidence of brain injury at autopsy. Our patient had a typical presentation of severe ARDS.[11] Survivors of acute, severe toxic exposures may have neurologic sequelae.[13]

Other, less dangerous, toxic gas exposures include NH_3, CO_2, CH_4, and CO. NH_3 is a soluble gas commonly found in confinement systems in concentrations that produce irritation of the respiratory tract. We have seen a lethal NH_3 exposure in an industrial accident where the victim could not readily escape and, as a result, developed progressive bronchiolitis obliterans. Because NH_3 is highly soluble, the conjunctiva and mucosal membranes are usually irritated enough to cause humans to leave the environment. Levels of NH_3 high enough to cause serious toxicity have not been recognized in confinement areas.

CO_2 could be toxic by high concentrations leading to hypoxia. Ventilation systems are usually adequate to prevent such CO_2 toxicity. CO_2 has a mild narcotic effect that may contribute to injury.[5]

CH_4 may exceed acceptable threshold levels[2] but is not considered highly toxic. Injury would most likely be due to hypoxia if CH_4 concentrations were high enough. CH_4 at levels of 12 to 15% may create an explosive hazard; heaters in confinement areas could act as an ignition source. CH_4 has a weak anesthetic effect[5] that may contribute to risks.

Toxic concentrations of CO may occur in manure pits, but CO has not been incriminated in reported injuries. Levels of CO have usually not been high enough to be associated with the sudden collapse commonly seen in manure pit accidents. The source of toxic concentrations of CO is more likely to be a combustion heater than the decaying organic matter.

Treatment of the toxic gas exposure from a manure pit involves removing the victim from the toxic environment and, if necessary, performing cardiopulmonary resuscitation. If pulmonary edema occurs, it should be treated as noncardiogenic pulmonary edema. Oxygen is commonly given although there is no proof of its efficacy. The use of amyl nitrite and sodium nitrite in the treatment of H_2S has been suggested,[13,16] but, again, there is little proof of the efficacy. Nitrites increase methemoglobin production which competes with the cytochrome oxidase for the free sulfide ion, creating a nontoxic sulfmethemoglobin complex which slowly releases the sulfide to the very active endogenous detoxification system.[13]

Prevention is critical because of the extreme toxicity of H_2S. Preventive measures are outlined in detail in a NIOSH publication;[6] basic preventative measures include:

1. Confinement buildings should be constructed to minimize exposure. Adequate ventilation is essential particularly when the pits are located under the confinement center.
2. Manure pits should be constructed to minimize the need for entry. Lids, valving systems, mixing systems, and discharge systems should be designed so that both entry into and time in the tank are minimized. Openings should be constructed and protected to prevent accidental entry (reported with children).
3. Workers should be educated about the dangers of, the

approach to, and the emergency treatment of toxic gas exposure. Promotional material for these systems should contain warnings so that buyers are aware of the risks. Such warnings may affect the number and kind of safety factors built into the system.

4. Signs in the appropriate language(s) warning of the hazards should be appropriately posted.
5. A self-contained protective breathing apparatus should be available at each site and worn on each entry.
6. Individuals entering a pit should wear a rescue harness so that they can be retrieved without the entry of the rescuer; a simple rope tied around the chest or waist may be adequate. A harness or rope should be readily available at every site.
7. Individuals should not enter pits without observation of a trained, physically capable rescuer.
8. Entry should be avoided after agitation of the slurry. When pits are agitated, animals should be removed from the area because, during those times, even good ventilation systems may be inadequate. Large losses of livestock in such situations have been reported.[2]

The U.S. Center for Disease Control has requested that accidents involving liquid manure systems be reported to them through state health departments. The address is Special Studies Branch, Chronic Diseases Division, Center for Environmental Health, Centers for Disease Control, Altanta, GA, 30333.

REFERENCES

1. **Donham, K.J., Rubino, M., Thedell, T.D., and Kammermeyer, J.,** Potential health hazards to agricultural workers in swine confinement buildings, *J. Occup. Med.*, 19, 383, 1977.
2. **Donham, K.J., Knapp, L.W., Monson, R., and Gustafson, K.,** Acute toxic exposure to gases from liquid manure, *J. Occup. Med.*, 24, 142, 1982.
3. **Morse, D.L., Woodbury, M.A., Rentmeester, K., and Farmer, D.,** Death caused by fermenting manure, *J.A.M.A.*, 245, 63, 1981.
4. **Stevens, A.J.,** Health risks associated with animal waste from intensive livestock units, *Ann. Ist. Super. Sanita,* 14(2), 221, 1978.
5. **Cockcroft, D.W. and Dosman, J.A.,** Editorial: respiratory health risks in farmers, *Ann. Intern. Med.*, 95, 380, 1981.
6. DHEW, Criteria for a Recommended Standard ... Occupational Exposure to Hydrogen Sulfide, DHEW (NIOSH) Publ. No. 77-158, Center for Disease Control, National Institute for Occupational Safety and Health, Public Health Service, U.S. Department of Health, Education and Welfare, U.S. Government Printing Office, Washington, D.C., May 1977.
7. Comments from the Editor, *Ann. Intern. Med.*, 95, 382, 1981.
8. **Berlin, C.M.,** Letter to the Editor: death from anoxia in an abandoned cesspool, *Ann. Intern. Med.*, 95, 387, 1981.
9. Deaths associated with liquid–manure systems — United States, *Morbidity Mortality Wkly. Rep.*, 30(13), 151, 1981.
10. **Hagley, S.R. and South, D.L.,** Fatal inhalation of liquid manure gas, *Med. J. Aust.*, 2, 459, 1983.
11. **Osbern, L.N. and Crapo, R.O.,** Dung lung: a report of toxic exposure to liquid manure, *Ann. Intern. Med.*, 95, 312, 1981.
12. **Kaye, S.,** *Handbook of Emergency Toxicology,* 4th ed., Charles C Thomas, Springfield, IL, 1980, 140.
13. **Gosselin, R.E., Smith, R.P., and Hodge, H.C.,** *Clinical Toxicology of Commercial Products,* 5th ed., Williams & Wilkins, Baltimore, 1983, 111.
14. **Beauchamp, R.O., Jr., Bus, J.S., Popp, J.A., Boreiko, C.J., and Andjelkovich, D.A.,** A critical review of the literature on hydrogen sulfide toxicity, *CRC Crit. Rev. Toxicol.*, 13, 25, 1984.
15. **Adelson, L. and Sunshine, I.,** Fatal hydrogen sulfide intoxication, *Arch. Pathol.*, 81, 375, 1966.
16. **Arena, J.M.,** *Poisoning: Toxicology, Symptoms, Treatments,* 4th ed., Charles C Thomas, Springfield, IL, 1979, 237.

Health Risks from Animal Confinement Units

J. A. Merchant and K. J. Donham

INTRODUCTION

Allergic rhinitis and asthma in animal handlers is well documented;[1–4] nonspecific airways obstruction from livestock or poultry exposure is less well defined. This has recently been brought to our attention by studies of workers in animal confinement buildings. Animal confinement technology, increasingly used in the U.S. and other industrialized countries, applies principals of industrial mass production to livestock production. Animals are raised from birth in closely confined quarters, which are often fully enclosed and designed principally for temperature and humidity control. Hence, exposure to gases from urine and feces degradation, animal dander, airborne feed dust, fungi, and bacteria and their products (especially endotoxin) are not controlled.

Respiratory symptoms consistent with airways obstruction are common among those working in these facilities.[5–11] Between 0.64 and 2.96 million workers and family members are occupationally exposed to poultry, swine, or beef confinement and processing operations in the U.S.[5–12] The economic advantages of this method of livestock production are resulting in increased animal confinement use. The large number of workers reporting acute respiratory symptoms with this type of exposure leads to public health concern. The possibility that this technology may cause chronic airways obstruction is concerning.

ENVIRONMENTAL EXPOSURES IN ANIMAL PROCESSING AND CONFINEMENT BUILDINGS

The slatted floor-pit system has been used extensively, especially in swine confinement units. This design is relatively inexpensive, requires little space, provides maximal retention of manure, requires little energy and maintenance for manure handling, and is more attractive in terms of meeting outdoor ambient air and water pollution limits.[12] Therefore, this type of design will continue to be used despite documentation of toxic exposures within these units.

Both toxic dusts and gases are found within confinement units. The dust, much of which is respirable,[13–17] arises principally from two sources: the animals (e.g., dried fecal material and animal dander[16]) and the animal feed which commonly contains various cereal grains, soybean, meat, and fish meal and additives such as antibiotics.[16,18] Mineral dust, insect parts, and pollen have also been reported.[15,16,19] The relatively warm and humid environment provides a rich culture medium for fungi and bacteria. Common fungi include species of *Pencillium, Fusarium, Aspergillus flavus, Scopolariopsis, Verticillium, Phycomycete,* and *Paecilomyces*.[3,15,16] Both Gram–positive and Gram–negative bacteria are common, and relatively high levels of endotoxin occur in both swine confinement and poultry processing facilities.[9,10,16,20–24] Thermophilic actinomycetes, *Thermoactinomyces candidus, Micropolyspora faeni,* and *Saccharomonospora viridis,* also occur in swine confinement facilities.[10,16,25] Total particulate concentrations range between 10 and 30 mg/m^3 in confinement operations.[9,10,12,15,16,18,24] Endotoxin levels have averaged about 11 µg/g of dust.[9,10,16,20–22,24]

Toxic gases are commonly found in these confinement units — particularly those with slatted floors. Ammonia (NH_3) in swine confinement units frequently exceeds its threshold limit value (TLV) (generally ranging from 20 to 40 ppm).[10,12,15,26] NH_3 levels vary between and within confinement units. Absorption of NH_3 onto respirable particulates in humid environments provides increases of the complexity of this occupational environment. Hydrogen sulfide (H_2S) is a second common toxic gas found in these pits. Very high levels of H_2S occur with liquid manure agitation, implicated in the deaths of several workers.[15,27,28] H_2S levels are normally relatively low (1 to 2 ppm).[15] Carbon dioxide (CO_2) may increase but rarely approaches its TLV. Carbon monoxide (CO), however, may be a problem. A common source of CO is a poorly ventilated propane heater.[29,31,33] Levels >20 ppm have been measured. Human health hazards have not been documented; however, animal toxicities from CO, deaths, and abortions have been documented. Under normal operating conditions, CO varies from 4 to 17 ppm, well below the TLV of 60 ppm.[15] Many other compounds, some of which are toxic, have been documented in swine confinement units[17] (Table 1).

HEALTH EFFECTS

The health effects which arise from exposure to animal confinement units have not been completely characterized. A number of pulmonary manifestations, including pulmonary edema, asthma, bronchitis, bronchiolitis, airways obstruction, and organic dust toxic syndrome, may arise from this type of exposure. Airways obstruction of uncertain mechanism and pathology is the most common health effect.[4,5,7–11,32] Several antigens associated with hypersensitivity pneumonitis are found in confinement environments, although this disease has not been confirmed in this environment. Episodes of a febrile illness suggesting organic dust toxic syndrome have been reported by 15% of workers. Individuals with massive exposures to H_2S frequently die from respiratory arrest and acute pulmonary edema.[27,28] The pathology and the future clinical course of those who survive such an exposure have not been documented.

TABLE 1
Compounds Identified in Swine Confinement Buildings[42–49]

Alcohols	Carbonyls	Sulfides
Methanol	Acetaldehyde	Dimethyl sulfide
Ethanol	Propronaldehyde	Diethyl sulfide
2-Propanol	*iso*-Butyraldehyde	Disulfides
n-Propanol	Hexanol	Esthers
n-Butanol	Acetone	Ethyl formate
iso-Butanol	Formaldehyde	Methyl acetate
iso-Pentanol	Heptaldehyde	*iso*-Propyl acetate
Acids	Valeraldehyde	*iso*-Butyl acetate
Butyric	Octaldehyde	*iso*-Propyl propionate
Propyl acetate	Decaldehyde	*n*-Butyl acetate
Acetic	Amines	Fixed gases
Propionic	Methylamine	Carbon monoxide
iso-Butyric	Ethylamine	Carbon dioxide
iso-Valeric	Trimethylamine	Methane
Disulfides	Triethylamine	Ammonia
	Mercaptans	
	Methylmercaptan	

Pulmonary Edema

Uncontrolled anaerobic and facultative microbial growth within liquid manure produces many gases as end products. The fixed gases methane (CH_4), NH_3, CO_2, and H_2S are the most important.[15] Agitation of liquid manure to break up solids prior to pumping releases high concentrations of H_2S (>500 ppm).[28] At least 16 deaths and 9 prolonged illnesses have been reported in workers acutely exposed to these toxic gases.[27,28,34–36] Pulmonary edema is frequently reported at autopsy; death certificates often indicate death by "asphyxiation".[27,28] Survivors may remain dyspneic for several weeks.[28] These cases, no doubt, underestimate those which have occurred in this setting.

Asthma

Asthma is common in animal confinement workers. Postal questionnaire surveys of Iowa swine confinement workers have shown that 25 to 35% report wheezing and/or chest tightness.[10,15] Twelve to 33% of workers report delayed symptoms, some of which are consistent with asthma.[10,12,13] Malaise and fever may accompany the delayed symptoms. Selection out of this industry among those workers most affected is common and consistent with selection effects among others exposed to organic dusts.[15] Elevated levels of IgE to hog pelt, hog urine, and hog serum and IgG to hog barn dust and hog feed have been reported among individual swine producers,[2,3,10] but are not thought to account for the high rate of respiratory symptoms observed among these workers.[37]

Bronchitis

Chronic cough and phlegm production occurs in 65 to 75% of Iowa swine confinement workers.[25,31,33,38] These symptoms are increased among cigarette smokers.[3,10,12] The prevalence of bronchitis is also associated with years of exposure.[3,12] Like asthma, persistent bronchitis is a common complaint among those who leave jobs in this industry. Pathological and physiologic studies are lacking in this common clinical syndrome among these workers.

Airways Obstruction

Across–shift decrements in expiratory flow rates, while statistically significant, are not large.[8,11,15,39] Reductions in forced expired volume in 1 s (FEV_1) and expiratory flow rates range from 5 to 10% among workers, tend to be less in volunteers, tend to be greater in winter (with higher exposure), and tend to be greater in smokers and those with long duration of exposure to animal confinement housing.[15,39] There is presently no evidence that these workers are at greater risk of irreversible airways obstruction than are nonfarm workers.[3,10,15,39] There is only one relatively small study; no large cross-sectional or prospective studies of lung function among these workers have yet been reported. The relatively recent introduction of swine confinement units (mean duration of exposure <10 years in Iowa) and the frequent outward migration of affected workers are additional limitations in such studies.

Organic Dust Toxic Syndrome

Hypersensitivity pneumonitis (HP) may occur among these workers, as fungi and thermophilic bacteria associated with HP in other settings are found in this environment, and precipitins have been documented.[3,15] Delayed malaise, cough, and fever, a syndrome similar to farmer's lung, are frequently reported.[15] HP associated with confinement housing has, however, not been described. Experimental data from year-long exposures of rabbits and guinea pigs in a commercial swine confinement unit resulted in frequent loss of cilia, squamous metaplasia, mucopurulent pneumonias, and a granulomatous type of pneumonia; the latter may be consistent with hypersensitivity pneumonitis or chronic infection.[2,15] Further studies of animal confinement workers are needed to determine whether HP occurs within this population.

HP–like symptoms are likely analogous to the organic dust toxic syndrome seen in cotton and grain workers. HP antigens are present in lower concentrations than in traditional HP environments.[19] Endotoxin levels, in the range of 0.1 to 0.5 mg/m^3, may be associated with such symptoms.[40]

PREVENTION

The key to prevention of respiratory disease among these workers is improved building design and management procedures. Increased general ventilation will reduce gas exposures, but it is limited in bringing dust levels to healthful levels. Additional dust control technology is needed, but proven, affordable systems are not available as yet, and therefore research and development in this area is badly needed.[50] Therefore, the use of respirators in these units is recommended until effective control measures are in use. It appears that the half–face respirator with ammonia cartridge with a dust prefilter or the positive–pressure full-face piece type of respirator is the most efficient and is usually tolerated well by these workers. However, respirator effectiveness and other intervention strategies await evaluation. Traditional medical surveillance with employee transfer within the industry does not appear to be a viable option.

REFERENCES

1. **Bar-Sela, S., Teichtahl, H., and Lutsky, I.,** Occupational asthma in poultry workers, *J. Allergy Clin. Immunol.,* 73, 271, 1984.
2. **Katila, M., Mantyjervi, R.A., and Ojanen, T.,** Sensitization against environmental antigens and respiratory symptoms in swine workers, *Br. J. Ind. Med.,* 38, 334, 1981.
3. **Harries, M.G. and Cromwell, O.,** Occupational asthma caused by allergy to pig's urine, *Br. Med. J.,* 284, 867, 1982.
4. **Donham, K.J., Haglind, P., Peterson, Y., and Rylander, R.,** Environmental and health studies of farm workers in Swedish swine confinement buildings, *Br. J. Ind. Med.,* in press.
5. **Donham, K.J., Rubino, M., Thedel, T.D., and Kammermeyer, J.,** Potential health hazards to agricultural workers in swine confinement buildings, *J. Occup. Med.,* 19, 383, 1977.
6. **doPico, G.,** Workgroup report on diseases, *Scand. J. Work Environ. Health,* in press.
7. **Donham, K.J., Zavala, D.D., and Merchant, J.A.,** Respiratory symptoms and lung function among workers in swine confinement buildings: a cross–sectional epidemiological study, *Arch. Environ. Health,* 39, 96, 1984.
8. **Donham, K.J., Zavala, D.D., and Merchant, J.A.,** Acute effects of the work environment on pulmonary functions of swine confinement workers, *Am. J. Ind. Med.,* 5, 367, 1984.
9. **Thelin, A., Tegler, O., and Rylander, R.,** Lung reactions during poultry handling related to dust and bacterial endotoxin levels, *Eur. J. Respir. Dis.,* 65, 266, 1984.
10. **Donham, K.J.,** Hazardous agents in agricultural dusts and methods of evaluation, *Am. J. Ind. Med.,* 10, 205, 1986.
11. **Haglind, P., Rylander, R., and Clark, C.S.,** Respiratory functions among workers in swine confinement buildings, in *Occupational Lung Disease,* Bernard, J. and Gee, L., Eds., Raven Press, New York, 1984, 238.
12. **Donham, K.J. and Gustafson, K.E.,** Human occupational hazards from swine confinement, in *Agricultural Respiratory Hazards: Annals of the American Conference of Governmental Industrial Hygiene,* Vol. 2, American Conference of Governmental Industrial Hygienists, Cincinnati, OH, 1982, 137.
13. **Curtis, S.E., Anderson, C.R., Simon, J., Jensen, A.H., Day, D.L., and Kelley, K.W.,** Effects of aerial ammonia, hydrogen sulfide and swine house dust on rate of body weight gain and respiratory tract structure in swine, *J. Anim. Sci.,* 41, 735, 1975.
14. **Bundy, D.S. and Hazen, T.E.,** Dust levels in swine confinement systems associated with different feeding methods, *Trans. Am. Soc. Agric. Eng.,* 18, 137, 1975.
15. **Donham, K.J.,** Respiratory Disease Hazards of Swine Confinement Workers, final report, NIOSH Grant NIH 1 RO1 OH 00825 (PB84241512-NTIS), National Institutes of Health, Atlanta, GA, 1982.
16. **Donham, K.J., Scallon, L., Popendorf, W.J., Treuhaft, M., and Roberts, R.,** Characterization of dusts collected from swine confinement buildings, *Am. Ind. Hyg. Assoc. J.,* in press.
17. **Nilsson, C.,** Dust investigations in pig houses, in Dust in Animal Houses, Proc. Int. Soc. of Animal Hygiene, Hilliger, H.G., Ed., March 13 and 14, Hanover, 1984, 31.
18. **Leistikow, B., Pettit, W., Donham, K., Merchant, J., and Popendorf, W.,** Respiratory risks in poultry handling, presented at Int. Symp. on Health and Safety in Agriculture, Saskatoon, Saskatchewan, Canada, October 8 to 11, 1985.
19. **Donham, K.,** Hazardous agents in agricultural dusts and methods of evaluation, *Scand. J. Work Environ. Health,* in press.
20. **Thedell, T.D., Mull, J.C., and Olenchock, S.A.,** A brief report of gram-negative bacterial endotoxin levels in airborne and settled dusts in animal confinement buildings, *Am. J. Ind. Med.,* 1, 3, 1980.
21. **Olenchock, S.A., Lenhart, S.W., and Mull, J.C.,** Occupational exposure to airborne endotoxins during poultry processing, *J. Toxicol. Environ. Health,* 9, 339, 1982.
22. **Clark, S., Rylander, R., and Larsson, L.,** Airborne bacteria, endotoxins and fungi in dust in poultry and swine confinement buildings, *Am. Ind. Hyg. Assoc. J.,* 7, 537, 1983.
23. **Lenhart, S.W.,** Sources of respiratory insult in the poultry processing industry, *Am. J. Ind. Med.,* 6, 89, 1984.
24. **Jones, W., Morring, K., Olenchock, S., Williams, T., and Hickey, J.,** Environmental study of poultry confinement buildings, *Am. Ind. Hyg. Assoc.,* 45, 760, 1984.
25. **Dutkiewicz, J.,** Exposure to dust-borne bacteria in agriculture. I. Environmental studies, *Arch. of Environ. Health,* 33, 250, 1978.
26. **Donham, K.J. and Popendorf, W.J.,** Ambient levels of selected gases inside swine confinement buildings, *Am. Ind. Hyg. Assoc. J.,* in press.
27. **Osbern, L.M.** and **Crapo, R.O.,** Dung lung: a report of toxic exposure to liquid manure, *Ann. Intern. Med.,* 95, 506, 1981.
28. **Donham, K.J., Knapp, L.W., Monson, R., and Gustafson, M.D.,** Acute toxic exposure to gases from liquid manure, *J. Occup. Med.,* 24, 142, 1982.
29. **Donham, K.J., Carson, T.L., and Adrian, B.R.,** Carboxyhemoglobin values in swine relative to carbon monoxide exposure: guidelines for monitoring animal and human health hazards in swine-confinement buildings, *Am. J. Vet. Res.,* 43, 813, 1982.
30. **Muehler, A.J.,** A research review of swine housing and waste management, Publ. No. AENG 873, Department of Agricultural Engineering, Urbana, University of Illinois Cooperative Extension Service, 1969.
31. **Carso, T.L. and Donham, K.J.,** Carbon monoxide abortion in Iowa swine, *Am. Assoc. Vet. Lab. Diagn.,* 21, 179, 1978.
32. **Donham, K.J.,** Respiratory hazards of livestock confinement workers, in Proc., 7th Int. Congr. of Rural Medicine, Salt Lake City, UT, 1978, 224.
33. **Carson, T.L., Donham, K.J., Dominick, M.A., et. al.,** Carbon monoxide induced abortion in swine: an update, *Am. Assoc. Vet. Lab. Diagn.,* 23, 59, 1980.
34. **Morse, D.L., Woodbury, M.A., Rentmeester, K., and Farmer, D.,** Death caused by fermenting manure, *J.A.M.A.,* 245, 63, 1981.
35. MMWR, Deaths associated with liquid–manure systems — United States, *Morbidity Mortality Wkly. Rep.,* 30, 152, 1981.
36. **Hockaday, T.,** The day Laird Schumm cheated death, reprinted from Country Guide, Winnipeg, Canada, by the National Safety Council, *Farm Saf. Rev.,* 34, 5, 1976.
37. **Matson, S.C., Swanson, M.L., Reed, C.E., and Yunginger, J.W.,** IgE and IgG immune mechanisms do not mediate occupational-related respiratory or systemic symptoms in hog farmers, *J. Allergy Clin. Immunol.,* 72, 299, 1983.
38. **Donham, K.J., Kammermeyer, J., and Mutel, C.,** Agricultural respiratory problems, a slide-tape presentation, National Medical Audiovisual Center, Washington, D.C., 1978.
39. **Donham, K.J.,** unpublished data.
40. **Rylander, R. and Vesterlund, J.,** Airborne endotoxins in various occupational environments, in *Endotoxins and Their Detection with the Limulus Amebocyte Lysate Assay,* Allan R. Liss, New York, 399.
41. **Barth, C.L. and Polkowski, L.B.,** Identifying odorous components of stored dairy manure (air pollution), *Trans. Am. Soc. Agric. Eng.,* 17, 737, 1974.
42. **Berthea, R.M. and Narayan, R.S.,** Identification of beef cattle feedlot odors, *Trans. Am. Soc. Agric. Eng.,* 15, 1135, 1972.
43. **Burnett, W.E.,** Air pollution from animal wastes — the determination of malodors by gas chromatographic and organoleptic techniques, *Envir. Sci. Technol.,* 3, 744, 1969.
44. **Hartung, L.D., Hammond, E.G., and Miner, J.R.,** Identification of carbonyl compounds in a swine-building atmosphere, in *Proc., Livestock Waste Management and Pollution Abatement,* American

Society of Agricultural Engineers, St. John, MI, 1971, 105.

45. **Andersen, P. and Schnheyder, H.,** Antibodies to hen and duck antigens in poultry workers, *Clin. Allergy,* 14, 421, 1984.
46. **Merkel, J.A., Hazen, T.E., and Miner, J.R.,** Identification of gases in a swine confinement building atmosphere, *Trans. Am. Soc. Agric. Eng.,* 12, 310, 1969.
47. **Miner, J.R. and Hazen T.E.,** Ammonia and amines: components of swine-building odor, *Trans. Am. Soc. Agric. Eng.,* 12, 722, 1969.
48. **White, R.K., Taiganides, E.P. and Cole, G.D.,** Chromatographic identification of malodors from dairy animal wastes, in *Proc., Livestock Waste Management and Pollution Abatement,* American Society of Agricultural Engineers, St. John, MI, 1971, 110.
49. **Donham, K.J., Yeggy, S., and Dague, R.,** Chemical and physical parameters of liquid manure from swine confinement facilities: health implications for workers, swine, and the environment, *Agric. Wastes,* 14, 97, 1985.
50. **Nilsson, C.,** Experiences with different methods of dust reduction in pig houses, in Dust in Animal Houses, Proc., Int. Soc. of Animal Hygiene, Hanover, March 13 and 14, 1984, 90.

Respiratory Risks in Poultry Farmers

B. Leistikow, W. Pettit, K. Donham, J. Merchant, and W. Popendorf

INTRODUCTION

Respiratory problems in poultry farmers[1] may be similar to those in swine producers where we have reported farmer respiratory symptoms[2] and function test alterations.[3] In this report, we outline results of a pilot survey in Iowa.

The U.S. has a large and changing poultry industry which had sales of $13 billion in 1980.[4–6] As a result of increasing poultry consumption, poultry now makes up approximately 30% of meats consumed,[7] and these changes have been associated with consolidation into fewer and larger facilities[7–9] with 20,000 farms representing less than 10% of the U.S. farms having poultry in 1982.[10] In these farms, workers are exposed full-time to environments inside poultry facilities.[11]

Poultry industry human respiratory pathogens are diverse and may be categorized in order of importance into allergenic, infectious, and toxic agents. Allergenic pathogens have been shown to cause both asthma and hypersensitivity pneumonitis in workers. Occupational asthma in two unrelated chicken raisers has been documented by their cutaneous hypersensitivity and FEV_1 (forced expired volume in 1 s) decrements on challenge testing with chicken feed[12] and Northern fowl (chicken) mite,[13] respectively. Other chicken workers with asthma[13] and turkey workers with respiratory symptoms[14] have been shown to have immediate cutaneous hypersensitivity or IgE to poultry antigens.

Hypersensitivity pneumonitis, with positive precipitins to poultry antigens and typical, delayed symptoms on challenge testing, has been documented in one chicken handler and one turkery handler. Other poultry workers have positive precipitins, a history of previous typical symptoms, or both.[14,15]

Infectious respiratory pathogens in the poultry industry include psittacosis, aspergillosis, histoplasmosis, and several other uncommon diseases. Psittacosis, caused by *Chlamydia psittaci,* is a disease consisting of fever, chills, myalgias, and headache with or without pulmonary infiltrates on X–ray.[16] It is believed to be spread to humans by infectious aerosols arising, for example, from evisceration during poultry processing.[17,18]

Toxic respiratory pathogens in the poultry industry appear to be largely limited to ammonia and endotoxin–containing dust. Hydrogen sulfide, methane, carbon dioxide, carbon monoxide, mercaptan, formaldehyde, and nitrogen dioxide levels have been found to be low or undetectable.[19-22] Ammonia levels appear to increase with duration since litter/manure removal, decreasing ventilation,[19] and with increasing bird age. Typical ammonia levels in occupied broiler and turkey houses range from 2 (new litter)[19] to 22 ppm[20] when buildings are open during warm weather. Levels of 15 to 80 ppm were measured in occupied broiler houses with doors and windows closed during cooler weather.[1] A Minnesota turkey house averaged 46 ppm of ammonia during the winter.[23] Peak winter ammonia concentrations exceeded 50 ppm, and may be over 100 ppm.[23] These ammonia peaks are usually caused by the rototilling (to loosen the litter) that was done in this facility.[24]

Dust and endotoxin levels have been measured in several different poultry operations (Table 1). Lower total dust (from about 1 to 5 mg/m^3) and endotoxin (500 ng/m^3 or less) levels have been found in layer houses even in winter,[25] and during routine feeding, watering, and adjusting in broiler houses with 25-d–old or younger birds.[1] Higher levels of total dust (6 mg/m^3 or greater) were noted in broiler houses with older birds, in

TABLE 1
Reported Dust and Endotoxin Levels in the Poultry Industry

Bird type	Activity	Bird age (weeks)	No. of measures (dust/endotoxin)	Total dust (mg/m^3)	Total endotoxin (ng/m^3)	Ref.
Broiler	Moving birds	2	1/1	7.8	8807	1
	[a]	4—6	6/3	8—25	37—3172	1
	Feed, water, adjustments	1—3.6	15/13	0—5.5	1—77	1
	Processing	—	56/49	6—29	360—2700	27
Layer	Unload cages	—	11/11	23	1090	26
	Load cages	—	6/6	28.1	450	26
	In winter	—	6/6	1—3.7	20—500	25

[a] Area samples.

a broiler house while moving 14-d–old birds,[1] when moving birds into and out of cages,[26] in the live broiler–handling area at a processing plant,[27] and in a turkey house.[28] Endotoxin levels were greater than 500 ng/m^3 while moving broilers (8807 ng/m^3),[1] moving layers out of their cages (1090 mg/m^3),[26] or processing live broilers (430 to 2150 ng/m^3).[27]

Acute and chronic symptoms and pulmonary function test changes have been related to poultry raising: acute symptoms such as cough, phlegm, dyspnea, wheezing, stuffy nose, or eye irritation,[1,14,26] as well as chronic symptoms such as chronic cough, chronic phlegm production, or persistent airway irritation and cough.[1,26]

Despite difficulties in significantly relating individuals' interview exposure histories to symptoms, and symptoms to pulmonary function decrements, significantly disproportionate (0.21 l) FEV_1 decrements have been noted in workers on chicken farms with high as opposed to low (<1.0 mcg/m^3) amounts of endotoxin.[26]

MATERIALS AND METHODS

The 579 people on the Iowa State University Poultry Extension Newsletter mailing list were mailed a pretested questionnaire in August of 1980, requesting information about their farm operations, symptoms while in their layer buildings, and health history. Dust measurements were made with a GCA MINIRAM Model PDM-3 which uses light scatter sensing to measure dust amounts, with preferential response to the particle size range of 0.1 to 10 μm..[29] Ammonia measurements were made using Draeger colorimetric tubes. Indoor and outdoor, dry and wet bulb temperatures were recorded.

RESULTS AND DISCUSSION

Adverse symptoms were acknowledged frequently by egg farm workers (Table 2). The frequency of the symptoms in the 156 egg farm workers as a group is quite similar, within 8 percentage points in every instance, to the symptom frequencies in the subset of 65 egg farmers working in 10,000+ bird operations. Each of the respiratory symptom frequencies in the egg farmers as a group, is lower, often significantly so, than in a group of Iowa swine confinement workers previously studied.[2] As in the swine confinement workers, cough was the single most frequent symptom in egg farm workers. However, in egg farm workers, burning or watering eyes was the next most common symptom. To quantify the pulmonary symptoms, a "Pulmonary SCORE" was calculated, which was determined by the number and frequency with which an individual experienced pulmonary symptoms. Table 3 indicates that the Pulmonary SCORE was positively correlated with the numbers of hours worked per week in poultry, indicating a crude dose–response relationship. Smoking history was also an important factor affecting the symptoms experienced. There was a positive correlation between present smokers or ex-smokers and the Pulmonary SCORE among both egg farm and swine farmers.

Forty-four poultry facilities were assessed by walkthrough evaluation. These included 10 turkey barns, 16 broiler barns, 10 layer barns, and 8 slaughterhouses (Tables 4 and 5). Environmental measurements are unavailable for three of the facilities, all slaughterhouses. Ammonia levels, measured during spring and summer, were below 13 ppm for all facilities except three. Those three facilities all had market-age meat birds, and 220 ppm of ammonia or less. Dust levels were high (6 to 20 mg/m^3) in unmisted broiler and turkey facilities with birds near market age (7– and 20–week–old birds, respectively). Dust levels were high during broiler loading, but only moderate during processing live birds (probably due to wet birds after rain). Barns with younger birds or misting had lower dust levels. Dust levels were uniformly low in layer barns.

Dust, endotoxin, and ammonia exposures, especially concentrated in moving or handling older birds, appear to be the most likely cause of poultry worker symptoms. In particular, higher poultry endotoxin exposures have been associated in one study with higher FEV_1 decrements through the workshift moving layers.[26] Confirmation of that finding and further epidemiologic study of poultry worker respiratory exposures and pulmonary function are needed.

TABLE 2
Symptoms of Poultry and Swine Confinement Farm Workers

	% of category reporting symptoms			
	Egg farmers[a]	Swine farmers[b]	X^2	p
Cough	48	67	15	<0.001
Sputum or phlegm	35	56	18	<0.001
Tightness of chest	22	36	10	<0.005
Shortness of breath	26	30	1.1	>0.25
Wheezing	27	27	1.1	>0.25
Burning or watering eyes	48	39	3.5	<0.07

[a] n = 156.
[b] n = 486.

TABLE 3
Factors Affecting Prevalence of Acute Pulmonary Symptoms in Egg Farm Workers

Positive correlation	Number of hours worked per week[a]
	Smoking history
No correlation	Age
	Number of layers raised
	Type of building

[a] Spearman correlation coefficient = 0.19; p <0.02.

TABLE 4
Poultry Farm Walkthrough Data, 1985

Bird type, location	No. of Barns	Temp gradient in–out (°F)[a]	Bird age (weeks)	Dust level (mg/m³ [b])	NH_3 (ppm)	Mist[c]
Turkeys						
California	3	0—2	14—20	0.20—0.57	3—20	Yes
Minnesota	1	24	8	0.9—1.65	11	No
Iowa, Minnesota	4	6—24	4—9	0.17—1.7	2—11	No
	1	8	12	2.0—3.3	10	No
	1	10	18	16	2	No
Broilers						
Arkansas	6	5—9	2—6	0—1.2	1—10	No
California	2	–15[a]	4—5	0.3—4.3	3—6	Yes
	2	–(13—20)	6	0.5—6.0	2—5	Yes
Other	1	18	3	1—4.1	12	No
	2	13	7	6.0—20	20	No
Other (breeding birds)	2	1—2	28—61	0.35—3.1	2—5	No
Layers						
Iowa, California	10	–6—29	27—68	34—1.6	1—12	No

[a] Negative values represent cooled buildings.
[b] Measured with a GCA MINIRAM light–scatter dust meter preferentially sensitive to 0.1 to 10 μm particles.
[c] Dampens and cools birds.

TABLE 5
Poultry Processing Live Handling Data, 1985

Bird type	ID	Relative humidity (%)	Birds/hour	No. of handlers	Dust (mg/m³ [a])	Birds (dry or wet)
Turkey						
Hens	1	60	1,700	6	5.0	Dry
	2	90	2,100	5	1.3	Wet
	3	60	2,900	6	3.5	Wet
Toms	1		2,100	11		
	2		2,400	8		
Broilers	1	60	5,000	4	3.0	Dry
	2	90	8,000	5	2	Wet
	3		17,000	6		

[a] Measured with a GCA MINIRAM light–scatter dust meter preferentially sensitive to 0.1-10–μm particles.

REFERENCES

1. **Williams, T.M., Hickey, J.L.S., Boehlecke, B.A., and Jones, W.G.,** HETA 83–195–1426 David Mayer poultry farm, Hobgood, North Carolina, NIOSH Health Hazard Evaluation Report, Centers for Disease Control, Public Health Service, U.S. Department of Health and Human Services, Atlanta, GA, 1983.
2. **Donham, K.J. and Gustafson, K.,** Human occupational hazards from swine confinement, *Ann. Am. Conf. Gov. Ind. Hyg.*, 2, 137, 1982.
3. **Donham, K.J., Zavala, D.C., and Merchant, J.,** Acute effects of the work environment on pulmonary functions of swine confinement workers, *Am. J. Ind. Med.*, 5, 367, 1984.
4. 1982 Census of Agriculture, U.S. Summary, Vol. 1, Part 51, Table 50, U.S. Department of Agriculture, Washington, D.C., 1984.
5. **U.S. Poultry Statistics by State: 1980,** Watt Publishing, Mount Morris, IL, 1982.
6. **Lasley, F.A.,** The U.S. Poultry Industry: Changing Economics and Structure, ERS, Agric. Econ. Rep. No. 502, U.S. Department of Agriculture, Washington, D.C., 1983.
7. **Anon.,** Continued growth ahead for U.S. further processors, *Turkey World,* p. 6, September 1984.
8. 1978 Census of Agriculture, U.S. Summary, Table 35, U.S. Department of Agriculture, Washington, D.C., 1980.
9. **Cobb, C.W. et al.,** Economic Indicators of the Farm Sector: Production and Efficiency Statistics, 1982, U.S. Department of Agriculture, Washington, D.C., 1983.
10. 1982 Census of Agriculture, U.S. Summary, Tables 19, 20, and 21, U.S. Department of Agriculture, Washington, D.C., 1984.
11. **Jester, R.C.,** Occupational health hazards on broiler farmers, *Poul. Dig.*, p. 248, June 1984.
12. **Paggiaro, P.L., Loi, A.M., and Toma, G.,** Bronchial asthma and dermatitis due to spiramycin in a chick breeder, *Clin. Allergy*, 9, 571, 1979.
13. **Bar-Sela, S., Teichtahl, H., and Lutsky, I.,** Occupational asthma in

poultry workers, *J. Allergy Clin. Immunol.*, 73, 271, 1984.

14. **Boyer, R.S. et al.,** Hypersensitivity lung disease in the turkey raising industry, *Am. Rev. Respir. Dis.*, 109, 630, 1974.
15. **Warren, C.P.W. and Tse, K.S.,** Extrinsic allergic alveolitis owing to hypersensitivity to chickens — significance of sputum precipitins, *Am. Rev. Respir. Dis.*, 109, 672, 1974.
16. **Anderson, D.C., Stoesz, P.A., and Kaufmann, A.F.,** Psittacosis outbreak in employees of a turkey-processing plant, *Am. J. Epidemiol.*, 107, 140, 1978.
17. **Palmer, S.R.,** Psittacosis in man — recent developments in the UK: a review, *J. R. Soc. Med.*, 75, 262, 1982.
18. **Potter, M.E., Kaufmann, A.K., and Plikaytis, B.D.,** Psittacosis in the United States, 1979, Morbid. Mort. Wkly. Rep., 32, 27SS, 1983.
19. **Jones, W., Morring, K., Olenchock, S.A., Williams, T., and Hickey, J.,** Environmental study of poultry confinement buildings, *Am. Ind. Hyg. Assoc. J.*, 45, 760, 1984.
20. **Janni, K.A., Redig, P.T., and Newman, J.,** Seasonal aerosol concentrations in turkey grower barns, Minnesota Turkey Research, 1984, Miscellaneous Publication 30, Agricultural Experimental Station, University of Minnesota, Minneapolis, MN, 1984.
21. **Balazova, G., Hluchan, E., and Gruss, K.,** An analysis of the working environment of a poultry farm (in Slovakian), *Prac. Lek.*, 19, 163, 1967.
22. **Van Der Haar, R., Waegemaekers, M., and Van Welie, R.,** Wageningen Gezondheidsleer, Rep. No. R-1984-177, *The Netherlands*, 1984.
23. **Janni, K.A. and Mulhausen, J.,** The physical environment in the turkey house, presented at Midwest Poultry Federation Convention, Minneapolis, MN, 1985.
24. **Redig, P.T.,** personal communication.
25. **Clark, S., Rylander, R., and Larsson, L.,** Airborne bacteria, endotoxin and fungi in dust in poultry and swine confinement buildings, *Am. Ind. Hyg. Assoc. J.*, 44, 537, 1983.
26. **Thelin, A., Tegler, O., and Rylander, R.,** Lung reactions during poultry handling related to dust and bacterial endotoxin levels, *Eur. J. Respir. Dis.*, 65, 266, 1984.
27. **Olenchock, S.A., Lenhart, S.W., and Mull, J.C.,** Occupational exposure to airborne endotoxins during poultry processing, *J. Toxicol. Environ. Health*, 9, 339, 1982.
28. **Cravens, R.L., Beaulieu, H.J., and Buchan, R.M.,** Characterization of the aerosol in turkey rearing confinements, *Am. Ind. Hyg. Assoc. J.*, 42, 315, 1981.
29. GCA MINIRAM Model PDM-3 Operations Manual, GCA Corporation, Technology Division, Bedford, MA, 1984.

Environmental and Health Studies of Workers in Swedish Swine Buildings

K. Donham, P. Haglind, R. Rylander, and Y. Peterson

The air environment inside modern intensive swine production facilities (confinement buildings) has been shown previously to be a potential health hazard for workers.[1–6] However, there have been very few studies linking both extensive environmental and worker health data. The present study was conducted to more clearly define the relationships between the air environment and the air environment in the swine confinement working environment.

Environmental analyses were performed on the air inside 30 swine buildings from 28 Swedish farms. Health data were taken on 57 persons working on these farms and on 55 matched controls. In addition, data were collected on the health of the pigs raised in these buildings and on the ventilation system and other engineering and management factors related to environmental control.

Environmental measurements are summarized in Table 1. Health data collected included an interview for occupational workers' history, health history, and disease symptoms. The medical questionnaire included acute respiratory symptoms[8] directly related to their work, plus the complete American Thoracic Society questionnaire.[9] Pulmonary function testing was done before and after a work period,[9,10] and a blood sample was collected to test for antibodies to suspected allergens in the work environments.[11]

Figure 1 summarizes some of the significant findings from the first part of the questionnaire interview. Bronchitis symptoms were very commonly reported, with 60% reporting cough and 40% reporting cough with phlegm.

Symptoms of bronchoconstriction were reported by 28% of the workers. Eye, nose, and throat irritation was also commonly reported (28, 41, and 34%, respectively). Many workers (21%) developed chest tightness symptoms over the work period. Only 7% of the workers reported chest tightness before, but 28% reported chest tightness after work. There was no significant difference between smokers and nonsmokers in development of chest tightness over the work shift.

Table 2 summarizes some of the significant findings from the ATS portion of the questionnaire. These results corroborate the bronchitic symptoms identified from the first part of the questionnaire. Cough (39%) and cough with phlegm (16%) were frequently reported by the swine farm group, while the respective symptoms were reported by 13 and 7% of the controls. Wheezing was reported by 21% of the swine group but only 13% of controls. Frequent chest colds (53%), work absence due to chest illness (30%), and history of pneumonia (16%) were all reported about twice as frequently in the swine worker group as compared to controls. Smoking appeared to be an additive or synergistic factor with work in swine buildings. Bronchitis symptoms were twice as prevalent in swine workers who smoked as in nonsmoking swine workers, and nine times as frequent compared to nonsmoking controls. Similar additive or synergistic effects of smoking and swine-building exposure were seen with frequent chest cold, work absence from chest illness, and a history of pneumonia.

TABLE 1
Environmental Measurements in Swine Stables from 28 Swedish Farms

Measurements	Mean values
Total dust[a]	
Area sampling	4.6 mg/m^3
Personal sampling	6.4 mg/m^3
Respirable dust[b]	
Area sampling	0.33 mg/m^3
Personal sampling	0.34 mg/m^3
Endotoxin	
Total dust[c]	
Area sampling	0.2 μgm/m^3
Personal sampling	0.23 μgm/m^3
Respirable dust	
Area sampling	0.12 μgm/m^3
Personal sampling	0.012 μgm/m^3
Microbial sampling	
Total microbes (viable + non–viable)[d]	18×10^6 Organisms/m^3
Total viable	
Nucleopore method[e]	1.4×10^6 cfu/m^3
Anderson method	4.5×10^5 cfu/m^3
Viable mold	
Nucleopore	3×10^4 cfu/m^3
Anderson method	2×10^4 cfu/m^3
Viable bacteria	
Nucleopore	$1.4 \text{x} 10^6$ cfu/m^3
Anderson method	$4.3 \text{x} 10^5$ cfu/m^3
Ammonia (area sampling,[e] human breathing zone)	9 ppm
Carbon dioxide (area sampling,[e] human breathing zone)	1.740 ppm

[a] Dust collected on 37–mm filters with personal sampling pumps with a flow of 1.9 1/m.
[b] Casella cyclone separator was used.
[c] Amebocyte Limulus assay.
[d] Microscopic examination of dust collected on 37–mm nucleopore filters with personal pumps with a 1–l/min flow.[9]
[e] Drager colorimetric tubes used.

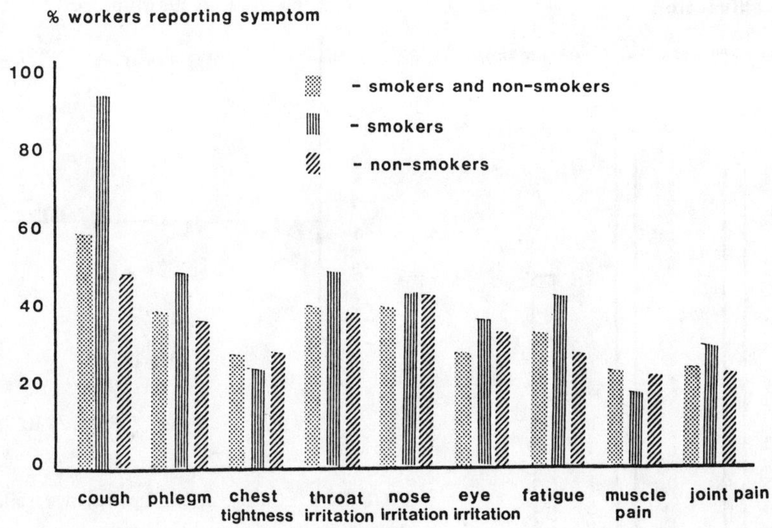

FIGURE 1. Acute symptoms among workers in swine confinement buildings which they attribute directly to their work.

TABLE 2
American Thoracic Society Respiratory Disease Questionnaire. Swine Confinement Worker and Controls[a]

Symptoms	All SWF[b]	All C[c]	Smokers		Nonsmokers	
			SWF (n = 16)	C (n = 14)	SWF (n = 41)	C (n = 41)
Cough	39[d]	13	63	29	29[e]	7
Cough with phlegm	16	7	25	21	12	2
Cough with phlegm (3 weeks)	7	4	6	14	15	2
Wheezing with a cold	21	13	25	14	20	12
Wheezing without a cold	12	16	31	21	5	15
Wheezing most of the time	4	2	13	0	0	2
Frequent chest colds	53[f]	20	50	21	54[e]	20
Off work from chest illness	30	18	31	14	29	20
History pneumonia	16	6	7	14	17[g]	2

[a] Results reported in percent of total responses.
[b] SWF = swine farm workers (n = 57).
[c] C = Controls (n = 55), farmers, not working with swine or poultry. Matched by age (±5 years), sex, and smoking status (±5 pack years). All cases reported as physician confirmed, and occurred during their active working age.
[d] $p < 0.01$
[e] $p < 0.05$
[f] $p < 0.001$
[g] $p > 0.05 < .1$

Baseline pulmonary function testing (Figure 2) revealed low FEF_{50} (forced expiratory flow) and FEF_{75} values (84 and 60%, respectively, of predicted values). There was little difference in these values between smokers and nonsmokers. A similar pattern was seen for the ratio of FEV_1 to forced vital capacity (FVC) (75%). The mean of baseline FVC and FEV_1 measurements was within 95% of predicted value.

Pulmonary function change over a work shift (Figure 3) resulted in significant decrements in FEF_{50} (6%) and FEF_{75} (4.3%) in smokers, but not in nonsmokers. There was only a slight mean decrement of FVC and FEV_1 over the work period (1.0%).

Correlations were made between change in pulmonary function over workshift and various environmental contaminants measured by the nucleopore method.[12] The strongest correlation was between endotoxin and fermentive bacteria (Table 3).

The results agree with previous studies in North America, in that swine confinement workers experience a significant amount of airway symptoms including bronchitis, hyperreactive airways, and mucous membrane irritation.[5,8] They also corroborate a previous study which demonstrated that swine workers have increased respiratory infections and other chest illness that results in significant increase in loss of work time.[4]

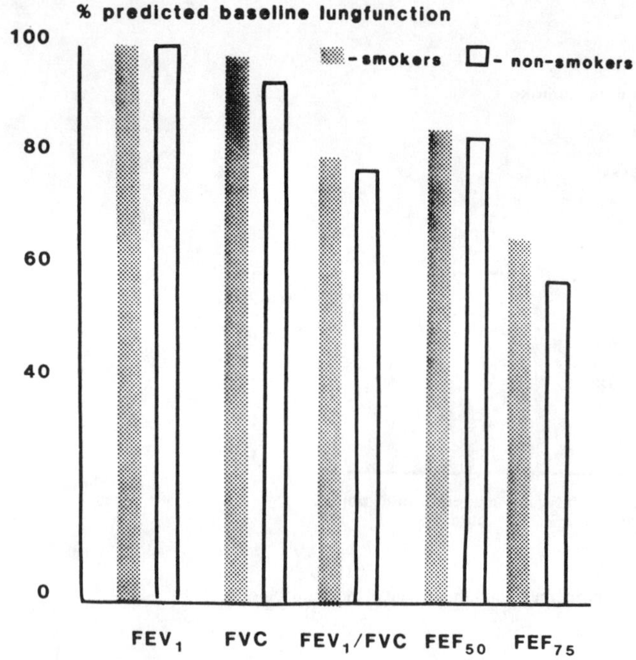

FIGURE 2. Baseline lung function of 57 swine confinement workers.

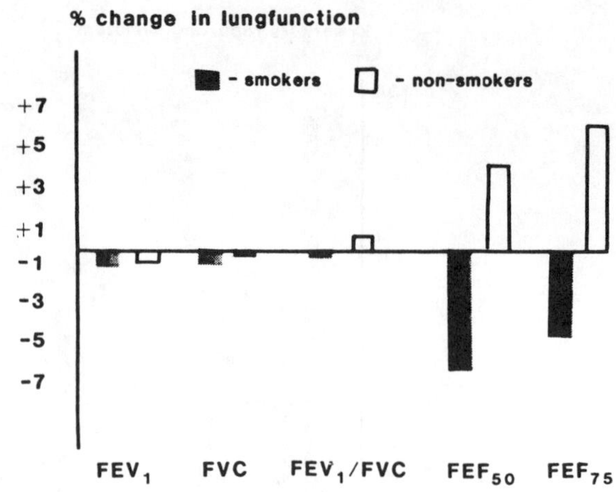

FIGURE 3. Change in lung function over work shift of 57 swine confinement workers.

Cigarette smoking, similar to our previous studies, seems to compound the health effects of this working environment.[4,5,8] The pulmonary function results indicate that the primary measurable physiologic effect is with decreased flow rates at 50 and 75% of lung volume. This suggests that this environment may be adversely affecting the peripheral airway, especially among smokers.

TABLE 3
Pearson Correlations between Levels of Air Contaminants in Swine Buildings and Changes in Pulmonary Function over Workshift

	FEV_1	FVC	FEF^{50}	FEF^{75}
Dust	–0.11	—	-0.14	–0.26
Endotoxin	–0.45[a]	–0.09	–0.23	–0.39[a]
Respirable microbes	–0.24	–0.01	–0.20	–0.31
Gram–negative bacteria	—	—	–0.22	–0.14
Enterococci	–0.29	–0.17	–0.26	–0.20
Fermentive bacteria	–0.30[a]	–0.53[a]	–0.85[a]	–0.57[a]
Mold	–0.44[a]	–0.30	–0.10	—

[a] $p < 0.05$.

REFERENCES

1. **Donham, K.J., Scallon, L., Popendorf, W.J., Treuhaft, M., and Roberts**, Characterization of dusts collected from swine confinement buildings, *Am. Ind. Hyg. Assoc. J.*, 47, 404, 1986.
2. **Donham, K.F. and Popendorf, W.**, Ambient levels of selected gases inside swine confinement buildings, *Am. Ind. Hyg. Assoc. J.*, 46, 658, 1985.
3. **Donham, K.J., Knapp, L.W., Monson, R., and Gustafsson, K.E.**, Acute toxic exposure to gases from liquid manure, *J. Occup. Med.*, 24, 9142, 1982.
4. **Donham, K.J., Zavala, D.C., and Merchant, J.A.**, Respiratory symptoms and lung function among workers in swine confinement buildings: a cross sectional epidemiological study, *Arch. Environ. Health*, 39, 96, 1984.
5. **Donham, K.J., Zavala, D.C., and Merchant, J.A.**, Acute effects of the work environment on pulmonary functions of swine confinement workers, *Am. J. Ind. Med.*, 5, 367, 1984.
6. **Attwood, P., Brouwer, R., Ruigewaard, P., et al.**, A study of the relationships between airborne contaminants and environmental factors in Dutch swine and confinement buildings, *Am. Ind. Hyg. Assoc. J.*, 48, 745, 1987.
7. **Holness, D.L., O'Blenis, E.L., Sass-Kortsak, A., Pilger, C., and Nethercott, J.R.**, Respiratory effects and dust exposures in hog confinement farming, *Am. J. Ind. Med.*, 11, 571, 1987.
8. **Donham, K.J. and Gustafsson, K.E.**, Human occupational hazards from swine confinement, *Ann. Am. Conf. Gov. Ind. Hyg.*, 2, 137, 1982.
9. **Ferris, B.G.**, Epidemiology standardization project, *Am. Rev. Respir. Dis.*, 118, 1, 1979.
10. **Gardner, R.M.**, ATS statement — snowbird workshop on standardization of spirometry, *Am. Rev. Respir. Dis.*, 119, 831, 1979.
11. **Belin, L., Eriksson, N., and Malmberg, P.**, Antibodies to microbial antigens in various farm populations, *Am. J. Ind. Med.*, 10, 1987.
12. **Palmgren, U., Strom, G., Malmberg, P., and Blomqvist, G.**, The nucleoporefilter method: a technique for enumeration of viable and non-viable microorganisms in highly contaminated environment, *Am. J. Ind. Med.*, 10, 1987.

Respiratory Status and Environmental Exposure of Hog Confinement and Control Farmers in Ontario

D. L. Holness and J. R. Nethercott

INTRODUCTION

The hog confinement process, whereby hogs are raised under carefully controlled and monitored conditions, has allowed substantially higher efficiency in modern pork production. While hog confinement has led to dramatic increases in productive capacity, limited information is available regarding the influence of this technique on the health of the hog farmers.

Farming has been associated with several respiratory diseases including farmer's lung, asthma and pulmonary edema, and bronchiolitis associated with exposure to nitrogen dioxide.[1–3] Fatalities have occurred due to high levels of toxic gases from hog manure.[4,5] Several studies have reported a higher prevalence of respiratory symptoms in hog confinement farmers.[6–9] A decrease in lung function over a 4–h period of exposure to dust was demonstrated in a group of hog farmers who had respiratory symptoms.[10]

The Ontario Pork Producer's Marketing Board has funded an initial investigation of the health status of Ontario hog farmers. The studies included the identification of those hazards to which hog farmers may be exposed during the normal farm operations by assessing dust and gas exposure in a small number of selected hog confinement farmers; the documentation of the health status of pork producers in Ontario by a questionnaire survey; and a more detailed study of the respiratory status of hog confinement farmers.[11] This report deals primarily with the detailed assessment.

A total of 96 farmers were examined in the detailed investigation. This included 53 hog confinement farmers and 43 control farmers who worked primarily on dairy and beef farms.[11]

FARMER AND FARM CHARACTERISTICS

Both the hog and control farmers had a mean age of 40 years. They were also similar with respect to height, weight, years of farming, and smoking history. Hog producers had marketed an average of 1310 hogs in the previous year and were hog farmers for a mean of 10.5 years. The hog farms used more low-moisture corn and commercial feeds. Most hog farms used antibiotics, whereas they were rarely used on the control farms. Both groups spent a similar amount of time feeding, but hog farmers used floor feeding more often than the control farmers. Hog farmers spent more time grinding feed.

ENVIRONMENTAL ASSESSMENT

The total TWA dust levels were 3.22 ± 3.10 and 1.24 ± 1.02 mg/m3 for 52 hog and 41 control farmers, respectively, while respirable dust levels were 0.41 ± 0.84 and 0.14 ± 0.14 mg/m^3. Floor feeding led to higher respirable dust and significantly higher total dust exposures. The use of high-moisture corn and the grinding of feed indoors were associated with significantly higher total dust levels.

The mean levels of carbon dioxide and ammonia were 2000 and 9 ppm, respectively. No hydrogen sulfide, carbon monoxide, or methane was detected.

SYMPTOMS

Cough, sputum, chronic bronchitis, and nasal complaints were significantly more frequent in hog farmers, while wheezing and flu–like symptoms tended (nonsignificantly) to be higher in hog farmers than in controls (Table 1). Cough, wheeze, flu–like symptoms, and shortness of breath were associated with farm work in both hog and control farmers. Cough, wheeze, and shortness of breath tended to be related to general barn conditions and not to a specific job, whereas flu–like symptoms were frequently associated with one specific job such as moving moldy corn or cleaning out silos or granaries. From the questionnaire (n = 759), nasal irritation, cough, wheeze, and dyspnea were more commonly related to feeding hogs than to working with liquid manure, whereas dizziness was more commonly associated with working with liquid manure (Table 2).

PULMONARY FUNCTION

The lung function results were similar for the hog farmers and control farmers. For the group as a whole, 8% of forced vital capacity (FVC) readings, 16% of forced expired volume

TABLE 1
Symptom Prevalence in Hog and Control Farmers

	Hog farmers	Control farmers	*p* value
Number	53	43	
Symptoms			
Cough	57%	21%	0.001
Sputum	32%	12%	0.023
Chronic bronchitis	26%	7%	0.021
Wheeze	26%	12%	0.077
Rhinitis	47%	21%	0.008
Flu–like complaints	21%	7%	0.057

Modified from Nethercott, J. R., *Am. J. Ind. Med.*, 11(5), 575, 1987.

TABLE 2
Percentage of Hog Farmers with Various Symptoms while Feeding Hogs and Working with Liquid Manure (from the Questionnaire Study)

Symptom	Feeding hogs	Working with liquid manure	*p* value[a]
Headache	7	8	0.540
Dizziness or lightheadedness	3	8	0.001
Runny or sore eyes	10	7	0.140
Sore throat	6	4	0.260
Stuffy nose or sneezing	30	7	0.001
Cough	24	9	0.001
Wheezing or tightness in chest	13	6	0.001
Shortness of breath	9	4	0.001
Nausea or vomiting	1	2	0.520

[a] *p* value obtained by Mantel–Haenszel procedure.

in 1 s (FEV1) values, and 18% of the FEV_1/FVC ratios were below their respective 95% confidence limits.

EXPOSURE–EFFECT RELATIONSHIPS

Exposure-effect relationships were examined in several ways. No significant relationships were found when exposure was categorized as high, medium, and low; when each individual's exposure was compared with their lung function; or when a cumulative dust exposure index, in which both the length of exposure and dust levels were included, was used.

DISCUSSION

Hog farmers had higher total and respirable dust exposures than the control farmers. Clark et al. found mean area total dust concentrations of 3.08 mg/m³ on eight hog confinement farms,[12] while Curtis et al. reported mean area total dust levels of 1.28 mg/m³ in five university hog gestation farms and 6.15 mg/m³ in five commercial hog production operations.[13] Dust-sampling methods in these studies differed from the present study in that high volume samplers were used for shorter sampling periods and only area, not personal, measurements were obtained.

Floor-feeding methods and indoor grinding appear to contribute to the hog farmers' higher dust exposure. Hog farmers used floor–feeding methods more frequently than control farmers. In hog farmers, floor feeding was associated with significantly greater total dust exposure and a tendency to higher respirable dust exposure than those using other methods. Total dust exposure for those hog farmers grinding indoors was significantly higher. Additionally, hog farmers spend significantly more time grinding feed than do control farmers, and indoor grinding tends to be more prevalent on hog farms. Thus, indoor grinding may also be an important factor elevating dust exposure of the hog farmers.

Respiratory symptoms were significantly more common in the hog farmers as compared to the control farmers; the symptom prevalences in both groups are similar to other studies.[4,8,14] In the present study, symptoms in both hog farmers and controls were significantly associated with workplace exposures. Flu–like symptoms, similar to those reported by others,[9] were associated with working with moldy or old feed. These episodes may be related to farmer's lung, silo filler's disease, or a reaction akin to mill fever or grain fever.[15–17] Possible causes include fungi, bacterial endotoxin, mycotoxin, or chemical components of the organic materials either by a hypersensitivity or direct toxic mechanism.

There were no significant differences between the lung function of hog and control farmers. However, the mean length of exposure to hog confinement farming of 10 years may be too short to detect any abnormalities. Eighteen percent of the farmers had an FEV_1/FVC less than 70%, similar to the 16% of grain workers in a study by do Pico et al.[17] The absence of a relationship between dust exposure and lung function may be because there is no relationship or because the exposure information is inappropriate. When dust levels vary significantly from day to day, the use of dust values obtained on 1 d of sampling will be inadequate to examine exposure-effect relationships. Alternatively, if a specific component of the dust (e.g., fungal spores, bacterial endotoxin) is responsible for the health effect and is variable in content from one dust sample to another, gravimetric dust measurements will not give a reliable estimation of exposure. It is also possible that the health effects are not caused by a component of the dust at all, but rather by a gas or vapor. Additionally, workers experiencing symptoms in the past may have altered their work practices or used respiratory protection to limit their exposure. Thus, farmers with high past exposure leading to health problems may currently have low exposures.

If the abnormality is the result of a hypersensitivity reaction, an exposure–effect relationship may not exist for the entire group, since only a small proportion of workers might be affected. Those affected would be expected to react to relatively low exposure, but no exposure–effect relationship might be evident in the analysis of the group as a whole.

In summary, the hog farmers studied had higher levels of total and respirable dust exposure than did the control farmers. Floor feeding and indoor feed grinding are contributors to the elevated dust exposure. The hog farmers reported more respiratory symptoms than did the control farmers; no differences in lung function were found between the two groups. No relationships between dust exposure levels or length of exposure and symptoms or lung function were evident.

REFERENCES

1. **Warren, C.P.W.,** Lung disease in farmers, *Can. Med. Assoc. J.*, 116, 391, 1977.
2. **Chan–Yeung, M., Ashley, M.J., and Grzybowski, S.,** Grain dust and the lungs, *Can. Med. Assoc. J.*, 118, 1271, 1978.
3. **Donham, K.J. and Gustafsson, K.E.,** Human occupational hazards from swine confinement, *Ann. Am. Conf. Gov. Ind. Hyg.*, 2, 137, 1982.
4. **Harries, M.G. and Cromwell, O.,** Occupational asthma caused by allergy to pigs' urine, *Br. Med. J.*, 284, 867, 1982.
5. **Donham, K.J., Knapp, L.W., Monson, R., and Gustafson, K.,** Acute toxic exposure to gases from liquid manure, *J. Occup. Med.*,

24, 142, 1982.

6. **Donham, K.J., Rubino, M., Thedell, T.D., and Kammermeyer, J.,** Potential health hazards to agricultural workers in swine confinement buildings, *J. Occup. Med.,* 19, 383, 1977.
7. **Katila, M.J., Mantyjarvi, R.A., and Ojanen, T.H.,** Sensitisation against environmental antigens and respiratory symptoms in swine workers, *Br. J. Ind. Med.,* 38, 334, 1981.
8. **Matson, S.C., Swanson, M.C., Reed, C.E., and Yunginger, J.W.,** IgE and IgG–immune mechanisms do not mediate occupation–related respiratory or systemic symptoms in hog farmers, *J. Allergy Clin. Immunol.,* 72, 299, 1983.
9. **Pratt, D.S. and May, J.J.,** Feed–associated respiratory illness in farmers, *Arch. Environ. Health,* 39, 43, 1984.
10. **Donham, K.J., Zavala, D.C., and Merchant, J.,** Acute effects of the work environment on pulmonary functions of swine confinement workers, *Am. J. Ind. Med.,* 5, 367, 1984.
11. **Holness, D.L., O'Blenis, E.L., Sass–Kortask, A., Pilger, C., and Nethercott, J.R.,** Respiratory effects and dust exposures in hog confinement farming, *Am. J. Ind. Med.,* 11, 571, 1987.
12. **Clark, S., Rylander, R., and Larsson, L.,** Airborne bacteria, endotoxin and fungi in dust in poultry and swine confinement buildings, *Am. Ind. Hyg. Assoc. J.,* 44, 537, 1983.
13. **Curtis, S.E., Drummond, J.G., Kelley, K.W., et al.,** Diurnal and annual fluctuations of aerial bacterial and dust levels in enclosed swine houses, *J. Anim. Sci.,* 41, 1502, 1975.
14. **Warren, C.P.W.,** Respiratory disorders in Manitoba cattle farmers, *Can. Med. Assoc. J.,* 125, 41, 1981.
15. **Holness, D.L., Taraschuk, I.G., and Goldstein, R.S.,** Acute exposure to cotton dust, a case of mill fever, *J.A.M.A.* , 247, 1602, 1982.
16. **do Pico, G.A., Reddan, W., Flaherty, D., et al.,** Respiratory abnormalities among grain workers, *Am. Rev. Respir. Dis.,* 115, 915, 1977.
17. **do Pico, G.A., Reddan, W.G., Tsiastis, A., Flaherty, D., Jacobs, S., and Rankin, J.,** Pulmonary reaction to grain dust and its constituents, *Chest,* 80, 57S, 1981.

Organic Dust Toxic Syndrome (Pulmonary Mycotoxicosis) — A Review of the Experience in Central Wisconsin

D. A. Emanuel, J. J. Marx, Jr., B. Ault, R. C. Roberts, M. J. Kryda, and M. W. Treuhaft

INTRODUCTION

Toxic reactions that were thought to be due to inhaled mycotoxins were first published in 1975, but were recognized at least 10 years earlier.[1] The illness produced symptoms that were similar to acute farmer's lung disease (FLD), but which were almost always associated with opening a "capped" silo. An ongoing study has included 198 individuals with similar histories; all were dairy farmers with exposure either to hay or corn silage frequently in July, August, and September. The disease entity was originally termed pulmonary mycotoxicosis, but at an international symposium in Sweden in 1985, the description organic dust toxic syndrome (ODTS) was applied.[2] The present report is a review of the clinical features, diagnostic studies, treatment, and follow–up of ODTS.

CLINICAL FINDINGS

One hundred and ninety–eight individuals with a mean age of 39 years (range 22 to 68) were seen with a similar history of exposure to either hay or corn silage, although 15 individuals had exposure to moldy grain (Table 1). Exposures occurred most often in July, August, and September as compared with January, February, and March for the patients with FLD (Table 2). There were 139 of 198 patients seen that had acute symptoms and/or a white blood cell count (WBC) >10,000. Complaints began most frequently 4 to 6 h after exposure, and, in decreasing order, were cough (54%), fever (51%), shortness of breath (51%), chills (25%), myalgia (33%), and malaise (20%). Physical examination revealed a paucity of signs, although 45 had occasional crackles at the lung bases. Increased WBC counts (10,000 to 26,000) occurred in 65 individuals seen within 12 to 24 h of their acute exposure; the majority of the WBCs were neutrophils. There was little or no evidence of eosinophilia.

Serum precipitin studies for *Micropolyspora faeni, Thermoactinomyces vulgaris, T. candidus, Saccharomonospora viridis,* plus five different strains of *Aspergillus fumigatus, A. sydowi, A. glaucus,* and *A. niger* were performed in all. Eighty–six of 198 patients had negative precipitins; 71 were positive to the thermophilic group and 37 were positive to the *Aspergillus* group. In the precipitin–positive group there were 59 individuals who had 79 exposures with acute symptoms of cough, shortness of breath, fever, chills, or malaise, and/or a WBC count >10,000. In the precipitin–negative group 64 patients had 71 exposures with acute symptoms and/or WBC count >10,000.

Pulmonary function studies in 33 patients revealed arterial hypoxemia in 10 and an abnormal DLCO in only 4 (measured by the single breath technique).

TABLE 1
Patients with History of or with Acute Illness[a]

	Precipitin negative	Precipitin positive			
	ODTS	ODTS	FLD with history of ODTS	FLD	Total
Number of patients	86	96	16	195	393
Years	1963—1984	1963—1981	1964—1981	1955—1984	—
Peak year	1980	1978	1979	1973	—
Percent of males	90%	93%	94%	83%	—
Age range	22—68	16—62	17—58	6—85	—
Median	39	32	37	51	—

[a] Marshfield Clinic, Marshfield, WI, 1955 to 1984.

TABLE 2
Number of cases in Each Category That Occurred in Each Month[a]

	Precipitin negative	Precipitin positive		
	ODTS	ODTS	FLD with history of ODTS	FLD
January	5	1	3	22
February	3	3	1	17
March	4	3	1	20
April	5	2	0	25
May	1	3	1	12
June	4	3	0	11
July	13	10	2	7
August	13	7	4	8
September	5	7	2	11
October	8	8	0	8
November	6	6	2	11
December	2	6	0	17

[a] Marshfield Clinic, Marshfield, WI, 1955 to 1984.

Bronchoalveolar lavage (BAL) on six patients demonstrated a distinctly different cell population than that seen in lavages from patients with FLD. Since most of the chest radiographs were unremarkable, an attempt was made to lavage a dependent portion of the lungs. The lavage fluid was immediately processed by centrifugation to separate the cells from the lavage fluid. The cell pellet was washed with buffered saline and the cells counted using a standard hemocytometer. Lymphocytes were identified as T cell or B cell by conventional immunofluorescent microscopy with commercially available antisera. The BAL demonstrated an increased cellularity which was predominantly neutrophilic leukocytes and alveolar macrophages.

Three patients were seen soon after symptoms developed following the opening of a filled silo. One was a laboratory worker with no previous significant exposure to dusty forage. Each of the silos had been filled at least 2 months previously. Each of these individuals agreed to multiple examinations. Thus, we had the opportunity to study the sequential changes in the content of BAL fluid during and following the clinical episode. Changes in the resident and transient cell populations in the lung, the microorganisms involved, as well as the development of immunologic sensitivity were measured. Table 3 outlines the salient laboratory findings during BAL of ODTS. In two of three patients the initial cell count was markedly higher than subsequent studies and higher than we have seen in FLD patients. The cell differentials showed a marked granulocytosis as compared to a lymphocytosis in FLD. Neither IgG nor IgA were elevated relative to the total protein as is seen in FLD. Specific antibody levels (measured by an enzyme immunoassay) to the usual organisms associated with FLD were not elevated in ODTS (Table 4). Complements (C4 and C3) were initially elevated but quickly returned to normal levels. Endotoxin (measured by the Limulus assay) was transiently present in the lavage fluid.

Significant numbers of microorganisms could be isolated from BAL fluids (Table 5). This was particularly true of the actinomycetes, commonly found in forage. It was not surprising that the numbers of colony forming units (CFUs) decreased as these were cleared from the lung, but even after 112 d, in the absence of any reexposure, significant numbers of thermophilic actinomycetes persisted. The types of organisms isolated are presented in Table 6.

Pulmonary biopsies, obtained on two of the more severe patients, revealed a multifocal acute process related to the terminal bronchioles, alveoli, and interstitial areas. Most of the bronchioles were filled with an exudate composed of neutrophils and histiocytes. The exudate extended into the surrounding interstititial tissue and into a number of the alveoli

TABLE 3
BAL in ODTS

Patient	Days following exposure	Total cells recovered ($\times 10^7$)	Lymph	PMN
MT	1	4.7	2	79
	6	4.2	12	8
	43	2.5	2	10
	112	2.4	7	2
HB	4	6.2	5	52
	54	2.3	3	7
PA	4	1.1	4	26
	53	1.9	10	10
Normals	—	0.6	62	8.5
FLD	?	2—3	10	2

TABLE 4
IgG Antibody in BAL Fluid following Acute ODTS

		Antigen			
Patient	Days following acute episode	*M. faeni*	*Tha. vulgaris*	*Tha. candidus*	*Asp. fumigatus*
MT	1[a]	0.52	0.10	0.03	0.62
	6	0.18	0.12	0.08	0.53
	43	0.56	0.28	0.12	0.39
	112	0.35	0.28	0.27	0.59
HB	4	0.09	0.16	0.00	0.61
	54	0.14	0.00	0.01	0.34
PA	4	0.17	0.07	0.02	0.66
	53	0.10	0.00	0.00	0.44

[a] ELISA value = corrected optical density.

TABLE 5
Microorganisms Isolated from BAL Fluid ODTS

Patient	Days following exposure	Total fungi	Mesophilic actinomycetes	Thermophilic actinomycetes	Mesophilic bacteria
MT	1	2×10^2	$>10^5$	$>10^5$	Not identifiable
	6	—	1.7×10^2	9.0×10^2	2.0×10^3
	43	2×10^1	8.0×10^1	1.5×10^3	1.5×10^3
	112			6.0×10^2	10^4
HB	4	9×10^1	10^2	2.0×10^1	1.1×10^2
	59	2×10^1	—	—	2.5×10^3
PA	4	1×10^1	2×10^1	5×10^2	1.7×10^3

TABLE 6
Microorganisms Recovered from BAL in ODTS (Formerly Pulmonary Mycotoxicosis)

Patient	Fungi	Actinomycetes
MT	*Penicillium* *Aspergillus nidulans* *Aspergillus* spp. *Mucorales*	*Micropolyspora faeni* *Thermoactinomyces candidus* *Thermomonospora* *Streptomyces* *Saccharomonospora viridis*
HB	*Penicillium* *Aspergillus*	*M. faeni* *T. candidus* *Streptomyces*
PA	Unidentified	*M. faeni* *T. candidus* *S. viridis* *Streptomyces*

with some consolidation. There were a few scattered plasma cells, chiefly in the interstitial tissue and about small blood vessels and bronchi. Many corpora amylacea present in the alveoli were more numerous in areas involved with exudate, but were also seen in areas where there was no inflammatory reaction. Methenamine silver stains revealed large numbers of fungal spores throughout the specimen, and culture of the biopsy material, blood agar at both 22 and 37°C, revealed an amazing array of organisms. At least five different species of fungi were noted, including *Fusarium* spp. and *Pencillium* spp.

Hospitalization was rarely required; most patients were successfully treated as outpatients. Most reported feeling ill for 3 to 5 d and not feeling well for another 5 d, but were able to return to farm duties during that period. Eleven patients were treated with short-duration steroid therapy because of severity of febrile reaction or suspected recurrence of FLD.

DISCUSSION

The silo provides an efficient method for storing and processing of various forages that are produced on the farm. In the early part of the growing season haylage (cut grasses) can be efficiently processed, while the late season crops, such as corn and soybeans, can be placed in a silo for use during the winter. We have speculated that the increase in ODTS in our farming population over the last few years is a manifestation of a change in farming techniques in our area. The use of large storage silos for chopped forage has been the recommended procedure for the last 10 years or so.

This technique, however, permits the accumulation of large numbers of fungal spores in the silage; subsequent disruption provides high exposures in a confined, poorly vented area. A large dust load appears to be important in the induction of this disorder. There do not appear to be any unique host factors which predispose to this acute illness; generally all individuals appropriately exposed become ill. A documented transition of ODTS to FLD has not been demonstrated in our population. In spite of this, previous epidemiologic surveys have noted increased reports of ODTS–like illnesses in farmers with precipitins as opposed to farmers without precipitins. The acute inflammation demonstrated in individuals with ODTS would present the opportunity for sensitization, since the available antigens are present for prolonged periods.

The acute inflammatory process appears to be in direct response to the inhaled dust and not a specific immunologic reaction.[3,4] As mentioned, all subjects were known to be serologically negative and were nonsmokers. None of these individuals developed serological reactivity, and in fact, there is an apparent complete return to normal cell characteristics of the BAL fluid.

The clinical presentation of this illness is similar to FLD, but the total clinical picture is different. ODTS is an acute illness which appears to follow exposure to large amounts of moldy silage. We have also seen this illness following exposure to moldy oats and moldy corn. Recurrences require a similar degree of exposure and affect all individuals exposed. Hypersensitivity lung disease or FLD, on the other hand, affects only selected individuals with recurrence requiring only minimal exposure. The absence of precipitins to the thermophiles also distinguishes ODTS from FLD. Perhaps most striking are the reports of miniepidemics in all individuals exposed to the moldy material. Some farmers that we have interviewed recall the typical flu-like illness following uncapping of their silos almost every year, but at no other time. Reports of the disease from Minnesota, North and South Dakota, Arkansas, Iowa, Indiana, New York, and Pennsylvania indicate that this exposure has a broad geographic distribution.

It is amazing that the chest radiograph does not show abnormality as a result of this dust inhalation in view of the fever, cough, and respiratory distress that is produced. Arterial desaturation, on occasion, can be demonstrated and pulmonary function studies occasionally show some reduction in diffusing capacity. Most often the measurement of ventilation was normal despite the respiratory distress.

Silo gases as a rule are not involved, since this is a danger only during the first 5 to 10 d after a silo has been filled. In most cases, the silo has been filled weeks to months before the acute exposure.

Acute FLD can also cause similar symptoms and a similar delayed onset of several hours. The absence of radiographic abnormalities, the lack of rales, and the lack of a diffusing abnormality in ODTS would indicate that this disease process is different. The failure to find precipitating antibodies in the majority of cases would also argue against this being FLD. On the other hand, the presence of precipitins does not necessarily mean that the patients did not have ODTS. Approximately 8 to 9% of all farmers do have precipitins.[4,5] A peak incidence of the disease occurring in July, August, September, and October is also different from that of FLD which occurs during December, January, February, March, and April. It is also significant that this disease can occur in miniepidemics and we have documented as many as five individuals becoming ill from a common exposure.

The report by Brinton et al.[6] emphasizes that this is a biological phenomenon not limited to the farm environment. Indeed, FLD disease rarely affects family members of coworkers. ODTS has also been thought to involve the activation of the alternative pathway of complement by organic dust.[7]

While we had originally described the illness as possibly due to mycotoxins, we have not been able to substantiate this. Mycotoxins do have the capability of inducing human disease,[8] and while ochratoxin and aflatoxin have been demonstrated in significant amounts in the inhaled moldy material, these have not been consistently present.

Olenchock et al. and Rylander[9–13] have recently proposed that endotoxins, present in the inhaled material in significant concentrations, can produce this reaction. Indeed, Rylander[13] has also advanced arguments favoring endotoxin as the cause of mill fever in cotton workers. Olenchock[9] demonstrated endotoxin levels in silos ranging from 8.8 to 87.3 ng/mg of dust sample.

While the exact etiology of the toxic lung reaction remains to be identified, this does seem to be a distinct disease entity, different than the FLD due to the thermophilic actinomycetes in the following respects: (1) miniepidemics do occur,[6] (2) all individuals exposed develop the illness, (3) a different age group is involved, (4) the illness occurs at a different time of year, (5) the histopathology is different, (6) BAL provides distinction in the cell population, and (7) serologic sensitivity apparently does not occur.

REFERENCES

1. **Emanuel, D.A., Wenzel, F.J., and Lawton, B.R.,** Pulmonary mycotoxicosis, *Chest,* 67, 293, 1975.
2. **Ryland, R.,** The international workshop on health effects of organic dust in the farm environment, Skokloster, Sweden, *Scand. J. Work Environ. Health,* in press.
3. **Marx, J.J., Arden–Jones, M.P., Treuhaft, M.W., Gray, R.L., Motszko, C.S., and Hahn, F.F.,** The pathogenic role of inhaled microbial material in pulmonary mycotoxicosis as demonstrated in an animal model, *Chest,* 80, 76S, 1981.
4. **Pratt, D.S. and May, J.J.,** Feed–associated respiratory illness of farmers, *Arch. Environ. Health,* 39, 43, 1984.
5. **Madsen, O., Kock, L.E., Wenzel, F.J., Robbins, J.L., and Schmidt, C.O.,** The prevalence of farmer's lung in an agricultural population, *Am. Rev. Respir. Dis.,* 113, 171, 1976.
6. **Brinton, W.T., Vastbinder, E.E., Greene, J.W., Marx, J.J., Hutcheson, R.H., and Schaffner, W.,** Outbreak of organic dust toxic syndrome in a college fraternity, *J.A.M.A.,* 258, 1210, 1987.
7. **Edwards, J.H., Baker, J.T., and Davies, B.H.,** Precipitin test negative farmer's lung activation of the alternative pathway of complement by moldy hay dusts, *Clin. Allergy,* 4, 379, 1974.
8. **Sansonov, P.F.,** Mycotoxicosis of man and agricultural animals, in *Kiev Izd Akad Nauk,* 6th ed., Bilai, Ukr SSR, 1960, 131.
9. **Olenchock, S.A., May, J.J., Pratt, D.S., and Morey, R.R.,** Endotoxins in the agriculture environment, *Scand. J. Work Environ. Health,* in press.
10. **Emanuel, D.A., Wenzel, F.J., Bowerman, C.I., and Lawton, B.R.,** Farmer's lung; clinical, pathologic and immunologic study of twenty–four patients, *Am. J. Med.,* 37, 392, 1964.
11. **Lacey, J. and Lacey, M.E.,** Spore concentrations in the air of farm buildings, *Trans. Br. Mycol. Soc.,* 47, 547, 1964.
12. **May, J.J., Pratt, D.S., Stallones, L., Morey, P.R., Olenchock, S.A., Deep, I.W., and Bennett, G.A.,** A study of silo openings: the work environment and its physiologic effects, *Scand. J. Work Environ. Health,* in press.
13. **Rylander, R.,** Bacterial toxins and the etiology of byssinosis, *Chest,* 79, 34S, 1981.

A Study of Dust Generated During Silo Opening and Its Physiologic Effects on Workers

J. J. May, D. S. Pratt, L. Stallones, P. R. Morey, S. A. Olenchock, I. W. Deep, and G. A. Bennett

INTRODUCTION

Most material stored in the common nonoxygen-limited silo is anaerobic, and thus protected from microbial contamination. The uppermost 1 to 3 ft is exposed to air and often heavily overgrown with microorganisms. Because of the size of the silo headspace and the aerodynamics of the silo chute, removal of this "cap" material can generate extremely large amounts of dust. Workers exposed to these conditions may subsequently experience high fever, myalgias, headaches, and chest tightness,[1,2] a problem currently referred to as the organic dust toxic syndrome (ODTS).[3] This study was undertaken to better define the work environment associated with this illness.

STUDY DESIGN

Five dairy silos in upstate New York were studied in October 1983. All were nonoxygen-limited, cement block structures from 40 to 60 ft in height and 12 to 18 ft in diameter. Four silos contained haylage (alfalfa, timothy, grass, etc.) which had been ensiled for 15 to 16 weeks. The remaining silo (#4) had been filled with corn silage approximately 3 weeks prior to the study. Four of the silos were enclosed at the base by a feed room connecting the silo to the barn. The other silo (#2) was free-standing, connected to the barn by a conveyor belt. (This feature of silo 2, as well as the windy conditions on the day of unloading, clearly influenced the levels of dust subsequently measured at its base.) In four of the silos (1, 2, 3, and 5), a plastic sheet had been placed over the silage at the completion of filling. Additional silage was then blown on top of this sheet in silos 1 and 2.

Eight silo workers and five farm wife controls were studied in relation to the uncapping of the silos. The workers were all male (mean age 32.3 years). One worker had a history of mild asthma (silo 3) and one (silo 4) had previously had ODTS on one occasion. Three workers were smokers. The controls had a mean age of 49 years, none smoked, and none had respiratory problems. Venous blood samples were obtained on subjects and controls prior to unloading and 6 h later. Spirometry was obtained on all workers prior to silo unloading, following completion of the job, and later at 6 h postcompletion.

DUST LEVELS

Respirable and total dusts were collected using two-piece, closed-face samplers (polyvinylchloride 37–mm diameter filter, 5 μm pore size). The respirable dust sampler eliminated particles larger than 10 μm, allowing 50% of particles with an aerodynamic diameter of 3.5 μm into the sampler. These samplers were worn by workers and also placed in the silo at the center of the headspace, near the chute, and at the base of the silo. Additional total dust estimation as well as particle sizing was carried out within the silo headspace with multistage cassette impactors and at the silo with Andersen Ambient Samplers.

Total dust in the five silos varied widely, as did dust levels at the silo base (Table 1). Silos 1 and 2 had mean dust levels of 113.4 and 20.2 mg/m^3, respectively, in the headspace near the chute, with 104 and 19.4 mg/m^3 near the center of the headspace. These levels are substantially higher than those

TABLE 1
Dust Levels within the Silo (mg/m^3)

Silo #		Personal sampler	Within silo near chute	Within silo near center	Silo base
1	Total[a]		100	86.8	129[b]
			86.2	91.8	
			154	101	
	Respirable[c]	24.6	22.1	24.8	—
		40.3			
2	Total[a]		20.8	13.5	6.8[b]
			19.6	25.2	3.8
	Respirable[c]	10.1	20.0	—	2.2
		10.5			
3	Total[a]		1.9	2.7	3.6[b]
			2.7	2.4	3.0
	Respirable[c]	1.4	0.9	1.4	0.5
4	Total[a]		3.2	1.8	7.1[b]
			3.1	1.4	6.0
	Respirable[c]	0.9	0.6	—	2.5
		0.3			
5	Total[a]		0.8	0.9	4.2[b]
			0.8	0.7	2.7
	Respirable[c]	0.8	0.2	0.5	0.7

[a] Total dust levels, unless otherwise indicated, measured with two–piece, closed–face cassette.
[b] Total dust level as measured with an Andersen Ambient Sampler.
[c] Respirable dust levels, measured with two–piece, closed–face cassette, personal and area samplers.

encountered in the remaining three silos. Results at the silo bases were also divergent, with 129 mg/m^3 for silo 1 and all other silos having levels <10 mg/m^3.

Respirable dust levels, as measured with personal and other two-piece, closed-face cassette samplers, also varied widely between silos (Table 1). Respirable dust levels in silos 1 and 2 were quite high (24 and 10 mg/m^3, respectively), while levels in silos 3, 4, and 5 were lower, ranging from 0.2 to 1.4 mg/m^3.

The size characteristics of the dust particles are expressed in terms of the mass median diameter (MMD). An MMD of 10, for example, indicates that 50% of the total weight of particles has an aerodynamic diameter of greater than 10 µm and the other 50% (by weight) of particles is smaller than 10 µm. Particle–sizing studies showed an MMD within two silo headspaces of 6.8 and 7.0 µm. The MMD of particles collected at the bases of four silos ranged from 4.5 to 7.2 µm with geometric standard deviations between 1.8 and 3.7. Thus, the majority of dust particles are sufficiently small to be respirable.

VIABLE ORGANISMS IN SILAGE AND DUST

The airborne concentration of mesophilic (25°C) fungi, thermophilic (45°C) fungi, and thermophilic (55°C) bacteria in the dust was estimated using all–glass impingers (AGI), which pulled the dust-laden air through 50 ml sterile water at 12.5 l/min. Immediately following sampling, AGI fluid was plated by serial dilution. Rose Bengal–Streptomycin plates were used for fungi and were incubated at 25 and 45°C. The tryptic soy agar media used for thermophilic bacteria was incubated at 55°C. Samples of cap silage were stored at –20°C, and subsequently streaked on potato dextrose agar (PDA). These samples were also subjected to surface sterilization to allow isolation of any internal fungi. Additionally, dust sampling filters were stored at –20°C, subsequently suspended in aqueous solutions, and cultured as above. Finally, fungi collected in the AGI and cultured on Rose Bengal–Streptomycin agar were subcultured on Sabouraud dextrose agar and later identified on PDA. Most samples were heavily contaminated with bacteria and isolation of fungi required the use of acidified PDA.

Concentrations of viable organisms in the sampled dust are reported in Table 2. Of note are the extremely high levels of all types of organisms encountered in the most dusty silo (silo 1). In particular, the levels of thermophilic bacteria (in excess of 4 billion/m^3) are among the highest ever recorded in agricultural dusts.[4]

Identification of the predominant fungi is summarized on Table 2. In some silos, particularly silo 1, the Zygomycetes *(Rhizopus* and *Mucor)* were predominant. In addition to fungi, substantial numbers of bacteria were present but not identified.

ENDOTOXIN LEVELS

Endotoxin concentration was measured using a nonpyrogenic aqueous extract of dust collected at the silo base with an Andersen sampler. In two silos (1 and 2) with high dust levels, samples were also analyzed from closed-face cassette samplers. In the case of respirable dust, samples from two-piece, open-face cassettes were assayed or the average of all Andersen samples less than 10 µm was used (silo base measurements). These dust extracts were assayed in duplicate by a spectrophotometric modification of the *Limulus* amebocyte lysate gel test.[5] The endotoxin level was multiplied by the corresponding dust level to determine the amount of total of respirable endotoxin per cubic meter.

Endotoxin levels at the base of all silos but #1 were low; the total level was 3.5 to 4.9 ng/m^3, with respirable levels of 0.4 to 5.5 ng/m^3. There was sufficient dust collected in cassette samplers from silos 1 and 2, and combined with the results of the area sampling, it becomes apparent that mean levels of endotoxin were quite high. In silo 1, total dust endotoxin levels averaged 8850 ng/m^3 with respirable levels of 1322 ng/m^3. These levels in silo 2 were 160 and 57 ng/m^3, respectively.

TABLE 2
Microorganisms in Silage and Silage Dust

	Silo 1	Silo 2	Silo 3	Silo 4	Silo 5
25°C fungi[a]	100	0.41	3.2	1.3	0.14
45°C fungi[a]	120	0.06	0.43	1.5	0.01
55°C bacteria[a]	4300	220	0.06	4.5	—
Organisms	*Rhizopus* *Mucor* *Aspergillus*	*Penicillium* *Mucor* *Aspergillus* *Fusarium* *Alternaria* *Nigrospora* *Absidia* *Circinella*	*Penicillium* *Mucor* *Aspergillus* *Fusarium* *Alternaria* *Geotrichum* *Cladosporium* *Monascus*	*Rhizopus* *Mucor* *Aspergillus* *Fusarium*	*Bryssochamys* *Absidia* *Aspergillus* *Trichoderma* *Alternaria*

[a] In millions of viable organisms per cubic meter. All dust specimens were collected in the headspace near the chute doorway.

MYCOTOXIN LEVELS

Samples of "cap" silage weighing approximately 500 g were frozen at –20°C. These samples were examined for trichothecenes and zearalenone. Methanol–water extracts of silage samples were analyzed by gas chromatography of the hepta–fluorobutyryl derivative and by gas chromatography–mass spectrometry.[6] In addition, skin tests were performed on ICR white mice using a 70% ethanol extract of each silage sample.

Analysis of samples of cap silage from the five silos revealed trace amounts of deoxynivalenol (0.1 to 0.2 ppm) in silo 4, which contained corn silage. The presence of this mycotoxin was confirmed by gas chromatography–mass spectrometry and also gave a positive skin test on ICR white mice. The remainder of the silage samples tested did not show evidence of mycotoxins.

PHYSIOLOGIC MEASUREMENTS

Venous white blood cells (WBC) were counted in the eight working subjects and in the five nonexposed controls. Comparison of baseline and 6–h postopening blood samples within each group showed no clear change for the control group ($p > 0.1$), but an apparent increase in WBC counts for the post-opening samples in the working subjects ($p = 0.04$) by the Mann–Whitney test.[7]

Functional levels of C3 and C4 were measured in silo workers and controls using a standard hemolytic assay.[8] Levels were measured in preunloading serum samples as well as samples obtained 6 h after completion of the job. The hemolytic titers for C3 and C4 showed reductions in the 6–h postwork levels compared to the baseline levels in the eight subjects. These changes, when compared to those in the control group, were not found to be significantly ($p < 0.05$) different with the *t* test. Microouchterlony testing revealed no sign of precipitating antibodies against an aqueous extract of matched silage in any of the eight subjects.

Baseline spirometry showed normal or supernormal function in all but two individuals (both were asymptomatic; one of these demonstrated a mild obstructive defect and the other showed mild restriction). When analyzed by *t* test, repeat spirometry showed no significant alterations (all *p* values ≥ 0.33) either immediately after completion of silo unloading or 6 h later.

The workers in both silo 1 and 2 commented on the dusty conditions. Those in silo 1 were forced to stop midway through the job and catch their breath for a few moments. Afterwards they complained of a foul taste in their mouths, burning in their throat, and some burning in their eyes. The workers in silo 2 noted the presence of the dust but voiced no particular symptoms related to it. At 6 h postexposure, both workers in silo 1 complained of feeling feverish. They experienced headache, malaise, and one individual complained of nausea. They experienced rises in WBC counts of 4500 and 4300 cells per cubic millimeter, respectively. By 24 h postexposure, however, both were asymptomatic.

COMMENTS

We have documented extremely high levels of dust, much of it respirable. This dust was shown to contain excessive numbers of viable organisms. Thermophilic bacteria, predominantly *Thermoactinomyces vulgaris,*[9] were particularly numerous, with viable levels in excess of 4 billion/m^3 in silo 1. Levels within silo 2 were only 10– to 20–fold lower.

Equally impressive are the levels of endotoxin found in the two different silos. Levels of 80 ng/m^3 in association with cotton dust appear to be sufficient to generate physiologic responses.[5] The amount of respirable endotoxin in silo 2 (50 ng/m^3) approached this level and respirable endotoxin in silo 1 (1.3 μg/m^3) was more than 15–fold higher. It should be recalled that these levels of endotoxin and organisms were encountered by individuals without respiratory protection who were exercising strenuously, usually with minute ventilations >20 l/min.

Other notable findings include the occurrence of leukocytosis and fever in silo 1 workers — a silo found to contain no measurable mycotoxins, thus casting doubt upon the role of mycotoxins in ODTS. The absence of precipitating antibodies vs. silage extract is in accordance with clinical observations.[1,2] Complement levels showed a nonsignificant fall, raising the question of possible reductions in complement levels that might have been demonstrated with a larger study or different sampling times.

Finally, and hopefully of some practical importance to farmers, we have documented wide variability in the amount of dust encountered while uncapping a dairy silo. Some of this seems to relate to the technique of capping used at the completion of filling. Our limited experience would suggest that the practice of placing a plastic sheet on top and blowing additional silage on this sheet should be reevaluated. This appears to promote drying after a period of microbial overgrowth. We have demonstrated that extremely high levels of dust can be generated in these situations, and that the dust contains high levels of thermophiles and endotoxin. It appears that some component(s) of this dust can provoke a leukocytosis and symptoms of ODTS in the absence of substantial pulmonary function changes. Although we are unaware of controlled studies, we believe that few would argue against our recommendation to farmers that appropriately fitted, NIOSH-approved respiratory protection be used in these situations.

REFERENCES

1. **Emanuel, D.A., Wenzel, F.J., and Lawton, B.R.,** Pulmonary mycotoxicosis, *Chest,* 67, 293, 1975.
2. **May, J.J., Stallones, L., Darrow, D., and Pratt, D.S.,** Organic dust toxicity (pulmonary mycotoxicosis) associated with silo unloading, *Thorax,* 41, 919, 1986.
3. **doPico, G.A.,** International Workshop on Health Effects of Organic Dusts in the Farm Environment, Skokloster, Sweden: 23—25 April 1985, report on diseases, *Am. J. Ind. Med.,* 10, 261, 1986.
4. **Lacey, J.,** The microflora of grain dusts, in *Occupational Pulmonary Disease,* Dosman, J.A. and Cotton, D.J., Eds., Academic Press, New York, 1980, 417.

5. **Castellan, R.M., Olenchock, S.A., Hankinson, J.L., Millner, P.D., Cooke, J.B., Bragg, C.K., Perkins, H.H., Jr., and Jacobs, R.R.,** Acute bronchoconstriction by cotton dust: dose–related responses to endotoxin and other dust factors, *Ann. Intern. Med.*, 101, 157, 1984.
6. **Bennett, G.A., Stubblefield, R.A., Shannon, G.M., and Shotwell, O.L.,** Gas chromatographic determination of deoxynivalenol in wheat, *J. Assoc. Off. Anal. Chem.*, 66, 1478, 1983.
7. **Mosteller, F. and Rourke, R.E.K.,** *Sturdy Statistics, Nonparametric and Order Statistics,* Addison-Wesley, Reading, MA, 1973, chap. 3.
8. **Olenchock, S.A., Mull, J.C., and Major, P.C.,** Extracts of airborne grain dusts activate alternative and classical complement pathways, *Ann. Allergy,* 44, 23, 1980.
9. **Labeda, D.,** personal communication.

Hypersensitivity Pneumonitis due to *Aspergillus fumigatus* in Papermill Workers

M. M. J. Kryda, M. A. Emanuel, J. J. Marx, Jr., and M. W. Treuhaft

Exposure to organic dusts in the papermaking industry has been known to cause hypersensitivity lung disease. Several different organisms have been isolated from the papermill environment and implicated as agents for sensitization and disease induction. This report focuses on cases of hypersensitivity pneumonitis caused by exposure to *Aspergillus fumigatus* in the workplace.

The index case was a 32–year–old man who presented with complaints of episodic cough, fever, chills, dizziness, and fatigue several hours after leaving work. Symptoms had been present for several years, generally occurring in the summer. Symptoms were worse after a vacation or several days off work. He was a chip unloader for 7 years and dated the onset of symptoms to the initiation of this job. Symptoms were worse when he was unloading "hot cars", boxcars which had been shipped from out–of–state suppliers of wood chips. Frequently, they would be on the railroad sidings and in the sun for in excess of 2 weeks. Upon opening these boxcars the patient would note tremendous amounts of heat and the presence of a large amount of greenish dust which became easily airborne upon attempting to unload the substance with a front-end loader. The patient had used various masks and respirators in an attempt to minimize his exposure. However, he noted the filter elements frequently became clogged and required changing too often. In addition, the heat in the work area made wearing the tight mask quite uncomfortable.

He was a 3–year ex–smoker of one pack per day for 6 years. There was no allergic history or asthma. He used an oil forced–air furnace with humidification, but did not note any symptoms at home nor worsening of symptoms in the winter. Physical examination was normal.

The white blood cell count was 18,000 (81 neutrophils, 6 bands, 8 lymphs, 2 monos, and 1 eosinophil). Chest radiographs demonstrated a reticulonodular infiltrate which had progressed over 4 years. Pulmonary function studies and arterial blood gases were normal.

Bronchoalveolar lavage contained 69% lymphocytes, 10% polymorphonuclear neutrophils, and 18% macrophages. There were strong serum precipitin reactions to multiple strains of *A. fumigatus* and to *A. niger* as well, but none to farmer's lung antigens or to maple bark stripper's antigen. *A. fumigatus* precipitins were also found in bronchoalveolar lavage fluid. Lung biopsy demonstrated chronic interstitial pneumonitis with bronchiolitis consistent with hypersensitivity pneumonitis. *A. fumigatus* was cultured from the patient's respiratory mask (95% of the microorganisms) and from the wood chips.

Four other workers with similar symptom complaints were identified. Three were chip unloaders, and one was a wood–room foreman. All had dyspnea, recurrent fever, and cough. The wood–room foreman had a lifelong history of asthma complicated by the development of bronchopulmonary aspergillosis. Specific *A. fumigatus* immunoglobulin levels demonstrated a clear discrimination between the individual with bronchopulmonary aspergillosis and the four with hypersensitivity pneumonitis secondary to *Aspergillus.*

The papermaking industry has a consistent requirement for substances which will provide raw fiber for manufacturing paper. Numerous substances are used, such as rags, waste products, and various lumbering products including wood chips. These chips are loaded into closed boxcars and may spend several weeks in transit to papermills. The high moisture content provides an excellent culture medium for *Aspergillus fumigatus* and other microorganisms, particularly in the summer. Steam and a greenish dust may emanate from these cars. Both papermill authorities and workers indicate that the hot cars are seasonal in character, occurring in the summer and early fall months.

Hypersensitivity pneumonitis due to *Cryptostroma corticale, Alternaria,* and thermophilic actinomycetes has been previously reported in the papermaking industry. This indicates frequent contamination of organic products with various organisms capable of inducing hypersensitivity pneumonitis; there may be elements specifically related to the environment of papermills which enhance susceptibility of sensitization.

In light of the above, negotiations were undertaken with the company involved and the related unionized work force. A specific epidemiologic study was presented and stressed; however, the company and work force did not wish to participate in any compulsory study. A compromise involving a survey of workers on a voluntary basis was performed. Fifty workers were given serologic testing on a voluntary basis. Of those workers, 24 had precipitins and 22 volunteered to be studied further. Of the 22, 4 had hypersensitivity pneumonitis, 1 had probable hypersensitivity pneumonitis (and declined further evaluation), and 3 had symptoms that might represent hypersensitivity pneumonitis.

As a result of this survey, the paper company involved instituted environmental controls and the requirement of effective and efficient respirators. No new additional cases of hypersensitivity pneumonitis have been identified.

Respiratory Disease Risk Relative to Farmer's Lung Disease Antibody Status

J. R. Guernsey, D. P. Morgan, J. J. Marx, E. P. Horvath, W. E. Pierce, and J. A. Merchant

INTRODUCTION

Diagnosis of farmer's lung disease (FLD) is usually based on a positive clinical and occupational history and is generally confirmed by the finding of circulating antibodies to FLD antigens. The presence of circulating antibodies is a sign of past exposure to FLD antigens;[1,2] most individuals with antibodies have no symptoms. The pathogenetic implications of a positive serology are uncertain.[3,4]

Evidence supporting a type III immune complex pathogenesis of FLD has been summarized by Parkes:[5] (1) large quantities of precipitating, complement–fixing, circulating antibodies are present in most patients with acute disease; (2) the symptoms of acute disease (fever, rigor, myalgia, breathlessness) occur 4 to 8 h postexposure; (3) bronchial provocation tests with inhaled extracts of the responsible agent reproduce all the features of acute disease in patients in 4 to 8 h, but not in controls; (4) intradermal tests result in an Arthus–type reaction at 4 to 8 h in symptomatic individuals. Long–term presence of circulating antibodies is correlated with a decrease in carbon monoxide diffusing capacity.[6]

There is increasing evidence for a role of cell–mediated hypersensitivity in FLD[7,8] and the acute and chronic pathologic changes are consistent with this. Organic dust antigens can also directly activate the alternate complement pathway.[9]

The organic dust toxic syndrome (ODTS) or "pulmonary mycotoxicosis", first described by Emanuel et al.,[10] produces similar symptoms to acute FLD, but the respiratory effects are completely reversible and precipitating FLD antibodies are not present. This illness also occurs with "silounloading" and is considered a condition distinct from FLD.[11]

To determine whether antibodies are associated with respiratory impairment, a population–based cohort analysis comparing similarly exposed groups with and without circulating antibodies is required. Such an analysis has not been reported. Serological studies in acute and chronic FLD have generally been limited to cross–sectional prevalence studies[12,13] and to longitudinal assessment of clinically selected cases.[14,15] Because of the complete resolution of symptoms in many cases of acute FLD, and reporting bias,[16] correlation of serological status with subsequent clinical and physiological findings has been difficult.

POPULATION STUDIES

A population–based study of 1444 farmers over the age of 18 living in a three–county area in central Wisconsin[17] showed that 147 had circulating antibodies by the double immunodiffusion assay method to at least one of the following antigens: *Micropolyspore faeni, Thermoactinomyces vulgaris, Saccharomonospora viridis, Aspergillus fumigatus, T. candidus,* and *T. sacchari.* These antibody–positive individuals were evaluated clinically and six had FLD for a case–prevalence rate of 4.2 per 1000 farmers.

This population was reexamined 7 years later. A positive cohort of 143 subjects with FLD antigens in the original study was matched with 143 FLD antigen–negative subjects. The study participation rate was 78% (224/286); there was no difference in sociodemographic characteristics between the 224 participants and the 42 located, living nonparticipants. Reasons for nonparticipation included "not interested" (31%), "too busy" (21%), "too ill" (n = 3, 7%; one of these three was classified as a FLD case in 1977), and "other" (41%) (e.g., "don't like to give blood" and "have already been tested for farmer's lung antibodies"). The deaths of five subjects in each cohort were not related to lung disease (n = 9) or to FLD antigen (n = 1, emphysema). The sociodemographic and farming characteristics of the two groups were similar (Table 1) except for the mean size of dairy herd ($p < 0.025$).

There was slight increase in risk associated with a positive antibody status for most respiratory symptoms (Table 2); this was significant for chronic phlegm, chronic cough, and dyspnea. When the study population was stratified by smoking status, the significant relationships for chronic cough and dyspnea disappear. Chronic phlegm remains significantly associated with a positive antibody status amongst smokers; the risk ratio is even larger in nonsmokers. However, tests for significance were not possible because of low numbers. The risk for smoking–related respiratory symptoms (Table 2) is stronger in the antibody–negative cohort. The consistency of this pattern suggests that smoking interacts with antibody status in the expression of certain respiratory symptoms. Mean pulmonary function test values by cohort smoking status are presented in Table 3. Lower percent predicted measures of pulmonary function caused by smoking are observed; there were no significant differences between antibody–positive and antibody–negative subgroups.

RELEVANCE

The observation that chronic cough, chronic phlegm, and dyspnea on exertion in 1984 are significantly associated with the presence of antibodies in 1977 is interesting and consistent with the literature on the disease progression of acute FLD.[1]

TABLE 1
Sociodemographic and Farming Characteristics of the Two Cohorts

	Positive cohort n = 112	Negative cohort n = 112
Sociodemographic		
Mean age	51.2	50.5
Female (%)	32	31
With high school education (%)	62	70
With second job other than farming (%)	17	27
Reporting having ever smoked		
In 1977 (%)	34	34
In 1984 (%)	29	37
Born in Wisconsin (%)	91	85
Farming in 1984 (%)	69	63
Dairy farmers in 1984 (%)	68	57
Farming		
Mean years farming	28.0	25.3
Mean hours farming per week	63.3	57.2
Mean total acres	292.4	249.8
Mean cultivated acres	207.9	183.4
Mean hay acres	88.2	77.9
Mean corn acres	77.9	66.8
Mean size dairy herd	101.1	76.5[a]

[a] $p < 0.025$.

TABLE 2
Risk Ratios for Respiratory Symptoms

	Positive cohort vs. negative cohort by smoking status			Smokers vs. nonsmokers by cohort		
	Smokers		Non-smokers		Positive cohort	Negative cohort
Symptom	cRR[a]	aRR[b]	aRR[b]	cRR[a]	aRR[b]	aRR[b]
Usual cough	1.33	1.12	1.98	2.11[c]	1.96	3.46[d]
Chronic cough	2.10[e]	1.95	2.99	2.81[c]	2.63[d]	4.04[e]
Usual phlegm	1.43	1.43	1.62	2.03[c]	1.99[e]	2.25[e]
Chronic phlegm	2.36[d]	2.17[e]	3.59	2.97[f]	2.79[f]	4.62
Wheeze with cold	0.78	0.76	0.90	2.13[f]	1.92[e]	2.25[f]
Wheeze apart from cold	1.00	1.24	1.11	2.02[e]	2.39	1.73
Breathless on exertion	1.57[e]	1.69	1.57	1.69[e]	1.70	1.59

[a] cRR = crude risk ratio.
[b] aRR = adjusted risk ratio.
[c] $p \leq 0.005$.
[d] $p \leq 0.01$.
[e] $p \leq 0.05$.
[f] $p \leq 0.0005$.

Emanuel et al.[13] found that chronic cough was the most constant symptom in a clinical series of 24 FLD patients. Exertional dyspnea was also common in this group. Bishop et al.[23] noted that shortness of breath was consistently found in all stages of disease in a follow–up study of 19 patients with

TABLE 3
Mean Percent Predicted Pulmonary Function Test Values by 1977 DID[a] Cohort and 1984 Smoking Status

	Positive cohort[c]		Negative cohort[c]	
Pulmonary function test[b]	Smokers (n = 33)	Nonsmokers (n = 79)	Smokers (n = 41)	Nonsmokers (n = 71)
FVC	102.2	103.1	99.9	103.2
FEV_1	94.9	102.2	94.0	102.8
FEV_1/FVC	88.9	96.5	91.4	96.6
MMFR	68.5	78.7	66.2	83.1
PEFR	101.7	107.3	95.5	105.6
FEF_{25}	85.8	96.0	80.4	94.2
FEF_{50}	66.0	75.2	62.9	79.0
FEF_{75}	39.7	46.1	38.4	48.1

[a] DID = double immunodiffusion.
[b] Expressed as percent predicted normal values according to Knudson/Cherniack.
[c] No significant differences were observed between antibody–positive vs. antibody-negative cohorts.

serology. The increased prevalence of dyspnea on exertion, but lack of pulmonary function changes associated with antibody status, was reported by Roberts et al.[3] in 1976. Hapke et al.[20] noted the increased prevalence of chronic cough and chronic phlegm in chronic FLD. Berrens[4] observed that in chronic FLD patients, the presenting symptoms at first suggested chronic obstructive lung disease. However, a careful history, radiologic abnormalities, and a combination of obstructive and restrictive ventilatory defects led to the diagnosis of chronic FLD. More recently, Braun et al.[6] noticed a high incidence of abnormalities usually ascribed to disease of the airways. They particularly emphasized the elevated prevalence of symptoms among nonsmokers compared to normal nonsmoking populations. Freedman and Ault[21] showed airways hyperresponsiveness was a common finding in FLD.

Dyspnea on exertion has been a prominent finding in a number of studies. Pepys[12] first presented data from a group of farmers with chronic, insidious–onset FLD where 99% complained of dyspnea and 88% complained of cough. In a biopsy study of six chronic FLD cases performed by Seal et al.,[14] in which all six were found to have interstitial fibrosis with lymphocytic infiltration, the prominent complaint was dyspnea on exertion. Lung function studies yielded a variable picture in the chronic stage.

Barbee et al.[15] followed up on a group of patients with acute FLD; 30% were experiencing dyspnea on exertion 3 to 6 years after the first episode. All symptomatic individuals had abnormal chest films at the end of the follow–up period. The most striking and consistent physiologic abnormality in symptomatic patients was a decreased carbon monoxide–diffusing capacity. Braun et al.[6] found that the most common symptoms at last follow–up of 141 cases of acute FLD were cough and dyspnea on exertion.

It has been reported that antibody–positive individuals are

more likely to be nonsmokers, suggesting a possible protective effect of smoking.[17] Braun et al.[6] pointed out that symptoms often associated with airways abnormalities were more prevalent than would be expected in a rural nonsmoking population. The elevated risks for chronic cough and chronic phlegm (2.99 and 3.59) for nonsmokers reflect the fact that, for this study, few antibody–negative nonsmokers were experiencing respiratory symptoms. However, although smoking may protect against exposure, it fails to protect against disease; risk ratios for respiratory symptoms were elevated for smokers in the negative cohort relative to nonsmokers in the negative cohort.

No evidence of restrictive or obstructive lung disease associated with antibody status was detected. The 7–year follow–up period may have been too short to observe significant changes. In support of this possibility, Braun et al.[6] noted that, though the mean length of disease was 14.8 years in a follow–up of acute cases, the duration was as great as 40 years in one individual. Hapke et al.,[20] finding that neither fibrotic narrowing of respiratory bronchioles in one patient nor hyperplasia of the mucous–secreting apparatus in another was reflected in lung function findings, concluded that in some patients with considerable disability, conventional pulmonary function tests may demonstrate little or no impairment. Braun et al.[6] stated that outcome of pulmonary function studies in FLD has not been well characterized. They observed that during the last follow–up visit, physiologic abnormalities were common, but severe abnormalities were present in only a small number. Thus, although we found no association between antibody status and decreased lung function, it is not possible to conclude that there was no association between antibody status and evidence of interstitial fibrosis; more detailed clinical studies are required.

There appears to be no increased risk of mortality associated with antibody status. The duration of follow–up in this study may not have been sufficient, however, to draw this conclusion. Further studies of this population may elucidate patterns of mortality associated with antibody status.

CONCLUSIONS

A number of recent exposure–specific studies of agricultural workers have demonstrated increased respiratory morbidity associated with agriculture. The common denominator is exposure to organic dust, which is a complex mixture of vegetable particles and fragments, microorganisms and their products, insects and insect fragments, feed additives, and avian and rodent proteins. While the vegetable dust itself is by far the most important exposure, individual circumstances of growing, storing, and using the product(s) from which the organic dust arises will influence the exposure(s). Yet epidemiological studies of FLD have focused primarily on clinical serology–positive cases with little attention to the broader question of nonspecific respiratory symptoms and functional changes associated with organic dust exposures. This view is supported by Madsen et al.,[22] who in a cross–sectional study of Utah dairy farmers (n=471), found that 14 or 3% have acute episodes thought to be consistent with FLD. However, only 2 of 14 were serology positive, suggesting some of these episodes might have been more consistent with ODTS. Their finding that 30% had spirometric evidence of obstructive airways disease raises the possibility that much of the epidemiologically detected functional changes and symptomatology associated with FLD is, in actual fact, associated with an airway response to organic dust. Subclinical FLD may, however, be developing in a subgroup of these subjects. Clinical investigations of this population may yield interesting information on the pathogenesis of this disease.

REFERENCES

1. **Salvaggio, J.D.,** Diagnostic significance of serum precipitins in hypersensitivity pneumonitis, *Chest,* 62, 242, 1972.
2. **Emanuel, D.A. and Kryda, M.J.,** Farmer's lung disease, *Clin. Rev. Allergy,* 1, 509, 1983.
3. **Roberts, R.C., Wenzel, F.J., and Emanuel, D.A.,** Precipitating antibodies in a midwest dairy farming population toward the antigens associated with farmer's lung disease, *J. Allergy Clin. Immunol.,* 57, 518, 1976.
4. **Berrens, L., deRidder, G., and deBoer, F.,** Longitudinal studies of immunological parameters in farmer's lung, *Scand. J. Respir. Dis.,* 58, 205, 1977.
5. **Parkes, W.R.,** *Occupational Lung Disorders,* 2nd ed., Butterworths, London, 1982.
6. **Braun, S.R., doPico, GA., Tsiatis, A., Horvath, E., Dickie, H.A., and Rankin, J.,** Farmer's lung disease: long term clinical and physiologic outcome, *Am. Rev. Respir. Dis.,* 119, 185, 1979.
7. **Reynolds, H.Y., Fulmer, J.D., Kazmierowski, J., Roberts, W.C., Frank, M.M., and Crystal, R.G.,** Analysis of cellular and protein content of bronchoalveolar lavage fluid from patients with idiopathic pulmonary fibrosis and chronic hypersensitivity pneumonitis, *J. Clin. Invest.,* 59, 165, 1977.
8. **Weinburger, S.E., Kelman, J.A., Elson, N.A., Young, R.C., Jr., Reynolds, H.Y., Fulmer, J.D., and Crystal, R.G.,** Bronchoalveolar lavage in interstitial lung disease, *Ann. Intern. Med.,* 89, 459, 1978.
9. **Edwards, J., Wagner, J.C., and Seal, R.M.E.,** Pulmonary responses to particulate materials capable of activating the alternate pathway of complement, *Clin. Allergy,* 6, 155, 1976.
10. **Emanuel, D.A., Wenzel, F.J., and Lawton, B.R.,** Pulmonary mycotoxicosis, *Chest,* 67, 293, 1975.
11. **Pratt, D.S. and May, J.J.,** Feed associated respiratory illness in farmers, *Arch. Environ. Health,* 39, 43, 1984.
12. **Pepys, J. and Jenkins, P.A.,** Precipitin (FLH) test in farmer's lung, *Thorax,* 20, 21, 1965.
13. **Emanuel, D.A., Wenzel, F.J., Bowerman, C.I., and Lawton, B.R.,** Farmer's lung: clinical, pathologic and immunologic study of twenty-four patients, *Am. J. Med.,* 37, 392, 1964.
14. **Seal, R.M.E., Hapke, E.J., Thomas, G.O., Meek, J.E., and Hayes, M.,** The pathology of the acute and chronic stages of farmer's lung, *Thorax,* 23, 469, 1968.
15. **Barbee, R.A., Callies, Q., Dickie, H.A., and Rankin, J.,** The long-term prognosis in farmer's lung, *Am. Rev. Respir. Dis.,* 97, 223, 1968.
16. **Smyth, J.T., Wenzel, J., Adleins, F.E., Lloyd, M., Moore, B., and McWhite, E.,** Farmer's lung in Devon, *Thorax,* 23, 469, 1975.
17. **Gruchow, H.W., Hoffman, R.G., Marx, J.J., Emanuel, D.A., and Rimm, A.A.,** Precipitating antibodies to farmer's lung antigens in a Wisconsin farming population, *Am. Rev. Respir. Dis.,* 124, 411, 1981.
18. **Ferris, B.,** Epidemiology standardization project, *Am. Rev. Respir. Dis.,* 119, 831, 1979.
19. **Marx, J.J. and Gray, R.L.,** Comparison of the enzyme linked

immunosorbent assay and double–immunodiffusion test for the detection and qualification of antibodies in farmer's lung disease, *J. Allergy Clin. Immunol.*, 70, 1982.

20. **Hapke, E.J., Seal, R.M.E., Thomas, G.O., Meek, J.C., and Hayes, M.,** Farmer's lung, *Thorax,* 23, 451, 1968.
21. **Freedman, P.M. and Ault, B.,** Bronchial hyperreactivity to methacholine in farmer's lung disease, *J. Allergy Clin. Immunol.*, 67, 59, 1981.
22. **Madsen, D., Klock, L.E., Wenzel, F.J., Robbins, J.L., and Schmidt, C.D.,** The prevalence of farmer's lung in an agricultural population, *Am. Rev. Respir. Dis.*, 113, 171, 1976.
23. **Bishop, J.M., Melnick, S.C., and Raine, J.,** Farmer's lung: studies of pulmonary function and etiology, *Q. J. Med.*, 32(127), 257, 1963.

The Prevalence of Respiratory Symptoms among Farmers and Ranchers in Southeastern South Dakota

E. H. Schlenker, G. R. Lenardson, C. McClain, E. Barnes, and R. R. Parry

The health hazards associated with farming include accidents, exposure to toxic gases and dusts in silos and confinement housing, and exposure to bacterial endotoxins, herbicides, and pesticides, to name a few.[1–7] Respiratory diseases found in farmers include occupational asthma related to animal and grain handling;[8–12] silo filler's disease, which occurs from inhalation of toxic oxides of nitrogen,[7] and farmer's lung, a hypersensitivity pneumonitis caused by thermophilic actinomycetes in moldy vegetable matter.[13,14] Some, for example, silo filler's disease or H_2S toxicity,[3,5] can cause death acutely. Others, for example, occupational asthma, are chronic and cause more morbidity than mortality. In addition to these relatively well–known disease entities, subclinical influenza–like or bronchitis–like symptoms may be intermittently debilitating.[7,9,15] Cigarette smoking and climatic factors may act in an additive or synergistic manner to exacerbate symptoms and findings.

Because surveys of the prevalence of respiratory disease among South Dakota farmers and ranches have not been previously documented, this pilot study was conducted in southeastern South Dakota. The ATS standardized questionnaire[16] was sent to 400 farm owners in Clay County, SD. Questions related to raising crops and animals, the use of masks, confinement housing, and use of pesticides and herbicides were also included. Data were analyzed using nonparametric tests such as the chi–square test and G test as well as the Student *t* test.

Of 176 respondents, 153 were born in South Dakota and 8 were women; the mean age was 48.4 years and the mean length of time worked on a farm or ranch was 33.4 years. Nine percent were current smokers, 35% were ex–smokers, and 56% were lifetime nonsmokers. The ex–smokers were significantly older (52.9 ± 12.0 years, $p \leq 0.01$) compared to smokers (43.3 ± 11.4) or nonsmokers (46.5 ± 13.3).

Nonfarm symptoms not associated directly with raising crops (hay, oats, alfalfa, soybeans, corn) and/or animals (dairy cattle, hogs, feeder pigs, beef cattle, etc.) were acknowledged by 71% of smokers, 30% of nonsmokers, and 34% of ex–smokers. Farm–related symptoms were reported by 55% of nonsmokers ($p < 0.01$ vs. "nonfarm symptoms"), by 64% of ex–smokers ($p < 0.01$), and by 71% of smokers. Nonsmokers and ex–smokers had less chronic cough, phlegm, cough and phlegm, and wheezing than did smokers (Table 1). Significant differences between the prevalence of symptoms related to farming were seen for shortness of breath, headache, "whistling" chest, and runny eyes related to crop raising in smokers vs. nonsmokers. Ex–smokers showed only an increased prevalence of shortness of breath related to crop raising, compared to nonsmokers.

The prevalence of doctor–confirmed asthma was 4.5%, chronic bronchitis 18%, emphysema 2.8%, and hay fever 11.4%.

Farmers with symptoms related to raising crops reported primarily runny eyes, cough, and runny nose, occurring during the summer (28%) and fall (31%). The most consistent symptom related to raising animals was cough (31%). Forty–three percent of respondents indicated that those symptoms associated with animal raising were equally bad all year, whereas 17.8% indicated summer and 14.4% indicated winter as "problem" seasons.

Of 138 farmers who raised hogs, 64 claimed that these animals were responsible for their symptoms, and of 29 farmers who raised feeder pigs, 34% claimed that these animals were responsible for their symptoms (Table 2). Only 4 of the 127 farmers who raised beef cattle reported that these animals were responsible for their symptoms. Comparison of symptom prevalence in hog and nonhog farmers revealed animal–related symptoms to be markedly greater in hog farmers. The prevalence of symptoms specifically related to raising crops was not different between the two groups, except runny eyes ($p < 0.01$).

Confinement housing of hogs, feeder, pigs and/or chickens was reported by 74 farmers. The proportion of farmers who reported symptoms associated with confinement housing (65%) was not different when compared to those who did not utilize this method (54%).

Masks were used by 82 respondents. Most farmers wore masks while cleaning out grain bins, and when using herbicides and pesticides (which 153 farmers used). Only 5 used masks in the hog house. Most masks were 3M paper or cloth "filter masks", although 5 farmers used cannisters and dust respirators. Only 27 farmers used masks the entire time they were performing a task. Most individuals who did not use a mask felt that the mask was uncomfortable, not immediately available, or that they did not need one.

This study, like others,[8,9] shows smoking to be an important determinant of "work–related" symptoms. Smokers reported a higher percentage of symptoms related to raising crops, but not to raising animals. There is a gradual increase in symptom prevalence related to raising crops in nonsmokers vs. ex-

TABLE 1
Percent of Smokers, Ex–Smokers, and Nonsmokers Reporting Symptoms

	Nonfarm symptoms				Farming symptoms						
	Chronic cough	Phlegm	Cough and phlegm	Wheezing during a cold	Runny nose	Shortness of breath	Headache	Cough	Whistling in chest	Runny eyes	Fever
Nonsmokers	9.2[a]	16.3[a]	8.2[a]	21.4[a]	12.2[b]	5.1	5.1	30.6	5.1	8.2	
n = 98					29.6[c]	8.2	13.3	20.4	5.1	17.3	5.1
Ex–smokers	9.8[a]	18[a]	9.8[a]	34.4[a]	19.7[b]	18	14.8	32.8	14.8	14.8	
n = 61					31.1[c]	23[d]	19.7	31.1	13	24.5	3.3
Smokers	47	58.8	35	82.3	17.6[b]	11.7	11.7	23.5	11.7	5.9	
n = 17					47[c]	41[a]	35[a]	41	29[a]	47[a]	11.7

[a] Percent is significantly (p <0.05) different from that of the smokers' corresponding value.
[b] Refers to symptoms associated with raising animals.
[c] Refers to symptoms associated with raising crops.
[d] Percent is significantly (p <0.02) different from the nonsmokers' corresponding value.

From *SD J. Med.*, 38(7), 5, 1985. With permission.

TABLE 2
Percent of Hog Farmers and Nonhog Farmers Reporting Symptoms

	Nonfarm symptoms					Farming symptoms						
	Chronic cough	Phlegm	Cough and phlegm	Wheezing during a cold	Cold goes to chest	Runny nose	Shortness of breath	Whistling in chest	Headache	Cough	Runny eyes	Fever
Nonhog farmers	5	13.5	8	21.6	27	2.7[a]	5.14	0	2.7	0	0	
n = 38						16.2[b]	13.5	8	10.8	24	16.2	5.4
Hog farmers	13	21	11.5	36	29	20[a,c]	11.5	12.3[c]	9	38.4[c]	13.8[c]	
n = 138						36[b]	18	11.5	19.6	26	24.6	4.3

[a] Refers to symptoms associated with raising animals.
[b] Refers to symptoms associated with raising crops.
[c] Indicates that the prevalence of the corresponding symptoms in hog farmers is significantly (p <0.01) greater than in nonhog farmers

From *SD J. Med.*, 38(7), 5, 1985. With permission.

smokers vs. smokers. Warren[17] found a similar pattern in a study of cattle farmers in Manitoba.

By contrast, Matson and co–workers[15] found a higher prevalence of symptomatology related to confinement housing in nonsmokers than in ex–smokers. The present study goes further than others in that we attempt to distinguish symptomatology related to the raising of crops from those related to raising animals. Respondents in the present study indicated that runny eyes, runny nose, and cough were most often associated with raising crops, whereas nonsmokers and ex–smokers reported cough as the most prevalent symptom. Bronchitis symptoms (cough and phlegm) have been noted in other studies of swine–confinement workers.[3,15]

Crop– and animal–related symptoms appeared to have different seasonal variations. Crop–related symptoms occurred more frequenty in summer and fall, as noted by others,[7] whereas animal–related symptoms occur equally all year, or are increased during the winter months[15] when CO, CO_2, H_2S, and respirable particles are higher in confinement housing.[3,5]

Hogs appear to contribute greatly to the animal–related symptomatology. Hogs were raised by 78% of the participants and of thoseparticipants reporting symptoms, most associated them with hogs. Also, most of the farmers using confinement houses had feeder pigs and/or hogs. Finally, hog farmers showed a significant increase in animal–related symptoms. The prevalence of symptoms was not as high as in other studies of hog farmers.[9,10,15] There was also no difference in the prevalence of symptomatology of individuals who raised animals in confinement, compared to those who did not — a finding different from an Iowa study.[9] This difference may relate to smaller–size hog farm operations in the present study. The average number of hogs raised by respondents in this study was 437 per year (range 2 to 4000); large confinement operations in Iowa may raise 1500 swine in a 6–month period.[18]

Masks may influence respiratory symptoms. Although almost 50% of farmers claimed to use masks, only 15% used masks all the time. Masks were used appropriately when cleaning out grain bins and handling chemicals, but only five individuals used masks in the hog house. The conclusions from this and other studies suggest that hog exposure contributes to respiratory dysfunction; yet protective measures either have not been found, or have not been used by farmers to decrease exposure. Further studies and education must be pursued to reduce the incidence of increased respiratory morbidity associated with farming.

REFERENCES

1. **de Haller, R. and Suter, F.,** *Aspergillosis and Farmer's Lung in Man and Animal,* Hans Huber, Bern, 1974.
2. **Morgan, W.K.C. and Seaton, A.,** *Occupational Lung Diseases,* W.B. Saunders, Philadelphia, 1975.
3. **Donham, K.J.,** Respiratory hazards of livestock confinement workers, in Proc. 7th Int. Congr. of Rural Medicine, Salt Lake City, UT, 1978, 17.
4. **Berry, G.,** *Statistical Anaylses in Occupational Lung Diseases,* Weill, H. and Turner–Warwick, M., Eds., Marcel Dekker, New York, 1981, 427.
5. **Donham, K.J., Knapp, L.W., Monson, R., and Gustafson, K.,** Acute toxic exposure to gases from liquid manure, *J. Occup. Med.,* 24, 142, 1982.
6. Hazards of farming, *Des Moines Register,* October 16, 1984.
7. **Pratt, D.S. and May, J.J.,** Feed–associated respiratory illness in farmers, *Arch. Environ. Health,* 39, 43, 1984.
8. **Babbott, F.L., Gump, D.W., Sylvester, P.L., MacPherson, B.V., and Holly, C.,** Respiratory symptoms and lung function in a sample of Vermont dairymen and industrial workers, *Am. J. Public Health,* 70, 241, 1980.
9. **Donham, K.J., Donald, C.Z., and Merchant, J.A.,** Respiratory symptoms and lung function among workers in swine confinement buildings: a cross–sectional epidemiology study, *Arch. Environ. Health,* 39, 96, 1984.
10. **Katila, M.–L., Mantyjarvi, R.A., and Ojanen, T.J.,** Sensitization against environmental antigens and respiratory symptoms in swine workers, *Br. J. Ind. Med.,* 38, 334, 1981.
11. **Gruchow, H.W., Hoffman, R.G., Marx, J.J., Emanuel, D.A., and Rimm, A.A.,** Precipitating antibodies to farmer's lung antigens in a Wisconsin farming population, *Am. Rev. Respir. Dis.,* 124, 411, 1981.
12. **Harries, M.G. and Cromwell, O.,** Occupational asthma caused by allergy to pig's urine, *Br. Med. J.,* 284, 867, 1982.
13. **Madsen, D., Klock, L.E., Wenzel, F.J., LaMar Robbins, J., and Schmidt, C.D.,** The prevalence of farmer's lung in an agricultural population, *Am. Rev. Respir. Dis.,* 113, 171, 1976.
14. **Mionkiare, S.,** Clinical aspects of farmer's lung: airway reactivity, treatment and prognosis, *Eur. J. Respir. Dis.,* 137, 1, 1984.
15. **Matson, S.C., Swanson, M.C., Reed, C.E., and Yunginger, J.W.,** IgE and IgG–immune mechanisms do not mediate occupation–related respiratory or systemic symptoms in hog farmers, *J. Allergy Clin. Immunol.,* 72, 299, 1983.
16. **Ferris, B.G.,** Epidemiology standardization project. II. Recommended respiratory disease questionnaires for use with adults and children in epidemiological research, *Am. Rev. Respir. Dis.,* 118 (Suppl.), 7, 1978.
17. **Warren, C.P.W.,** Respiratory disorders in Manitoba cattle farmers, *Can. Med. Assoc. J.,* 125, 41, 1981.
18. **Mueller, A.G. and Kesler, R.P.,** Costs of confinement hog systems, Vet. Ext. Newsl., (Iowa State Univ., Ames), 192, 3044, 1976.

Pi Phenotype and the Development of Air Flow Obstruction in Saskatchewan Agricultural Workers

S. L. Horne

Because we are located in a province where a large portion of the population is engaged in agricultural activity, we have wondered if a particular genetic constitution makes agricultural workers more or less susceptible than nonagricultural workers to the development of air flow obstruction or emphysema. This paper describes two studies which we have carried out which suggest that the answer to this question is yes.

We have chosen to look at a particular genetic system which in the past 15 years has become well known (perhaps even infamous) for its association with emphysema.[1] This system, the protease inhibitor, or Pi system, is responsible for the production of the serum protein, α_1–antitrypsin (α_1AT). Utilizing differences in concentration and/or electrophoretic mobility, over 30 alleles of the Pi locus have been described. At least five of these (M1, M2, M3, S, and Z) are found in polymorphic frequencies. Some, most notably Z, which is found in about 3% of Saskatchewan residents, are associated with varying degrees of reduction in α_1AT level. Others, such as the F, I, N, and P variants, have no known physiological differences, yet tend, along with Z, to show up more frequently than expected in series of COPD patients. The Z allele is thought to contribute to an increased risk of air flow obstruction, because its low levels of α_1AT do not inhibit elastase sufficiently to prevent leukocyte elastase from attacking elastin fibers in the lungs. Repeated destruction of elastin is thought to eventually lead to the development of emphysema.[2]

In our first study, we used a consecutive series of male emphysema patients seen in our respiratory medicine clinic over a period of 2 years to determine if different Pi phenotypes and occupations were represented equally among the patients in the same proportions as the general population. Those patients with nonreversible air flow obstruction as well as reduced diffusing capacity for carbon monoxide and radiographic evidence of pulmonary changes consistent with emphysema were included in this analysis. Smoking history and occupation were obtained from each patient's chart. Patients were categorized as (1) farmer or grain worker, (2) other blue collar, and (3) white collar or professional workers. All subjects were current or former smokers. α_1AT phenotyping was carried out using isoelectric focusing on Servalyte gels at pH 4 to 5.[3]

Our respiratory medicine unit is a referral center for the entire province of Saskatchewan, so we then estimated the chance for any male Saskatchewan resident over 45 years of age to have been seen during a 2–year period in our unit as an emphysema patient. We estimated the number of individuals in the province in each of six phenotypic and occupational groups, determined the number of patients in each group, and divided to find the proportion of men in each group who were patients.

Estimates of the number of farmers, and other blue and white collar workers were based on questionnaires from a province–wide survey of a comparable age group. Approximately 50% were farmers, 25% other blue collar, and 25% white collar. The estimated number of smokers was based on a study of 2000 farmers and 1600 residents of the town of Humboldt categorized by employment. Seventy percent of farmers were smokers or ex–smokers. Eighty–one percent of other blue collar workers and 73% of white collar workers were smokers or ex–smokers. Frequency of α_1AT variants was obtained from a total of 2800 unselected grainworkers,[4] university students, and Humboldt residents.[5] M and S alleles were considered nonvariants. Four percent of individuals had one or two variant alleles, F, I, N, P, Z, etc. The number of males over age 45 in Saskatchewan was obtained from the 1981 Canada Census.[6]

The estimated numbers of ever–smoking Saskatchewan males categorized by employment and α_1AT type are shown in Table 1 along with the number from each group whom we actually saw as patients. Also shown are the proportions, which ranged from a low of 0.00040 to a high of 0.00522. A

TABLE 1
Proportion of Smoking Male Saskatchewan Residents over 45 Years with Emphysema Who Were Seen in Our Respiratory Medicine Unit

		Proportion		
	No. patients	Estimated no. in population	Seen as patients	Relative risk
Farmer variant	6	1,988	0.00302	7.55
Farmer nonvariant	42	47,712	0.00088	2.2
Other blue collar variant	6	1,150	0.00522	13.05
Other blue collar nonvariant	41	27,605	0.00148	3.70
White collar variant	0	1,037	<0.00096	<2.4
White collar nonvariant	10	24,878	0.00040	1
Total	105	105,370	0.00101	—

chi–square test indicated that patients were not drawn with equal frequency from the six groups.

Relative risks were calculated by dividing all values by 0.00040 and are illustrated in Figure 1. Nonvariant farmers had a relative risk of 2.2, while other blue collar workers had a risk of 3.7. Addition of a variant phenotype increased these relative risks to 7.5 and 13, respectively. Excluding group F, the risk for each group is significantly different from every other group with the exception of comparisons between groups C and D and groups D and E.

Since all our patients were smokers, the lowest risk group which we examined was smoking nonvariant white collar workers. Using this group as a reference, farming increased risk by a factor of 2.2, and blue collar work increased risk by 3.7. Having a Pi variant further increased risk to farmers and blue collar workers by 3.4. Pi variant had less effect on white collar workers (<2.4). Thus, at least among smokers, farming increases risk, blue collar work increases risk more than farming, and a Pi variant increases risk to farmers and blue collar workers more than to white collar workers.

This study was carried out using emphysema patients. They comprise a specially selected group. Because they have emphysema, they must also have a concentration of factors, probably in addition to smoking, which have contributed to the development of the disease. Most of these other contributing factors and their interactions, whether genetic or environmental, have not yet been identified even though their existence is strongly suggested.

In contrast to this design, in our second study we worked with a population selected for one genetic factor, moderate α_1AT deficiency, and one environmental factor, heavy exposure to grain dust. All other contributory factors should have been present only as they exist in a random population. We then compared the differences in symptoms, pulmonary function, and chest radiographs between two groups which differed only by the one genetic factor, α_1AT deficiency, and which had similar environmental exposure to grain dust.[7]

We found no differences in the prevalence of reported

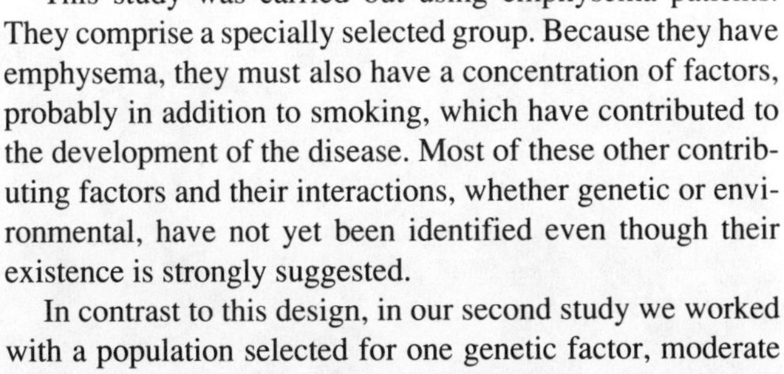

FIGURE 1. Relative risk for smokers only of being seen in our Respiratory Medicine Unit with emphysema according to their type of employment and Pi type. White collar workers have an assigned risk of one. All other groups are expressed as multiples of the white collar workers.

TABLE 2
Mean Pulmonary Function Values (±1 SD) for 28 Matched MZ and M Grainworkers

	MZ	M	*p*
FEV_1 (% expected)	94 ± 15	104 ± 12	<0.05
FEV_1/FVC (%)	76 ± 11	80 ± 7	<0.05
MMFR (% expected)	90 ± 36	107 ± 31	<0.05
V_{max50} (% expected)	102 ± 37	123 ± 33	<0.05

symptoms: cough, sputum, wheeze, or dyspnea between the two groups.

Using a paired *t* test, the MZ grain workers had significantly poorer mean values for forced expired volume in 1 s, forced vital capacity, maximum midexpiratory flow rate (FEV_1, FEV_1/FVC, MMFR), and V_{max50} (Table 2). In 11 other tests, including lung volumes, diffusing capacity, single breath nitrogen, and airways reactivity, the MZ workers had poorer mean values in 9, while in 2, the mean values were equal.

There was a threefold greater prevalence of radiographic abnormalities in the Pi MZ workers, with 35% of the MZ men showing abnormalities compared with only 11% of the Pi M men. The chi–square test for paired observations was, however, nonsignificant ($0.10 < p < 0.25$).

We established a cutoff point below which individuals were considered to have air flow obstruction. This cutoff point was jointly $FEV_1/FVC \times 100 < 70$ and FEV_1 <85% of predicted. We then calculated the odds ratio for the MZ as compared with the M men and found it to be 3, the same as that reported by Cox et al.[8] and similar to the 3.4 and 3.5 odds ratios which we calculated for our farming and blue collar patients previously described. While not significant in the grain workers, in the patient groups, ratios of 3.4 and 3.5 were significant ($p < 0.01$).

Thus, in a group of agricultural workers, namely, grain handlers, it appears that the MZ phenotype alone contributes a threefold increase in risk of air flow obstruction. As well, variant phenotypes appear to contribute a 3.5–fold increase in risk of emphysema to farmers. So, in both farmers and grain workers, variant phenotypes appear to increase the chance of developing air flow obstruction or emphysema approximately threefold. As well, Pi variants may have less effect in white collar workers than in farmers or other blue collar workers.

We have considered here only one genetic system previously implicated as a risk factor in the development of emphysema.

We and other groups have identified other genetic factors which may make contributions to risk of emphysema of equal or greater magnitude than α_1AT deficiency. These include type A blood, nonsecretor status, and absence of Lewis antigen.[9,10] It will be important to compare the contribution to risk of these systems in the various occupational groups to determine if other genetic systems interact with specific types of employment as the Pi system appears to do.

REFERENCES

1. **Kueppers, F.,** Inherited differences in alpha$_1$–antitrypsin, in *Genetic Determinants of Pulmonary Disease, Lung Biology in Health and Disease,* Vol. 11, Litwin, S.D., Ed., Marcel Dekker, New York, 1978, 23.
2. **Janoff, A.,** Proteases and lung injury: state–of–the–art minireview, *Chest,* 83 (Suppl.), 54S, 1983.
3. **Horne, S.L., Tennent, R., Lovegrove, A., Dosman, J.A., and Cockcroft, D.W.,** Pi type MZ and an increased risk of pneumonia, *Clin. Invest. Med.,* 7, 85, 1984.
4. **Horne, S.L., Clarke, M.E., and Barnett, G.D.,** Alpha$_1$–antitrypsin deficiency: its effect on pulmonary function in Saskatchewan grain buyers, in *Occupational Pulmonary Disease: Focus on Grain Dust and Health,* Dosman, J.A. and Cotton, D.J., Eds., Academic Press, New York, 1980, 347.
5. **Horne, S.L., Junk, G.J., Cockcroft, D.W., Lovegrove, A., and Dosman, J.A.,** Pi MZ phenotype and an increased prevalence of reported psoriasis in a community survey, *Hum. Hered.,* 36, 266, 1986.
6. Canada Census, Vol. 2, 93–967–93–969, 1981.
7. **Horne, S.L., Tennent, R.K., Cockcroft, D.W., Cotton, D.J., and Dosman, J.A.,** Pulmonary function in Pi, M, and MZ grainworkers, *Chest,* 89, 795, 1986.
8. **Cox, D.W., Hoeppner, V.H., and Levison, H.,** Protease inhibitors in patients with chronic obstructive pulmonary disease: the alpha–1–antitrypsin heterozygote controversy, *Am. Rev. Respir. Dis.,* 113, 601, 1976.
9. **Cohen, B.H.,** Chronic obstructive pulmonary disease: a challenge in genetic epidemiology, *Am. J. Epidemiol.,* 112, 274, 1980.
10. **Horne, S.L., Cockcroft, D.W., Lovegrove, A., and Dosman, J.A.,** ABO, Lewis and secretor status and relative incidence of air flow obstruction, *Dis. Markers,* 3, 55, 1985.

Acute Disorders Attributable to Exposure to Grain in Farmers

C. P. W. Warren, V. Holford–Strevens, M. Cheang, and J. Manfreda

Epidemiological studies of grain elevator workers have revealed a high prevalence of acute symptoms by the workers when at work.[1–5] Asthma can be produced by grain dust or its constituents and inhalation challenge in the laboratory produces acute falls in airflow.[6–8] Grain dust is complex and contains many active and allergenic materials.[9]

This paper will describe the frequency, and related factors, of acute symptoms experienced by farmers when handling grain. The data were obtained from a cross–sectional study of an adult rural Manitoba population, the details of which have been published more fully.[10]

All permanent residents aged 20 to 65 years on the 1st of December 1978 in two rural municipalities, Dufferin and Roland, and one rural town, Carman, were the subjects. The subjects were identified from the population registry of the Manitoba Health Services Commission. A questionnaire consisting of the NHLBI respiratory questionnaire supplemented by questions on occupation, farming practice, and symptoms related to farming was administered. The acute symptoms related to grain dust exposure are the subject of this report. Lung function tests were performed with a rolling seal spirometer and recorded on analog tape which was analyzed with a Hewlett–Packard computer. Normal predicted values were generated from the measurements of lung function in asymptomatic, disease–free, nonsmoking participants in the study. Allergy tests were performed using the prick method to six common and five grain–related allergens as reported elsewhere.[11] Total IgE levels, specific IgE by RAST against certain allergens, and specific IgG by ELISA against a few antigens were measured.

The few female subjects exposed to grain were omitted from this analysis. Male subjects were classified into 445 current farmers, 242 former farmers, and the 257 who had never farmed. Their characteristics are summarized in Table 1. The nonfarmers had been hardly involved in growing grain and only 5% had ever shoveled grain. Thus, this paper will devote itself to acute disorders from exposure to grain in current and former farmers. Current farmers spent an average of 60 d a year exposed to grain vs. only 26 d per year for former farmers. Grain was grown on 80 and 90% of farms in the preceding 1 and 5 years, respectively. Of current farmers 29% were smokers, 23% ex–smokers, and 43% nonsmokers, while of former farmers 37% were smokers, 32% ex–smokers, and 29% were nonsmokers (the balance were pipe/cigar smokers).

Forty–six percent of subjects reacted to at least one allergen with a 2–mm wheal. House dust most often produced a reaction. There was no association between size or frequency of skin reactions and smoking habits. Overall there was no difference in size or frequency of allergy reactions or total IgE levels between current and former farmers.

Table 2 indicates the nature and frequency of acute grain dust–related respiratory symptoms. Shortness of breath (SOB) was most frequent and occurred in 41% of current and 27% of former farmers. It developed within 1 h in 84% of instances and lasted up to 4 h in 78%. Cough, wheezing, and grain fever occurred in 8 to 23%. Grain fever occurred least frequently; half of the affected had had only one attack but a few had had many attacks. Fever was accompanied by dyspnea in 42%. Although a large number of farmers had no symptoms it should be noted that 41% of current and 21% of former farmers had had two or more symptoms. Common factor loading of symptoms was studied by Boolean factor analysis to reduce the six binary variables of presence or absence of a symptom into a smaller set of factors. Only wheezing and shortness of breath were linked. Age, duration of farming, and history of grain dust exposure were similar in symptomatic and asymptomatic farmers. Current farmers

TABLE 1
The Personal and Farming Habits of the Male Subjects

	Current farmer	Former farmer	Never farmed
Number	445	242	237
Mean age (years)	42	50	38
Mean size of farm (HA)	264	179	—
Years of farming	22 ± 14	19 ± 15	—
Ever–grown grain	98%	97%	—
Days worked with grain/year[a]	60 ± 41	26 ± 47	—
Ever–shoveled grain	97%	97%	5%

[a] Mean ± standard deviation.

TABLE 2
The Frequency (%) of Acute Symptoms on Exposure to Grain Dust in Male Farmers

	Current	Former
Cough	22	14
Wheeze	23	12
SOB	41	27
Grain fever	15	8
Stuffy nose	18	10
Skin itch	26	16

who developed shortness of breath with grain worked with it more days per year than those who did not.

We found that about 20% developed shortness of breath frequently when exposed to grain and a further 20% did this occasionally. Smoking, past and present, was associated with shortness of breath frequently and grain fever was associated with nonsmoking; shortness of breath was associated with 3 mm or larger allergy skin reactions; grain fever was not (Table 3). Total IgE was significantly higher in current and former farmers for cough and in the latter for wheeze and shortness of breath. Allergic reactions, in general, rather than any specific allergic reactions were related to symptoms.

Barley, followed by oats, was blamed most often as the grain provoking shortness of breath. Most farmers reacted to several grains. Grain–fever was not associated with specific grains, but rather with moldy grain.

The mean lung function, expressed as percent predicted, was compared in the current and former farmers for those with or without acute grain–related symptoms for non–, ex–, and current smokers. For those who developed grain– related shortness of breath frequently and for wheezing (Table 4, current farmers) the lung function was worse in all three categories of smokers.

A positive RAST was less frequent than skin reactions against the same allergens and ELISA detection of farmer's lung antibodies was uncommon. Neither correlated with any particular symptom or abnormal lung function.

TABLE 3
Frequency (%) of Acute Shortness of Breath and Grain Fever in Farmers Related to Smoking and Strong (>3 mm) Skin Reactions

	Smoking		Allergy reaction	
	Nonsmokers	Ever smokers	No reaction	3 mm
SOB frequently	13	19[a]	14	24[a]
SOB occasionally	19	19	20	17
Grain fever	17	9	12	16

[a] $p < 0.05$.

TABLE 4
The Relationship between Lung Function (Mean % Predicted) in Current Farmers With (+) or Without (–) Grain–Related Wheezing in Nonsmokers, Ex–Smokers, and Current Smokers

	Smoking					
	Nonsmokers		Ex–smokers		Smokers	
	–	+	–	+	–	+
FVC	101	97	102	93[a]	100	97
FEV_1	100	93[a]	99	87[a]	94	89[a]
FEV_1/FVC	100	97[a]	97	94	95	92

[a] $p < 0.05$.

Our study found that acute respiratory symptoms develop frequently in farmers on exposure to grain dust. This equates with the report of disorders in farmers harvesting grain.[12] Cross–sectional studies of acute symptoms in population studies are methodologically difficult. We had to rely on recollection of acute symptoms and this may have been biased. Moreover, we related acute symptoms, experienced in the past, with lung function and allergy tests done at the time of the survey. It is with some reservation that one draws conclusions of the temporal relationship of shortness of breath and wheezing with allergy tests and lung function. On the other hand, the advantage of such a survey was the possibility of obtaining information on symptoms from a large number of individuals, in our case, from a total adult rural community including those who had withdrawn from farming and exposure to grain dust prior to the survey.

Studies of farming environments and work–related disorders are handicapped by the difficulty of obtaining adequate estimates of exposure. We had to depend on the rather crude indices of duration and intensity of exposure such as average number of days spent each year in handling grain.

In our study farmers had a lower frequency of acute symptoms than grain elevator workers; this could be due to differences in the questionnaire or in exposure to grain dust. The farmer's exposure is intermittent compared to elevator workers. In the study the two symptoms of wheezing and shortness of breath were associated. Shortness of breath occurring frequently was associated with smoking. Wheeze and, to a lesser extent, shortness of breath were associated with allergy to common allergens and a raised total IgE. These findings suggest these symptoms are asthmatic and more common in atopics. Interestingly, these symptoms were not related to allergy to grain– associated allergens. It is possible that the skin reactivity simply identified grain farmers with airway hyperresponsiveness as it does in elevator workers. In studies of grain elevator workers there are fewer atopics than in controls,[2,3,5] suggesting the selection of nonatopics into elevators. This did not appear to be so for farmers who had the same allergy tests and total IgE as nonfarmers.

The acute symptoms of cough and shortness of breath were associated with smoking and this suggests an interaction. Such an interaction has been observed for chronic symptoms and abnormal lung function in elevator workers.[2–5] Smoking may increase the susceptibility of an individual to react to grain dust.

Acute changes in lung function in grain elevator workers have been found during a work shift,[13,14] and workers in whom this occurs are more prone to declining lung function over the years.[15] This association may be evident in this study by the finding that farmers who report wheezing and shortness of breath on exposure to grain had worse lung function at the time of the study. The study was done in winter so that recent exposure to grain was minimal and lung function was not affected by acute changes.

The frequency of grain fever in grain elevator workers varies from 0 to 32% between studies.[1–5] This may reflect differences in questioning or in exposure to type of grain. The frequency in this study was in the middle of the reported rates. Grain fever differed from other acute symptoms in that it was not associated with skin reactivity or total IgE, but was weakly associated with nonsmoking. Although about half of those with grain fever also developed shortness of breath, cluster analysis suggests this is coincidental. It is likely that grain fever is a systemic response to the dust, perhaps due to its endotoxin content;[16] it leaves no residual effects.

In conclusion, farmers are affected acutely by grain dust and should use dust control measures and masks to reduce these acute effects and perhaps prevent chronic disorders.

REFERENCES

1. **Tse, K.S., Warren, C.P.W., Janusz, M., McCarthy, D.S., and Cherniack, R.M.,** Respiratory abnormalities in workers exposed to grain dust, *Arch. Environ. Health,* 27, 74, 1968.
2. **Broder, I., Mintz, S., Hutcheon, M., Corey, P., Silverman, F., Davies, G., Lleznoff, A., Peress, L., and Thomas, P.,** Comparison of respiratory variables in grain elevator workers and civic workers in Thunder Bay, Canada, *Am. Rev. Respir. Dis.,* 119, 193, 1979.
3. **Chan–Yeung, M., Schulzer, M., MacLean, L., Dorken, E., and Grzybowski, S.,** Epidemiologic health survey of grain elevator workers in British Columbia, *Am. Rev. Respir. Dis.,* 121, 392, 1980.
4. **Cotton, D.J., Graham, B.L., Li, K.Y.R., Froh, F., Barnett, G.D., and Dosman, J.A.,** Effects of grain dust exposure on respiratory symptoms and lung function, *J. Occup. Med.,* 25, 131, 1983.
5. **do Pico, G., Reddan, W., Tsiaatis, A., Peters, M.E., and Rankin, J.,** Epidemiologic study of clinical and physiologic parameters in grain handlers of the northern United States, *Am. Rev. Respir. Dis.,* 130, 759, 1984.
6. **Warren, C.P.W., Cherniack, R.M., and Tse, K.S.,** Hypersensitivity reactions to grain dust, *J. Allergy Clin. Immunol.,* 53, 139, 1974.
7. **Chan–Yeung, M., Wong, R., and Maclean, L.,** Respiratory abnormalities among grain elevator workers, *Chest,* 73, 461, 1979.
8. **do Pico, G.A., Jacobs, S., Flaherty, D., and Rankin, J.,** Pulmonary reactions to durum wheat, *Chest,* 81, 55, 1982.
9. **Manfreda, J. and Warren, C.P.W.,** The effects of grain dust on health, *Rev. Environmental Health,* 4, 239, 1984.
10. **Manfreda, J., Holdford–Strevens, V., Cheang, M., and Warren, C.P.W.,** Acute symtoms following exposure to grain dust in farming, *Environ. Health Persp.,* 66, 73, 1966.
11. **Warren, C.P.W., Holford–Strevens, V., Wong, J., and Warren, C.P.W.,** The relationship between smoking and total immunoglobulin E levels, *J. Allergy Clin. Immunol.,* 69, 370, 1982.
12. **Darke, C.S., Knowelden, J., Lacey, J.,** and **Ward, A.M.,** Respiratory disease of workers harvesting grain, *Thorax,* 31, 294, 1976.
13. **Corey, P., Hutcheon, M., Broder, I., and Mintz, S.,** Grain elevator workers show work–related pulmonary function changes and dose–effect relationships with dust exposure, *Br. J. Ind. Med.,* 39, 330, 1982.
14. **do Pico, G.A., Reddan, W., Anderson, S., Flaherty, D., and Smalley, E.,** Acute effects of grain dust exposure during a work shift, *Am. Rev. Respir. Dis.,* 128, 399, 1983.
15. **Tabona, M., Chan–Yeung, M., Enarson, D., MacLean, L., Dorken, E., and Shulzer, M.,** Host factors affecting longitudinal decline in lung spirometry among grain elevator workers, *Chest,* 85, 782, 1984.
16. **Dutkiewicz, J.,** Exposure to airborne bacteria in agriculture. I. Environmental studies, *Arch. Environ. Health,* 33, 250, 1978.

Grain Dust and Health

Overview of Adverse Pulmonary Effects of Grain Dust

I. Broder

INTRODUCTION

Grain dust has been recognized since the 18th century to have a strongly irritating effect on the skin, eyes, nose, and lungs.[1] Over the past 50 years, there have been a number of reports which have characterized many different aspects of the adverse pulmonary effects of grain dust exposure.

CROSS–SECTIONAL STUDIES

This review will start with a consideration of cross–sectional studies of the health of grain handlers which included a control population and were relatively general in focus.[2–8] The results of each of these cross–sectional studies were compared by determining the percent difference in symptom prevalence and pulmonary function between the grain worker and control populations. This was done within each study by subtracting symptom prevalence in the controls from that in the grain workers, and also by obtaining the difference between the mean of pulmonary function variables for the two groups and expressing this as a percent of the mean of that of the grain handlers.

Respiratory Symptoms

Respiratory symptoms were found to occur in an excess of about 5 to 15% among nonsmoking grain handlers (Table 1). A consistently greater excess was reported in a study in which the questionnaires were self–completed.[7] All of the excesses were statistically significant in the latter study, while only cough was present in significant excess in one of the others. Among the currently smoking grain handlers the excess of symptoms was similar to that among the nonsmokers (Table 2). These excesses were variably statistically significant for cough, sputum, wheeze, and shortness of breath. Generally, the prevalence of symptoms observed among the grain handlers approximated that observed in studies which lacked controls.[9–13]

Pulmonary Function

The most consistent difference in the pulmonary function results of the grain handlers relative to the controls was a small decrease in the forced expired volume in 1 s (FEV_1) (Tables 1 and 2). This difference was independent of smoking and in two instances was statistically significant.

The excess of cough, sputum, wheeze, and shortness of breath among grain handlers, as well as the decrease in their FEV_1, was consistent with the interpretation that the long–term exposure to grain dust caused chronic bronchitis. This was confirmed by a study designed to compare the prevalence of chronic bronchitis between a population of nonsmoking grain handlers and controls.[14]

Interaction of Grain Dust and Smoking

Although the effects of grain dust on pulmonary symptoms and function tended to be similar in both smokers and nonsmokers (Tables 1 and 2), studies conducted in Saskatchewan showed that grain workers who were employed for 5 to 10 years or less had significant impairment of small airway function if they smoked, but not if they were nonsmokers.[6,15] This was not observed in longer duration grain handlers or in controls matched for age and smoking.

Chest X–Rays

Chest X–rays were examined in three of the studies,[2,3,7] and were found to show no differences between grain handlers and controls. However, grain dust is known to contain about 5% silica,[4,16,17] and there have been reports of silicosis in grain handlers.[16,18,19]

TABLE 1
Percent Difference in Symptom Prevalence[a] or in Pulmonary Function[b] between Grain and Control Workers

	Nonsmokers				
	Broder[3]	**Chan–Yeung[4]**	**Cotton[6]**	**DoPico[7]**	**Herbert[8]**
No. grain/control	65/40	130/40	195/195	65/63	16/16
Cough	+14[c]	+6[d]	+6	na[e]	+5
Sputum	+12	+11	+11	+21[c]	na
Bronchitis	+6	na	+11	+25[c]	na
Wheeze	–2	–3	+17[c]	+27[c]	+5
Dyspnea	+4	+3	na	+26[c]	+5
FVC	–2	–5[c]	+1	na	+4
FEV_1	–5	–4[c]	0	na	–13

[a] Symptom prevalence in grain handlers minus that in controls.
[b] Pulmonary function of grain handlers minus that in controls, expressed as a percent of that of grain handlers.
[c] Difference between grain and control workers has p value ≤0.05 by chi–square or t test.
[d] Prevalence of symptoms was estimated from bar graphs in this report.
[e] na indicates that a variable is not available.

TABLE 2
Percent Difference in Symptom Prevalence[a] or in Pulmonary Function[b] between Grain and Control Workers

	Current Smokers						
	Becklake[2]	Broder[3]	Chan–Yeung[4]	Cotton[6]	DoPico[7]	Herbert[8]	Patel[5]
No. grain/control	69/23	231/86	300/42	195/195	153/106	20/20	47/15
Cough	+39[c]	+7	+14[d]	+12	na[e]	na	na
Sputum	+12	0	+1	na	+20[c]	na	na
Bronchitis	+20	+7	na	+13	+27[c]	–5	na
Wheeze	+24[c]	0	+6	–2	+22[c]	–20	na
Dyspnea	+12	–2	+4	na	+12[c]	+40[c]	na
FVC	–10[c]	–6[c]	–3	+1	na	+2	+6
FEV_1	–9	–5[c]	–2	0	na	–10	+3

[a] Symptom prevalence in grain handlers minus that in controls.
[b] Pulmonary function of grain handlers minus that in controls, expressed as a percent of that of grain handlers.
[c] Difference between grain and control workers has p value ≤0.05 by chi–square or t test.
[d] Prevalence of symptoms was estimated from bar graphs in this report.
[e] na indicates that a variable is not available.

Grain Fever

Grain fever was reported by 5 to 30% of the grain handlers in several of the controlled[2,7,8] and other studies of grain handlers.[9,11,12,20] This problem was not clearly related to the presence of immediate wheal and flare reactions or serum precipitins to grain antigens, or to the occurrence of pulmonary function or serum complement changes following exposure to grain dust.[2,7,8,10,20–25] Also, grain fever has been shown to occur even on the initial exposure to grain dust in individuals who have not previously been employed as grain handlers.[22,23,26] The development of grain fever appears to be directly related to the level of grain dust exposure, since the symptoms have been observed where the concentration of dust was high,[2,7,9,26] but not in elevators having relatively low dust levels.[3,4] The requirement for high dust levels would be consistent with bacterial endotoxin being the cause of grain fever, since this material is a known pyrogen,[27] but is present in only relatively low concentration in grain dust.[28]

LONGITUDINAL STUDIES

Longitudinal studies have been performed in populations of grain handlers and control workers in both Vancouver and Thunder Bay, over a period of 2.5 to 3 years.[29,30] The Vancouver population also was reexamined after 6 years, but without the controls.[31]

Longitudinal Changes in Respiratory Symptoms and Pulmonary Function

The results of the initial longitudinal studies were similar, showing relatively minor changes in respiratory symptoms among both the grain handlers and controls between the two surveys.[29,30] There were only small decreases in the pulmonary function of the grain workers at the second time point, which tended to be in slight excess of the controls (Table 3), especially among the older age groups in the Vancouver study. However, the changes among the grain handlers in both studies were smaller than seen in two contemporary longitudinally studied general populations surveyed in other communities.[32,33] The annual pulmonary function changes among the Vancouver grain handlers were smaller when examined over 6 years than over the initial 2.5 years.[31] There were no significant associations found in either the Vancouver or Thunder Bay studies between the degree of change in pulmonary function of the grain handlers and the presence either of respiratory symptoms or of positive allergy skin test reactions at the first survey.[29,30]

Healthy–Worker Effect

The reparticipation rate of these longitudinal studies was about 70% over a 2.5– to 3–year interval.[29,30] The Thunder Bay results showed that approximately 10% of the original populations of grain workers and controls were not available to participate in the second survey because they had quit their respective jobs during the interval following the first.[30] The grain handlers who were known to have quit were found to have a significantly higher prevalence of cough and shortness of breath at the time of the first survey, relative to those who remained employed, despite being younger and having a shorter duration of employment. No comparable differences were seen in a similar comparison made among the control population. The healthy–worker effect has undoubtedly led to an underestimate of the adverse health effects of employment as a grain handler.

The bias resulting from the healthy–worker effect was likely avoided in a study in which grain handlers and control workers in Thunder Bay were examined on two occasions, the first soon after being hired and the second approximately 2.5 months later.[34] No workers were known to have quit for reasons of health during this interval. The grain and civic workers were comparable in pulmonary symptoms at the time

TABLE 3
Change in Pulmonary Function of Grain Handler and Control Populations Expressed in Cm^3/Year

	Nonsmokers						Current smokers					
	Chan–Yeung[29] [a]		Broder[30]		Bossé[32]	Tashkin[33]	Chan–Yeung[29]		Broder[30]		Bossé[32]	Tashkin[33]
	G	C	G	C	C	C	G	C	G	C	C	C
No.	81	37	44	22	398	414	152	23	141	46	354	278
FVC	–25	+14	–37	–17	–68	–60	–39	+13	–40	–40	–91	–64
FEV_1	–16	+31	–7	–10	–61	–56	–39	–5	–3	+17	–78	–70
FEF_{50}			–97	–50		–96			–70	+43		–158
FEF_{75}			–53	–47		–60			–60	+47		–80
MMF	–126	+21				–86	–142	–41				–122

[a] Values were pooled from age–specific tables in this report.

of starting their jobs. Subsequently, the grain handlers developed a marked increase in cough and sputum while the controls remained unchanged. The excess of symptoms in the grain workers at the end of the observation period was greater than previously observed in a cross–sectional study within the same populations.[3]

PREDICTABILITY OF LONGITUDINAL CHANGE IN PULMONARY FUNCTION

The Vancouver study demonstrated a direct relationship between longitudinal decrease in pulmonary function and both decrease in pulmonary function over a day or week of work as measured at the first cross–sectional examination, and methacholine responsiveness determined at the 6–year follow–up examination.[29,31] Thus, the presence of an acute work–related decrease in pulmonary function may prove to be a useful approach for predicting longitudinal loss of pulmonary function in grain handlers.

EXPOSURE–RESPONSE RELATIONSHIPS

Duration of Employment as Exposure Variable

In several studies, years of employment as a grain handler were found to correlate positively with the prevalence of pulmonary symptoms and negatively with pulmonary function.[6,7,14] However, other studies reported either that duration of employment varied positively with pulmonary function in some workers,[35] or that there was no association with pulmonary symptoms or function.[3–5,8] The absence of a consistent relationship between years of employment as a grain handler and the occurrence of adverse pulmonary effects may be due to a variable influence of the healthy–worker effect.

Grain Dust Levels as Exposure Variable

Studies of exposure–response relationships have been performed in Thunder Bay[35] and Superior–Duluth,[25] using dust measurements as the exposure variable.

Acute Exposure–Response Relationships

The acute association was examined in the Thunder Bay study by obtaining the slope of the relationship for each worker, between his change in pulmonary function on each of several days that this was measured and his exposure to dust on the same day. The results showed a negative slope for the FEF_{50} (forced expiratory flow) in 74% of the men, which was significantly more prevalent than could be expected to occur by chance. This relationship was seen only with the respirable fraction of grain dust, and was independent not only of mask wearing, but also of age, duration of employment, pack years of cigarette smoking, and baseline level of FEF_{50}. No significant relationships were seen with the forced vital capacity (FVC), FEV_1 and FEF_{25}.

The Superior–Duluth study examined the acute exposure–response relationship, both with respect to pulmonary symptoms which occurred during the work day and to the change in pulmonary function over the same day. Each worker contributed only single data points and no adjustment was made for mask wearing. Workers who complained of one or more of cough, sputum, shortness of breath or wheezing were reported to have a significantly higher exposure to total grain dust (4.1 ± 8.1 mg/m^3) than those who had no work–related symptoms (2.1 ± 4.5 mg/m^3). Also, a significantly negative relationship was observed between total grain dust exposure and change over the day in FVC, FEF_{50}, and FEF_{25}. The pulmonary function relationships were not significantly influenced by age, height, or smoking status.

Chronic Exposure–Response Relationships

The Thunder Bay study also attempted to examine the chronic exposure–response relationship. This was done by assuming that the pulmonary function results obtained on the morning of day 1 of a week of work represented the baseline for each worker and that the average of his dust measurements over the subsequent week represented his usual dust exposure. The analysis showed that there was a statistically significant inverse relationship between respirable dust exposure and the baseline FEV_1, FEF_{50}, and FEF_{25}.[35] This was seen only among workers who did not wear a mask and did not occur with the nonrespirable dust fraction. Nevertheless, these results may be at least in part indicative of an acute exposure–response relationship, since on day 1 of a work week the grain handlers could be anticipated to demonstrate some residual changes in their pulmonary function from the acute effects of the week before.

NATURE OF PULMONARY FUNCTION CHANGES

Cross–sectional Studies

The cross–sectional studies reviewed above, as well as others, variably demonstrated that grain workers had a decrease in their FEV_1 and flow rates relative to their control populations.[2–4,6–8,14] Compartments of lung volume also have been found to be diminished, including the vital capacity, functional residual capacity, residual volume, and total lung capacity (TLC).[2,3,7,36,37] In addition, the shorter–duration grain handlers especially showed a smaller change in their flow rate when breathing helium relative to air, and had a greater increase in nitrogen concentration during phase 3 of the single–breath nitrogen test.[6,7,15] The airway resistance of grain handlers was variably increased,[36,37] and their carbon monoxide diffusing capacity was diminished, but not when adjusted for the TLC.[2,3,7] The changes in the grain handlers relative to the controls were generally small and were only variably found to be statistically significant in the studies in which they were measured. These observations made in cross–sectional studies of grain handlers likely represent the combined effects of acute and chronic exposure to grain dust.

Acute Studies

Similar pulmonary function changes have been found to occur under direct observation, in short–term studies of individuals or small groups who were examined over a day or several days of grain handling, or during experimental exposure to grain dust or extracts of grain dust or mites.[20,22–26,35,38–41] An additional finding obtained in the acute studies was a decrease in static lung compliance.[23,41] The decrease in compliance was not apparent in one of the original reports when the volume coordinate utilized was the vital capacity,[41] but was observed when the TLC was substituted.[35]

Chronic Changes

Grain handlers who were temporarily laid off in Thunder Bay, or who worked through a period of reduced elevator activity, have been found to show improvement in their FEV_1, FVC, and flow rate after a delay of about 1 month.[42] These results are suggestive of a chronic pulmonary function change, since the acute effects of grain dust tend to reverse over several days.[26] If so, the chronic changes in FVC, FEV_1, and flow rate were similar to those of the acute changes. Whether the chronic and acute changes are similar in other tests of pulmonary function is not known.

Interpretation of Pulmonary Function Changes

The presence of airflow limitation at both high and low lung volume, increased airway resistance, and changes both in phase 3 of the single–breath nitrogen test and in flow characteristics with helium are all indicative of the presence of obstruction at the level of both the large and small airways. The reduction of lung volume additionally may be compatible with obstruction, since this has been reported with small airway constriction in animals.[43] Similarly, small airway constriction has been observed to be accompanied by pressure–volume changes in man.[44] This interpretation would be further consistent with the known tendency of small inhaled particles to be deposited at the bifurcation of alveolar ducts.[45] Nevertheless, it is equally probable that the reported volume restriction and pressure–volume change may be at least in part due to an effect of grain dust at the alveolar level, of a magnitude sufficiently small for the diffusing capacity to remain unaffected. This is supported by observations made in animal models of grain dust exposure, where a granulomatous pneumonitis has been reported in both rats and rabbits,[46,47] although not in mice.[48]

Reversibility of Pulmonary Function Changes

The reversibility of the acute changes in pulmonary function is well defined.[22–26,39–41] Also, the results obtained in the Thunder Bay study of laid–off grain handlers indicated that the chronic pulmonary function changes are at least partially reversible in workers employed for an average of less than 5 years.[42]

Allergic and Nonallergic Subjects

Several of the foregoing acute studies were based on subjects who were allergic to grain dust or grain mites,[22,24,39–41,49] while other studies included control subjects who had not previously been exposed to grain dust, but who responded to relatively high concentrations of grain dust in a manner comparable to the allergic ones.[20,23,26] Moreover, the onset of the pulmonary function changes was relatively immediate in both the allergic and nonallergic subjects, and both returned to baseline over a period ranging from several hours to a few days. The two groups were also similar in demonstrating a late bronchoconstrictor response.[23,39,41] Thus, the acute pulmonary function changes which result from exposure to grain dust appear to be similar whether they occurred in allergic or nonallergic individuals.

Nonallergic Airway Responsiveness

Grain handlers are further known to show an increase in their airway responsiveness to histamine and methacholine.[24,31,36,37,39,41] This is known to show a direct relationship retrospectively with their change in pulmonary function measured over a week of work,[29,31] and also with symptoms of bronchitis related, at least in part, to smoking.[37,39,50] Increased nonallergic airway responsiveness did not correlate with the atopic status of the grain handlers, with the presence of a positive skin test or positive serum precipitin reactions to antigens present in grain dust, or with the presence of a positive or a negative inhalation–challenge response to grain dust or grain dust extract.[24,36,37,39,50] The increased nonallergic airway responsiveness of grain handlers is likely a similar phenomenon to that seen following bronchoconstriction induced by irritants and respiratory infection,[51,52] as well as in atopic individuals following an episode of allergic bronchoconstriction.[53,54] This would imply that the increased airway responsiveness of grain handlers is a secondary phenomenon and perhaps may demonstrate, rather than predict, that a given worker likely experiences pulmonary effects from continuing exposure to grain dust.

MECHANISM OF ADVERSE PULMONARY EFFECTS OF GRAIN DUST

Type 1 Allergy

Some grain handlers have been diagnosed as having allergic asthma from inhaling grain dust or one of its constituents.[19,22,24,39–41,49,55–58] This may be variably associated with rhinitis, conjunctivitis, urticaria, and other features of a type 1 allergic reaction.

However, it appears unlikely that type 1 allergy accounts for a significant proportion of the adverse pulmonary manifestations observed in grain handlers, since even the prevalence of asthma in this occupation has been found not to exceed that of control populations.[3,6,14] Also, several epidemiological studies have demonstrated that grain handlers have a lower than expected frequency of features of atopy, suggesting that individuals who are predisposed tend either not to become employed in this work or to quit soon after starting.[2,3,36,37] Moreover, a number of studies have shown that the presence of positive skin reactions in populations of grain handlers correlated poorly with the presence of respiratory symptoms, pulmonary function status, and change in pulmonary function over time.[2,3,5,20,21,24,29–31,50]

A number of epidemiologic studies have demonstrated that up to 50% or more of grain handlers complain of work–related respiratory symptoms.[2,3,7,8,12] The high association of symptoms with grain handling is suggestive of an irritant effect, and is supported by the observation that people exposed to grain dust for the first time developed symptoms and pulmonary function changes which did not differ from those of grain handlers.[20,23,26]

Type 3 Allergy

A number of reports have dealt with the possibility that the adverse pulmonary effects of grain dust may involve a type 3 reaction, since this mechanism is at least partially the basis of farmer's lung.[59] A case of grain weevil asthma has been reported in which immediate and late skin test and inhalation bronchial responses were observed.[58] The late bronchial response was associated with fever, basal lung crepitations, and a decrease in the carbon monoxide diffusing capacity, which did not improve following the inhalation of a bronchodilator aerosol. Although this one case is suggestive of a combined type 1 and type 3 pulmonary response to grain weevil antigen, late skin responses have not generally been observed in studies of grain handlers.[22,39] There have been reports of late bronchial responses following inhalation challenge of symptomatic grain handlers with grain dust or extracts of grain dust or grain mite.[22–24,39,41] These responses were associated with fever, elevated circulating white blood cell counts, and impairment of carbon monoxide diffusing capacity. However, the same features also have been observed in control subjects who were ostensibly being exposed to grain dust for the first time.[23] Late bronchial responses in grain handlers have not been associated with evidence of complement activation in serum,[20,23,41] although this negative finding does not exclude the possibility of complement being consumed locally in the lung. Similarly, grain handlers have shown no increase in the prevalence of serum precipitins to grain antigens, and when present, these have not correlated well with the presence of respiratory symptoms or pulmonary function changes.[2,22,23,39,50] Finally, the chest X–rays obtained in epidemiological studies of grain handlers have not suggested that hypersensitivity pneumonitis is a prevalent problem in this occupational group.[2,3,7] Thus, there is little to suggest the common involvement of a type 3 reaction in the adverse pulmonary effects of grain dust.

Other Possible Mechanisms of Adverse Pulmonary Effects of Grain Dust

The adverse pulmonary effects of grain dust appear to involve an inflammatory response, as suggested by the fever and elevated circulating white blood cell level of the acute reaction,[20,22,23,25,41] and the elevated serum α_1–antitrypsin level of the chronic state.[50,60] Grain dust and grain dust extract are known to cause a nonimmunologic activation of complement[28] which results in the formation of complement–derived inflammatory cell chemotactic factors.[61–63] One of these factors has been demonstrated to cause an interstitial pneumonitis when introduced into the lungs of rabbits.[61] Thus, activation of complement by grain dust inhaled into the lung may, at least in part, explain the basis of the adverse pulmonary effects of grain dust.

SYNOPSIS

The adverse pulmonary effects of chronic exposure to grain dust are associated predominantly with the clinical features of chronic bronchitis, but occasionally cases of allergic grain dust asthma also occur. The magnitude of these problems appear to be only modest, as indicated by the results of both cross–sectional and longitudinal studies. However, studies of this type tend to underestimate the adverse effects of grain dust, since most are likely biased by the healthy–worker effect. The pulmonary function changes which result from grain dust exposure are both acute and chronic. The acute changes in pulmonary function become apparent within minutes of the initial exposure to grain dust and are reversible over hours or days, while the chronic effects are at least partially reversible over a period of months. The acute changes demonstrate features of obstruction at the level of both the large and small airways, as well as being consistent with the presence of a pulmonary–restrictive defect. The chronic changes are not well characterized. In individual workers, the chronic pulmonary function changes examined prospectively over time vary directly with their previously defined acute decrease over a day or a week of work. Exposure–response relationships are demonstrable between total and respirable dust exposure and both acute pulmonary symptoms and function changes, as well as possibly with chronic pulmonary function changes. Grain handlers show increased nonspecific bronchial reactivity, which is likely a secondary phenomenon demonstrating that a given worker experiences pulmonary effects from continuing exposure to grain dust. The adverse pulmonary effects of grain dust are most likely due to an irritant effect, perhaps involving a complement–mediated inflammatory response. Only isolated instances may be explained by a type 1 or type 3 allergic mechanism.

REFERENCES

1. **Ramazzini, B.,** *Demorbis Artificiam Diatribia (Disease of Workers),* Wright, W.C., transl., University of Chicago, Chicago, 1940.
2. **Becklake, M.R., Jodoin, G., Lefort, L., Rose, B., Mandl, M., and Fraser, R.G.,** A respiratory health study of grainhandlers in St. Lawrence River ports, in *Occupational Pulmonary Disease: Focus on Grain Dust and Health,* Dosman, J. and Cotton, D.J., Eds., Academic Press, New York, 1980, 239.
3. **Broder, I., Mintz, S., Hutcheon, M., et al.,** Comparison of respiratory variables in grain elevator workers and civic outside workers of Thunder Bay, Canada, *Am. Rev. Respir. Dis.,* 119, 193, 1979.
4. **Chan–Yeung, M., Schulzer, M., MacLean, Dorken, E., and Grzybowski, S.,** Epidemiology health survey of grain elevator workers in British Columbia, *Am. Rev. Resp. Dis.,* 121, 329, 1980.
5. **Patel, K.R., Symington, I.S., Pollock, R., and Shaw, A.,** A pulmonary survey of grain handlers in the West of Scotland, *Clin. Allergy,* 11, 121, 1981.
6. **Cotton, D.J., Graham, B.L., Li, K.Y.R., Froh, F., Barnett, G.D., and Dosman, J.A.,** Effects of grain dust exposure and smoking on respiratory symptoms and lung function, *J. Occup. Med.,* 25, 131, 1983.
7. **DoPico, G.A., Reddan, W., Tsiatis, A., Peters, M.E., and Rankin, J.,** Epidemiologic study of clinical and physiologic parameters in grain handlers of northern United States, *Am. Rev. Respir. Dis.,* 130, 759, 1984.
8. **Herbert, F.A., Woytowich, V., Schram, E., and Baldwin, D.,** Respiratory profiles of grain handlers and sedentary workers, *C.M.A.J.,* 125, 46, 1981.
9. **Williams, N., Skoulas, A., and Merriman, J.E.,** Exposure to grain dust. I. A survey of the effects, *J. Occup. Med.,* 6, 319, 1964.
10. **Skoulas, A., Williams, N., and Merriman, J.E.,** Exposure to grain dust. II. A clinical study of the effects, *J. Occup. Med.,* 61, 659, 1964.
11. **Kleinfeld, M., Messite, J., Swencicki, R.E., and Shapiro, J.,** A clinical and physiologic study of grain handlers, *Arch. Environ. Health,* 16, 380, 1968.
12. **Tse, K.S., Warren, P., Janusz, M., McCarthy, D.S., and Cherniack, R.M.,** Respiratory abnormalities in workers exposed to grain dust, *Arch. Environ. Health,* 27, 74, 1973.
13. **Sheridan, D., Deutscher, C.L., Tan, L., et al.,** The relationship between exposure to cereal grain dust and pulmonary function in grain workers, in *Occupational Pulmonary Disease: Focus on Grain Dust and Health,* Dosman, J. and Cotton, D.J., Eds., Academic Press, New York, 1980, 229.
14. **Dosman, J.A., Cotton, D.J., Graham, B.L., Li, K.Y.R., Froh, F., and Barnett, G.D.,** Chronic bronchitis and decreased forced expiratory flow rates in lifetime nonsmoking grain workers, *Am. Rev. Respir. Dis.,* 121, 11, 1980.
15. **Cotton, D.J., Graham, B.L., Li, K.Y.R., Froh, F., Barnett, G.D., and Dosman, J.A.,** Effects of smoking and occupational exposure on peripheral airway function in young cereal grain workers, *Am. Rev. Respir. Dis.,* 126, 660, 1982.
16. **Dunner, L., Hermon, R., and Bagnall, D.J.T.,** Pneumoconiosis in dockers dealing with grain and seeds, *Br. J. Radiol.,* 19, 506, 1946.
17. **Farant, J.P. and Moore, C.F.,** Dust exposure in the Canadian grain industry, *Am. Ind. Hyg. Assoc. J.,* 39, 177, 1978.
18. **Heatley, T.F., Kohn, D., and Rex, C.R.,** A case of silicosis caused by wheat dust, *J.A.M.A.,* 124, 980, 1944.
19. **Cohen, V.D. and Osgood, H.,** Disability due to inhalation of grain dust, *J. Allergy,* 24, 193, 1953.
20. **Cockcroft, A.E., McDermott, M., Edwards, J.H., and McCarthy, P.,** Grain exposure — symptoms and lung function, *Eur. J. Respir. Dis.,* 64, 189, 1983.
21. **DoPico, G.A., Reddan, W., Flaherty, D., et al.,** Respiratory abnormalities among grain handlers, *Am. Rev. Respir. Dis.,* 115, 915, 1977.
22. **Warren, P., Cherniack, R.M., and Tse, K.S.,** Hypersensitivity reactions to grain dust, *J. Allergy Clin. Immunol.,* 53, 139, 1974.
23. **DoPico, G.A., Flaherty, D., Bhansali, P., and Chavaje, N.,** Grain fever syndrome induced by inhalation of airborne grain dust, *J. Allergy Clin. Immunol.,* 69, 435, 1982.
24. **DoPico, G.A., Jacobs, S., Flaherty, D., and Rankin, J.,** Pulmonary reaction to durum wheat, a constituent of grain dust, *Chest,* 81, 55, 1982.
25. **DoPico, G.A., Reddan, W., Anderson, S., Flaherty, D., and Smalley, E.,** Acute effects of grain dust exposure during a work shift, *Am. Rev. Respir. Dis.,* 128, 399, 1983.
26. **McCarthy, P.E., Cockcroft, A.E., and McDermott, M.,** Lung function after exposure to barley dust, *Br. J. Ind. Med.,* 42, 106, 1985.
27. **Atkins, E., Feldman, J.D., Francis, L., and Hursh, E.,** Studies on the mechanism of fever accompanying delayed hypersensitivity. The role of the sensitized lymphocyte, *J. Exp. Med.,* 135, 1113, 1972.
28. **Olenchock, S.A. and Major, P.C.,** Immunological studies of grain dust, *J. Environ. Pathol. Toxicol.,* 2, 9, 1979.
29. **Chan–Yeung, M., Schulzer, M., MacLean, L., et al.,** A follow–up study of the grain elevator workers in the Port of Vancouver, *Arch. Environ. Health,* 36, 75, 1981.
30. **Broder, I., Corey, P., Davies, G., et al.,** Longitudinal study of grain elevator and control workers with demonstration of healthy worker effect, *J. Occup. Med.,* 27, 873, 1985.
31. **Tabona, M., Chan–Yeung, M., Enarson, D., MacLean, L., Dorken, E., and Schulzer, M.,** Host factors affecting longitudinal decline in lung spirometry among grain elevator workers, *Chest,* 85, 782, 1984.
32. **Bossé, R., Sparrow, D., Rose, C.L., and Weiss, S.T.,** Longitudinal effect of age and smoking cessation on pulmonary function, *Am. Rev. Respir. Dis.,* 123, 378, 1981.
33. **Tashkin, D.P., Clark, V.A., Coulson, A.H., et al.,** The UCLA population studies of chronic obstructive respiratory disease, *Am. Rev. Respir. Dis.,* 130, 707, 1984.
34. **Broder, I., Hutcheon, M.A., Mintz, S., et al.,** Changes in respiratory variables of grain handlers and civic workers during their initial months of employment, *J. Ind. Med.,* 41, 94, 1984.
35. **Corey, P., Hutcheon, M., Broder, I., et al.,** Grain elevator workers show work–related pulmonary function changes and dose–effect relationships with dust exposure, *Br. J. Ind. Med.,* 39, 330, 1982.

36. **Gerrard, J.W., Mink, J., Cheung, S.S.C., Tan, L.K.T., and Dosman, J.A.,** Nonsmoking grain handlers in Saskatchewan: airway reactivity and allergic status, *J. Occup. Med.*, 21, 342, 1979.
37. **Mink, J.T., Gerrard, J.W., Cockcroft, D.W., Cotton, D.J., and Dosman, J.A.,** Increased bronchial reactivity to inhaled histamine in nonsmoking grain workers with normal lung function, *Chest,* 77, 28, 1980.
38. **Gandevia, B. and Ritchie, B.,** Relevance of respiratory symptoms and signs to ventilatory capacity changes after exposure to grain dust and phosphate rock dust, *Br. J. Ind. Med.*, 23, 181, 1966.
39. **Chan–Yeung, M., Wong, R., and MacLean, L.,** Respiratory abnormalities among grain elevator workers, *Chest,* 75, 461, 1979.
40. **Warren, C.P.W., Holford–Strevens, V., and Sinha, R.N.,** Sensitization in a grain handler to the storage mite lepidoglyphus destructor (Schrank), *Ann. Allergy,* 50, 30, 1983.
41. **Davies, R.J., Green, M., and Schofield, N.M.,** Recurrent nocturnal asthma after exposure to grain dust, *Am. Rev. Respir. Dis.*, 114, 1011, 1976.
42. **Broder, I., Mintz, S., Hutcheon, M.A., Corey, P.N., and Kuzyk, J.,** Effect of layoff and rehire on respiratory variables of grain elevator workers, *Am. Rev. Respir. Dis.*, 122, 601, 1980.
43. **Woolcock, A.J., Macklem, P.J., Hogg, J.C., et al.,** Effect of vagal stimulation on central and peripheral airways in dogs, *J. Appl. Physiol.*, 26, 806, 1969.
44. **Michaels, R., Sigurdson, M., Thurlbeck, S., and Cherniack, R.,** Elastic recoil of the lung in cigarette smokers: the effect of nebulized bronchodilator and cessation of smoking, *Am. Rev. Respir. Dis.*, 119, 707, 1979.
45. **Brody, A.R., Hill, L.H., and Warheit, D.B.,** Induction of early alveolar injury by inhaled asbestos and silica, *Fed. Proc.*, 44, 2596, 1985.
46. **Friborsky, V., Ulrich, L., Tesik, L., and Malik, E.,** Morphological changes of rat lung after exposure to plant dust, *Acta Morphol. Acad. Sci. Hung.*, 20, 191, 1972.
47. **Stepner, N., Broder, I., and Baumal, R.,** Animal model of grain worker's lung, *Environ. Health Perspect.*, 66, 31, 1986.
48. **Armanious, M., Cotton, D., Wright, J., Dosman, J., and Carr, I.,** Grain dust and alveolar macrophages: an experimental study of the effects of grain dust on the mouse lung, *J. Pathol.*, 136, 265, 1982.
49. **Ingram, C.G., Jeffrey, I.G., Symington, I.S., and Cuthbert, O.G.,** Bronchial provocation studies in farmers allergic to storage mites, *Lancet,* 2, 1330, 1979.
50. **Broder, I., Davies, G., Hutcheon, M., et al.,** Variables of pulmonary allergy and inflammation in grain elevator workers, *J. Occup. Med.*, 25, 43, 1983.
51. **Empey, D.W., Laitinen, L.A., Jacobs, L., et al.,** Mechanisms of bronchial hyperreactivity in normal subjects after upper respiratory tract infections, *Am. Rev. Respir. Dis.*, 113, 131, 1976.
52. **Golden, J.A., Nadel, J.A., and Boushey, H.A.,** Bronchial hyperirritability in healthy subjects after exposure to ozone, *Am. Rev. Respir. Dis.*, 118, 287, 1978.
53. **Cartier, A., Thomson, N.C., Frith, P.A., Roberts, R., and Hargreave, F.E.,** Allergen–induced increase in bronchial responsiveness to histamine: relationship to the late asthmatic response and change in airway caliber, *J. Allergy Clin. Immunol.*, 70, 170, 1982.
54. **Boulet, L.P., Cartier, A., Thomson, N.C., et al.,** Asthma and increases in nonallergic bronchial responsiveness from seasonal pollen exposure, *J. Allergy Clin. Immunol.*, 71, 399, 1983.
55. **Duke, W.W.,** Wheat hairs and dust as a common cause of asthma among workers in wheat flour mills, *J.A.M.A.*, 105, 97, 1935.
56. **Harris, L.H.,** Allergy to grain dusts and smuts, *J. Allergy,* 10, 327, 1938/1939.
57. **Ordman, D.,** Cereal grain dusts as a cause of respiratory allergy in South Africa, *S.Afr. Med. J.*, p. 784, August 2, 1958.
58. **Lunn, J.A. and Hughes, D.T.D.,** Pulmonary hypersensitivity to the grain weevil, *Br. J. Ind. Med.*, 24/25, 158, 1967/1968.
59. **Schatz, M., Patterson, R., and Fink, J.,** Immunopathogenesis of hypersensitivity pneumonitis, *J. Allergy Clin. Immunol.*, 60, 27, 1977.
60. **Ashley, M.J., Corey, P., and Chan–Yeung, M.,** Smoking, dust exposure, and serum α_1–antitrypsin, *Am. Rev. Respir. Dis.*, 121, 783, 1980.
61. **Shaw, J.O., Henson, P.M., Henson, J., and Webster, R.O.,** Lung inflammation induced by complement–derived chemotactic factors in the alveolus, *Lab. Invest.*, 42, 547, 1980.
62. **Olenchock, S.A., Mull, J.C., and Major, P.C.,** Extracts of airborne grain dusts activate alternative and classical complement pathway, *Ann. Allergy,* 44, 23, 1980.
63. **Olenchock, S.A., Mull, J.C., and Major, P.C.,** Complement activation by commercial allergen extracts of cereal grain, *Clin. Allergy,* 10, 395, 1980.

A Study of the Pulmonary Histopathologic Effects of Chronic Grain Dust Exposure in Rabbits

N. Stepner, I. Broder, and R. Baumal

INTRODUCTION

We have previously developed an animal model of grain worker's lung in the rabbit, and described the results of light microscopic examination of the lung tissue.[1] The following report deals with the use of scanning and transmission electron microscopy in assessing the alveolar and interstitial response to chronic grain dust exposure in this model.

METHODS

Rabbits

New Zealand white rabbits weighing 2 kg were obtained in a single batch from one supplier. The controls were housed continuously in the animal quarters, while the experimentals were moved into a separate room on the days when they were exposed to grain dust.

Dust Exposure Chambers

Two exposure chambers were utilized which consisted of a 1 m^3 working area supplied with grain dust from a fluidized bed, as described in detail previously.[2,3] The delivery rate was adjusted to maintain a dust level of approximately 20 mg/m^3 in the chambers.

Grain Dust

We used a single batch of grain dust obtained from the air exhaust collectors of two large grain elevators in Thunder Bay, Canada which handle wheat predominantly and to a lesser extent oats, rye, and barley. The level of grain dust in each chamber was measured using a Bendix 44 pump and a Millipore filter cassette holding a 37–mm polytetrafluoroethylene filter (Millipore, Flouropore, 1.0 µm pore size).[4] The airborne dust in the chambers was further characterized on several occasions using an Andersen cascade impactor.[5]

Experimental Design

This study included 47 rabbits; 23 were controls and 24 were experimentals. Both groups were divided into five subgroups containing four or more animals each. The experimental animals were exposed to 20 mg/m^3 of grain dust for 7 hr/day, 5 day/week, for a duration of 1, 3, 6, 12, or 24 weeks. Exposures were staggered so that all rabbits were approximately the same age at the time of sacrifice. This study was approved by the Animal Care Committee of the University of Toronto.

Bronchoalveolar Lavage

The rabbits were anesthetized with an injection of sodium nembutal (50 mg/kg), exsanquinated, and bronchoalveolar lavage[6,7] was then performed. A random sample of lavage cells was processed for scanning electron microscopy following standard methods.

Histopathological Procedures

Following the bronchoalveolar lavage, a water manometer was connected to the proximal end of the tracheal cannula and the lungs were inflated to 20 cm H_20. The lungs were perfused free of blood with cold heparinized saline via a cannula placed in the pulmonary artery.[8] The lungs were kept at a constant volume and fixed by *in situ* perfusion via the pulmonary artery with cold universal fixative (4% formaldehyde–1% glutaraldehyde, pH 7.2).[9] Sections containing both central and peripheral regions of lung were stained with hematoxylin–eosin and cut at a thickness of 5 µm.

Electron Microscopy Procedures

Standard methods were used to postfix, dehydrate, stain, or gold coat the tissue sections for electron microscopy.[10,11] A JOEL JSM–35 scanning electron microscope and a Phillips 201 transmission electron microscope were utilized.

RESULTS

Examination of Grain Dust from Exposure Chamber

The mean dust level was 21 mg/m^3, range 15 to 29 mg/m^3. The aerodynamic size distribution of the dust demonstrated that about 50% of the particles were between 3.3 and 5.8 µm, 4% were <3.3 µm in size, and the remainder were larger.

Identification of Grain Dust Particles in the Lungs

Dust particles were not identified by light microscopy of bronchoalveolar lavage cells, or in the stained tissue sections. Scanning electron microscopy occasionally demonstrated particles with adherent macrophages in the lavage fluid; grain dust material was also frequently observed either as individual particles or as clumps associated with strands of airway secretions and entwined macrophages on the ciliated surface of the conducting airways. No dust particles were observed on the epithelial surface of alveoli or at alveolar duct bifurcations. No comparable positive observations were made in samples obtained from the control animals.

Light and Electron Microscopic Examination of the Lung Tissue

All experimental animals demonstrated histopathological changes after 5 d of exposure to grain dust.[1] The extent of the tissue involvement reached a peak at 3 weeks, diminished at 6 and 12 weeks, and then increased again at 24 weeks. The changes consisted of a granulomatous interstitial pneumonitis which centred on the membranous and respiratory bronchioles and extended to the alveolar regions. The interstitial pneumonitis was patchy but widespread, and characterized by consolidation with vascular congestion and diffuse alveolar septal thickening. Consolidation was associated with an interstitial infiltrate composed predominantly of macrophages, epithelioid cells, lymphocytes, and giant cells which often formed granulomata. Neutrophils and eosinophils were present, moreso within the initial 3 to 6 weeks. There was also extensive cellular exudation into the alveolar spaces and conducting airways, comparable to the tissue infiltrate. Control rabbit's lungs were normal by light microscopy.

Scanning electron microscopy of lung tissue from exposed animals showed that the ciliated surfaces of the secondary bronchioles and distally were covered with patches and clumps of "hairy" cells, most of which were macrophages. A few smaller "hairy" cells were present as well and were likely polymorphonuclear leukocytes. The alveoli were generally filled with similar cells and the septa were thickened (Plate 1A*). Comparable sections from control animals showed the ciliated surfaces of small bronchioles and the epithelial surfaces of alveoli to be clear, with the septa of normal thickness (Plate 1B*).

Transmission electron microscopy, performed on lung tissue of exposed rabbits from sites free of significant granulomata, showed alveolar septa thickened with edema and increased collagen deposition (Plate 2A*). Interstitial neutrophils, lymphocytes, plasma cells, and macrophages were present and the numbers of fibroblasts and type II alveolar epithelial cells per unit area were increased. Intraalveolar macrophages were also present. The type I alveolar epithelial cells, the basement membrane, and the endothelium were all normal. These ultrastructural changes were not evident in comparable sections from control animals (Plate 2B*). All of the findings described were observed in animals which had been exposed to grain dust for only 5 d.

DISCUSSION

The daily and weekly exposure intervals utilized in the foregoing study were selected to represent those of grain terminal elevator workers. The concentration of dust employed was two–fold greater than the federally regulated level in Canada,[12] but equal or higher ambient levels are often encountered.[13–15]

Over 50% of the dust was <5.8 μm in size; therefore, a significant amount was capable of being inhaled to the alveolar regions of the experimental animals. According to Brody et al.,[16] it is the presence of respirable dust in the terminal regions of the lungs which is important in initiating the early inflammatory response as seen in the lungs of animals exposed to asbestos and silica. We did not observe the grain dust particles in this explicit area, although this was the center of the inflammatory response.

The interstitial response seen in the lungs of the exposed animals was similar in histological appearance to a type 4 reaction[17] and also to a hypersensitivity pneumonitis,[18] but the rapidity of onset seems to weigh strongly against the involvement of an immunological mechanism. There was no indication, even after 5 d of exposure, of an eosinophilic response suggestive of a type 1 allergic reaction.[19] Also, the acute inflammation and hemorrhagic necrosis of a type 3 reaction was not seen at any time.[20]

We and others have hypothesized that the adverse pulmonary effects of inhaled grain dust may be due to the known complement–activating properties of this material.[21,22] Nevertheless, the histopathologic changes seen in our model did not resemble the acute changes described when a complement–derived chemotactic factor was instilled intratracheally in rabbits.[23] This difference does not necessarily exclude the possible role of complement in the response to grain dust, since a chronic, diffuse effect of low levels of complement activation may differ from that of a single instillation of a large bolus. The response to grain dust may be conditioned additionally by the simultaneous presence of particulates.

REFERENCES

1. **Stepner, N., Broder, I., and Baumal, R.,** Animal model of grain worker's lung, *Environ. Health Perspect.*, 66, 31, 1986.
2. **Fisher, A.M. and Christie, H.,** A dust dispersal machine for inhalation experiments, *Br. J. Ind Med.*, 19, 236, 1962.
3. **Nelson, N.M., Rajhans, G.S., Morton, S., and Brown, J.R.,** Silica flour exposure in Ontario, *Am. Ind. Hyg. Assoc.*, 39, 261, 1978.
4. **Corey, P., Hutcheon, M., Broder, I., and Mintz, S.,** Grain elevator workers show work–related pulmonary function changes and dose–effect relationships with dust exposure, *Br. J. Ind. Med.*, 39, 330, 1982.
5. **Rajhans, G.S.,** Inertial and gravitational collectors, in *Air Sampling Instruments for Evaluation of Atmospheric Contaminants*, 5th ed., American Conference of Governmental Industrial Hygienists, Cincinnati, OH, 1978, 0.
6. **Brain, J.D. and Corkery, G.C.,** The effect of increased particles on the endocytosis of radiocolloids by pulmonary macrophages *in vivo:* competitive and toxic effects, in *Inhaled Particles IV*, Walton, W.H., Ed., Pergamon Press, New York, 1977, 551.
7. **Brain, J.D. and Kavet, R.I.,** Phagocytosis: quantification of rates and intercellular heterogeneity, *J. Appl. Phys.*, 42, 432, 1977.
8. **Gil, J. and Weibel, E.R.,** Improvements in demonstration of lining layer of lung alveoli by electron microscopy, *Respir. Phys.*, 8, 13, 1969/1970.
9. **McDowell, E.M. and Trump, B.F.,** Histological fixatives suitable for diagnostic light and electron microscopy, *Arch. Pathol. Lab. Med.*, 100, 405, 1976.
10. **Pease, D.C.,** *Histological Methods for Electron Microscopy*, 2nd ed., Academic Press, New York, 1964.
11. **Johari, O. and DeNee, P.B.,** Handling, mounting and examination of particles for scanning electron microscopy, in *Scanning Electron Microscopy*, Johari, O., Ed., ITTRI, Chicago, 1972, 249.
12. Labour Canada, Guidelines for Environmental and Medical Surveil-

* Plates 1 and 2 appear on pages 106 to 109.

lance Programs in Industry, Occupational Health and Safety Branch, File No. 895–7–11, January 26, 1978.

13. **DoPico, G.A., Reddan, W., Anderson, S., Flaherty, D., and Smalley, E.,** Acute effects of grain dust exposure during a work shift, *Am. Rev. Respir. Dis.,* 128, 399, 1983.
14. **Chan–Yeung, M., Schulzer, M., Maclean, L., et al.,** A follow–up study of the grain elevators workers in the port of Vancouver, *Arch. Environ. Health,* 36, 75, 1981.
15. **Yoshida, K. and Maybank, J.,** Physical and environmental characteristics of grain dust, in *Occupational Pulmonary Disease: Focus on Grain Dust and Health,* Dosman, J.A. and Cotton, D.J., Eds., Academic Press, New York, 1980, 441.
16. **Brody, A.E., Hill, L.H., and Warheit, D.B.,** Induction of early alveolar injury by inhaled asbestos and silica, *Fed. Proc.,* 44, 2596, 1984.
17. **Hay, J.B.,** Delayed (cellular) hypersensitivity, in *Inflammation, Immunity and Hypersensitivity,* 2nd ed., Movat, H.Z., Ed., Harper & Row, New York, 1979, 271.
18. **Roberts, R.C. and Moore, V.L.,** Immunopathogenesis of hypersensitivity pneumonitis, *Am. Rev. Respir. Dis.,* 116, 1075, 1977.
19. **Baumal, R., Hooi, C., McAvoy, D., and Broder, I.,** Time course of morphometric changes after acute allergic bronchoconstriction in the guinea pig, *Int. Arch. Allergy Appl. Immunol.,* 71, 131, 1983.
20. **Kaplan, R.L., Schocket, A.L., King, T.E., Maulitz, R.M., Good, J.T., Sanford, R.E., and Sahn, S.A.,** A model of immune complex–mediated pleuropulmonary injury, *Am. J. Pathol.,* 100, 115, 1980.
21. **Olenchock, S.A. and Major, P.C.,** Immunological studies of grain dust, *J. Environ. Pathol. Toxicol.,* 2, 9, 1979.
22. **Broder, I., Davies, G., Hutcheon, M., Leznoff, A., Mintz, S., Thomas, P., and Corey, P.,** Variables of pulmonary allergy and inflammation in grain elevator workers, *J. Occup. Med.,* 25, 43, 1983.
23. **Shaw, J.O., Henson, P.M., Henson, J., and Webster, R.O.,** Lung inflammation induced by complement–derived chemotactic fragments in the alveolus, *Lab. Invest.,* 42, 547, 1980.

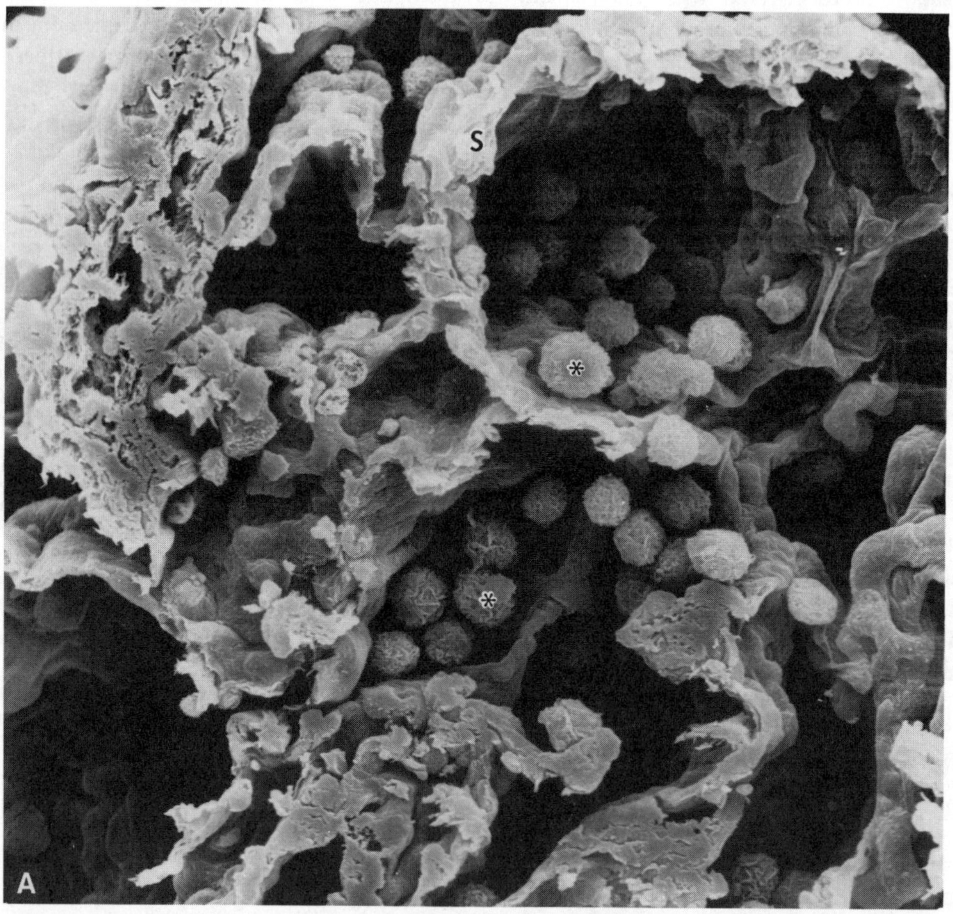

Plate 1A

PLATE 1. Scanning electron microscopy of alveolar regions. (Original magnification × 320.) (A) Photomicrograph from a 1–week exposed animal illustrating the extent of inflammatory cell infiltration (*) into the luminal surface of the alveoli. Septa (S) appear thickened. (B) Comparable section from a control animal showing the absence of intraalveolar cells (*) and normal–appearing septa (S).

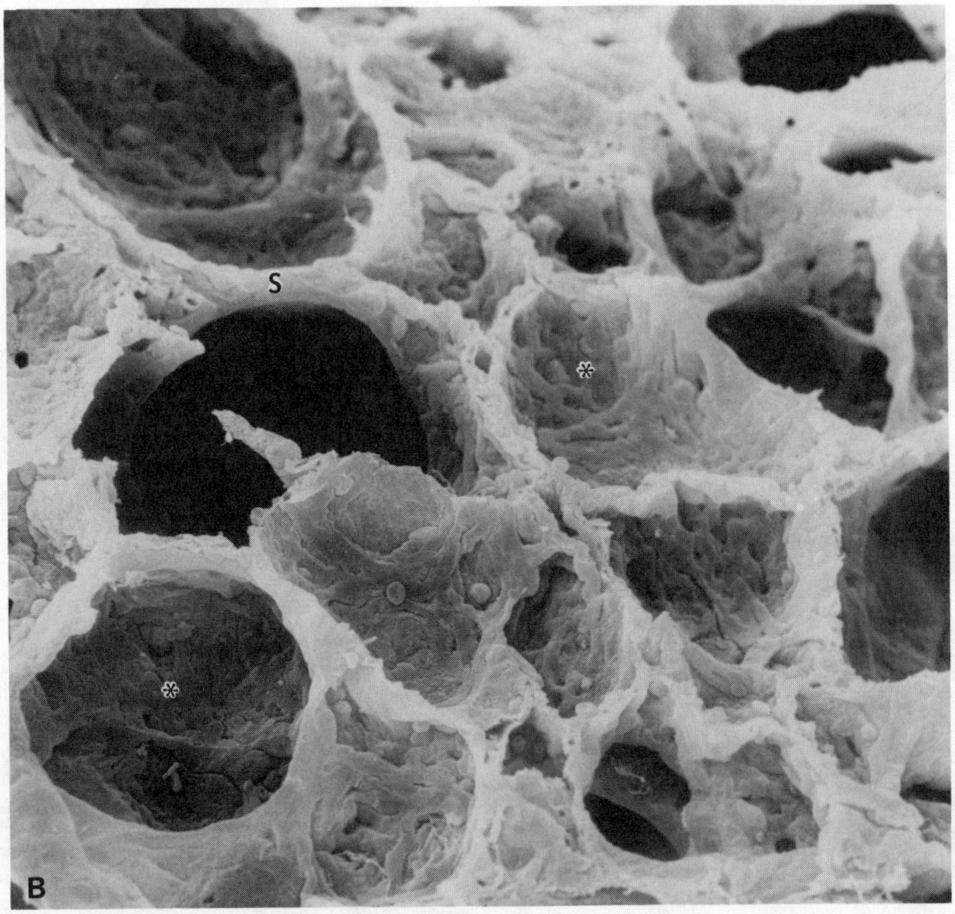

Plate 1B

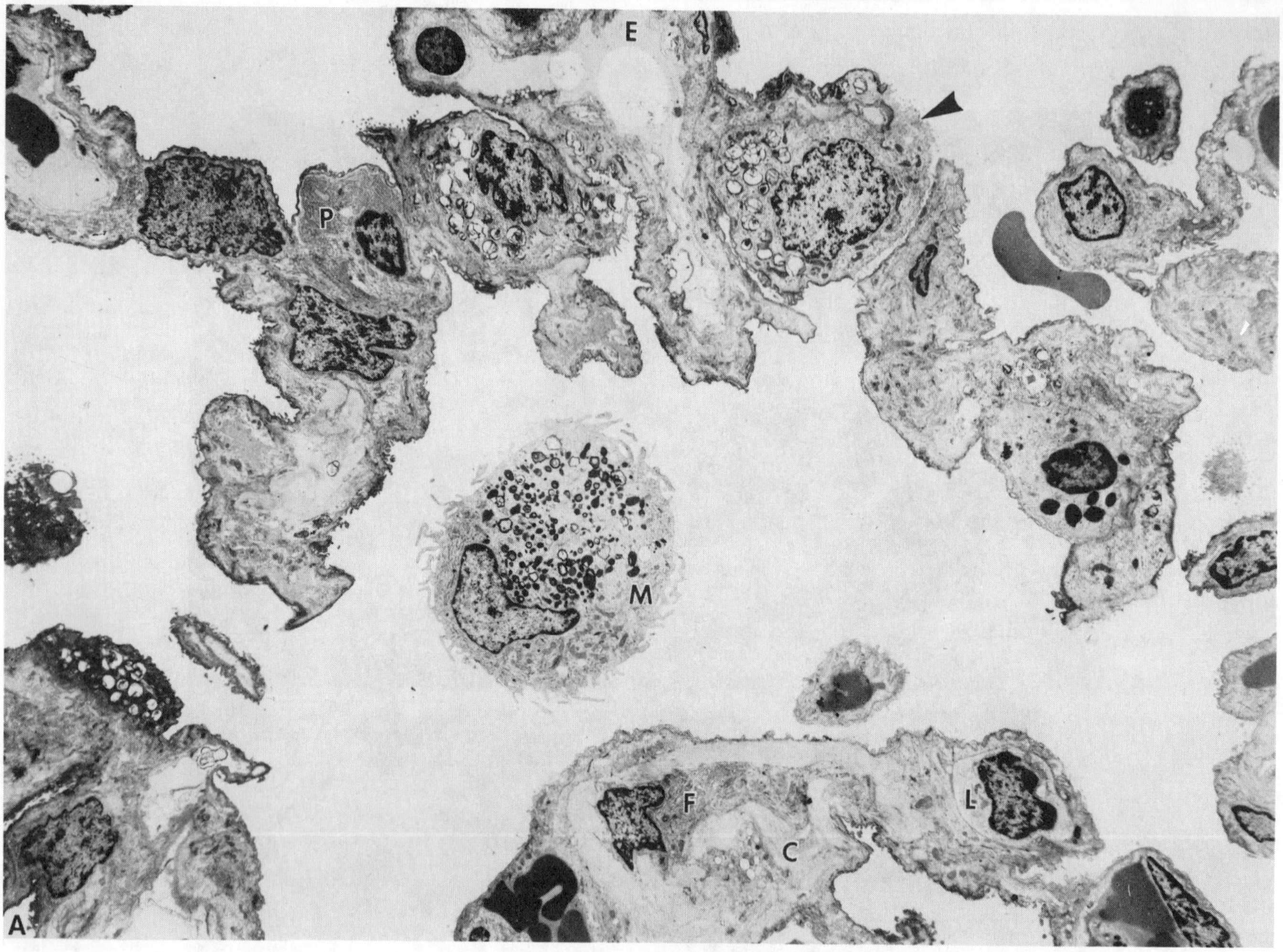

Plate 2A

PLATE 2 Transmission electron microscopy illustrating the ultrastructure of comparable alveolar areas to those shown in Plate 1. (Original magnification × 1500.) (A) Photomicrograph from 1-week exposed animal showing septa to be thickened due to the accumulation of interstitial edema (E), collagen fibers (C), fibroblasts (F), lymphocytes (L), and plasma cells (P). Intraalveolar macrophages (M) are present and there is an increase in the number of type II alveolar epithelial cells (arrow head). (B) The alveolar septa in the lungs of the control rabbits appeared normal in thickness and showed no increase in collagen or cellular constituents.

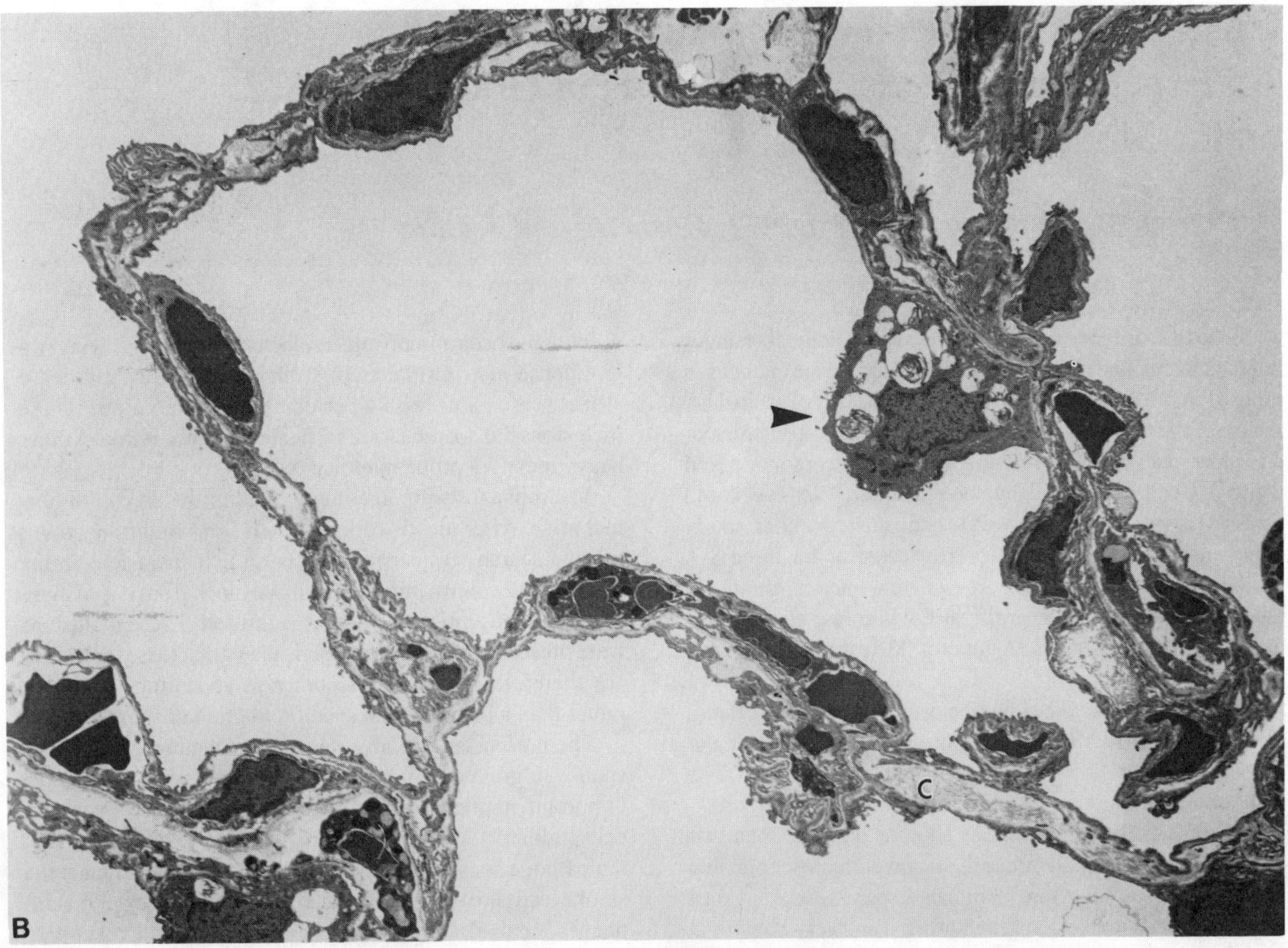

Plate 2B

Grain Dust and Alveolar Macrophages: An Experimental Study of the Effects of Grain Dust on the Mouse Lung*

D. Cotton, M. Armanious, J. Wright, J. Dosman, and J. Carr

Since little information exists on the pathological changes induced by inhaled grain dust, we have studied the changes in mouse lungs exposed to high atmospheric levels of grain dust. An experimental dust–generating chamber was constructed[1] and atmospheric dust levels in the chamber were monitored using a Bendix personal cyclone sampler.[2] Six–week–old Swiss white mice were exposed to grain dust for either a daily exposure of 8 h, 5 d/week, for 16 weeks, or for three 8–h periods per day for 2 to 28 d. At each time interval four to six animals were exposed to grain dust, while four to six littermates served as controls. A total of 33 test and 34 control animals were studied.

Total dust levels in the chamber were highest after 30 min of exposure (5400 ± 700 mg/m^3; mean ± 1 SD), decreasing to 1900 ± 800 mg/m^3 at 4 h and to 1200 ± 500 mg/m^3 at 8 h. The respirable fraction was 4% of the total dust.

The lungs of control animals showed no more than two intraalveolar macrophages in a high–power microscopic field and usually less than that. Animals exposed to grain dust showed a consistent lesion minimal at 5 d and very obvious at 10, 20, and 28 d. A few polymorphs were present in the lumina of small bronchi. Alveolar sacs and alveoli showed clusters of eosinophilic macrophages with a few polymorphs (Plate 1**). This phenomenon was well marked in all exposed animals and best seen at 80 d. These cells were PAS positive and did not show stainable iron. There was no consistent change in bronchi and no evidence of fibrosis. The difference between control and test animals was clear–cut and was not quantitated.

Thick plastic sections showed the macrophages described above to contain inclusions, sometimes elongated, staining densely but not metachromatically with toluidine blue. By transmission electron microscopy, large macrophages containing prominent inclusions were seen throughout the course of the experiment. These were most numerous and easily found at 80 d and were studied in detail at the time. Apart from the usual ultrastructural features of macrophages, these cells contained numerous elongated inclusions, numbering a mean of 30 in 30 cell profiles studied (Plate 2**). The inclusions were up to 2 µm in length, and occasionally longer. An inclusion was bound by smooth membrane and comprised one to four, and occasionally more, electron–dense rods lying in a less dense matrix (Plate 2). At higher magnification dense striae were visible with a periodicity of 4 to 6.5 nm. These inclusions did not contain significant amounts of iron or other heavy metal by probe microanalysis.

Macrophages were uncommon within the alveoli of normal mice. After identification of inclusions in dust–exposed mice, a search was carried out for similar inclusions in the tissues of 12 control mice. Only in one block from one of these animals were similar inclusions identified. These inclusions were present in only four cells. It is likely that these inclusions are therefore a normal phenomenon appearing to excess, rather than a phenomenon specific to grain dust exposure.

The presence of an alveolar macrophage response in response to grain dust inhalation in exposed mice could have important implications if this phenomenon also occurs in occupationally exposed workers. One of the mechanisms in the pathogenesis of emphysema in smokers is that cigarette smoke leads not only to increased numbers of alveolar macrophages,[3] but also to increased production and release of proteolytic enzymes from these alveolar macrophages[4] which may lead to pulmonary autodigestion. By similar mechanisms, chronic grain dust exposure could lead to the development of emphysema, particularly in smoking grain workers[5] or in workers genetically deficient in α_1–antitrypsin.

The prominent macrophage inclusions are similar to those which have been demonstrated in mouse marrow macrophages.[6] Organized paracrystalline inclusions are a feature of macrophages in several species including man.[7] The elongated inclusions presently seen seem to be peculiar to the mouse and many either represent secretory inclusions[6] or stored noniron–containing red cell residue.[8] It seems rather more likely that they are secretory inclusions, since large numbers of red blood cells were not a feature of the process at any point and since no evidence of sequential breakdown of red blood cells was seen. It is quite clear that these inclusions are not in any way specifically induced by exposure to grain dust, since they are found very rarely and in small numbers in macrophages in the lungs of mice after nitrogen dioxide inhalation.[9]

* Previously published in expanded from (*J. Pathol.*, 136, 265, 1982).

** Plates 1 and 2 appear on pages 111 and 112.

REFERENCES

1. **Crider, W.L., Barkley, N.P., and Strung, A.A.,** Dry–powder aerosol dispersing device with long–time output stability, *Rev. Sci. Instrum.,* 39, 152, 1968.
2. **Beechmans, J.M.,** Analysis of the cyclone in a size selective aerosol sampler, in *Aerosol Measurement,* Lundgren, Ed., 1979, 56.
3. **Finley, T.N., Swenson, E.W., Curran, W.S., Huber, G.L., and Ladman, A.J.,** Bronchopulmonary lavage in normal subjects and patients with obstructive lung disease, *Ann. Intern. Med.,* 87, 784, 1980.
4. **Harris, J.O., Olsen, G.N., Castle, J.R., and Maloney, J.S.,** Comparison of proteolytic enzyme activity in pulmonary alveolar macrophages and blood leukocytes in smokers and non–smokers, *Am. Rev. Respir. Dis.,* 111, 579, 1975.
5. **Cotton, D.J., Graham, B.L., Li, K.Y.R., Froh, F., Barnett, G.D., and Dosman, J.A.,** Effects of grain dust exposure and smoking on respiratory symptoms and lung functions, *J. Occup. Med.,* 25, 131, 1983.
6. **Hudson, G.,** Crystalloid material in macrophages of mouse bone marrow, *Acta Anat.,* 71, 100, 1968.
7. **Carr, I. and Morris, P.,** The fine structure of human macrophage granules in sarcoidosis, *J. Pathol.,* 122, 29, 1977.
8. **Simon, G.T.,** Morphology in relation to function, in *The Reticuloendothelial System,* Vol. 1, Carr, L. and Dames, W. Th., Eds., Plenum Press, New York, 1980, 469.
9. **Ranga, V. and Kleinerman, J.,** Lung injury and repair in the blotchy mouse, effects of nitrogen dioxide inhalation, *Am. Rev. Respir. Dis.,* 123, 90, 1981.

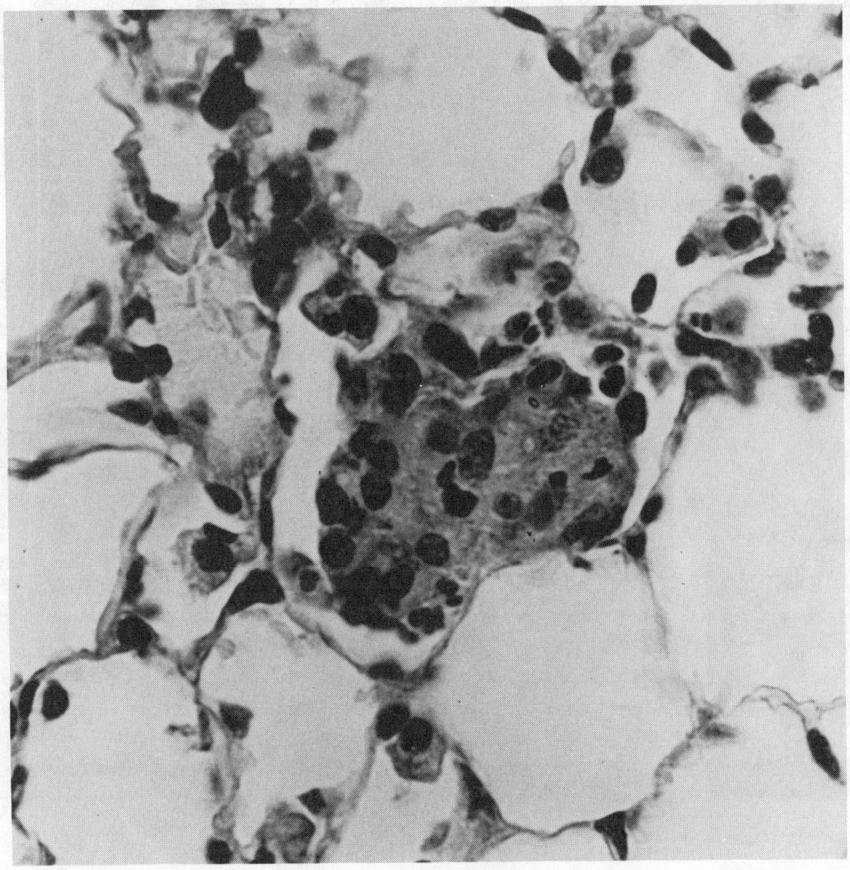

PLATE 1. Mouse lung exposed to grain dust for 28 d showing a cluster of macrophages in an alveolus. (Hematoxylin and eosin; magnification × 350.)

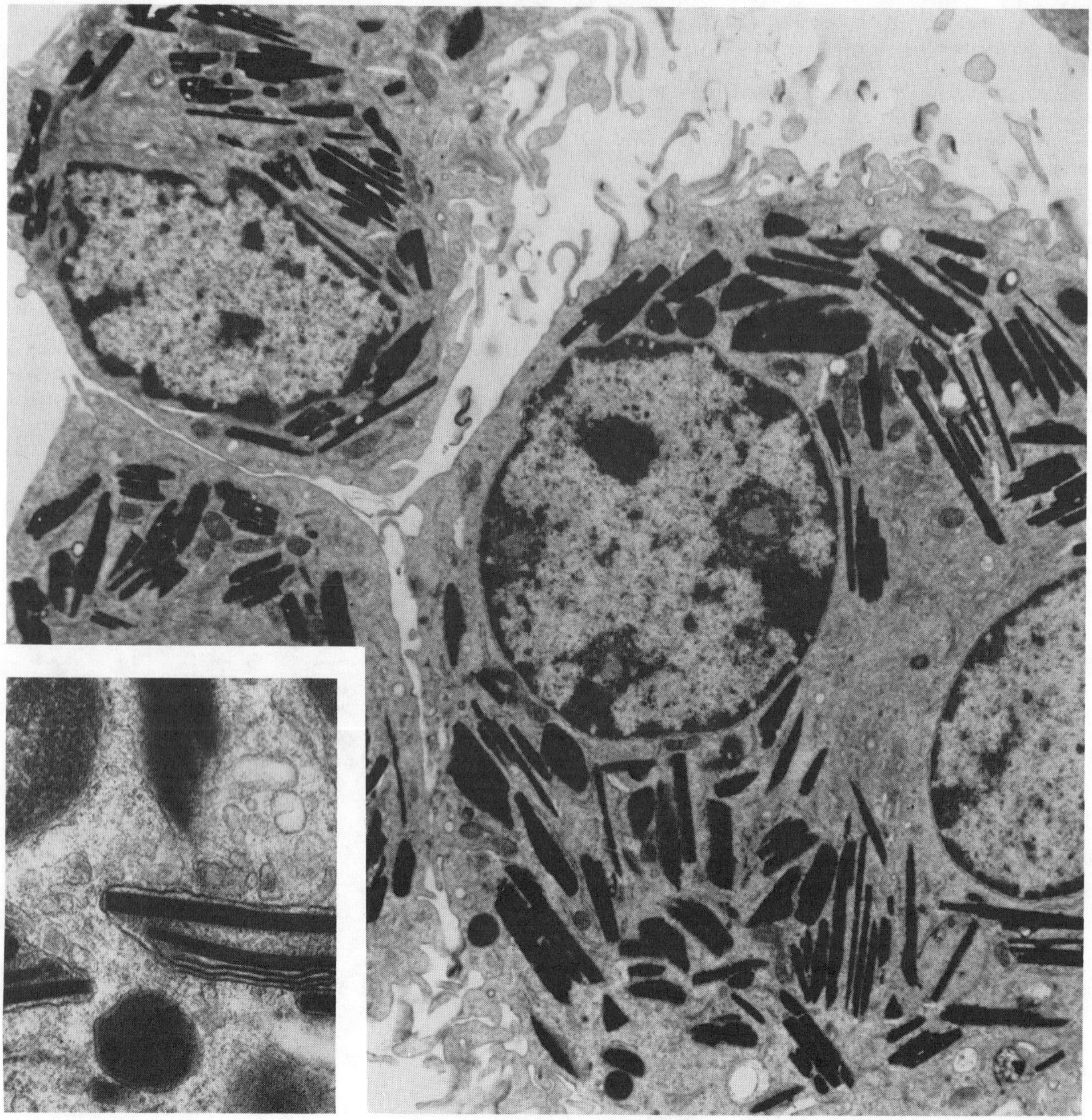

PLATE 2 A cluster of macrophages similar to those seen in Plate 1 showing numerous electron-dense inclusions. (Magnification × 8500.) (Inset) Detail of electron–dense inclusions. The inclusions are membrane bound and contain several elongated rods. (Magnification × 46,000.)

Cellular Immune Reactions to Grain Dust and Extracts of Grain Dusts

D. M. Lewis and S. A. Olenchock

INTRODUCTION

Exposures to grain dusts occur in a variety of agricultural settings ranging from the harvesting of the grain at the farm, to the storage and transport of grain, to the processing of grain for foodstuffs. The deleterious effects of grain dusts on lung function are well documented and have prompted some investigators to suggest that they should no longer be regarded as merely nuisance dusts.[1] We have been investigating the mechanisms by which grain dusts induce pulmonary reactions, with particular reference to their effects on the immune system. In previous studies, extracts of grain dusts have been shown to activate complement[2] and thereby generate the inflammatory peptides C3a and C5a. While complement activation can account for part of the inflammatory reaction seen following inhalation of grain dusts, this mechanism is not sufficient to explain all of the signs and symptoms of inflammation associated with such exposure.

Recent studies have suggested that certain substances, termed "polyclonal cell activators", may be important in the pathogenesis of inflammatory diseases of the lung.[3] Polyclonal cell activators are substances which can stimulate cells of the immune system by immunologically nonspecific mechanisms, resulting in the activation of much larger cell populations and, thus, the generation of greater amounts of inflammatory mediators than would occur through antigenic stimulation. It has been shown that inhalation of such substances can induce pulmonary damage.[4,5]

In light of these findings, we investigated whether extracts of grain dusts contain such polyclonal activators. Specifically, we tested the ability these extracts have to stimulate lymphocytes to proliferate and macrophages to produce interleukin 1 (IL–1) in culture.

RESULTS AND DISCUSSION

In order to determine if the grain would induce inflammatory reaction in rat lungs, we exposed rats to 1.0 mg spring wheat dust suspended in 0.25 ml saline by intratracheal instillation. Twenty–four hours after instillation of the grain dust suspension, we found that the total white cell count of the peripheral blood had not significantly changed, but that there was a statistically significant (p <0.001) increase in the relative number of neutrophils in the blood (Table 1). The cell population obtained by bronchoalveolar lavage showed even greater changes. As shown in Table 1, there was a significant increase in the percentage of neutrophils found in the lavage fluid. The results indicate that grain dust can induce an acute inflammatory reaction in the rat lung.

Aqueous extracts[2] of grain dust samples were tested for their mitogenic potential of inducing lymphocyte proliferation, using rat spleen cells. The spleen cells from normal healthy rats were cultured in RPMI–1640 media containing 10% fetal calf serum for 4 d. Extracts of the grain dust were added to the culture on day 1, and the proliferative response to the extracts was monitored. As shown in Figure 1, there was

TABLE 1
Cell Counts from Rats Exposed to a Suspension of Spring Wheat Dust by IT Instillation

	Blood cell counts	Lavage cell counts	
Treatment	Neutrophils (%)	Total	Neutrophils (%)
Spring wheat	40 ± 2.7[a]	44.0 ± 6[b]	55.4 ± 1.5[c]
Saline	26 ± 2.2	13.6 ± 1.5	3.6 ± 0.9

[a] Percent neutrophils in peripheral blood, mean ± SD, n = 5.
[b] Total number of cells obtained by pulmonary lavage, mean ± SD × 10^6, n = 10.
[c] Percent neutrophils in lavage samples, mean ± SD, n = 10.

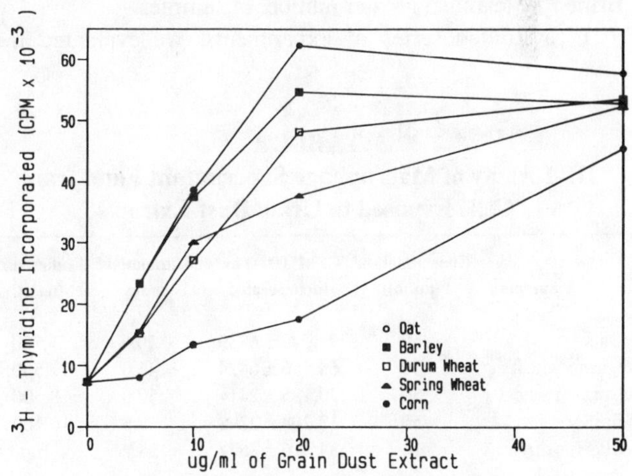

FIGURE 1. Concentration–response curves for mitogenic activity of aqueous extracts of airborne dusts for rat splenic mononuclear cells as measured by tritiated thymidine incorporation. Extracts were lyophilized, and the resulting residue dissolved in culture media at the concentrations indicated. Each concentration of the extract was tested in replicates of four, and the experiment repeated at least twice for each extract. Rat splenic mononuclear cells were recovered from a ficol–hypaque gradient, washed, and resuspended in RPMI–1640 media containing 10% fetal calf serum and antibiotics to a concentration of 6 × 10^6/ml. (Adapted from Olenchock, S. A., Lewis, D. M., and Mull, *Environ. Health Perspect.*, 66, 119, 1986.)

a concentration– dependent proliferative response seen with each of the extracts. Extracts of barley dust were the most potent, while corn dust extracts were the least potent. In all cases the responses to the grain dust extracts were significantly less than were obtained with known mitogens (concanavalin A and phytohemagglutinin) and were most consistent when cells were cultured at relatively high cell densities (6×10^6 cells per milliliter). These data suggest that grain dust extracts contain substances that are mitogenic for rat lymphocytes. The possibility exists that the responses observed were antigenic–, rather than mitogenic–induced responses. Randomly selected rat serum samples were negative for antibodies to the extract by counterimmunoelectrophoresis (data not shown), which would indicate that these animals had not been immunized to the grain dusts, but we cannot formally exclude the possibility that we observed an antigenically specific cellular immune reaction to the dust extracts.

To further investigate the effects of these extracts on the cellular immune system, alveolar macrophages from normal healthy rats were cultured with the extract, and the production of IL–1 was monitored. The macrophages were cultured for 24 h with the various grain dust extracts (50 μg/ml), and the culture fluid was tested for IL–1 content using a thymocyte proliferation assay.[6] All the extracts tested induced significant IL–1 production (Table 2). Because bacterial endotoxin is a known potent IL–1 stimulator and a contaminant of the grain dust extracts, the endotoxin content of the extracts was determined. There was no apparent correlation with the endotoxin content of the samples and the IL–1–inducing activity (Table 2), suggesting the extracts contain other substances that can induce IL–1 production. This observation needs to be confirmed by testing a greater number of samples.

In a limited series of experiments, we evaluated the IL–1–inducing activity of grain dust extracts *in vivo*. Rats were exposed to either spring wheat or oat dust extracts by intratracheal instillation, and pulmonary lavage was performed 24 h later. The lavage fluid was concentrated five– to tenfold and tested for IL–1 activity, but none of the samples showed any IL–1 activity. The cells obtained from the lavage were cultured overnight in media alone, and culture supernatant fluid was tested for IL–1 activity. As shown in Table 3, the macrophages obtained from rats exposed to the grain extracts *in vivo* did produce IL–1 when cultured *in vitro*.

The demonstration of the IL–1–inducing activity of these extracts both *in vitro* and *in vivo* is noteworthy because of the numerous signs and symptoms of inflammation that are mediated by IL–1. The effects mediated by IL–1 include fever induction,[6] sleep promotion,[7] neutrophil release from bone marrow,[8] fibroblast proliferation,[9] protein catabolism in muscle cells,[10] and acute– phase protein production by the liver.[11] Many of these effects are observed in grain fever.

TABLE 2
IL–1 Assay of Macrophage Supernatant Fluid from Cells Exposed to Grain Dust Extracts

Sample	Concentration[a] (μg/ml)	CPM[3] [H]–Thy[b] incorporated	Stimulation[c] index	Endotoxin[d] (ng/ml)
Oats	50	71,029 ± 6839	50.6	1,700
Spring wheat	50	64,556 ± 6424	46.0	11,500
Durham wheat	50	70,185 ± 2414	50.0	8,500
Barley	50	12,208 ± 2298	8.6	10,100
LPS (positive cont.)	10	54,316 ± 2674	38.7	nd
Media (negative cont.)	—	1,404 ± 120	—	nd

[a] Concentration of extract added to macrophage cultures expressed as μg/ml of dissolved solids.
[b] Counts per minute of tritiated thymidine incorporated by thymocytes exposed to macrophage culture supernatants. Each sample tested in replicates of four.
[c] Stimulation index is the fold increase over negative control.
[d] Endotoxin concentration in ng/ml as determined by the limulus lysate assay.

TABLE 3
IL–1 Production by Alveolar Macrophages Exposed to Grain Dust *In Vivo*

Sample	Number of cells[a] recovered × 10[6]	CPM[3] [H]–Thy[b] incorporated	Stiumulation index
Oats	10.0	3,221 ± 507	5.9
Spring wheat	13.0	8,810 ± 1,099	16.0
Saline	6.0	546 ± 99	1.1
Controls[d]			
Oats		20,437 ± 940	39.7
Spring wheat		18,586 ± 767	36.1
LPS (positive cont.)		19,860 ± 2,927	38.6
Media (negative cont.)		515 ± 92	—

[a] Average number of cells recovered by pulmonary lavage.
[b] Counts per minute of tritiated thymidine incorporated by thymocytes exposed to macrophage culture supernatants.
[c] Stimulation index is the fold increase over negative control.
[d] Control cultures are normal rat alveolar macrophages exposed to the agents *in vitro* for 24 h.

CONCLUSION

The results of this study provide evidence that extracts of grain dust contain substances which can stimulate the cells of the immune system in an immunologically nonspecific manner. These findings, together with the complement–activating potential of the grain dust extracts, indicate that grain dusts may stimulate inflammatory reactions by a variety of mechanisms. In addition, it appears that the various grain dusts differ in their inflammatory potential. Detailed investigation of the mechanisms by which these dusts stimulate an inflammatory reaction may provide a rational means of determining the inflammatory potential of a given grain dust, and thus an estimate of which dusts are most likely to cause pulmonary reaction in the workplace.

REFERENCES

1. **Chan–Yeung, M., Schulzer, M., MacLean, L., Dorken, E., Tan, F., Lam, S., Enarson, D., and Grzybowski, S.,** A followup study of the grain elevator workers in the Port of Vancouver, *Arch. Environ. Health,* 36, 75, 1981.
2. **Olenchock, S.A., Mull, J.C., and Major, P.C.,** Extracts of airborne grain dust activate alternative and classicial complement pathways, *Ann. Allergy,* 44, 23, 1980.
3. **Willoughby, W.F., Willoughby, J.B., and Gerberick, G.F.,** Polyclonal activators in pulmonary immune disease, *Clin. Rev. Allergy,* 3, 197, 1985.
4. **DeMaria, T.F. and Burrell, R.,** Effects of inhaled endotoxin–containing bacteria, *Environ. Res.,* 23, 87, 1980.
5. **Willoughby, W.F., Willoughby, J.B., Cantrell, R.B., and Wheelis, R.,** *In vivo* responses to inhaled proteins. II. Cell–mediated and immune complex mediated lung injury, *Lab. Invest.,* 40, 399, 1979.
6. **Lachman, L.B., Page, S.O., and Metzgar, R.S.,** Purification of human interleukin 1, *J. Supramol. Struct.,* 13, 457, 1980.
7. **Murphy, P.A., Simon, P.L., and Willoughby, W.F.,** Exogenous pyrogens made by rabbit peritoneal exudate cells are identical with lymphocyte– activating factors made by rabbit alveolar macrophages, *J. Immunol.,* 124, 2498, 1980.
8. **Krueger, J.M., Walter, J., Dinarello, C.A., Wolfe, S.M., and Chedid, L.,** Sleep–promoting effects of endogenous pyrogen (interleukin 1), *Am. J. Physiol.,* 246, R994, 1984.
9. **Kampschmidt, R.F.,** Leukocytic endogenous mediator/endogenous pryogen, in *The Physiologic and Metabolic Responses of the Host to Infection and Inflammation,* Powanda, M.C. and Canoico, P.G., Eds., Elsevier/North–Holland, Amsterdam, 1981, 55.
10. **Schmidt, J.A., Mizel, S.B., Cohen, D., and Green, I.,** Interleukin 1, a potent regulator of fibroblast proliferation, *J. Immunol.,* 128, 2177, 1982.
11. **Baracos, V., Rodemann, H.P., Dinarello, C.A., and Goldberg, A.L.,** Stimulation of muscle protein degradation and prostaglandin E_2 release by leukocyte pyrogen (interleukin 1), *N. Engl. J. Med.,* 308, 553, 1983.
12. **Selinger, M.J., McAdam, K.P., Kaplan, M.M., Sipe, J.D., Vogel, S.N., and Rosensteich, D.L.,** Monokine induced synthesis of serum amyloid A protein by hepatocytes, *Nature,* 285, 498, 1980.
13. **Olenchock, S.A., Lewis, D.M., and Mull,** *Environ. Health Perspect.,* 66, 119, 1986.

Activation of Complement by Silage

S. A. Olenchock, J. J. May, D. S. Pratt, D. M. Lewis, J. C. Mull, and L. Stallones

INTRODUCTION

The potential for adverse respiratory effects due to inhalation of airborne agricultural dusts is well documented. Respiratory or systemic responses associated with farm–related dusts occur worldwide in such diseases as hypersensitivity pneumonitis,[1,2] allergy,[3] asthma,[4,5] chronic bronchitis,[6] and grain fever syndrome.[7] The dusts related to these various adverse responses have been shown to activate complement as a potential disease mechanism that can account for many of the signs and symptoms observed in the workers. Farm–related products such as grain and grain dusts,[8–10] moldy hay dust,[11] rice,[12] sugarcane,[13] and microorganisms[14] activate complement, generating biologically active complement fragments. Consequently, complement activation may result in the initiation of or contribution to the pathophysiology related to agricultural dust–exposed workers at various stages of production, from the farm to the port grain terminal.

Recently, an acute febrile illness of short duration was identified in farmers who were unloading the material in their silos.[15] This entity is considered as an "organic dust toxic syndrome", and it is to the role of silage in this illness that this study is directed. Specifically, the interaction of silage with human serum complement is investigated in order to explore one disease mechanism which could be occurring in silo unloaders. This paper reports the *in vitro* activation of human serum complement by aqueous extracts of silage. Additionally, *in vivo* changes in hemolytic complement components C3 and C4 will be reported for farmers who unloaded silos as part of their normal work routine.

COMPLEMENT ACTIVATION *IN VITRO*

Extracts of each of the four different silage samples consumed normal human hemolytic complement component C3 in a dose–dependent fashion. Figure 1 illustrates the dose relationships for the silage extracts. As the amount of silage extract increased, greater consumption of hemolytic C3 was observed. There appeared to be a difference in the relative complement–activity rankings of the silages. For example, the silage from farm A showed the greatest C3 consumption at all four data points (37.9 to 85.1% consumption). At the opposite extreme, the extract of silage from farm D was the least active against C3, although it was still of considerable reactivity (0.0 to 46.0% consumption). Calculation of the relative complement–activity rankings was performed by probit analysis and the rankings were reported in terms of the concentrations which consume 50% of the available serum C3H50 U per milliliter. As suggested by Figure 1, the extract of silage from farm A was the most active. Only 9.97 μl of extract was required to consume 50% of the available C3H50 in 0.5 ml of pooled normal human serum. The activities of extracts from farms B, C, and D were 37.50, 46.33, and 144.52 μl/0.5 ml, respectively. Of interest, the Gram–negative bacterial endotoxin concentrations of the silage extracts ranked similarly. Silages from farms A and B contained the highest levels of endotoxin contamination, 31.35 and 31.97 ng/mg, respectively. The silage from farm C contained 6.32 ng/mg, and the sample from farm D contained 0.06 ng/mg. Examination of aliquots of the extract–treated sera by two–dimensional electrophoresis revealed that C3 was converted to the active fragment C3b in a dose–dependent fashion. Immunochemical conversion of C3 was observed as an increase in the C3b arc with corresponding decrease in the native C3 arc (data not shown). Treatment with only saline showed little or no nonspecific (background) conversion of C3 to the active C3b. When the normal human serum is pretreated with EDTA which blocks both classical and alternative complement pathway activation, no conversion of C3 to C3b is observed with any dose of extract. Prior chelation of the sera with EGTA, which blocks the classical pathway activation but does not block the Ca–independent activation of C3 by the alternative pathway, resulted in the conversion of

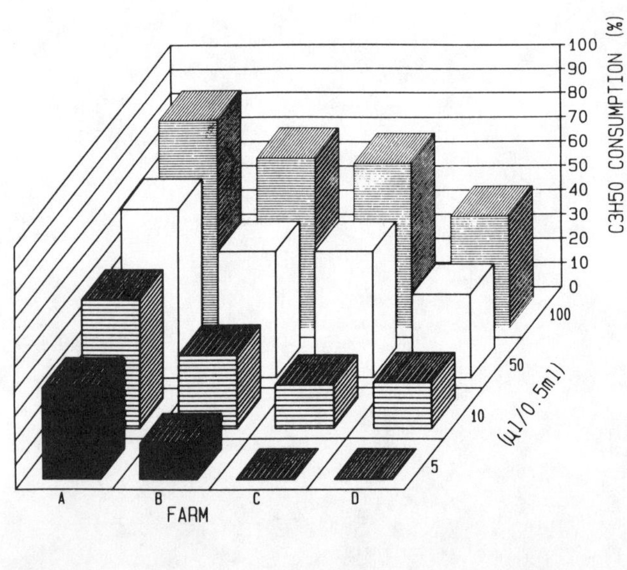

FIGURE 1. Consumption of hemolytic complement component C3 in pooled normal human serum which was reacted *in vitro* with different quantities of aqueous extracts of silage samples from four different farms.

C3 to C3b. Unchelated sera supported conversion as shown previously for extracts of airborne grain dusts.[9]

COMPLEMENT CHANGES *IN VIVO*

As a result of actual workday practices of unloading their silos, nine individuals were exposed to airborne silage dusts; three became ill and complained of headache, malaise, and low–grade fever as described elsewhere.[15] Table 1 shows the demographic characteristics of these individuals. Control individuals were the wives of the farmers, which explains the sex differences in the study populations. Additionally, the mean age of the controls was higher than the test group. Figure 2 shows the mean preexposure concentrations of hemolytic C3 in the sera of the study groups. Mean concentrations in the sera at 6 h postexposure are shown, also. No differences in the mean (±SEM) serum concentrations of the control group were noted between pre– and postexposure times (8500 ± 569 vs. 8500 ± 436 U/ml). However, the mean group concentration decreased (pre– to postexposure) approximately 500 C3H50 U/ml for both the whole exposed group (6933 ± 614 to 6422 ± 500 U/ml) and for those who became ill (7267 ± 939 to 6733 ± 809 U/ml). These changes, however, were not statistically significant at $p < 0.05$. Hemolytic C4 concentrations increased significantly ($p < 0.03$) from 49,767 ± 12,682 to 53133 ± 12971 U/ml during the 6–h period in the unexposed control group (Figure 3). During that same period, a nonsignificant decrease in hemolytic C4 was observed in the exposed group (33,689 ± 5011 to 31,244 ± 3937 U/ml). A significant ($p < 0.04$) decrease in the serum C4 levels was found in the ill group pre– to postexposure (27,167 ± 3815 to 24,167 ± 3396 U/ml).

TABLE 1
Description of Study Groups

Group	Number of participants	Sex	Mean age (years ± SE)
Total exposed	9	Male	33.4 ± 4.2
Ill[a]	3	Male	32.0 ± 4.4
Controls	4	Female	48.5 ± 4.4

[a] Ill subjects within the total exposed group.

DISCUSSION

Silage joins the list of other farm–related products such as grains, moldy hay, rice, sugarcane, and associated microorganisms which can activate human serum complement. Aqueous extracts of silage samples from four different farms in New York State consume human hemolytic C3 in a dose–dependent fashion *in vitro* Immunochemical studies show that native C3 is converted to active C3b, and that the alternative complement pathway is activated. Complement activation and the generation of active complement fragments can initiate or contribute to respiratory or systemic signs and symptoms of disease, either at the site of complement activation or as soluble mediators of inflammation.[16] Because complement components are present in human airways,[17–19] the potential exists for the direct interaction of inhaled silage dusts, represented in this study by aqueous extracts, and complement. Although the *in vitro* data suggest that the antibody–independent alternative complement pathway is activated by extracts of silage, the experimental design did not negate the potential role of classical pathway involvement as well. Gram–negative bacterial endotoxins, which can activate both the classical and alternative complement pathways,[20] were natural contaminants of the silage. Those samples which were the greatest activators of C3 (farms A and B) contained the highest concentrations of endotoxins. Likewise, the least complement–reactive sample (from farm D) contained the

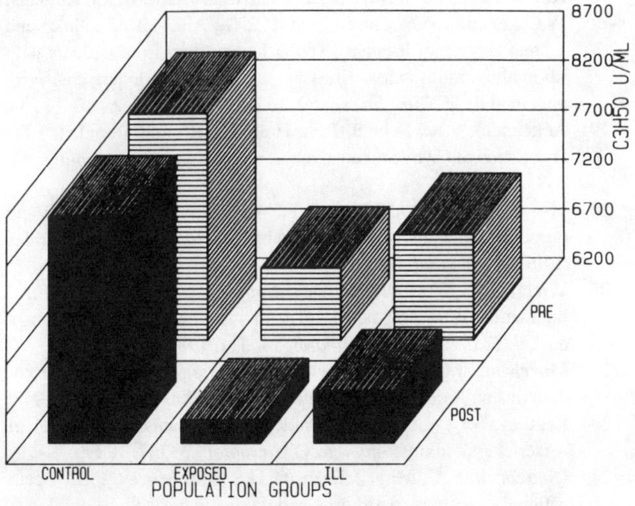

FIGURE 2. Levels of hemolytic complement component C3 in sera of control, exposed, and ill subjects before (pre–) and 6 h after (post–) silo unloading.

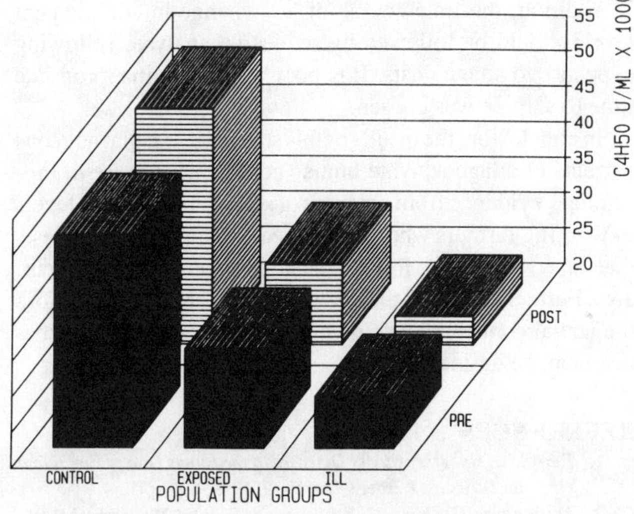

FIGURE 3. Concentrations of hemolytic complement component C4 in sera of control, exposed, and ill subjects before (pre–) and 6 h after (post–) silo unloading.

least amount of endotoxins. The potential contribution of endotoxins in the observed consumption of human C3 must be considered, therefore. Additionally, other as yet unidentified complement–activating components in the silage extracts are likely to be present as well.[21] In addition to their activity on human complement, however, endotoxins may induce other inflammatory reactions by interacting directly with both humoral and cellular components in the lung, blood, and other tissues.[22,23]

During the course of their regular job of unloading the farm silo, farmers are exposed to airborne dusts which contain Gram–negative bacterial endotoxins[24] as well as viable Gram–negative bacteria, fungi, and thermophilic bacteria.[25] When one considers the relatively uncharacterized nature of airborne silage dusts, the full extent of the toxic and irritating respiratory insults is not known. Individuals whom we studied were exposed to the unloading operations for approximately 45 min. Sera were collected both before and 6 h after the work. Although the study and control groups are admittedly small, some interesting observations can be made from the data. Changes in the concentrations of serum hemolytic C3 and C4 occurred in the test groups which did not occur in the control group. Serum C3 levels decreased as a result of exposure, but remained unchanged in the control group during the same time period. When considered in light of the *in vitro* data, one could suggest that C3 was activated *in vivo* by the alternative pathway. Control C4 levels increased during the 6–h period, but the C4 concentrations in the exposed and ill groups decreased. These data suggest that the classical complement pathway was activated as well. Additionally, the possibility exists that C4 is activated directly by a component in the silage.[9,21] Although statistical comparisons were made on the changes in C3 and C4 concentrations, it is apparent that the numbers in the study groups were not of sufficient size to form the basis of a definitive statement concerning the *in vivo* changes in serum complement after exposure to silage dusts. In addition, the time course of any change in complement levels should be followed by sequential analyses following exposure to silage dusts. It is possible that an inappropriate time to sample was chosen.

In conclusion, the results of this study suggest that aqueous extracts of silage activate human complement *in vitro*. Preliminary evidence from analyses of serum complement levels in working farmers who are exposed to silage dusts suggests that *in vivo* changes in functional complement may occur, also. Further studies of the *in vivo* changes are required before the actual contribution of this inflammatory disease mechanism can be defined.

REFERENCES

1. **Pepys, J.,** *Hypersensitivity Diseases of the Lungs Due to Fungi and Organic Dusts,* S. Karger, New York, 1969, 69.
2. **Dutkiewicz, J., Kus, L., Dutkiewicz, E., and Warren, C.P.W.,** Hypersensitivity pneumonitis in grain farmers due to sensitization to *Erwinia herbicola, Ann. Allergy,* 54, 65, 1985.
3. **Warren, P., Cherniack, R.M., and Tse, K.S.,** Hypersensitivity reactions to grain dust, *J. Allergy Clin. Immunol.,* 53, 139, 1974.
4. **Davies, R.J., Green, M., and McC. Schofield, N.,** Recurrent nocturnal asthma after exposure to grain dust, *Am. Rev. Respir. Dis.,* 114, 1011, 1976.
5. **Chan–Yeung, M., Wong, R., and MacLean, L.,** Respiratory abnormalities among grain elevator workers, *Chest,* 75, 461, 1979.
6. **Dosman, J.A., Cotton, D.J., Graham, B.L., Li, K.Y.R., Froh, F., and Barnett, G.D.,** Chronic bronchitis and decreased forced expiratory flow rates in nonsmoking grain workers, *Am. Rev. Respir. Dis.,* 121, 11, 1980.
7. **DoPico, G.A., Flaherty, D., Bhansali, P., and Chavaje, N.,** Grain fever syndrome induced by inhalation of airborne grain dust, *J. Allergy Clin. Immunol.,* 69, 435, 1982.
8. **Olenchock, S.A., Mull, J.C., Major, P.C., Gladish, M.E., Peach, M.J., III, Pearson, D.J., Elliott, J.A., and Mentech, M.S.,** Activation of the alternative pathway of complement by grain, *Clin. Allergy,* 8, 125, 1978.
9. **Olenchock, S.A., Mull, J.C., and Major, P.C.,** Extracts of airborne grain dusts activate alternative and classical complement pathways, *Ann. Allergy,* 44, 23, 1980.
10. **Broder, I. and McAvoy, D.,** Characterization of precipitation reaction between grain dust and normal human serum and comparison of reactive and nonreactive grain handlers, *Clin. Immunol. Immunopathol.,* 21, 141, 1981.
11. **Edwards, J.H.,** A quantitative study on the activation of the alternative pathway of complement by mouldy hay dust and thermophilic actinomycetes, *Clin. Allergy,* 6, 19, 1976.
12. **Strunk, R.C., Pinnas, J.L., John, T.J., Hansen, R.C., and Blazovich, J.L.,** Rice hypersensitivity associated with serum complement depression, *Clin. Allergy,* 8, 51, 1978.
13. **Li, X. and Vogt, W.,** Activation of the classical complement pathway by a polysaccharide from sugar cane, *Immunopharmacology,* 5, 31, 1982.
14. **Marx, J.J., Jr. and Flaherty, D.K.,** Activation of the complement sequence by extracts of bacteria and fungi associated with hypersensitivity pneumonitis, *J. Allergy Clin. Immunol.,* 57, 328, 1976.
15. **Pratt, D.S. and May, J.J.,** Feed–associated respiratory illness in farmers, *Arch. Environ. Health,* 39, 43, 1984.
16. **Brown, A.J. and Frank, M.M.,** Complement activation, *Immunol. Today,* 2, 129, 1981.
17. **Robertson, J., Caldwell, J.R., and Waldman, R.H.,** Evidence for the presence of components of the alternative (properdin) pathway of complement activation in respiratory secretions, *J. Immunol.,* 117, 900, 1976.
18. **Reynolds, H.Y., Fulmer, J.D., Kazmierowski, J.A., Roberts, W.C., Frank, M.M., and Crystal, R.G.,** Analysis of cellular and protein content of bronchoalveolar lavage fluid from patients with idiopathic pulmonary fibrosis and chronic hypersensitivity pneumonitis, *J. Clin. Invest.,* 59, 165, 1977.
19. **Ackerman, S.K., Friend, P.S., Hoidal, J.R., and Douglas, S.D.,** Production of C2 by human alveolar macrophages, *Immunology,* 35, 369, 1978.
20. **Morrison, D.C. and Kline, L.F.,** Activation of the classical and properdin pathways of complement by bacterial lipopolysaccharides (LPS), *J. Immunol.,* 118, 362, 1977.
21. **Wirtz, G.H., Westfall, S.S., Davidson, B.J., and Olenchock, S.A.,** Interactions of complement with an extract of airborne spring wheat dust, *J. Toxicol. Environ. Health,* 14, 511, 1984.
22. **Morrison, D.C. and Ulevitch, R.J.,** The effects of bacterial endotoxins on host mediation systems, *Am. J. Pathol.,* 93, 527, 1978.
23. **Bradley, S.G.,** Cellular and molecular mechanisms of action of bacterial endotoxins, *Annu. Rev. Microbiol.,* 33, 67, 1979.
24. **Olenchock, S.A., May, J.J., Pratt, D.S., and Morey, P.R.,** Occupational exposures to airborne endotoxins in agriculture, in *Detection of Bacterial Endotoxins with the Limulus Amebocyte Lysate Test,* Watson, S.W., Levin, J., and Novitsky, T.J., Eds., Alan R. Liss, New York, 1987, 475.
25. **Sorenson, W.G.,** personal communication.

Possible Sources of Complement–Fixing Activity in Grain Dust: Grain Dust May Contain a Tannin–Like Material Which Fixes Complement

D. Skea and I. Broder

INTRODUCTION

Grain elevator operators collectively develop respiratory symptoms and pulmonary function changes which are indicative of obstructive and restrictive ventilatory defects.[1–4] The available evidence suggests that the mechanism of this condition is nonimmunologic.[5–9] The possibility that grain dust is capable of inducing inflammation was suggested by the demonstration that grain dust activates the complement system in normal human serum *in vitro*.[10–13]

Grain dust is a heterogeneous mixture of particles derived from grain and contaminants such as fungal organisms, their metabolites and spores,[14] bacteria and bacterial products including endotoxins,[15] insects and mites and products and particles thereof[16], agricultural chemicals such as pesticides,[17] and such inorganic substances as soil and silica dust.[18] Many grain components as well as many of the contaminating components of grain dust have the potential to activate the complement system. It is the purpose of this paper to review the possible sources of complement–fixing activity in grain dust.

POSSIBLE SOURCES OF COMPLEMENT–FIXING ACTIVITY IN GRAIN

Massey et al.[19] demonstrated that the gluten fraction of wheat fixes complement. Furthermore, Olenchock et al.[20] and Broder and McAvoy[21] have demonstrated that extracts of crushed, clean kernels of grain have complement–fixing activity. These studies suggest that at least part of the complement–fixing activity of grain dust is due to a component or components of the grain itself. Possible sources of complement–fixing activity in grain include proteolytic enzymes, lectins, and tannins.

Proteolytic Enzymes

Since the complement proteins act by proteolytic cleavage, it is conceivable that proteolysis by an unrelated enzyme could activate the complement cascade if the cleavage occurred in the correct location.[22] Alternatively, random proteolytic digestion of the complement proteins would result in a loss of complement activity which, in a complement–fixation assay, might appear to be due to complement activation. The possible presence of proteolytic activity in grain dust is supported by the demonstration of a papain–like enzyme in extracts of moldy hay dust.[23] This activity, however, appeared to be distinct from the complement–fixing activity of the moldy hay dust in terms of molecular weight. Furthermore, in a different study, soluble wheat gluten which was boiled and autoclaved for the purpose of destroying enzyme activity still fixed complement.[19] These studies did not support a proteolytic or other enzyme activity as the source of complement–fixing activity in moldy hay dust or wheat gluten. Thus, the available evidence does not suggest that a similar activity is responsible for the complement– fixing activity of grain dust.

Lectins

Lectins are plant proteins or glycoproteins with two or more binding sites which interact specifically with carbohydrates.[24] The immunoglobulins and many of the complement proteins are glycoproteins. Thus, it is possible that lectins may activate complement by nonspecific aggregration of IgG or IgM, or by direct interaction with the carbohydrate moieties of the complement proteins. There is no evidence, however, that lectins fix complement. In fact, Broder and McAvoy[21] found that wheat germ lectin failed to fix complement at concentrations which caused IgG precipitation and hemagglutination. Thus, there is no evidence to support a lectin or lectins as the complement–fixing source in grain dust.

Other Proteins or Nucleic Acids

The complement–fixing activity of grain dust is resistant to degradation by trypsin, chymotrypsin and Staphylococcal protease, DNase, and RNase.[25] These results suggest that the complement–fixing activity of grain dust is not due to a protein.

Tannins

Tannins are a heterogeneous group of naturally occurring, high molecular weight (500 to 5000) plant compounds which are structurally based on a phenolic subunit.[26] Tannins are classified into two broad categories. The hydrolyzable tannins contain carbohydrate residues sterically linked to phenolic acids such as gallic or ellagic acid. The condensed tannins are comprised of polymerized flavanoids and are not broken down by hydrolysis. An important property of tannins is their ability to cross–link proteins. The presence of tannins has been demonstrated in sorghum grain and barley,[27] and, thus, it is possible that tannins may be present in grain dust.

Wirtz et al.[25] demonstrated that periodate oxidation decreased, but did not completely abolish the complement–fixing activity of grain dust extract. Oxidation by periodate results in the cleavage of carbon–carbon bonds between adjacent hydroxyl groups and, thus, is relatively specific for the degradation of carbohydrate.[28] This study suggests that carbohydrate plays a role in the fixation of complement by grain dust. On this basis, Wirtz et al.[25] postulated that the complement–fixing activity of grain dust was due to a polysaccharide. However, an alternate explanation is that this activity is due to hydrolyzable tannins. Wirtz et al.[25] further demonstrated by Sephadex G–200 gel filtration chromatography that the molecular weight of the complement–fixing component of grain dust was greater than 200,000. The molecular weight of tannins is less than this. Nevertheless, this does not exclude the possibility that hydrolyzable tannins are the complement–fixing component of grain dust. Tannins are known to form supramolecular complexes with proteins, and these complexes would not be dissociated under the conditions of gel filtration.

Studies conducted in our laboratory have demonstrated that the biological activities of grain dust extract resemble those of tannic acid. Tannic acid is a hydrolyzable tannin of molecular weight 1700 and is isolated from Near East gall nuts. Both grain dust extract and tannic acid fix complement and interact with IgG and albumin.[29] These activities are stable to heating at 100°C and are removed from solution by adsorption with hide powder,[29] which is a property of tannins.[30,31]

Methanol is the conventional solvent employed by agricultural biochemists in the initial extraction step of the purification of tannin from sorghum grain.[32] We found that grain dust extract prepared in methanol, dried, and reconstituted in aqueous solution fixed complement.[29] This activity was heat stable and was removed from solution by adsorption with hide powder.[31] Adsorption with PVP did not affect the complement–fixing activity of this preparation.[29]

Thus, there are similarities between tannic acid and grain dust extract, consistent with the possibility that the complement–fixing activity of grain dust is due to a tannin or tannins.

POSSIBLE SOURCES OF COMPLEMENT–FIXING ACTIVITY AMONG THE CONTAMINANTS OF GRAIN DUST

Fungal Organisms and Bacteria

Grain dust is known to contain many species of fungi and bacteria. Among these, *Micropolyspora faeni*, *Thermoactinomyces vulgaris,* and the staphylococcal product protein A are known to activate complement.[33,34] However, the similar complement–fixing activity of extracts prepared either from crude grain dust or from clean, crushed grain kernels makes it unlikely that such contaminants contribute importantly to the complement–fixing activity of grain dust.[21]

Gram–Negative Bacterial Endotoxins

Endotoxins are complex lipopolysaccharides which activate both the alternate and classical pathways of complement.[35] The presence of endotoxin has been demonstrated in grain dust,[15] but the levels were found not to correlate with that of complement–fixing activity.[36] Also, the major portion of the complement–fixing activity of grain dust extract was shown not to coelute with endotoxin on Sepharose 4B gel filtration chromatography.[25] Thus, the complement–fixing activity of grain dust cannot be easily explained by the presence of endotoxin.

CONCLUSIONS

There are several components of grain dust which may potentially contribute to its complement–fixing activity, including both grain constituents such as tannins,[29] and contaminating materials such as fungal organisms,[14] bacteria,[34] and endotoxin.[15,35] The available evidence suggests that a grain–derived tannin likely accounts for most of this activity.

REFERENCES

1. **Broder, I., Mintz, S., Silverman, F., et al.,** A comparison of respiratory parameters in grain handlers and civic outside workers in Thunder Bay, in *Occupational Pulmonary Disease: Focus on Grain Dust and Health,* Dosman, J. A. and Cotton, D. J., Eds., Academic Press, New York, 1980, 317.
2. **Becklake, M.R., Jodoin, G., Lefort, L., Rose, B., Mandl, M., and Fraser, R.G.,** A respiratory health study of grain handlers in St. Lawrence River ports, in *Occupational Pulmonary Disease: Focus on Grain Dust and Health,* Dosman, J. A. and Cotton, D. J., Eds., Academic Press, New York, 1980, 239.
3. **Chan–Yeung, M., Wong, R., and Maclean, L.,** Respiratory abnormalities among grain elevators workers, *Chest,* 75, 461, 1979.
4. **Corey, P., Hutcheon, M., Broder, I., et al.,** Grain elevator workers show work–related pulmonary function changes and dose–effect relationships with dust exposure, *Br. J. Ind. Med.,* 39, 330, 1982.
5. **Gerrard, J.W., Mink, J., Cheung, S., Tan, L., and Dosman, J.A.,** Non–smoking grain handlers in Saskatchewan: airways reactivity and atopic status, *J. Occup. Med.,* 21, 342, 1979.
6. **Gryzbowski, S., Chan–Yeung, M., and Ashley, M.J.,** Atopy and grain dust exposure, in *Occupational Pulmonary Disease: Focus on Grain Dust and Health,* Dosman, J. A. and Cotton, D. J., Eds., Academic Press, Academic Press Inc., New York, 1980, 257.
7. **Broder, I., Davies, G., Hutcheon, M., et al.,** Variables of pulmonary allergy and inflammation in grain elevator workers, *J. Occup. Med.,* 25, 43, 1983.
8. **DoPico, G.A., Reddan, W., Flaherty, D., et al.,** Respiratory abnormalities among grain handlers. A clinical, physiologic and immunologic study, *Am. Rev. Respir. Dis.,* 115, 915, 1977.
9. **Chan–Yeung, M., Schulzer, M., Maclean, L., Dorken, E., and Gryzbowski, S.,** Epidemiologic health survey of grain elevator workers in British Columbia, *Am. Rev. Respir. Dis.,* 121, 329, 1980.
10. **Olenchock, S.A., Mull, J.C., and Major, P.C.,** Conversion of C3 and consumption of hemolytic complement by extracts of airborne grain dust, *Chest,* 75 (Suppl.), 225, 1979.
11. **Olenchock, S.A., Mull, J.C., and Major, P.C.,** Activation of complement and generation of chemotactic factors by commerical grain extracts, *Am. Rev. Respir. Dis.,* 119 (Abstr.), 229, 1979.
12. **Olenchock, S.A., Mull, J.C., and Major, P.C.,** Extracts of airborne grain dusts activate alternative and classical complement pathways, *Ann. Allergy,* 44, 23, 1980.
13. **Olenchock, S.A., Mull, J.C., and Major, P.C.,** Complement activation by commercial allergen extracts of cereal grains, *Clin. Allergy,* 10, 395, 1980.
14. **Wallace, H.A.H.,** Fungi and other organisms associated with stored grain, in *Grain Storage: Part of a System,* Sinha, R.N. and Muir, W.E., Eds., AVI Publishing, Westport, CT, 1973, 71.

15. **Deluccia, A.J., Godshall, M.A., and Palmgreen, M.S.,** Gram negative bacterial endotoxins in grain elevator dusts, *Am. Ind. Hyg. Assoc. J.*, 45, 336, 1984.
16. **Dykstra, W.W.,** Rodents in stored grain, in *Grain Storage: Part of a System,* Sinha, R.N. and Muir, W.E., Eds., AVI Publishing, Westport, CT, 1973, 181.
17. **Bond, E.J.,** Chemical control of stored grain insects and mites, in *Grain Storage: Part of a System,* Sinha, R.N. and Muir, W.E., Eds., AVI Publishing, Westport, CT, 1973, 137.
18. **Williams, N., Skoulas, A., and Merriman, J.E.,** Exposure to grain dust. I. A survey of the effects, *J. Occup. Med.*, 6, 319, 1976.
19. **Massey, A., Capner, P.M., and Mowbray, J.F.,** Activation of the alternate pathway by gluten, *Immunology,* 33, 339, 1977.
20. **Olenchock, S.A., Mull, J.C., Major, P.C., et al.,** Activation of the alternate pathway of complement by grain, *Clin. Allergy,* 8, 125, 1978.
21. **Broder, I. and McAvoy, D.,** Characterization of precipitation between grain dust and normal human serum and comparison of reactive and non–reactive grain handlers, *Clin. Immunol. Immunopathol.*, 21, 141, 1981.
22. **Marx, J.J., Nanboren, M., and Moore, V.L.,** Activation of the complement cascade by a purified component of pigeon dropping extract, *Clin. Immunol. Immunopathol.*, 8, 487, 1977.
23. **Edwards, J.H.,** The effect of soluble material from moldy hay dust on the complement system, *Immunology,* 37, 91, 1979.
24. **Allen, A.K., Neuberger, A., and Sharon, N.,** The purification, composition and specificity of wheat–germ agglutinin, *Biochem. J.*, 131, 155, 1973.
25. **Wirtz, G.H. Westfall, S.S., Davidson, B.J., and Olenchock, S.A.,** Interactions of complement with an extract of airborne spring wheat dust, *J. Toxicol. Environ. Health,* 14, 511, 1984.
26. **Walker, J.R.L.,** *The Biology of Plant Phenolics,* Edward Arnold, London, 1975, 36.
27. **Bate–Smith, E.C. and Rasper, V.,** Tannins of grain sorghum, *J. Food Sci.*, 34, 203, 1969.
28. **Morrison, R.T. and Boyd, R.N.,** *Organic Chemistry,* 3rd ed., Allyn and Bacon, Newton, MA, 1976, 538.
29. **Skea, D.L.,** *Characterization of the Biological Activity of Grain Dust Relative to Harmful Effects on the Lung,* Thesis, University of Toronto, Toronto, Canada, 1985.
30. **Greifeneder, J.,** personal communication.
31. **Anderson, R.A. and Sowers, J.A.,** Optimum conditions for bonding of plant phenolics to insoluble polyvinylpyrrolidone, *Phytochemistry,* 7, 293, 1968.
32. **Hagerman, A.E. and Butler, L.G.,** Determination of tannin in tannin–protein precipitates, *J. Agric. Food Chem.*, 28, 944, 1980.
33. **Edwards, J.H.,** A quantitative study on the activation of the alternate pathway of complement by moldy hay dust and thermophillic actinomycetes, *Clin. Allergy,* 6, 19, 1976.
34. **Stalenheim, G. and Castesson, S.,** Protein A from *Staphylococcus aureus.* Conversion of complement factor C3 by aggregrates between IgG and protein A, *FEBS, Lett.*, 14, 79, 1971.
35. **Morrison, D.C. and Kline, L.F.,** Activation of the classical and properdin pathways of complement by bacterial lipopolysaccharides (LPS), *J. Immun.*, 118, 362, 1977.
36. **Olenchock, S.A. and Major, P.C.,** Immunological studies of grain dust, *J. Environ. Pathol.*, 2, 9, 1979.

Effect of Short–Term Exposure to Grain Dust in Workers and Volunteers

G. A. doPico

To study the physiological characteristics of the respiratory and systemic (grain fever) reaction to occupational grain dust exposure and the pathogenic role of complement activation, 12 subjects underwent short–term exposure to airborne grain dust in an environmental chamber. Our aim was to simulate an environment that the grain workers considered most likely to provoke the symptoms of grain fever. Exposures lasted 1 to 2 h depending on tolerance. The mean dust concentration reached 80 mg/m3, range 23 to 164 mg/m^3. The grain workers were exposed to a mean of 50 ± 29 mg/m3 and the controls to a mean concentration of 132 ± 34 mg/m^3.

Grain fever is a short–term flu–like syndrome that develops during and/or after occupational exposure to high concentrations of grain dust. Epidemiologic studies report grain fever in 6 to 32% of grain handlers.[1–4] We found that 32% (99 of 310) of grain workers[1] reported one or more episodes of grain fever in their work life. Although some workers reported only an isolated episode after their first heavy exposure, 16 workers related over 100 episodes in their work life. In another study, we recognize that facial warmth, fever, or a "chilly" sensation felt by workers during or after a work shift were seldom associated with fever, i.e., body temperature >37.8°C.[7]

We have also demonstrated by bronchoprovocation that solutions of airborne grain dust or durum wheat can provoke immediate and late asthmatic responses.[5] Grain dust contains a wide variety of organic constituents, e.g., grain, fungi, bacteria, and endotoxin, that could be potentially responsible for the workers' respiratory and systemic responses to grain dust. Because of its clinical resemblance to hypersensitivity pneumonitis, grain fever has been hypothesized to be a type III allergic reaction[8] to fungi contained in the dust. Physiologic or radiologic evidence of pneumonitis with grain fever has not been demonstrated. However, like the extracts of organisms associated with hypersensitivity pneumonitis,[9] airborne grain dust also activates complement by the alternate pathway in *in vitro* studies.[10]

We studied six asymptomatic healthy nonsmokers who had never been exposed to grain dust and six grain handlers with histories of recurrent work–related respiratory and systemic symptoms.

Atopic status and hypersensitivity to grain dust antigens were determined by skin prick test and serum precipitin tests. Airway hyperresponsiveness was determined using dilution increments of methacholine and PC_{20} was calculated.

On a control day without exposure to grain dust and before and at regular intervals after exposure, we documented symptoms, oral temperature, blood leukocyte count, total C_3 level, activation of complement by alternate and classical pathway, and pulmonary function tests. Static and dynamic compliance and pulmonary and respiratory resistance could be measured in 9 of the 12 subjects before, immediately after, and 3 to 4 h after exposure. The frequency (ΔRL 0 to 112 bpm) and gas dependence (ΔR_L air–$He0_2$) of resistance were measured by both the esophageal balloon method and the forced oscillations technique.

All subjects developed symptoms. The most common were flu–like illness, facial warmth, headache, cough, chest tightness, and throat burning (Table 1). Symptoms improved in 24 h in all but one who required treatment with bronchodilators and aspirin. Temperature above 37.8°C was noticed in four grain workers and two controls. The maximum elevation in temperature was noticed 2 to 4 h after exposure. Ten of the subjects developed leukocytosis with a relative increase in neutrophils reaching a maximum level of 11,700 to 24,300 4 to 8 h after challenge. In some, leukocytosis persisted for >24 h.

Only one subject showed a significant change in C3 complement (above 42 mg).

Nine of the 12 subjects developed airflow obstruction detectable by forced expired volume in 1 s (FEV_1) changes, but all had evidence of significantly decreased airflow by other tests such as FEF_{50} or FEF_{75} (forced expiratory flow) (Table 1).

The airways response occurred in subjects with and without preexisting airways obstruction, with or without airway hyperresponsiveness, as well as in atopic and nonatopic subjects, and in skin reactors and nonreactors to grain dust antigens.

FEV_1 and forced vital capacity (FVC) % change (mean ± SD) from baseline during a control day and on a dust challenge day in the 12 subjects are shown in Figure 1. There were significant changes immediately after removal from the chamber and 2 and 4 h later. Greater decrements were seen in FEF50, FEF_{75}, and $FEF_{25–75}$. The diffusing capacity increased after exposure.

The pulmonary resistance increased by 140 ± 38%. FRC showed a small but significant mean increase of 18% over baseline. Static compliance did not change significantly, whereas dynamic compliance did (–21%) (Figure 2). The decrement in FEV_1, FEF_{50}, and FEF_{75} from the control day were 16, 26, and 21%, respectively, after exposure with a tendency to return to baseline, but remaining lower than the baseline 2 or 3 h after exposure.

Immediately after exposure, compliance became fre-

TABLE 1
Results of Selected Parameters

No.	Skin tests: CAA	Skin tests: Grain dust	Respirable dust level (mg/m^3)	Temp (C°): Pre	Temp (C°): Post	WBC n × 1000: Pre	WBC n × 1000: Post	Largest postchallenge percent change from baseline: $FEV_{1.0}$	FVC	MMF	V_{max50}	V_{max75}	DLCO
							Grain Workers						
I	–	–	86.7	36.8	38.0[a]	7.7	14.3[a]	–3	–6	4	–25[a]	–6	+14
II	–	–	44	36.7	37.2	6.4	11.7[a]	–35[a]	–73[a]	–55[a]	–40[a]	–12	–5
III	–	–	86.7	36.6	39.0[a]	5.8	24.3[a]	–12	–3	–31[a]	–29[a]	–41[a]	+16
IV	–	–	23	36.8	38.4[a]	6.2	21.0[a]	–27[a]	–16	–43[a]	–29[a]	–51[a]	–17[b]
V	+	+	23	36.3	36.9	5.3	8.6	–27[a]	–29[a]	–55[a]	–36[a]	–50[a]	–21[b]
VI	+	–	38.6	36.8	38.9[a]	8.5	17.0[a]	–25[a]	–15	–40[a]	–44[a]	–38[a]	+17
							Controls						
VII	+	–	164	36.7	36.4	7.1	8.8	–23[a]	–11	–36[a]	–37[a]	–35[a]	+7
VIII	–	–	143	36.7	37.8[a]	5.7	24.3[a]	–37[a]	–6	–33[a]	–37[a]	–35[a]	–2
IX	+	–	90.1	37.0	37.2	9.9	19.0[a]	–27[a]	–13[a]	–58[a]	–54[a]	–53[a]	+14
X	+	+	164	36.8	36.9	6.8	14.6[a]	–11	–11	–25[a]	–29[a]	–21	–13
XI	+	–	143	36.9	38.4[a]	5.5	21.2[a]	–23[a]	–22[a]	–18	–10	–8	–6
XII	–	–	90.1	36.7	37.4	7.4	22.0[a]	50[a]	–39[a]	–70[a]	–67[a]	26[a]	+15

Note: CAA = common allergens; + = atopy = two or more positive reactions to ragweed, grass, tree, alternaria, cat hair, or feathers; airborne grain dust extract; + = wheal ≥3 mm; temp = temperature; WBC = leukocyte count.

[a] Considered significantly changed from baseline.
[b] Similar decrement measured on control day.

quency dependent (–41 ± 8% ΔC_{DYN} 0 to 112 bpm) and the resistance showed a greater gas density dependence (–31 ± 5% ΔR_L air–HeO_2) when compared to baseline, as is expected with actual airways constriction. Two or 3 h later we found a significant decrease in the resistance from 140 to 61% of baseline and a loss of gas density dependence of resistance (–10% ΔRL air–HeO_2), but a persistence in the frequency dependence of compliance (–34 ± 10% ΔCD_{YN} 0 to 112 bpm).

In summary, these studies demonstrate that inhalation of a high concentration of grain dust for 1 to 3 h can induce the grain fever syndrome, leukocytosis, respiratory symptoms, and airways constriction in both previously exposed grain workers and previously unexposed controls.

Airways constriction was predominantly central immediately after an acute exposure, as indicated by a fall in FEV_1, FEF_{50}, FEF_{75}, an increase in FRC and pulmonary resistance, the development of frequency dependence of compliance, and of gas density dependence of pulmonary resistance with no change in the static compliance.

With time, the site of constriction appeared to change to more peripheral sites as suggested by the persistence of frequency, persistence of dynamic compliance, loss of gas density dependence of resistance, and decrease in the magnitude of the resistance elevation.

At these high dust concentrations, grain fever and respiratory symptoms occur independent of previous exposure history, atopic status, preexisting airways obstruction, or airway hyperresponsiveness.

There was no evidence of allergic alveolitis, complement activation, or relation to the presence of serum precipitins.

Grain fever syndrome is probably due to toxins contained in grain dust and/or nonspecific mediator release after interaction between grain dust and lung cells. No evidence to support the type III allergic reaction hypothesis was found.

At high dust concentrations the airways constriction seemed to be triggered by an irritative or inflammatory reaction to components of grain dust, including toxins. Previous allergic sensitization does not appear to be playing a major role.

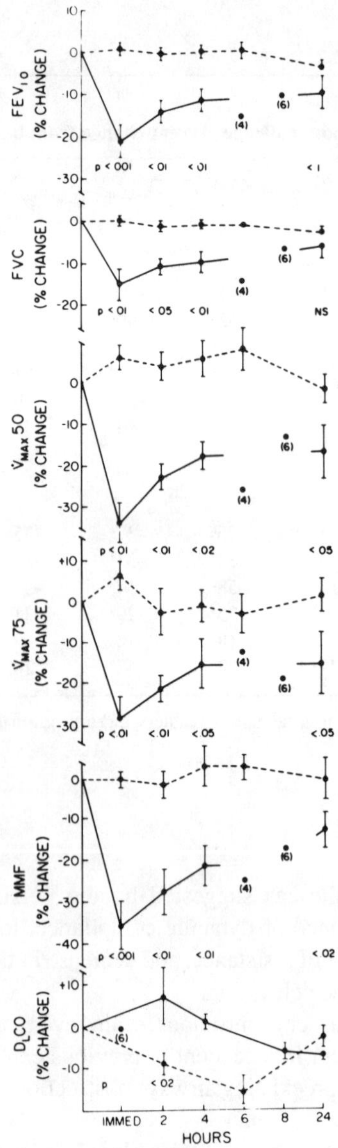

FIGURE 1. Percent change in lung function (vertical axis) vs. time (horizontal axis) on control day (dotted lines) and following grain dust exposure (solid lines).

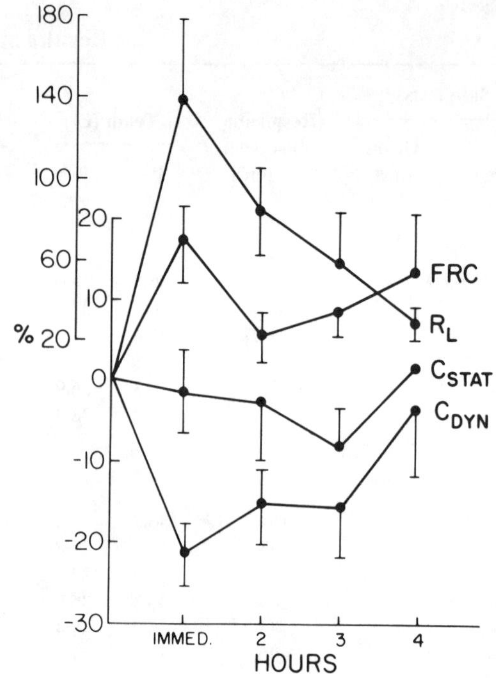

FIGURE 2. Percent change in pulmonary resistance (vertical axis top scale) and FRC, dynamic compliance, and static compliance (vertical axis lower scale) vs. time (horizontal axis) following grain dust exposure.

REFERENCES

1. **doPico, G.A., Reddan, W., Tsiatis, A., Peters, M.E., and Rankin, J.,** Epidemiological study of clinical and physiological parameters in grain handlers of northern United States, *Am. Rev. Respir. Dis.*, 130, 759, 1984.
2. **Williams, N., Skoulas, A., and Merriman, J.E.,** Exposure to grain dust, I. Survey of the effects, *J. Occup. Med.*, 6, 319, 1964.
3. **doPico, G.A., Reddan, W., Flaherty, D., Tsiatis, A., Peters, M., Rao, P., and Rankin, J.,** Respiratory abnormalities among grain handlers, *Am. Rev. Respir. Dis.*, 225, 915, 1977.
4. **Becklake, M.R.,** Grain dust and health: state of the art, in *Occupational Pulmonary Disease: Focus on Grain Dust and Health,* Dosman, J.A., and Cotton, D.J., Eds., Academic Press, New York, 1980, 189.
5. **doPico, G.A., Jacobs, S., Flaherty, D., and Rankin, J.,** Pulmonary reaction to durum wheat—a constituent of grain dust, *Chest,* 81, 55, 1982.
6. **doPico, G.A., Flaherty, D., Bhansali, P., and Chavaje, N.,** Grain fever induced by inhalation challenge—clinico–physiological and immunological characteristics, *J. Allergy Clin. Immunol.*, 69, 435, 1982.
7. **doPico, G.A., Reddan, W., Anderson, S., Flaherty, D., and Smalley, E.,** Acute effects of grain dust exposure during a work shift, *Am. Rev. Respir. Dis.*, 128, 399, 1983.
8. **Cotton, D.J. and Dosman, J.A.,** Grain dust and health. I. Host factors, *Ann. Intern. Med.*, 88, 840, 1978.
9. **Marx, J. and Flaherty, D.K.,** Activation of the complement sequency by extracts of bacteria and fungi associated with hypersensitivity pneumonitis, *J. Allergy Clin. Immunol.*, 57, 328, 1976.
10. **Olenchock, S.A., Mull, J.C., Gladish, M.E., Peach, M.S., III, and Major, P.C.,** Airborne grain dust activation of complement, in *Occupational Pulmonary Disease: Focus on Grain Dust and Health,* Dosman, J.A., and Cotton, D.J., Eds., Academic Press, New York, 1980, 263.

Mycotoxins in Grains and Dust

D. Abramson

Fungi impact upon our health and welfare in both beneficial and harmful ways. Commercial fungal fermentation processes are efficient and economical sources of antibiotics, other medicinal products, and various foodstuffs. In contrast, other fungal processes have been responsible for widespread illness and death in human populations,[1] as well as for occupational health problems. Examples of fungal–related disease include hypersensitivity reactions, infections (mycoses), and mycotoxicoses. Hypersensitivity reactions include farmer's lung disease (extrinsic alveolitis). Mycoses result from the invasion of the tissues by living fungi; examples include histoplasmosis and candidiasis. Mycotoxicoses, on the other hand, arise from the toxicity of certain metabolites (mycotoxins) produced by fungi; examples include ergotism, yellow rice disease, alimentary toxic aleukia, and aflatoxicosis.

Toxigenic fungi are prolific and widespread in the air and soil throughout the world. Consequently, many field crops, stored products, and agricultural commodities already contain spores. When environmental conditions are favorable, certain species of fungus will start growing, and some strains will produce mycotoxins.

When field crops are subject to preharvest infection by toxigenic fungi, mycotoxin formation is difficult to prevent. In certain bioclimatic zones, the invasive action of toxigenic field fungi is facilitated by sustained favorable conditions of temperature and moisture. In the future, this situation may be improved by breeding resistant cultivars, or by introducing new crops into the affected locales. In many agricultural areas, however, mycotoxin contamination is usually absent before storage, and arises from the action of toxigenic postharvest fungi. Controlled storage atmospheres and antifungal chemical agents have been investigated as methods of mycotoxin prevention, but it appears that mycotoxins can be minimized economically, mainly by proper storage practices based upon an understanding of storage ecosystems.[2]

In storage, crop spoilage and fungal growth result from the interaction of several factors. At the onset of storage, crops are infected by spores from other grain and dockage, as well as from the air. Mechanical damage due to harvesting or insect action disrupts the outer seed coat and facilitates infection. Insects, through their movements, may also act as vectors, spreading the inoculum. Fungal growth follows, depending on available oxygen, moisture, heat, and time. Heat and moisture may be augmented by storage–insect presence, and by the metabolism of the fungi themselves, starting a cyclical process that often results in so–called hot spots. Sufficient moisture to start this process may come from rain seepage, melting of wind–driving snow, or moisture migration and condensation on the walls of the storage structure. The subsequent formation of the mycotoxin(s) depends on the presence and predominance of toxigenic fungal strains, sufficient heat for further fungal development, a substrate of suitable chemical composition, and enough time for fungal metabolism to enter into its toxin–producing phase.

Mycotoxins can remain in the substrate long after their producing fungi have died. Most mycotoxins are not chemically labile, and many are heat stable up to their melting points. Methods of chemical decontamination have been studied;[3] ammonia, for example, has shown promise in the detoxification of grain containing aflatoxins. More aggressive chemical agents have been tried with less success, including chlorine, ethylene oxide, ozone and sulfur dioxide gases, and various alkaline and oxidizing liquids. Physical methods of decontamination, such as flotation, heating, radiation, and milling, have also been investigated.

The biological effects of the mycotoxins have been extensively investigated. The discovery in the early 1960s of the hepatotoxic aflatoxins, and of their carcinogenicity and widespread occurrence, has resulted in their being the most intensively studied. The LD_{50} values vary with species, age, sex, and route, but for comparative purposes are in the range of 7.2 to 17.9 mg/kg p.o. in rats.[4] The aflatoxins are rare in Canadian agriculture; of greater relevance are the nephrotoxins ochratoxin A and citrinin, the hepatotoxin and carcinogen[5] sterigmatocystin, and the trichothecenes vomitoxin (deoxynivalenol), diacetoxysciropenol, and T–2 toxin (Table 1). The published LD_{50} values[4] for these mycotoxins depend on species and route: ochratoxin A 22 mg/kg, rat p.o.; sterigmatocystin 166 mg/kg, rat p.o. and 60 mg/kg, i.p.; citrinin 67 mg/kg, rat i.p. and 35 mg/kg, mouse i.p.; vomitoxin 70 mg/kg, mouse i.p.; diacetoxyscirpenol 23 mg/kg, mouse i.p.; and T–2 toxin 4 mg/kg, rat p.o. and 3 mg/kg, mouse i.p.

The evidence of the effects of mycotoxins is necessarily indirect in most studies of human mycotoxicosis.[1,6] Although certain ergot alkaloids have been characterized as producers of ergotism, other causative associations are less definite. For example, T–2 toxin and citreoviridin have been linked to past outbreaks of alimentary toxic aleukia and yellow rice disease only by retrospective inference. More recently, attempts have been made to link dietary ochratoxin content to the occurrence of Balkan endemic nephropathy.[7] In other studies, dietary aflatoxins have been implicated as hepatocarcinogens in Africa and southeast Asia, and have been associated with the occurrence of Reye's syndrome.[1]

Although evidence of hazards associated with inhalation of mycotoxins is scarce, three reports in the literature impli-

TABLE 1
Mycotoxins and Associated Fungi in Western Canadian Grain

Commodity	Mycotoxin (max levels, μg/kg)	Associated fungi[a]	Sample source or features	History of storage	Ref.
Wheat	Ochratoxin A (27,000) Citrinin (80,000)	*P. viridicatum* *P. cyclopium*	Associated with human lung problems	Unknown, but stored damp	13
Rye	Ochratoxin A (720) Citrinin (960)	*P. viridicatum*	Associated with human lung problems	Unknown, but stored damp	13
Oat	Ochratoxin A (3,000) Citrinin (10,000)	*P. viridicatum*	Associated with human lung problems	Unknown, but stored damp	13
Wheat and wild oats	Ochratoxin A (70)	*P. viridicatum* *P. cyclopium*	Associated with human lung problems	Unknown, but stored damp	13
Oats and barley	Ochratoxin A (22,000) Citrinin (60,000)	*P. viridicatum*	Associated with human lung problems	Unknown, but stored damp	13
Wheat	Sterigmatocystin (300)	*P. viridicatum* *A. versicolor*	Associated with human lung problems	Unknown, but stored damp	13
Wheat	Ochratoxin A (700)	*P. cyclopium*	Granary storage study	21% MC, 20 weeks	14
Barley	Ochratoxin A (3,900)	*P. cyclopium*	Granary storage study	21% MC, 20 weeks	14
Oats and barley	T–2 toxin (2,500) Diacetoxyscirpenol (800) Ochratoxin A (380)	*P. cyclopium* *Fusarium* spp.	Mycotoxicosis	Bin inundated with rain seepage	15
Barley	Ochratoxin A (48)	*P. cyclopium*	Mycotoxicosis	High moisture storage	15
Wheat oats and barley	Ochratoxin A (5,900)	*P. cyclopium* *A. ochraceus*	Moldiness	Unknown	15
Barley	Sterigmatocystin (800)	*A. versicolor*	Granary storage study	16% MC, 66 weeks	16
Barley	Sterigmatocystin (3,450) Ochratoxin A (240)	*P. cyclopium* *A. versicolor*	Granary storage study	20% MC, 66 weeks	16
Corn	Ochratoxin A (3,600)	*P. cyclopium*	Granary storage study	21% MC, 8 weeks	17
Wheat	Citrinin (4,400) Ochratoxin A (1,100)	*P. cyclopium*	Mycotoxicosis	Unknown, but stored damp	18
Barley	Ochratoxin A (57)	*P. cyclopium*	Mycotoxicosis	Unknown	18
Wheat and barley	Vomitoxin (95)	*Fusarium* spp.	Mycotoxicosis	Unknown	18
Wheat	Vomitoxin (12,600)	*Fusarium* spp.	Mycotoxicosis	Unknown	19

[a] *Aspergillus ochraceus* Wilhelm; *A. versicolor* (Vuill.) Tiraboschi; *Fusarium* spp. not determined; *Penicillium cyclopium* = *P. verrucosum* var. *cyclopium* (Westling) Samson, Stolk & Hadlok; *P. viridicatum* = *P. verrucosum* (Dierckx) var. *verrucosum*.

cate aflatoxins as a hazard in the dust form.[8] In the first occurrence, a chemical engineer, aged 68, contracted bronchial carcinoma after working for 3 months on methods for sterilizing aflatoxin–contaminated peanut meals. Upon his death some months later, thin–layer chromatography indicated what appeared to be aflatoxin B_1 in samples of lung tissue. In a second case, a biochemistry professor and a graduate student both contracted colon cancer after purifying aflatoxins by thin–layer chromatography. The illness was linked to inhalation of aflatoxin–containing silica gel dust scraped off the thin–layer plates in an unventilated environment. In a third instance, an epidemiological study was undertaken at a peanut–cleaning and oil–pressing operation. Air samples taken at the site showed dust levels of 250 to 410 μg/kg (ppb) aflatoxins. Although the sample size was small (60 to 70) and the workers admittedly consumed peanuts on the job, the rate of multiple cancer and liver cancer disease taken together was more than three times that of the matched control group.

In the light of these implications, it is not surprising that studies on mycotoxins in grain dust so far have focused on the aflatoxins. Some of these results are summarized in Table 2. Three studies were undertaken to investigate the potential hazard from inhalation of aflatoxins in airborne dust to agricultural workers, livestock farmers, and elevator operators while handling contaminated corn.[8–10] The various air–sampling devices included high–volume total samplers, battery–powered personal samplers, and four–stage sampling particle separators. Chemical analysis methods were modified to accommodate reduced sample sizes. Maximum aflatoxin levels of 52,200 μg/kg dust, corresponding to 1020 ng/m^3 air, were observed in the personal sampler of a combine operator standing outside the combine cab and downwind during transfer of contaminated corn from combine to truck. This dust was thought to have a high conidia content; previous aflatoxin assays indicated levels in *Aspergillus flavus* conidia as high as 152,000 μg/kg. Although no consistent relationships were found between the aflatoxin concentration in the airborne dust and bulk corn, nor between the aflatoxin concentration and different particle sizes, it is evident that contaminated airborne dust is associated with the contaminated commodity.

TABLE 2
Levels of Aflatoxins in Corn and Associated Airborne Dusts during Harvesting and Handling

Operation	Alfatoxins assayed	Aflatoxin concentration[a] (μg/kg) (ppb) Dust	Corn	Ref.
Corn unbinning and rebinning	B_1	10—182	232	8
	B_2	1—42	18	—
Pouring ground corn through divider	B_1	160—4,380	—	8
	B_2	15—548	—	—
	$B_1 + B_2$	—	2,250	
Harvest				
Georgia farm #1	$B_1 + B_2$	ND—124	18	9
Georgia farm #3	$B_1 + B_2$	3—11	406	9
Grain elevator	$B_1 + B_2$	ND—543	ND—388	9
Harvest, Georgia farm	Total	281—52,200	1,640	10
Grinding corn and feeding to livestock	Total	ND—43,700	1,240—1,850	10
Unloading farm trucks at elevator	Total	ND—3,270	ND—288	10
Unloading rail cars and trucks from elevator	Total	ND—2,430	ND—273	10

[a] ND = none detected by assay procedure specified.

Although aflatoxin levels were found to be high in some locations, the exposure to aflatoxins would not last the entire day.[11] The estimated inhaled volume of air during an 8–h work shift would not consequently be very useful in calculating the daily exposure. In addition, aflatoxin levels in the grain being handled would vary throughout the day, and from year to year in the crop. Furthermore, constant exposure of the workers to the aflatoxin–contaminated dusts would be unlikely, as most are seasonal workers. Despite these mitigating factors, the issue and use of comfortable, effective dust masks are highly recommended when handling contaminated grain.

Little is known regarding the absorption rates or biochemical activation reactions of aflatoxins and other mycotoxins in lung tissue. Ingestion of mycotoxins in the diet, and subsequent dilution in the stomach with other foods presents a lower concentration of the toxin than found at sites in the respiratory system after inhalation of contaminated dusts.[8] Furthermore, the distribution of mycotoxins among airborne particles is probably not uniform, and dusts of high–average concentration would present some extremely contaminated particles at deposition sites. The risks from inhalation are therefore quite different from the risks of ingestion, and accurate assessment of the former would not be possible solely on toxicological evidence of the latter.

In western Canadian grain, ochratoxin A, sterigmatocystin, and citrinin appear to be the main mycotoxin contaminants (Table 1), although vomitoxin has recently appeared in isolated wheat– and corn–growing areas. Grain samples from storage bins containing ochratoxin A, sterigmatocystin, and citrinin have been associated with lung problems in farmers and elevator operators.[13] Before the hazards from inhalation of these mycotoxins in grain dust can be determined, investigations in at least three areas appear warranted: first, measurement of these mycotoxins in samples of airborne dust generated during the handling of contaminated grain; second, animal inhalation studies with contaminated dusts; and third, studies on respiratory routes of exposure to these mycotoxins. Meanwhile, grain contaminated with mycotoxins should be handled with caution.

REFERENCES

1. **Wyllie, T.D. and Morehouse, L.G., Eds.,** *Mycotoxic Fungi, Mycotoxins, Mycotoxicoses,* Vol. 3, Marcel Dekker, New York, 1978.
2. **Abramson, D. and Mills, J.T.,** Mycotoxin production during storage, in *Mycotoxins: A Canadian Perspective,* Scott, P.M., Trenholm, H.L., and Sutton, M.D., Eds., NRCC Publications, Ottawa, 1985, 69.
3. **Young, J.C.,** Decontamination, in *Mycotoxins: A Canadian Perspective,* Scott, P.M., Trenholm, H.L., and Sutton, M.D., Eds., NRCC Publications, Ottawa, 1985, 69.
4. **Cole, R.J. and Cox, R.H.,** *Handbook of Toxic Fungal Metabolites,* Academic Press, Toronto, 1981.
5. **Purchase, I.F.H. and Van Der Watt, J.J.,** Carcinogenicity of sterigmatocystin, *Food Cosmet. Toxicol.,* 8, 289, 1970.
6. **Shank, R.C., Ed.,** *Mycotoxins and N–Nitroso Compounds: Environmental Risks,* Vol. 1 and 2, CRC Press, Boca Raton, FL, 1981.
7. **Steyn, P.S.,** Ochratoxins and related dihydroisocoumarins, in *Mycotoxins—Production, Isolation, Separation and Purification,* Betina, V., Ed., Elsevier, Amsterdam, 1984, 183.
8. **Burg, W.R., Shotwell, O.L., and Salzman, B.E.,** Measurements of airborne aflatoxins during the handling of contaminated corn, *Am. Ind. Hyg. Assoc. J.,* 42, 1, 1981.
9. **Burg, W.R., Shotwell, O.L., and Salzman, B.E.,** Measurements of airborne aflatoxins during the handling of 1979 contaminated corn, *Am. Ind. Hyg. Assoc. J.,* 43, 580, 1982.
10. **Burg, W.R. and Shotwell, O.L.,** Aflatoxin levels in airborne dust from contaminated corn during harvest and at an elevator in 1980, *J. Assoc. Off. Anal. Chem.,* 67, 309, 1984.
11. **Anon.,** Aflatoxin and airborne dusts from contaminated corn, *Grain Qual. Newslett.,* 7, 5, 1985.
12. **Ehrlich, K.C. and Lee, L.S.,** Mycotoxins in grain dust: method for analysis of aflatoxins, ochratoxin A, zearalenone, vomitoxin and secalonic acid D, *J. Assoc. Off. Anal. Chem.,* 67, 963, 1984.
13. **Scott, P.M., Van Walbeek, M., Kennedy, B., and Anyeti, D.,** Mycotoxins (ochratoxin A, citrinin and sterigmatocystin) and toxigenic fungi in grains and agricultural products, *J. Agric. Food Chem.,* 20, 1103, 1972.
14. **Abramson, D., Sinha, R.N., and Mills, J.T.,** Mycotoxin and odor formation in moist cereal grains during granary storage, *Cereal Chem.,* 57, 346, 1980.
15. **Abramson, D., Mills, J.T., and Boycott, B.R.,** Mycotoxins and mycoflora in animal feedstuffs in western Canada, *Can. J. Comp. Med.,* 47, 23, 1983.
16. **Abramson, D., Sinha, R.N., and Mills, J.T.,** Mycotoxin and odor formation in barley stored at 16 and 20% moisture in Manitoba, *Cereal Chem.,* 60, 350, 1983.
17. **Abramson, D., Sinha, R.N., and Mills, J.T.,** Quality changes and mycotoxin formation in granary–stored corn at 16 and 21% moisture content, *Sci. Aliment.,* 5, 653, 1985.
18. **Abramson, D. and Mills, J. T.,** unpublished.
19. **Clear, R.M. and Abramson, D.,** Occurrence of fusarium head blight and deoxynivalenol (vomitoxin) in two samples of Manitoba wheat in 1984, *Can. Plant Dis. Surv.,* 66, 9, 1986.

Explosion Hazards in Facilities Handling Agricultural Products

C. W. Kauffman

INTRODUCTION

Within the last decade in the U.S. there have been 200 explosion incidents in facilities handling agricultural products; these resulted in 139 deaths and 422 injuries. In a single 8–d period centered on Christmas 1977, there were five explosions resulting in 59 deaths and 47 injuries. Subsequent to this particularly disasterous week, the U.S. Department of Labor issued a Hazard Alert,[1] the U.S. Department of Agriculture issued a Task Force Report,[2] the National Academy of Sciences issued four reports from the Panel on the Causes and Prevention of Grain Elevator Explosions,[3–6] and the U.S. Department of Health and Human Services issued a Criteria Document,[7] all of which pertained to the prevention of dust explosions within facilities handling agricultural products. This information provides a substantial background for the Proposed Rulemaking, Grain Handling Facilities; Safety Hazards, U.S. Department of Labor, Occupational Safety and Health Administration, 29CFR Parts 1910 and 1917 on which hearings were held during the summer of 1984. All of the preceding clearly substantiate that the technical circumstances leading to explosions in agricultural product–handling facilities are clearly understood and that measures may be undertaken to prevent the future occurrence of these explosions.

BASES FOR DUST EXPLOSIONS

Everyone is familiar with the fire triangle—oxidizer, fuel, and ignition. When these three components come together under appropriate conditions, a fire will result. The removal of any one of the three components prevents the beginning of a fire or allows it to be extinguished.

A similar set of necessary components also exists for a dust explosion. These are represented in the explosion pentagon—oxidizer, fuel, ignition, mixing, and confinement—which is recognized to also contain the three components of the fire triangle. In addition to fire, a dust explosion also generates missiles and blast waves. Because of the five components, a dust explosion should be easier to prevent than a fire, but because of its rapid occurrence, it is almost impossible to extinguish once it has begun.

For both fires and explosions in the situation considered here, the components may be restricted to those which occur within grain–handling facilities. The oxidizer is atmospheric air. Grain or seed is the fuel for fires and the particles generated or manufactured from them are the fuel for explosions. Ignition is provided by flames, mechanical or electrical devices, or hot work. The confinement is provided by building structures, material– handling systems, or manufacturing equipment. The mixing can be provided by mechanical agitation during handling or manufacture or be an inherent part of the explosion process itself.

A differentiation between primary and secondary dust explosion can be made based upon the time of the introduction of the ignition source into the pentagon. If the mixing of the dust and air occurs prior to its introduction, a primary explosion occurs. If the mixing occurs as a consequence of an existing combustion process which itself acts as the ignition source, a secondary explosion results.

PREVENTION OF DUST EXPLOSIONS

The component of the pentagon which should receive major attention is the fuel, which consists of small dust particles generated by mechanical abrasion during handling or by manufacturing processes. It is the uncontrolled accumulation of this dust in layers throughout the facility over periods of time ranging from hours to weeks which supports the secondary explosions that account for the majority of the human and property damage. This layered dust can be dealt with in three different ways. Gentler handling of the material, which implies slower speed or well–engineered equipment, leads to less abrasion and less dust generation. A tight materials–handling system allows less dust to escape, while pneumatic pickups contain the dust at transfer points. Blow down, vacuuming, and sweeping can remove previously settled dust.

Inside of processing or conveying equipment, suspended dust may be minimized by first dealing with initially low–dust–content materials. Second, the design of the equipment may affect the dispersion of the dust into the air. Finally, pneumatic systems may be used to remove suspensions of dust.

It takes a relatively small amount of dust to support an explosion. For many agricultural dusts a hazardous suspension is approximately 50 g/m^3; and, if a uniform mixture of this dust settles in a 2.43–cm ceiling volume, it will produce a layer 0.4 mm thick on the floor.

Another component which should receive significant attention is the ignition source. Traditionally, this is where the maximum effort has been made with regard to the prevention of explosions. Ignition sources, though, are difficult to deal with in that they may, for example, be basically uncontrollable, e.g., lightning. They may appear very quickly, too quickly to be assessed and made harmless; an electrical or mechanical component may fail in a fraction of a second.

To a large degree the prevention of ignition sources depends upon four items. First, equipment must be procured which is suitable for use in a hazardous environment. Close attention must be paid to applicable codes. Second, the approved equipment must be installed in the specified manner. Faulty installation may defeat the safety features. Third, the equipment must be maintained to keep the performance within original specifications. Last, the performance of all equipment must be continually monitored to see if it is performing to specifications, and if it requires maintenance.

Handbook values given for the ignition energy of suspended dust clouds are as low as tens of millijoules. Actual failed electrical or mechanical devices are capable of providing at least hundreds of joules.

The mixing component has already been mentioned to some degree during the discussion of suspended dust. Efforts should be made not to suspend dust into air. This may require slower speed or better–designed processing equipment and the elimination of free–flowing material and large material drops.

The elimination, dilution, or modification of air on a continuing basis in agricultural product facilities do not seem to be reasonable methods to control the fourth component of the explosion pentagon at this time. Already, deaths occur in such facilities arising from confined space entry. However, it may be a possibility on an intermittent basis, through the use of packaged explosion suppression devices which discharge into small, well–defined volumes after being triggered by a developing explosion. The major considerations in the employment of these devices are their acquisition cost, their degree of sophistication with regard to installation and future maintenance, and that they deal with perhaps a minor, but already– existing explosion.

The final component to be considered is confinement, which results when there is inadequate venting. This means that equipment designed to contain manufacturing processes or the material transport system and structures for the storage of materials and the exclusion of weather do not support internal pressurization. Provisions for the noncontainment of pressure are most easily considered in preconstruction stages, but some retrofitting of existing equipment and structures is possible. Structures used in agricultural product–handling facilities may be quite large, and at this time existing venting data have not been verified for these volumes. Certainly, venting requirements should not be taken as any less than those given by linear interpolation of the existing data. While venting is certainly detrimental to the propagation of an existing explosion, an explosion has not been prevented. In addition, the venting process itself creates jets of high–pressure, high–temperature gases which themselves can cause significant damage.

ACTUAL DUST EXPLOSIONS

During the tenure of the National Academy of Sciences Panel on the Causes and Prevention of Grain Elevator Explosions, on–site investigations were conducted for 14 different explosion incidents.[8] In all cases except one, a satisfactory scenario of events leading up to the accident was constructed. The ignition source and the location of the primary explosion were identified, as well as the path of and the availability of the fuel to support the secondary explosion. In the one case, the ignition source was not identified, but subsequent events were successfully described.

In three of the last five fatal incidents, all of which occurred in 1984, a satisfactory scenario describing the explosion seems to have been developed. At the Lake Lilian Farmers Coop, where two were killed and four injured, the ignition source resulted from the structural failure of a feed–grinding mill, and the primary explosion occurred in the product–conveying system into which the mill discharged. The explosion broke out of this system and sufficient layered dust was present to support a secondary explosion which caused little structural damage to the facility, as there were a sufficient number of windows present to allow for venting. At the Blue Valley Coop, where two were killed and none injured, the ignition source was an overheated bearing on one of the bucket elevators. The primary explosion occurred in the concrete legwells of the bucket elevators, but it is reported that there were no secondary explosions because of the design of the elevator and the level of housekeeping within it. At the Con Agra Agribasics Feed Mill, where 3 were killed and 11 injured, corn was being ground and classified at the time of the explosion. Although it is clear that the primary explosion initiated in the product–handling system, the ignition source could not be identified. Two of the components in use, the grinder and the classifier, had been cited as the ignition source in other recent explosions. The level of housekeeping was poor and secondary explosions occurred which resulted in massive structural damage because of inadequate venting.

DUST COMBUSTION RESEARCH

The explosive nature of dust and products based upon agricultural materials has been well known for well over half a century.[9] Extensive investigations and considerable testing have been undertaken and reported for such materials.[10–13] At this time there appears to be sufficient technical knowledge available in order to prevent entire facilities from exploding. However, some interesting scientific details yet remain to be understood. In many situations smouldering combustion in a dust heap which is subsequently dispersed provides the primary explosion. Some data have recently been reported[14] concerning the ignition requirements and rates of smouldering for starch–based dusts, and this has been used to develop and support an elementary theory. With regard to a suspended dust cloud, the severity of a resulting explosion has been characterized traditionally by the maximum pressure reached and the maximum rate of pressure rise achieved. However, both of these measured values are apparatus dependent as opposed to a fundamental parameter such as the burning velocity. Measurements are now available[15,16] of the laminar as well as the turbulent burning velocity and values are in the vicinity of 1 m/s. The thickness of a dust layer required to support a secondary explosion has been traditionally determined from tests concerning the explosion characteristics of suspended dust clouds. Data are now being collected in an apparatus which accurately duplicates a secondary explosion,[15,17] and results indicate that perhaps a layer as thick as 1.2 mm in a 2.43–m ceiling enclosure is required. Additionally, the secondary explosion is seen to be greatly affected by properties which determine the entrainment characteristics of the dust.

CONCLUSIONS

In facilities handling agricultural products the chief reason for the occurrence of damaging explosions is the large distributed accumulations of fuel in the form of dust layers throughout the facility. Mechanical and electrical ignition sources are plentiful and energetic, mainly resulting from the improper use of equipment and the failure to maintain and monitor it. Dust explosions are not inevitable!

REFERENCES

1. **Bingham, E.**, Grain Elevator Industry Hazard Alert, Occupational Safety and Health Administration, Washington, D.C., January 5, 1978.
2. Prevention of Dust Explosions in Grain Elevators — An Achievable Goal, A Task Force Report, U.S. Department of Agriculture, Washington, D.C., 1978.
3. The Investigation of Grain Elevator Explosions, Panel on Causes and Prevention of Grain Elevator Explosions, NMAB 367–1, National Materials Advisory Board, National Academy of Sciences, Washington, D.C., September 1980.
4. Prevention of Grain Elevator and Mill Explosions, Panel on Causes and Prevention of Grain Elevator Explosions, NMAB 367–2, National Materials Advisory Board, National Academy of Sciences, Washington, D.C., 1982.
5. Pneumatic Dust Control in Grain Elevators — Guidelines for Design Operation and Maintenance, Panel on Causes and Prevention of Grain Elevator Explosions, NMAB 367–3, National Materials Advisory Board, National Academy of Sciences, Washington, D.C., 1982.
6. Guidelines for the Investigation of Grain Dust Explosions, Panel on Causes and Prevention of Grain Elevator Explosions, NMAB 367–4, National Materials Advisory Board, National Academy of Sciences, Washington, D.C., 1983.
7. Occupational Safety in Grain Elevators and Feed Mills, DHHS (NIOSH) Publ. No. 83–126, U.S. Department of Health and Human Services, Washington, D.C., September 1983.
8. **Kauffman, C.W. and Hubbard, R.F.**, An Investigation of Fourteen Grain Elevator Explosions Occurring between January 1979 and April 1981, Directorate of Safety Standards Program, Occupational Safety and Health Administration, May 1984.
9. **Price, D.J. and Brown, H.H.**, Dust Explosions — Theory and Nature of, Phenomena, Causes and Methods of Prevention, National Fire Protection Association, Quincey, MA, 1923.
10. **Bartknecht, W.**, *Explosions, Course Prevention Protection,* Springer–Verlag, Berlin, 1981.
11. **Field, P.**, *Dust Explosions, Handbook of Powder Technology,* Vol. 4, Elsevier, New York, 1982.
12. **Cross, J. and Farrer, D.**, *Dust Explosions,* Plenum Press, New York, 1982.
13. **Nagy, J. and Verakis, H.C.**, *Development and Control of Dust Explosions,* Marcel Dekker, New York, 1983.
14. **Leisch, S.O., Kauffman, C.W., and Sichel, M.**, Smoldering combustion in horizontal dust layers, in 20th Int. Symp. on Combustion, University of Michigan, Ann Arbor, August 1984.
15. **Kauffman, C.W., Srinath, S.R., Tezok, F.I., Nicholls, J.A., and Sichel, M.**, Turbulent and accelerating dust flames, in 20th Int. Sym. on Combustion, University of Michigan, Ann Arbor, August 1984.
16. **Tezok, F., Kauffman, C.W., Sichel, M., and Nicholls, J.A.**, Turbulent burning velocity measurements for dust/air mixtures in a constant volume spherical bomb, in 10th Int. Colloquium on Dynamics of Explosions and Reactive Systems, University of California, Berkeley, August 1985.
17. **Srinath, S.R., Kauffman, C.W., Nicholls, J.A., and Sichel, M.**, Flame propagation due to layered combustible dusts, in 10th Int. Colloquium on Dynamics of Explosions and Reactive Systems, University of California, Berkeley, August 1985.

Prospective Changes in Lung Function in Grain Elevator Workers in Large Terminals in Vancouver

M. Chan–Yeung and D. A. Enarson

INTRODUCTION

Some cross–sectional studies of grain workers have shown a high prevalence of respiratory symptoms and lung function abnormalities compared with control populations.[1–6] Little, however, is known about the longitudinal decline in lung function among grain workers. In 1975 we began a prospective comparative study of the respiratory health of grain workers in the four large terminal elevators in the Port of Vancouver using a group of civic workers as controls to determine whether exposure to grain dust affects respiratory health. We found that at the beginning of the study in 1975, grain workers had an excess of respiratory symptoms and a lower level of lung function compared with civic workers.[4] During the first 3–year follow–up period, we found that the annual decline in lung function was greater among grain workers than among civic workers.[7] In this report we present the results of changes in lung function and respiratory symptoms in grain workers compared with those in civic workers over a period of 6 years.

SURVEY OF B.C. GRAIN HANDLERS

The first survey was carried out in 1975 on 642 (587) grain workers and 136 civic workers. The second study was performed in 1978 on 584 (396) grain workers and 111 civic workers. The third one was done in 1981 on 671 (323) grain workers and 66 civic workers. For each study the participation rate was 94 to 95%. Bracketed numbers are the white male grain workers who took part in the prospective longitudinal study. During the third study only 323 (55%) of the original grain worker cohort and 66 (48.5%) of the original civic worker cohort remained. Data on white male workers who did not change their smoking habits over the period of study are presented in this report.

Subject Characteristics

Characteristics of the workers in the two groups in the prospective study are shown in Table 1. The grain workers were significantly younger and had been employed for a shorter period of time than the civic workers. There were significantly less nonsmokers and ex–smokers and atopic subjects among the grain workers than among the civic workers.

Respiratory Symptoms

Table 2 shows a comparison of the prevalence of respiratory symptoms, assessed on a modified MRC questionnaire,[8] in the two groups (still employed in 1981) at the time of the initial survey in 1975 and in 1981, and the incidence of new symptoms in the period of study. As can be seen, the prevalence of wheeze in the absence of colds was lower in the grain handlers at the beginning of the study. All other symptoms were more common in the grain handlers. Over the period of follow–up, cough and phlegm production appeared at a similar rate in the two groups. Wheeze, although initially less common, appeared at a higher rate in grain handlers over the period of study, so that by 1981 it was more common than in

TABLE 1
Characteristics (in 1975) of Workers[a] in the Longitudinal Study — 1975 to 1981

N	Grain (267)	Civic (63)	*p* value
Age (in years/mean ± SD)	38.7 ± 11.0	43.7 ± 9.9	0.001[b]
Height in cm (mean ± SD)	176.8 ± 6.9	177.6 ± 6.6	0.395[b]
Duration of employment in months (mean ± SD)	109.1 ± 83.2	144.7 ± 116.8	0.025[b]
Smoking habit (%)			
Nonsmokers	25.8	33.3	—
Ex–smoker	31.8	46.0	—
Current smoker	42.3	20.6	0-006[c]
Atopy (%)	16.9	32.3	0.011[c]

[a] White males who did not change their smoking habits.
[b] Probability from *t* test.
[c] Probability from chi–square.

TABLE 2
Chest Symptoms (%) in Participants of Longitudinal Study (1975—1981), Adjusted for Smoking

	Prevalence (%)				Incidence (%)	
	1975		1981		(1975—1981)	
Symptom	Grain	Civic	Grain	Civic	Grain	Civic
Cough	27	20	24	26	13	17
Phlegm	26	18	28	20	9	14
Dyspnea	31[a]	18	33[a]	14	21	9
Wheeze	9[a]	16	14	18	19	12

[a] $p < 0.05$ by chi–square for grain workers vs. civic workers.

the civic workers. Dyspnea remained more common in grain handlers over the whole period of study.

Changes in Lung Function

The annual decline in lung function in the two groups was compared by analysis of covariance to adjust for differences in age and smoking habits. Among nonsmokers, the grain workers showed a significantly greater decline in forced expired volume in 1 s (FEV_1) compared to civic workers (–31.2 ± 14.3 vs. +4.1 ± 11.2 ml, $p = 0.012$). The rates of decline in FEV_1 among smokers were not significantly different between the two work groups (–35.9 ± 4.5 vs. –31.2 ± 5.6 ml, $p = 0.684$).

Host Factors Affecting Longitudinal Decline in Lung Function in Grain Workers

We attempted to identify host factors that affect the longitudinal decline in lung function. These factors include: age, smoking, atopic status, presence or absence of respiratory symptoms, presence or absence of lung disease, work week change in lung function during the first study (1975), presence of airway hyperresponsiveness to methacholine assessed by a standardized methacholine inhalation test[9] at the third study (1981), initial expiratory flow rate measurements (FEV_1, forced vital capacity [FVC], forced expiratory flow [$FEF_{25-75\%}$]) and initial α_1–antitrypsin level. The results have already been published in another communication[10] and they are summarized in Table 3.

Age was found to be one of the strongest determinants of decline in flow rates; the older the worker, the greater the annual decline. There was no difference in flow rate decline related to smoking habit. The work week change in spirometry during the first study correlated positively and strongly with the annual decline in expiratory flow rates. Grain workers who had airway hyperresponsiveness (PC_{20} <8 mg/ml) in the third study experienced a significantly greater decline in FEV_1 and $FEF_{25-75\%}$.

The atopic status of the workers, the presence or absence of respiratory symptoms, or lung disease (history of asthma and chronic bronchitis) and α_1–antitrypsin had no influence on the longitudinal decline in lung function. The initial lung function also affects the subsequent decline in lung function; the greater the initial lung function, the greater the decline.

Dust Concentration and Longitudinal Decline in Lung Function

In an attempt to identify factors in the workplace which might influence the longitudinal decline in lung function, we have used a "nested" case–control model. The results of this study have already been published in detail elsewhere.[11] In this study "cases" were defined as those 10% of studied grain workers with the worse trend in FEV_1 over the study period (n = 27). Each of these was matched with two "controls" identified as grain workers of the same age with the same smoking habit who had the best trend in FEV_1 over the same period (n = 54).

TABLE 3
Estimates of Pulmonary Function Decline from Multiple Regression Analysis

	FEV_1		FVC		$FEF_{25-75\%}$	
	Coeff[a]	*p* value	Coeff[a]	*p* value	Coeff[a]	*p* value
Initial PF[b] (+10%)	–8.00	0.0118	–7.11	0.0752	–12.47	0.0003
Age (+10 years)	–10.45	0.0096	–6.48	0.1556	–26.36	0.0006
PC_{20} <8 mg/ml	–40.63	0.0003	–26.75	0.0245	–69.59	0.0017
Work week change (–100 ml)	–8.90	0.0000	–9.95	0.0000	7.89	0.0000
Ex–smoker	12.83	0.2449	5.91	0.6370	12.95	0.5334
Current smoker	–0.37	0.9722	1.55	0.8984	–29.26	0.1505
Atopy	7.85	0.4835	6.62	0.6023	8.95	0.6750
α_1–Antitrypsin (+1 mg)	0.0002	0.9748	0.04	0.6639	0.05	0.7632

[a] Coeff = regression coefficient.
[b] PF = pulmonary function.

"Cases" and "controls" were successfully matched for age (40 ± 11 vs. 40 ± 11 years) and smoking (26 ± 7 vs. 21 ± 7 cigarettes per day in 18 smokers, and 16 ± 7 vs. 20 ± 10 in 9 ex–smokers), and were similar in height (178 ± 7 vs. 176 ± 7 cm), atopy prevalence (15 vs. 11%), and initial FEV_1 (3806 ± 935 vs. 3624 ± 808 ml). The cases demonstrated a striking decrement in FEV_1, averaging more than 100 ml/year (–682 ± 272 ml/6 yr) whereas the controls demonstrated a net rise in FEV_1 (+108 ± 163 ml) over the period. We attempted to relate the likelihood of being a case to particular jobs within the grain elevators. We divided the jobs into three broad categories according to levels of exposure: cleaners and sweepers were likely to have the highest exposure; maintenance personnel were likely to have intermittent high exposure; and others who worked away from the actual grain–handling process and were likely to have a lower level of exposure to grain dust. We found that cases were found more frequently than controls to be cleaners and sweepers, while the controls were more likely to be in the "other" job category.

We also found that there was a difference in the distribution of cases and controls among the four elevators. Two elevators (A and B) were associated with cases and two were not (C and D). Table 4 shows the distribution of cases and controls and the odds ratio (OR) of being a case and the total dust concentration for each of the six job categories in the elevators. Cleaners/sweepers and maintenance personnel in elevators A and B were more likely to be cases (OR of 6 and 3, respectively). The mean total dust levels associated with these job categories were high (32 ± 36 and 12 ± 16 mg/m^3, respectively). None of the other job categories was more likely to be associated with cases (OR of 1 or less).

RELEVANCE

In our first prevalence study in 1975 we found that grain workers had significantly lower FEV_1 and FVC compared to civic workers.[7] In our first follow–up study 3 years after the

TABLE 4
Distribution of Cases and Controls, Odds Ratio for Being a Case, and Mean Total Dust Levels according to Job Category and Elevator Site

	Elevators A and B			Elevators C and D		
	Cleaner sweeper	Maintenance	Other	Cleaner sweeper	Maintenance	Other
Controls (n)	2	4	11	4	12	21
Cases (n)	6	5	3	2	5	6
Odds ratio	6.0	3.0	0.4	1.0	0.8	0.4
Total Dust Levels (mg/m³)						
Maximum	144	48	26	72	24	17
Mean ± SD	32 ± 36	12 ± 16	7 ± 7	6 ± 10	6 ± 8	3 ± 3
#	33	14	23	53	15	37

initial study we found that grain workers had a more rapid decline in lung function compared to the same group of controls. This observation was confirmed in the 6–year follow–up study in that the annual decline in lung function was greater for grain workers than controls. The effect of grain dust exposure on lung function was particularly evident among nonsmokers.

We have attempted to answer the question, "Who is most likely to suffer respiratory impairment in grain workers and why?" We found that the effects are more striking in older grain workers. This does not necessarily imply that age is a risk factor, but may merely indicate more prolonged duration of exposure, or exposure to higher dust levels in the past. Since age and duration of exposure are closely correlated, it is difficult to separate them.

As expected, smoking was found to correlate with decrement in lung function in civic workers. In grain workers, smoking has very little effect on the decrement in lung function. We have shown that among non–smokers, grain workers have significantly more rapid decline in function compared to civic workers. Among current smokers, however, the differences between grain workers and civic workers was not as great as that for nonsmokers and, indeed, was not statistically significant. This could be interpreted in several ways. It could be said that among smokers, the effect of grain dust on the lungs is not as great as among nonsmokers. Conversely, one could argue that among those exposed to grain dust, the effect of smoking is not as great. Whatever the explanation, it seems reasonable to conclude that the effects of grain dust exposure and smoking are not potentiating and may, in fact, be competitive.

We have again shown, as in our first follow–up study, that acute response during a work week at the first study, in 1975, is associated with the rate of change of lung function over the entire period of 1975 to 1981. It is not possible to state whether the acute response over a work week represents a specific response to grain dust or nonspecific bronchial reactivity. We have demonstrated that those with bronchial reactivity to methacholine in 1981 had experienced a significantly greater decline in lung function in the preceding 6 years. These individuals, however, did not correspond to those showing a decline in lung function over a work week in 1975. Each of these characteristics was associated with an adverse outcome in lung function, but the two are unrelated.

Certain characteristics were not related to the trend in lung function over the 6 years. These included atopy, respiratory symptoms, and α_1–antitrypsin level. Some of these observations must be qualified. For example, the presence of respiratory symptoms was determined at the initial study and no attempt was made to determine how long the symptoms had been present or whether there had been a change in the complaint over the period of study. In the study of change in lung function in relation to initial lung function level, we found that grain workers with the best lung function initially had a worse decline in lung function. A possible explanation is the "middling tendency" or "regression to the mean", in which variation tends to be reduced by repeated measurements.

Using a "nested" case–control model, we investigated factors in the working environment and their association with rapid decline in lung function in grain workers. The group of workers most likely to demonstrate this decline in FEV_1 (more than 100 ml/year) was exposed to high levels of total dust (around 15 mg/m³), as compared with those most likely to have a favorable outcome who were exposed to much lower mean levels (just over 5 mg/m³). There was considerable variation in the dust levels in the two groups. Moreover, the data must be interpreted with caution, as the odds ratios indicate the likelihood of experiencing a severely adverse trend and not just any health effects.

In conclusion, we have found that grain workers had a more rapid decline in lung function over a 6–year period than civic workers. The host factors adversely affecting the decline in lung function in grain workers include age, acute work–week change in lung function during the initial study, and presence of airway hyperresponsiveness during the third study. We have demonstrated a dose–response relationship between total dust exposure and impairment of lung function with a cohort. The higher the dust exposure, the greater the likelihood of rapid decline in lung function in grain workers. The findings of this study emphasize the serious outcome of exposure to high levels of grain dust, and support proposals recommending a lower limit of dust exposure in grain–handling operations rather than the current standard.

REFERENCES

1. **Williams, N., Skoulas, A., and Merriman, J.E.,** Exposure to grain dust. I. A study of the effects, *J. Occup. Med.*, 6, 319, 1964.
2. **doPico, G.A., Reddan, W., Flaherty, D., Tsiatis, A., Peters, M., Rao, P., et al.,** Respiratory abnormalities among grain handlers: a clinical physiologic and immunologic study, *Am. Rev. Respir. Dis.*, 115, 915, 1977.
3. **Becklake, M.R., Jodoin, G., Utz, G., Lefort, L., Rose, B., and Fraser, R.B.,** A health study of grain handlers in St. Lawrence River Ports, *Am. Rev. Respir. Dis.*, 115 (Abstr.), 200, 1977.
4. **Chan–Yeung, M., Schulzer, M., MacLean, L., Dorken, E., and Grzybowski, S.,** Epidemiologic health survey of grain elevator workers in British Columbia, *Am. Rev. Respir. Dis.*, 121, 329, 1980.
5. **Broder, L., Mintz, S., Hutcheon, M., Corey, P., Silverman, F., Davies, G., et al.,** Comparison of respiratory variable in grain elevator workers and civic outside workers in Thunder Bay, Canada, *Am. Rev. Respir. Dis.*, 119, 193, 1979.
6. **Dosman, J.A., Cotton, D.J., Graham, B.L., Li, K.Y., Froh, F., and Barnett, D.,** Chronic bronchitis and decreased forced expiratory flow rate in lifetime non–smoking grain workers, *Am. Rev. Respir. Dis.*, 121, 11, 1980.
7. **Chan–Yeung, M., Schulzer, M., MacLean, L., Dorken, E., Tan, F., Lam, S., et al.,** A follow–up study of grain elevator workers in the Port of Vancouver, *Arch. Environ. Health*, 36, 75, 1981.
8. Medical Research Council's Committee on the Aetiology of Chronic Bronchitis, Standardized questionnaires on respiratory symptoms, *Br. Med. J.*, p. 1665, 1980.
9. **Lam, S., Wong, R., and Yeung, M.,** Nonspecific bronchial reactivity in occupational asthma, *J. Allergy Clin. Immunol.*, 63, 28, 1979.
10. **Tabona, M., Chan–Yeung, M., Enarson, D., MacLean, L., Dorken, E., and Schulzer, M.,** Host factors affecting longitudinal decline in lung function among grain elevator workers, *Chest*, 35, 782, 1984.
11. **Enarson, D.A., Vedal, S., and Chan–Yeung, M.,** Rapid decline in FEV_1 in grain handlers: relation to level of dust exposure, *Am. Rev. Respir. Dis.*, in press.

Respiratory Profiles of Grain Handlers and Sedentary Workers*

F. A. Herbert, V. Woytowich, E. Schram, and D. Baldwin

During 1978 we studied 71 men employed at three similar inland grain terminals in Alberta. Previous studies by others on total dust levels in these terminals had revealed concentrations ranging from less than 10 to 128 mg/m^3 in personal samples taken from workers during an 8–h shift.[1] These studies found that all persons in the plant working in any capacity shared in this exposure. A report of the employment records of the workers at these terminals in the period 1958 to 1978 was provided by their government employer.

A control group of 120 sedentary male employees of similar age, background, and geographic area, but unexposed to grain dust, was obtained from two departments of the federal civil service in Edmonton. Each grain handler was paired with a control subject by age and smoking history prior to the calculation of the results.

The clinical characteristics of the grain handlers and the matched control subjects, grouped according to smoking history, are shown in Table 1. Eight young men who had smoked only briefly, and their matched counterparts, were excluded to prevent dilution of the group of smokers.

A comparison of respiratory complaints made between all the grain handlers and controls by chi–square analysis indicated a significant difference between the two groups (p <0.05). Detailed analysis indicated that this difference was due to a significantly greater prevalence (p <0.01) of grade 1 dyspnea (dyspnea when hurrying on the level or walking up an incline) in the grain handlers who smoked compared with their controls.

A respiratory illness lasting at least 1 week every 3 years was more common among the controls than the grain handlers who were ex–smokers. However, overall, respiratory symptoms were more common in smokers than in nonsmokers or ex–smokers, but not significantly so.

A history of flu–like illness associated with grain dust exposure, interpreted by us as "grain fever", was noted in 14% of the 63 grain handlers, all of whom were smokers or ex–smokers. Ocular irritation was noted by six grain handlers and one control subject (not a significant difference). Dermatitis of all types was reported with equal frequency by the grain handlers and control subjects.

The pulmonary function data are shown in Table 2. There was no significant difference between the grain handlers who were nonsmokers and their controls, but the grain handlers who were smokers had a significantly lower mean forced expired volume in 1 s per forced vital capacity (FEV_1/FVC) than their controls (p <0.05); the frequency distribution showed that 15 of the 20 grain handlers had values below 70%, compared with 10 of the 20 control subjects. The mean FEV_1 was lower, but not significantly so, in the grain handlers who were ex–smokers, compared with their controls. The data for the workers at terminal 1 were not significantly different on the two widely separated occasions on which they were studied, when the dust levels differed greatly.

When age was standardized, a simple linear regression analysis did not show a deterioration of pulmonary function (a reduction in the FEV_1/FVC) in any of the grain handler groups with increasing duration of exposure to grain dust.

The FEV_1 values were highest in the nonsmokers and lowest in the smokers among both the grain handlers and the control subjects (p <0.05).

The immunologic data are detailed in Table 3. Skin testing had positive results with any of the antigens in 12 grain handlers (18% of the 66 tested), only 2 of whom had positive reactions to barley, wheat, or oat dust. Among these 12 men, 2 complained of wheezing when in contact with dust only, and 5 had a family history of asthma or allergic rhinitis. All 12 had an FEV_1 that was 80% of the predicted value. Thus, skin testing did not identify these men as being at high risk, and this technique will not likely become a useful means of preemployment screening.

The reasons for end of employment at the three grain terminals during the period 1958 to 1978 are summarized in Table 4. Of the 64 men, 32 left because of mandatory retirement at the age of 65 years, 8 took an early retirement apparently unrelated to illness, and 4 transferred to similar work in other grain terminals. Of the remaining 20, 1 man died abruptly, 2 left at the age of 60 years because of chronic obstructive pulmonary disease, and 17 resigned for personal reasons.

Among the nonsmokers in this study, the grain handlers were not significantly different from the controls in the frequency of their clinical complaints or in their spirometric results, although if the groups had been large the intergroup differences in spirometric values might have been significant.

The grain handlers in our study who were smokers had significantly more complaints of dyspnea and poor pulmo-

* Previously published in expanded form (*C.M.A.J.*, 125, 46, 1981).

TABLE 1
Clinical Data for 63 Grain Handlers and 63 Matched Control Subjects

	Mean ± 1 SD					
	Nonsmokers		Smokers		Ex–smokers	
Variable	Grain handlers (n = 16)	Control subjects (n = 16)	Grain handlers (n = 20)	Control subjects (n = 20)	Grain handlers (n = 27)	Control subjects (n = 27)
Age (years)	40 ± 14	44 ± 13	51 ± 10	52 ± 8	48 ± 10	45 ± 12
Years employed	10.9 ± 10.5	—	14.7 ± 9.5	—	17.2 ± 11.4	—
Pack years of smoking	—	—	36 ± 19	37 ± 16	—	—
	No. of Subjects					
Cough and/or sputum for ≥3 months/years for 2 years	4	3	13	14	6	3
Grade 1 dyspnea[a]	2	1	10[b]	2	8	4
Wheeze any time	3	2	8	12	4	6
Wheeze and cough with exposure to grain dust	4	—	12	—	2	—
Nasal stuffiness	5	3	10	11	9	6
"Grain fever"[c]	0	—	2	—	7	—
Flu–like illness[d]	7	6	7	7	11	16
Respiratory illness for ≥1 weeks every 3 years	1	0	4	3	0	5
Rhonchi	1	0	6	1	0	0

[a] Dyspnea when hurrying on the level or walking up an incline.
[b] Difference from control group significant at $p < 0.01$.
[c] Flu–like illness associated with grain dust exposure.
[d] Malaise, chills, muscle aches, and fever.

From Herbert, F. A. et al., *Can. Med. Assoc. J.*, July 1, 125, 1981. With permission.

TABLE 2
Pulmonary Function Data for the 63 Grain Handlers and 63 Control Subjects

	Mean ± 1 SD					
	Nonsmokers		Smokers		Ex–smokers	
Variable	Grain handlers	Control subjects	Grain handlers	Control subjects	Grain handlers	Control subjects
Forced vital capacity (FVC), % of predicted	105 ± 16	101 ± 37	104 ± 11	102 ± 16	100 ± 21	107 ± 12
Forced expiratory volume in 1 s (FEV_1), % of predicted	97 ± 15	110 ± 12	82 ± 18	90 ± 22	85 ± 31	101 ± 13
FEV_1/FVC	74 ± 8	79 ± 8	60 ± 12[a]	70 ± 9	69 ± 13	75 ± 9

[a] Difference from control group significant at $p < 0.05$.

From Herbert, F. A. et al., *Can. Med. Assoc. J.*, July 1, 125, 1981. With permission.

nary function than their controls. Others have reported that smoking in grain handlers produces airways obstruction in excess of that expected from smoking alone.[2,3] Our data support the concept that there may be an interaction between the effects of inhaled dust and smoke that causes a deterioration of lung function, which is greater than that due to smoking alone.

REFERENCES

1. **Farani, J.P. and Moore, C.P.**, Grain dust exposures in the five inland Canadian Government elevators, Rep. 75 SR–27, Occupational Health Unit, Medical Services Branch, Department of National Health and Welfare, Ottawa, 1975, 13.
2. **doPico, G., Reddan, W., Tsiatis, A., and Rankin, J.**, A comparison of pulmonary function parameters in grain handlers and city workers, *Am. Rev. Respir. Dis.*, 119 (Suppl. 4, Abstra.), 213, 1979.
3. **Dosman, J.A.**, Chronic obstructive airways disease and smoking in grain workers (E), *Ann. Intern. Med.*, 87, 784, 1977.

TABLE 3
Immunologic Data for the Grain Handlers and Control Subjects

Finding	No. (and %) Grain handlers	Control subjects	
Positive results of skin testing			
With any antigen	12/66 (18)	—	—
With wheat, barley, or oats	2/66 (3)	—	—
Family history of asthma or allergic rhinitis	18/71 (25)	21/71	(30)

From Herbert, F. A. et al., *Can. Med. Assoc. J.*, July 1, 125, 1981. With permission.

TABLE 4
Reasons for End of Employment at the Three Grain Terminals in 1958—1978

	Retirement				
Terminal	Mandatory	Elective[a]	Transfer	Sudden death	Resignation
I	7	8	0	1	6
II	15	2	2	0	9
III	10	0	2	0	2
Total	32	10	4	1	17

[a] After approximately 30 years of service; in two cases because of chronic obstructive lung disease.

From Herbert, D. A. et al., *Can. Med. Assoc. J.*, July 1, 125, 1981. With permission.

Effects of Grain Dust Exposure and Smoking on Respiratory Symptoms and Lung Function*

D. J. Cotton, B. L. Graham, K. L. R. Li, G. D. Barnett, and J. A. Dosman

Although numerous studies have documented changes in respiratory symptoms and pulmonary function in grain workers, only recently have attempts been made to assess the separate and combined effects of occupational grain dust exposure and cigarette smoking in these workers.[1–8] In the present study we have assessed both the separate and combined effects of cigarette smoking and grain dust exposure on respiratory symptoms and pulmonary function in grain elevator workers in the province of Saskatchewan.

All male Caucasian grain elevator workers in the province of Saskatchewan answered a standardized questionnaire and performed a series of pulmonary function tests. We identified 195 lifetime nonsmoking (<400 cigarettes) grain workers and individually matched each nonsmoking grain worker to a smoking grain worker of similar age (the lesser of ±3 years or 10% of age) and years of employment in the grain elevator (the lesser of ±2 years or 10% of exposure years). From the same geographic location, we matched each grain worker to a nonsmoking or smoking control subject of the same age and number of cigarette pack years for smokers (the lesser of ±2 years or 10% of pack years). Community control subjects were excluded if they had previously been grain elevator workers, had been farmers in the last 5 years, or had professional occupations (lawyer, dentist, physician, veterinarian). Both grain workers and controls were excluded if they had ever worked in a mine, but none were excluded because of symptoms of chest disease. All workers and controls were male Caucasians and were gainfully employed. Thus, 195 matched quartets were formed, each with two grain workers (a nonsmoker and a smoker) and two community controls (also a nonsmoker and a smoker).

We defined chronic bronchitis by a positive response to any one or more of the following three questions from a modified British MRC questionnaire:[5,9] (1) "Do you usually bring up phlegm from the chest first thing in the morning? (not from the back of your nose; count swallowed phlegm from the chest)"; (2) "Do you usually bring up phlegm from your chest at other times during the day or night?"; (3) "Do you bring up phlegm like this as much as twice a day, 4 or more days of the week?"; as well as a positive response to the following question, "Do you bring up phlegm from the chest on most days for as much as three months of the year?", and a response of for "two years or more" on response to the following question: "For how many years have you raised phlegm from your chest?". In assessing the prevalence of chronic bronchitis, we considered only men who had been exposed to grain dust for ≥2 years and their control subjects (n = 154 in each group).

Pulmonary function studies were undertaken in a mobile van using methods previously described.[2] We obtained three acceptable forced expiratory maneuvers while breathing air and three acceptable forced expiratory maneuvers following three vital capacity (VC) inhalations of a mixture of 80% helium and 20% oxygen (He–0_2). From each forced expiratory maneuver breathing air we measured the forced expiratory volume in 1 s (FEV_1), the forced vital capacity (FVC), the $FEV_1/FVC \times 100$, the maximal midexpiratory flow rate (MMFR), and the maximal expiratory flow rate at 50 and 25% FVC ($\dot{V}_{max50}$ and $\dot{V}_{max25}$, respectively). We also measured the percent increase in the maximal expiratory flow breathing He–0_2 compared with air at 50% FVC ($\Delta\dot{V}_{max50}$).[9] The slope of phase III of the alveolar plateau ($\Delta N_2/l$) of the single–breath nitrogen test was also recorded using methods previously described.[10]

For purposes of statistical analysis, we chose the largest single value for FEV_1, FVC, MMFR, $\dot{V}_{max50}$ and $\dot{V}_{max25}$ from the acceptable forced expiratory maneuvers whose FVCs were within ±5% and obtained $FEV_1/FVC \times 100$ from the ratio of the largest FEV_1 and FVC. $\Delta\dot{V}_{max50}$ was calculated from the largest $\dot{V}_{max50}$ breathing air and the largest $\dot{V}_{max50}$ breathing helium. $\Delta N_2/l$ was the mean value of three single–breath nitrogen tests whose VCs agreed within ±5%.

The $FEV_1/FVC \times 100$, $\dot{V}_{max50}$ and $\dot{V}_{max25}$ were expressed as percent expected values.[11] Expected values were obtained from Morris et al.[12] for MMFR, Dosman et al.[9] for $\dot{V}_{max50}$ and Dosman et al.[13] for $\Delta N_2/l$. Percent expected values were compared between groups using an analysis of variance procedure for 2 × 2 factorial design[14] while blocking on quartets. With this analysis an independent assessment of the effects of smoking and grain dust exposure can be determined as well as detection of interaction between smoking and grain dust exposure. The prevalence of chronic bronchitis from the questionnaire was analyzed separately for the effects of grain dust and for the effects of smoking using the Mantel–Haenszel procedure.[15]

In addition to examining the symptoms and pulmonary

* Previously published in expanded form (*J. Occup. Med.*, 25, 131, 1983).

function changes of the total group (n = 195 quartets), we also examined two exposure categories classified according to the employment history of the nonsmoking grain workers in each quartet, as follows: (1) ≤5 years exposure (n = 94 quartets), and (2) >20 years exposure (n = 24 quartets).

We also sampled the airborne grain dust in 63 of the elevators in which the workers were employed using area samplers[16] and measured dust concentration in the breathing zone of 123 employees during an 8–h shift using personal samplers (Bendix).[17]

Table 1 outlines the anthropomorphic characteristics of the nonsmoking and smoking grain workers and their matched community controls. For both the entire sample (n = 195 in each group) and for the two grain dust exposure categories there were no differences in the matched variables, i.e., age, years of employment in grain elevators (for grain workers), and pack years of smoking (for smokers) among the groups. However, for the total group, nonsmoking grain workers were heavier than nonsmoking community controls (p <0.05), and smokers in the greater–than–20–years exposure category were slightly shorter than nonsmokers (p <005).

Table 2 shows the overall results (grain workers, controls, smokers and nonsmokers) for the FEV_1, $FEV_1/FVC \times 100$, MMFR, V_{max50}, $\dot{V}_{max25}$, $\Delta\dot{V}_{max50}$, and $\Delta N_2/l$. For both the total group and in the higher exposure category, MMFR and $\dot{V}_{max25}$ were lower in grain workers than in controls. In the lower exposure category, $FEV_1/FVC \times 100$ and $\dot{V}_{max25}$ were lower in grain workers than in controls (Table 2). In smokers vs. nonsmokers, FEV_1/FVC was lower for the lower exposure category (p <0.05); and FEV_1/FVC, MMFR, $\dot{V}_{max50}$ and $\dot{V}_{max25}$ were all lower in both the total group and the higher exposure category (Table 2). There was no evidence for an interaction between grain dust and smoking in the total group (Figure 1).

Smokers had a higher $\Delta N_2/l$ than nonsmokers for the total group and for smokers matched to the higher exposure category, but there were no overall effects of grain dust in these groups (Figure 2). However, in the lower exposure category, for $\Delta N_2/l$ there was an interaction between smoking and grain dust exposure: $\Delta N_2/l$ was higher in smoking grain workers compared with both smoking controls and nonsmoking grain workers. An interaction similar to that observed for $\Delta N_2/l$ was also seen for $\Delta\dot{V}_{max50}$. In the lower exposure category (Figure 3), the smoking grain workers had a lower $\Delta\dot{V}_{max50}$ than either smoking controls or nonsmoking grain workers.

The symptom of chronic bronchitis (Figure 4) was greater in grain workers vs. controls (p <0.01) for both the total group as well as in the older exposure category. Chronic bronchitis was also greater in smokers vs. nonsmokers for the lower and higher exposure categories as well as for the total group (Figure 4).

The concentration of general work–area dust in grain elevators depended on the operation of dust control equipment (cyclone) in the elevators. In 54% of the elevators, control equipment was in operation. In these elevators the concentration of the total dust in the general work area was 6.1 ± 9.2 mg/m^3 (mean ± 1 SD) and in the breathing zone (8–h average) was 1.9 ± 3.2 mg/m^3. In 46% of elevators no dust control equipment was in operation. In these elevators the total dust concentration in the general work area was 13.6 ± 22.1 mg/m^3 and the 8–h average concentration in the breathing zone was

TABLE 1
Anthropomorphic Data[a]

Exposure category: Employment (years)	Smoking category	No. Controls	No. Grain workers	Age (years) Controls	Age (years) Grain workers	Employment (years) Controls	Employment (years) Grain workers	Smoking (pack years) Controls	Smoking (pack years) Grain workers	Height (cm) Controls	Height (cm) Grain workers	Weight (kg) Controls	Weight (kg) Grain workers
Total	Nonsmoker	195	195	30.6 ±11.0	30.6 ±11.1	—	8.2 ±7.7	—	—	174.7 ±6.5	174.5 ±5.9	78.9 ±12.0	82.2[b] ±11.9
	Smoker	195	195	30.7 ±11.0	30.7 ±11.1	—	8.1 ±7.7	13.2 ±11.2	13.1 ±11.0	174.3 ±6.1	173.5 ±5.8	80.7 ±14.7	79.9 ±13.0
<5	Nonsmoker	94	94	23.6 ±6.1	23.5 ±5.9	—	2.4 ±1.4	—	—	174.7 ±6.2	174.6 ±6.3	76.0 ±11.1	78.9 ±12.5
	Smoker	94	94	23.7 ±6.0	23.5 ±5.7	—	2.4 ±1.4	6.7 ±5.5	6.7 ±5.5	175.3 ±5.7	174.4 ±6.2	80.4 ±15.3	79.6 ±12.7
>20	Nonsmoker	24	24	51.4 ±6.0	51.5 ±5.9	—	24.6 ±3.6	—	—	175.5 ±5.9	174.9 ±5.2	82.7 ±10.8	86.8 ±8.5
	Smoker	24	24	51.3 ±6.1	51.4 ±5.4	—	24.4 ±4.1	30.8 ±11.9	30.6 ±11.3	173.0 ±7.2	172.1 ±5.4[c]	80.6 ±12.3	82.4 ±14.4

[a] Mean values ± 1 SD for each group.
[b] Significant difference between nonsmoking grain workers and nonsmoking community controls (p <0.05).
[c] Significant overall differences between smokers and nonsmokers (p <0.05).

TABLE 2
Pulmonary Function Studies (% Expected) of Combined Groups (Mean ± 1 SD)

Exposure category (years)	Group	No.	FEV_1	$FEV_1/FVC \times 100$	MMFR	$\dot{V}_{max50}$	$\dot{V}_{max25}$	$\Delta\dot{V}_{max50}$	$\Delta N_2/1$
Total	Grain workers	390	109.5 ± 13.4	94.4 ± 9.0	92.5 ± 25.4[a]	105.4 ± 30.5	95.4 ± 36.5[a]	117.2 ± 85.7	110.0 ± 56.8
	Controls	390	109.7 ± 15.2	95.6 ± 9.4	97.0 ± 30.9	109.5 ± 36.5	102.0 ± 40.8	121.5 ± 65.3	108.0 ± 52.3
	Nonsmokers	390	111.0 ± 14.4	96.1 ± 9.0	98.6 ± 28.2	111.0 ± 33.9	103.9 ± 39.0	124.5 ± 79.4	91.6 ± 42.2
	Smokers	390	108.2 ± 14.1[b]	93.9 ± 9.3[b]	90.8 ± 28.0[b]	103.9 ± 33.1[b]	93.6 ± 38.0[b]	114.3 ± 72.5	126.4 ± 59.8[b]
<5	Grain workers	188	108.3 ± 13.1	94.6 ± 8.2[a]	93.9 ± 24.1	105.3 ± 28.6	97.4 ± 33.4[a]	116.2 ± 80.9[c]	110.1 ± 47.9[d]
	Controls	188	110.0 ± 13.6	96.2 ± 7.7	98.0 ± 25.1	109.8 ± 30.3	104.7 ± 36.0	125.1 ± 54.7	105.6 ± 36.9
	Nonsmokers	188	109.9 ± 13.5‘	96.3 ± 7.8	98.1 ± 25.0	109.3 ± 30.7	104.0 ± 35.4	128.6 ± 70.7	98.6 ± 35.8
	Smokers	188	108.3 ± 13.2	94.5 ± 8.1[b]	93.9 ± 241	105.8 ± 28.2	98.1 ± 34.2	112.6 ± 66.8	117.1 ± 46.9
<20	Grain workers	48	107.1 ± 15.3	91.9 ± 9.8	80.1 ± 25.2[a]	94.3 ± 31.8	76.8 ± 31.5[a]	117.7 ± 105.6	126.6 ± 96.2
	Controls	48	111.1 ± 20.0	93.1 ± 12.1	94.8 ± 42.9	107.1 ± 51.2	95.8 ± 47.7	115.0 ± 69.1	125.6 ± 90.8
	Nonsmokers	48	113.9 ± 17.5	94.9 ± 9.2	96.3 ± 35.9	109.4 ± 40.8	96.5 ± 42.7	120.4 ± 87.4	84.5 ± 75.2
	Smokers	48	104.3 ± 16.9[b]	90.3± 12.2[e]	78.6 ± 33.7[e]	92.0 ± 43.5[e]	76.1 ± 37.6[b]	112.3 ± 90.9	167.8 ± 91.0[b]

[a] Indicates a significant difference between grain workers and controls ($p < 0.05$).
[b] Significant ($p < 0.01$) difference between smokers and nonsmokers.
[c] Significant interaction between smoking and grain dust exposure ($p < 0.05$).
[d] Significant interaction between smoking and grain dust exposure ($p < 0.01$).
[e] Significant ($p < 0.05$) differences between smokers and nonsmokers.

6.2 ± 14.3 mg/m^3. The average respirable fraction of dust in the breathing zone was 28 ± 25% for workers in all elevators. The total dust concentration in the general work area exceeded 10 mg/m^3 in 24% of elevators without dust control equipment and in 15% of elevators with dust control equipment in operation.

In this study we have found that grain dust exposure is associated with evidence of airways obstruction and an increased prevalence of respiratory symptoms, a finding that supports two recent short longitudinal studies[5,6] and confirms our previous findings[2] that nonsmoking grain workers have a greater prevalence of respiratory symptoms and decreased expiratory flow rates compared with nonsmoking community controls, particularly in the lower and higher exposure categories.

We found small but significant reductions in FEV_1/FVC and $\dot{V}_{max25}$ in grain workers employed less than 5 years. This may relate to the fact that the most recently hired workers (job class: "helpers") might be heavily exposed to grain dust, especially compared to older employees (job class: "agents"). Alternatively, the least–exposed worker population may be less influenced by selection. Young workers who are most affected by grain dust exposure may subsequently leave this employment and, thus, would not be studied in a population of workers with a longer employment history.[3] Despite evidence for mild airflow obstruction in young grain workers vs. community controls, we observed that not a single nonsmoking grain worker reported a previous physician's diagnosis of asthma, whereas the prevalence in nonsmoking community controls was 6%. This suggests another selection process, possibly related to preemployment pulmonary health, that would favour better lung function and lower symptom prevalences in nonsmoking workers. It has been previously suggested[1] that asthmatics seldom commence employment in grain elevators, presumably because grain dust exposure increases their asthmatic symptoms.

We also observed that grain workers in the higher exposure category had lower forced exhaled flow rates and a greater prevalence of chronic bronchitis than community controls of similar age, suggesting that cumulative grain dust exposure adversely affects the lung. By the magnitude of the differences observed, the data suggested that in this group of subjects, both grain dust (a mean exposure of 24.5 years) and smoking (a mean of 30.7 pack years) alter lung function to a similar degree, and no significant interaction between smoking and grain dust was observed. However, separating acute and chronic effects in this study is not possible. The effect in men with high exposure years is most likely minimized by the fact that older grain workers may represent a survivor population, the more severely affected workers having already left the workplace. Reductions in forced exhaled flow rates in older grain handlers in St. Lawrence River ports[4] have also been previously reported.

In the total group of grain workers, the increase in the respiratory symptom of chronic bronchitis clearly differentiated this group from community controls. The prevalence of this symptom was not significantly higher in grain workers vs. controls for the younger exposure category, but was increased in the older exposure category. This finding supports previous reports comparing a small group of smoking grain workers with smoking port workers occupationally unexposed to grain dust,[3] as well as the experience in British Columbia comparing grain workers with civic workers[5] and studies in nonsmoking grain workers.[2] The increased prevalence of chronic bronchi-

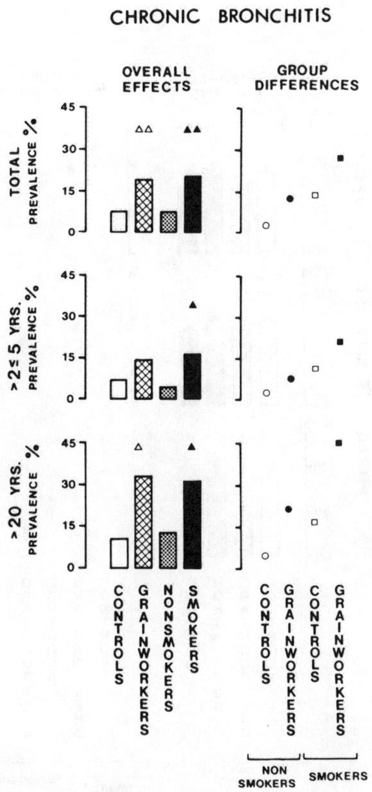

FIGURE 1. Mean percent expected[12] values (±SEM) for MMFR. Total group (top panel: overall effects, n = 390; group differences, n = 195) is divided into two exposure categories (lower panels) consisting of workers exposed equal to or less than 5 years (overall effect, n = 188; group differences, n = 94) and greater than 20 years (overall effects, n = 48; group differences, n = 24). For graphs on left side, bars indicate mean values for overall effects (controls, open bars; grain workers, cross–hatched bars; nonsmokers, stippled bars; and smokers, black bars). For graphs on right side, symbols (open and solid circles, open and solid squares) indicate group mean values (±1 SEM) (open circle and square for nonsmoking and smoking controls, respectively; solid circle and square for nonsmoking and smoking grain workers, respectively). Solid triangle and paired solid triangles indicate significant overall differences between smokers and nonsmokers ($p < 0.05$ and $p < 0.01$, respectively). Open triangle and paired open triangles indicate significant overall differences between grain workers and controls ($p < 0.05$ and $p < 0.01$, respectively). In subsequent figures the dagger and double dagger indicate significant interaction between smoking and grain dust exposure ($p < 0.05$ and $p < 0.01$ respectively).

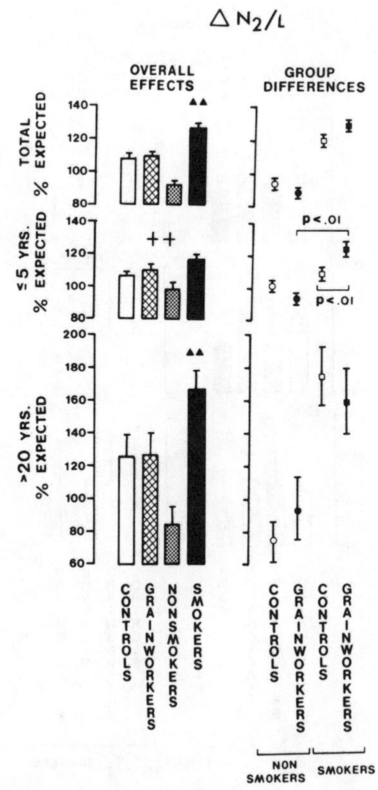

FIGURE 2. Mean percent expected[13] values (±1 SEM) for $\Delta N_2/l$ (see Figure 1 for details).

tis in grain workers compared with community controls suggests that exposure to grain dust causes effects on the tracheobronchial tree that are separate from the effects expected by exposure to cigarettes. Fletcher and Peto[18] have recently suggested that cigarette smokers suffer from both airflow obstruction and mucus hypersecretion and that these may represent unrelated disease processes. Although chronic airflow obstruction has been associated with respiratory disability and compensation in smoking grain workers,[19] we found more dramatic effects of grain dust exposure on respiratory symptoms in grain workers. The significance of the increase in chronic bronchitis in workers in terms of eventual respiratory disability remains uncertain, but the nuisance and discomfort of this symptom for workers must also be considered.

Smoking grain workers in the lower exposure category had a greater $\Delta N_2/l$ and a lower $\Delta \dot{V}_{max50}$ than either smoking controls of similar age and pack years of cigarette exposure or nonsmoking grain workers. This suggests that the combination of cigarette smoking and occupational exposure in these cereal grain workers leads to obstruction in peripheral airways,[10] a phenomenon not associated with either the matched groups of nonsmoking grain workers or the smoking community controls. While the significance of peripheral airways obstruction in these smoking workers remains uncertain, such functional abnormalities correlate well with inflammatory changes in the small airways in cigarette smokers.[20] The fact that these changes can be found in smoking grain workers who are young (age 23.5 ± 5.7 years), but are not observed in smoking community controls, suggests that the addition of occupational exposure to grain dust in smokers may accelerate the pathological changes in the peripheral airways of the lung in these workers. Recent animal experiments indicate that acute exposures to high atmospheric concentrations of crude grain dust primarily cause a proliferation of macrophages at the alveolar level[21] similar to the effects of instillation of carbon particles or the acute effects of cigarette inhalation.

The environmental survey carried out during the same period as the pulmonary function and questionnaire survey indicated that the concentration of total dust was usually less than the 10–mg/m^3 threshold limit set by Labour Canada.[22] Despite this, we were able to demonstrate changes in lung

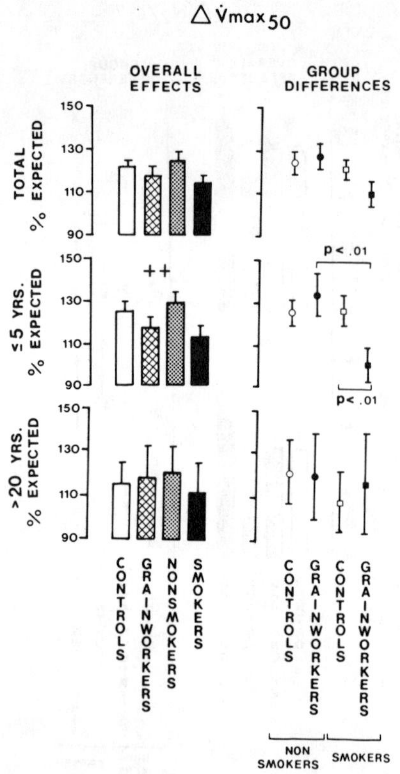

FIGURE 3. Mean percent expected[9] values (±1 SEM) for $\Delta\dot{V}_{max50}$ (see Figure 1 for details).

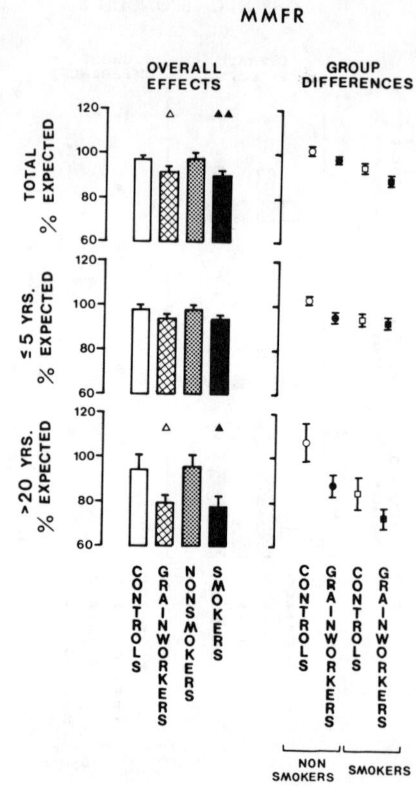

FIGURE 4. Prevalence of chronic bronchitis (%) defined as chronic phlegm production for at least 3 months/years for the previous 2 years. Younger category of workers was altered to include only workers employed greater than 2 years, but less 5 five years (see Figure 1 for details).

function and respiratory symptoms in grain workers in both low– and high–exposure categories. The environmental measurements only reflect current exposure levels, and cannot be used to assess cumulative dust exposure. Nevertheless, the demonstration of grain dust effects at these low current dust levels suggests that the present threshold limit value of 10 mg/m^3 for total grain dust may require reevaluation in the future.

In summary, decreases in lung function and increased respiratory symptoms in grain elevator workers suggest that the lung is being adversely affected by occupational exposure to grain dust. The separate effects of grain dust exposure and smoking appear to be similar and their combined effects to be additive. However, the effects of both smoking and grain dust exposure may have been affected by the multiple factors that determine initial and continued exposure to both grain dust and cigarette smoke. Clearly, a prospective study is necessary to determine more accurately the effects of grain dust exposure.

REFERENCES

1. **Mink, J.T., Gerrard, J.W., Cockcroft, D.W., Cotton, D.J., and Dosman, J.A.,** Increased nonspecific bronchial reactivity to inhaled histamine in nonsmoking grain workers with normal lung function, *Chest,* 77, 28, 1980.
2. **Dosman, J.A., Cotton, D.J., Graham, B.L., Li, K.Y.R., Froh, F., and Barnett, G.D.,** Chronic bronchitis and decreased forced expiratory flow rates in lifetime nonsmoking grain workers, *Am. Rev. Respir. Dis.,* 121, 11, 1980.
3. **Becklake, M.R., Jodoin, G., Lefort, L., Rose, B., Mandl, M., and Fraser, R.G.,** A respiratory health study of grainhandlers in St. Lawrence River ports, in *Occupational Pulmonary Disease: Focus on Grain Dust and Health,* Dosman, J.A. and Cotton, D.J., Eds., Academic Press, New York, 1980, 239.
4. **Broder, I., Mintz, S., Hutcheon, M., Corey, P., Silverman, F., Davies, G., Leznoff, F., Peress, L., and Thomas, P.,** Comparison of respiratory variables in grain elevator workers and civic outside workers of Thunder Bay, Canada, *Am. Rev. Respir. Dis.,* 119, 193, 1979.
5. **Chan–Yeung, M., Schulzer, M., MacLean, L., Dorken, E., and Grzybowski, S.,** Epidemiological health survey of grain elevator workers in British Columbia, *Am. Rev. Respir. Dis.,* 121, 329, 1980.
6. **Broder, I., Mintz, S., Hutcheon, M.A., Corey, P., and Kuzyk, J.,** Effect of layoff and rehire on respiratory variables in grain workers, *Am. Rev. Respir. Dis.,* 122, 601, 1980.
7. **Weil, H. and Diem, J.,** Relationships between cigarette smoking and occupational pulmonary disease, in *Occupational Pulmonary Disease: Focus on Grain Dust and Health,* Dosman, J.A. and Cotton, D.J., Eds., Academic Press, New York, 1980, 65.
8. **Elmes, P.C.,** Relative importance of cigarette smoking in occupational lung disease, *Br. J. Ind. Med.,* 38, 1, 1981.
9. **Dosman, J.A., Bode, F., Urbanetti, J., Martin, R., and Macklem, P.T.,** The use of a helium–oxygen mixture during maximum expiratory flow to demonstrate obstruction in small airways in smokers, *J. Clin. Invest.,* 55, 1090, 1975.
10. **Cotton, D.J., Graham, B.L., Li, K.–Y.R., Froh, F., Barnett, G.D., and Dosman, J.A.,** Effects of smoking and grain dust exposure in young workers, *Am. Rev. Respir. Dis.,* 126, 660, 1982.

11. **Schoenberg, J.B., Beck, D.J., and Bouhuys, A.,** Growth and decay of pulmonary function in healthy blacks and whites, *Respir. Physiol.*, 33, 367, 1978.
12. **Morris, J.F., Koski, A., and Johnson, L.C.,** Spirometric standards for healthy nonsmoking adults, *Am. Rev. Respir. Dis.*, 103, 57, 1971.
13. **Dosman, J.A., Cotton, D.J., Graham, B.L., Hall, D., Li, R., Froh, F., and Barnett, G.D.,** Sensitivity and specificity of early diagnostic tests of lung function in smokers, *Chest,* 79, 6, 1981.
14. **Steel, R.G.D. and Torrie, J.H.,** *Principles and Procedures of Statistics,* McGraw–Hill, New York, 1960.
15. **Mantel, N. and Haenszel, W.,** Statistical aspects of analysis of data from retrospective studies of disease, *J. Natl. Cancer Inst.*, 22, 719, 1959.
16. **Yoshida, K. and Maybank, J.,** Physical and environmental characteristics of grain dust, in *Occupational Pulmonary Disease: Focus on Grain Dust and Health,* Dosman, J.A. and Cotton, D.J., Eds., Academic Press, New York, 1980, 441.
17. **Farant J.-P. and Moore, C.F.,** Dust exposures in the Canadian grain industry, *Am. Ind. Hyg. Assoc. J.*, 39, 177, 1978.
18. **Fletcher, C. and Peto, R.,** The natural history of chronic airflow obstruction, *Br. Med. J.*, 1, 1645, 1977.
19. **Elliott, G.F.,** Problems related to compensations, in *Occupational Pulmonary Disease: Focus on Grain Dust and Health,* Dosman, J.A. and Cotton, D.J., Eds., Academic Press, New York, 1980, 581.
20. **Cosio, M., Ghezzo, M.S.C., Hogg, J.C., Corbin, R., Loveland, M., Dosman, J., and Macklem, P.T.,** The relations between structural changes in small airways and pulmonary function tests, *N. Engl. J. Med.*, 298, 1277, 1978.
21. **Armanious, M., Cotton, D.J., Wright, J., Dosman, J.A., and Carr, I.,** Grain dust and alveolar macrophages: an experimental study of the effects of grain dust on the mouse lung, *J. Pathol.*, 136, 265, 1982.
22. Labour Canada 1978, Guidelines for Environmental and Medical Surveillance Program in the Grain Industry, Ottawa Occupational Safety and Health Branch File No. 895–7–11, January 1978.

Some Biological Components of Grain and Forage Dusts

J. D. Smith

INTRODUCTION

Dust components, especially those of biological origin, originating from the production and processing of grain and fodder, may be deleterious to human and animal health. Primary plant pathogens and pests degrade plant tissues, facilitating colonization by secondary organisms. Plant tissues break down, fragment, and become comminuted and dispersed during harvesting, wetting, drying, and processing. Mineral and biological topsoil components and fragments of associated flora and fauna are dust constituents. Pollen, fungus spores, and animal remains are often allergenic. Pesticide residues of uncertain toxicity may be present and some inorganic soil or plant components, for example, small silica particles, are potential pulmonary carcinogens. The different dust components may act synergistically to produce ill health.[1,2]

The main components of field dust are (1) plant cells and fragments from the crop or weeds, including pollen; (2) soil particles, e.g., clay, silica, and degraded plant material (humus); (3) bacteria, pathogenic and nonpathogenic, from plants and animals; (4) actinomycetes from all hosts and habitats; (5) fungi—plant and animal pathogens and saprobes; (6) protozoa and nematodes—small forms may be airborne; (7) insects and mites from plants and animals; and (8) other animal remains, including feces of birds and small animals.

DEVELOPMENT OF MICROORGANISMS AND PESTS IN STORED CROPS

Although each crop species has its particular range of field pathogens and primary animal pests, they may have secondary organisms in common which colonize or consume the harvested tissues. A reduction in numbers of many of the pathogenic and saprobic species of the field microflora and in the microfauna occurs when grain or fodder is stored. However, some wet–stored, mycotoxin–producing *Fusarium* spp. may continue growth when storage temperatures are low and humidities high.[3] Some microorganisms with resistant life forms may endure long periods of dry storage, but some insects and mites are very sensitive to desiccation and die. Storage fungi of the genera *Aspergillus* and *Penicillium* originate in the field in small numbers and contaminate the grain.[4] Grain pests are harbored in remnants of the previously stored crop, in outdoor spillage, or move in from natural habitats. In stored grain, species of *Aspergillus* and *Penicillium* are usually the most important contaminants and degraders. They may also be allergenic and produce mycotoxins. They can affect great bulks of grain and contribute much to store dustiness following wet harvests. At temperatures below 5°C and grain moisture below 11% little mold activity is likely, although some species may grow very slowly below –7°C. Insects will raise the temperature when grain is stored above 18°C, but will cease activity at about 0°C, whereas mites may remain active at 0°C and can survive at lower temperatures.[5,6]

Xerophilic fungi, such as species in the *A. restrictus* and *A. glaucus* groups, colonize stored, tough grain first and produce moisture for less xerophilic species in the *A. versicolor* group and the actinomycete *Streptomyces griseus*. With increases in temperature and moisture, *A. fumigatus, Streptomyces* spp., *Microplyspora faeni,* and *Thermoactinomyces vulgaris* take over. Heating in large bulks of stored, damp grain is often associated with heavy infestations of insects and mites which thrive in condensed moisture on the outside or in wet patches deep in the grain mass. More heat is generated than can be dissipated and, successively, mesophilic, thermophilic, and extremely thermophilic fungi, actinomycetes, and bacteria develop.[3,6,7]

The activities of grain insects, mites, fungi, actinomycetes, and bacteria are interlocked in a food economy through a succession of energy transfers (Figure 1). Rodents, birds, and primary consumer insects and mites cause seed injury and fungal infection. These insects and mites produce feces and food debris, improving the C/N ratio, favoring the second–level feeders which, in turn, leave debris suitable for the scavenging third–level consumers. In this complex cycling and recycling of food materials, grain and animal debris, microbial spores, and structures are produced, with accumulated toxins and allergens.[3]

Fodder is produced from a wide range of forage species. Some saprobes and secondary consumers may be common to forage and grain plants. If fodder is made rapidly and stored dry, few saprobes will colonize it. At less than 18% moisture there is no heating or molding, but if baled at above 30% moisture content it may heat spontaneously. Between 40 and 60% moisture and 40 to 60°C the allergenic thermophilic actinomycetes *M. faeni, Saccharomonospora viridis,* and *T. vulgaris,* implicated in farmer's lung, and fungi of the allergenic, pathogenic, and mycotoxigenic *A. fumigatus* group proliferate.[8]

MYCOTOXIGENIC FUNGI IN GRAIN AND FODDER DUSTS

Species of field and storage fungi in the genera *Fusarium, Aspergillus,* and *Penicillium,* in particular, are potential producers of toxic metabolites, e.g., trichothecenes, aflatoxins,

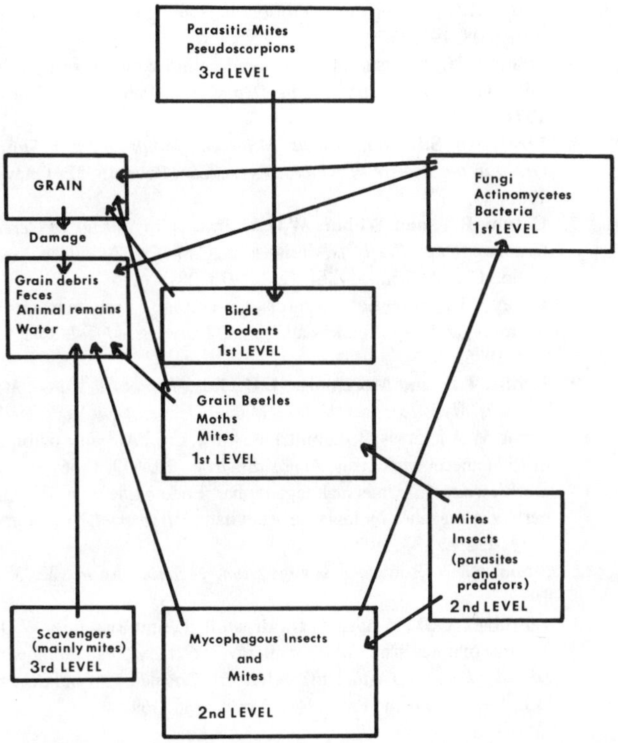

FIGURE 1. Food economy in a grain bulk (After Sinha, R. N., *Ann. Technol. Agric.*, 22, 351, 1973.)

and ochratoxins. While ingestion is the usual route of uptake by humans and animals, the possibility of dermal uptake seems feasible and some of the mycotoxins are potent carcinogens.[9] Recent studies have shown that long–term inhalation of spores of *Stachybotrys atra,* a straw saprobe, may result in intoxication.[10] The possibility of synergistic or adjuvant effects between allergens, pathogens, and mycotoxins seems worthwhile pursuing.

BACTERIA IN GRAIN DUST

Many bacterial species have been recorded as pathogens, epiphytes, or saprobes of grain or fodder crops. Bacteria seem to be less important as spoilage organisms of grain than fungi or actinomycetes, except in the last stages of heating. Spores of the yellow pigmented strains of *Erwinia herbicola* are predominant among the Gram–negative airborne bacteria, particularly in grain–processing plants. Some isolates of these produce potent endotoxins.[11]

MITES IN GRAIN AND FODDER DUSTS

Mites are ubiquitous on grain and fodder crops in the field. Hay stores may show heavy infestations. More than 40 species have been found in stored grain, where *Aspergillus siro* and *A. ferris* are important species. Some species show a preference to particular grain and storage fungi species and storage conditions.[12] Their remains, which are known allergens, are often present in considerable amounts in grain and fodder dusts. *Siteroptes graminum,* associated with the silvertop disease of many grass species, may be the principal agent in the spread of spores of *Fusarium poae,* a potentially mycotoxigenic fungus and a weak pathogen on grasses and cereals. Skin tests have shown allergenic responses to the house dust mite, *Dermatophagoides pteronyssinus,* and to *Acaris siro, Tyrophagus putrescens, Glycyphagus domesticus,* and *G. destructor* from hay.[13]

PLANT PARTS IN GRAIN DUST

Most of the organic (combustible) grain dust in elevator grain–cleaning operations of wheat is derived from the fibrous cells of the pericarp and trichomes (epidermal hairs) abraded in the cleaning process.[2] Readily identifiable shrunken pollen cells which are potential allergens, and outer cells of the pollen sacs, are present also. The trichomes from wheat, 100 to 1500 μm in length, are mainly from the brush end of the grain. They are brittle and fragment transversely and longitudinally into variably sized pieces; the finer ones may then act in similar fashion to asbestos fibers. Organic dust particles frequently support the growth of fungi, yeasts, and bacteria. Fungal hyphae are sometimes noted infesting the trichomes, whose sharp points could perhaps act as "microhypodermic needles", facilitating the entry of allergens and toxins into the upper respiratory tract.

DISCUSSION

Although the most significant environmental factors determining the succession of the wide range of organisms in stored crops are moisture and temperature, substrate and pH also influence this. It has proved difficult to predict which species will become dominant from an *in vitro* study of their interactions. Enzymic heating of grain or the subsequent heating brought about by high insect or mite populations influences its moisture content and the subsequent succession of storage microorganisms, insects, and mites in an interlocking food economy. Both grain and fodders are abundant sources of a wide range of storage mites, their remains, and excreta, some with specific allergic responses. Some mites show feeding preferences with specific fungi. The possible synergistic or adjuvant properties of the favored fungus to the mite in causing allergies need testing. Mites and actinomycetes appear to play a greater role in the degradation of stored fodders and the production of allergies, such as farmer's lung, than in grain, yet respiratory problems associated with hay handling are receiving less attention in North America than elsewhere.

Although many storage fungal species of grain and fodders are known to be potentially mycotoxigenic, little is known about the health hazards associated with the inhalation of mycotoxins, especially in low concentrations, especially on the adjuvant or synergistic effects of mycotoxins with allergens and pathogens in the dust of stored crops.

Many bacterial species occur as plant pathogens, epiphytes, or saprobes of grains and fodders. Possibly they have received less attention as causes of health problems of workers because they have not generally been considered as important economically as other grain organisms or pests in grain deterioration. The potent endotoxins produced by some yellow pigmented isolates of *Erwinia herbicola,* common in the air of grain–processing plants, could possibly enhance the immunological responses of antigens from actinomycetes and fungi implicated in farmer's lung.

The sharp–pointed fragmented trichomes abraded from wheat grains during cleaning have been considered as irritants of the upper respiratory tract. This possibility could be examined by respiratory challenge studies. The finer fragments of these are of respirable size and the associated fungi, bacteria, and yeasts and the pollen in the grain dust are potential allergens.

REFERENCES

1. **Abraham, J.,** Pulmonary disease among farm workers, *Ann. Am. Conf. Gov. Ind. Hyg.,* 2, 293, 1982.
2. **Yoshida, K. and Maybank, J.,** Physical and environmental characteristics of grain dust, in *Occupational Pulmonary Disease: Focus on Grain Dust and Health,* Dosman, J. and Cotton, D., Eds., Academic Press, New York, 1980, 441.
3. **Sturz, A.V. and Johnson, H.W.,** Early colonization of the ears of wheat and barley by *Fusarium poae, Can. J. Plant Pathol.,* 5, 107, 1983.
4. **Sinha, R.N.,** Ecology of microflora in stored grain, *Ann. Technol. Agric.,* 28, 191, 1979.
5. **Sinha, R.N.,** Spoilage of farm stored grain by molds, insects and mites in Western Canada, *Can. Dep. Agric.,* Publ. 1437, Ottawa, 1971.
6. **Loschiavo, S.R.,** Ed., *Insects, Mites and Molds in Farm Stored Grain in the Prairie Provinces,* Agric. Can. Publ. 1595E, Ottawa, 1983, 31.
7. **Cotton, R.T. and Wilbur, W.D.A.,** Insects, in *Storage of Cereal Grain and Their Products,* Christensen, C.M., Ed., American Association Cereal Chemistry, St. Paul, 1978, 281.
8. **Lacey, J.,** Potential hazards to animals and man from micro–organisms in fodders and grain, *Trans. Br. Mycol. Soc.,* 65, 171, 1975.
9. **Wyllie, T.D. and Morehouse, L.G.,** Eds., *Mycotoxic Fungi, Mycotoxins, Mycotoxicoses,* Vol. 1., Marcel Dekker, New York, 1977.
10. **Croft, W.A., Jarvis, B.B., and Yatawara, C.S.,** Airborne outbreak of trichothecene toxicosis, *Atmos. Environ.,* 20, 549, 1986.
11. **Dutkiewitz, J.,** Untersuchungen uber Endotoxine von Erwinia herbicola und ihre biologische Activitat, *Z. Bacteriol Hyg. Abt. 1 Orig. A,* 236, 487, 1976.
12. **Sinha, R.N.,** Ecology of storage, *Ann. Technol., Agric.,* 22, 351, 1973.
13. **Cuithbert, O.D., Wraith, D.G., Brostoff, J., and Brighton, W.D.,** The role of mites in hay and dust allergy, in *Occupational Pulmonary Disease: Focus on Grain Dust and Health,* Dosman, J.A. and Cotton, D.J., Eds., Academic Press, New York, 1980, 469.

The Threshold Limit Value for Grain Dust

R. S. Ratney

The establishment of a threshold limit value (TLV) for grain dust or, for that matter, for any toxicant in the workplace is essentially an exercise in the evaluation of dose–response data. Doses and exposures may be created in carefully controlled animal or human experiments or may be the results of "natural experiments" in which people are exposed in the normal course of their work. In either case it is necessary to identify clearly the health effects of concern and to determine the incidence of adverse health outcomes produced by different levels of exposure (including none). In many cases it is possible to observe an exposure level below which identical health outcomes are observed for exposed and nonexposed subjects. This is the TLV.

The most prominent health effect of grain dust, the one most obvious to workers and most familiar to their physicians, is an acute irritation of the upper respiratory tract, eyes, and occasionally the skin. Cough, wheezing, breathlessness, inflamed and watery eyes, and intense itching of the skin have been described. In 1713 Ramazinni was the first to publish a description of these symptoms, and in the succeeding centuries case reports and epidemiologic studies have identified and, to some extent, quantified the contributions of grain species, fungal and microbial toxins, smoking, and airborne dust concentrations to these acute health effects. A gradual decline in pulmonary function apparently occurs with repeated exposure to grain dust. Careful studies, particularly in Canada, have explored the effects of smoking, atopy, layoff, and dust concentration on this decline.

In the past two decades worker exposures to airborne contamination have been measured in programs monitoring total and respirable dust as well as fungal and microbial spores and pesticides.[1–6] Although there has been wide variation between studies, the relative dustiness of different operations within the grain–handling industry has been consistent, independent of the country or the method of measurement. A few reports have been published[1,7,8] in which health effects and dust exposures were compared.

These studies suggest that grain dust is more than a nuisance dust. For purposes of establishing TLV, the American Conference of Governmental Hygienists notes that "nuisance dusts have a long history of little adverse effect on lungs and do not produce significant organic disease or toxic effect when exposures are kept under reasonable control."[9] A TLV of 10 mg/m^3 is assigned to nuisance dusts, although there is no implication that this is some sort of boundary between safe and unsafe conditions. The TLV for grain dust should thus be below 10 mg/m^3. doPico et al.[1] showed that workers with exposures below 4–mg/m^3 demonstrated a prevalence of respiratory symptoms similar to control subjects. A TLV of 4 mg/m^3 has thus been recommended and placed on a "Notice of Intended Changes" to elicit new data and comment. The proposal of a 4 mg/m^3 TLV has encouraged the submission of a number of comments primarily critical of the doPico et al. study on perceived statistical and methodological grounds. None of the other published reports suggests that a different value should be chosen.

Despite all of the medical and exposure data available, the establishment of an appropriate TLV for grain dust has not been easy. Two characteristics of the published data make the construction of dose–response curves difficult. First, exposure measurements have generally not been made for the individual workers whose health status was being determined, and second, the time spans for which exposures were determined did not match the time needed for grain dust to cause the health effects being observed.

The first problem is more or less self–evident. It is usually difficult or impossible to determine the actual contaminant exposure of every subject in an epidemiological study. Generally, representative samples are taken assuming (or perhaps hoping) that all workers with the same job title or working in the same facility have the same exposures. It is recognized that there will be a distribution of actual exposures in a group, but the measurements actually taken represent the best estimate of group average and, in any case, are usually the only measurements available.

The second problem has been discussed by Roach,[10] but it has not been widely recognized even among professional industrial hygienists. Many substances, grain dust among them, produce a health effect whose onset and decay closely follow changes in airborne contamination. For instance, if a worker in a nonirritating environment is suddenly subject to a high grain–dust concentration, because of his entry into a dusty workspace or because the process produces a brief spurt of dust, he will rapidly experience the characteristic irritation. As the dust concentration declines, the irritation will wane. A device which collects an integrated dust sample will average the high–concentration spikes with the lower exposures during the rest of the day and will provide a measure of exposure which is not at all characteristic of the conditions which caused the irritation. If the worker is asked at the end of the day about his subjective experience, he will not usually report some average level of irritation, but only whether irritation was experienced at some time during the day. In order to determine the actual dust concentration which causes irritation, it would be necessary to use a continuous recording instrument or to take a large number of short–term samples.

The appropriate exposure measure likely determining chronic decline in pulmonary function is cumulative dose (mg/m^3 × year). This is impossible to determine for an individual with any degree of completeness. Rather, what is done is to measure current exposures and to estimate previous dust concentrations using what are thought to be reasonable assumptions. This process introduces a large variance into the estimate of integrated dose which may overwhelm the other sources of variance used in the construction of dose–response relationships. One cannot do anything about measuring past exposures, but one should subject any stated assumptions about them to a searching analysis of their validity and also seek out unstated assumptions that may have been used.

These difficulties make it impossible to establish a TLV that is completely valid and could withstand the most searching criticism. However, a TLV is not expected to have the same precision as a physical constant such as the speed of light, and with full recognition of the inaccuracies of estimating exposures and disease prevalences it is still possible to derive a best estimate for a level which would minimize or prevent adverse health effects.

As stated earlier, grain dust is clearly not a nuisance dust and, thus, it would be inappropriate to set a TLV of 10 mg/m^3. This still leaves a wide range of possible values. The work of doPico[1] suggests that 4 mg/m^3 is without adverse effects. Enarson, et al.[7] present evidence that workers will not experience severe chronic decrements in lung function at exposures below 5 mg/m^3. In both studies one could find fault with the exposure characterization and the epidemiology, but it is significant that both reached similar conclusions on the basis of different health effects in different groups of workers in different geographical areas. In the absence of any contrary evidence, 4 or 5 mg/m^3 has to be considered as the best estimate of the appropriate TLV.

There has been some uncertainty as to whether grain dust exposures should be measured in terms of total dust concentrations or of the respirable mass fraction. Some investigators have used one, some the other, and some have measured both total and respirable dust. Respirable dust measurements would seem to be the most relevant to health effects in the gas exchange region of the lung. This does not take into consideration the physical characteristics of grain dust and of size–selective sampling devices. Ideally, one would like to be able to sample only that fraction of the airborne dust which would penetrate the upper airways and reach the alveolar spaces. The cyclones and elutriators which have been developed to accomplish this purpose have generally been found to mimic fairly closely the particle–size selection characteristics of the human lung when tested with mineral dusts or unit density polystyrene spheres. Grain dust particles are different from mineral dusts in that their particle densities are much lower, they develop electrostatic charges more readily, and absorb signficant quantities of water in humid atmospheres. Cyclones and elutriators differ from lung airways in that they operate at ambient humidities rather than the 100% humidity in the lung, can maintain persistent electrostatic charges, and have different airflow characteristics. It is thus questionable whether the currently used devices have the same size–selection characteristics as the human lung. On the other hand, total dust measurements are accurate indicators of environmental conditions and no assumptions have to be made relating measured dust concentrations and actual lung doses. Accordingly, for the present, it is recommended that worker exposures be measured in terms of total rather than respirable dust.

Grain, of course, is not a single entity. Wheat, corn, barley, rye, oats, rice, and millet are grown, stored, and shipped as bulk commodities throughout the world. A given facility may handle several different grains at different seasons and on occasion may deal with nongrain agricultural commodities such as soybeans. Most of the published epidemiological investigations do not identify the species and variety of grain to which their study subjects were exposed. In the terminal elevators of the Great Lakes and Pacific Northwest ports, the major grain is wheat, but this is not clearly identified in the major Canadian and U.S. studies.

Strictly speaking, the studies of doPico et al.[1] and Enarson et al.[7] should only be used to support a TLV for wheat dust and not for all grains, since it was presumably wheat that was being handled in the elevators where their test subjects worked. This would leave open to question the appropriate control level to be used in the large number of facilities throughout the world where other grains are processed. However, even though dose–response data have not been published for grains other than wheat, there have been reports that other grains are qualitatively similar. Cockcroft et al.,[11] in a study of British dock workers, showed that barley dust is particularly irritating and described symptoms that are similar to those experienced by workers exposed to wheat dust. Darke et al.[12] studied symptom prevalences and pulmonary function changes in grain harvesters and stated that barley, wheat, and oat dusts all caused similar respiratory distress. On this basis, it is recommended that these three grain dusts be covered by the same TLV.

REFERENCES

1. **doPico, G.A., Reddan, W., Anderson, S., Flaherty, D., and Smalley, E.,** Acute effects of grain dust exposure during a work shift, *Am. Rev. Respir. Dis.*, 128, 399, 1983.
2. **Farant, J.P. and Moore, C.F.,** Dust exposures in the Canadian grain industry, *Am. Ind. Hyg. Assoc. J.*, 39, 177, 1978.
3. **Schrag, K.R. and Eng, P.,** Dust concentrations in Alberta's grain industries, *Am. Ind. Hyg. Assoc. J.*, 33, 100, 1972.
4. **Machida, T.,** Atmospheric grain dust contamination survey at a country elevator, *Sci. Rep. Fac. Agric. (Ibaraki Univ.)*, No. 25 (Abstr.), 81, 1977.
5. **Molocznik, A.,** Assessment of air pollution with dust in the working medium of grain mills and elevators, *Biul. Sluzby Sanit. Epidemiol. Wojewodztwa Katowickiego*, 15 (Abstr.), 153, 1971.
6. **Molocznik, A.,** Estimation of dustiness of working places connected with the type of grain processing, *Rocz. Nauk Roln. Ser. C*, 75 (Abstr.), 223, 1981.
7. **Enarson, D.A., Vedel, S., and Chan–Yeung, M.,** Rapid decline in FEV_1 in grainhandlers: relation to level of dust exposure, *Am. Rev. Respir. Dis.*, 132, 814, 1985.
8. **Corey, P., Hutcheon, M., Broder, I., and Mintz, S.,** Grain elevator workers show work–related pulmonary function changes and

dose–effect relationships with dust exposure, *Br. J. Ind. Med.*, 39, 330, 1982.

9. American Conference Government Industrial Hygienists, Threshold Limit Values for Chemical Substances and Physical Agents in the Work Environment and Biological Exposure Indices with Intended Changes for 1984—1985.
10. **Roach, S.A.,** A more rational basis for air sampling programs, *Am. Ind. Hyg. Assoc. J.*, 27, 1, 1966.
11. **Cockcroft, A.E., McDermott, M., Edwards, J.H., and McCarthy, P.,** Grain exposure: symptoms and lung function, *Eur. J. Respir. Dis.*, 64, 189, 1983.
12. **Darke, C.D., Knowelden, J., Lacey, J., and Milford, W.A.,** Respiratory disease of workers harvesting grain, *Thorax,* 31, 294, 1976.

Occupational Asthma

Occupational Asthma: An Overview

J. Pepys

INTRODUCTION

Occupational asthma has become, only recently (in 1981), a prescribed industrial disease in the U.K., vs. 1961 in Germany and 1965 in France, in spite of the frequent invocation of Ramazzini's dictum, "What is your occupation?" Seven categories of causes are listed at present in the U.K.: (1) isocyanates; (2) complex platinum salts; (3) acid anhydride and polyamine hardening agents; (4) fumes from the use of pine resin, colophony in the bonding of soldering flux; (5) proteolytic enzymes — used in the manufacture of "biological" washing powders, and in baking, brewing, fish, silk, and leather industries; (6) animals or insects used for research, education, or in laboratories; (7) dusts arising from milling or handling flour and from harvesting, drying, and transporting grain. In relation to agriculture, those categories dealing with agents of organic origin are more readily appreciated, but the chemical agents listed are models for the possible role of chemicals as a whole. A comprehensive review of occupational asthma in general and related to the above list is available in a multiauthor text.[1]

In view of the common prevalence of bronchial disorders, the diagnosis of occupational asthma has to be based on: (1) definitive criteria for the diagnosis of asthma; (2) evidence for an occupational causal relationship; and (3) identification of the specific immunological causal agent and distinction from nonspecific effects. A limited number of the many facets of these questions which appear to be of particular interest will be discussed.

ASTHMA AND PATTERNS OF ASTHMATIC RESPONSE

Asthma can be defined as airway hyperresponsiveness to specific immunological, i.e., allergic, or nonspecific, nonimmunological agents. The distinguishing feature of allergy is that in sensitized subjects very small and, often, minute doses can elicit asthmatic responses, in contrast to the larger doses needed to produce responses by nonimmunological mechanisms in either sensitized or nonsensitized subjects.

Evidence that the asthma is related to the occupation can be shown by self–recording of peakflow rates at frequent intervals during the working day over sufficiently long periods to cover times at and away from work.[2]

Occupational–type provocation tests have shown that there are several patterns of asthmatic response in terms of time, duration, severity, and response to therapeutic agents, which may be present alone or in various combinations.[3,4] These are immediate or nonimmediate, the latter comprising three patterns. The first of these occurs between 1 and 5 h after the test; the second and most common starts after several hours, is maximal at 5 to 8 h, and resolves within a day; and the third and latest is a nocturnal recurrent response, appearing about 16 h after the test and in which circadian rhythms play a part. The recurrent response, capable of appearing nightly for many nights after the single small challenge required for allergic reaction, is important clinically because it may obscure the diagnosis of occupational asthma by persisting over long periods when away from work. Improvement of asthma when off work is one of the criteria used as evidence for an occupational relationship. It is necessary to ensure that a recurrent response is not occurring at the time of provocation testing; this could cause false–positive results to an unrelated agent under test.

In addition to these provoked patterns of response, it is common for subjects with occupational asthma to describe a period during which exercise–induced bronchospasm was present, followed later by asthma coming on in relation to work without exercise, often in the evening at the end of a day's work. This is consistent with a nonimmediate pattern of response and suggests, also, that in the early stages of sensitization, airway hyperresponsiveness was developing and was manifest by the effects of exercise.

FACTORS INFLUENCING RESPIRATORY ALLERGY

Three main factors are involved in the development of respiratory allergy to occupational or nonoccupational agents: (1) immunological reactivity of the subject; (2) the nature and source of the causal agents; and (3) the circumstances of exposure.

Immunological Reactivity of The Subject

The most important immunological aspect in relation to sensitization to organic agents and to a number of low–molecular–weight chemicals is the presence of an atopic constitution. Atopy is defined here as the capacity to produce specific IgE antibody in response to ordinary exposure to common environmental allergens. This can readily be determined by prick skin tests with a battery of three or four relevant allergens. Between 30 and 40% of asymptomatic populations belong in this group and may show associated decreased β–adrenergic and increased α–adrenergic and cholinergic activity.[5] They can be regarded as primed to react to specific and/or nonspecific stimuli. Such subjects can become sensitized to certain occupational allergens more rapidly and in a higher proportion than those classified as nonatopic. In certain industries preemployment screening for atopy in this way is used to select employees and thereby reduce the rate of sensitization. There is, however, the problem of whether to exclude so many subjects in order to prevent sensitization and respiratory symptoms in only a few subjects as, for example, in workers handling experimental animals. Control of exposure to prevent sensitization is desirable, but not readily feasible in agriculture. Guidance on career selection by atopy screening has a place in this context.

The respiratory symptoms in atopics associated with IgE sensitization are rhinitis and asthma. By contrast, the production of extrinsic allergic alveolitis by inhaled organic dusts is a feature of nonatopic subjects.

The roles in occupational asthma of other classes of antibody capable of mediating type I reactions and of precipitins which could mediate type III reactions have yet to be determined.

Nature and Source of the Causal Agents

Allergy to organic environmental allergens such as pollens, mold spores, and organic dusts is common and the sources known. It has become evident in recent years that important sources of allergens are the feces and urine of animals, mites, and insects. These may be as, or more, important clinically than the dander allergens, for example, that are commonly used for investigation.

Excretory Allergens

The urine of mice (Figure 1) and rats, rabbits, and guinea pigs contains low–molecular–weight potent allergens.[6] In the case of the mouse these are present in the urine of the male and not the female. Allergens are also present in the urine of domestic animals and in the pig. It is probable that further study will widen the range of sources of urinary allergens. The presence of allergens mainly, in the fecal pellets of the house dust allergy mite genus, *Dermatophagoides,* suggests that similar studies should be made with storage mites common in agriculture and known to cause asthma. The allergens in

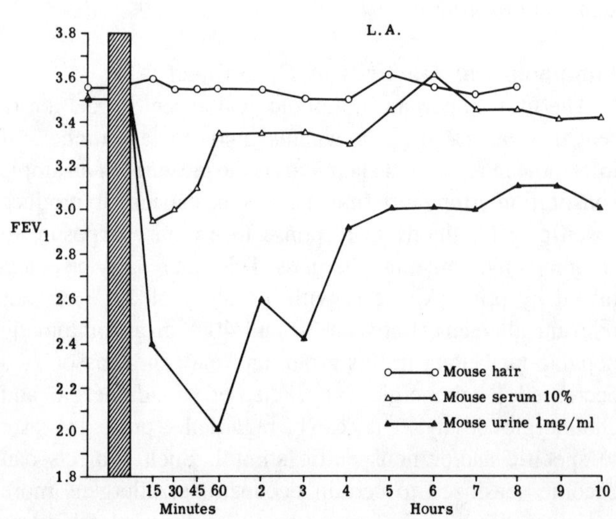

FIGURE 1. Bronchial provocation test. Immediate asthmatic response to mouse urine (excretory) allergen.

locust allergy are also found in the excretory tract and other insects, such as cockroaches, may fall into the same category. Avian serum proteins in the feces are an established cause of extrinsic allergic alveolitis; their role in asthma due to avian species has not, apparently, been studied.

The excretory allergens so far identified are specific and potent; their control and avoidance, where possible, are indicated.

Emanations

"Wet" Allergens

Emanations from maggots have been found to cause asthma and have been termed "wet" allergens.

Fume Emanations — Wood Products

The fumes from pyrolyzed pine resin, colophony, used as bonding for soft solders in the electronics industry etc. are a common cause of asthma found to affect about 20% of exposed workers.[2] The main allergen appears to be abietic acid. Affected workers can get asthma from exposure to pine trees, attributable to their resinous emanations. In one case, an asthmatic subject not exposed to colophony fumes developed asthma in a pine forest and had an asthmatic response to the occupational–type inhalation test with pyrolyzed abietic acid. Colophony is widely used in many forms, as a cold or hot adhesive, and may be present in cigarette tobacco, printed material, cosmetic preparations, etc. It is not known how sensitized subjects would react to pine sawdust or other pine products.

An organic acid, plicatic acid, has been identified as an allergen in Western red cedar.[7] This, together with the finding of abietic acid as a colophony allergen, suggests that other types of wood may have similar causal mechanisms. While cessation of work exposure may be followed by return to normal — as with Western red cedar[8] — avoidance of the relevant allergen or sensitizing chemical in the nonoccupational environment may not always be possible.

Low–Molecular–Weight Chemicals

Occupational asthma due to low–molecular–weight chemicals, organic and inorganic, is now being increasingly recognized and confirmed by occupational– type tests with minute amounts, far below the level of any nonspecific, nonimmunological effect, and by immunological findings. In many workers, for example, IgE and other antibodies to inorganic agents have been found, such as the complex halide salts of platinum (Figure 2)[9] and, in a single case so investigated, to nickel sulfate.[10] IgE and other antibodies to isocyanates, acid anhydrides, and other protein reactive chemicals have been found,[11] and asthmatic responses are shown by provocation tests (Figure 3).

In sensitization to low–molecular–weight chemicals, antibodies may be produced depending upon the methods, procedures, and conditions used against the hapten, the hapten–carrier conjugate needed for sensitization, or against the modified carrier protein itself. IgE antibodies have been

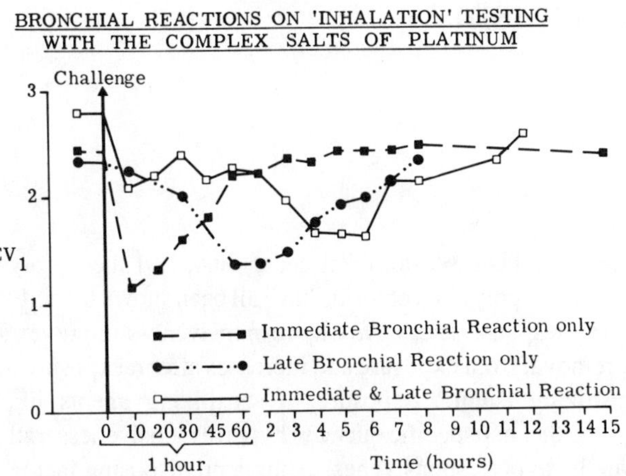

FIGURE 2. Bronchial provocation, occupational–type test. Patterns of immediate, late, and dual responses to complex halide salts of platinum in different sensitized subjects.

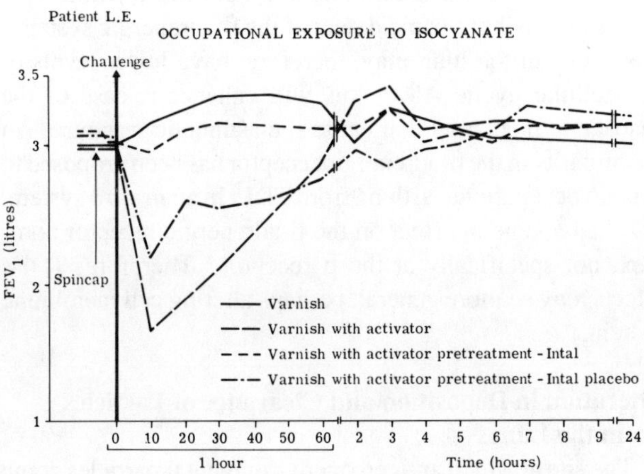

FIGURE 3. Bronchial provocation, occupational–type test. Immediate asthmatic response to toluene di–isocyanate (polyurethane varnish activator). Typical blocking of reaction by Intal (cromolyn sodium) and not by lactose placebo.

found against new antigenic determinants (NADs) produced in human serum albumin, for example, by chemical haptens. Serological tests have shown that similar or identical NADs may be produced by related or unrelated chemicals. This raises the possibility that in a subject sensitized by one chemical and with the production of NADs, allergic reactions may be produced by an unrelated chemical capable of modifying the human serum albumin so as to produce identical or similar NADs. The whole phenomenon, yet to be tested clinically, could be termed, "extrinsically determined — auto — allergy", resulting from exposure either to the actual sensitizing agent or to an unrelated chemical.[12]

Circumstances of Exposure

The circumstances of exposure influence sensitization and elicitation of responses in sensitized subjects. The effects of intensity and duration of exposure on occupational sensitization are known mainly for the highly potent agents, such as the halide salts of platinum. In this case and in general, atopics are sensitized more rapidly and by small amounts as befits their immunological constitution.

In sensitized subjects intervals between exposures can be relevant, with the tendency for reactions to occur, sometimes for the first time, after such an interval. A more regular effect is shown in humidifier fever and byssinosis where the bronchial response appears after each weekend off work.

CONCLUSIONS

Recognition of the modes of clinical onset and the patterns of asthmatic response elicited by occupational–type provocation tests is essential for the understanding of occupational asthma. Armed with this knowledge, history taking can be very revealing. Insight into this, under ordinary working conditions, often can be obtained by monitoring peak flow measurements made at frequent intervals over periods of several weeks. The responses to inhaled bronchodilators during this time may be helpful in deciding whether the asthmatic pattern at the time is of an immediate type, which responds well, or of the nonimmediate type which, as a rule, responds poorly. This information can be a guide to identifying causal exposures and is also of value for clinical management.

REFERENCES

1. **Pepys, J.,** Occupational respiratory allergy, in *Clinics in Immunology and Allergy,* Vol. 4 (No. 1), W. B. Saunders, London, 1984.
2. **Burge, P.S.,** Occupational asthma, rhinitis and alveolitis due to colophony, in *Clinics in Immunology and Allergy*, Vol. 4 (No. 1), Pepys, J., Ed., W.B. Saunders, London, 1984, 55.
3. **Pepys, J. and Hutchcroft, B.J.,** Bronchial provocation tests in etiologic diagnosis and analysis of asthma, *Am. Rev. Respir. Dis.,* 112, 829, 1975.
4. **Pepys, J.,** Clinical and therapeutic significance of patterns of allergic reactions of the lungs to extrinsic agents, *Am. Rev. Respir. Dis.,* 116, 573, 1977.
5. **Kaliner, M.A.,** Mast–cell derived mediators and bronchial asthma, in *Airway Reactivity,* Hargreave, F.E., Ed., Astra Pharmaceuticals, Canada, 1980, 175.
6. **Longbottom, J.L.,** Occupational allergy to animal allergens, in *Clinics in Immunology and Allergy,* Vol. 4 (No. 1), Pepys, J., Ed., W.B. Saunders, London, 1984, 19.
7. **Chan–Yeung, M., Giclas, P.C., and Henson, P.C.M.,** Activation of complement by plicatic acid, the chemical compound responsible for asthma due to Western red cedar (*Thuja plicata*), *J. Allergy Clin. Immunol.,* 65, 333, 1980.
8. **Chan–Yeung, M., Lam, S., and Koener, S.,** Clinical features and natural history of occupational asthma due to Western red cedar *(Thuja plicata), Am. J. Med.,* 72, 411, 1982.
9. **Pepys, J.,** Occupational allergy due to platinum complex salts, in *Clinics in Immunology and Allergy,* Vol. 4 (No. 1), Pepys, J., Ed., W.B. Saunders, London, 1984, 131.
10. **Malo, J.–L., Cartier, A., Doepner, M., Nieboer, E., Evans, S.L., and Dolovich, J.,** Occupational asthma caused by nickel sulfate, *J. Allergy. Clin. Immunol.,* 69, 55, 1982.
11. **Bernstein, I.L. and Berstein, D.I.,** Respiratory allergy to synthetic resins, in *Clinics in Immunology and Allergy,* Vol. 4 (No. 1), Pepys, J., Ed., W.B. Saunders, London, 1984, 83.
12. **Pepys, J.,** Auto–antibodies induced by extrinsic low molecular weight chemicals (EDAA), in *Proc. 12th Int. Congr. of Allergology and Clinical Immunology,* Reed, C.W., Ed., in preparation.

Occupational Asthma: Pathogenic Mechanisms

S. M. Brooks

DEFINITION

Occupational asthma is a disorder in which there is generalized, usually reversible airway obstruction caused by the inhalation of substances or materials that a worker manufactures, uses directly, or that are incidentally present at the work site. This encompasses a variety of airway disorders of different mechanisms, both immunologic and nonimmunologic.

PREDISPOSING FACTORS

Atopy imparts a greater risk of sensitization and development of asthma in selected occupations. The nature of the exposure to the offending agent (e.g., intermittent or continuous) may be a determinant. A significant relationship has been noted between exposure to isocyanate spills and the presence of clinical asthma and chronic bronchitis,[1] suggesting intermittent high-concentration exposure may be important. A relationship between the intensity of Western red cedar dust exposure and prevalence of occupational asthma has also been observed; sawyers, packers, and splitters — those with the dustiest jobs — showed the highest prevalence.[2]

MECHANISMS TO EXPLAIN OCCUPATIONAL ASTHMA

Immunologic

Immunologic mechanisms operate in many cases of occupational asthma, affecting primarily atopic individuals. Allergy to certain industrial materials — the enzymes of *Bacillus subtilis*, platinum salts, animal danders, locusts, or gum acacia — is more common in atopic than in nonatopic subjects.[1] By contrast, however, asthma caused by Western red cedar and toluene diisocyanate (TDI) seems to develop in more nonatopic persons. Antibodies other than IgE may contribute to tissue changes in some chronic allergic reactions; however, the relation to acute asthmatic responses is uncertain.

Airway Smooth Muscle Hyperresponsiveness

An important feature of most cases of occupational asthma is the increased responsiveness of the airways to physical, chemical, and pharmacologic stimuli. The mechanism(s) of airway hyperresponsiveness is (are) not known. Possible mechanisms include decrease in baseline airway caliber; increased responsiveness of the smooth muscle itself; changes induced by component(s) of inflammation; abnormality in the autonomic nervous control of the smooth muscle; and increase in the accessibility of the stimuli to target cells.[3]

Induced increases in airway responsiveness, associated with late but not early asthmatic responses, are important in the pathogenesis of occupational asthma. Dimethyl ethanolamine, TDI, Western red cedar dust, and the pyrolysis products of polyvinyl chloride have all been shown to increase airway responsiveness. Airway responsiveness improves after removal from exposure and increases after reexposure to a sensitizing agent.[5] Exposure to sensitizing agents likely causes the nonspecific airway hyperresponsiveness rather than the hyperresponsiveness being a predisposing factor.

Intracellular Nucleotides

The β-adrenergic receptor is closely associated with adenylate cyclase, a plasma membrane indirectly controlling active secretion, transport, and carbohydrate storage. Asthmatics may respond abnormally to β-adrenergic-stimulating agents or may have some defect of the β-adrenergic system.[6] Mast cells in the lung may, therefore, have lower levels of intracellular cyclic AMP; this will enhance release of the vasoactive mediators that cause the asthmatic response. An abnormality in the β-adrenergic receptor has been proposed to explain occupational asthma from TDI.[7] In an *in vitro* system, TDI had a general effect on the β-adrenergic receptor complex, not specifically at the β receptor.[8] Therefore, TDI's effects may be more general, perhaps altering cell membrane protein.

Alteration in Deposition and Clearance of Particles From the Lungs

The site at which antigen or environmental particles come into contact with responding tissue may be an important determinant of the nature and severity of the asthmatic response. Airflow obstruction, either reversible or fixed, leads to more central deposition of all inhaled particles. This may alter both immunological and nonimmunological responses. Impaired pulmonary defenses by cigarette smoke, air pollution, chemicals, or dust may reduce clearance and thus lead to altered responses.

Airway Inflammation

Airway hyperresponsiveness may depend on the development of airway inflammation. Ozone-induced increase in airway responsiveness to acetylcholine in dogs is associated with neutrophilic inflammation.[9]

Epithelial Damage

Airway epithelial damage may be important in occupational asthma. Environmental agents can increase the permeability of the airway epithelium to macromolecular substances by alterations in tight junctions.[10] Reversible airway epithelial damage may result in transient airway hyperresponsiveness due to sensitized subepithelial vagal sensory nerve endings, leading to exaggerated cough reflex and bronchoconstriction. Increased permeability may allow greater access to bronchial smooth muscle, resulting in a direct effect by the environmental agent on muscle or on mast cells enhancing airway contraction or mediator release in both.

EXAMPLES OF OCCUPATIONAL ASTHMA

Substances of Animal Origin

Occupational asthma may be caused by IgE directed against animal proteins, as, for example, in shepherds, farmers, jockeys, laboratory and research technicians, animal handlers, and veterinarians. Animal hair, dander, and epidermal squamae can be the causal agents; urinary, fecal, or salivary proteins may be more important. In addition, mites, other small insects, and mold in the work environment may be implicated.

IgE-mediated occupational asthma to proteolytic enzymes occurs among workers preparing, handling, or packing the enzyme. Adequate dust suppression can reduce sensitization as there is a good relationship between sensitization, and level of exposure.[11] Preventive measures to reduce disease prevalence include enclosing the dust sources, increasing plant ventilation rates, and vacuum cleaning dust instead of sweeping.

Insects such as mushroom fly, aphid, bedbug, locust, bee, cockroach, housefly, moth, daphnia, Mexican bean weevil, and sewage filter fly may cause occupational asthma in entomologists, beekeepers, laboratory workers, and mushroom workers.[1] Sensitizing materials include scales, hairs, and excreta. Asthma from trypsin and from primitive organisms living in the sea that attach to the surface of oyster shells has been reported. The latter occurs in seasonal workers who crush oyster shells to obtain the meat.

Substances of Plant Origin

Substances of plant origin are probably the most commonly reported causes of reversible airway disease in industry. This includes IgE sensitization to grain, flour, maiko, mold, castor beans, green coffee beans, garlic, and kapok, sensitization of uncertain mechanism to low-molecular-weight chemicals, e.g., plicatic acid in Western red cedar,[12] and nonimmunological responses to various fiber dusts (cotton, hemp, etc.).

IgE-mediated grain allergy may be seen in millers, bakers, and farm workers handling grain. The specific antigen(s) responsible is(are) not known, and may include components of the cereal grain, parasitic fungi (smut, rust), saprophytes *(Aspergillus, Alternaria),* wheat weevil, mites, or Gram-negative bacteria.[13] Exposure to grain dust in lifetime nonsmoking grain workers was associated with an increased prevalence of chronic bronchitis and mild airflow obstruction.[14] An additive but not potentiating effect of smoking and grain dust exposure has been demonstrated.[15]

Byssinosis, generally not considered occupational asthma, is a disease characterized by nonimmunologic airflow obstruction caused by the dusts of cotton, flax, jute, sisal, or soft hemp. The symptom-complex prevalence relates to dust levels and particularly the coarse protein particles, rather than the mineral or cellulose portion of the cotton. The workers who appear to be at greatest risk for disease are those employed in the initial processing stages (i.e., carding) of cotton, flax, jute, sisal, or soft hemp. Mechanisms suggested to explain byssinosis include: (1) a nonimmunologic mast cell-mediator release; (2) an endotoxin-like action from the Gram-negative bacterial *(Enterobacter)* comtamination; and (3) an allergic mechanism.

Substances of Chemical Origin

A variety of chemicals, both simple and complex, are associated with occupational asthma. In general, they include chemicals with molecular weights of less than 1000 (micromolecules), such as chloramine, ethylenediamine, paraphenylenediamine, formaldehyde, and sulfathiazole, and other substances such as pesticides, chromium, vanadium, platinum, tannic acid, aliphatic polyamines, sulfonechloramides, amprolium hydrochloride, polyglycine acid chloride, spiramycin, phenylglycerine, phthalic anhydride (also trimellitic, tetrachlorophthalic, and hexahydrophthalic anhydride), epoxy resins, soldering fluxes, and isocyanates.[1]

Isocyanates are widely used in the production of polyurethane foam and in the manufacture of plastics, foam surface-coating elastomers, adhesives, and fibers. Occupational exposure to TDI, by far the most important isocyanate commercially and toxicologically, may occur after accidental leakage, during production, spraying, or molding operations, disposal of TDI waste, welding of polyurethane-covered wires, burning of polyurethane products, or the use of polyurethane floor varnish with a TDI activator. Isocyanate exposure from autobody paints and fillers is also common.

The mechanism(s) of TDI-asthma is(are) not certain; both immunologic and nonimmunologic mechanisms may be involved. TDI leads to sensitization in about 5% of workers; the sensitization is presumed immunologic despite difficulty in identifying IgE antibodies or other immune mechanism(s). High exposure may lead to nonimmunologic airway damage and continuous exposure may lead to accelerated lung function decline. As in plicatic acid sensitization, early environmental control may result in complete remission, but prolonged exposure after development of symptoms of occupational asthma may result in permanent airway hyperresponsiveness (i.e., asthma). In our study of seven confirmed TDI-asthmatics (mean age 42.6 years), my colleagues and I observed the disease to last as long as 12 years after exposure ended.

Phthalic anhydride is a reagent in the manufacture of a variety of industrial products including plastics, epoxy resins, and paints. The prevalence of sensitization to phthalic anhydride is unknown; however, asthma in exposed workers is an important toxic effect. IgE and IgG antibodies have been identified to phthalic acid-human serum albumin conjugates. Trimellitic anhydride (TMA) also may cause IgE-mediated asthma as well as symptoms simulating allergic alveolitis. Extensive, more serious infiltrative pulmonary processes associated with anemia after exposure to TMA fumes (rather than the dust) have been reported.

Chromium, a potent sensitizer used in the manufacture of pigments and in tanning, has been reported to cause bronchial asthma. The hexavalent compound is the most active one chemically and the most widely encountered in occupational health problems. Complex salts of platinum, used in the manufacture of catalysts, electroplating, production of fluorescent screens, and occupational groups such as jewelers and platinum refinery workers, can cause a syndrome of asthma, urticaria, and rhinitis known as platinosis.[13] Specific IgE antibodies to platinum salts and also heat-stable, short-term sensitizing (STS) antibodies, presumably STS-IgG, have been identified. Persistence of asthma, elevated levels of IgE, and hyperresponsive airways have been reported in workers who have not been exposed for several years. Cold air challenge tests were positive in 11% of the 107 current and 31% of 29 terminated platinum workers; more than one third of the current workers with positive cold air responses and one half of the terminated employees with positive cold air responses had positive prick skin reactions to platinum salt.

A VARIANT OF OCCUPATIONAL ASTHMA: REACTIVE AIRWAYS DYSFUNCTION SYNDROME (RADS)

Over the past several years, a clinical entity in which a single exposure to a very high concentration of an irritant results in a syndrome simulating asthma has been identified.[16] This syndrome, which is characterized by cough, dyspnea, and wheezing, has been termed reactive airways dysfunction syndrome, or RADS, because of the characteristic hyperresponsive airways.[16] In early cases, underlying bronchial asthma with previously unmanifested hyperresponsive airways was implicated with the agent causing irritant-induced bronchospasm. The persistence of asthma and airway hyperresponsiveness could not be explained. Overlooked sensitizing-agent exposure and coincidental development of bronchial asthma were other possibilities. Finally, there is the possibility that high concentrations of irritant agents, smoke, gases, and fumes, actually damaged the airways, resulting in bronchiolitis obliterans. A characteristic of RADS is that it simulates asthma. It is probable that many of the complaints (cough, wheezing, chest tightness) develop because of the clinically manifested hyperresponsive airways. Symptoms developed acutely after a specific incident in which there was an unusually high exposure. The duration of exposure averaged 3.2 h (range <15 min to 12 h). In fact, in about 40% of the subjects, the exposure was <1 h; it was ≤2 h in about 70%. There was an average 4.5-h interval between exposure and onset of symptoms; in 40% symptoms developed immediately, and in most, symptoms developed by 5 h. RADS persisted for 3 or more years in over 70% of the subjects; the mean duration of RADS was 2.4 years. A variety of jobs and offending agents were reported; in all cases the exposure was to unusually high levels of irritant, with most occurring after an accident or improper work practices. The pathogenesis of RADS probably involves nonimmunologic inflammatory mechanisms.

REFERENCES

1. **Brooks, S.M.,** Occupational asthma, in *Bronchial Asthma: Mechanisms and Therapeutics,* Weiss, E.B., Segal, M.S., and Stein, M., Eds., Little, Brown, Toronto, 1985, 461.
2. **Brooks, S.M., Edwards, J.J., Jr., and Edwards, F.H.,** An epidemiologic study of workers exposed to Western red cedar and other wood dusts, *Chest,* 80, 30S, 1981.
3. **Boushey, H.A., Holtzman, M.J., Sheller, J.R., and Nadel, J.A.,** Bronchial hyperreactivity, *Am. Rev. Respir. Dis.,* 121, 389, 1980.
4. **Cockcroft, D.W., Ruffin, R.E., Dolovich, J., and Hargreave, F.E.,** Allergen-induced increase in nonallergic bronchial reactivity, *Clin. Allergy,* 7, 503, 1977.
5. **Lam, S., Wong, R., and Yeung, M.,** Nonspecific bronchial reactivity in occupational asthma, *J. Allergy Clin. Immunol.,* 63, 28, 1979.
6. **Austen, K.F.,** A review of immunological, biochemical and pharmacological factors in the release of chemical mediators from human lung, in *Asthma: Physiology, Immunopharmacology and Treatment* (2nd Int. Symp.), Austen, K.F. and Lichtenstein, L.M., Eds., Academic Press, New York, 1973, 109.
7. **Butcher, B.T., Salvaggio, J.E., Weill, H., and Ziskind, M.M.,** Toluene diisocyanate (TDI) pulmonary disease: immunologic and inhalation challenge studies, *J. Allergy Clin. Immunol.,* 58, 89, 1976.
8. **McKay, R.T. and Brooks, S.M.,** Effect of toluene diisocyanate on beta adrenergic receptor function, *Am. Rev. Respir. Dis.,* 128, 50, 1983.
9. **Holtzman, M.J., Fabbri, L.M., O'Byrne, P.M., Gold, R.D., Aizawa, H., Walters, E.H., Alpert, S.E., and Nadel, J.A.,** Importance of airway inflammation for hyperresponsiveness induced by ozone, *Am. Rev. Respir. Dis.,* 127, 686, 1983.
10. **Simani, A.S., Inoue, S., and Hogg, J.C.,** Penetration of the respiratory epithelium of guinea pigs following exposure to cigarette smoke, *Lab. Invest.,* 31, 75, 1974.
11. **Newhouse, M.L., Tagg, B., and Pocock, S.J.,** An epidemiological study of workers producing enzyme washing powders, *Lancet,* 1, 689, 1970.
12. **Chan-Yeung, M., Barton, G.M., MacLean, L., and Grzybowski, S.,** Occupational asthma and rhinitis due to Western red cedar *(Thuja plicata), Am. Rev. Respir. Dis.,* 108, 1094, 1973.
13. **Pepys, J., and Davies, R.J.,** Occupational asthma, in *Allergy: Principles and Practice,* Vol. 2, Middleton, E.J., Reed, C.E., and Ellis, E.F., Eds., C.V. Mosby, St. Louis, 1978, 812.
14. **Dosman, J.A., Cotton, D.J., Graham, B.L., Li, R.K.Y., Froh, F., and Barnett, G.D.,** Chronic bronchitis and decreased forced expiratory flow rates in lifetime nonsmoking grain workers, *Am. Rev. Respir. Dis.,* 121, 11, 1980.
15. **Chan-Yeung, M., Schulzer, M., MacLean, L., Dorken, E., and Grzybowski, S.,** Epidemiologic health survey of grain elevator workers in British Columbia, *Am. Rev. Respir. Dis.,* 121, 329, 1980.
16. **Brooks, S.M., Weiss, M.A., and Bernstein, I.L.,** Reactive airways dysfunction syndrome (RADS), *Chest,* 88, 376, 1985.

Occupational Asthma: Nonallergic Airway Responsiveness

D. W. Cockcroft

INTRODUCTION

Virtually all currently symptomatic asthmatics have hyperresponsiveness of the tracheobronchial smooth muscle to nonallergic or nonspecific stimuli such as histamine, cholinergic agonists, exercise, or cold air.[1] This so-called nonallergic airway hyperresponsiveness (hereafter referred to as airway hyperresponsiveness) can be easily demonstrated in the laboratory with standardized histamine or methacholine inhalation tests.[2] Measurement of airway (hyper) responsiveness is useful in the diagnosis of bronchial asthma, in the assessment of its severity, and in monitoring of treatment. In addition, it has recently been recognized that serial measurements of airway responsiveness to histamine or methacholine may be used to provide strong evidence for recent exposure to either an allergenic or a low-molecular-weight sensitizing agent.[3]

Occupational asthma, asthma *caused* by exposure to gases, dusts, fumes or vapors encountered in the workplace,[4] is usually due to sensitization. Allergens, haptens, or low-molecular-weight, nonhaptenic chemicals may be the cause. The two most common low-molecular-weight sensitizing chemicals in the latter category are plicatic acid (in Western red cedar) and the aromatic diisocyanates. Sensitivity to these is difficult to document as there are no simple cutaneous or blood (RAST) tests of sensitivity. Sensitivity must, therefore, be confirmed by tedious, controlled, occupational-challenge tests[5] which may be difficult to interpret and are associated with significant patient morbidity. Serial measurements of airway responsiveness and peak flow rates, at work and away from work, may be useful in confirming a diagnosis of occupational asthma caused by a sensitizing agent.[3]

FACTORS INFLUENCING THE LEVEL OF AIRWAY RESPONSIVENESS

Although airway hyperresponsiveness is a characteristic feature of current asthma, the level of airway responsiveness will vary from time to time depending upon recent exposures. The largest increases in airway responsiveness appear following exposure to a clinically relevant sensitizing agent, either an allergen[6] or a low-molecular-weight chemical.[3,8] Increases in airway responsiveness are strongly associated with the late asthmatic response,[6-8] a feature of about 90% of positive, single-exposure challenges with low-molecular-weight sensitizing chemicals. The increases in airway responsiveness seen following allergen or low-molecular-weight sensitizing chemical exposure are large, last for several days or longer, and are associated with increased symptoms of asthma.[6,7]

Exposure to allergens under natural circumstances[9] or to low-molecular- weight chemicals in the workplace[3] also increases airway responsiveness. Likewise, environmental avoidance of allergens[10] or low-molecular-weight sensitizers[7] results in improvement in airway responsiveness. Airway responsiveness may improve into the normal range once a sensitizing agent has been removed from the asthmatic's environment. This may occur within 2 or 3 d in occupational asthma of recent onset,[11] but more commonly occurs over a period of several weeks or months in the established occupational or allergic asthmatic.[7,10] About 50% of occupational asthmatics appear to develop permanent, and likely acquired, airway hyperresponsiveness as a result of their occupational exposure; the major determinant of this appears to be duration of exposure after onset of symptoms.[12]

The precise mechanism and pathophysiology of the late asthmatic responses and induced increases in airway responsiveness remain uncertain. However, animal studies and bronchoalveolar lavage studies suggest that airway inflammation with eosinophils and/or neutrophils may be important.

Sensitizing allergens/chemicals are likely the only asthma-triggering factors which will produce such profound changes in airway responsiveness. Viral respiratory tract infections may produce inflammation and significant increases in airway responsiveness;[13] this has not been well studied in asthmatics, and may not occur as often or as severely as was once speculated. Most other asthma-triggering factors produce predominantly bronchospasm (the ease with which they do so being symptomatic of the current level of airway responsiveness) and not changes in airway responsiveness. In particular, inhaled irritant gases probably act primarily through a bronchospastic route, although gases such as ozone in moderately high concentrations are recognized to produce small, short-lived increases in airway responsiveness to which tolerance is rapidly acquired.[14]

For these reasons, large increases in airway responsiveness, occurring in the absence of any other possible explanation such as viral infection or large reduction in baseline airway caliber, signify recent exposure to a sensitizing agent, either allergen or low-molecular-weight chemical.

CLINICAL VALUE OF AIRWAY RESPONSIVENESS MEASUREMENTS IN OCCUPATIONAL ASTHMA

Diagnosis of Asthma

Measurements of airway responsiveness are said to be most valuable in the diagnosis of asthma when symptoms are atypical and lung function is normal.[1] This is common particularly in early occupational asthma where atypical symptoms (e.g., nocturnal cough) and normal lung function may be seen. As with other types of asthma, demonstration of airway hyperresponsiveness may be used to confirm a diagnosis of asthma. It is important to recall that airway responsiveness may return to normal in as little as 2 or 3 d away from work,[11] and that a normal measurement of airway responsiveness when the subject is away from work does not exclude a diagnosis of occupational asthma.

Inferring Sensitization

A second, more important use of airway responsiveness determinations is the differentiation of asthma caused by a sensitizing agent from irritant/exercise-induced bronchospasm in a worker who has coincidentally developed airway hyperresponsiveness. The former is "occupational asthma", while the latter is not. Differentiating these two is a common problem with important medical and medico-legal/compensation implications. The occupational asthmatic should change jobs in order to lessen the likelihood of developing permanent and severe airway responsiveness, and he should receive full compensation; early diagnosis is essential. By contrast, the subject with irritant-related bronchospasm in the workplace is less eligible for compensation and, if symptoms permit, may continue to work at the same job with appropriate protective measurements such as reduced exposure and protective face gear.

Serial measurements of airway responsiveness to histamine and home/work monitoring of peak flow rates may be safely and conveniently used as an alternative to occupational challenge tests to confirm occupational asthma caused by a low-molecular-weight sensitizing agent. Peak flow rates can be monitored with portable peak flow meters and are recorded along with symptoms and medication use in relation to work, noting exposure to any suspect processes or chemicals.[3,15] The pattern of peak flow rates associated with occupational asthma is shown in Figure 1.[15] Occupational exposure produces a gradual fall in peak flow rates, which becomes progressively more marked and fails to recover to baseline. Likewise, withdrawal from exposure and institution of therapy results in a gradual improvement in peak flow rates. By contrast, Figure 2 shows the result of daily monitoring of peak flow rates in a subject who proved not to have occupational asthma. This asthmatic subject has an exaggerated diurnal variation in flow rates but no work-related change. The major problem in this approach is patient compliance, which is often so poor as to lead to uninterpretable studies.

Serial measurements of histamine and methacholine PC_{20} in relation to work and work-free periods are probably more useful. A significant, reproducible, otherwise unexplained work-related fall in histamine or methacholine PC_{20} is virtually diagnostic of sensitization to a chemical (or allergen) at work.[3] Although the definition of "a significant change" has not been agreed upon, as a rough guideline, a fall in PC_{20} in excess of fourfold (two concentrations) is likely to be significant.[2] Although a greater number of lab visits will be required

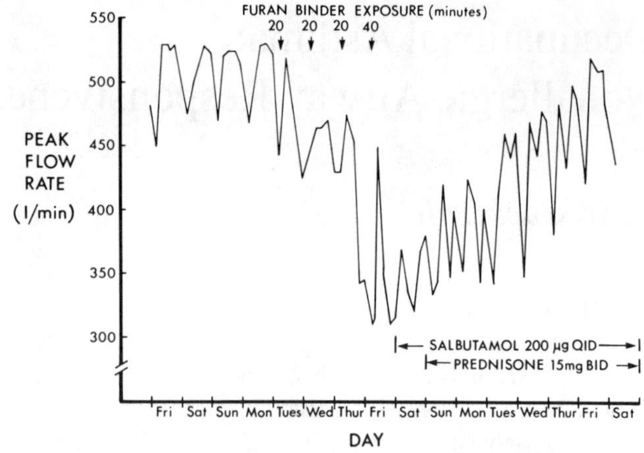

FIGURE 1. Peak flow rates (l/min) in a subject with occupational asthma caused by a furan-based binder system is plotted on the vertical axis against time on the horizontal axis. (Reproduced from Cockcroft, D.W., Cartier, A., Jones, G., Tarlo, S.M., Dolovich, J., and Hargreave, F.E., *J. Allergy Clin. Immunol.*, 66, 458, 1980. With permission.)

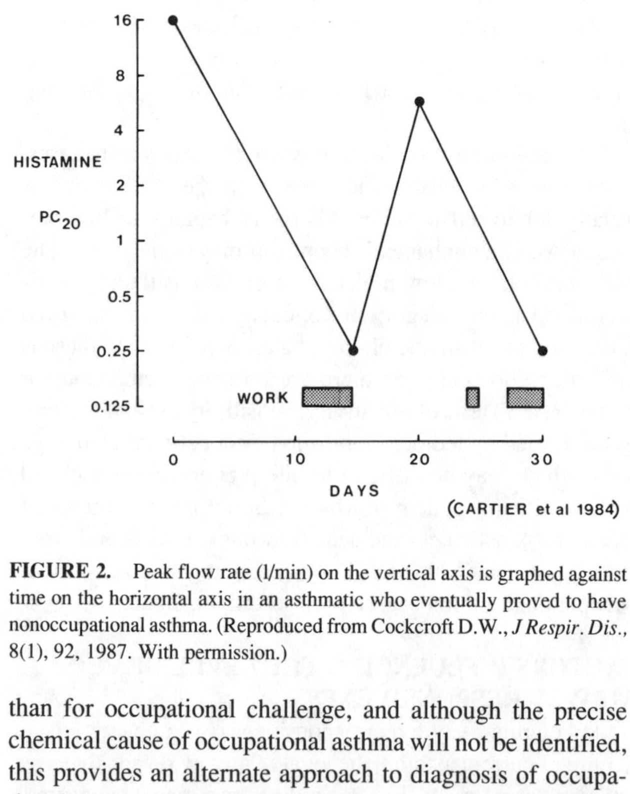

FIGURE 2. Peak flow rate (l/min) on the vertical axis is graphed against time on the horizontal axis in an asthmatic who eventually proved to have nonoccupational asthma. (Reproduced from Cockcroft D.W., *J Respir. Dis.*, 8(1), 92, 1987. With permission.)

than for occupational challenge, and although the precise chemical cause of occupational asthma will not be identified, this provides an alternate approach to diagnosis of occupational asthma, for example, when occupational challenges cannot be performed.

Figure 3 shows an example of a worker who demonstrated on two occasions a 32- to 64-fold reduction in histamine PC_{20} after several days at work. Although the precise cause of occupational asthma was not identified, this reproducible occupation-induced increase in airway responsiveness was considered diagnostic of occupational asthma caused by sensitization.[3] Figure 4 shows an example of a negative study; this asthmatic had an equivocal history of occupational asthma and airway hyperresponsiveness to inhaled histamine was the same when at work as when not at work.

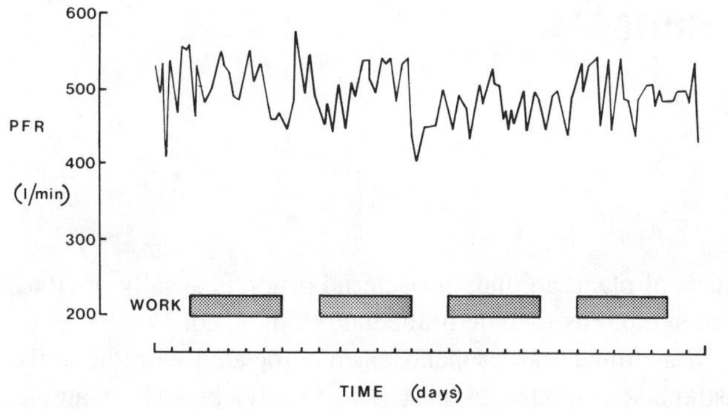

FIGURE 3. Histamine PC_{20} (mg/ml) is plotted on the vertical axis against time on the horizontal axis with work periods noted in the shaded areas. This subject proved to have occupational asthma, although the chemical could not be identified. (Reproduced from Cartier, A., Pineau, L., and Mallo, J.L., *Clin. Allergy,* 14, 193, 1984. With permission.)

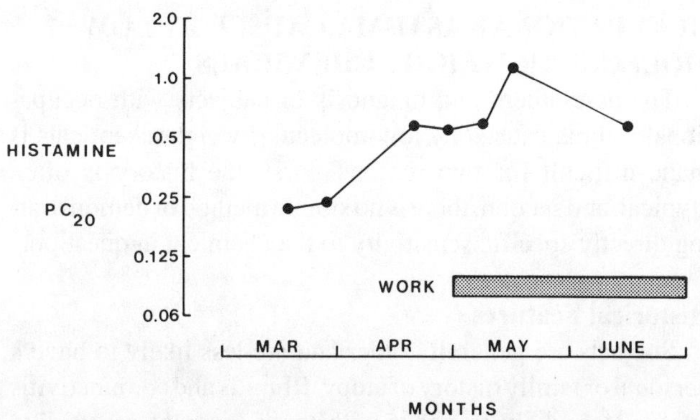

FIGURE 4. Histamine PC_{20} (mg/ml) on the vertical axis is plotted against time on the horizontal axis with work periods noted in the shaded areas. This patient eventually proved to have asthma which was thought not to be caused by his work.

The "sensitivity" and "specificity" of this approach for diagnosing occupational asthma remain to be thoroughly investigated. There is, however, no doubt that reproducible work-induced increase in airway responsiveness is essentially the definition of occupational asthma.[4] Therefore, once a "significant change" can be defined (probably on a subject-to-subject basis), false positives are unlikely; i.e., reproducible work-related fall in histamine PC_{20} is a diagnostic test with a high specificity and high-positive predictive value. Until more detailed studies are available, the sensitivity and negative predictive value of the test, both of which require knowledge of the "false negatives", cannot be determined.

Follow-Up

Finally, measurements of airway responsiveness may be used in both short- and long-term follow-up of the occupational asthmatic. This may be of value in determining the adequacy of environmental control, which is, of course, the only important therapeutic maneuver and which is mandatory in the management of occupational asthma. Follow-up studies may also help determine long-term disability and role of compensation in long-term disability in such asthmatics.

CONCLUSION

In conclusion, airway responsiveness measurements may be useful in occupational asthma in three areas. First, airway responsiveness measurements may aid in the confirmation of asthma; however, airway responsiveness may rapidly return to normal when away from the sensitizing agent. Second, reproducible work-induced increases in airway responsiveness can be used to infer a diagnosis of occupational asthma, particularly in conjunction with a characteristic pattern of self-monitored peak flow rates. This approach may be more convenient in certain circumstances than occupational challenges in the laboratory. Finally, measurements of airway responsiveness are of value in the follow-up studies in the occupational asthmatic.

REFERENCES

1. **Hargreave, F.E., Ryan, G., Thomson, N.C., O'Byrne, P.M., Latimer, K., Juniper, E.F., and Dolovich, J.,** Bronchial responsiveness to histamine or methacholine in asthma: measurement and clinical significance, *J. Allergy Clin. Immunol.,* 68, 347, 1981.
2. **Juniper, E.F., Frith, P.A., Dunnett, C., Cockcroft, D.W., and Hargreave, F.E.,** Reproducibility and comparison of responses to inhaled histamine and methacholine, *Thorax,* 33, 705, 1978.
3. **Cartier, A., Pineau, L., and Malo, J.L.,** Monitoring of maximum expiratory peak flow rates and histamine inhalation tests in the investigation of occupational asthma, *Clin. Allergy,* 14, 193, 1984.
4. **Newman Taylor, A.J.,** Occupational asthma, *Thorax,* 35, 241, 1980.
5. **Pepys, J. and Hutchcroft, B.J.,** Bronchial provocation tests in etiologic diagnosis and analysis of asthma, *Am. Rev. Respir. Dis.,* 112, 829, 1975.
6. **Cartier, A., Thomson, N.C., Frith, P.A., Roberts, R., and Hargreave, F.E.,** Allergen-induced increase in bronchial responsiveness to histamine: relationship to the late asthmatic response and change in airway caliber, *J. Allergy Clin. Immunol.,* 70, 170, 1982.
7. **Lam, S., Wong, R., and Yeung, M.,** Nonspecific bronchial reactivity in occupational asthma, *J. Allergy Clin. Immunol.,* 63, 28, 1979.
8. **Mapp, C.E., DiGiacomo, G.R., Omini, C., Broseghini, C., and Fabbri, L.M.,** Late, but not early, asthmatic reactions induced by toluene-diisocyanate are associated with increased airway responsiveness to methacholine, *Eur. J. Respir. Dis.,* 69, 276, 1986.
9. **Boulet, L.P., Cartier, A., Thomson, N.C., Roberts, R.S., Dolovich, J., and Hargreave, F.E.,** Asthma and increases in nonallergic bronchial responsiveness from seasonal pollen exposure, *J. Allergy Clin. Immunol.,* 71, 399, 1983.
10. **Platts-Mills, T.A.E., Mitchell, E.B., Nock, P., Tovey, E.R., Moszoro, H., and Wilkins, S.R.,** Reduction of bronchial hyperreactivity during prolonged allergen avoidance, *Lancet,* 2, 675, 1982.
11. **Cockcroft, D.W. and Mink, J.T.,** Isocyanate-induced asthma in an automobile spray painter, *Can. Med. Assoc. J.,* 121, 602, 1979.
12. **Chan-Yeung, M., Lam, S., and Koener, S.,** Clinical features and natural history of occupational asthma due to Western red cedar *(Thuja plicata), Am. J. Med.,* 72, 411, 1982.
13. **Empey, D.W., Laitinen, L.A., Jacobs, L., Gold, W.M., and Nadel, J.A.,** Mechanisms of bronchial hyperreactivity in normal subjects after upper respiratory tract infection, *Am. Rev. Respir. Dis.,* 113, 131, 1976.
14. **Boushey, H.A. and Holtzman, M.J.,** Experimental airway inflammation and hyperreactivity, *Am. Rev. Respir. Dis.,* 131, 312, 1985.
15. **Cockcroft, D.W., Cartier, A., Jones, G., Tarlo, S.M., Dolovich, J., and Hargreave, F.E.,** Asthma caused by occupational exposure to a furan-based binder system, *J. Allergy Clin. Immunol.,* 66, 458, 1980.
16. **Cockcroft, D.W.,** Keys to early recognition of Occupational asthma, *J. Respir. Dis.,* 8(1), 92, 1987.

Occupational Asthma: Patient Assessment

D. W. Cockcroft

INTRODUCTION

Occupational asthma is asthma (reversible intrathoracic airflow obstruction with airway hyperresponsiveness) which is caused by exposure to gases, vapors, or dusts encountered in the workplace.[1] Since financially important decisions result from the diagnosis of "occupational asthma", it is imperative that the diagnosis be made as objectively as possible. Once the diagnosis of asthma has been confirmed, a topic which will not be covered in this review, the following modalities are useful in the evaluation of the occupational or suspected occupational asthmatic:

I. Accurate documentation of the history
   A. With regard to subject and symptoms
   B. With regard to exposure(s) in the environment

II. Objective demonstration of occupational increase in asthma severity
   A. Serial measurements of ambulatory peak flow rate (PFR)
   B. Serial measurements of nonallergic airway responsiveness

III. Objective demonstration of sensitivity to the agent in question
   A. Skin testing
   B. RAST
   C. Bronchial provocation testing

NONIMMUNOLOGIC OCCUPATIONAL ASTHMA

Nonimmunologic (pharmacologic, inflammatory) mechanisms account for only a small number both of causes and cases of occupational asthma. Since sensitization is not a feature of this type of occupational asthma, the "dose" of the causative agent is generally larger, often the result of an "industrial accident". An accurate history is the only means available for diagnosis of these types of asthma. The event is often well recalled and sometimes well documented; more than one person may have been affected. Management, both in terms of prevention and in terms of ongoing employment of the affected asthmatic, should be directed at the environment rather than the subject(s).

IgE SENSITIZATION TO HIGH-MOLECULAR-WEIGHT COMPOUNDS

Occupational asthma caused by IgE directed against a high-molecular- weight chemical (plant, animal, bacteria) encountered in the workplace can be diagnosed by an accurate history followed by documentation of sensitivity by skin testing. The history is usually typical; exposure to organic dusts of plant, animal, or bacterial origin is usually obvious and symptoms include immediate rhinitis, conjunctivitis, as well as immediate bronchospasm associated with the early asthmatic response. A past (and family) history of atopic disease is common and these patients are often young. Following accurate documentation of asthma and of the environmental exposure, specific sensitivity can be documented by allergen prick skin testing with the relevant allergen extract with demonstration of the immediate wheal and flare response characteristic of IgE-mediated sensitivity.

OCCUPATIONAL ASTHMA CAUSED BY LOW-MOLECULAR-WEIGHT CHEMICALS

The assessment and diagnosis of subjects with occupational asthma caused by low-molecular-weight chemicals is made difficult for two reasons. First, the history is often atypical, and second, there is no simple method of demonstrating directly specific sensitivity to the chemical in question.

Historical Features

Subjects are generally older and are less likely to have a personal or family history of atopy. Rhinitis and conjunctivitis appear less likely to occur with these types of sensitivity. Challenge tests with relatively high-concentration, short-duration exposures in the laboratory produce late responses (3 to 12 h) in 90% of subjects (about half isolated and half as part of a dual response). Nonallergic airway responsiveness to histamine and methacholine (the fundamental abnormality in asthma) will increase for several days in association with the late response.[2,3] The clinical correlates of this are that work-induced symptoms may occur mainly in the evening and may persist long after exposure has ceased.

The "classic" history would initially involve mild, perhaps atypical, symptoms occurring in the evenings during the work week; often these are progressive throughout the work week. Symptoms will eventually become more troublesome and take more days away from work to resolve. Symptoms will appear at work due to development of early responses and/or due to irritant effects on the (newly) hyperresponsive airways. Nocturnal and persistent weekend symptoms may direct patient and physician away from the work place. Intermittent and/or unrecognized exposure(s) may further confuse the issue.

The major diagnostic difficulty is in differentiating occupational asthma from preexisting or coincidentally acquired asthma with work-related symptoms due to exertion and/or irritants at work. Paradoxically, irritant-induced symptoms may be more closely related to work. However, crescendo work-week symptoms and major weekend/vacation resolution should not occur; this is due to lack of significant changes in airway responsiveness.

Demonstration of Occupational Increase in Asthma Severity

The pathophysiological features noted above, particularly the progressive increase in airway responsiveness,[2,3] can be used as a diagnostic tool. Serial measurements of peak flow and of histamine or methacholine airway responsiveness may be used to infer exposure to a sensitizing agent,[4] based on the hypothesis that large increases in airway responsiveness occur only after such exposures (and possibly after viral respiratory infection, but this is easily ruled out). A characteristic pattern of progressive fall in peak flow with gradual postexposure recovery and reproducible work-related increase in airway responsiveness which is otherwise unexplained can be used to diagnose occupational asthma.[4] However, the sensitizing chemical will not be identified.

Demonstration of Sensitization

Objective demonstration of sensitivity to low-molecular-weight compounds, in IgE-hapten-mediated cases, can be done by skin testing or RAST. Examples include platinum salts, nickel salts, and acid anhydrides. Reagents for these tests, generally conjugated with human serum albumin, may have limited availability.

Unfortunately, most low-molecular-weight sensitizing chemicals produce sensitivity by obscure but, presumably, immunologic mechanisms; i.e., IgE has yet to be clearly implicated (e.g., isocyanates, Western red cedar). Occupational challenge tests remain the only objective method for demonstrating sensitivity. Although sophisticated challenge methods are available, for diagnostic purposes challenges which mimic occupational exposure in the lab (painting, soldering, sanding, sawing, etc.) are acceptable.[5] These tests must be controlled with an equally irritating chemical or substance, and should be carried out single-blind. A minimum of 2 8- to 10-h days in the lab are required; typical results in Western red cedar vs. spruce (control) challenges over 2 d include early asthmatic response, dual asthmatic response, and isolated late asthmatic response as shown in Figure 1. Increases in airway responsiveness to histamine or methacholine[2-4] should probably be sought routinely.

Occupational challenges are tedious and potentially dangerous. Physician attendance and resuscitation facilities must be available. Some centers routinely recommend hospitalization for the duration of such challenges.

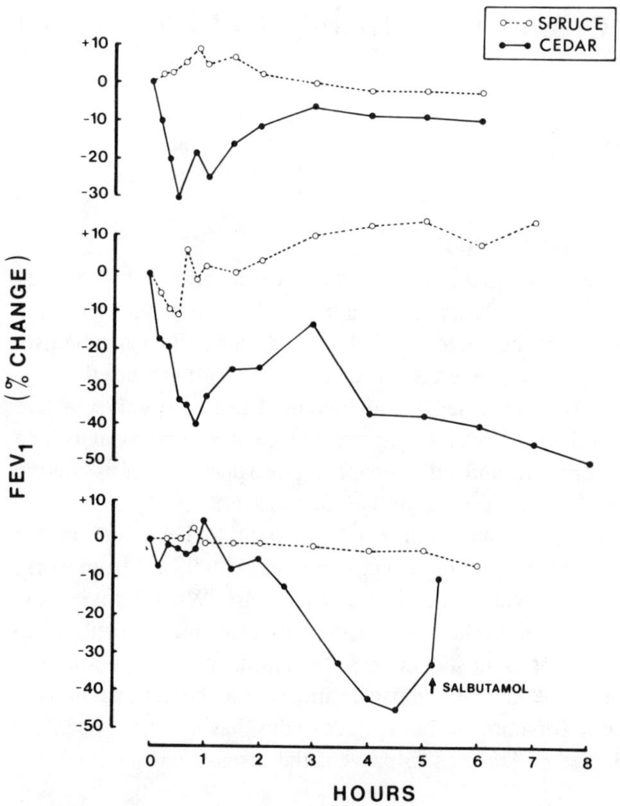

FIGURE 1. Bronchial responses to Western red cedar (spruce control) showing early, dual, and late responses (top to bottom).

REFERENCES

1. **Newman Taylor, A.J.,** Occupational asthma, *Thorax,* 35, 241, 1980.
2. **Lam, S., Wong, R., and Yeung, M.,** Nonspecific bronchial reactivity in occupational asthma, *J. Allergy Clin. Immunol.,* 63, 28, 1979.
3. **Cockcroft, D.W.,** Mechanism of perennial allergic asthma, *Lancet,* 2, 253, 1983.
4. **Cartier, A., Pineau, L., and Malo, J.L.,** Monitoring of maximum expiratory peak flow rates and histamine inhalation tests in the investigation of occupational asthma, *Clin. Allergy,* 14, 193, 1984.
5. **Pepys, J. and Hutchcroft, B.J.,** Bronchial provocation tests in etiologic diagnosis and analysis of asthma, *Am. Rev. Respir. Dis.,* 112, 829, 1975.

Follow-Up Studies In Occupational Asthma

A. Cartier

INTRODUCTION

Occupational asthma is defined as asthma that is induced by a specific "sensitizing" agent at work. It excludes asthma that is exacerbated at work through an irritant mechanism (e.g., cold air, exercise, strong odors). With increased recognition of the disease and increased use of highly reactive chemicals in industry, the prevalence of occupational asthma is increasing and will probably replace pneumoconioses as the most common occupational lung disease.

Occupational asthma was thought to be a self-limited disease with complete recovery after withdrawal from work; it is now evident that this is not the case. We will review the outcome of workers with occupational asthma and attempt to identify prognostic factors, to outline how these subjects should be managed, and to define criteria of functional impairment for compensation needs; this has recently been reviewed.[1] Measures to prevent the disease will also be discussed.

PROGNOSIS OF OCCUPATIONAL ASTHMA

Once the diagnosis of occupational asthma has been confirmed, two alternatives are left to the worker: leave the workplace or continue to work. What is the outcome for each alternative chosen by the patient?

Follow-Up of Workers Withdrawn from Work

Several studies have addressed the outcome of workers with occupational asthma who left work. In 1972 Pain and Symons[2] studied six workers with red cedar- or toluene diisocyanate (TDI)-asthma removed from work for 1 to 36 months. Although three workers were asymptomatic, all showed an increase in airway responsiveness to methacholine. In 1975 Adams[3] found a higher incidence of chest symptoms among workers with presumed TDI-asthma who had left work for 2 to 11 years. Moller et al.[4] studied 12 workers with a history suggestive of TDI-asthma removed from work for a mean of 4.5 years; 7 had a positive challenge to TDI, 6 with persistent increase in airway responsiveness to methacholine, and 1 with "normal" airway responsiveness. Paggiaro et al.[5] reevaluated 12 workers with proved occupational asthma to TDI after a mean of 2 years away from work; 8 had persistent asthma while 7 had increased airway responses to bethanechol.

Chan-Yeung et al.[6] examined the outcome of workers with occupational asthma to Western red cedar. Of 75 workers removed from work for 1 to 9 years, half became asymptomatic; the rest had persistent asthma and needed regular asthma medication including corticosteroids in some. Persistence of asthma was associated with a longer duration of exposure after onset of symptoms and before withdrawal from work. Airway hyperresponsiveness to methacholine, initially present in all, was greater in those who remained symptomatic. On follow-up, 20 of 26 asymptomatic workers had improved airway responsiveness to methacholine, only 8 of these to normal. By contrast, 7 symptomatic workers had airway responsiveness to methacholine which increased (n = 4) or remained unchanged (n = 3). Workers with persistent symptoms who remained symptomatic had lower initial spirometry with less improvement on follow-up compared to those who became asymptomatic.

Finally, Burge[7] has shown that of 20 workers with occupational asthma due to colophony who had left work for 12 and 45 months, only 2 became asymptomatic, whereas 7 still had increased responsiveness to histamine. The fact that 13 workers still had chest symptoms with normal airway responsiveness to histamine suggests either occasional colophony or pine resin exposure, or symptoms not due to asthma (chronic bronchitis, malingering, etc.). These studies thus demonstrate that occupational asthma to TDI, red cedar, and colophony can lead to persistent asthma.

We have extended these results to a variety of etiologic agents.[8] Two groups of occupational asthmatics were evaluated: 31 snowcrab-processing workers and 32 workers with occupational asthma due to various other agents (isocyanates in 10, red cedar in 6, flour in 5, wood dust in 4, antibiotics in 2, and cereals, tea powder, soldering products, and freon in 1 subject each). Workers with occupational asthma due to various agents were older and duration of follow-up, total duration of exposure, and duration of exposure after onset were longer (Figure 1). Initial and follow-up forced expired volume in 1 s (FEV_1) were lower, suggesting more severe asthma than crab workers, possibly due to longer exposure to the sensitizing agent. In crab-processing workers (Figure 2), 12 subjects who remained symptomatic at follow-up had worked for a longer period of time after onset of asthma. The symptomatic workers required only intermittent or regular bronchodilators, and had lower initial and follow-up spirometry. Workers with occupational asthma due to various other agents and who required regular medication had a longer duration of exposure after onset and a lower FEV_1 on initial and follow-up survey (Figure 3). Airway responsiveness, assessed by the concentration of histamine producing a 20% fall in FEV_1 (PC_{20}), is shown in Figures 4 and 5. Snow crab asthmatics who became asymptomatic had improved airway responsiveness; two subjects returned to normal (Figure 4). Similar, less marked changes occurred in symptomatic work-

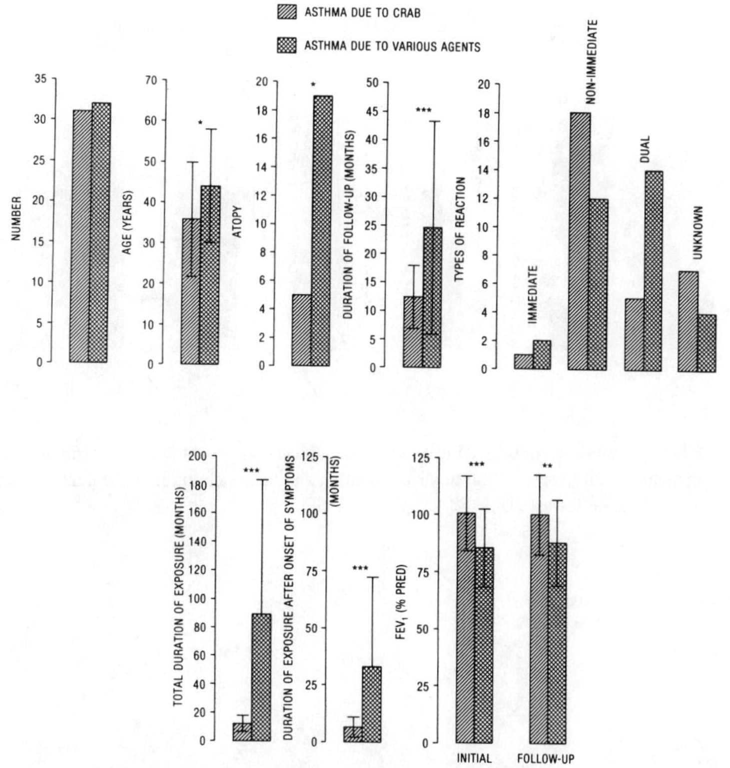

FIGURE 1. Anthropometric, clinical, and physiological results in workers with occupational asthma due to crab and to various agents. Comparison made by unpaired *t*-test. *, $p < 0.05$; **, $p < 0.01$; ***, $p < 0.001$. Bars represent 1 SD.

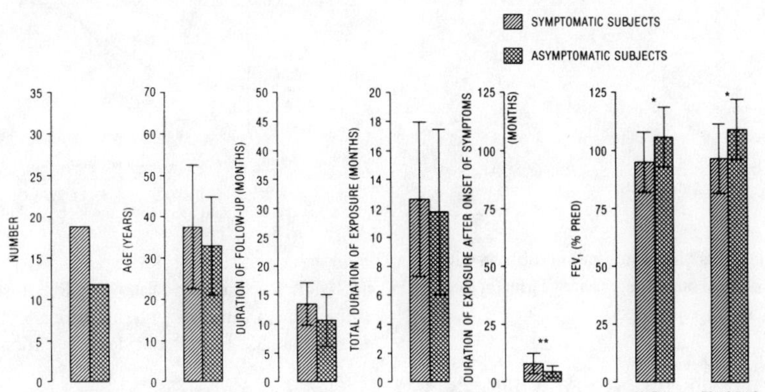

FIGURE 2. Results according to the outcome in subjects with asthma due to crab, dividing them into symptomatic and asymptomatic subjects. Same level of significance for asterisks. Bars represent 1 SD.

ers. In subjects with asthma due to various other agents, there was no change in PC_{20} in either treatment group (Figure 5). Some showed significant improvement in airway responsiveness; only one returned to normal. Our results are similar to those of Chan-Yeung et al. in red cedar asthma.

In summary, the prognosis of workers with occupational asthma who leave work is related to the duration of exposure after onset of symptoms, the presence of airway obstruction, and the level of nonallergic airway responsiveness at the time of diagnosis. Occupational asthma can cause persistent asthma after withdrawal from work.

Asymptomatic subjects often have persistent airway hyperresponsiveness, a prevalence higher than would be expected in a normal population.[9] We could thus postulate

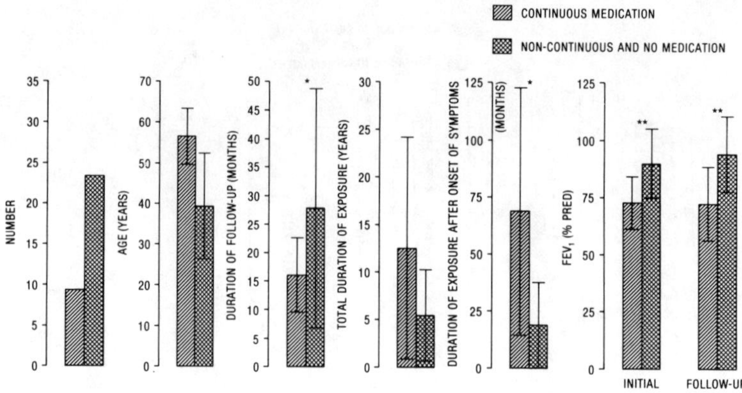

FIGURE 3. Results according to the outcome in subjects with asthma due to various agents, dividing them into subjects on continuous medication and on noncontinuous medication. Same level of significance for asterisks. Bars represent 1 SD.

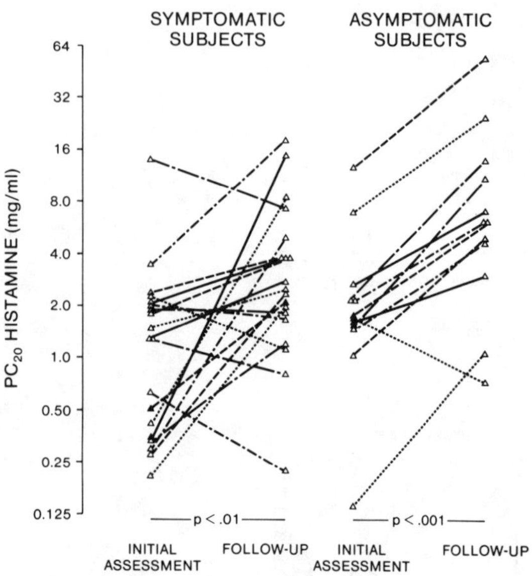

FIGURE 4. Initial and follow-up PC_{20} histamine (mg/ml) in subjects with asthma due to crab according to their clinical outcome. A change greater than 3.2-fold is considered significant in any given subject.[19]

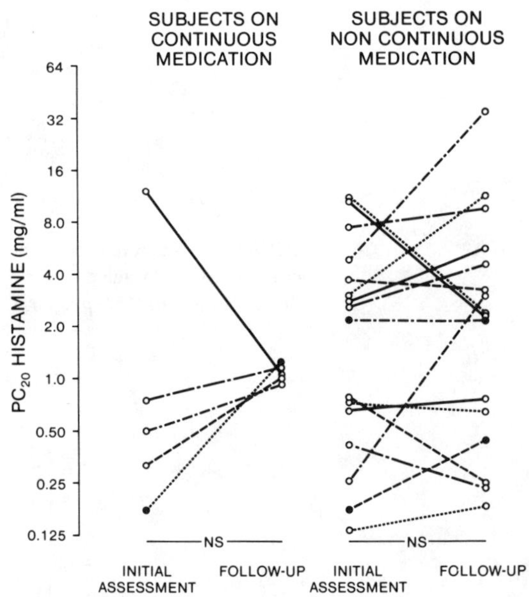

FIGURE 5. Initial and follow-up PC_{20} histamine (mg/ml) in subjects with asthma due to various agents, according to their clinical outcome. A change greater than 3.2-fold is considered significant in any given subject.[19]

that these findings are "abnormal" and reflect airway damage. These studies stress the importance of introducing health programs and regular surveys in selected industries to ensure early diagnosis and removal of affected individuals.

Follow-Up of Workers Still Exposed

Few studies have addressed outcome of workers with occupational asthma who had continued exposure to the sensitizing agent. Chan-Yeung et al.[6] reevaluated 50 workers who continued exposure to red cedar for a mean of 34.7 months after the initial assessment. The FEV_1 and forced vital capacity (FVC) decreased in these workers and the methacholine PC_{20} decreased in 9 and remained unchanged in 13 of the 22 tested. Burge[7] showed that workers with occupational asthma to colophony, who were moved away from colophony but within their original factory, did not change their spirometry. Most showed improved histamine responsiveness, but only three out of nine returned to normal. Finally, Paggiaro et al.[5] reported that continued TDI exposure in 15 workers with TDI-asthma led to deterioration of airflow obstruction and further increase in airway responsiveness to bethanechol.

These data again stress the importance of removing subjects from work once the diagnosis of occupational asthma is made in order to prevent worsening of the asthmatic condition.

COMPENSATING OCCUPATIONAL ASTHMA

We now recognize that occupational asthma can lead to a permanent impairment of pulmonary function tests with airflow obstruction and/or airway hyperresponsiveness. This implies that it should be a compensable disease. Persistent asthma seen in workers with occupational asthma after they have left work cannot be interpreted as the unmasking of an individual predisposition to develop late onset asthma.[1] In Britain,[10] France,[11] and all Canadian provinces, occupational asthma is recognized as a compensable disease. Compensation responsibilities should include, first, job rehabilitation and, second, determination of compensation for any permanent disability.

Physicians should advise governmental agencies as regards to the type of job that should be offered to the worker. Exposure to the sensitizing agent or cross-reacting products (e.g., various isocyanates) must be strictly avoided even in minute amounts. Exposure to irritants such as fumes, cold air, strong odors, or jobs requiring strenuous efforts should be avoided if the worker has significantly heightened nonallergic airway responsiveness.

Disability is defined[12] as an inability to carry out a specific task or job, and impairment is an abnormality of physiologic function that persists after treatment. Physicians can assess the respiratory impairment of an occupational asthmatic, whereas determination of disability is usually the prerogative of a nonmedical committee. Indeed, disability implies that factors other than respiratory impairment (e.g., age, intelligence, education) are taken into account. Furthermore, a worker might have no functional respiratory impairment assessed by spirometry and airway responsiveness, yet still be partially disabled as he cannot be reexposed to the agent to which he has been sensitized without risk of recurrent asthma.[4] Persistent heightened nonallergic airway responsiveness in subjects sensitized to common inhaled allergens can lead to asthma upon their (nonoccupational) contact, a condition which would not have been otherwise clinically relevant.

When and how can we assess functional impairment in these workers?

Unlike pneumoconioses, which are irreversible, impairment associated with occupational asthma may be transient with complete recovery in some. The time course of recovery after cessation of exposure is not well known and may be very short[13] or take months.[14] This should be kept in mind. In Quebec, impairment is currently determined 2 years after leaving work, and reassessed 3 years later and considered definitive at that time.

In contrast to pneumoconioses for which impairment is evaluated by chest radiograph, lung volumes, pulmonary mechanics, carbon monoxide diffusing capacity, and blood gas analysis at rest and on exercise, impairment due to asthma is better evaluated by two functional criteria: spirometry before and after bronchodilators and the level of airway responsiveness to histamine or methacholine.[4,11] The persistent need for medication might be used as an additional criterion.

Questionnaires are not very useful in evaluation of functional impairment. Even if a validated questionnaire were available, it would still always be subjective and possibly biased. An example of questionnaire bias is the poor correlation between suggestive history and confirmation of occupational asthma in snowcrab-processing workers.[14] Questionnaires are used mainly to confirm the results of our tests and to try to evaluate disability as well as degree of control of symptoms, and the "minimum" medication requirement which may be a better reflection of asthma severity than is reduced spirometry. Spirometry before and after a bronchodilator is essential. It is possible, based on predicted values, to establish a standardized scale of impairment; however, spirometry may be normal in asthmatics who will develop symptoms on exposure to cold air, exercise, dust, fumes, etc. Such bronchial lability can be assessed by the level of airway responsiveness. Airway response to histamine/methacholine correlates with minimum medication requirement, ease of inducton of exercise/stress-induced bronchospasm, diurnal fluctuations of peak expiratory flow rates, response to bronchodilator, and baseline airway caliber.[15]

If workers continue exposure, protective respirators and medication are usually used. Masks are often ineffective because of poor fit and poor compliance; the minute exposure required to induce symptoms cannot be prevented. Prophylactic use of sodium cromoglycate or corticosteroids might control symptoms of asthma, but does not seem to prevent the deterioration of airway function.[6] Subjects who continue to work in the same environment despite sensitization require careful follow-up including spirometry and assessment of airway responsiveness.

PREVENTION OF OCCUPATIONAL ASTHMA

In order to prevent occupational asthma, consideration should be given to decrease the exposure to known offending agents and to set up health programs in high-risk industries.[16] Exposure to known offending agents can rarely be totally avoided; however, it can be significantly decreased by improving ventilation, isolating production of hazardous materials, preventing accidental high exposure, and using respirators. Reduction of exposure has led to a marked (but not complete) reduction in sensitization in the enzyme detergent industry.[17] Thus, maintenance of threshold limit values may reduce, but not entirely prevent, occupational asthma.[18] Substitution of the offending agent by less harmful material or a different formulation (e.g., encapsulation) of the product may prove to be useful. Respirators, especially those using an airline system with complete protective clothing, may effectively prevent occupational asthma.

Health programs in the industries are also needed. Employer and employee education on the risk associated with any job and the importance of reducing exposure should be stressed. Preemployment screening for atopy, clinical asthma, or subclinical airway hyperresponsiveness may be useful. However, atopy is a risk factor for occupational asthma in only a few instances (e.g., platinum salts, *B. subtilis* enzymes, animals) and does not apply to isocyanates or red cedar asthma. There are no published prospective data indicating that increased airway responsiveness is a risk factor.

Antismoking programs should be encouraged, although

smoking is relevant in only rare instances.[14] Finally, periodic medical examinations should be carried out for early identification of workers who develop asthma or other evidence of sensitization (e.g., rhinitis, conjunctivitis, urticaria). Prospective studies are needed to determine if histamine/methacholine inhalation tests, skin tests, and *in vitro* antibody tests will allow early detection of sensitization before the occurrence of (symptomatic) occupational asthma.

CONCLUSION

With the increased prevalence of occupational asthma, employers and Workmen's Compensation Boards should concentrate their efforts on prevention by: (1) reducing exposure to sensitizing agents; (2) identifying risk factors and indicators that will allow us to remove workers before they develop asthma; (3) establishing health programs in industries in order to facilitate early diagnosis; and (4) removing affected workers early.

REFERENCES

1. **Yeung, M. and Grzybowski, S.,** Prognosis in occupational asthma, *Thorax,* 40, 241, 1985.
2. **Pain, M.C.F. and Symons, H.S.,** Bronchial reactivity in occupational asthma, *Med. J. Aust.,* 1, 522, 1972.
3. **Adams, W.G.,** Long term effects on the health of man engaged in the manufacture of toluene diisocyanate, *Br. J. Ind. Med.,* 32, 72, 1975.
4. **Moller, D.R., McKay, R.T., Bernstein, I.L., and Brooks, S.M.,** Long term follow-up of workers with TDI asthma, *Am. Rev. Respir. Dis.,* 129, A159, 1984.
5. **Paggiaro, P.L., Loi, A.M., Rossi, O., Ferrante, B., Pardi, F., Roselli, M.G., and Baschieri, L.,** Follow-up study of patients with respiratory disease due to toluene diisocyanate (TDI), *Clin. Allergy,* 14, 463, 1984.
6. **Chan-Yeung, M., Lam, S., and Koener, S.,** Clinical features and natural history of occupational asthma due to Western red cedar *(Thuja plicata), Am. J. Med.,* 72, 411, 1982.
7. **Burge, P.S.,** Occupational asthma in electronics workers caused by colophony fumes: follow-up of affected workers, *Thorax,* 37, 348, 1982.
8. **Hudson, P., Pineau, L., Cartier, A., and Malo, J.L.,** Follow-up of occupational asthma due to various agents, *J. Allergy Clin. Immunol.,* 73, 174, 1984.
9. **Malo, J.L., Pineau, L., Cartier, A., and Martin, R.R.,** Reference values of the provocative concentrations of methacholine that cause 6% and 20% changes in forced expiratory volume in one second in a normal population, *Am. Rev. Respir. Dis.,* 128, 8, 1983.
10. **Hendrick, D.J. and Fabbri, L.,** Compensating occupational asthma, *Thorax,* 36, 881, 1981.
11. **Gervais, P.,** Le pneumologue et le medecin du travail devant le risque immuno-allergologigue en milieu professionnel, *Rev. Fr. Mal. Respir.,* 11, 409, 1983.
12. **Morgan, W.K.C.,** Pulmonary disability and impairment. Can't work? Won't work?, *ATS News,* 8, 30, 1982.
13. **Cockcroft, D.W. and Mink, J.T.,** Isocyanate-induced asthma in an automobile spray painter, *Can. Med. Assoc. J.,* 121, 602, 1979.
14. **Cartier, A., Malo, J.L., Forest, F., Lafrance, M., Pineau, L., St-Aubin, J.J., and Dubois, J.Y.,** Occupational asthma in snow crab-processing workers, *J. Allergy Clin. Immunol.,* 74, 261, 1984.
15. **Hargreave, F.E., Ryan, G., Thomson, N.C., O'Byrne, P.M., Latimer, K., Juniper, E.F., and Dolovich, J.,** Bronchial responsiveness to histamine or methacholine in asthma: measurement and clinical significance, *J. Allergy Clin. Immunol.,* 68, 347, 1981.
16. **Parkes, W.R.,** *Occupational Lung Disorders,* Butterworths, London, 1982.
17. **Juniper, C.P., How, M.J., Goodwin, B.F.J., and Kinshott, A.K.,** *Bacillus subtilis* enzymes: a 7-year clinical epidemiological and immunological study of an industrial allergen, *J. Soc. Occup. Med.,* 27, 3, 1987.
18. **Bernstein, I.L.,** Occupational asthma, *Clin. Chest Med.,* 2, 255, 1981.
19. **Dehaut, P., Rachiele, A., Martin, R.R., and Malo, J.L.,** Histamine dose-response curves in asthma: reproducibility and sensitivity of different indices to assess response, *Thorax,* 38, 516, 1983.

Grain Dust Asthma, Does It Exist?

M. Chan-Yeung

INTRODUCTION

In 1713 Ramazzini[1] the "Father of Industrial Medicine", described in an article entitled "Diseases of Sifters and Measurers of Grain"... "all kinds of grain and in particular wheat, whether they are stored in pits or trenches or in granaries and barns, always have mixed in with them a very fine dust....Hence, whenever it is necessary to sift wheat and barley or other kinds of grain to be ground in the mill or to measure it when corn merchants convey it hither and thither, the men who sift and measure are so plagued by this kind of dust that when the work is finished they heap a thousand curses on their calling. The throat, lungs and eyes are keenly aware of serious damage; the throat is choken and dried up with dust, the pulmonary passages become coated with crust formed by dust, and the result is a dry and obstinate cough. The eyes are much inflamed and watery and almost all who make a living by sifting or measuring grain are short of breath and cachectic and rarely reach old age." This passage has been quoted extensively and has been interpreted by many that the entity grain dust asthma was being described. When one reads it carefully one wonders whether this is so. "Short of breath, cachectic, and rarely reach old age" could be tuberculosis.

IS THERE SUCH AN ENTITY AS GRAIN DUST ASTHMA?

In this presentation, asthma will be defined in the conventional sense as variable airflow obstruction associated with nonspecific airway hyperresponsiveness.

In 1974 Warren and co-workers[2] performed allergy and respiratory investigations on 15 grain workers with respiratory symptoms and 5 controls. Inhalation provocation tests with extracts of grain dust were carried out. Eight out of 15 grain workers developed asthmatic responses after challenge; 7 of 8 had a positive immediate skin test to crude grain dust extract. Immediate, late, or dual asthmatic responses were described. Systemic symptoms (fever and leukocytosis) occurred in 6 grain workers and in 2 controls. The controls did not have respiratory symptoms or lung function changes after challenge.

In 1979 our group[3] studied 22 grain workers with respiratory symptoms for lung function abnormalities and 11 asymptomatic grain workers with normal lung function as controls. We used grain dust as well as grain dust extract for challenge. Six out of 22 grain workers developed asthmatic responses after challenge, 3 with isolated immediate response and 3 with dual responses. Systemic symptoms and leukocytosis were not found. None of the control subjects developed any reaction. The immediate asthmatic response was inhibited by pretreatment with disodium cromoglycate, while the late response was partially inhibited by treatment with beclomethasone diproprionate before and at intervals after challenge. Grain workers who had positive asthmatic response on challenge had a marked degree of nonspecific airway hyperresponsiveness (methacholine $PC_{20} = 0.36 \pm 0.34$ mg/ml); those with negative challenge but respiratory symptoms and/or lung function abnormality had some degree of airway hyperresponsiveness (methacholine $PC_{20} = 1.87 \pm 6.88$ mg/ml). The controls had no airway hyperresponsiveness. Moreover, those who had a positive grain dust challenge tended to have a higher eosinophil count than those who did not react (244 and 156/cu mm, respectively). None of our workers had a positive skin test to the crude grain dust extract.

The above two studies suggest that grain dust asthma exists and may be due to immunologic mechanisms.

IS GRAIN DUST ASTHMA AN ALLERGIC REACTION?

With the exception of Warren and co-workers,[2] who found a good correlation between positive inhalation tests and immediate skin reaction to grain dust extract, other investigators have demonstrated no such correlation. Moreover, specific antibodies have not been found in the sera of patients with positive inhalation challenge.[3,4] One problem is the difficulty in identifying the responsible allergen in the grain dust. The composition of grain dust is complex, and includes various cereal grains, fungi, bacteria, insects, mites, fumigants, etc. Studies to identify the allergen(s) have been difficult. Two studies have demonstrated specific agents in the grain dust. Davies et al.[5] reported a farmer who developed recurrent nocturnal asthma after inhalation challenge with grain dust. Examination of the grain samples together with the skin and serological tests suggested that the grain mite, *Glycyphagus destructor*, might be an important allergen. Another study was conducted by doPico and co-workers,[6] who performed inhalation tests on 11 grain workers using extracts of durum wheat, durum wheat airborne dust, and grain insects and mites. Five out of 11 workers showed a positive response to durum wheat extract (1 isolated immediate and 4 isolated late response). Only one subject had a positive bronchial response to durum wheat dust extract and none reacted to mites or insect extracts. No parenchymal or systemic reaction was found. There are no other studies that have successfully identified the allergen(s).

WHAT IS THE PREVALENCE OF GRAIN DUST ASTHMA?

There is virtually no information on the prevalence of grain dust asthma. It has been suggested that asthmatics are likely to seek other employment and leave the industry early. The results of our own prevalence study in Vancouver supported this notion. The prevalences of atopy (immediate skin reaction to one or more common allergens) and current asthma (affirmative answers to: [1] have you ever had asthma?; [2] was it diagnosed by a physician?; and [3] do you still have it?) among 669 male grain workers in the Terminal Elevators in Vancouver and 560 male office workers were 17.3 vs. 31.3%, respectively, for atopy (p <0.05), and 2.4 vs. 2.7%, respectively, for asthma even though the grain workers were significantly younger (38.5 vs. 43.6 years, p <0.05). Only 11 (1.6%) out of 16 asthmatic grain workers claimed that their asthma started after employment. The prevalence of grain dust asthma amongst elevator employees is, indeed, very low.

ACUTE REVERSIBLE AIRWAY OBSTRUCTION IN GRAIN WORKERS

There are now a number of studies[7-10] showing that some grain workers demonstrate an acute shift change in lung function. The first, conducted by Gandevia and Ritchie in 1966[7] in 24 subjects, showed that a mean change in forced expired volume in 1 s (FEV_1) after exposure to wheat dust was –141 ml over a single exposure (half an hour) and –291 ml over the whole shift. The comparable figures following exposure to phosphate rock dust were –43 and –130 ml, respectively. During our initial cross-sectional study[8] we measured spirometry on Monday morning (48 h away from work), Monday afternoon (at the end of one work shift), and Friday afternoon (after 5 d of work) on grain workers and controls (noncedar sawmill workers). We found a slight, but statistically significant, drop in mean FEV_1 and forced vital capacity (FVC) among grain workers after one work shift. The readings on Friday afternoon were significantly lower than Monday morning readings, but not different from Monday afternoon readings. Among sawmill workers there was a slight, but statistically significant, increase in FEV_1 and FVC after one work shift and one work week. In our study, 3.9% of grain workers had a decrease in FEV_1 of greater than 10% over one work shift.[8]

WHAT IS THE MECHANISM OF ACUTE SHIFT CHANGE IN LUNG FUNCTION?

Is it due to grain dust-induced allergic asthmatic response or to nonallergic bronchoconstriction? The acute shift change in lung function was examined carefully by Corey and coworkers.[9] They measured lung function pre- and postshift on Monday, Wednesday, and Friday in 34 grain workers. They found that the acute shift changes in $\dot{V}_{max50}$ were evident on Monday and Wednesday. There was a progressive drop in preshift $\dot{V}_{max50}$ from Monday to Friday. By Friday afternoon there was some recovery. This finding is a little different than, but not unlike, those found in byssinosis. The other important observation was that the acute shift change was dependent on the degree of exposure. The higher the exposure to respirable dust, the greater the acute shift change in lung function. They calculated that in 50% of workers, a decrease of 923 ml/s in $\dot{V}_{max50}$ was induced for each 1-mg/m^3 increase of respirable dust.

doPico et al.[10] studied 248 grainhandlers and 192 civic workers before and after 8-h work shifts. They found significant (p <0.05) differences in pre-post shift changes in FVC and $\dot{V}_{max50}$ among grain workers, but not in civic workers. Eleven percent of the grain workers had a greater than 10% fall in FEV_1 after one work shift. In grain workers they found a significant negative relationship between total dust levels and the changes in the above lung function parameters. There was also a dose-response relationship between total dust concentration and pre-post shift increase in leukocyte count. The work shift changes in lung function were not related to age, atopic status, or smoking.

The acute shift change in lung function in grain workers appears be an obstructive defect, directly related to the level of dust exposure, unrelated to age, smoking, duration of employment, atopic status, or skin test reactivity to grain dust antigens, and associated with an increase in leukocyte count. These characteristics suggest that in most cases the acute shift change is not a type I immediate hypersensitivity reaction but a nonallergic response to some constituent(s) of grain dust.

EVIDENCE OF NONALLERGIC BRONCHOCONSTRICTION

Extracts of grain dust have been shown by Olenchock and colleagues[11] to activate both the alternative and classical complement pathways *in vitro* with the release of complement fragments such as C3a and C5a. Components of both the alternative and classical pathways are present in human airways;[12] the potential for direct activation of both pathways after inhalation of grain dust therefore exists, either directly by the dust or as rapidly extracted material. This may be a cause of bronchoconstriction.

Warren and Holford-Stevens[13] demonstrated that extracts of grain dust released histamine from peritoneal mast cells of rats. We also showed that extracts of airborne grain dust released histamine from guinea pig lung fragment in a dose-response manner. These findings suggest that grain dust is capable of inducing bronchoconstriction by releasing histamine from the mast cells in the airways, a finding similar to cotton dust.[14]

SIGNIFICANCE OF ACUTE SHIFT CHANGE IN LUNG FUNCTION

A prospective study was carried out on the grain workers in Vancouver using civic worker controls. During the initial cross-sectional study in 1975 we measured pre- and postshift lung function. In 1978 we performed a 3-year follow-up and found that grain workers had a significantly greater decline in lung function compared to civic workers.[15] We studied the factors affecting the longitudinal decline in the lung function. Among the factors studied, acute shift change in lung function was found to be predictive of longitudinal decline. Those with the greatest acute drop in lung function over one work week

TABLE 1
Annual Change in Spirometry (1975—1981) According to Change in Spirometry over One Work Week (1975) in Grainhandlers by Smoking Habit

Smoking habit	Change in spirometry over a work week			
	>+4%	+4% to –4%	>–4%	p[a]
Nonsmokers				
FEV_1 (ml)	–74 ± 205	–17 ± 43	–54 ± 66	0.167
FVC (ml)	–36 ± 56	–13 ± 100	–52 ± 68	0.013
$FEF_{25-75\%}$ (ml/s)	–21 ± 114	–90 ± 186	–137 ± 116	0.001
Ex-smokers				
FEV_1 (ml)	–20 ± 47	–19 ± 44	–36 ± 56	0.004
FVC (ml)	+26 ± 56	–10 ± 47	–32 ± 44	0.000
$FEF_{25-75\%}$ (ml/s)	–28 ± 96	–93 ± 137	–118 ± 96	0.002
Current smokers				
FEV_1 (ml)	+8 ± 50	–33 ± 46	–83 ± 59	0.000
FVC (ml)	+21 ± 46	–5 ± 56	–74± 80	0.000
$FEF_{25-75\%}$ (ml/s)	–73 ± 107	–101 ± 123	–151 ± 80	0.000

[a] From multiple regression analysis, values are mean ± SD.

had the greatest longitudinal (3-year) decline in lung function.[15] This finding was subsequently confirmed in our 6-year follow-up study (Table 1).[16] Since the acute shift change in lung function is dose related, the greater longitudinal decline in lung function in those with greater acute shift change is probably a reflection of greater exposure rather than of individual susceptibility. This finding, if confirmed by other investigators, may have an important bearing on the determination of the maximal allowable concentration of grain dust.

CONCLUSION

In summary, this author believes that grain dust asthma does exist though it is uncommon. During the past 15 years, only two grain workers were referred to me with suspected grain dust asthma. It is unknown at present whether this is a truly allergic phenomenon with the allergen(s) not yet identified. The prevalence of acute reversible bronchoconstriction over one work shift varies from 4 to 11% depending on the level of dust concentration. The degree of acute shift change in lung function is directly correlated with the level of dust exposure. There is evidence to suggest that the acute reversible bronchoconstriction is due to nonallergic mechanisms different from the usual asthma manifestation. We also found that the acute work week change in lung function is an important determinant of subsequent longitudinal decline in lung function. There are considerable similarities between the syndrome of acute bronchoconstriction due to grain dust exposure, grain fever syndrome, and bysinnosis.

REFERENCES

1. **Ramazzini, B.,** *Demorbis Artificium Diatriba* (1713), W. Care Wright, W., Trans., University of Chicago Press, Chicago, 1940.
2. **Warren, P., Cherniack, R.M., and Tse, K.S.,** Hypersensitivity reactions to grain dust, *J. Allergy Clin. Immunol.*, 53, 139, 1974.
3. **Chan-Yeung, M., Wong, R., and MacLean, L.,** Respiratory abnormalities among grain elevator workers, *Chest,* 75, 461, 1979.
4. **Broder, I., Davies, G., Hutcheon, M., Leznoff, A., Mintz, S., Thomas, P., and Corey, P.,** Variables of pulmonary allergy and inflammation in grain elevator workers, *J. Occup. Med.,* 25, 43, 1983.
5. **Davies, R.J., Green, M., and Schofield, N.McC.,** Recurrent nocturnal asthma after exposure to grain dust, *Am. Rev. Respir. Dis.,* 114, 1011, 1976.
6. **doPico, G.A., Jacobs, S., Flaherty, D., and Rankin, J.,** Pulmonary reaction to durum wheat: a constituent of grain dust, *Chest,* 81, 55, 1982.
7. **Gandevia, B. and Ritchie, B.,** Relevance of respiratory symptoms and signs to ventilatory capacity changes after exposure to grain dust and phosphate rock dust, *Br. J. Ind. Med.,* 23, 181, 1966.
8. **Chan-Yeung, M., Schulzer, M., MacLean, L., Dorken, E., and Grzybowski, S.,** Epidemiologic health survey of grain elevator workers in British Columbia, *Am. Rev. Respir. Dis.,* 121, 329, 1980.
9. **Corey, P., Hutcheon, M., Broder, I., and Mintz, S.,** Grain elevator workers show work-related pulmonary function changes and dose-effect relationships with dust exposure, *Br. J. Ind. Med.,* 39, 330, 1982.
10. **doPico, G.A., Reddan, W., Anderson, S., Flaherty, D., and Smalley, E.,** Acute effects of grain dust exposure during a work shift, *Am. Rev. Respir. Dis.,* 128, 399, 1983.
11. **Olenchock, S.A., Mull, J.C., and Major, P.C.,** Extracts of airborne grain dusts activate alternative and classical complement pathways, *Ann. Allergy,* 44, 23, 1980.
12. **Robertson, J., Caldwell, J.R., Castle, J.R., and Waldman, R.H.,** Evidence of the presence of components of the alternative (properdin) pathway of complement activation in respiratory secretions, *J. Immunol.,* 117, 900, 1976.
13. **Warren, C.P.W. and Holford-Stevens, V.,** Induction of histamine release *in vitro* from rat peritoneal mast cells by extracts of grain and grain dust, presented in the New Orleans Conf. on Agricultural Dusts, New Orleans, 1984.
14. **Douglas, J.S., Zuskin, E., and Bouhuys, A.,** Relationship between *in vivo* bronchospasm induced by textile dust extracts and *in vitro* histamine release from pig lung, *Am. Rev. Respir. Dis.,* 109, 712, 1974.
15. **Chan-Yeung, M., Schulzer, M., MacLean, L., Dorken, E., Tan, F., Lam, S., Enarson, D., and Grzybowski, S.,** A follow-up study of the grain elevator workers in the port of Vancouver, *Arch. Environ. Health,* 36, 75, 1981.
16. **Tabona, M., Chan-Yeung, M., Enarson, D., MacLean, L., Dorken, E., and Schulzer, M.,** Host factors affecting longitudinal decline in lung spirometry among grain elevator workers, *Chest,* 85, 782, 1984.

Environmental Assessment and Health Promotion

Grain Dust Assessment in Country Elevators

K. Yoshida

INTRODUCTION

Grain dust is known as organic or viable dusts in the work environment. However, the characteristics of the viable fraction of grain dust have been studied only for grain worker health protection. Physicochemical properties of grain dust as nonviable particles require further investigation from the point of view of controlling worker inhalation exposure. This paper is intended to present characteristics of nonviable grain dust in conjunction with problems associated with the assessment of airborne contamination in country grain elevators.

MECHANISMS OF DUST GENERATION

A typical country elevator consists of a truck scale, a receiving hopper, a bucket elevator, storage bins, a hopper scale, and various chutes for transferring grain to and from the bins and for loading railcars or truck trailers. Some elevators are equipped with grain-cleaning facilities.[1]

Dust is emitted whenever grain is moved from one location to another. Newly developed elevators employ closed-duct systems and ventilation is provided at each transfer point. The suction is generated by a train of dust control cyclones, located outside of the building. However, primary sources of dust generation are at the receiving and shipping hoppers that still require sufficient ventilation.[2] Prior to the use of cyclones, the ventilation stack at the top of the penthouse was used to eliminate dust in the penthouse. Dust emission from the rooftop vent contributes to the dustfall intensity of the surrounding community.[3]

PHYSICAL CHARACTERISTICS

Particle Size Distribution

Emission sources of inhalable grain dust are numerous; therefore, the mode of emission is multimodal. The particle size distribution usually takes the form of a bimodal function. One type of fine-particle spectrum originated from the process of condensation such as exhaust emission of a diesel engine in a delivery or shipping truck, while another type of coarser particle was produced by the mechanical disturbance of the grain. Although they are bimodal, the distribution curves take log-normal function.[3]

Particles generated by condensation are usually submicron in diameter and remain within the diffusional size range. On the other hand, dust particles produced by mechanical disturbances are coarser and remain within the inertial size range from a few microns to hundreds of microns in diameter. These two groups of dust behave in a distinctive manner in an aerodynamic field such as the human respiratory system.

In the analyses of particle size spectra, the following statistical parameters are determined: mass medium aerodynamic equivalent diameter (MMAED or MMD), which is the diameter of a particle with unit density and divides the total mass of dust within a sample into two equal parts, one half of the mass smaller than MMD and one half larger. The cumulative mass of dust is plotted against the particle aerodynamic diameter. An intercept of this curve-linear graph at 50% cumulative mass yields MMD.

The uniformity of particle size distribution is expressed by the geometric standard deviation (sigma-g). The value of the particle diameter at 84.6% intercept of the cumulative mass on the curve divided by MMD (50%) gives the value of sigma-g. If the value approaches unity, the degree of particle size distribution is said to be very homogeneous; however, the range of sigma-g with grain dust fluctuates widely. Usually, for mechanically dispersed dusts the values of sigma-g range from 2 to 3.[3]

Respirable Dust Concept

Potential health hazards associated with the inhalation of an aerosol may be evaluated with the provision of the following information: (1) the total quantity of material which can be absorbed from the aerosol, (2) the primary site of deposition and the distribution of the absorbed particles among them, (3) the significant elimination rates from the primary sites of deposition, (4) the secondary sites of deposition for material dissolved from the primary sites, and (5) the biological effects caused by absorbed material deposited on, or passing through, the various organs.[4]

"Respirable dust" is that which is deposited in the nonciliated portion of the lung. This application of the concept of respirable dust and concomitant selective sampling was intended only for "insoluble" particles which exhibit prolonged retention in the lung. It was not intended to include dusts which have an appreciable solubility in body fluids and those which are primarily chemical intoxicants. According to the alveolar deposition curve adopted by the American Conference of Governmental Industrial Hygienists (ACGIH), the particles below the cutoff MMD of 3.5 µm at 50% collection efficiency of a size-selective sampler may be considered respirable dust.[4] A comparison of standards for respirability has been discussed.[5] The respirable mass fraction of dust is determined graphically on a particle size distribution curve by intercepting it at 3.5 µm of the particle diameter. The fraction of respirable mass in the country elevator ranged from 10 to 40% of the total airborne dust mass.[3]

Morphology

Grain dust is a mixture of constituents varying with the type of grain handled; therefore, the morphology of dust particles is complex. Particles of grain dust may be grouped into the following shapes: spheres mainly composed of starch particles, spherical caps of broken starch or soil particles, cubes of soil particles, broken crop fibers, rods of broken and split fibers, disks of soil or broken grain particles, and cones of the tips of trichome particles or quartz.

Respirable particles with different geometries behave distinctively in inhalation and deposition. Therefore, an ordinary aerodynamic diameter (equivalent to unit density particle) cannot be used in assessing the respirable mass fraction of dust. The diameter must be multiplied by an aerodynamic shape factor or a product of the shape factor. Sample calculations of inhalation characteristics of grain dust by employing aerodynamic shape factors (volume shape factor, projected area diameter resistance shape factor, and fiber diameter resistance shape factor) were attempted.[6] The shape factor is one of the contributing parameters in inhalation exposure of dust particles containing respirable quartz.[7] Studies on shape factors were conducted for highly toxic dusts such as uranium ore dust; however, relatively little is known for grain dust.

The apex or tip of a wheat kernel contains a brush composed of many hairs that are extensions of epidermal cells. Some hairs are 1 mm long, but most are 500 μm, with some as short as 120 μm. The hairs are bulbous at the base and taper to the tip, and are straight or slightly curved. This morphology, microscopically, is described as "trichome".[8] Trichome particles may be split into a respirable size (like asbestos particles) of 10 μm long and 2 μm in optical diameter with an aspect ratio of 5. These respirable particles may not be cleared easily from the site of deposition due to predominant penetration rather than interception or sedimentation. The presence of trichome particles in grain dust was confirmed by using a scanning electron microscope.[3]

CHEMICAL COMPOSITION

Water-Insoluble Fraction

Grain dust is considered to be predominantly organic; however, it contains a substantial amount of mineral dust, mainly from the farm soil.[9] The content of the water-insoluble fraction in airborne dust inside the work area of a country elevator ranged from 55 to 65% of the total mass of the dust. The water-insoluble fraction of dust in a certain size range may be considered theoretically inhalable dust.

Combustible Fraction

The organic mass of grain dust may be one of the major contributors to dust explosions in grain elevators. The combustible fraction of grain dust inside a country elevator ranged from 35 to 45% by mass.[3] The ash content of grain dust in terminal elevators was also investigated.[10]

Trace Elements

The chemical composition of grain dust collected in several country elevators appeared similar in concentrations and generally compatible with the terrestrial abundance of each element. The elements with concentrations higher than their terrestrial abundance are aluminum, manganese, titanium, vanadium, gold, and silver.[3] The high concentration of aluminum and titanium might have originated from the equipment used for harvesting and hauling the grains. The higher concentration of manganese and vanadium may have been contributed by the soil where the grain was harvested, or by fuel combustion in a diesel engine during harvesting or hauling. The high concentration of gold and silver might be the residue of pesticides.

Quartz Content

The mass concentration of respirable quartz was correlated with specific grain processed in a given type of elevator. The quartz content in grain dust (6.5% of the total mass of dust[10]) in country elevators ranged considerably lower than that of terminal elevators. The origin and nature of quartz particles in grain dust are still unclear; however, three possible sources may be postulated: (1) soil particles picked up while harvesting grain, (2) inclusion of quartz-containing soil particles in the growing kernel of grains, or (3) the quartz cells naturally developed on the plant tissue of crops for protection against disease or insect infestation. The appreciable presence of quartz (or free silica) in grain dust has been recognized. However, there are no reports of silicosis among chronically exposed workers. No abnormalities were found which could be attributed to grain dust exposure and, in particular, to its quartz fraction.[11]

METHOD OF GRAIN DUST MEASUREMENT

Dust-Sampling Practices

Methods of general work-area sampling include the following three primary practices: (1) determination of the total mass concentration of dust by using a group of stationary samples at representative locations in the work area, (2) determination of particle size distribution (size-selective sampling) by using a cascade impactor or other size-selective sampler such as cyclones, and (3) grab sampling of a large quantity of airborne dust, using a high volume sampler or an industrial vacuum cleaner in order to produce a sufficient mass of samples for chemical or microscopic analyses.[2]

Particle Size Analysis

For size-selective sampling of airborne dust, a cascade impactor with 6 or 11 stages may be used and must be operated at a constant flow rate. The mass of dust particles deposited on each substrate is gravimetrically determined and a computer graphics program processes the data to obtain statistical parameters.[12]

Particle Morphology

The shape of grain dust may be investigated under an optical microscope of the polarized light type, while identifying quartz particles that exhibit birefringency under the polarized light. For detailed investigation, the use of a scanning electron microscope (SEM) coupled with an energy-dispersing X-ray analyzer (EDA) is a standard practice.[13]

Respirable Dust

In the assessment of worker exposure to grain dust, the target agent is its respirable fraction. There are three conflicting areas that make realistic assessment of worker exposure difficult: the methodology of sampling and analyses, factors affecting the exposure index, and the adequacy of instrumentation.

One crucial factor in grain dust sampling is that the cutoff efficiency of the commonly used 10-mm nylon cyclone as a size-selective (or respirable) sampler has not been studied for organic dusts of wider size spectrum and of various shape factors. A discussion presented in a comparison of performance between open-face and closed filter holders with dusts of wider size spectrum suggests that the closed holder with a 4-mm inlet orifice may act almost as a size-selective sampler against large particles. Therefore, the closed-face sample may tend to underestimate the total dust concentration.[14] The same argument may be applicable to the collection efficiency of an inlet orifice (4-mm rectangular) of a 10-mm nylon cyclone.[15]

Respirable Quartz

Dust particles collected on a membrane filter of a 10-mm nylon cyclone are regarded as respirable dust. The dust sample then will be analyzed for free silica or quartz content by using infrared spectrophotometry[16,17] or by using X-ray diffraction spectrophotometry.[18] For the determination of respirable quartz content, it is necessary to collect a large mass of airborne dust on a filter. Usually, in personal sampling for the time-weighted average (minimum of 15 min for 32 samplings), it is impossible to collect a sufficient dust mass under the normal work environment of a country elevator. The respirable quartz, therefore, may be quantified only in grab samples.[3]

Chemical Elements

The chemical elements contained in grain dust may be determined either by using neutron activation and gamma spectrometry or by the EDA system of the SEM. The neutron activation method provides concentration of a wide spectrum of elements contained in a mass of dust (on a filter paper), while the EDA method can detect chemical elements in a single particle. For dust samples collected in a general area, the use of neutron activation analysis provides easier sample preparation and a quick result.[3]

CONCLUSIONS

1. Grain dust has a complex composition and its mechanism of inhalation exposure is not fully understood.
2. While the exposure indices of various job categories are established in large terminal elevators, an adequate method of dust assessment is still needed for country elevators.
3. Dust-sampling instruments that are capable of performing for particles of a wide spectrum are needed to precisely determine worker exposure to respirable dust.
4. Further studies are needed of the characteristics of inhalation of quartz particles in grain dust.

REFERENCES

1. **Schrag, K.R.,** Dust concentrations in Alberta's grain industries, *Am. Ind. Hyg. Assoc. J.*, 33, 100, 1972.
2. **Yoshida, K.,** On the nature of environmental grain dust pollution on the Canadian prairie, *Air Pollut. Control Assoc.*, No. 77, 13P, 1977.
3. **Yoshida, K.,** Physical and environmental characteristics of grain dust, in *Occupational Pulmonary Disease: Focus on Grain Dust and Health*, Dosman, J.D., and Cotton, D.J., Eds., Academic Press, New York, 1980, 441.
4. **Lippmann, M. and Harriw, W.B.,** Size-selective samplers for estimating respirable dust concentrations, *Health Phys.*, 9, 155, 1962.
5. **Lippmann, M.,** Size-selective health hazards sampling, in *Air Sampling Instruments for Evaluation of Atmospheric Contaminants*, H1-19, American Conference of Governmental Industrial Hygienists, Cincinnati, OH, 1983.
6. **Yoshida, K.,** Dynamic shape factor determinations of respirable grain dust, *Proc. 10th Aerosol Tech. Meet.*, p. 11P, 1977.
7. **Kotrappa, P.,** Shape factors for quartz aerosol in respirable size range, *Aerosol Sci.*, 2, 353, 1971.
8. **Bradbury, D. and MacMasters, M.M.,** Structure of the mature wheat kernal, I. Gross anatomy and relationship of parts. II. Microscopic structure of pericap, seed coat, and other coverings of the endosperm and germ of hard red winter wheat, *Cereal Chem.*, 33, 329, 1956.
9. **Omura, T. and Nogawa, K.,** Respiratory function abnormalities in workers exposed to diatomaceous earth dust, *Jpn J. Ind. Health*, 20, 254, 1978.
10. **Farant, J.P. and Moore, C.F.,** Dust exposures in the Canadian grain industry, *Am. Ind. Hyg. Assoc. J.*, 39, 177, 1978.
11. **Yoshida, K.,** Design of sampling and analytical techniques for respirable grain dust — for correlation with a clinical study of pulmonary disease characteristics among grain handlers, *SRC Rep.*, P77-2, 1977.
12. **Yoshida, K. and Wallace K.,** Size analysis of suspended dust by Sierra cascade impactor using computer graphic method, *SRC Rep.*, P77-6, 1977.
13. **Denee, P.B.,** Mine dust characterization using the scanning electron microscope, *Am. Ind. Hyg. Assoc. J.*, 33, 654, 1972.
14. **Beaulieu, H.J.,** A comparison of aerosol sampling techniques, "open versus closed-face" filter cassettes, *Amer. Ind. Hyg. Assoc. J.*, 41, 758, 1980.
15. **Yoshida, K.,** Aerodynamic parameter determination of inlet orifice of CEA alpha dosimeter, *Proc. Int. Conf. Occup. Rad. Safety in Mining*, 1984.
16. **Dodgson, J. and Whitaker, W.,** The determination of quartz in respirable dust samples by infrared spectrophotometry. I. The potassium bromide disc method, *Ann. Occup. Hyg.*, 16, 373, 1973.
17. **Dodgson, J. and Whitaker, W.,** The determination of quartz in respirable dust samples by infrared spectrophotometry. II. The direct analysis of quartz deposited on filters, *Ann. Occup. Hyg.*, 16, 389, 1973.
18. American Conference of Governmental Industrial Hygienists, Round table discussion on analytical techniques for quartz, American Conference of Governmental Industrial Hygienists, Cincinnati, OH, 1973.

Assessment of Dust Nature and Levels in the Grain Industry

J.-P. Farant

INTRODUCTION

Several epidemiological studies of populations exposed to grain dust have been conducted in the Canadian grain industry[1-4] and elsewhere[5,6] during the past 20 or more years. In most studies the prevalence of certain symptoms and the results of lung function tests were largely similar. However, as for similar studies of other organic dusts, these investigations have not identified causative agent(s) responsible and the mechanisms whereby grain dust affects the respiratory system. In this presentation, a review is made of the nature and airborne concentrations of potential causative agents as well as the impact that modernization of certain elevator operations may have had on these.

NATURE OF GRAIN DUST

Generation

Grain dust is produced by the abrasion and attrition of grain kernels each time the grain mass is handled and the extent of its production is a function of the degree of processing. It has been estimated that approximately 3 to 4 lb of dust is produced per ton of grain processed through a typical elevator system.[7] This amount will vary somewhat with the type, source, and quality of grain handled, previous handling, conveying equipment characteristics, and the degree of attrition inherent in the operation.

Composition

Although the prime constituents of grain dust are derived from the parent grain and grain plant, this dust is a rather complex mixture which contains minerals such as quartz and mica, wild seeds, pollen, fungal hyphae, spores, mycotoxins, insect parts, animal and insect fecal material, plant and animal hairs, bacteria, various pesticides, and a host of other biological contaminants too numerous to be listed here. Efforts have been made in recent years to further characterize grain dust components to the extent of other organic dusts such as cotton.[8,9] Results of these investigations have been presented elsewhere in the symposium[10-14] and there would be no purpose in reiterating them here. The main thrust of this research appears to be concerned with the microbial and mycological agents in grain dust.

Particle Size Distribition

In an earlier publication,[15] we had reported that grain dust particles, similar to other organic and inorganic dusts, varied considerably in size distribution. Although the results presented were only preliminary, they had indicated that particle size distribution in a grain dust cloud was strongly related to the type of grain handled. Thus, the respirable dust mass samples with a two-stage personal sampler, a 10-mm nylon cyclone, was characteristically 17% for barley, 18% for wheat, and only 6.9% for corn dust. These results apparently support most grain handlers' complaints of personal discomfort which were greatest for barley and wheat and least for corn dust, which is generally considered to be a "heavy" dust. To our knowledge, no further efforts have been made to investigate the particle size distribution in dust clouds generated during the handling of these or other grains. Several portable, personal instruments are presently commercially available for this type of investigation. These, such as the Anderson minisampler, can be used with battery-operated, personal-type portable air pumps and could afford data on the particle size distribution for different dust clouds in a variety of work situations involving grain dust exposure.

Mineral Constituents

To date, attention on the mineral content of grain dust has focused mainly on its quartz content. In a previous publication we had reported that quartz accounted for some 10 to 20% of the inorganic material found in respirable grain dust from different sources. The amount of inorganic material and its associated quartz content apparently varied according to the type of respirable grain dust, barley generally being highest with $43.4 \pm 5.3\%$ (inorganic material) and $6.1 \pm 3.8\%$ (quartz). Conversely, rapeseed dust contained the least with 21.2 ± 2.2 and $4.4 \pm 3.9\%$, respectively. One factor appeared foremost in distinguishing grains processed at different types of elevators, that is, grain cleanliness. Thus, grain received at country grain elevators is relatively soiled and the mean quartz content of its respirable dust was correspondingly highest. Conversely, transfer elevators usually handle only cleaned grains and the respirable dust in these elevators have a relatively low inorganic (19.3%) and quartz (2.1%) content. It is interesting to note that respirable corn dust contained relatively low amounts of quartz (1.2%).

The origin and nature of quartz in grain dust are still relatively unknown. We had earlier postulated that an important source for this mineral would be quartz cells found in *Gramina* genera of grasses such as wheat, barley, oats, and other grains.[16] Other important sources include soil contamination of harvested crops and the inclusion of quartz-containing soil particles on the growing kernels of grains. The apparent absence of abnormalities to date in the X-rays of grain workers would suggest that the quartz found in grain dust apparently does not cause silicosis.

Mycological Constituents

The microflora of grain dust evolves through harvesting, storage, and handling and is a reflection of the microflora of the grain at any one time. In an earlier report, Lacey[18] had described the fungi found in grain dust at harvest, during early stages of storage of dry grain, and during the later stages of storage of moist grain and of accumulated dust deposits in elevators. He demonstrated that the species of fungi found can vary markedly, the predominant types reflecting the predominant microflora of the grain, itself a reflection of the conditions of ripening and storage. It was also noted that many of these microorganisms recorded from grain are well-known allergens such as *Cladosporium, Alternaria, Penicillium, Ustilago, Aspergillus, Mucor,* and others.

We have described[18] the occurrence of fungi in country, terminal, and transfer elevators. Thirty species of fungi, mainly psychotolerant thermophiles or mesophiles, belonging to some 15 genera were identified. These are listed in Table 1. *P. expansum,* a storage fungi, was by far the most predominant species, and was observed in 69% of the samples collected. *A. repens* and *Hormodendrum cladosporioides* were also very prevalent and were found in 45% of the samples. *Alternaria* was observed in 29% of the grain dust samples. Most other fungi in Table 1 occurred rather sporadically. All of the fungi identified were among a list of some 43 genera and 102 species of fungi isolated by Machacek et al.[19] during their investigation of seed-borne fungi in various grains in Canadian grain elevators. Wallace and Sinha[20] found similar fungi in hot spots in silo-stored grains on the Canadian Prairies.

We also determined the airborne concentrations of some six morphologically identifiable spore types *(Ustilago, Aspergillus, Mucor, Hormodendrum, Alternaria, Epicoccum)* in Canadian grain elevator environments.[18] Dust samples were collected on open-faced membrane filter samplers, portions of which were then clarified and examined microscopically at 450 × magnification. Many air samples collected in this fashion were discarded because of overloading by extraneous dust particles. Presently, we are investigating the possibility of circumventing this serious limitation by sampling the dust cloud with a specially arranged two-stage (10-mm nylon cyclone) sampler, which preferentially collects fungal spores and particles <10 μm in diameter. Preliminary results have been very encouraging.

The occurrences of fungal spores in dust samples collected between July and September in various elevators during the handling of the new season's grain is shown in Table 2. The air spora found is one in which the fungi associated with grain at harvest is predominant. Concentrations varied between 0.01 to 958 x 10^6 spores per cubic meter, with considerably more spores in the air of non-ventilated than of ventilated elevators. *Ustilago* spores were by far the most prevalent and were found in 70% of the samples collected. The abundance of this spore type, *Ustilago,* probably reflects the extent to which the grain crop had been infected by this plant pathogen at the time of sampling. *Aspergillus, Mucor,* and *Cladosporium* spores were found in most samples collected. *Epicoccum* spores were rarely noted.

Lacey[18] has studied dust deposits in elevators in Manitoba where respiratory abnormalities have occurred in workers. He found both field and storage fungi as well as bacteria and actinomycetes. Storage fungi were usually far more abundant than field fungi, while bacteria were both ubiquitous and numerous. Actinomycetes species usually associated with farmer's lung, although present in many samples, were rarely numerous.

TABLE 1
Species of Fungus Identified in Airborne Grain Dust Samples Collected in Terminal, Transfer, and Country Grain Elevators

Mucor racemosus	*Rhizopus nigricans*
Paecilomyces sp.	*Cephalosporium* sp.
Aureobasidium sp.	*Scopulariopsis* sp.
Isaria sp.	*Botrytis* sp.
Hormodendrum cladosporioides	*H. macrocarpium*
Epicoccum sp.	*Alternaria* sp.
Yeast sp.	*Aspergillus repens*
A. fumigatus	*A. flavus*
A. niger	*A. versicolor*
Eurotium amstelodami	*Penicillium expansum*
P. terrestre	*P. commune*
P. janthinellum	*P. duclauxi*
P. oxalicum	*P. decumbens*
P. cyclopium	*P. chrysogenum*
P. granulatum	

TOTAL AND RESPIRABLE GRAIN DUST CONCENTRATIONS

Dusty conditions in grain elevator work areas can be attributed to the type and degree of activity in progress; the extent of enclosure provided at transfer points where grain falls freely from one conveying system to another; the extent, efficiency, and upkeep of exhaust ventilation provided at transfer points; and work and housekeeping practices. Dust exposures are also markedly subject to seasonal variations in grain throughput in an elevator.

TABLE 2
Airborne Spore Concentrations Found in Various Types of Canadian Grain Elevators[18]

Elevator type	Range of airborne spore concentrations (10^6 spores/m^3)
Country/ventilated	0.04—38.1
Country/nonventilated	0.01—358
Terminal	0.21—78.2
Storage	5.97—163

We have previously reported results for the measurement of workshift total and respirable grain dust concentrations collected in 8 terminal, 9 transfer, and 14 country grain elevators.[15] These results are summarized in Table 3.

Not unexpectedly, the highest dust concentrations were generally found in elevators with either no ventilation systems or with poorly maintained systems. Grain dust exposures were higher during the performance of housekeeping and maintenance work in elevators. This result was also not unexpected as workers usually performed these chores with air hoses or brooms, and thus generated thick dust clouds to which they were exposed for the duration of the operation.

We have recently been requested to repeat our investigation of total and respirable dust exposure in transfer elevators that had been studied some 10 years ago. Significant changes have taken place in these elevators since the summer surveys of 1975. For instance, certain jobs such as weighmen have been eliminated through automation and computerization of the grain-weighing operation. Other operations have also undergone drastic changes. In addition, the workforce in all three elevators has been markedly reduced. Nevertheless, several operations have remained relatively unchanged and, in some instances, the same person was performing the same function at the same work station. This offered an opportunity to perform a comparison of grain dust exposures then and now. Table 4 shows the results obtained whenever such a comparison was possible, that is, whenever virtually the same job was performed in a relatively unchanged work environment and at the same time of year. According to the results shown, in most instances, grain handler exposure to grain dust, both respirable and total, has been reduced significantly since the first survey. This overall decrease in exposure can be attributed neither to different levels of activity at the time of the investigation nor to significantly different sampling procedures, since these were virtually the same in both instances. The decrease noted can be attributed to the provision of enhanced exhaust ventilation at the grain transfer points in the elevators and to grain-handler awareness of the risks of exposure to grain dust.

TABLE 3
Total and Respirable (Full Workshift) Grain Dust Concentrations Measured in Canadian Elevators

Worksite/ sample type and number	Total dust concentrations (mg/m^3)	Respirable dust concentrations (mg/m^3)
Country elevators (14)		
Personal samples (30)	0.41—225	0.10—97.4
General area (123)	0.04—41.5	ND[a]—3.35
Transfer/terminal elevators (19)		
Personal samples (434)	0.41—412	0.12—47.5
General area (320)	0.18—781	ND—76.3

[a] Not detectable.

TABLE 4
Comparison of Grain Dust Exposure Results Obtained in Transfer Elevators during the Summers of 1975 and 1985

Job title	Work station	Average[a] total dust concentration (mg/m^3) 1975	1985	Average[a] respirable dust concentration (mg/m^3) 1975	1985
Distributor	Bin floor	10.48	7.57	1.23	0.86
Feed tender	Ground floor	17.1	0.59	1.20	0.21
Towerman	Gallery tower	1.71	1.14	0.82	0.38
Feed tender	Shipping floor	14.1	9.73	0.77	1.40
Feed tender	Receiving floor	1.41	3.16	0.61	0.22
Conveyorman	Receiving floor	10.1	6.73	1.51	0.63

[a] Geometric average.

CHANGES IN GRAIN INDUSTRY OPERATIONS AND THEIR IMPACT ON GRAIN HANDLER EXPOSURES TO GRAIN DUST

Several operations in the grain industry are readily amenable to automation and some, such as the weighing of grain, have been computerized in several elevators. In fact, some modern elevators are fully automated and require but a few employees to operate them. Thus, it is conceivable that in a very few years, grain handling as we know it now will largely have disappeared as an occupation, at least in most large terminal elevators. Although it is entirely possible that a relatively small number of individuals will be exposed to grain dust in the near future, it is ironic that the most highly exposed workers will still be with us, namely, the grain handler who performs daily maintenance tasks such as sweeping, cleaning silos, and other housekeeping chores. For these individuals, however small their numbers, it is imperative that we ensure that their exposure to grain dust will not affect their health and well-being.

REFERENCES

1. **Skoulas, A., Williams, N., and Merriman, J.E.,** Exposure to grain dusts. II. A clinical study of the effects, *J. Occup. Med.*, 6, 359, 1964.
2. **Tse, K.S., Warren, P., Janusz, J., McCarthy, D.S., and Cherniak, R.M.,** Respiratory abnormalities in workers exposed to grain dust, *Arch. Environ. Health,* 27, 74, 1973.
3. **Ross, T. and Smith, J.,** Lung Function in 948 Grain Elevator Workers, paper presented at the Ann. Meet. Ontario Thoracic Society, October 1974.
4. **Becklake, M.R.,** personal communication, 1971.
5. **Kleinfeld, M., Messite, J., Swencili, R.E., and Shapiro, J.A.,** Clinical and physiological study of grain handlers, *Arch. Environ. Health,* 16, 380, 1969.
6. **Smith, A.R., Greenberg, L., and Siegel, W.,** Respiratory diseases among grain handlers, *Ind. Bull.,* 20, 35, 1941.
7. **Thimsen, D.J. and Aften, P.W.,** A proposed design for grain elevator dust collection, *J. Air Pollut. Control Assoc.,* 18, 739, 1968.
8. **Brown, D.F., McCall, E.R., Picollo, B., and Gripp, V.W.,** Survey of effects on variety and growing location of cotton on cardroom dust composition, *Am. Ind. Hyg. Assoc. J.,* 38, 107, 1977.

9. **Wakelyn, P.J., Greenblatt, G.A., Brown, D.F., and Tripp, V.W.,** Chemical properties of cotton dust, *Am. Ind. Hyg. Assoc. J.*, 37, 22, 1976.
10. **Skea, D.,** Grain dust contains a tannin-like material which fixes complement, in Int. Symp. Health and Safety in Agriculture, Saskatoon, Saskatchewan, October 8 to 11, 1985.
11. **Palmgren, M.S.,** Fluctuating occurrence and concentrations of mycotoxins in grain dust, in Int. Symp. on Health and Safety in Agriculture, Saskatoon, Saskatchewan, October 8 to 11, 1985.
12. **Smith, J.D.,** Some biological components of grain and fodder dusts, in Int. Symp. on Health and Safety in Agriculture, Saskatoon, Saskatchewan, October 8 to 11, 1985.
13. **Abrahamson, D.,** Mycotoxins in grain and dust, in Int. Symp. on Health and Safety in Agriculture, Saskatoon, Saskatchewan, October 8 to 11, 1985.
14. **Morey, P.R.,** Dust and microbial exposures during silo unloading, in Int. Symp. on Health and Safety in Agriculture, Saskatoon, Saskatchewan, October 8 to 11, 1985.
15. **Farant, J.-P. and Moore, C.F.,** Dust exposures in the Canadian grain industry, in *Occupational Pulmonary: Disease Focus on Grain Dust and Health,* 1980, 417.
16. **Blackman, E.,** Observations on the development of the silica cells of the leaf sheath of wheat (triticum aistiuum), *Can. J. Bot.*, 47, 827, 1969.
17. **Farant, J.-P., Moore, C.F., and Windish, J.P.,** Airborne fungal spores in Canadian grain elevators, Health & Welfare Canada Rep. No. HHAW73-2, 1973.
18. **Lacey, J.,** Microflora in grain elevators, *Occupational Pulmonary Disease: Focus on Grain Dust and Health,* 1980, 543.
19. **Machacek, J.E., Cherewick, W.J., Meed, H.W., and Broadfoot, W.C.,** A study of some seed-borne diseases of cereals in Canada, 11 kinds of fungi and prevalence of disease in cereal seed, *Sci. Agric.*, 31, 193, 1951.
20. **Wallace, K.A.H. and Sinha, R.N.,** Fungi associated with hot spots in farm stored grain, *Can. J. Plant Sci.*, 42, 130, 1962.

Dust and Microbial Exposures During Silo Unloading

P. R. Morey, J. J. May, D. S. Pratt, L. Stallones, S. A. Olenchock, and W. G. Sorenson

INTRODUCTION

An acute febrile illness may develop 4 to 12 h after the farmer is exposed to microbially contaminated silage dust. The illness, which may last for 2 to 5 d, is characterized by fatigue, fever, muscle aches, cough, and shortness of breath. Affected farmers do not appear to develop permanent scarring changes in the lung, and can continue to work on the farm for years so long as care is taken to avoid subsequent massive exposures.[1,2] This illness, which appears to be distinct from farmer's lung,[1] was originally described as "silo unloaders syndrome"[2] and is now referred to in the more general context of "organic dust toxic syndrome" (ODTS).[3]

We have found that ODTS commonly occurs on dairy farms in New York State when the dry uppermost silage in concrete-staved silos is removed after weeks to months of storage.[2] It is the usual practice in this region at the time of harvest to fill silos with hay (grass, legumes) or chopped corn. Moisture content of the added green material is sufficiently great (about 60%) so that it packs tightly, and anaerobic conditions develop 0.5 to 1 m beneath the surface. This underlying material undergoes fermentation with the associated production of preservatives such as lactic and propionic acids. Vegetable material in the uppermost cap of silage remains under largely aerobic conditions. Because of heavy fungal contamination, this silage is unfit for livestock feed, and it must be removed manually before the top-unloader machine characteristic of these silos is lowered onto the silage surface to recover the fermented material. ODTS is associated with the brief, intense exposure to dusts from the cap of spoiled silage. This study investigated the amount of dust and the levels of airborne microorganisms and endotoxins generated during silo unloading.

SILAGE DUSTS

Five silos were studied. The condition of the silage cap varied. In the first two a plastic sheet covered the ensiled forage. An additional 1/2 to 1 m harvested material had then been deposited on top of the sheet. This cap material was dry and permeated with mycelium. In the last three, silage was either left uncovered or it was protected by a plastic sheet with no additional covering forage. In these silos the uppermost 1/3 to 1 m of silage was moldy; this spoiled silage was, however, uniformly moist or wet.

Levels of total and respirable dusts for locations within and at the base of silos are given in Table 1 (see Reference 4 for sampling and analytical procedures utilized). Highest levels of dust were found when dry as compared to moist or wet silage was unloaded. Total and respirable dust levels exceeded 100 and 20 mg/m^3, respectively, in silo 1 where the moldy hay being removed was very dry. For silo 2, somewhat lower dust concentrations were found at its base as compared to within-silo measurements. In this instance, the discharge chute terminated in the outdoor environment and not in a feed shed as in the other four silos. Gusty winds, which were blowing across the outdoor sampling site at the base of silo 2, rapidly dispersed dust clouds generated as spoiled silage was unloaded. In silos 3, 4, and 5 total and respirable dust concentrations never exceeded 4 and 2 mg/m^3, respectively. For reasons unknown, somewhat higher dust levels than expected were found at the base of silos 4 and 5.

Total dust levels for area samples both near the discharge chute and near the silo center were, for the most part, similar (e.g., for silo 1, arithmetic mean of center samples = 104 mg/m^3, arithmetic mean of chute samples = 113 mg/m^3). Dust levels within the working space of tower silos studied thus appear fairly homogeneous at the time when spoiled cap material was manually unloaded. It was also apparent that higher dust levels occur in those silos where the silage to be discarded is dry as opposed to being moist or wet.

Personal respirable dust exposures (data not shown) during

TABLE 1
Dust Levels (mg/m^3) Both within and at the Base of Silos during Removal of Moldy Cap Silage

		Location		
Silo	Type of dust	Within silo near discharge chute doorway	Within silo near its center	At silo base near discharge chute
1	Total	113[a]	104	129
	Respirable	22.1	24.8	—
2	Total	20.2	19.4	5.1
	Respirable	12.0	—	2.2
3	Total	2.3	2.6	2.8
	Respirable	0.9	1.4	0.5
4	Total	3.2	1.5	6.4
	Respirable	0.6	—	2.5
5	Total	0.8	0.6	3.2
	Respirable	0.2	0.5	0.7

[a] Each entry is an average of two to four samples.

silage removal were also determined (see Reference 4 for details). The time required to unload moldy silage (generally about 0.75 h) is affected by several variables including the depth of spoilage, the number of farmers at work, and the intensity of their work activity. In general, the levels of respirable dusts collected within silos by area and personal sampling are similar (e.g., for silo 2 arithmetic mean of area samples = 12.0 mg/m^3; arithmetic mean of personal samples = 10.3 mg/m^3).

The aerodynamic mass median diameter of dusts collected within silos was about 7 μm, while that present at the base of silos ranged from 4.5 to 7.2 μm.[4] Approximately 60 and 65 to 85% of the dusts collected within the silo and near the discharge chute at the silo base, respectively, were <10 μm aerodynamic size. Thus, most of the dust present at both locations is respirable.

BIOAEROSOLS IN SILAGE DUST

The concentration of viable microorganisms collected by the all-glass impinger (AGI) and the filter cassette sampler during silage unloading is given in Table 2. Dusts generated within silo 1, which was characterized by a very dry cap silage, were most heavily contaminated with microorganisms. Mesophilic and thermotolerant fungi in numbers within the range of 1 to 2 × 10^8/m^3 were detected by both impinger and filter samplers. The concentration of airborne thermophilic bacteria was approximately 4 × 10^9/m^3. In silo 2, which was also characterized by a dry cap silage, airborne thermophilic bacteria exceeded 10^7/m^3. By contrast, 25 and 45°C fungi were present in the same silo at levels that did not exceed 10^6/m^3. In silos 3, 4, and 5, which were characterized by cap silage that was moist or wet, levels of microorganisms did not exceed 10^7/m^3 during unloading.

TABLE 2
Levels of Airborne Microorganisms Collected by AGIs and Filter Cassette Samplers within Silos during Removal of Moldy Cap Silage[a]

Silo	Type of sampler	25°C fungi	45°C fungi	55°C bacteria
		Microorganisms in millions/m^3		
1	AGI	100	120	4300
	Filter[b]	140	160	3500
2	AGI	0.41	0.06	220
	Filter[b]	0.13	0.03	16
3	AGI	3.2	0.43	0.06
	Filter	0.5	0.20	0.34
4	AGI	1.3	1.5	4.5
	Filter	5.5	6.1	1.1
5[c]	AGI	0.14	0.014	—
	Filter	0.01	0.007	1.1

[a] All glass impingers and filter cassette samplers located within silos near discharge chute doorway.
[b] Silo 1: Gram-negative bacteria present at 4.4 million/m^3; silo 2: Gram-negative bacteria present, but not countable because of excessive spreading of colonies.
[c] Silo 5: concentration of 25 and 45°C fungi at silo base near discharge chute equals 1.1 and 1.2 millions/m^3, respectively.

The concentration of airborne 25 and 45°C fungi was quantified near the discharge chute of silo 5 (within the feed shed) by use of the AGI (Table 2, footnote c). While the levels of these fungi in the feed shed exceeded 10^6/m^3, the same organisms were present at levels one to two orders of magnitude lower within the silo. Although we did not carry out sampling for bioaerosols outside of silos (with exception of silo 5), it is apparent that significant exposure to viable microorganisms (Table 2) and dusts (Table 1) occurs in the vicinity of the feed shed during unloading. In addition, the mean particle size of silage dusts near the discharge chute appears to be somewhat less than that characteristic within silos (data not shown). Thus, a greater fraction of silo dust in the feed shed may be respirable, possibly because of the breakup of vegetable matter as it tumbles down the discharge chute.

Eluate from filter samples was analyzed for the presence of Gram-negative bacteria. These bacteria were detected only in the dusts collected at silos 1 and 2, and were quantifiable only in the former silo where their concentration was estimated to be 4 x 10^6/m^3 (Table 2, footnote b). Respirable and total dusts from polyvinylchloride filters from which the gravimetric data in Table 1 were constructed were also analyzed for the presence of endotoxin. As was found for Gram-negative bacteria, endotoxins were present in reportable quantities in silage dusts from silos 1 and 2 (data not shown). Levels of endotoxin in dusts in silos 1 and 2 exceeded 1.0 and 0.05 μg/m^3, respectively. Because only meager amounts of dusts were found in silos 3, 4, and 5, we could not accurately estimate endotoxin levels in these dusts.

During the course of these studies we made several technical observations that may be useful to others conducting bioaerosol sampling in the agricultural environment. One inherent difficulty associated with using the AGI in a rural environment is the necessity of plating out the collected bioaerosol soon (probably within 6 h) after sampling. In this study, because of the proximity of the microbiology laboratory to the dairy farms where sampling was carried out, collected bioaerosol from the AGI was processed within 3 h. The portion of the collected bioaerosol not used in the initial analysis was refrigerated (4°C) for 3 d and subsequently reanalyzed for thermophilic microorganisms by serial dilution plating (Table 3). Levels of 45°C fungi and 55°C bacteria as determined from collected bioaerosol that had been refrigerated were 1.3 × 10^8 and 2.8 × 10^9/m^3, respectively, which was not appreciably different from the initial analysis. Thus, when using the AGI, it is appropriate to hold collected bioaerosol after sampling at refrigerator temperatures for periods of up to 3 d without sample deterioration at least for thermophilic microorganisms. This is an important consideration when the laboratory and proper incubation facilities are remote from the site where sampling is carried out.

We have observed during these studies that the filter

TABLE 3
Levels of Thermotolerant and Thermophilic Microorganisms Released during Silage Unloading Based on Processing of AGI-Collected Bioaerosol, Some of Which Was Refrigerated for 3 d[a]

	Microorganisms in millions/m^3	
Collected bioaerosol	**45°C fungi**	**55°C bacteria**
Processed within 3 h at 25°C	120	4300
Held at 25°C for up to 3 h; subsequently held at 4°C for 3 d and then processed	130	2800

[a] All glass impingers located in silo near discharge chute doorway.

cassette sampler gave results similar to the AGI for microorganisms collected during silo unloading (Table 2). It should be noted that after collection we held filter cassette samplers at 4°C for a period of 3 weeks prior to laboratory analysis. The stability of the collected viable particulate on cellulose ester filters, the ease of operating the sampling train associated with this sampler, and the portability and intrinsically safe nature of the pumps that can be used with filter cassettes are all compelling reasons to suggest this sampler as a standard instrument for very dusty agricultural environments such as silos and grain elevators.

DISCUSSION

The clinical responses of farm workers unloading moldy cap silage are described in other papers.[2] Based on environmental measurements presented here it may be concluded that a significant health hazard due to bioaerosol exposure may exist when moldy silage is unloaded from tower silos. Thus, in silo 1 the level of 55°C microorganisms exceeded the $10^9/m^3$ threshold that is thought to trigger febrile symptoms due to short-term exposures in agricultural environments.[5] In addition, levels of endotoxin in total dust in silos 1 and 2 equaled or exceeded the suggested threshold of 0.08 $\mu g/m^3$ of vertically elutriated endotoxin associated with acute pulmonary function responses in cotton mill workers who smoke.[6]

In the silos that we studied it was clear that the highest dust and bioaerosol exposures occurred in those instances where the protective plastic sheet used to cover silage was itself covered by an additional layer of vegetable material. This additional blown-on material supports an intense microbial growth and undergoes considerable drying. Since the purpose of the layer of added silage is to prevent movement of the plastic sheet, it would be far better to hold the latter in place with objects such as chain or pieces of discarded tires. In this manner exposure to excessive levels of dust and bioaerosol can be minimized as was the case for silos 3, 4, and 5, where spoiled silage to be unloaded was moist or wet.

SUMMARY

Dairy farmers who remove moldy hay and corn from tower silos are exposed to a very dusty atmosphere and as a result may develop an acute febrile illness referred to as ODTS. In one silo where the uppermost 1 m of silage was very dry and moldy, two farmers worked for 0.75 h to remove this material. During this time the respirable and total dust levels within the silo were about 25 and 100 mg/m^3, respectively. Concentrations of mesophilic fungi, thermotolerant fungi, thermophilic bacteria, and Gram-negative bacteria in the dust were approximately 1×10^8, 1×10^8, 4×10^9, and $4 \times 10^6/m^3$, respectively. Levels of endotoxin in total and respirable dusts within this silo exceeded 1 $\mu g/m^3$. In silos at other farms where the condition of the spoiled uppermost silage was moist or wet, dust levels during unloading were lower, and airborne fungi and bacteria were present in numbers 2 to 4 log orders below that encountered when dry material was removed.

REFERENCES

1. **Emanuel, D.A., Wenzel, F.J., and Lawton, B.R.,** Pulmonary mycotoxicosis, *Chest,* 67, 293, 1975.
2. **Pratt, D.S. and May, J.J.,** Feed-associated respiratory illness in farmers, *Arch. Environ. Health,* 39, 43, 1984.
3. **Donham, K.J. and Rylander, R.,** Epilogue: health effects of organic dusts in the farm environment, *Am. J. Ind. Med.,* 10, 339, 1986.
4. **Morey, P.R., Kullman, G., and Patnode, R.,** Silo Unloader's Syndrome, NIOSH Technical Assistance Rep. 84-015, U.S. Department of Health and Human Services, Morgantown, WV, 1984.
5. **Malmberg, P., Palmgren, U., and Rask-Anderson, A.,** Relationship between symptoms and exposure to mold dust in Swedish farmers, *Am. J. Ind. Med.,* 10, 316, 1986.
6. **Haglind, P. and Rylander, R.,** Exposure to cotton dust in an experimental cardroom, *Br. J. Ind. Med.,* 41, 340, 1984.

Environmental Assessment of Respiratory Hazards in the Poultry Industry

K. L. Morring

INTRODUCTION

In 1982 the U.S. poultry industry produced 70 billion eggs, 4.2 billion broilers, and 165 million turkeys;[1] this involved 114,000 farms[2] and about 106,000 poultry-processing workers.[3] The total number of workers involved in poultry production can be estimated to exceed 200,000.

Poultry workers are exposed to a variety of airborne dusts, gases, microorganisms, and endotoxins. The prevalence of respiratory disease has not been assessed in these workers, but there have been many case reports of disease associated with the poultry environment. High prevalences of respiratory symptoms, signs of airway obstruction, and immunological reactions to poultry dust have been documented.[4-9] Suggested causes of asthma and hypersensitivity pneumonitis symptoms include endotoxin, Gram-negative bacteria, and allergens.[10-16] A feed additive (antibiotic) and poultry mites can cause occupational asthma in some poultry workers.[17,18] Epidemics of psittacosis have also been documented.[19,20]

Poultry production primarily consists of broilers, eggs, and turkeys. Fertilized eggs are transported to the hatchery, and after hatching, the birds are vaccinated and sexed. For broilers, chicks are raised to maturity in confinement houses, usually 6 to 8 weeks, then sent to the processing plant. Both sexes (in separate houses) can be reared as broilers. Female chicks are also reared as egg producers or layers. These chicks (called pullets) are reared in confinement until they approach egg production age, then they are placed in cages and produce eggs for about 40 weeks. Other poultry, such as turkeys and ducks, are managed in a similar fashion to broilers.

The following is a description of respiratory hazards found in the different segments of the poultry industry.

HATCHERIES

Little information has been published on the environmental conditions in poultry hatcheries. Effective sanitary measures have long been required to reduce chick mortality. Formaldehyde, a suspected carcinogen, is often used in hatcheries as a disinfectant; however, levels have not been assessed. Exposure to vaccines is also a potential hazard in hatcheries, but no case reports of adverse health effects have been documented.

Airborne dusts and gases are not present in concentrations that adversely affect the hatching chicks (and, presumably, hatchery workers), but there has been concern over airborne microorganisms. In four commercial hatcheries, the lowest airborne microbial concentrations (about 1000 colony-forming units [CFU] of bacteria per cubic meter and 200 CFU of fungi per cubic meter) was associated with the lowest chick mortality; the highest chick mortality rate was found in the hatchery with airborne bacterial and fungal levels of about 8500 and 430 CFU/m_3, respectively.[21] High microbial counts were associated with newly hatched chicks and their hatching debris.[22] Microorganisms contaminated the hatcher environment by adhering to fluff and down from newly hatched chicks. Concentrations in the hatcheries fluctuated with the degree of activity around the emerging chicks. This was also demonstrated in a turkey hatchery where airborne bacterial levels increased from 700 to 16,000 CFU/m^3 during the removal of poults from the hatchers.[23]

EGG PRODUCTION

In commercial egg production, adult hens or layers are kept in pens or in suspended wire cages. There is less dust with cage management as opposed to litter management because cages restrict the birds' movement and eliminate litter usage. Airborne dust levels ranged from 2 to 7 mg/m^3 total dust in facilities where the birds were caged.[2-4] During loading and unloading cages, total dust concentrations ranged from 5.6 to 34.6 mg/m^3.[10] Endotoxin levels during these operations varied from 0.13 to 1.42 $\mu g/m^3$. In Swedish confinement houses and egg rooms, total dust levels averaged 2.3 mg/m^3.[25] Levels of airborne microorganisms in these facilities were from 210,000 to 680,000 (median = 420,000) bacteria CFU/m^3 and 200 to 1900 (median = 500) fungi CFU/m^3. Gram-negative bacteria were found in concentrations of 14,000 to 560,000 (median = 41,000) CFU/m^3 and endotoxin levels averaged 0.3 $\mu g/m^3$.

Ammonia (NH_3), released during microbial degradation of manure, is common in layer houses. NH_3 levels have ranged from 1 to 27 ppm in the summer and 1 to 34 ppm in the winter.[26] Levels are a little higher in the winter because ventilation is decreased to conserve heat. In Hawaiian layer houses during the winter, levels of 5 to 140 ppm were measured. These levels were detected just prior to scheduled manure removal and factors such as high ambient temperatures and relative humidity influenced NH_3 release.

BROILER PRODUCTION

Most of the environmental sampling has been focused on the broiler industry's confinement houses and processing plants. Airborne dusts, gases, microorganisms, and endotoxins have been quantified in these facilities because the worst conditions seem to occur here, and there have been many case reports of illness associated with these facilties.

The process of raising chicks to market weight is frequently contracted with local farmers who supply confinement houses, labor, equipment, and utilities, while the company provides chicks, feed, heating fuel, transportation, and professional consultation. A contract grower often manages 50,000 to 100,000 birds simultaneously and spends on average 12 h/week in each house.[6]

Confinement houses are long, narrow buildings with windows along both sides. Fans and curtains can be adjusted to provide ventilation in the houses. Litter such as wood shavings, straw, or peanut hulls covers the floor to absorb excretions. Litter is often used for several flocks before it is discarded. This practice increases the potential for disease spread and can contribute to high concentrations of airborne dusts and gases.[27]

Gases such as NH_3, CO_2, CO, and others can be present in sufficient concentrations to affect the birds and poultry workers. The sources of CO_2 include animal respiration, manure fermentation, and gas flame heaters. Faulty incubation devices can produce CO. Other gases such as H_2S, CH_4, and SO_2 and oxides of nitrogen can be generated from manure decomposition and heating fuel combustion. Formaldehyde may be present if it is used to disinfect the house between flocks. Harmful levels of gases can build up in confinement houses, especially when ventilation is reduced to avoid excessive heat loss. Moisture may condense on walls and other cold surfaces in poorly insulated houses and this results in wet litter which favors the release of NH_3.[27] Other factors such as ambient temperature, relative humidity, litter pH, and method of litter management also affect NH_3 concentrations.[28]

NH_3 concentrations ranged from 10 to 40 ppm in Swedish broiler houses where higher concentrations were associated with older birds.[24] This relationship was reversed in a recent study; NH_3 concentrations in the buildings that housed 7- and 30-d-old birds averaged 75 and 11 ppm, respectively.[29] In an unventilated section of this house, 170 ppm NH_3 was documented. Old or used litter was used in both houses. Concentrations ranged from 2 to 66 ppm in broiler houses where litter was changed annually.[26] CO_2 concentrations have ranged from 500 to 1000 ppm, while CO, H_2S, NO_2, NO_x, CH_4, mercaptans, formaldehyde, and hydrocarbons were all below the limit of detection.[29]

Litter can be a major source of airborne dust and microorganisms; the quantity of dust varied with the type of litter and increased with litter age.[30] Older birds are associated with higher airborne dust, microorganisms, and endotoxin concentrations.[24,29,31] This may be due to increased size and activity of the birds. Skin debris or dander, feed particles, mold spores, bacteria, feather fragments, litter fragments, and crystalline silica are the major constituents of confinement house dust.[32,33] Levels of airborne dust in these broiler confinement houses in North Carolina ranged from 2.0 to 9.32 mg/m^3 total dust and 0.08 to 0.48 mg/m^3 respirable dust.[29] In studies of turkey confinement houses, total dust concentrations ranged from 9.6 to 12.4 mg/m^3.[32,34]

Several researchers have documented airborne bacterial levels ranging from 12,500 to 7,750,000 CFU/m^3 and fungal concentrations of 2500 to 23,000 CFU/m.[29,31,35] One source of microorganisms in confinement houses is the litter itself; new peanut hull litter contained more viable fungi than old litter.[36] Fresh wood chips had higher viable bacteria and fungi concentrations than old litter (120,000 vs. 74,000 bacterial CFU/m^3 and 23,000 vs. 2500 fungal CFU/m^3).[29]

Gram-negative bacterial endotoxin concentrations in settled dust averaged 11.4 ng/mg, whereas airborne total dust yielded concentrations averaging 6.4 to 15 ng/mg (0.02 to 0.06 μg/m^3).[29,37]

PROCESSING PLANTS

To facilitate broiler transport to processing plants, catching crews are sent to the farm to place the birds in crates. Catching crews enter the confinement houses at night, as the birds will not move around when there is little light. The highest concentrations of total dust and NH_3 were measured during the catching operations where total dust ranged from 9 to 17 mg/m^3 and 40 to 50 ppm NH_3 detected.[24]

In a shackling room, birds are removed from crates one at a time and hung by their feet in shackles. Particles of feathers, skin, fecal matter, and litter become airborne as the birds flap their wings and struggle. Airborne dust, microbes, and endotoxin levels can be quite high here. Total and respirable dust levels ranged from about 7 to 24 and 0.4 to 2.6 mg/m^3, respectively.[38,39] One study reported average total dust concentration of 56 mg/m^3 in a shackling room before the ventilation system was redesigned.[40] Bacterial concentrations ranging from 19,000 to 650,000 CFU/m^3 and fungal concentrations averaging 2765 CFU/m^3 have been reported.[31,41,42] In a duck processing plant, 1,061,000 CFU/m^3 microorganisms (bacteria and fungi) were found in the shackling area.[31] The range of airborne endotoxin levels in a shackling room was 0.4 to 2.1 μg/m^3.[39]

Once the birds are killed and plucked, dust is no longer a problem in the processing plant. Airborne microbial levels were highest in the dressing room (83,400 CFU/m^3), with diminishing numbers in shackling, eviscerating, and holding rooms (about 22,500, 9800, and 3400 CFU/m^3, respectively).[42]

NH_3 and CO_2 may also be present, but the sources are much different than in the confinement house. NH_3 is used in the refrigerating units; workers may be exposed to NH_3 from leaks in the refrigeration system. CO_2 evolves from the sublimation of dry ice, used in great quantities to cool the fresh meat. At one processing plant, NH_3 could not be detected in the plant, but CO_2 exceeded 3000 ppm around the wrapping and packaging area.[38]

Concentrations of dust, gases, microorganisms and endotoxins and related occupational standards are presented in Tables 1 to 4 respectively. Concentrations of some contaminants are usually found below Occupational Safety and Health Administration's (OSHA) permissible exposure limits or National Institute of Occupational Safety and Health's (NIOSH) recommended levels (Tables 1 and 2). OSHA stan-

TABLE 1
Dust Concentrations in the Poultry Industry (Range Expressed Unless Otherwise Noted)

	Total (mg/m^3)	Respirable dust (mg/m^3)	Ref.
Egg production (caged layers)	2—7		24
	1.1—3.7		25
Loading/unloading cages	5.6—34.6[a]		10
Confinement houses (on litter)	12—17		24
	2.0—9.3[a]	0.08—0.48[a]	29
Turkey house	12.4[b]		32
	9.6[b]		34
Catching crew in broiler house	9.17		24
Shackling area in processing plant	7.1—18.5	1.1—2.6	38
	6.4—24.1	0.4—1.6	39
	55.6[b]		40
OSHA permissible exposure (TWA)	15	5	

[a] Range of means.
[b] Mean value.

TABLE 2
NH_3 and CO_2 Levels in the Poultry Industry (Range Expressed Unless Otherwise Noted)

	NH_3 (ppm)	CO_2 (ppm)	Ref.
Egg production (caged layers)	5—15		31
In summer	1—27		26
In winter	1—34		
Natural ventilation	5—140		
Confinement houses (on litter)	10—40	500—1,000	29
	9—75		
In unventilated area	170		
In summer	1—28		26
In winter	2—66		
Catching crew in broiler houses	40—50		31
Processing plant			
Shackling area	<5		38
Holding area	<5	>3,000[a]	
OSHA permissible exposure limit (TWA)	50	5,000	
NIOSH recommended standard (TWA)	50	10,000	

[a] Exceeded the limit of detection.

TABLE 3
Concentrations of Airborne Viable Microorganisms in the Poultry Industry (Mean Values Expressed Unless Otherwise Noted)

	Bacteria (CFU/m^3)	Fungi (CFU/m^3)	Both (CFU/m^3)	Ref.
Hatchery	1,053—9,823[a]	212—4,876[a]		21
			225,500	
Turkey poult removal				
Before	1,555[b]	417[b]		23
During	15,974[b]	855[b]		
After	742[b]	509[b]		
Egg production (caged layers)	290,000—680,000	200—1,300		25
Egg room	120,000—210,000	160—1,900		
Confinement houses (on litter)			172,800—751,000[b]	31
New litter	120,000[b]	23,000		29
Old litter	74,000[b]	2,500[b]		
Processing plants				
Shackling area	19,790	2,757		42
	360,000—650,000[b]			41
Duck shackling area			1,061,100	31
Dressing	77,748	5,680		42
Eviscerating	8,128	1,661		
Holding	2,191	1,237	198,400	

[a] Range of means.
[b] Actual values.

dards and NIOSH recommended levels are not designed to protect the hypersensitive individual. There are no standards, required or recommended, for microorganisms and endotoxins, although guidelines for exposure to airborne endotoxins have been proposed.[43] It was suggested that airborne endotoxin concentrations exceeding 0.5 μg/m^3 can produce a reaction in humans.

It is evident from the data that the poultry workers' environment is laden with a variety of airborne contaminants. Many case reports of respiratory diseases associated with these substances have been documented, yet the prevalence of acute and chronic lung dysfunction has not been determined. Health effects from exposure to contaminants at these concentrations and their interactions, as well as the prevalence of

TABLE 4
Airborne Endotoxin Concentrations in the Poultry Industry (Mean Values)

	Endotoxin (μg/m^3)	Ref.
Egg production (caged layers)	0.13—0.50	25
Egg room	0.12	
Loading/unloading cages	0.13—1.42	10
Confinement houses (on litter)		
In total dust	0.02—0.06	29
In respirable dust	0.005—0.01	
In settled dust	11.5[a]	37
Shackling area of processing plant		
In total dust	0.6—0.9	39
In respirable dust	0.03—0.04	

[a] Units = μg/g.

hypersensitive reactions among poultry workers, should be documented.

REFERENCES

1. Statistical Abstract of the United States 1985, 105th ed., Bureau of the Census, U.S. Department of Commerce, Washington, D.C., 1985, 671.
2. *1982 Census of Agriculture*, Vol. 1, Geographic Area Series, Part 51, U.S. Summary and State Data, Bureau of the Census, U.S. Department of Commerce, Washington, D.C., 1984, 11.
3. *1982 Census of Manufacturers*, Preliminary Report, Industry Series, Poultry Dressing Plants' Poultry and Egg Processing (Industries 2016 and 2017), Bureau of the Census, U.S. Department of Commerce, Washington, D.C., 1984.
4. **Bar-Sela, S., Teichtahl, H., and Lutsky, I.,** Occupational asthma in poultry workers, *J. Allergy Clin. Immunol.*, 73, 271, 1984.
5. **Stewart, X., Kapp, J., Digiovanna, A., et al.,** A retrospective study of the effect of occupational farmland chicken dust and smoking on forced expiratory flow rates and the incidence of cough, wheezing, shortness of breath, and chronic bronchitis, *Respir. Care*, 29 (Abstr.), 1033, 1984.
6. Health Hazard Evaluation Report: David Mayer Poultry Farm, HETA, 83-195-1426, National Institute of Occupational Safety and Health, U.S. Department of Health and Human Services, Washington, D.C., 1984.
7. **Wuthe, H., Bergmann, K.C., and Vogel, J.,** Frequency of lung function disturbances and immunological status of industrial poultry farmers, *Eur. J. Respir. Dis.*, 62, 38, 1982.
8. **Petro, W., Bergmann, K.C., Heinze, R., et al.,** Long-term occupational inhalation of organic dust — effect on pulmonary function, *Int. Arch. Occup. Environ. Health*, 42, 119, 1978.
9. **Elman, A.J., Tebo, T., Fink, J.N., and Barboriak, J.J.,** Reactions of poultry farmers against chicken antigens, *Arch. Environ. Health*, 17, 98, 1968.
10. **Thelin, A., Tegler, O., and Rylander, R.,** Lung reactions during poultry handling related to dust and bacterial endotoxin levels, *Environ. J. Respir. Dis.*, 65, 266, 1984.
11. **Stahuljak-Beritic, D., Dimov, D., Buthovic, D., and Stilinovic, L.,** Lung function and immunological changes in poultry breeders, *Int. Arch. Occup. Environ. Health*, 40, 131, 1977.
12. **Boyer, R.S., Klock, L.E., Schmidt, C.D., et al.,** Hypersensitivity lung disease in the turkey raising industry, *Am. Rev. Resp. Dis.*, 109, 630, 1974.
13. **Warren, C.P.W. and Tse, K.S.,** Extrinsic allergic alveolitis owing to hypersensitivity to chickens — significance of sputum precipitins, *Am. Rev. Respir. Dis.*, 109, 672, 1974.
14. **Avila, R.,** Extrinsic allergic alveolitis in workers exposed to fish meal and poultry, *Clin. Allergy*, 1, 343, 1971.
15. **Korn, D.S., Florman, A.L., and Gribetz, I.,** Recurrent pneumonitis with hypersensitivity to hen litter, *J.A.M.A.*, 205, 114, 1968.
16. **Adams, C.T.B.,** Farmer's lung, *Br. Med. J.*, 1, 114, 1966.
17. **Paggiaro, P.L., Loi, A.M., and Toma, G.,** Bronchial asthma and dermatitis due to spiramycin in a chick feeder, *Clin. Allergy*, 9, 571, 1979.
18. **Lutsky, I., Teichtahl, H., and Bar-Sela, S.,** Occupational asthma due to poultry mites, *J. Allergy Clin. Immunol.*, 73, 56, 1984.
19. **Andrews, B.E., Major, R., and Palmer, S.R.,** Ornithosis in poultry workers, *Lancet*, p. 632, March 21, 1981.
20. **Durfee, P.T., Pullen, M.M., Currier, R.W., and Parker, R.L.,** Human psittacosis associated with commercial processing of turkeys, *J. Am. Vet. Med. Assoc.*, 167, 804, 1975.
21. **Gentry, R.F., Mitrovic, M., and Bubash, G.R.,** Application of Andersen sampler in hatchery sanitation, *Poult. Sci.*, 41, 794, 1962.
22. **Magwood, S.E.,** Studies in hatcherty sanitation. I. Fluctuations in microbial counts of air in poultry hatcheries, *Poult. Sci.*, 43, 441, 1964.
23. **Gardner, F.A., Golan, F.A., and Denton, J.H.,** Microbiological parameters associated with hatching turkeys, *Poult. Sci.*, 59, 966, 1980.
24. **Carlsson, B.O.,** Environmental hygiene problems associated with broiler production, *Nord. Hyg. Tidskr.*, 52, 22, 1971.
25. **Clark, S., Rylander, R., and Larsson, L.,** Airborne bacteria, endotoxin and fungi in dust in poultry and swine confinement buildings, *Am. Ind. Hyg. Assoc. J.*, 44, 537, 1983.
26. **Nakaue, H.S., Cain, J.R., and Quarles, C.L.,** Poultry house gases, in *Poultry Environmental Quality and Production*, Western Regional Research Project W-136, Colorado State University Experimental Station, Fort Collins, CO, 1980, 20.
27. **Carlile, F.S.,** Ammonia in poultry houses: a literature review, *Worlds Poult. Sci. J.*, 40, 99, 1984.
28. **Elliot, H.A. and Collins, N.A.,** Factors affecting ammonia release in broiler houses, *Trans. ASAE*, 25, 413, 1982.
29. **Jones, W., Morring, K., Olenchock, S.A., Williams, T., and Hickey, J.,** Environmental study of poultry confinement buildings, *Am. Ind. Hyg. Assoc. J.*, 45, 760, 1984.
30. **Lillie, R.J.,** Air pollutants affecting the performance of domestic animals: a literature review. USDA ARS Agriculture Handbook 380, U.S. Department of Agriculture, Washington, D.C., 1970.
31. **Dutkiewicz, J.,** Exposure to dust-borne bacteria in agriculture. I. Environmental studies, *Arch. Environ. Health*, 33, 250, 1978.
32. **Cravens, R.L., Beaulieu, H.J., and Buchan, R.M.,** Characterization of the aerosol in turkey rearing confinements, *Am. Ind. Hyg. Assoc. J.*, 42, 315, 1981.
33. **Koon, J., Howes, J.R., Grub, W., and Rollo, C.A.,** Poultry dust: origin and composition, *Agric. Eng.*, 44, 608, 1963.
34. **Parker, S., Beaulieu, H., and Pacheco, D.,** Attempts at aerosol control in a turkey rearing confinement with negative air ionization, *Am. Ind. Hyg. Assoc. J.*, 44, 911, 1983.
35. **Sauter, E.A., Petersen, C.F., Steele, E.E., and Parkinson, J.F.,** The airborne microflora of poultry houses, *Poult. Sci.*, 60, 569, 1981.
36. **Arafa, A.S., Miles, R.D., Harmes, R.H., et al.,** Fungistatic compounds in broiler production. IV. Microbiological and physical characteristics at litter, *Poult. Sci.*, 58, 1462, 1979.
37. **Thedel, T.D., Mull, J.C., and Olenchock, S.A.,** A brief report of gram-negative bacterial endotoxin levels in airborne and settled dusts in animal confinement buildings, *Am. J. Ind. Med.*, 1, 3, 1980.
38. Health Hazard Evaluation Report Perdue, Inc. HETA 84, National Institute of Occupational Safety and Health, U.S. Department of Health and Human Services, 1985, 280.
39. **Olenchock, S.A., Lenhart, S.W., and Mull, J.C.,** Occupational exposure to airborne endotoxins during poultry processing, *J. Toxicol. Environ. Health*, 9, 339, 1982.
40. **Wesley, R.L. and Hughes, H.A.,** Dust control at the line station in poultry processing plants, *Poult. Sci.*, 62, 772, 1983.

41. **Lenhart, S.W., Olenchock, S.A., and Cole, E.C.,** Viable sampling for airborne bacteria in a poultry processing plant, *J. Toxicol. Environ. Health,* 10, 613, 1982.

42. **Kotula, A.W. and Kinner, J.A.,** Airborne microorganisms in broiler processing plants, *Appl. Microbiol.,* 12, 179, 1964.

43. **Rylander, R. and Morey, P.,** Airborne endotoxin in industries processing vegetable fibers, *Am. Ind. Hyg. Assoc. J.,* 43, 811, 1982.

NIOSH Nontextile Cotton Industry Study: An Industrial Hygiene Overview

J. N. Zey, G. M. Piacitelli, W. G. Jones, and J. L. Clere

INTRODUCTION

From 1976 to 1982 the National Institute for Occupational Safety and Health (NIOSH) conducted a cross-sectional environmental/medical study of the nontextile cotton industry. The primary objectives of the environmental component were measurement of airborne cotton dust concentrations using the vertical elutriator cotton dust sampler and assigning corresponding time-weighted average exposures to individual employees.

This paper is a summary of two previously published articles which discuss the design and implementation of the study, environmental sampling results, and problems encountered in using the vertical elutriator.[1,2] Detailed accounts of the medical and environmental investigations of each segment are presented in five separate reports.[3-7]

Occupational exposure criteria for the nontextile industry have been in a state of transition for several years. In 1978 the Occupational Safety and Health Administration (OSHA) established a permissible exposure limit of 500 μg/m^3 of air as measured with the vertical elutriator.[8] This standard was subsequently stayed, and in 1983 OSHA proposed a revised standard exempting all nontextile industries except waste utilization plants, which were to be covered under the old 1.0-mg/m^3 total dust standard.[9,10] Prior to the nontextile cotton industry study, the NIOSH recommended standard for cotton dust exposure was 200 μg/m^3. This level was based primarily on studies conducted in the cotton textile industry.[11] This was a principal consideration in the decision to conduct an evaluation of the nontextile industry. During testimony at the OSHA public hearing in September 1983, NIOSH recommended that dust levels be reduced to the lowest level feasible in the nontextile industry.[12]

The five segments of the nontextile industry evaluated were (1) cotton gins (n = 35), (2) cotton compress warehouses (n = 13), (3) cotton classing offices (n = 13), (4) cottonseed oil mills (n = 18), and (5) cotton waste utilization plants (n = 13).

These industries process cotton and cotton by-products and, as such, form a periphery to the cotton textile industry. Seed cotton is collected in the field using mechanical harvesters and then transported to the gin. After initial cleaning, cotton fibers (called lint) are separated from the seed in a "gin stand". Lint undergoes additional cleaning and tempering and is then packaged into bales. Bales of lint are transported to a warehouse for storage, prior to marketing. Each warehouse studied had a compressing operation to produce a smaller "high-density" bale and, thus, increase both storage and transportation capacity. Samples from each bale of cotton are collected at either the gin or warehouse and forwarded to a cotton classing office. These facilities, operated by the U.S. Department of Agriculture, grade each sample based on the length and maturity of the fibers, in addition to the overall color, preparation, and trash content of the sample.

Cottonseed oil mills receive cottonseed from the gin. After cleaning, short fibers (called linters) still adhering to the seed are removed and baled. Linters are sold to various markets including waste plants.

Cotton waste utilization plants receive waste materials from textile mills, cotton gins, and cottonseed oil mills. These materials are cleaned by manual and machine processing. Different waste materials are usually blended and sold for use in products such as bedding.

METHODS

Airborne cotton dust levels were measured using the Lumsden-Lynch vertical elutriator operating at a flow rate of 7.4 ± 0.2 l of air per minute.[13,14] The flow rate was controlled with a critical orifice which was recalibrated after about ten plant studies. Preweighed polyvinyl chloride filters in open-face cassettes were used to collect the cotton dust samples. All samples were hand carried to the NIOSH laboratory in Morgantown, WV for gravimetric analysis. Flow rate checks using a rotometer were conducted at the end of each plant survey to determine if any samplers had material plugging the critical orifice. The number of vertical elutriators used to assess separate work areas depended on the number of employees and size of the area. A minimum of two samplers were used in any area where employees spent a significant period of time. Elutriators were most often supported on a portable tripod stand with the separation chamber inlet at the approximate breathing-zone height (about 1.5 m) of employees in the area. Sampling was conducted to coincide with initial and follow-up medical tests (pulmonary function and questionnaires) performed on individual employees.

To assess dust exposures for individual employees, the NIOSH investigators monitored the employees' location throughout the sampling period. Information collected on medically tested employees included identification, total time spent in each work area, and the appropriate vertical elutriators to be used in calculating a dose. By combining average dust levels for corresponding work areas, a time-weighted average was established for each employee.[1,2]

RESULTS AND DISCUSSION

Figure 1 presents the range of mean plant dust concentrations for each segment. For example, of the 35 cotton gins studied, the lowest mean concentration for all elutriated dust samples collected at one plant was 101 µg/m^3, and the highest mean concentration was 2050 µg/m^3. The corresponding values for the other four segments were 39 to 831 µg/m^3 for cotton compress warehouses, 81 to 376 µg/m^3 for cotton-classing offices, 502 to 2040 µg/m^3 for cottonseed oil mills, and 237 to 3970 µg/m^3 for cotton waste utilization facilities.

Figure 2 presents the average of the mean dust concentrations for all plants within each segment. For example, the average of 35 separate plant mean concentrations was 521 µg/m^3 for cotton gins. Average concentrations for the other segments were 229 µg/m^3 for compress warehouses, 165 µg/m^3 for classing offices, 1010 µg/m^3 for cottonseed oil mills, and 1390 µg/m^3 for waste utilization plants. Approximately half (54%) of the 92 plants studied had mean dust concentrations of 500 µg/m^3 or less. Of the 42 plants with a mean dust concentration above 500 µg/m^3, all but one were in the ginning, cottonseed oil, and waste utilization segments.[1]

Factors affecting airborne dust levels in cotton-processing facilities have been previously discussed.[15-17] Factors affect-

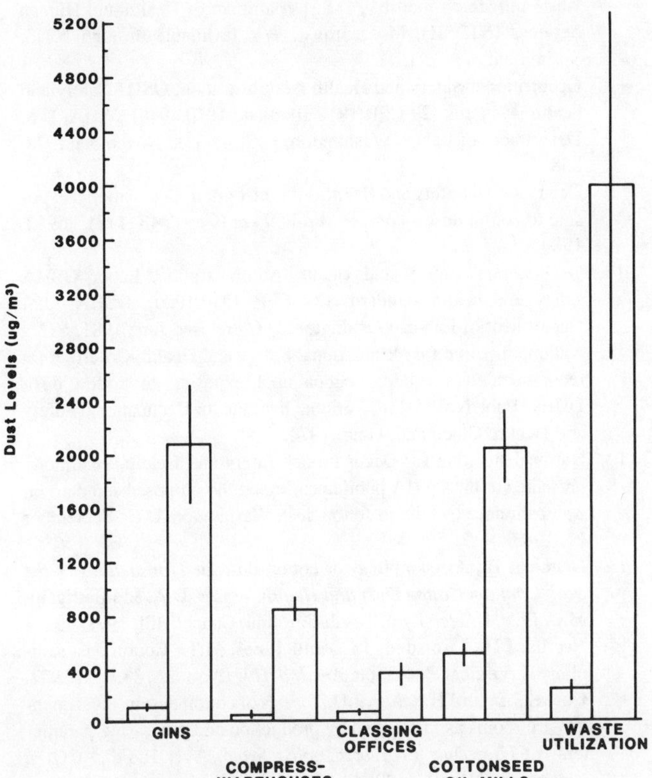

FIGURE 1. High and low mean plant dust levels in each segment (center line indicates standard error).

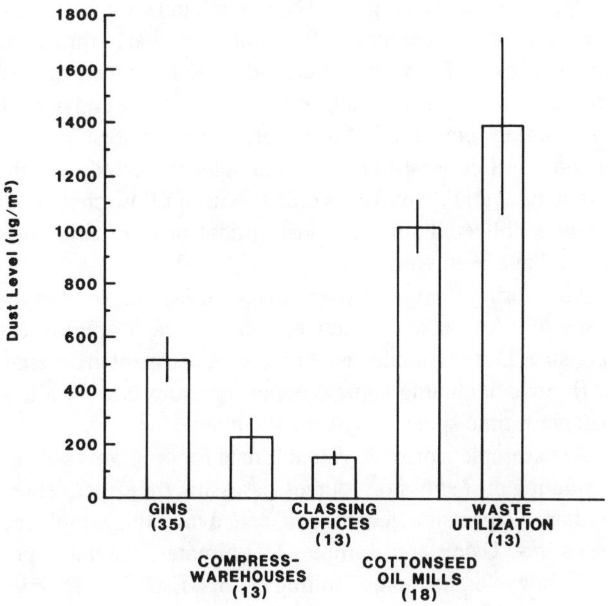

FIGURE 2. Average elutriated mean plant dust levels by segment (center line indicates standard error). ( ) = number of plants.

ing the intra- and intersegment variability in the dust concentrations obtained in this study include the type of material (waste material vs. cotton) processed, the type of processing (machine vs. manual) conducted, maintenance of processing equipment, availability and maintenance of dust suppression equipment, and the cotton-harvesting method used. The cotton-classing office segment was the only one in which all facilities had installed dust suppression equipment. Processing at the facilities involves primarily manual manipulation of cotton samples. Additionally, all facilities have temperature- and humidity-controlled environments, and the facility layouts are very similar. These factors contribute to this segment having the lowest overall mean dust values.

Among the other segments, waste utilization had the highest overall mean dust values. This segment had very little similarity among the 13 plants evaluated. A few waste plants had extensive dust suppression equipment and good maintenance programs. Other plants had no dust suppression equipment and minimal maintenance. Some facilities utilized a lot of manual processing, while others had very little. This segment also processes waste material which contains more trash (dirt, vegetable matter) than the cotton samples processed in the classing offices.

Close monitoring of the samplers and employees enabled the NIOSH investigators to detect and modify many potential problems. Sampler problems included blown motor fuses, particularly in plants with high dust levels. Needle valves had to be adjusted periodically to maintain correct gauge pressure. In facilities with very old electrical equipment, circuit breakers occasionally tripped. In one instance, all vertical elutriators had to be moved during the plant lunch break because the work force started a different operation in a separate area after lunch.

The portable stands proved beneficial, but their use necessitated securing the sampler/stand unit to prevent it from being knocked over. Flow rate checks detected plugged critical orifices in a few instances. Checking flows at the end of each plant survey prevented the loss of many samples. When a plugged orifice was found, only samples collected with the elutriator at that plant were voided. Without flow checks, all samples collected since the last critical oriface calibration would have been voided.

Monitoring the employees' whereabouts enabled the investigators to make a better estimate of each employee's exposure. Dose estimates accounted for time spent in separate work areas including some outdoor locations and, in a few instances, time spent away from the plant.

Area samplers provide valuable data for programs such as evaluating dust emission sources, assessing the effectiveness of dust suppression techniques, and evaluating employee exposures. Using area samples to estimate personal exposures, however, has some limiting factors. Establishing time-weighted average exposures with the vertical elutriator, or any area sampler, should work best in facilities with relatively consistent dust levels throughout each work area and employees who remain in one work area during their shift. The feasibility of estimating accurate doses decreases as dust level variability and employee mobility increase. Of the five segments studied, cotton-classing offices had the best conditions for assessing employee exposures using area samplers. Elutriated dust levels were fairly consistent in the various work areas. Additionally, the majority of employees remained in one area throughout the work shift. These conditions were encountered in all 13 plants. Some plants in the other segments had similar conditions, but no other segment had as much consistency between the separate facilities. Many facilities had both variable dust concentrations in individual work areas and employees who worked in multiple areas. Dose estimation was more difficult in these facilities.

Some work locations, such as the seed tunnel in a cottonseed oil mill, are prohibitive for using a sampler as large as the vertical elutriator. Other locations such as the yard area in a cotton gin are prohibitive due to the need for line power.

CONCLUSION

Variability in elutriated dust samples was obtained within and between the five segments of the nontextile cotton industry studied. Facilities with the higher dust levels were most often those processing lower-quality materials and which had poor maintenance programs and no dust suppression equipment.

The vertical elutriator has been used to sample cotton dust since the early 1970s. Much of the early work was conducted in the cotton textile industry. This study exhibited some problems in using the elutriator in the nontextile industries. Some sampling locations require a more portable and/or a battery-operated sampler. Assessing personal cotton dust exposures for employees who work in multiple work areas is very difficult when using stationary area samplers that measure full-shift dust concentrations. Portable battery-operated area samplers would be beneficial, but a practical, personal cotton dust sampler would solve most, if not all, of the problems encountered.

REFERENCES

1. **Zey, J.N., Piacitelli, G.M., and Jones, W.G.,** Characterizations of airborne cotton dust concentrations in the non-textile cotton industry, *Am. Ind. Hyg. Assoc. J.*, 45(8), 538, 1984.
2. **Zey, J.N.,** Industry surveys conducted by NIOSH, *Ann. of the American Conf. of Governmental Industrial Hygienists,* Vol. 2, Kelley, W.D., Ed., ACGIH, Cincinnati, OH, 1982, 37.
3. National Institute for Occupational Safety and Health, Respiratory disorders and dust exposure in sectors of the cotton industry of the United States. I. Cotton gins, U.S. Department of Health and Human Services (NIOSH), Morgantown, WV (available through NTIS, Springfield, VA 22161).
4. National Institute for Occupational Safety and Health, Respiratory disorders and dust exposures in sectors of the cotton industry of the United States. II. Cotton warehouse compressors, U.S. Department of Health and Human Services (NIOSH), Morgantown, WV (available through NTIS, Springfield, Virginia 22161).
5. National Institute for Occupational Safety and Health, Respiratory disorders and dust exposures in sectors of the cotton industry of the United States. III. Cottonseed oil mills, U.S. Department of Health and Human Services (NIOSH), Morgantown, WV (available through NTIS, Springfield, VA 22161).
6. National Institute for Occupational Safety and Health, Respiratory disorders and dust exposures in sectors of the cotton industry of the United States. IV. Cotton classing offices, U.S. Department of Health and Human Services (NIOSH), Morgantown, WV (available through NTIS, Springfield, VA 22161).
7. National Institute for Occupational Safety and Health, Characterization of byssinosis and other pulmonary abnormalities in the cotton waste utilization industry, U.S. Department of Health and Human Services (NIOSH), Morgantown, WV (available through NTIS, Springfield, VA 22161).
8. Occupational Safety and Health Administration, OSHA safety and health standards (29 CFR 1910.1043 and 1910.1046) OSHA, U.S. Department of Labor, Washington, D.C., revised November 1978, 658.
9. Occupational Safety and Health Administration, Occupational exposure to cotton dust — proposed rule. *Fed. Regist.*, 48 (113), 26962, 1983.
10. Occupational Safety and Health Administration, Labor OSHA safety and health standards (29 CFR 1910.1000), OSHA, U.S. Department of Labor, Washington, D.C., revised June 1981, 632.
11. National Insitute for Occupational Safety and Health, Criteria for a recommended standard: occupational exposure to cotton dust, DHEW Publ. No. 75-118, National Institute for Occupational Safety and Health, Cincinnati, Ohio, 1974.
12. National Institute for Occupational Safety and Health, Testimony presented at the OSHA public hearing on the proposed standard on occupational exposure to cotton dust, Washington, D.C., September 21, 1983.
13. **Lynch, J.R.,** Air sampling for cotton dust, in *Transactions of the Natl. Conf. on Cotton Dust and Health,* Fraser, D.A. and Battigelli, M.C. Eds., University of North Carolina, Chapel Hill, 1971, 33.
14. **Neefus, J.D., Lumsden, J.C., and Jones, M.T.,** Cotton dust sampling. II. Vertical elutriation, *Am. Ind. Hyg. Assoc. J.*, 38, 394, 1977.
15. **Cocke, J.B. and Hatcher, J.D.,** Levels of cotton dust in experimental card rooms as influenced by production and processing parameters, in *Cotton Dust, Proc. of a Topical Symp.*, Ayer, H. Ed., ACGIH, Cincinnati, OH, 1975, 395.
16. **Cocke, J.B., Wesley, R.A., and Kirk, I.W.,** Harvesting and processing factors affecting cotton-dust levels in the card room, ARS-USDA Marketing Research Rep. No. 1065, 1977.
17. **Wesley, R.A. and Wall, J.A.,** Levels and chemical composition of cotton gin dust, *Am. Ind. Hyg. Assoc. J.*, 39, 962, 1978.

Respiratory Protection for Use Against Organic Dusts

S. W. Lenhart and L. D. Reed

INTRODUCTION

Workers comprising the agricultural sector of the American workforce were described during the keynote address of a 1981 symposium on agriculture respiratory diseases as the largest population at risk of exposure to occupational hazards of the lung.[1] Agricultural workers encounter risks of respiratory disease through exposure to a variety of living and nonliving airborne contaminants;[2] and, because of the health risks associated with excessive inhalation of certain organic dusts,[3] minimizing the risks associated with overexposure is certainly desirable. For those job situations where exposure to organic dusts cannot be adequately reduced through engineering and/or administrative measures, as is generally the case for a large percentage of agricultural activities, respirators offer the only measure for reducing potentially hazardous exposures to levels considered to be safe. However, two important questions which should be asked are (1) what evidence is currently available to make a respirator selection for protection against organic dusts? and (2) what additional information is needed from future research activities to more completely ensure the maintenance of health of those agricultural workers who wear respirators?

The purposes of this paper are to attempt to answer these questions by summarizing current knowledge regarding the performance of air-purifying respirators as protection against organic dusts, and by describing National Institute of Occupational Safety and Health (NIOSH)-funded field research in agricultural environments designed to evaluate dose-response relationships in conjunction with respirator usage.

BACKGROUND

The ability of negative-pressure respirators to provide adequate protection against organic dusts has been questioned.[4-7] In a report of a clinical study of 12 individuals with pigeon breeders' disease, the health of a patient who continued his exposure to organic dusts with the use of a respirator as the only protective measure deteriorated into apparent permanent pulmonary injury.[4] Unfortunately, no information was given regarding exposure levels or the extent to which the subject wore his respirator. Pigeon breeders' disease belongs to the group of hypersensitivity pneumonitides that includes farmer's lung.

The first modern report of hypersensitivity pneumonitis appeared in 1932 when farmer's lung was described as an acute pneumonitis occurring in workers exposed to moldy hay.[8] The most frequently cited etiologic agents of this disease are the spores of certain thermophilic actinomycetes. In a report on farmer's lung resulting from a joint meeting of the British Medical Research Council and the Ministry of Agriculture, Fisheries and Food, the opinion was offered that those farmers who had experienced acute attacks of the disease and had become sensitized to the causative spores would gain little protection from wearing a respirator.[5] However, the same report contained the statement that healthy people could expect worthwhile protection through the use of the same respirator type. One author has expressed his view, without benefit of experimentation or reference, that air-purifying respirators would be of little practical value in preventing farmer's lung principally because of the extremely small size of the spores of thermophilic actinomycetes.[6]

Enzyme detergent lung is the name given to another of the many diseases resulting from the inhalation of organic dusts. In a report of an investigation of chest illness among workers in a factory using a preparation containing proteolytic enzyme derived from *Bacillus subtilis* in the manufacture of detergent products, half-mask air-purifying respirators were described as insufficient protection.[7] As in the case with previous reports, no data were provided to support opinion.

Support for respirators as appropriate protection against the inhalation of organic dusts is derived solely from the results of clinical reports.[9-12] In one report, a patient who had experienced many episodes with symptoms of pigeon breeders' disease was reportedly able to handle his racing pigeons without apparent difficulty while wearing a respirator.[9]

Another report which related exposures of organic dusts from pigeons described a laboratory worker who had suffered acute attacks of purulent rhinitis, chills, fever up to 104°F (40°C), striking malaise, and fatigue. After air-purifying respirators were provided, this individual was able to work without difficulty, even in situations which had previously made him acutely ill.[10] Success using a respirator for this laboratory situation is contradicted by clinical evidence which suggested that the use of helmeted, powered, air-purifying respirators did little to prevent or delay the progression of laboratory animal allergy.[11]

During a clinical assessment using inhalation provocation tests, six subjects known to have hypersensitivity pneumonitis underwent identical paired inhalation tests of 20 min to 5 h duration using natural, but unmeasured, methods of antigen exposure.[12] For one test the subjects wore half-mask air-purifying respirators with electrostatically charged, resin-impregnated wool filters. These respirators were reported to have successfully protected four of the subjects from any suspicion of a positive response, while the remaining two subjects reported symptoms of minimal severity. Unprotected challenges provoked late responses of mild to moderate severity characteristic of hypersensitivity pneumonitis. Levels of exposure were not measured to confirm the similarity of challenges between the two trials.

During another inhalation provocation trial, an individual known to suffer from farmer's lung worked for 1 h in a shed forking moldy hay while wearing a half-mask respirator with high-efficiency filters.[13] Air samples showed an ambient spore concentration of 7.3×10^7 m^{-3}. A sample taken from inside the respiratory facepiece showed a spore concentration of 7×10^5 m^{-3}. This represents a penetration of 1%, suggesting the achievement of a workplace protection factor of 100. The subject remained asymptomatic for 2 d after the first trial. A second trial in the same shed containing a spore concentration similar to that measured during the first trial lasted for only 3 min. While working without benefit of a respirator, the subject began to cough severely and mild dyspnea developed; symptoms persisted through the next day.

The results of laboratory experiments have provided additional information related to the performance of respirators as protection against organic dusts.

Respirator filters of different rated efficiencies have been found through laboratory experimentation to permit a mean penetration of actinomycete spores ranging from 44% for nuisance dust filters to 1% and less for filters used with both negative-pressure and powered air-purifying respirators.[14]

The overall penetration of dust respirators by *B. subtilis* was evaluated with the aid of a specially designed mouth collector during controlled exposure periods of 10 min.[15] The mouth collector consisted of a rubber mouthpiece containing a metal cartridge packed with absorbent cotton, which was aseptically removed from the cartridge after each trial and placed in sterile water for assay. Each test subject was required to inhale through his mouth and exhale through his nose so that only inspiratory air would be sampled; a breathing rate of 10 l/min^{-1} was assumed. Penetrations of 3% or less were reported.

While the necessity for assuming an inspiratory flow rate and dependence upon cooperation from test subjects restricts the workplace application of mouth collectors,[16] this technique has an immediately obvious advantage over a respirator evaluation technique used several years ago which required sacrifice of the test subject.[17] During these experiments rabbits were fitted with respirators in order to measure the penetration of pathogenic tubercle bacilli. After an animal was exposed for 30 min to contaminated air, it was sacrificed and its lungs were removed for quantitative culturing. An average penetration of 23% was reported, but direct extrapolation of the data to human beings was cautioned.[17]

DISCUSSION

Reviewing the preceding assemblage of conflicting information regarding the use of air-purifying respirators as protection against organic dusts leads to the following general observation. The pulmonary response or nonresponse of a respirator wearer working in an environment containing organic dust depends upon certain factors of the respirator, such as facepiece seal and filter efficiency; the airborne concentration and mass median diameter of contaminants; and the existing health status of the respirator wearer. Obviously, these variables influence not only organic dust exposure, but also, in varying degrees, the pulmonary response of any respirator wearer working in any contaminated environment.

Because differences exist among the various classes of respirators with regard to their protective capabilities, respirators are assigned protection factors as a guidance for their selection when contaminant concentrations have been measured.[18] It is generally assumed that a properly used and properly maintained respirator will provide a level of protection equal to or greater than its assigned protection factor[19] for the vast majority of wearers.[20] Given a known contaminant concentration, assigned protection factors can be useful for indirectly predicting the exposure concentration expected inside a respirator wearer's facepiece. For example, the concentration of bacterial endotoxin during certain poultry-processing activities reportedly ranges between 130 and 2150 $ng.m^{-3}$.[21,22] In order to reduce these endotoxin levels solely with the use of respirators to less than 500 ngm^{-3}, the amount presumed to cause human reactions,[23] a half-mask air-purifying respirator could be selected. This type of respirator has an assigned protection factor of 10.[18,24,25] Suppose, on the other hand, that reduction of the aeromicrobiological content found during the same poultry-processing activities (as high as 6.5×10^5 colony-forming units [CFU] m^{-3} [26]) to levels normally found in urban atmospheres (between 125 and 5×10^3 CFU m^{-3} [27-31]) was desired. In this case, selection of a respirator with an assigned protection factor of 100 or greater would be more appropriate.

Historically, assigned protection factors have been based on laboratory tests in which a group of respirator wearers perform a specific regimen of head and body movements while in a test chamber containing a challenge aerosol. More recently, workplace protection factors[19] have been used to determine an assigned protection factor for a particular type of respirator. Assigned protection factors based on workplace measurements have not always resulted in agreement with those factors resulting from laboratory tests. For instance, half-mask powered air-purifying respirators with high-efficiency filters have been assigned protection factors of 500,[32] 1000,[24] and 3000.[25] However, the results of a study to determine workplace protection factors for such a device, when used in a primary lead smelter, suggested an assigned protection factor of not greater than 50.[20]

More research is needed regarding in-facepiece exposures experienced by agricultural workers who wear respirators while working in environments containing airborne organic dusts, and the extent of any acute or chronic effects that such exposures have upon their respiratory systems. Better insight concerning dose-response relationships, together with workplace protection factor measurements for various types of respirators, will serve to more completely ensure the maintenance of health of agricultural workers.

Because of the health benefits to be derived by agricultural workers from the use of properly selected respirators, NIOSH has funded a 3-year research grant with the objective of developing a model respirator program for controlling exposures to organic dusts. This objective will be accomplished through research involving agricultural workers from the grain, poultry, and swine confinement industries. Determinations of workplace protection factors for three classes of respirators will be combined with information concerning assessments of pulmonary function and respiratory symptoms over work shifts both with and without the use of respirators. The project was scheduled for completion in late 1986.

CONCLUSIONS AND RECOMMENDATIONS

Conflicting reports have been published concerning the use of air-purifying respirators as a means of protecting both sensitized and healthy individuals against overexposure to organic dusts. The confusion associated with these reports could be due to improper respirator selection initially, since exposure concentrations were seldom reported. In addition, none of the cited reports contained information concerning user training or fit testing prior to respirator usage. Also missing from these reports is information regarding the duration of respirator usage while test subjects were in contaminated environments. Because of the uncertainties associated with the use of respirators as protection against organic dusts, selection of the most practical respirator with the highest assigned protection factor is recommended.

Listed in Table 1 are the current NIOSH recommendations for the assigned protection factors of respirators used against particulate exposures. To select an appropriate respirator, one must estimate the level of exposure and the toxicity of the contaminant, as well as determine the job requirements and the advantages and limitations of the respirator.

Because the use of a supplied air respirator and the use of a self-contained breathing apparatus are often not feasible for the majority of agricultural tasks involving organic dust exposures, selection of a powered air-purifying respirator with high-efficiency filters and tight-fitting facepiece is recommended. Because the potential for eye irritation has been demonstrated during certain exposures to organic dusts,[33-36] the use of tight-fitting goggles when half-masks are worn or the use of a full facepiece may be necessary. While a full facepiece respirator with a high-efficiency filter has the same assigned protection factor as the recommended powered respirator, use of this nonpowered respirator is advised only when quantitative fit-testing procedures are available.

The greater cost of powered air-purifying respirators relative to the cost of nonpowered air-purifying respirators may be objectionable to some. However, downgrading of protection, such as to half-mask or disposable respirators, is advised only after sufficient environmental and medical evidence has been collected to clearly demonstrate that the use of a respirator with less protective capability will reduce exposures to safe levels. Regardless of the type of respirator selected, only those respirators with NIOSH/MSHA approval should be used.

TABLE 1
Assigned Protection Factor Classifications of Respirators for Protection against Particulate Exposures

Assigned protection factor	Type of respirator
5	Single-use or quarter mask respirator
10	Any air-purifying half-mask respirator including disposable equipped with any type of particulate filter[a]
25	Any powered air-purifying respirator equipped with a hood or helmet and any type of particulate filter[a]
50	Any air-purifying full facepiece respirator equipped with a high-efficiency filter
	Any powered air-purifying respirator equipped with a tight-fitting facepiece and high efficiency filters[a]
1,000	Any supplied air respirator equipped with a half-mask and operated in a pressure demand or other positive pressure mode
2,000	Any supplied air respirator equipped with a full facepiece and operated in a pressure demand or other positive pressure mode
10,000	Any self-contained-type respirator equipped with a full facepiece and operated in a pressure demand or other positive pressure mode

[a] Assigned protection factors based upon the field evaluation of respirator workplace protection factors.

It is important to emphasize that the preceding recommendations are only for the control of exposures to organic dusts. These recommendations do not apply to environments where the potential exists for overexposure to toxic gases or vapors, or to environments which may be oxygen deficient.

REFERENCES

1. **Merchant, J.A.,** Keynote address, in *Agricultural Respiratory Hazards,* American Conference of Governmental Industrial Hygienists, Cincinnati, 1982, 5.
2. **Popendorf, W., Donham, K.J., Easton, D.N., and Silk, J.,** A synopsis of agricultural respiratory hazards, *Am. Ind. Hyg. Assoc. J.,* 46, 154, 1985.
3. **Rylander, R.,** Organic dusts and lung reactions — exposure characteristics and mechanisms for disease, *Scand. J. Work Environ. Health,* 11, 199, 1985.
4. **Fink, J.N., Sosman, A.J., Barboriak, J.J., Schlueter, D.P., and Holmes, R.A.,** Pigeon breeders' disease: a clinical study of a hypersensitivity pneumonitis, *Ann. Intern. Med.,* 68, 1205, 1968.
5. **Davies, C.N.,** Farmer's lung, *Ann. Occup. Hyg.,* 11, 377, 1968.
6. **Flindt, M.L.H.,** Pulmonary disease due to inhalation of derivatives of bacillus subtilis containing proteolytic enzyme, *Lancet,* 1, 1177, 1969.
7. **Crofton, J.W. and Douglas, A.C.,** *Respiratory Diseases,* F.A. Davis, Philadelphia, 1969, 504.
8. **Campbell, J.M.,** Acute symptoms following work with hay, *Br. Med. J.,* 2, 1143, 1932.
9. **Reed, C.E., Sosman, A., and Barbee, R.A.,** Pigeon-breeders' lung:

a newly observed interstitial pulmonary disease, *J.A.M.A.*, 193, 81, 1965.

10. **Siegal, F.P. and Quellette, J.J.,** Protecting pigeon-handlers, *Lancet*, 1, 733, 1969.
11. **Slovak, A.J.M., Orr, R.G., and Teasdale, E.L.,** Efficacy of the helmet respirator in occupational asthma due to laboratory animal allergy, *Am. Ind. Hyg. Assoc. J.*, 46, 11, 1985.
12. **Hendrick, D.J., Marshall, R., Faux, J.A., and Krall, J.M.,** Protective value of dust respirators in extrinsic allergic alveolitis: clinical assessment using inhalation provocation tests, *Thorax*, 36, 917, 1981.
13. **Gourley, C.A. and Braiwood, G.D.,** The use of dust respirators in the prevention of recurrence of farmer's lung, *J. Soc. Occup. Med.*, 21, 93, 1971.
14. **Lacey, J., Nabb, S., and Webster, B.T.,** Retention of actinomycete spores by respirator filters, *Ann. Occup. Hyg.*, 25, 351, 1982.
15. **Guyton, H.G., Buchanan, L.M., and Lense, F.T.,** Evaluation of respiratory protection of contagion masks, *Appl. Environ. Microbiol.*, 4, 141, 1956.
16. **Guyton, H.G., Mick, C.E., Decker, H.M., and Burgess, W.A.,** Techniques for evaluating biological penetration of respiratory masks on human subjects, *Am. Ind. Hyg. Assoc. J.*, 28, 462, 1967.
17. **Abramson, S.,** Experimental airborne infection, *Am. Rev. Respir. Dis.*, 73, 315, 1956.
18. National Institute for Occupational Safety and Health, NIOSH Respirator Decision Logic, NIOSH Publ. in press. 1985.
19. **Guy, H.P.,** Respirator performance terminology, *Am. Ind. Hyg. Assoc. J.*, 46(5), B22, 1985.
20. **Lenhart, S.W. and Campbell, D.L.,** Assigned protection factors for two respirator types based upon workplace performance testing, *Ann. Occup. Hyg.*, 28, 173, 1984.
21. **Thelin, A., Tegler, O., and Rylander, R.,** Lung reactions during poultry handling related to dust and bacterial endotoxin levels, *Eur. J. Respir. Dis.*, 65, 266, 1984.
22. **Olenchock, S.A., Lenhart, S.W., and Mull, J.C.,** Occupational exposure to airborne endotoxins during poultry processing, *J. Toxicol. Environ. Health*, 9, 339, 1982.
23. **Rylander, R. and Morey, P.,** Airborne endotoxin in industries processing vegetable fibers, *Am. Ind. Hyg. Assoc. J.*, 43, 811, 1982.
24. **Pritchard, J.A.,** *A Guide to Industrial Respiratory Protection,* NIOSH Publ. No. 76-189, National Institute for Occupational Safety and Health, Cincinnati, 1976, 144.
25. American National Standards Institute, *American National Standard Practises for Respiratory Protection,* ANSI Z88.2-1980, American National Standards Institute, New York, 1980, 21.
26. **Lenhart, S.W. and Olenchock, S.A.,** Sources of respiratory insult in the poultry processing industry, *Am. J. Ind. Med.*, 6, 89, 1984.
27. **Wright, T.J., Greene, V.W., and Paulus, H.J.,** Viable microorganisms in an urban atmosphere, *J. Air Pollut. Control Assoc.*, 19, 337, 1969.
28. **Lee, R.E., Jr., Harris, K., and Akland, G.,** Relationship between viable bacteria and air pollutants in an urban atmosphere, *Am. Ind. Hyg. Assoc. J.*, 34, 164, 1973.
29. **Mancinelli, R.L. and Shulls, W.A.,** Airborne bacteria in an urban environment, *Appl. Environ. Microbiol.*, 35, 1095, 1978.
30. **Bovallius, A., Bucht, B., Roffey, R., and Anas, P.,** Three-year investigation of the natural airborne bacterial flora at four localities in Sweden, *Appl. Environ. Microbiol.*, 35, 1095, 1978.
31. **Jones, B.L. and Cookson, J.T.,** Natural atmospheric microbial conditions in a typical suburban area, *Appl. Environ. Microbiol.*, 45, 919, 1983.
32. British Standards Institution, *Recommendations for the Selection, Use and Maintenance of Respiratory Protective Equipment,* BS 4275, British Standards Institution, London, 1974.
33. **Williams, N., Skoulas, A., and Merriman, J.E.,** Exposure to grain dust. I. A survey of the effects, *J. Occup. Med.*, 6, 319, 1964.
34. **Mattsby, I. and Rylander, R.,** Clinical and immunological findings in workers exposed to sewage dust, *J. Occup. Med.*, 20, 690, 1978.
35. **doPico, G.A., Tsiatis, A., Peters, M.E., Rao, P., and Rankin, J.,** Respiratory abnormalities among grain handlers, in *Occupational Pulmonary Disease,* Academic Press, New York, 1980, 213.
36. **Flaherty, D.,** Mechanisms of host response to grain dust, in *Agricultural Respiratory Hazards,* American Conference of Governmental Industrial Hygienists, Cincinnati, 1982, 198.

Single-Cell Model Systems in Toxicology

M. I. C. Gyongyossy-Issa and G. G. Khachatourians

Protection against toxic agents comes with understanding of the agent's mechanism of action. As with any pharmacologically active compound, the chemistry of the molecule determines the kinds of interactions that can take place at the single-cell level. The chemical nature of the compound and its route of entry effectively predetermine the access to, concentration of, and persistence in tissues. Using cellular model systems, interactions at a single-cell level can be determined and employed to guide extrapolation towards the effect of the agent at tissue, organ, or entire organism levels.

With this approach, we have been studying the mechanism of action of a family of sesquiterpenoid compounds, the 12,13-epoxytrichothecenes.[1] The trichothecene family of mycotoxins has been implicated as the causative agent of a large number of human and animal diseases. T-2 toxin (Figure 1), one of the most potent toxins of this group, is associated with toxicoses resulting from exposure to contaminated agricultural products.[2] The ensuing intoxications are characterized by common symptoms such as feed refusal, vomiting, diarrhea, distal paresis, and leukopenia.[1,2]

It is generally agreed that T-2 toxin has a radiomimetic effect, preferentially destroying actively multiplying cells.[2-4] It is in this context that the immune system has been shown to be a prime toxin target.[5] However, toxicity to many other cell types has also been demonstrated.[1-5] In fact, it appears that actively multiplying cells are not a preferred toxin target, but only constitute a target where the toxic effect is easily and rapidly detectable.

Current evidence suggests that the pathologic effects of T-2 toxin are due to the inhibition of major cellular biochemical functions.[1-6] In a variety of eukaryotic cells, the synthesis of DNA, RNA, and protein,[1,4,6,7] and in some cases lipid,[8] is demonstrably inhibited by T-2 toxin. In yeast, mitochondrial function is also impaired.[9] While inhibition of any of these functions is potentially cytotoxic, neither the exact sequence of events leading to cell death nor the first critical step in this sequence has been established. Using the single-cell model system of mouse splenic lymphocytes, we confirmed that the synthetic apparatus of DNA, RNA, and protein does indeed constitute intracellular target sites for T-2 toxin. In addition, we showed that these three synthetic systems are inhibited by the same threshold toxin dose.[6] The fact that all three types of macromolecular synthesis are affected equally, by an identical dose of T-2 toxin which also corresponds to the interaction of 10^5 T-2 toxin molecules with the same number of lymphocyte T-2 toxin receptors, suggests that full receptor occupation may be an early stage in the expression of T-2 toxin toxicity.[6,10] Thus, there appears to be a relationship between intracellular and extracellular events.

Xenobiotics' access to intracellular target sites is regulated by the cell membrane. Interaction of T-2 toxin with membrane receptors also indicates that the cell membrane may play a critical role in the modulation of T-2 toxin toxicity. From a chemical standpoint, T-2 toxin is potentially a membrane-interactive molecule: amphiphilic molecules, including some terpenes,[11] are known to be able to interact intimately with bilayer membranes and cause a range of effects from membrane expansion, osmotic protection, or membrane disintegration.[12] As a sesquiterpenoid[7] and an amphiphile,[13] T-2 toxin (Figure 1) could potentially intercalate into the hydrophobic, phospholipid bilayer of cell membranes. Our observations show that from an aqueous solution, T-2 toxin partitions preferentially into an organic phase such as octanol (Figure 2).[14] Similarly, T-2 may partition into the organic phase of the cell membrane. Preliminary experiments have shown that purified lymphocyte and erythrocyte membranes do absorb T-2 toxin rapidly from an aqueous dispension (Table 1). In an intact-cell system, using mouse lymphocytes as a model, it was demonstrated that competition for cellular binding sites by (^{3}H)-T-2 and unlabelled toxin takes place over a 500-fold range of unlabeled toxin concentration.[10] (^{3}H)-T-2 binding is not decreased to less than 40%. One possible explanation for this phenomenon implies the presence of many, low-affinity binding sites on lymphocyte membranes, possibly an expression of T-2 partitioning into the phospholipid bilayer. These observations provide circumstantial evidence of an intimate T-2 toxin-membrane relationship.

The importance of the cell membrane in T-2 toxin toxicity has been invoked by several authors,[4,7,15] and indirect evidence of the toxin-membrane interaction has been presented. The impairment of a range of leukocyte functions,[16] inhibition of macrophage phagocytosis,[17] as well as erythrocyte hemolysis[18,19] can all be ascribed to membrane perturbations by the toxin.

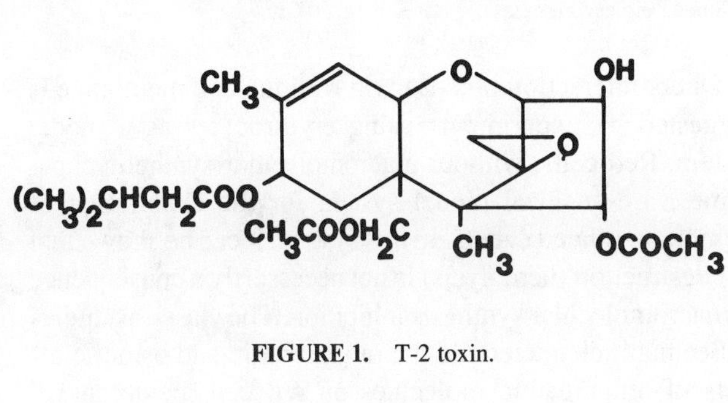

FIGURE 1. T-2 toxin.

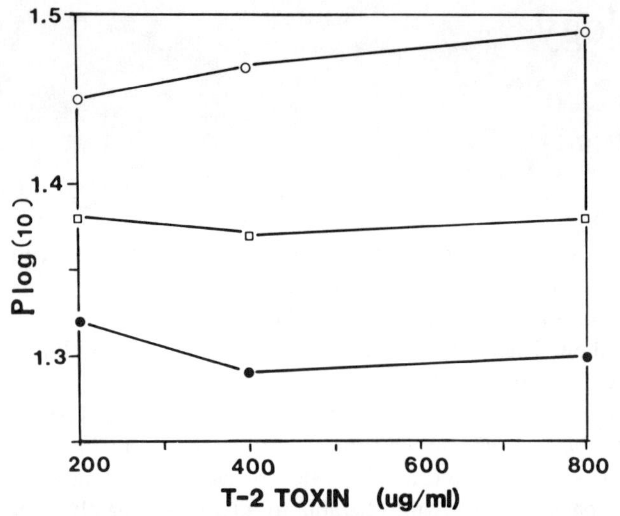

FIGURE 2. The octanol-water partition coefficient of T-2 toxin. Octanol, 2 (○), 5 (❑), and 10 (●) µl was used as the organic phase above T-2 toxin dispersed in 125 µl of veronal-acetate buffer, 5 mM, pH 7.0. Starting T-2 toxin concentration in the veronal-acetate buffer is given on the x axis. The log of the partition coefficient, P, was calculated[14] from $[\text{T-2 toxin}_{oct}]/[\text{T-2 toxin}_{buffer}] = P$, and is expressed on the y axis.

TABLE 1
Time Course of T-2 Toxin Binding to Lymphocyte Membranes, Lymphocytes, and Erythrocytes

	Specific (^{3}H)-T-2 toxin uptake (cpm)[a]		Hemolysis protection (%)[b]
Time (min)	Lymphocyte membranes[c]	Lymphocytes[c]	Erythrocytes[d]
1	<10	<10	—
2	—	30	16
3	—	60	—
6	<10	155	9
10	<10	340	—
20	—	360	−8
30	—	350	—
60	230	—	−10
200	—	210	−9

[a] Specific uptake = (binding of (^{3}H)-T-2 toxin – binding of (^{3}H)T-2 toxin in the presence of 50 µg/ml unlabeled T-2 toxin.[10]
[b] Percent reduction of hemolysis in the presence of 200 µg/ml T-2 toxin.[23]
[c] Mouse spleen cells.
[d] Guinea pig erythrocytes.

Direct interaction of T-2 toxin with the cell membrane is suggested by experiments using erythrocytes as a model system. Red cells, without macromolecular synthetic capabilities, are an ideal model system for the exploration of membrane-related events. In this system, it can be shown that cell destruction (hemolysis) is not necessarily a consequence of macromolecular synthesis inhibition. The classical studies of Seeman[12] characterized the morphologic and osmotic effects of amphipathic molecules on red cell membranes.[20] Subsequently, the "bilayer couple" hypothesis ascribed these changes to the unilateral expansion of one membrane leaflet of the bilayer.[21] Prompted by the sesquiterpenoid and amphiphilic structure of T-2 toxin, we carried out experiments intended to document a similar behavior for T-2 toxin. As other amphiphilic molecules that intercalate into the exterior membrane leaflet, T-2 toxin causes echinocyte transformation[22] and osmotic protection of red cells (Table 2).[23] Both these effects are rapid and can be observed as early as 2 min after contact, and are reminiscent of the rapid toxin uptake by intact lymphocytes. Apparently, the toxin does not remain sequestered in the exterior half of the membrane bilayer. With time, as the toxin molecules equilibrate across the membrane, echinocyte morphology returns to discocyte and, subsequently, after 3 h, stomatocytes appear (Table 2).[23] In the presence of high toxin concentrations (>400 mg/10^8 cells) the morphologic sequence is initial echinocytosis, followed by spheroechinocytosis and hemolysis.[19] The appearance of echinocytes can be correlated with initial osmotic protection followed by rapid lysis. Ektacytometer studies have confirmed the results of osmotic protection by demonstrating increased red cell deformability in buffers of relatively low osmotic strength (Figure 3).[24] This shift of the deformability profile is also correlated with concurrent echinocytosis and is characteristic of a change in the cell's surface-to-volume ratio. Such a change can be brought about by either an increase in membrane area, by cellular dehydration, or both.[25] Ektacytometer studies were also carried out on ATP-depleted cells or in the presence of EDTA. Similar T-2-induced deformability changes were observed in these metabolically depleted cells, indicating that the toxin was not acting via a mechanism involving initial metabolic depletion which would subsequently result in echinocytosis, but was interacting directly with the cell membrane.[24] This is the first demonstration of a direct amphiphilic mechanism of action for this toxin. However, in view of the toxin's chemical structure, other possible

TABLE 2
Time Course of Erythrocyte[a] Morphology Changes in Response to T-2 Toxin

Time (min)	Sample	Echinocytes (%)	Discocytes (%)	Stomatocytes (%)
2	T-2[b]	92	8	0
	Cont[c]	56	44	0
5	T-2	88	12	0
	Cont	—	—	—
10	T-2	93	7	0
	Cont	—	—	—
20	T-2	86	14	0
	Cont	49	51	0
60	T-2	89	10	1
	Cont	60	40	0
320	T-2	33	60	6
	Cont	48	52	0

[a] Guinea pig erythrocytes.[23]
[b] T-2 toxin at 200 µg/ml final concentration.
[c] T-2 toxin diluent (ethanol) added without toxin.

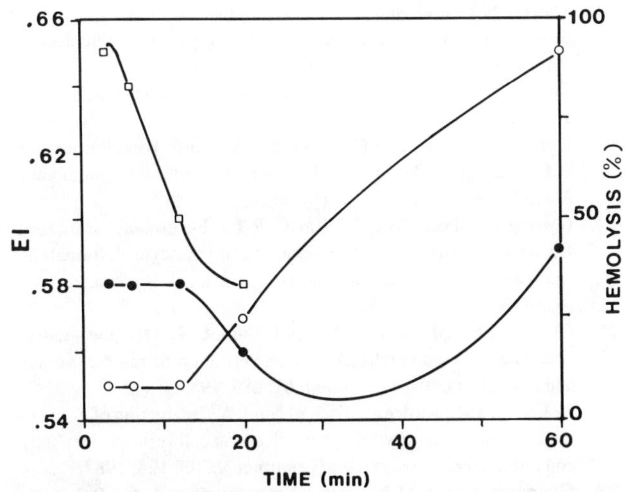

FIGURE 3. The effect of T-2 toxin on red cell deformability and hemolysis. The time course (min; x axis) of the change of human erythrocyte deformability in the presence of T-2 toxin (❑) or diluent (●) is shown. The y axis indicates the elongation index at the point of maximum deformation, EI_{max}. EI_{max} was measured by ektacytometry[23,25] and calculated as the ratio of two axes of a red cell deformed under a shear stress of 159 dyn/cm^2. A corresponding cell population was concurrently assessed for lysis as indicated by hemoglobin release. Percent T-2-specific hemolysis (i.e., background substracted) is indicated by the (❍) symbols and on the y axis.

mechanisms involving free radicals[18] or binding to SH groups[7] may also be acting, probably concurrently, on the cell membrane.

So far, we have discussed a potential mechanism of action of T-2 toxin which involves not only macromolecular synthesis inhibition, but also target cell membranes, and we have established that a membrane-interactive step is both possible and probable in T-2 toxicity. Where there are two interacting components — in this case, the toxin and the cell membrane — the components' individual contributions must be assessed. The cell membrane, too, must influence the degree to which T-2 toxin can interact with it. Recently, DeLoach and co-workers have demonstrated permeability and entrapment aspects of T-2 interaction with bovine carrier erythrocytes.[26]

To determine the influence of the membrane on the expression of T-2 toxicity, red cells of different compositions were tested as to their toxin susceptibility. It was found that on an equal molecule/cell basis, red cells from different species had widely different susceptibilities to the toxin.[27] Sheep erythrocytes were indifferent to the toxin, while porcine red cells showed osmotic protection and human and bovine cells were lysed (Table 3). This observation indicates that not all cell types are equally susceptible to T-2 toxin and that membrane composition, most importantly phosphatidyl choline,[28] is a critical factor in toxin susceptibility.

This line of reasoning is supported by studies on yeast membranes, where known membrane-modulating agents alter the susceptibility of the organism to T-2 toxin.[29] Recent experiments indicate that the membrane-bound hormone loading of red cells as a consequence of the estrous cycle may

TABLE 3
Dose-Dependent, Species-Specific Responses of Erythrocytes to T-2 Toxin

	T-2 toxin concentration (μg/ml)			
Erythrocytes[a]	**0**	**50**	**100**	**200**
Ovine	23.0	19.5	22.0	22.2
Human	28.9	28.1	28.8	38.0
Bovine	19.1	21.0	20.9	46.6
Porcine	28.2	26.2	25.8	23.2

[a] 2.5×10^7 erythrocytes per milliliter; 10 h incubation at 37°C.[26]

also influence T-2 toxin-membrane related effects such as cell deformability.

The single-cell model system in toxicology can be used to demonstrate individual aspects of a toxin's mechanism of action. In the case of T-2 toxin, the observations can guide further studies and extrapolation towards understanding the physicochemical factors that can influence such things as toxin localization or the modulation of the toxic effect that results from encounters with cells having different membrane compositions. Eventually, such model systems can be exploited to test potential means of interference, or protection against this toxin and others of similar chemistry.

REFERENCES

1. **Ueno, Y.,** Trichothecene mycotoxins. Mycology, chemistry and toxicology, in *Advances in Nutritional Research,* Vol. 3, Draper, H.H., Ed., Plenum Press, New York, 1980, 301.
2. **Bamburg, J.R.,** The biological activities and detection of naturally occurring 12,12-epoxy-9-trichothecenes, *Clin. Toxicol.,* 5, 495, 1972.
3. **Bamburg, J.R. and Strong, F.M.,** 12,13-Epoxytrichothecenes, in *Microbial Toxins,* Vol. 7, Kadis, S., Gieglerand, A., and Ajl, S.J., Eds., Academic Press, New York, 1971, 207.
4. **Ueno, Y., Nakajima, M., Sakai, K., Ishii, K., Sato, N., and Shimada, N.,** Comparative toxicology of trichothecene mycotoxins: inhibition of protein synthesis in animal cells, *J. Biochem.,* 74, 285, 1973.
5. **Lafarge-Frayssinet, C., Decloitre, F., Mousset, S., Martin, M., and Frayssinet, C.,** Induction of DNA single-strand breaks by T-2 toxin, a trichothecene metabolite of Fusarium. Effect on lymphoid organs and liver, *Mutat. Res.,* 88, 115, 1981.
6. **Gyongyossy-Issa, M.I.C. and Khachatourians, G.G.,** Interaction of T-2 toxin and murine lymphocytes and the demonstration of a threshold effect on macromolecular synthesis, *Biochim. Biophys. Acta,* 844, 167, 1985.
7. **Ueno, Y.,** Mode of action of trichothecenes, *Pure Appl. Chem.,* 49, 1737, 1977.
8. **Suneja, S.K., Ram, G.G., and Wagle, D.S.,** Effect of T-2 toxin administration to rats on liquid metabolism in liver, *Toxicol. Lett.,* 22, 113, 1984.
9. **Schappert, K.T. and Khachatourians, G.G.,** Abstract, *Can. Fed. Biol. Soc. Proc.,* 27, PA112, 1984.
10. **Gyongyossy-Issa, M.I.C. and Khachatourians, G.G.,** Interaction of T-2 toxin with murine lymphocytes, *Biochim. Biophys. Acta,* 803, 197, 1984.
11. **Zanker, K.S. and Blumel, G.,** Terpene-induced lowering of surface-tension in vitro: a rationale for surfactant substitution, *Res. Exp. Med.* (Berlin), 182, 33, 1983.

12. **Seeman, P.,** The membrane action of anesthetics and tranquilizers, *Pharmacol. Rev.,* 24, 583, 1972.
13. **Gyongyossy-Issa, M.I.C., Christie, E.J., and Khachatourians, G.G.,** Charge-shift electrophoretic behaviour of T-2 toxin in agarose gels, *Appl. Environ. Microbiol.,* 47, 1182, 1984.
14. **Leo, A., Hansch, C., and Elkins, D.,** Partition coefficients and their uses, *Chem. Rev.,* 71, 537, 1971.
15. **Ueno, Y. and Matsumoto, H.,** Mode of action of trichothecene mycotoxin, *Proc. 1st Intersec. Congr. I.A.M.S. (Tokyo),* 4, 315, 1974.
16. **Yarom, R., Sherman, Y., More, R., Ginsburg, I., Borinski, R., and Yagen, B.,** T-2 toxin effect on bacterial infection and leukocyte functions, *Toxicol. Appl. Pharmacol.,* 75, 60, 1984.
17. **Gerberick, G.F., Sorensen, W.G., and Lewis, D.D.,** Toxicity of T-2 toxin, a fusarium mycotoxin to alveolar macrophage *in vitro, Environ. Res.,* 33, 246, 1984.
18. **Segal, R., Milo-Goldzweig, I., Joffe, A.Z., and Yagen, B.,** Trichothecene-induced hemolysis. I. The hemolytic activity of T-2 toxin, *Toxicol. Appl. Pharmacol.,* 70, 343, 1983.
19. **Gyongyossy-Issa, M.I.C., Khanna, V., and Khachatourians, G.G.,** Characterization of hemolysis by T-2 toxin, *Biochim. Biophys. Acta,* 838, 252, 1985.
20. **Seeman, P., Kwant, W.P., Sauka, T., and Argent, W.,** Membrane expansion of intact erythrocytes by anesthetics, *Biochim. Biophys. Acta,* 183, 490, 1969.
21. **Sheetz, M.P. and Singer, S.J.,** Biological membranes as bilayer couples. A molecular mechanism of drug-erythrocyte interactions, *Proc. Natl. Acad. Sci. U.S.A.,* 71, 4457, 1974.
22. **Bessis, M.,** *Blood Smears Reinterpreted,* Springer International, 1977, 25.
23. **Gyongyossy-Issa, M.I.C., Khanna, V., and Khachatourians, G.G.,** Changes induced by T-2 toxin in the erythrocyte membrane, *Food Chem. Toxicol.,* 24, 311, 1986.
24. **Gyongyossy-Issa, M.I.C., Card, R.T., Fergusson, D.J., and Khachatourians, G.G.,** Pre-hemolytic erythrocyte deformability changes caused by trichothecene T-2 toxin: an ektacytometer study, *Blood Cells,* 12, 393, 1986.
25. **Clark, M.R., Mohandas, N., and Shohet, S.,** Osmotic gradient ektacytometry: comprehensive characterisation of red cell volume and surface maintenance, *Blood,* 61, 889, 1983.
26. **DeLoach, J.R., Andrews, K., and Nagi, A.,** Interaction of T-2 toxin with bovine carrier erythrocytes: effects on cell lysis, permeability, and entrapment, *Toxicol. Appl. Pharmacol.,* 88, 123, 1987.
27. **Gyongyossy-Issa, M.I.C. and Khachatourians, G.G.,** Role of the red cell membrane in T-2 toxin induced hemolysis, manuscript in preparation.
28. **DeLoach, J.R. and Mollenhauer, H.H.,** Interaction of T-2 toxin with erythrocytes, *Fed. Proc.,* 46, 1409, 1987.
29. **Schappert, K.T. and Khachatourians, G.G.,** Influence of the membrane on T-2 toxin toxicity in *Saccharomyces, Appl. Environ. Microbiol.,* 47, 681, 1984.

A Bioassay Based on Yeast Cytotoxicity for the Detection of Trichothecene Mycotoxins of Agricultural Importance

G. G. Khachatourians, K. T. Schappert, and H. A. Koshinsky

INTRODUCTION

The presence of mycotoxins in agriculturally important commodities can create a significant health and safety problem. The dimensions of this problem are not only relevant to the environment, but also to the consumption of such material. Analytical chemistry is the most suitable means for detecting the permissive levels of such toxins in a sample. However, these analytical assays are not convenient to do especially at the locations where an immediate reading and "early warning" is required.

There are several systems for testing mycotoxins.[1-8] These are shown in Table 1. It is clear that the higher eukaryotic test systems, while showing reactivity toward mycotoxins, are not able to identify the specific toxins. They are also quite cumbersome and require sophisticated procedures. The yeast bioassay described here is an economical and sensitive bioassay which could be done in most laboratories and/or in the environment. As such the bioassay described here can serve as an "early warning" system for the detection of the trichothecene family of mycotoxins.

DEFINITIONS AND DELIMITATIONS

Many fungi produce primary and secondary metabolites which could affect the sensitivity and the final outcome of a microbial toxicity biassay. In addition, any original material of soil or plant origin containing herbicide and pesticide residues could seriously affect such bioassay systems. Thus, the specific bioassay should be carefully defined and its limits

TABLE 1
A Selective Example of Bioassays for the Detection of Trichothecene Mycotoxins

Indicator organism	Trichothecene detected	Toxicity indicator	Sensitivity measurement: Concentration (μg)	Sensitivity measurement: Time (h)	Ref.
Tetrahymena pyriformis	Diacetoxyscirpenol	Growth inhibition	0.04	8	1
	Trichothecin	Growth inhibition	0.48	8	
Daphnia magna	Diacetoxyscirpenol	LD_{50}	1.2	24	
	Trichothecin	LD_{50}	8.0	24	
Artemia salina	T-2 toxin	Decreased motility	0.25	24	2
	T-2 toxin	Mortality	0.2	16	3
	Diacetoxyscirpenol	Mortality	0.28	24	1
	Trichothecin	Mortality	1.16	24	
Mouse	Diacetoxyscirpenol	Refusal to drink trichothecene in water	2		4
	T-2 toxin	Refusal to drink trichothecene in water	2		
	Vomitoxin	Refusal to drink trichothecene in water	2		
Rat	T-2 toxin	Skin redness	0.05—0.1	24	5
	T-2 toxin	Skin redness	2	48	6
Kluyveromyces fragilis	T-2 toxin	Growth inhibition	0.2	20	7
	Verrucarin A	Growth inhibition	0.12	20	8
	Roridin A	Growth inhibition	0.5	20	
	T-2 toxin	Growth inhibition	1.0	20	
	Diacetoxyscirpenol	Growth inhibition	10.0	20	
	HT-2 toxin	Growth inhibition	13.3	20	

known. Ultimately, if routine long-term screening of agricultural material with known herbicide and insecticide residues is required, indicator organisms resistant to herbicides and insecticides could be employed for bioassays. Here we will consider the most important component of the bioassay, the microorganisms used. An indicator organism is used primarily for identifying the presence of mycotoxins and monitoring organisms are used for quantification of mycotoxins, in well-defined environments containing other organisms.

The above organisms must be maintained and periodically tested to ensure purity. After a longitudinal study of many yeasts,[9-11] we arrived at the yeasts *Saccharomyces cerevisiae* GR262 and *Kluyveromyces fragilis* GK1005. They have ideal determinants for our bioassay. For example, they are easily grown, maintained, transferred, and distributed. They can also be used for tests under field conditions. They yield characteristic growth on solid media which, in the presence of mycotoxins, show definite zones of growth retardation. The indicator yeasts used in the bioassay have yielded highly reproducible and consistent readings with a standard error of no greater than 10% after at least 500 different bioassays. The limit of detection of the disk-type bioassay has been between 0.1 to 0.2 μg per disk.

The largest limitation for the bioassay is the absence of a guarantee that it would result in a complete answer without any false-positive or false-negative data. Therefore, comprehensive study of multiple toxins and synergistic effects of other environmental xenobiotics is needed. Nonetheless, we are satisfied with the initial bioassay for its use as an "early warning" system and hope that by using biotechnological approaches one could further enhance the bioassay system.

BIOASSAYS

Two yeasts, *S. cerevisiae* GR262 and *K. fragilis* GK1005, were selected for the standardized bioassay. In addition, strains of *S. cerevisiae* HK1 and HK8 could be used for preferentially testing or monitoring of specific mycotoxins (see below).

Exponentially growing cells of either *S. cerevisiae* or *K. fragilis* were standardized to an absorbance of 0.3 at 610 nm. Sterile cotton swabs were dipped into the standardized inoculums and used to swab the surface of solidifed medium in petri plates.

The trichothechenes were dissolved in 95% ethanol.

The medium consisted of yeast extract (1%, w/v), peptone (1%, w/v), and dextrose (2%, w/v). This was solidified with the addition of 1.75% (w/v) agar. Sterile filter paper disks (12.5 mm in diameter) were impregnated with the desired concentration of a trichothecene. The total volume per disk was brought to 50 μl with ethanol.

The zone of growth inhibition was identified by a disk-type bioassay which has been described fully elsewhere,[7] and in brief is as follows. A toxin-impregnated disk was aseptically transferred to the center of an inoculated petri plate. The plate was then inverted and incubated at 30°C *(S. cerevisiae)* or 35°C *(K. fragilis)*. The zones of growth inhibition were measured after 20 to 48 h.

The minimal inhibitory concentration (MIC) and effective concentration$_{50}$ (EC$_{50}$) values were determined in a broth-type bioassay.[12] Serial twofold dilutions of toxins were prepared in liquid medium in test tubes. The tubes were inoculated from an overnight culture of *K. fragilis* at a final concentration of 1% (v/v) and incubated at 35°C for 20 h. The growth that occurred in each tube was measured at 610 nm in a spectrophotometer.

We have previously shown that carbohydrate source and membrane-perturbing agents in the growth medium enhanced the sensitivity of *S. carlsbergensis* and *S. cerevisiae*.[9,10] We tested the influence of the same two parameters on the sensitivity of *K. fragilis* to T-2 toxin using a disk-type of bioassay. These results are shown in Table 2. In the absence of cetyltrimethylammonium bromide (CTAB) the sensitivity of *K. fragilis* to T-2 toxin was the greatest with glycerol as the carbohydrate source in the medium. However, in the presence of CTAB the sensitivity of *K. fragilis* to T-2 toxin was greatest when galactose was the carbohydrate source in the growth medium.

In the experiments that follow, galactose and CTAB were incorporated in a standardized disk type of bioassay and used to test the sensitivity of *K. fragilis* to various concentrations of T-2 toxin (Table 3). The degree of growth inhibition caused by T-2 toxin was concentration dependent.

The MIC and EC$_{50}$ values of five trichothecenes were determined using the broth-type bioassay. The results are

TABLE 2
The Influence of Carbohydrate in Growth Medium and CTAB on Growth Inhibition Induced by T-2 Toxin (50 μg/Disk) on *K. fragilis*

	Diameter of growth inhibition (mm)	
Carbohydrate	Without CTAB	With 0.5 μg/ml CTAB
Glucose	9.0	11.5
Glycerol	11.5	14.0
Lactose	7.0	14.5
Galactose	8.5	15.5

TABLE 3
The Influence of Various Concentrations of T-2 Toxin on Growth Inhibition of *K. fragilis*

T-2 toxin (μg/disk)	Diameter of growth inhibition (mm)
0.0	0.0
0.1	0.0
0.2	6.2
0.4	10.8
0.6	14.5
0.8	16.8
1.0	20.2

shown in Table 4. The growth of *K. fragilis* had greatest sensitivity to verrucarin A and the least sensitivity to HT-2 toxin. The MIC and EC_{50} values for verrucarin A were 0.12 and 0.03 µg/ml, respectively.

The specificity of the bioassay could be increased by strain selection of genetic manipulation. We have isolated T-2 toxin-resistant mutants (T-2^R) or *S. cerevisiae* GR262. The T-2^R strains HK1 and HK8 were tested for T-2 toxin sensitivity and roridin sensitivity (Table 5). These two mutants had a similar degree of resistance to T-2 toxin. However, strain HK1 was cross-resistant to roridin A, whereas strain HK8 was sensitive to roridin A. Therefore, it was possible to distinguish between these two trichothecenes based on the differential growth inhibition caused by these two trichothecenes.

EXAMINATION OF FUSARIOTOXIN FROM FUSARIA ISOLATED IN THAILAND

Although it is difficult to prove the presence of a specific fusariotoxin in samples of agricultural commodities, the techniques described here have been used for such a demonstration. Recent studies in our laboratory[14] have demonstrated the usefulness of the above bioassay for the direct determination of T-2 toxin in rye. Moldy rye samples were prepared as described.[13] The results of an experiment in which the moldy rye (0.25 mg) was incorporated into agar plugs are described here. The bioassay used experimentally infected samples and for positive controls included T-2 toxin-spiked rye samples.

TABLE 4
MIC and EC_{50} Values of Five Trichothecenes Using *K. fragilis* as the Indicator Organism

Trichothecene	MIC (µg/ml)	EC_{50} (µg/ml)
Verrucarin A	0.12	0.03
Roridin A	0.50	0.05
T-2 toxin	1.00	0.10
Diacetoxyscirpenol	10.00	1.00
HT-2 toxin	>13.26	4.00

TABLE 5
The Selective Sensitivity of T-2 Toxin-Resistant Mutants of *S. cerevisiae* to Roridin A

S. cerevisiae strain	Diameter of growth inhibition (mm)	
	T-2 toxin	Roridin A
GR262 (T-2^S)	9.5	16.7
HK 1 (T-2^R)	1.3	6.2
HK8 (T-2^R)	1.6	16.8

The mold-free rye did not inhibit the growth of *K. fragilis*. Rye samples spiked with 0.5, 1, and 2 µg of pure T-2 toxin per sample showed a concentration- dependent zone of growth inhibition of *K. fragilis*. Rye spiked with 0.5 µg of pure T-2 toxin did not inhibit the growth of the yeast. A rye sample spiked with 1 and 2 µg of pure T-2 toxin gave zones of growth inhibition of 4 and 13 mm, respectively. A rye sample infected with *Fusarium semitectum* S125-C2 did not inhibit the growth of *K. fragilis*. This rye sample contained T-2 toxin less than the detection limits of the bioassay. A rye sample infected with a known T-2 toxin-producing strain of *F. sporotrichioides* ATCC 48020 gave a zone of growth inhibition of 34 mm. The results of the bioassay are in full agreement with a skin bioassay and an analytical assay.[13] Clearly, our bioassay needs to be extended to other agriculturally important material in areas of the world where known incidents of mycotoxicosis are prevalent.

PROBLEMS CONCERNING CRITERIA AND STANDARDS

The standards and criteria used in this bioassay were established from the general knowledge of the mechanism of action of fusariotoxins in yeast[7-11] and ecological safety rules. The problems and criteria concerning the bioassay could be divided into four groups.

Specificity

The general bioassay is specific for inhibition of the yeast growth by a number of cytotoxic agents. In order for this test to respond to specific toxins, we must generate mutant cell lines which would respond to an individual class of toxins while showing tolerance or resistance to others. Isolation of such mutants and their biotechnological manipulations would enhance the specificity of this bioassay.

Sensitivity

The use of standards on a routine basis should allow the establishment of a threshold and sensitivity of a given bioassay under different laboratory/field conditions.

Altered Zones of Growth Inhibition

The general measure of cytotoxicity depends on the appearance of the zones in growth inhibition. Occasionally, these zones could be masked by background growth of, for example, a mixed population of yeast and contaminating bacteria. The timing of the incubation period, substrates used in the plate assay, and culture purity are very important. Strict adherence to aseptic techniques and/or the inclusion of antibiotics in the growth medium would suppress bacterial contamination.

Technical

While the routine performance of such bioassays does not require specific technical training, it is important that the nature of the bioassay and the constraints associated with the test be known to the monitoring person. These bioassays should be confined to the laboratory setting where the field sample is brought in and tested. Handling would be easier and the parameters of the bioassay would be more controlled. Some training of the laboratory worker in microbiology and toxicology would be of use.

CONCLUSIONS

The main objective in devising a bioassay is to screen large numbers of samples suspected of containing fungal toxins. Any bioassay yielding a positive result could indicate the presence of potentially hazardous mycotoxins for human and animal populations. However, there are numerous fungal metabolites which could interfere with such bioassays yielding false-positive results. The standardized bioassay can be improved in many ways to eliminate false-positive data. It is after such an improvement that the bioassay can become an effective "early warning" system.

Considering the costliness of analytical tools such as the mass spectrometer, it is clear that such *in vitro* bioassays have a place in routine screening programs. We are currently exploring the possibility that this bioassay can be used for the assessment of mycotoxins involved in health and safety in agriculture.

REFERENCES

1. **Beijl, J., Dive, D., and van Petegham, C.,** Comparison of some bioassay methods for mycotoxin studies, *Environ. Pollut. (A)*, 26, 173, 1981.
2. **Schmidt, R.,** Optical motility test for the detection of trichothecenes using brine shrimps, *Mycotoxin Res.*, 1, 25, 1985.
3. **Harwig, J. and Scott, P.M.,** Brine shrimp (*Artemia salina* L.) larvae as a screening system for fungal toxins, *Appl. Microbiol.*, 21, 1011, 1971.
4. **Burmeister, H.R., Vesonder, R.F., and Kwolek, W.F.,** Mouse bioassay for *Fusarium* metabolites: rejection or acceptance when dissolved in drinking water, *Appl. Environ. Microbiol.*, 39, 957, 1980.
5. **Wei, R.-D., Smalley, E.B., and Strong, F.M.,** Improved skin test for detection of T-2 toxin, *Appl. Microbiol.*, 23, 1029, 1972.
6. **Hayes, M.A. and Schiefer, H.B.,** Quantitative and morphological aspects of cutaneous irritation by trichothecene mycotoxins, *Food Cosmet. Toxicol.*, 17, 611, 1979.
7. **Schappert, K.T. and Khachatourians, G.G.,** A yeast bioassay for T-2 toxin, *J. Microbiol. Methods*, 3, 43, 1984.
8. **Schappert, K.T., Koshinsky, H.A., and Khachatourians, G.G.,** Growth inhibition of yeast by T-2 toxin, HT-2 toxin, T-2 triol, T-2 tetraol diacetoxyscirpenol, verrucarol, verrucarin A and roridin A mycotoxins, *J. Am. Coll. Toxicol.*, 5, 181, 1986.
9. **Schappert, K.T. and Khachatourians, G.G.,** Effects of fusariotoxin T-2 on *Saccharomyces cerevisiae* and *Saccharomyces carlsbergensis, Appl. Environ. Microbiol.*, 45, 862, 1983.
10. **Schappert, K.T. and Khachatourians, G.G.,** Influence of the membrane on T-2 toxin toxicity in *Saccharomyces* spp., *Appl. Environ. Microbiol.*, 47, 681, 1984.
11. **Sukroongreung, S., Schappert, K.T., and Khachatourians, G.G.,** Survey of sensitivity of twelve yeast genera toward T-2 toxin, *Appl. Environ. Microbiol.*, 48, 416, 1984.
12. **Washington, J.A. and Sutter, V.L.,** Dilution susceptibility test: agar and macro-broth dilution procedures, in *Manual of Clinical Microbiology*, 3rd ed., Lennete, E.H., Balows, A., Hausler, W.J., Jr., and Truant, J.P., Eds., American Society for Microbiology, Washington, D.C., 1983, 453.
13. **Sukroongreung, S., Schiefer, H.B., and Smith, J.D.,** Comparison of potential toxigenicity of *Fusarium* spp. from Southeast Asia and North American, *J. Sci. Soc. Thailand*, 11, 31, 1985.
14. **Schappert, K.T., Sukroongreung, S., and Khachatourians, G.G.,** unpublished data.

A Health Surveillance and Health Promotion Program for Grain Workers

M. Reddy

INTRODUCTION

Several different approaches are required to make medical surveillance in the grain industry a positive program. The primary focus must be to improve the work environment by reducing dust to the lowest reasonable level below the threshold limit value (TLV) of 10 mg m^3.[1] Environmental monitoring and dust collection systems will not be addressed in this presentation.

Workers exposed to grain dust have a higher prevalence of respiratory symptoms,[2,3] and are also believed to be at a higher risk of developing chronic obstructive airways disease (COAD). However, there is no clear definition of disease in these workers.[4] This results in ill-defined parameters to monitor. The following variable factors may complicate health surveillance: (1) cigarette smoking is a major contributor to both respiratory symptoms and airway obstruction. Smoking combined with grain dust exposure may in certain age groups have a synergistic effect;[4] (2) grain dust is complex, made up of many substances in varying amounts;[5] (3) dust exposure varies considerably according to peak work periods, types of grain, weather conditions, use of personal respiratory protection, and the efficiency of the elevator's dust control system; (4) there is currently no accurate method to relate grain dust exposure (dose) to response; (5) obesity may affect pulmonary function; (6) some healthy men have forced expired volume in 1 s/forced vital capacity (FEV_1/FVC) ratios less than 70% as a result of very large FVCs, thus falsely appearing to have a below-average pulmonary function; (7) years of exposure (both to grain dust and cigarette smoking); (8) age; and (9) allergies.

Understandably, then, medical surveillance of grain workers is not as straightforward as monitoring other occupational exposures, where cause and effect are more clearly defined.

MEDICAL SURVEILLANCE PROGRAM

Medical surveillance of grain workers in Canada has been required under the Canada Labour Code since 1978.[1] Saskatchewan Wheat Pool has been monitoring the health of their grain workers since 1973.

New employees are required to have a physical examination, chest radiograph, and pulmonary function test prior to commencing work in the grain industry.[1] This provides baseline data and identifies employees at risk. For example, men who suffer from chronic bronchitis, asthma, or emphysema should be discouraged from working in the grain industry. Cigarette smokers are also considered at risk and they are informed by letter of this risk and smoking cessation is encouraged.

Once every 3 years grain workers are required to have a pulmonary function test, chest X-ray, and complete a medical surveillance questionnaire.[1] This province-wide testing is carried out by the Saskatchewan Lung Association from a mobile unit. Quality control of testing is under the direction of the Section of Pulmonary Medicine, University of Saskatchewan. The program is designed to meet the objectives of the Employees Health Service, while at the same time accommodating ongoing research.

Results of medical surveillance are forwarded to Saskatchewan Wheat Pool Employees Health Services for evaluation and comparison with previous test results. These records are maintained under strict confidentiality. The evaluation process takes into consideration, where possible, the many factors reviewed earlier. Medical surveillance is not a diagnostic service; at best, we can identify those employees who appear to have comprised lung function and/or report significant respiratory symptoms and then refer them for further medical evaluation.

Medical surveillance without a mechanism for follow-up serves little purpose for improving the health of grain workers. At Saskatchewan Wheat Pool, our Employees Health Service arranges for follow-up examinations with chest specialists after telephone consultation with the worker's family physician. On written consent from the employee, the chest specialist forwards his reports to both the Employees Health Service and the employee's family physician. Diagnoses determined from follow-up examinations have included mild to moderate COAD, early emphysema, chronic bronchitis, asthma, and hyperreactivity of airways hyperresponsiveness; some employees are found to be normal on retest.

Frequency of follow-up varies according to the degree of impairment and symptoms. We have approximately 20 employees on a yearly follow-up program. Occasionally, a chest specialist will recommend to an employee that he should "get out of the grain industry". These workers may feel quite well and are not physically disabled from doing their job. The actual decision to leave the industry is one that the employee himself must make. Most grain workers will respond to this advice and leave the industry within 1 or 2 years or as soon as alternate employment has been found. Grain workers in Saskatchewan usually qualify for income assistance and vocational rehabilitation training program from the Workers' Compensation Board. Assistance from the Workers' Compensation Board is very important as it enables men to leave the industry before their health is severely impaired.

Results of medical surveillance and follow-up examinations are reported to the employee and to Labour Canada. Labour Canada guidelines also recommend that the employer receive a report indicating that the employee is either "fit to work in the grain industry", or "follow-up is recommended". Our Employees Health Service has discontinued this practice to ensure a greater degree of confidentiality for the employee. Management is provided with group statistics only.

Employees receive a complete report of their pulmonary function test along with a full explanation of what the tests actually measure. Smokers are encouraged to quit and all employees are encouraged to wear respiratory protection. In addition to written reports, we often telephone employees to discuss their medical surveillance results. This personal contact has proven to be very beneficial in determining if follow-up is required and it also provides a good public relations function by letting employees know that we are actively monitoring their respiratory health.

Reports to Labour Canada are coded to ensure employee confidentiality. Labour Canada is provided with lung function and chest X-ray results, reported symptoms, and smoking history. The employee is graded according to his FEV_1/FVC ratio, which we find can result in a technical misrepresentation of health status. Approximately 17% of employees with a reduced FEV_1/FVC ratio have FVCs greater than 110% of predicted values and a normal FEV_1.

The smoking habits of our employees understandably are of major concern. Several studies report that smoking is a major contributing factor in the prevalence of airway disease[4,6] in grain workers. Although a higher percentage of our grain workers smoke compared to the Canadian average of 33%, there continues to be some improvement noted in these numbers (Figure 1).

Prevalence of respiratory symptoms are much higher among our grain workers who smoke (Figure 2). However, the prevalence of symptoms overall is far less than what has been reported by various epidemiologic surveys,[7,8] that between 33 and 50% of grain workers have chronic productive cough. This has not been our experience with Saskatchewan Wheat Pool grain workers.

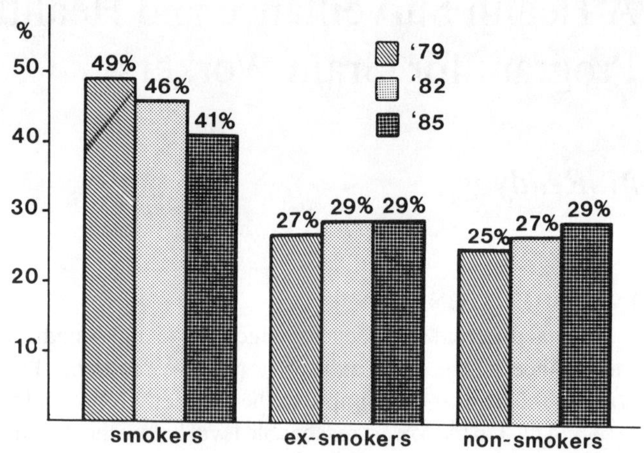

FIGURE 1. Changes in smoking habits of Saskatchewan Wheat Pool grain workers.

The prevalence of respiratory symptoms in Saskatchewan Wheat Pool grain workers has actually decreased since the last medical surveillance (Figure 3). This could be attributed to several factors: (1) decreased number of smokers; (2) improved dust collection systems; and (3) a more responsible attitude on the part of the worker with respect to wearing personal respiratory protection.

Table 1 illustrates the actual number reporting chronic or occasional symptoms in the 1985 medical surveillance. Chronic symptoms were not recorded separately in previous surveillance years; therefore, a comparison of chronic symptoms alone is not available.

Our statistics indicate that nonsmoking grain workers by

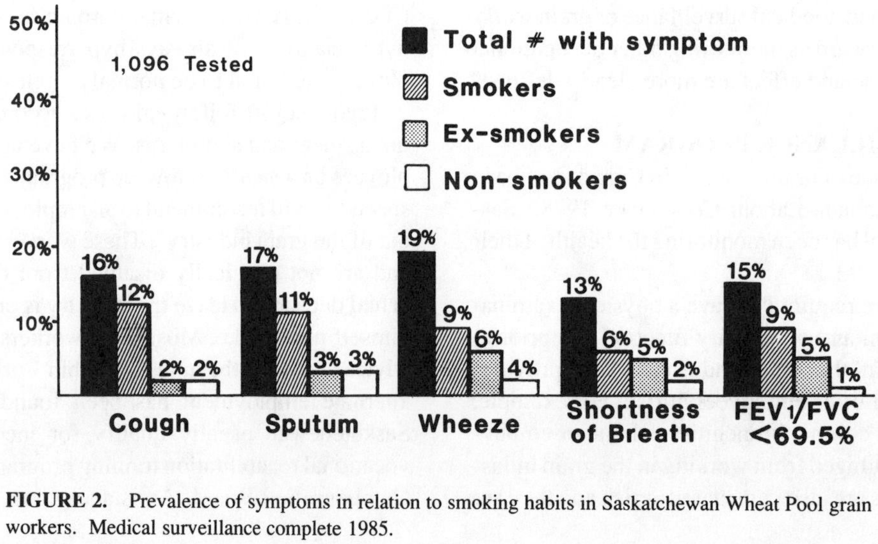

FIGURE 2. Prevalence of symptoms in relation to smoking habits in Saskatchewan Wheat Pool grain workers. Medical surveillance complete 1985.

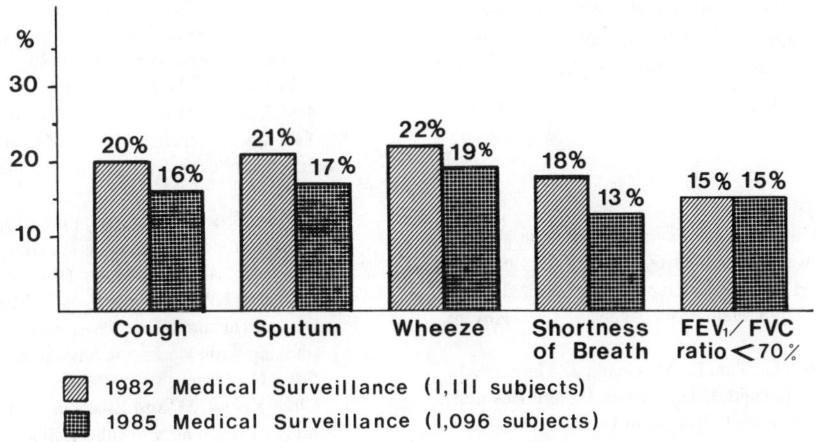

FIGURE 3. Changes in prevalence of respiratory symptoms of Saskatchewan Wheat Pool grain workers.

TABLE 1
Chronic and Occasional Symptoms

	Total number	Smoker	Ex-smoker	Nonsmoker
Tested	1096	451 (41%)	318 (29%)	327 (30%)
Cough (chronic)	139 (12.6%)	103 (9.4%)	18 (1.6%)	18 (1.6%)
Cough (occasional)	34 (3.1%)	24 (2.1%)	5 (0.5%)	5 (0.5%)
Sputum (chronic)	149 (13.5%)	98 (8.9%)	30 (2.7%)	21 (1.9%)
Sputum (occasional)	39 (3.5%)	26 (2.3%)	6 (0.5%)	7 (0.6%)
Wheeze (chronic)	28 (2.5%)	13 (1.1%)	9 (0.8%)	6 (0.5%)
Wheeze (occasional)	185 (16.8%)	92 (8.3%)	57 (5.2%)	36 (3.2%)
Shortness of breath (all)	145 (13.2%)	69 (6.3%)	59 (5.3%)	17 (1.5%)
$FEV_1/FVC \times 100 =$ less than 69.5%	169 (15.4%)	97 (8.8%)	61 (5.5%)	11 (1.0%)
Referral for follow-up	44 (4.0%)	24 (2.1%)	18 (1.6%)	2 (0.18%)

comparison infrequently develop respiratory disease. In response to medical surveillance conducted in 1984 and 1985, 43 employees have been referred to chest specialists for follow-up. In this group, we found only *two* lifetime nonsmokers. Studies by doPico et al.[9] and Gerrard et al.[10] support our observation.

Results from research studies[4,6,11] as well as our own statistical analysis prompted us to try and reduce the smoking habit among our grain workers. This, we believed, would greatly reduce respiratory symptoms and lower the risk of more serious respiratory diseases.

In 1982 we developed a "Respiratory Health Management Seminar" for our country elevator employees. Although our primary objective was to encourage smoking cessation, our approach was to promote all aspects of health maintenance (Figure 4). Specific topics addressed during the seminar were respiratory diseases found in the grain industry; smoking and exposure to grain dust; review of Saskatchewan Wheat Pool Medical Surveillance program, objectives and statistical results; film presentation on the respiratory system; explanation of lung function test; Workers' Compensation Board and rehabilitation of the grain worker with COAD; other lifestyle factors (i.e., weight control, exercise, alcohol abuse, and stress); dust control and dust monitoring; and personal respiratory protection.

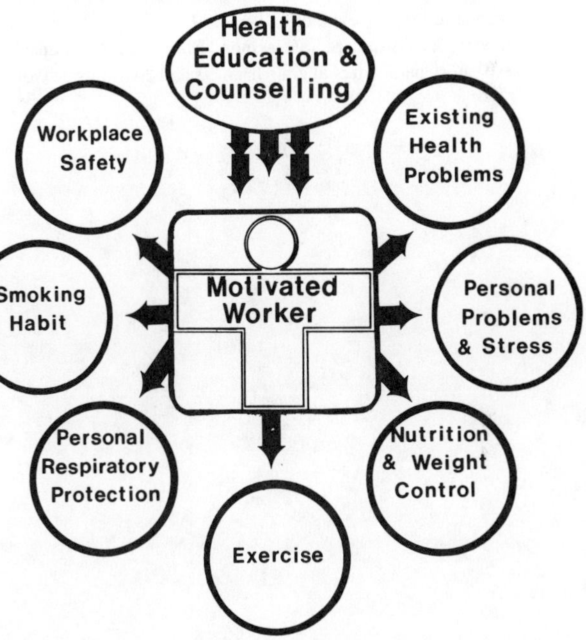

FIGURE 4. A health promotion model for grain workers.

Employees also had the opportunity to meet individually with the nurse to view and discuss their medical surveillance results. Employees' evaluations of the 85 seminars that have been presented to date have been extremely positive. Health promotion is an ongoing process. Respiratory health seminars are presented for new workers and health bulletins, stop smoking programs, individual letters and telephone calls are directed toward promoting employee health.

CONCLUSION

Effective medical surveillance is much more than a test

every 3 years. Follow-up examinations are necessary to ensure early diagnosis and appropriate treatment. In addition, health and safety education will encourage grain workers to accept a greater degree of responsibility for protecting their own health.

REFERENCES

1. Labour Canada, Guidelines for an Environmental and Medical Surveillance Program in the Grain Industry, January 15, 1981.
2. **Warren, C.P.W. and Manfreda, J.,** Respiratory Abnormalities in Grain Workers 1971—77: A Preliminary Analysis, University of Manitoba, Winnipeg.
3. **Sheridan, D., Deutscher, C., Tan, L., Maybank, J., Gerrard, J., Horne, S., Yoshida, K., Barnett, G.D., Cotton, D., and Dosman, J.A.,** The Relationship Between Cereal Grain Dust and Pulmonary Function in Grain Workers, University of Saskatchewan, Saskatoon.
4. **Dosman, J.A.,** Chronic obstructive pulmonary disease and smoking in grain workers, *Ann. Intern. Med.,* 87, 784, 1977.
5. **Yoshida, K. and Maybank, J.,** Physical and environmental characteristics of grain dust, in *Occupational Pulmonary Disease: Focus on Grain Dust and Health,* Dosman, J.A. and Cotton, D.J., Eds., Academic Press, New York, 1980, 399.
6. **Broder, I., Mintz, S., Silverman, F., et al.,** A comparison of respiratory parameters in grain handlers and civic outside workers in Thunder Bay, Canada, in *Occupational Pulmonary Disease: Focus on Grain Dust and Health,* Dosman, J.A. and Cotton, D.J., Eds., Academic Press, New York, 1980.
7. **Williams, N., Skoulas, A., and Merriman, J.E.,** Exposure to grain dust, I. A survey of the effects, *J. Occup. Med.,* 6, 319, 1964.
8. **Tse, K.S., Warren, P., Janusz, M., et al.,** Respiratory abnormalities in workers exposed to grain dust, *Arch. Environ. Health,* 27, 74, 1974.
9. **doPico, G.A., Reddan, W., Flaherty, D., et al.,** Respiratory abnormalities among grain handlers: a clinical, physiologic and immunologic study, *Am. Rev. Respir. Dis.,* 115, 915, 1977.
10. **Gerrard, J.W., Cheung, S.-S.C., Mink, J., Tan, L.K.-T., Cotton, D., and Dosman, J.A.,** Characteristic of airways reactivity in nonsmoking grain workers in Saskatchewan, *Ann. R. Coll. Phys. Surg. Can.,* 11 (Abstr.), 89, 1978.
11. **Chan Yeung, M. and Enarson, D.A.,** A prospective: comparative study of respiratory health of civic workers and grain handlers in Vancouver, Progress Report, Respiratory Division, Vancouver General Hospital, University of British Columbia, Vancouver, April 1983.
12. **Atherley, G.,** Editorial, Rights vs Rights (experts), in *At The Centre,* 8(2), July 1985.
13. Worksite Health Promotion and Human Resources: Program Summary at a Conf., Washington, D.C., October 11, 1983.
14. **Dosman, J.A. and Cotton, D.J.,** *Occupational Pulmonary Disease: Focus on Grain Dust and Health,* Academic Press, New York, in press.

Toxicology and Health Effects of Pesticides

Registration of Pesticides in Canada Under the Pest Control Products Act

S. W. Ormrod

BACKGROUND

The Pest Control Products Act, the federal legislation under which all pesticides are registered in Canada, is administered by Agriculture Canada. The purpose of this statute is to assure safety, merit, and value. The regulatory process is designed to achieve this fundamental objective through presale assessment and registration.

Safe and safety are the key words, or cornerstone, of the statute. The context of these fundamental principles is broadened in the regulations to focus attention specifically on human health and environmental and aquatic impact.

The registration process has traditionally operated on the principle of "guilty until proven innocent". Concrete information is needed to allow a determination of safety; absence of data indicating risk or hazard is not justification to register a product.

The statute provides three basic regulatory options: research permit, temporary registration, and full registration. This stepwise approach is followed with the newest active ingredients. A research permit provides the opportunity for limited scale testing, such as small plots and limited field demonstrations. This first step can be initiated with a minimum of data on hand. Temporary registration, by comparison, requires substantially more information. Basically, this interim step is used to provide an opportunity for confirmatory studies to answer final, outstanding questions. Ultimately, the product moves forward to full registration status. Older products are reviewed and updated through the reevaluation authorities of the regulations. Some current examples include phenoxy herbicides, fumigants, and chlorophenols. A reevaluation may be called, at the discretion of the Minister of Agriculture, when it is felt that the facts at hand no longer demonstrate safety. From an operational viewpoint, our reevaluation system is usually in step with the U.S., in part, because the reevaluation ranking scheme — which weighs such factors as age and quality of the data base, and quantity of product used — is similar in both countries.

Perhaps one of the major differences between the Canadian and American systems lies in the cancellation and suspension authorities held by the Minister of Agriculture. In Canada, these authorities are very broad and can be directly implemented with only limited procedural constraints. Cancellation stops sale at all levels of trade, while suspension allows stocks remaining at dealer and distributor levels to be sold.

The Pesticides Directorate of Agriculture Canada is the working level body accountable for day-to-day administration of the statute. In this role, the Directorate is primarily responsible for establishing which pesticides are and are not registered and used. As a result, it also is a target for much public pressure and criticism.

REGULATORY PROCESS

The Canadian regulatory process involves a team approach. Key players at the federal level are the Departments of Agriculture, Health, Environment, and Fisheries and Oceans. Provincial colleagues in all related disciplines, including Agriculture, Health, and Environment, are first-string members of the Canadian regulatory team. Certainly, it is a complicated process which, in recent times, has been played out in a fish bowl environment. Nevertheless, the system does provide for best-balanced decisions, as all points of view are brought to the table for consideration. The process, and the players involved, entail built-in checks and balances that are inherent in the structure of the institutions involved.

Although not without problems, the system does provide a fundamentally sound framework for decision making. Ultimately, the Federal Minister of Agriculture bears responsibility for the Pest Control Products Act, which means that his department is continuously called upon to make judgments and be judged by others on the basis of the decisions that are made.

The registration system is a science-driven process based primarily upon biology, chemistry, toxicology, and agronomy. It has often been claimed, with some justification, that the Canadian regulatory system is the toughest in the world. On the other hand, there are also critics at the opposite end of the scale who view our process as not sufficiently intensive.

We are sometimes pressured to relax the intensity of the science reviews that are conducted, but we have constantly resisted this type of suggestion. We believe that the quality and integrity of the science reviews conducted are the backbone and key strength of the Canadian process. Rather than erode this critical strength, we feel that the facts emerging from the science review process should be the key elements weighed in the subsequent decision-making process.

KEY QUESTIONS

Obviously, assurance of safety is an unimpeachable objective. Much more controversial is the question of what information is required to adequately demonstrate safety. This question continues to generate debate in the general context as well as in relation to specific chemicals or situations.

We attempt to define the ground rules by means of registration guidelines. We have, we believe, a good working document relative to health and safety guidelines and a draft paper on environmental fate. Since the statute and regulations provide only very general direction, these guidelines are critically important as they set out the required areas of study, such as: detailed product chemistry, performance, occupational exposure, nontarget organisms, and persistence, retention, and movement of residues.

DECISION MAKING

Recognizing the complexity of the system and the number of players involved, the obvious question is how does the decision-making process work? In practice, it functions at a number of levels, including *ad hoc* interdepartmental discussions to resolve differing viewpoints and perceptions.

Recognizing the divergent basic mandates held by the key federal agencies, differing points of view are virtually inherent in the system. As a result, there are negotiations, discussions, and decision-making conferences on specific products or problems.

In addition, there are regular, formal interdepartmental meetings to define operating procedures and principles and to maintain communication and understanding.

Traditionally, the decision-making process has operated on the basis of direct, regulatory recommendations from consultant departments. More recently, there has been a shift towards presentation of a more detailed account of the science review conducted, with only a general commentary on the preferred regulatory option. This approach provides more freedom in selection of regulatory avenues, but also carries additional responsibilities. As the regulatory authority, Agriculture Canada — quite properly — holds the final decision on registration. In making this decision the department must largely assume the responsibility which is inherent with authority.

Much stronger efforts are being made in the areas of consultation and communication. Traditionally, supporting data have usually indicated absence of health or environmental impact, fostering the perception, quite reasonably, that the product is risk free or completely "safe". Advances in science and technology have led to more precise observations which may indicate potential risk or possible hazard. This trend is likely to increase in the future, and the possibility for a "perfect 10" will become more and more remote.

As Agriculture Canada takes a more aggressive stand in pesticide regulatory decision making, it also assumes an obligation to share with provincial colleagues, user associations and the public at large, the nature of the data base supporting a product. This approach is widely and successfully used in the U.S. and is consistent with overall efforts within the department. Some specific examples include extensive dialogue and information exchanges that have occurred in relation to registration of: insecticides for wheat midge to accommodate an anticipated insect outbreak in 1984; and herbicides for triazine-tolerant rapeseed to accommodate the introduction of a new rapeseed cultivar that will tolerate residual levels of a previously used herbicide in soil.

The Pesticides Call Line, recently established by Agriculture Canada on a trial basis, and the annual Farm Letter on pesticides represent two more specific examples of this kind of initiative. Similar efforts are also being made at working and/or operational levels; for example, explanatory background letters to user organizations and provincial officials.

We see a critical need for pest control products, frequently already in use in other areas of the world, that cannot fully meet Canadian standards in the areas of health and safety or environmental impact. Recognizing this, we expect that Agriculture Canada will be obliged to continue to make pesticide regulatory decisions that are not universally acceptable. At the same time, increasing efforts are being made to provide technical background and regulatory rationale in support of registration decisions. Hopefully, this communication effort will provide an opportunity for improved understanding and informed decision making at all levels, from provincial regulatory systems to municipalities, user organizations, and, ultimately, to individual users.

Agriculture Canada appreciates the provincial contribution to pest management and pesticide control within that jurisdiction. While we feel an obligation to establish an overall regulatory framework through registration, we also recognize and respect the provincial interest in fine tuning the system by means of user training, use permits, maintenance of sale records by vendors, and other regulatory strategies. Indeed, not all provinces wish to provide for use all federally registered materials within their boundaries.

Traditionally, risk/benefit judgments have been made in pesticide regulation. More recently, there has been a growing interest in risk/benefit analysis. Although these two approaches are similar, the subtle differences involved are very important.

Although the decision may be correct, the judgment approach is seen to be arbitrary, incomprehensible, and without justification that can be readily understood or communicated. Risk/benefit analysis is unlikely to substantially change judgments that are made; however, it does provide a tool to explain the rationale used in arriving at the decision. When supported by a complementary communications effort, risk/benefit analysis can assist in improving education and understanding, which are fundamental elements of progress.

Many of the key changes in Agriculture Canada's approach to pesticide regulation have emerged in response to the Salter Report on Consultation in the Assessment and Registration of Pesticides (1984). Very briefly, this independent study made three basic recommendations to the department: (1) establishment of an Information Secretariat to provide information on pesticides to all sectors of the public through mechanisms such as backgrounders, publication of regulatory rationale, and a toll-free, call-line service; (2) establishment of an independent Pest Management Advisory Board (PMAB) to advise the Minister on broad policy questions and issues of special concern. The PMAB is not seen as an independent evaluator of regulatory decisions in either a peer review or a review board context. It is not intended to second guess or pass judgment on rational regulatory decisions based on good science; (3) strengthening communications and liaison with provinces to support the provincial role in pesticide regulation.

One of the first steps in implementing the Salter Report recommendations was the establishment of the Pesticides Call Line. This new communications vehicle has been operated since April 1, 1985. During this period, the Call Line has fielded a broad range of telephone inquiries from across Canada. The response to this type of information services seems to be very favorable.

The chairman of the Pest Management Advisory Board has been appointed, and the full complement of board members has been named.

Substantial, additional resources have been directed to pesticide regulation within Agriculture Canada and the stature of the regulatory unit has been raised within the organizational hierarchy. The intention is to maintain the scientific strengths that already exist, while improving our performance in broader consultation and communication processes.

FUTURE CHALLENGES

In addition to strengthening the traditional operating process, the Canadian regulatory community faces a number of important issues that, while not uniquely Canadian, are nevertheless very challenging. Foremost among these are data ownership, water contamination, and biotechnology.

For several years Agriculture Canada has operated on the basis of a Product Specific Registration Policy. This approach ties each individually formulated product to the unique data package supporting the technical active ingredient from which it was made. Traditionally, the Canadian system had operated on a generic basis where all sources of active ingredients were considered equivalent and identical. As a result of our experience with microcontaminants, particularly in the phenoxy herbicides and trifluralin, it became apparent that this traditional approach was too simplistic.

For all practical purposes, our current position on data ownership is equivalent to that in the U.S.: a manufacturer of the active ingredient must have his own basic data to obtain registration. There is, however, a great deal of discussion at the international level regarding some additional flexibility. We are participating in these discussions and have specific proposals under review by Canadian industry and other interested parties.

As we become more active in efforts to monitor low level contaminants in water, we are uncovering new challenges. For example, we are seeing the same basic picture on aldicarb and alachlor in Canada that has been observed in the U.S. Water contamination, and how to deal with it rationally, is likely to be one of the major issues of the 1980s.

In the area of biotechnology, we are very much at the front end of the learning curve. As this new field develops, we will be obliged to develop applicable regulatory standards. Usually, regulation is lagging behind innovation. This is particularly true for biotechnology in the area of pesticides.

Assessment of Pesticide Toxicity: Regulatory Viewpoints

L. Ritter

In Canada, the primary federal legislative means of controlling the availability, sale, and use of pesticides is the Pest Control Products Act (PCP Act). It is administered by the Department of Agriculture. The Act is designed to ensure that no person shall store, display, distribute, or use a pesticide under conditions that are unsafe to human or animal health, or that will adversely affect the environment. It also ensures that such products are effective for their intended purposes.

The federal departments primarily involved in this pesticide review process are Agriculture Canada, Fisheries and Oceans, Environment Canada and Health and Welfare Canada (Figure 1).

Agriculture Canada, in the Pesticides Directorate, receives the application for registration of the pesticide and is responsible for the evaluation process and the coordination of reviews from the other agencies. The Research Branch of this department contributes by developing information regarding the effectiveness, crop safety, crop residue, and soil residues of these new products. Undesirable levels of these chemicals must naturally be avoided in the crops, soil, water, etc.

Fisheries and Oceans Canada and Environment Canada are concerned with the effects of pesticides on fisheries, aquatic ecosystems, wildlife, etc., as well as the efficiency of these products in controlling forest insects and disease, and the prevention of contamination of the environment and the adequacy of disposal instructions provided on the labels.

Health and Welfare Canada is concerned with the health aspects of the people involved in the manufacture, handling, or application of pesticides, as well as anyone else who may otherwise become exposed to the products. This department is also responsible for ensuring that the use of pesticides on food crops will not result in residues that may be hazardous to human health.

Both Canada and the U.S.A. conduct their own review of basic data and arrive at independent conclusions. Registration in Canada requires full evaluation of the complete data package.

Registered products are subject to reevaluation in the light of new information during the life of the product (Section 19 PCP Regulations). Once a product has been registered and becomes established in pest control practices in agriculture or elsewhere, the scope of any reevaluation process is expanded to include provincial agencies. Reevaluation involving long-registered products is a longer and more complex procedure because of the essential participation of provincial advisors.

For "domestic class" products, the eligibility levels on the noted toxicological parameters are such that inadvertent exposure by swallowing or dermal contact is not likely to result in severe effects. Environmental consequences of the use of domestic class pesticides must be slight and the disposal of the container and product by means of regular garbage collection must be without appreciable risk. While the PCP Regulations do not specify the kinds of sale outlets through which domestic class products may be marketed, some provinces control this marketing aspect by licensing requirements of dealers.

"Commercial class" pest control products are those that are registered for use in commercial operations such as agriculture, forestry, industry, and similar situations. The toxicological parameters used to judge the eligibility of a pesticide for the commercial class allow for the inclusion of more toxic

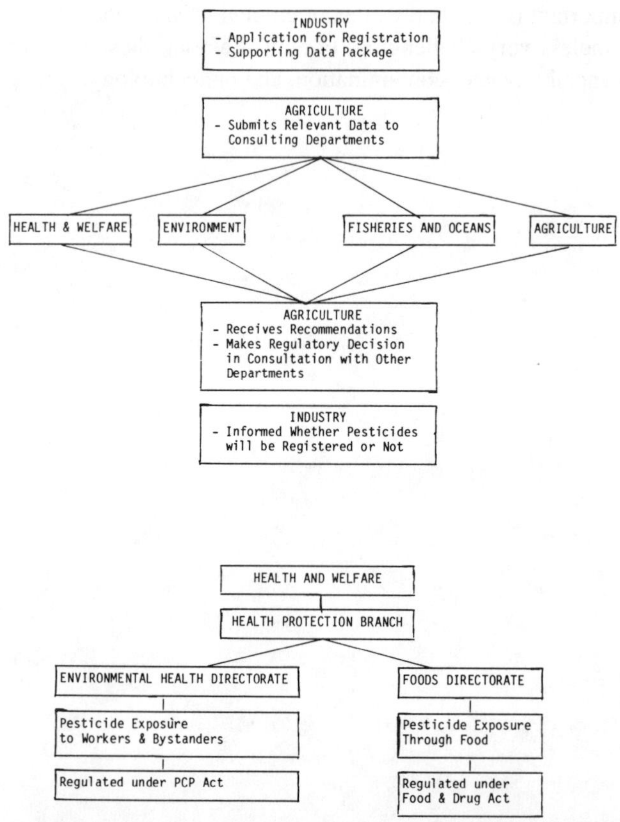

FIGURE 1. Pesticide submissions and registration in Canada.

products than in the domestic class. It is expected that commercial users are better equipped and more knowledgeable in safe practices than the private user. Further, most poisonings occur with children and imprudent storage of containers around the home; this situation does not normally occur in commercial operations.

"Restricted class" pest control products are really commercial products for which several regulatory instructions or limitations have been imposed. There is a point where label instructions on storage, handling, and use of a potentially hazardous product become inadequate to convey the need for unusual precautions in order to prevent detrimental effects to human health or environmental quality. The classification "restricted" in itself provides an alert value to the purchaser and user, but in each case the nature of the restriction is printed on the product label. The onus for compliance with the described restriction falls upon the user or the person in whose charge the product is found.

Certain products are restricted in use to persons who, by training or authority, are qualified to use them. The reason for restriction may be due to human health hazard, environmental concerns, etc.

Products registered for use in aquatic or forestry situations are classed as restricted. This limitation is imposed without regard to toxicological characteristics of the product so registered. The reason for this arbitrary assignment is that generally both the aquatic and forestry environments are in a natural state with a full complement of flora and fauna, some of which may be sensitive, and good operating practice must be ensured. Conversely, most agricultural situations involve greatly modified and simplified environments that can tolerate greater latitudes in pesticide use.

Aquatic and forestry control products may also be regulated at the provincial level by means of permit requirements under provincial legislation.

In addition to the classification scheme for a pest control product that is largely dependent upon toxicological parameters, there is a coordinated precautionary labeling scheme that is also dependent upon toxicological parameters. Precautionary labeling consists of series of statements, signs, and symbols that inform the purchaser and user on the nature and degree of hazard inherent in a particular product.

The precautionary labeling scheme for pest control products is described in the "Registration Guidelines", referred to earlier, that are included in the Registration Kit. The scheme recognizes four levels of risk potential represented in descending order of hazard by the words "DANGER", "WARNING", and "CAUTION", with the fourth category requiring no designation. The cautionary words are accompanied by international traffic symbols, again in descending order, consisting of the octagon stop sign, the diamond caution sign, and the inverted triangular yield sign.

Within the encircling stop, caution, and yield signs is a symbol that indicates the kind of risk potential. Four kinds of risk are recognized: the skull and crossbones for poisoning by ingestion, dermal, or inhalation routes; a flame symbol in respect to flammability; an exploding sphere in respect to explosivity; and a skeletonized hand in respect to corrosivity.

Pest control products are registered for sale and use by a comprehensive review process by various government departments. The information reviewed by these departments is, of course, provided by the manufacturer of the particular pesticide, and industry states that it takes 5 to 7 years to develop at a cost of up to $20 million. This review process is a means of demonstrating the effectiveness of the product and its safety to man and the environment, before it is available on the market.

In order for the safety of a product to be assessed, toxicity testing must be carried out. There are several kinds of studies performed: (1) acute studies (single exposure), (2) short-term studies (multiple exposure — 10% life span), (3) long-term studies (multiple exposure — life span), as well as (4) special studies (Table 1).

The parameter LD_{50} is frequently used in assessing the acute hazard of a compound. This refers to the dosage at which 50% of the animals die after being given a single dose. It is, however, also important to know what the long-term effects might be. Toxicity testing is conducted on more than one species of animal, usually rodents and dogs. Man is considered to be the most sensitive species for the purpose of hazard evaluation. One can see, however, the difficulty in testing this, as well as in extrapolating animal experiments to fit the human model.

It is essential that we have some indication as to the amount of product to which the user might be exposed. Currently, exposure models are being developed in the Pesticide Divi-

TABLE 1
Tests Required for Registration

Acute Toxicity (Technical and Formulations)
LD_{50} — oral, dermal, inhalation
Irritation — dermal, eye
Sensitization
Delayed neurotoxicity
Short-Term Toxicity (Technical and Formulations)
90-d oral (rat and dog)
12-month oral (dog)
90-d dermal (when indicated)
90-d inhalation (when indicated)
Delayed neurotoxicity (if acute test positive)
Long-Term and Special (Technical)
Chronic feeding (rat)
Oncogenicity (rat and mouse)
Pharmacokinetic (appropriate routes)
Mutagenicity
Teratology
Multigeneration reproduction
Exposure

sion to predict the exposure. These models show promise of being extremely useful for the registration process.

Protective clothing, including respirators, is only useful when worn correctly. Research is currently being done in many places concerning the design and comfort of clothing. There is evidence to show that unless the worker is going to be drenched during application, cotton or cotton/synthetic derivatives afford equal protection to rubberized protective suits.

Gloves are one of the most widely misused pieces of protective clothing. Regardless of the material, a glove replaced on a hand that has been contaminated may only serve to increase the dermal penetration. Studies done in California have shown that gloves can significantly reduce hand exposure when properly worn.

Respirators are uncomfortable. That is one of their major disadvantages. Also, many users appear to be unaware of the necessity of ensuring a good fit. Beards and respirators are not compatible. A new respirator designed primarily for dust situations may soon be available with a chemical cartridge. This unit incorporates a face shield with gasket attached to a hard hat. Air is pumped over the face by a small battery-driven pump. The two important features of this unit are comfort and coverage of the complete facial area, since facial exposure can be quite significant.

One of the greatest challenges facing all of us is to convince the user to use protective equipment properly. Another is to provide equipment that is both effective and comfortable.

In order to make a hazard assessment, it is essential that we have information on both the toxicity and exposure. These two pieces of information can be correlated in the following way. Negligible daily exposure (NDE) is assessed for animal toxicity studies: NDE = NOEL (no observed effect level)/safety factor. Theoretical maximum daily exposure (TDE) can be assessed from exposure studies. If TDE is smaller than the NDE there would be no apparent hazard for using the product as directed.

A similar calculation is used to set tolerances for food use. Negligible daily intake (NDI) = NOEL/safety factor and theoretical maximum daily intake (MDI) = maximum residue × consumption. The product is acceptable for food use if TDI is smaller than NDI.

In summary, the various steps involved in the registration of pesticides have been discussed and the relationship between the predictive toxicology as determined through animal testing and exposure estimates for estimating hazard has been shown.

The Public Role in Pesticide Registration

L. Salter

INTRODUCTION

A major change has recently been made in the assessment and registration of pesticides in Canada. It was instigated at the demand of advocate and industry groups, and at the insistence of the provincial authorities. Its purpose was to avoid the confrontational relations that had characterized pesticide registration for many years, and to help resolve public controversies over specific pesticides. The change was also designed to deal with complaints that the previous process had made informed participation difficult.

This change will be successful in dealing with complaints and resolving controversies only if it is accompanied by a sound understanding of the role of the public and the media in such controversies. The purpose of this paper is to provide the basis for such an understanding. It is based on the study conducted for Agriculture Canada that resulted in the changes to the registration process,[1] and a 2-year study of science and public policy, which included research on two standard setting organizations and four chemical controversies in Canada.[2]

THE RELATIONSHIP BETWEEN SCIENCE AND PUBLIC POLICY

Before analyzing the role of the public and media, it is necessary to have a picture of the relationship between science and public policy that both seek to influence. There are two possible ways to conceive of this relationship. One relies upon formal risk assessment procedures and the design of institutions related to them. In this view, the relationship is a rational one, and thus this approach can be called "rationalist". According to the rationalist perspective, a scientific assessment should precede a policy review and be conducted according to the canons of the scientific community. A policy assessment should focus on questions of values and the allocation of costs and benefits.

The other approach questions the rationality of any risk assessment and, more particularly, of chemical controversies. It conceives of risk assessment as involving mixed disputes, disputes that are simultaneously truth-seeking and justice-seeking in orientation. I will call this approach "dispute-centered", because it begins with a sociological analysis of actual chemical controversies, analyzing them as particular kinds of disputes. The emphasis in the dispute-centered approach is upon identifying the problems in resolving disputes, and seeking for ways to handle them. Usually this involves a case-by-case attempt to prevent and resolve controversies, rather than the adoption of formal procedures for the assessment of particular chemicals. I will deal with each approach in turn, first explaining how the relationship between science and public policy is conceived in it, and then exploring how it locates the role of the public and media groups in chemical controversies.

In the rationalist approach, it is possible to make decisions on the basis of scientific information about the toxicity of a particular chemical and to relay those judgments to policy makers and regulators.* Any one chemical might be more or less toxic, and the information to judge its toxicity is likely to be inconclusive or incomplete. The job of the regulatory scientist or the expert committee is to make judgments in the face of uncertain, inconclusive, or conflicting data on the basis of the best available data. Once these judgments are made, it is the task of the government official or commissioner of a public inquiry to evaluate the costs and benefits of each policy initiative, and to reach conclusions about the most practical and beneficial ones.

Within this rationalist perspective, neither the public nor the media has a role to play in the first stage of the assessment, the scientific assessment. Both laypeople and media lack the background expertise to make the necessary judgments, or even to interpret the information with adequate regard to the tenets of scientific inquiry. The public has a role to play, however, in the second stage of the assessment, since this assessment includes an evaluation of the value choices inherent in different public policies. Members of the public can indicate whether they are likely to suffer from the implementation of particular regulations — or their lack of implementation. They can provide guidance to regulators in the same manner that they provide assistance to legislators in determining the appropriate priorities for policies and allocation of resources. This role is important, because regulations are no different from governmental policies generally, and because members of the public are likely to know of specific costs and benefits that otherwise might not be identified by the regulators.

From the rationalist perspective, the role of the media in the relationship between science and policy is less clear. Like the members of the public, the media have no role to play in a scientific assessment, except to acclimatize members of the public to the methods of scientific reasoning and to provide enough background information so the public will know what is to be decided. Since only a few media devote significant resources to scientific reportage, media coverage of scientific

* The term "toxicity" is used to mean all aspects of chemical activity likely to cause ill effects on human health and safety.

assessments is often inadequate. As a result, media coverage is likely to generate controversy rather than informed opinion. Indeed, the existence of public controversy is often taken as a signal that the media have failed to provide adequate and evenhanded scientific information, and that they have inflamed an ill-informed public as a consequence.

The media's role in the policy debate is more complicated. On one hand, informed public debate about policy options requires dissemination of information through the media, and possibly some coverage of the different perspectives in debate. On the other hand, providing this background coverage about policy options is likely to promote conflict or controversy, simply because a news story is often written to focus on conflict and controversy. From a rationalist perspective, then, the media should play an essential role in generating informed public debate about policy options, but in practice the media often contribute to the failure of the debate and, as a result, to public controversy.

If the rationalist perspective focuses on the negative aspects of public controversy in the relationship between science and public policy, the alternative perspective takes controversy for granted. In the dispute-centered approach, controversy instigates a review of particular chemicals — often a necessary review — and infuses all of the judgments that are made as a result. There are several reasons why the actual process of risk assessment should be characterized as dispute centered. First, public debate does not take place according to the rules and procedures set by administrators. There are usually several trajectories of decision making in any chemical assessment, and several locations for public controversy over a specific chemical. Controversies are often unpredictable with respect to their location, participants or content.

Second, the articulation of scientific issues in the context of a situation where interests are being pursued vigorously means that scientific information will be used as a strategic resource by all parties. For example, when an inquiry is commissioned, the companies and their legal counsel have a duty to act in the inquiry in terms of their legal interests and to ensure that issues of liability and negligence are dealt with in a manner favorable to themselves. Similarly, government officials must defend past departmental or regulatory practices, even if they later agree to their revision. Only the lay participant can discuss the matter at hand without reference to political or economic interests, and these participants are likely to have the least access to resources for an independent scientific assessment. Most often, laypeople are dependent upon the work done by the companies or government on their behalf, and their participation is limited to a critique of the material before the inquiry. In such a situation, participation by all parties is likely to be interest based and defensive in tone.

Third, the dispute-centered perspective also lays stress on the way language is used in scientific controversies. Statements made in a scientific community about the status of particular scientific evidence have quite a different meaning when addressed to a lay audience or an audience of policy makers. For example, it would be scientifically accurate and unexceptional for a group of expert scientists to indicate in a particular situation that they cannot pinpoint the source of pesticide residues with any degree of certainty. To make the same statement in a regulatory context is to imply that no regulatory or legal action should be taken because a specific problem has not yet been identified. In disputes about chemical toxicity, often statements have a different significance depending upon the audience being addressed. The duality of language confounds the rationality of the risk assessment process.

The dispute-centered perspective is based on a perception of chemical controversies as positive developments in their own right, and as catalysts for the review of particular chemicals. From this perspective, each dispute should be seen as containing elements of truth-seeking (a scientific orientation) and justice-seeking (a value debate), but the two aspects of the dispute cannot be separated easily nor considered sequentially. The existence of controversy means that the various interest groups each will seek platforms for their views, including that of the various expert committees.

Expert committees are commissioned to deal with the truth-seeking or scientific aspects of a chemical assessment. From a dispute-centered perspective, these expert committees are not very successful in keeping the two spheres of assessment — scientific and regulatory — separate. Even though they intend their work to be scientific, expert committee members are engaged in deliberating the regulatory options because they must make judgments in the face of insufficient information and consider how these judgments will affect an industry and the public. The mixed character of their deliberations is furthered by their use of language, which is itself infused with interests. The terms they use have strategic importance in the hands of those who use their reports; their scientific statements have regulatory and legal implications when they are directed to a nonscientific audience.

In a dispute-centered approach, the role of the public is an important one. First, as has been noted, the public is mainly responsible for instigating the review of particular chemicals. Even though the public perceptions of risk may be inaccurate, these perceptions result in resources being allocated for chemical reassessment, and for a higher priority being placed upon the evaluation of specific chemicals. Second, public concern is likely to influence the scope of any research undertaken or commissioned as a result of the controversy. Third, the public functions as if it were an interest group, and its interest is reflected along with others at every stage of the assessment. Finally, members of the public are likely to have information about specific problems related to chemical use that have not come to the attention of regulators. The public inquiry that often accompanies one or both stages of a risk assessment is designed to ferret out these problems, even if it does so in an unsystematic manner.

What is the role of media, as seen from a dispute-centered approach? The answer is a complex one and can only be derived from the study of specific disputes. It is to this research that I now turn.

THE ROLE OF THE MEDIA IN A DISPUTE-CENTERED APPROACH

In the rationalist model, media coverage was intended to provide the basis for informed public opinion, but more often it resulted in public controversy. In contrast, in the dispute-centered approach, public controversy and media coverage are independent variables. Of course, the generation of controversy is often preceded by media coverage. What was surprising, and important, from the research that we have conducted on chemical controversies is how small an amount of media coverage is required to instigate public or regulatory concern about a specific chemical. In one of our case studies, two articles in a newspaper from a prairie province were sufficient to provoke a major response on the part of government, industry, and advocate groups.[3] Media do provide coverage of chemical accidents, of conferences and scientific reports, but this coverage does not, in itself, generate controversy. Indeed, given the narrative structure of the average news story, some conflict must normally be present for a conference or scientific report to be reported at all. Reportage of conflict is not sufficient to provoke widespread public concern, since almost all news stories focus on conflict.

In our research, we found that the usual development of controversy was as follows. Media coverage was instigated when the quasi-scientific press — *Science, New York Times,* etc. — published reports from either scientific conferences or from advocate groups or the courts. Alternatively, the mass media reported on specific regulatory, legal, or legislative decisions, or on catastrophes. This initial coverage was often of a limited nature, and seldom focused on or furthered controversy.

However limited, the initial coverage was sufficient to have some important effects. First, the well-established advocate groups read such reports and used the initial media coverage as a signal that they should develop a posture and related arguments with respect to the issue being discussed. Second, various interest groups used the media coverage as a signal that action was necessary to protect the interests of their members. Finally, government departments and regulators, which normally monitored media coverage, used it as an indication of when special attention was required.

It was the actions of these three groups — the advocates, the interest groups, and various government departments — that called a controversy into being. In response to the initial media coverage, each group positioned itself both in terms of its own past actions and in terms of any arguments it might have been required to provide. The groups also positioned themselves in relationship to each other and, in doing so, provided the basis for conflict and controversy among themselves. In other words, controversy was instigated when various groups began to conceive of themselves as protagonists and act accordingly. Controversy was furthered as each group developed not only arguments, but also a defense of its position in relation to others. This controversy was heightened immediately if there was more than one group involved from each sector. For example, the controversy was furthered if one advocate group took a position that another found troublesome. It was furthered if more than one interest group became involved, or if two or more government departments or agencies shared jurisdiction and were in conflict with each other.

We observed that this activity by the various groups in relation to each other was usually sufficient to generate more extensive media coverage. Each group brought the news to the media, feeding reporters with information, examples of scientific reports, and arguments in defense of their position. As is typical in news reporting, a great deal of this unsolicited information was used as news. As a result, new *ad hoc* advocate groups became vocal, and a more formal response from the various interest groups was forthcoming. The pressure from the advocate sector, and the formal participation of various interest groups was usually enough to provoke a series of government initiatives designed to resolve the now-significant controversy. These initiatives included special committees or consultations, public hearings, special research projects, and even full-scale inquiries, each of which provided a platform and an opportunity for the public airing of the controversy. At this stage of the controversy, news did not have to be "fed" to the media, but resulted from the public hearings, release of documents, statements by government officials and advocate groups, and committee reports themselves. These are "news-staging" events that are almost invariably reported in terms of the disputes they reflect.

In summary, seen from a dispute-centered approach and on the basis of our research, we would argue that media coverage was an important component of a chemical controversy, and indeed also of risk assessment. Media coverage was also not necessarily accurate coverage, and it did tend to concentrate on the conflicts among the various protagonists in the debate about particular chemicals. At the same time, it would be wrong to conclude that controversy was the result of media coverage, or even of irresponsible media coverage. Media coverage played an important, but not very controversial role in instigating the concern of advocate, interest, and government groups. The media subsequently reported upon conflicts between or within each of these groups, and was a repository for the arguments advanced by each group in defending its position. The media played their most important role in the last stages of a public controversy, when *ad hoc* groups became involved and when inquiries or expert committees were commissioned and reported their findings. Just as the inquiries provided a platform for the expression of conflicting views, the media became a platform for the protagonists in the controversy.

SCIENCE AND PUBLIC CONTROVERSY

If one adopts the dispute-centered approach, it is possible to identify the character of the relationship between science and public controversy. For example, as a controversy develops, the scientific assessment becomes increasingly infused with adversarial and interest-based conflicts. The various groups take a position in the debate in reference to each other.

Their arguments are developed to respond to the positions of their protagonists. In the process, their view of science is altered. Scientific information and research become instruments of argumentation within and between groups. Studies are cited to prove that another group's position is wrong. Groups seek out evidence from the disciplinary literature to support their views. In reviewing material they may have evaluated earlier, they draw a more conclusive and coherent picture, one that is consistent with the main thrust of their argument. If an inquiry is held, scientists are used by one party or another as expert witnesses. In such cases, scientists are interviewed by the interested parties before they are asked to testify, and the scientists who do testify are unlikely to muse about their contradictory findings.

The development of controversy also has an effect upon how scientific information is reported by the media. Once groups have become positioned in an adversarial debate, they offer to the media both information and a coherent picture of how that information should be understood. With the information and the coherent picture, reporters also get an adversarial perspective on the issues involved. Having no basis to question the information they are given, most reporters print the adversarial information as if it were based exclusively in science. If these reporters are then told that there is another side to the story, they print it, but describe the scientific debate in terms of its controversies rather than its research. The final result is media coverage that does provoke more public controversy, without necessarily furthering the level of public sophistication about the scientific issues involved.

There is another aspect to the relationship between science and public controversy. This was most evident in our research when we found that the debates about different chemicals took on a "set piece" appearance. In other words, the same arguments were made in a predictable fashion when different chemicals were being evaluated through public hearings. In order to deal with this phenomenon, we introduced a new concept, the "political chemical".[4] A political chemical is one that has become a lodestone for a more general debate about chemical toxicity or about the regulatory process. When chemicals become political chemicals, it is very difficult to advance specific evidence about their toxicity without, at the same time, engaging in a more general discussion about how toxicity should be regulated.

In the assessment of political chemicals, the same themes reoccurred. For example, questions about the suitability of different species were raised and debated in its most general terms. The options for regulation were presented as a series of clear-cut choices, in the same manner as the scientific material was seen to answer questions in unambiguous ways. Scientific and regulatory judgments were often combined, as the inconclusiveness of data was used as an introduction to a discussion about various regulatory options. Questions about the orientation of regulation and the trustworthiness of regulators were raised as part of the scientific assessment. At least some participants claimed that the issue to be determined was one of values or ethics, not science.

The result of the creation of political chemicals was a debate that could have occurred about any chemical, regardless of its particular mechanism of action within the human body. Risk estimates were commonly used, not so much to demonstrate how much concern was justified about a specific chemical, but as a way of comparing one chemical to another in a chemically dependent world. Recommendations emerging from the assessment of a political chemical were usually focused on regulatory procedures, or procedures for risk assessment, rather than on the specific conditions under which the chemical might be registered or not. The general tenor of the discussion, and its focus on regulation, compensated for the gaps in data that were available for assessment.

We found that political chemicals were created by a number of factors. First, they were the result of serendipity, the chance of occurrence of some factor that caused a chemical to be suitable for more general discussion. Second, jurisdictional complexity had an important effect in creating a political chemical, because disputes between different departments or levels of government compounded the issues to be determined, and underscored problems arising from the allocation of regulatory responsibility. Third, the involvement of several different advocate or interest groups increased the likelihood of a political chemical, as the groups used their different perspectives on regulation as a means of differentiating their positions. Finally, political chemicals were created when the issues about toxicity were relatively simple ones, when, for example, a chemical was clearly dangerous or had been in use for a considerable period of time. The unambiguous toxicity made it possible for laypeople to avoid highly technical debates about scientific information. At the same time, it made it possible for the manufacturers or users of the chemical to stress the inevitability of risk in a chemically dependent world. The revelation that a chemical in use for some time might be dangerous gave rise to questions about why the danger had not been identified earlier.

Political chemicals are significant for the pesticide registration process, because once they are created they alter the nature of the assessment. A number of factors having little relevance to the specific chemical at hand are brought into the debate, and they demand a response from the regulatory authorities. Moreover, both the public and the media play a more significant role in the assessment of political chemicals than they would otherwise, since regulatory issues are much more accessible to the laypublic, and much more conducive to extensive media coverage than scientific debates. At the same time, it is important to note that the media do not create political chemicals, and that their creation is often at the instigation of both interest and advocate groups. In effect, the creation of a political chemical gives rise to many strategic opportunities for all the interest and advocate groups to influence public policy. The result is not always a negative one, although it does deflect attention from the specific chemical under assessment.

THE PESTICIDE REGISTRATION PROCESS

In the past few years, major changes have been introduced by Agriculture Canada to deal with pesticides. Several are particularly relevant to this analysis. First, Agriculture Canada has set up an information secretariat, published background information on specific pesticides, and instituted a "call line" as means of providing information about pesticides in a timely manner. Second, Agriculture Canada has increased the resources for and profile of its pesticide registration activities, and established a mechanism to handle controversies over specific chemicals. In addition to these formal procedural and administrative changes, Agriculture Canada, along with several other departments handling decisions about potentially dangerous chemicals, has adopted formal risk assessment procedures. Finally, great interest has been shown in various government departments in problems of "risk communication". In light of the analysis provided above, how should one evaluate the potential for success of these changes?

It is too soon to judge the success of Agriculture Canada's information program, but the scope of its initiatives with respect to the provision of information is impressive. Advocate and interest groups and provincial officials can no longer claim to be "left in the dark" about the reasons for registration decisions. The heightened profile of the registration activities with Agriculture Canada responds to well-defined needs for a speedier and more rational registration process. Ironically because it adds new regulatory officials, it has resulted in less red tape and fewer instances of arbitrary decision making. The conflict-resolution approach adopted by Agriculture Canada is still being tested, but some of the problems remain because of the jurisdictional complexity of the relationship between assessment — a Federal responsibility involving several departments, registration — the responsibility of Agriculture Canada, and implementation — the responsibility of provincial authorities.

In spite of its interest in and adoption of risk assessment procedures, there is little evidence that government officials in Canada have adopted the rationalist perspective. In part, the jurisdictional complexity of decisions about pesticides in Canada precludes any simple, sequential process of risk assessment from being implemented. Too many departments within government must be consulted before a specific decision is taken. Instead, departmental officials seem to have opted for the dispute-centered approach. As a result, the new initiatives are designed to support the public's role as catalyst for chemical reassessment and to provide information before groups have adopted a stance in relation to particular chemicals.

At the same time, some officials are expressing interest in risk communication, an approach that often — although not always — is associated with a rationalist perspective. From a rationalist perspective, it is to be expected that risks will be inaccurately perceived, for the public is an outsider to the assessment process and the media are likely to be a negative influence in providing accurate information. The role of the risk communicator is to monitor media coverage so as to prevent public controversy, and the public perceptions of risk so as to correct them. From the rationalist perspective, public perceptions of risk constitute a problem to be resolved, and controversy exerts a negative influence upon regulatory and policy decisions. Risk communications need not be undertaken from a rationalist perspective, of course. If it is undertaken with an adequate understanding of chemical disputes and of the role of both the public and the media in those disputes, it is simply another term for increasing the capacity of the regulatory process to be responsive to the demands of its many publics.

REFERENCES

1. **Salter, L. and Leiss, W.,** Consultation in the Assessment and Registration of Pesticides, Agriculture Canada, Ottawa, 1984.
2. **Salter, L., Leiss, W., and Levy, E.,** *Mandated Science: Science and Scientists in the Creation of Standards,* Reidel, Amsterdam, 1987.
3. **Salter, L.,** Observation on the politics of risk assessment: the Captan case, *Can. Public Policy,* 11, 64, 1985.
4. **Salter, L.,** *Mandated Science,* chap. 5.

Magnitude and Source of Airborne Residues of Herbicides in Saskatchewan

R. Grover

INTRODUCTION

In the past three decades, the use of herbicides in the Canadian Prairies has increased over eight-fold, from less than 2 million kilograms in the late 1950s to over 17 million kilograms annually by the early 1980s. Herbicide use patterns in western Canada for the year 1984 show Saskatchewan as the major user of these herbicides (Table 1). The greater part of this use occurs from mid-June to mid-July of each year as postemergence treatments to control emerged weeds in crops. A significant portion of wild oat herbicides, such as triallate and trifluralin, are also applied as preemergence, soil-incorporated treatments and these are applied in early to late May of each year.

Herbicides, as well as other pesticides, can enter the atmosphere by a variety of routes such as droplet drift during application, postapplication vaporization from target areas, such as soils, crops/weeds, water bodies, etc., and as "blow-off" of treated soil particles, especially during dust storms (Figure 1).[1] This presentation summarizes our current knowledge of airborne residues of herbicides in the prairie atmosphere, especially that in Saskatchewan, and discusses some of the major inputs and/or sources.

TABLE 1
Herbicides Used for Agricultural Weed Control in Western Canada and in Saskatchewan (1984)[a]

Herbicides	Western Canada (kg)	Saskatchewan (kg)
Broad leaved weeds		
2,4-D (ester and amine)	3,431,300	2,092,200
MCPA (ester/amine/salt)	1,542,600	534,600
Bromoxynil/dicamba	2,209,200	1,018,500
Other herbicides[b]	1,570,100	525,500
Total	8,753,200	4,170,800
Grassy weeds		
Wild oat herbicides[c]	8,455,400	3,564,700
Grand total	17,208,600	7,735,500

[a] 1984 Herbicide Use Survey, Statistics Section, Manitoba Department of Agriculture.
[b] Dichloroprop, mecoprop, linuron, glyphosate, propanil, picloram, atrazine, etc.
[c] Asulam, barban, EPTC, flamprop-methyl, diclofop-methyl, difenzoquat, triallate, trifluralin, etc.

LEVELS OF AIRBORNE RESIDUES OF HERBICIDES IN SASKATCHEWAN

Air monitoring studies have established the presence of airborne residues of herbicides, such as 2,4-dichlorophenoxyacetic acid (2,4-D) esters, triallate, and trifluralin, both during and immediately following the spraying season.[2-4] Levels of 2,4-D butyl esters varied from <10 to 13,500 ng/m^3 in samples collected at various sites in Saskatchewan during the period 1966—1975, with confirmation of residues by various techniques in 330 of 1350 samples collected (Table 2). The low-volatile *iso*-octyl esters were also confirmed in several samples, with amounts ranging up to 590 ng/m^3. Most of these residues were detected in samples collected during and immediately following the application season, with negligible or no residues being detected thereafter.

On the other hand, airborne residues of herbicides, such as triallate and trifluralin, were detected in the air not only during the spraying season, but also throughout the growing season (Table 2, Figure 2). Airborne residues of triallate and trifluralin reached a maximum of 198 and 62 ng/m^3, respectively, when monitored at three sites in Saskatchewan during the 1978—1982 period.

Preliminary data on the monitoring of dust and rainfall samples collected within an intensive herbicide-use watershed in the vicinity of Regina indicated that dust and rain may be major carriers of herbicides in the prairies.[5] Herbicides, such as 2,4-D, dicamba, bromoxynil, diclofop, and triallate, were detected both in dust and rainfall samples at levels ranging from <0.1 to 1.5 ppb levels (Table 3).

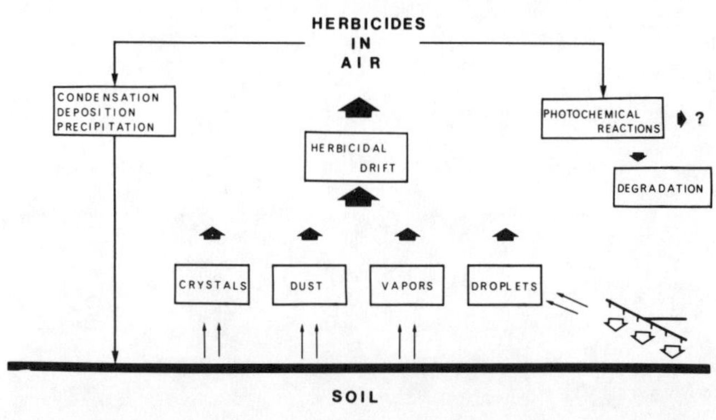

FIGURE 1. A schematic representation of herbicidal interactions in the environment, with emphasis on entry mechanisms.[1]

TABLE 2
Levels of Airborne Herbicide Residues at Various Locations in Saskatchewan[2,4]

Number of Locations	Number of Samples	Period	Concentration ranges (ng/m³)
		2,4-D-Butyl Esters	
9	1,235	1966—1975	10—13,500 (330)[a]
		2,4-D-*iso*-Octyl Esters	
4	558	1966—1975	50—590 (22)
		Triallate	
3	974	1978—1982	1—198 (826)
		Trifluralin	
Regina	239	1980—1981	1—2 (16)
Melfort	358	1980—1982	1—62 (283)

[a] Numbers over detection limits; confirmed by dual column, dual detector, or mass detector techniques.

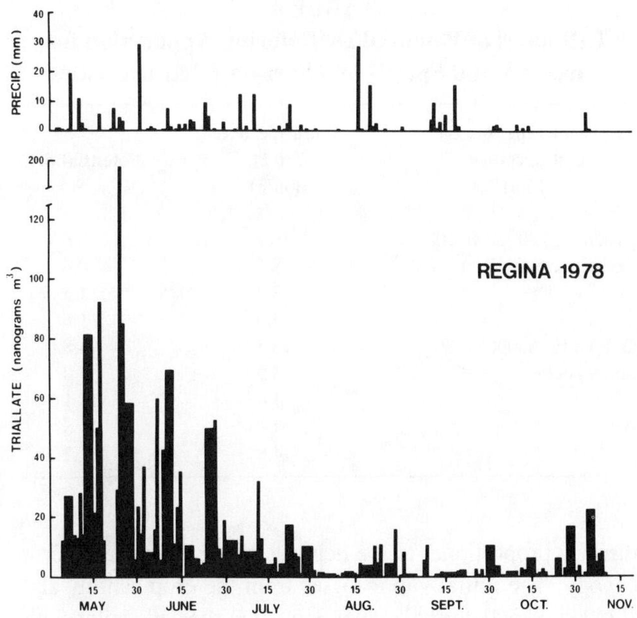

FIGURE 2. Airborne residues of triallate during the 1978 growing season at Regina, SK.

TABLE 3
Residues of Herbicides in Spring Runoff (Snowmelt), and Wet and Dry Precipitation in Samples Collected at Several Locations within a Small Agricultural Watershed near Regina, SK[5]

Sample type	Duration of sampling	ppb Dicamba	Bromoxynil	2,4-D	Trifluralin	Triallate	Diclofop
Spring runoff	March 27—April 4	<0.100 (25)[a]	<0.100 (25)	<0.100 (25)	<0.100 (25)	<0.100 (20)	<0.100 (25)
Precipitation — wet	May 7—September 4	<0.100 (16) 0.133—0.473 (6)	<0.100 (19) 0.227—0.518 (3)	<0.100 (15) 0.195—1.561 (7)	<0.100 (22)	<0.100 (15) 0.129—0.578 (7)	<0.100 (17) 0.111—0.315 (5)
Precipitation — dry	July 3—August 28	<0.100 (4) 0.100—0.272	<0.100 (7) 0.314—0.380	<0.100 (6) 0.243—0.437	<0.100 (10)	<0.100 (10)	<0.100 (4) 0.141—1.045

[a] Figures in parentheses denote numbers of samples analyzed.

SOURCES AND MECHANISMS OF HERBICIDE INPUT INTO THE ATMOSPHERE

Herbicide input into the atmosphere from agricultural use consists of (1) drift losses during application, and (2) postapplication losses, such as vapor losses from leaf and soil surfaces, and wind erosion losses.

Application Inputs

In Saskatchewan, over 90% of the herbicides are applied by ground rigs. However, drift losses, even from these ground driven systems, varied from <1% to as high as 20%, depending on (1) type of rig and the associated hardware, such as nozzles, (2) operation conditions, such as spray pressure, spray volume, etc., and (3) environmental conditions, especially wind speed.[6] Use of conventional tractor-pulled ground rigs equipped with low-pressure nozzles maintained spray drift below 1% under most operational and wind conditions (Table 4). Application drift from Spray coupes, a self-driven fast speed ground rig, varied from 5 to 15% of the amount applied, depending on the wind conditions. On the other hand, drift losses from aerial application were even higher.

Postapplication Inputs

Vapor Losses from Plant Surfaces

Several herbicides, such as 2,4-D, MCPA, dicamba, bromoxynil, etc., are applied as postemergence treatments. In such cases, major deposits occur on the vegetation canopy which includes both the crop and the competing weeds. It has been established under controlled laboratory conditions that the magnitude of vapor losses from leaf surfaces can be similar to that from glass surfaces.[7] Recent field studies, using 2,4-D *iso*-octyl ester on wheat, have demonstrated significant vapor losses under field conditions from deposits on the vegetation or crop canopy.[8] About 20% of the 2,4-D ester was lost to the atmosphere as vapor over a 5-d period following application (Figure 3). The daily accumulative losses were

TABLE 4
Estimates of Potential Drift during Application for Various Wind Speeds and Sprayer Characteristics[6]

Sprayer type, nozzle, and pressure (kPa)	Wind speed at 2 m ht (km/h)	Potential drift (%)
Ground rig 8002 LP, 100	16.4	0.6
Gound rig 8002, 210	8.5	0.6
	13.7	1.2
	22.0	1.6
Ground rig 650067, 280	16.5	4.8
Spray coupe	3.9	5.0
	21.6	15.1
Aircraft	7.9	14.8
	14.8	31.3

directly proportional to the deposits of the ester on the crop canopy. The initial vapor losses from the crop canopy also showed typical diurnal cycles, similar to the daily solar cycle (Figure 3). Two distinct processes were recognized in this study. The first process was the initial vaporization of the herbicide ester mass from the leaf surfaces, while the second one was the vertical transport of this vapor mass away from the leaf surfaces. Surface temperature was recognized as the controlling factor in the former process and, hence, the diurnal cycles, whereas turbulence was responsible for the vertical transport away from the surfaces.

Vapor Losses from Soils

Herbicides are also applied directly to the soil as preemergence treatments and either left exposed on bare soil or incorporated into the top 5 to 10 cm of the soil layer. A significant proportion of postemergence treatments also deposits on bare areas between crop rows or may reach the soil after application by "wash-off" of residues from leaf surfaces during a rainfall event.

The dominant factors governing the volatilization of herbicides from the soil surfaces are the magnitude of their adsorption or "binding" onto soil colloids, and their solubility in soil water. Field vapor flux estimates of herbicides such as 2,4-D ester deposited on soil surfaces have shown that there was a complete suppression of vapor flux during the solar noon (early afternoon) when soil surfaces became extremely dry, which occurs commonly under Saskatchewan conditions.[8] These studies also showed that there were sudden increases in vapor flux of the ester following rainfall or even following dew deposition on dry soil surfaces in the early hours of the morning.

Herbicides such as triallate and trifluralin are incorporated

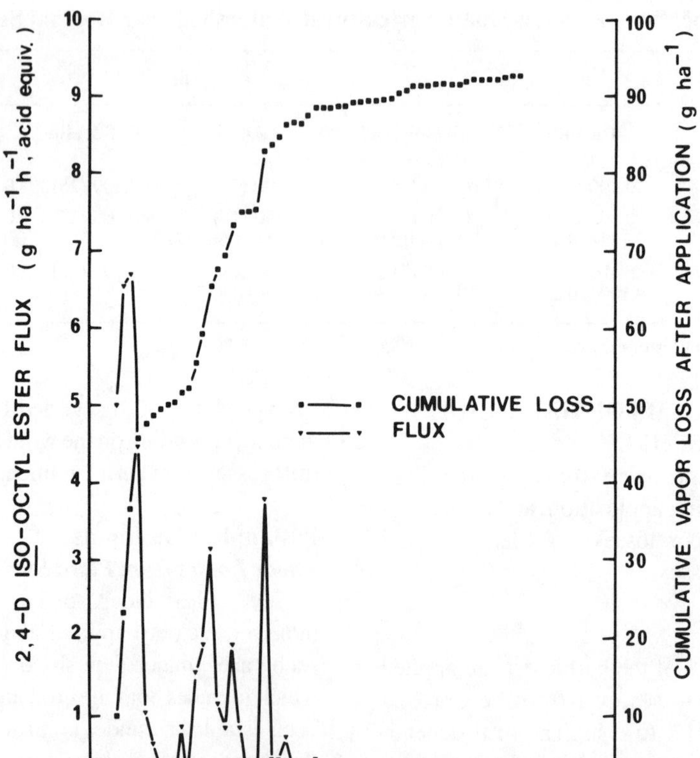

FIGURE 3. Volatilization rate and cumulative losses of 2,4-D *iso*-octyl ester over 5 d after application.

into the soil to reduce vapor losses by distribution or exposure to a comparatively large soil mass and, thus, to increase their longevity and presumably also their efficacy. However, even after incorporation, vapor losses of herbicides such as triallate and trifluralin were shown to be in the 15 to 20% range over a growing season.[2] When soil surfaces were relatively moist following their application, the initial vapor losses were rapid, and >50% of the total vapor losses occurred within a few days after application. Once the surface layers of the soil were depleted of the herbicide(s), subsequent losses were usually low but continuous over the entire growing season. In the latter case, adsorption to soil and water solubility was the major determinant factor in their upward transport to soil surfaces and subsequent vaporization.

In conclusion, it may be stated that readily detectable levels of airborne residues of herbicides, such as 2,4-D esters, triallate, trifluralin, etc., were present in air samples collected during and following their application in the prairies. Very little is known as to their ultimate fate in the atmosphere. Some of these residues are no doubt transported outside the region with perhaps appreciable portions reentering the soil/water bodies in the region via the deposition of dust and/or rainfall. The phasing out of the high volatile butyl esters of 2,4-D in the 1970s has greatly diminished the airborne residues of this herbicide. Finally, it may be stated that the current toxicological techniques are unable to demonstrate the effects, if any, of annual exposure over long periods of time to airborne residues of herbicides at levels present in the Canadian prairies.

REFERENCES

1. **Grover, R.,** Herbicide entry into the atmospheric environment, *Chem. Can.,* 26, 36, 1974.
2. **Grover, R.,** Transport of wild oat herbicides in the environment, in *Wild Oat Symp.,* Smith, A.E., Ed., Proc. Canadian Plains Research Center, Regina, Saskatchewan, 1983, 119.
3. **Grover, R., Kerr, L.A., and Khan, S.U.,** Multidetector gas chromatographic determination and confirmation of airborne triallate residues in Saskatchewan, *J. Agric. Food Chem.,* 29, 1082, 1981.
4. **Grover, R., Kerr, L.A., Wallace, K., Yoshida, K., and Maybank, J.,** Residues of 2,4-D in air samples from Saskatchewan, 1966—1975, *J. Environ. Sci. Health,* B11, 331, 1976.
5. **Grover, R., Kerr, L.A., Waite, D., and Sommerstad, H.,** South Saskatchewan Watershed Study, unpublished data, 1984.
6. **Maybank, J., Yoshida, K., and Grover, R.,** Spray drift from agricultural pesticide applications, *A.P.C.A. J.,* 28, 1009, 1978.
7. **Quee Hee, S.S. and Sutherland, R.G.,** Volatilization of butyl esters of 2,4-D from Pyrex and leaves, *Weed Sci.,* 23, 119, 1972.
8. **Grover, R., Shewchuk, S.R., Cessna, A.J., Smith, A.E., and Hunter, J.H.,** Fate of 2,4-D *iso*-octyl ester after application to a wheat field, *J. Environ. Qual.,* 14, 203, 1985.

Immunotoxicity of 2,4-D

B. R. Blakley

INTRODUCTION

Herbicides such as 2,4-dichlorophenoxyacetic acid (2,4-D) are used in large quantities throughout the world to control broad leaf weeds. Accidental exposure of livestock or occupational exposure of man to 2,4-D or its derivatives may occur, although levels exceeding 300 mg/kg in livestock[1,2] and 67 mg/kg in man[3] are usually required to produce clinical manifestations of toxicity. This relatively low toxicity is observed with 2,4-D compounds even though it is well absorbed from oral, respiratory, or dermal routes of exposure.[4,5] It is thought that rapid urinary excretion accounts for the low toxicity.[4] The potential for toxicity may be enhanced if highly volatile butylesters of 2,4-D are utilized, since greater chemical drift and environmental contamination may occur with these 2,4-D derivatives. "Flu"-like symptoms have often been reported by farmers subsequent to spraying with these derivatives. Some of these symptoms may involve an immunological component. Studies with 2,4-D derivatives have not investigated their impact on the immune system.

The purpose of this study was to evaluate the effects of oral exposure to the *n*-butylester of 2,4-D on antibody production and lymphocyte mitogen responses in mice at several exposure levels. Details of the 2,4-D ester preparation, animal exposure to the *n*-butylester of 2,4-D, and the immunological studies including lymphocyte mitogen and antibody production have been reported previously.[6]

RESULTS

Clinical and Pathological Effects

Symptoms of depression were observed for 24 h in mice administered 200 mg/kg 2,4-D (butylester), although no evidence of myotonia or gastrointestinal disturbances was seen. For mice in the other acute treatment groups, 0, 50, and 100 mg/kg appeared to be clinically normal. Evidence of perivascular edema and ganglial cell necrosis in the brain and spinal cord was seen in tissues from mice exposed to 200 mg/kg 2,4-D (butylester). In the subacute study, clinical symptoms or histological evidence of toxicity were not observed. Weight gains were not affected during the 3-week treatment period.

Antibody Production

Acute oral 2,4-D ester exposure stimulated the primary immune response against SRBC. All levels of 2,4-D ester exposure produced immunostimulation ($p = 0.003$). The maximum immunostimulatory effect was observed in the 100-mg/kg treatment group (240% of the corresponding untreated group). The primary immune response against SRBC was not altered by subacute 2,4-D ester exposure.

Mitogen Studies

Lymphocyte mitogen responses were altered by acute 2,4-D ester exposure. Acute exposure produced a dose-dependent enhancement of DNA synthesis induced by lipopolysaccharide (LPS), a B-lymphocyte mitogen ($p = 0.023$), but had a small dose-dependent stimulatory effect on T-lymphyocte mitogen responses induced by concanavalin A (Con A). This reflects the weak mitogenic activity of 2,4-D.

Subacute 2,4-D ester exposure does not produce a similar dose-dependent enhancement of DNA synthesis. The lymphocyte responses indicated that treatment effects were observed with the LPS ($p = 0.003$) and the no-mitogen ($p = 0.007$) groups. Maximum responses, enhancement in each instance, were observed in the intermediate exposure groups. No enhancement was observed in the high exposure group (100 mg/kg). Since a linear dose-dependent response was not observed, the responses were not evaluated using linear regression analysis. Subacute treatment effects were not seen in the lymphocyte studies with Con A ($p = 0.11$), indicating that 2,4-D did not alter T-lymphocyte blastogenesis.

DISCUSSION

Antibody production and B-lymphocyte mitogen responses were enhanced by acute 2,4-D butylester exposure. No significant effects were associated with T-lymphocytes, although the 2,4-D butylester appeared to produce weak, nonspecific, mitogen stimulation. Similar effects were not observed following subacute exposure. Antibody production was not affected, although enhancement of B-lymphocyte mitogen responses was seen at intermediate dosages of 2,4-D.

Only one study has examined functional alterations in the immune system following 2,4-D exposure.[7] The study indicated that 2,4-D hypersensitivity is not due to increased IgE production or altered delayed-type hypersensitivity reactions. It is more probable that 2,4-D may be functioning as a hapten in hypersensitivity reactions. Several 2,4-D derivatives have been associated with contact dermatitis, a hypersensitivity reaction,[8] which may involve an immunologic component.

Moderately high, acute 2,4-D ester exposure was necessary to enhance the immune responses. The highest exposure, 200 mg/kg, which produced clinical manifestations and histopathological changes, is less than the acute oral LD_{50} for rats, 1500 mg/kg,[9] but is likely to be well in excess of typical occupational exposures encountered by man. The FAO/WHO Allowable Daily Intake (ADI) in man is 0 to 3 mg/kg.[9] Clinical manifestations of toxicity in man begin to appear at dosages exceeding 67 mg/kg.[3] Subacute 2,4-D ester exposure which produced minimal immune alterations and no clinical or pathological manifestations also exceeded the accepted ADI recommendations for man. Since immune alterations were only observed at relatively high exposures the toxicological significance of pure 2,4-D-induced immune alteration should not be a major concern.

REFERENCES

1. **Bjorklund, N.E. and Erne, K.,** Toxicological studies of phenoxyacetic herbicides in animals, *Acta Vet. Scand.*, 7, 364, 1966.
2. **Buck, W.B., Osweiler, G.D., and Van Gelder, G.A.,** *Clinical and Veterinary Toxicology*, 2nd ed., Kendall/Hunt Publishing, Dubuque, IA, 1976.
3. **Seabury, J.H.,** Toxicity of 2,4-dichlorophenoxyacetic acid for man and dog, *Arch. Environ. Health*, 7, 202, 1963.
4. **Kolmodin-Hedman, B. and Erne, K.,** Estimation of occupational exposure to phenoxy acids (2,4-D and 2,4,5-T), *Arch. Toxicol.*, Suppl. 4, 318, 1980.
5. **Sauerhoff, M.V., Braun, W.H., Blau, G.E., and Gehring, P.J.,** The fate of 2,4-dichlorophenoxyacetic acid (2,4-D) following oral administration to man, *Toxicology*, 8, 8, 1977.
6. **Blakley, B.R.,** The effect of oral exposure to the n-butylester of 2,4-dichlorophenoxyacetic acid on the immune response in mice, *Int. J. Immunopharmacol.*, 8, 93, 1986.
7. **Cushman, J.R. and Street, J.C.,** Allergic hypersensitivity to the herbicide 2,4-D in Balb/c mice, *J. Toxicol. Environ. Health*, 10, 729, 1982.
8. U.S. Environment Protection Agency, Summary of Reported Pesticide Incidents Involving 2,4-D, Pesticide Incident Monitoring Service Report No. 283, U.S. Environment Protection Agency, Washington, D.C., 1980.
9. **Mercier, M.,** *Criteria (Dose/Effect Relationship) for Organochlorine Pesticides*, Commission of the European Communities, Pergamon Press, Toronto, 1981.

Kinetics and Components Involved in the Adsorption of 2,4-D by Soils

P. M. Huang and R. Grover

INTRODUCTION

There has been an increase in public concern over the use of pesticides in general, and of 2,4-dichlorophenoxyacetic acid (2,4-D) in particular. 2,4-D is the major phenoxy herbicide used in Canada.[15] It is used to control broad-leaved weeds in cereal crops, public parks, and on private lawns. Because of its widespread use and its possible toxicological effects, extensive reviews of the environmental and human health criteria for 2,4-D exist.[5,6,14,20]

2,4-D is known to be readily degraded by soil microorganisms.[11] However, the effect of soil components and environmental factors on the kinetics and mechanisms of the breakdown of the herbicide is still obscure. Furthermore, knowledge of how a series of soil materials governs the binding of 2,4-D still remains to be advanced. The ability of soil components to bind the herbicide may have an enormous impact on its transformations and degradation in soil environments.

The objective of this study was to investigate the rates of the adsorption of 2,4-D by selected Saskatchewan soils and the relative importance of soil components involved in the reaction.

MATERIALS AND METHODS

Soil Properties

Eight Saskatchewan soils, which were collected from Ap horizons, were used. The soils were sieved to pass a 0.5-mm screen. The pH values of the soils were determined at the soil-to-water ratio of 1:1. Organic matter was determined by the Wakley-Black method.[10] The clay contents of the soils were determined by the pipette method.[4] The amounts of the extractable Al and Fe oxides were determined by three methods, namely, the dithionite-citrate-bicarbonate (DCB) treatment,[9] ammonium oxalate treatment,[17] and sodium pyrophosphate treatment.[1]

Adsorption of 2,4-D by Soils

One gram of each soil was suspended in 10 ml of aqueous solution containing ^{14}C-labeled 2,4-dichlorophenoxy (2-^{14}C) acetic acid (2,4-D) at the initial concentration of 0.208 μg/ml. The suspension was equilibrated for 24 h in a constant-temperature bath with gentle agitation at 25 ± 0.2°C. At the end of the equilibration period, the suspension was centrifuged at 10,000 rpm for 10 min. 2,4-D remaining in solution was determined by liquid scintillation counting techniques using Beckman LS9000. The amounts of 2,4-D adsorbed by the soils were calculated by the difference in 2,4-D concentrations before and after equilibration with the soils. The increase in the adsorption of 2,4-D by the soils after a 24-h reaction period was negligible.

Time Function of 2,4-D Adsorption by Soils

One gram each of selected soils (Asquith, Balgonie, Melfort, Nobleford, Regina, Success) was suspended in 10 ml of aqueous solution of ^{14}C-labeled 2,4-D at the initial concentration of 0.208 μg/ml for 0.5, 1, 2, 4, 8, 16 and 24 h. The suspensions were kept in a constant-temperature bath at 25 ± 0.2°C. The concentration of 2,4-D remaining in solution at each reaction period was determined in the supernatant as described previously.

Adsorption of 2,4-D by Soils as Influenced by Acidification by Organic Acids

One gram each of Battrum and Nobleford soils was added to 10 mL of aqueous solution containing ^{14}C-labeled 2,4-D at 0.208 μg/ml and citric or malic acid at 10^{-2} *M* for 24 h. The suspensions were equilibrated in a constant-temperature bath at 25 ± 0.2°C. At the end of the reaction period, the pH of each suspension was measured. The equilibrium concentration of 2,4-D remaining in solution was assayed as described in the previous section.

All the determinations carried out in the present study were duplicated.

RESULTS AND DISCUSSION

The selected properties of the soils studied and the amounts of 2,4-D adsorbed by the soils at equilibrium are presented in Table 1. The ability of the soils to adsorb 2,4-D was most significantly correlated with their amounts of organic matter and the pyrophosphate-extractable Al and Fe (Table 2). The metal oxides extracted by the prophosphate treatment are bound to organic components[1] and are short-range ordered (poorly crystalline) in nature. These sesquioxidic components evidently possess reactive sites to bind 2,4-D introduced to the systems. Hydrolytic precipitation products of Al and Fe are constantly interacting with natural organics in soils.[8,18] The hydrolytic reactions of Al and Fe are very substantially perturbed by organic components of soils, leading to the formation of organically bound and short-range-ordered metal oxides. In soil environments, between 52 and 98% of the organic carbon is associated with the clay fraction.[7] It is likely that most of the remaining organic carbon is linked to metal oxides and hydroxides.[16] Therefore, the organo-mineral

TABLE 1
Selected Properties of the Soils Studied

Soil	pH	Organic matter (%)	Clay (%)	Extractable Al (ppm) Oxalate	DCB[a]	Pyrophosphate	Extractable Fe (ppm) Oxalate	DCB	Pyrophosphate	2,4-D adsorbed by soil at equilibrium (ng/g)
Asquith	7.6	1.67	11	600	1350	62	1000	3130	50	36
Balgonie	6.2	6.00	27	1400	1730	307	2600	6450	359	181
Battrum	7.6	3.31	42	1300	2030	105	1900	6870	74	77
Melfort	6.1	9.06	49	2200	2210	466	3600	9490	508	441
Nobleford	7.8	3.07	22	1200	1680	105	1300	4880	44	92
Regina	7.7	3.06	71	2700	2260	84	2900	7950	84	88
Success	7.7	2.86	29	1200	1420	70	1700	4880	67	49
Swift current	6.0	2.85	22	800	1790	198	1900	5360	201	66

[a] Dithionite-citrate-bicarbonate.

complexes are clearly largely responsible for the retention of 2,4-D by the soils (Table 2).

The adsorption of 2,4-D by the soils followed multiple first-order kinetics (both the fast and slow reactions). The rate constants of the fast and slow adsorption reactions of 2,4-D by the soils ranged from 4.8 to 122.7 and from 0 to 2.30×10^{-3} h^{-1}, respectively (Table 3). The adsorption of 2,4-D by the soils was virtually completed in a few hours. Thus, the fast reaction was largely responsible for the retention of 2,4-D by the soils. Among the soils studied, Melfort soils had the highest rate of the adsorption and Success soil the lowest during the fast adsorption stage. The rate constants of the adsorption of 2,4-D by these soils differed by about 26 times. This difference in the kinetics of the adsorption is attributed to the nature and surface reactivity of soil components. The bearing of the adsorption kinetics of 2,4-D upon its transformations in the environment merits further attention.

Acidification of the soils by organic acids such as citric and malic acids resulted in a great enhancement of the adsorption of 2,4-D by the soils (Table 4). The amounts of 2,4-D adsorbed were increased by as high as 4.2 times upon acidification. Low-molecular-weight organic acids certainly can degrade soil minerals through the attack of the dissociated protons and organic ligands.[19] The Al and Fe released from soil minerals would undergo hydrolytic reactions and/or complexation with organic ligands, leading to the formation of short-range-ordered sesquioxides[8,18] which are apparently responsible for the enhancement in the adsorption of 2,4-D by the soils upon the addition of the organic acids (Table 4). The competition between 2,4-D and citric and malic acids upon dissociation of hydrogen ions for the adsorption sites of the soils is evidently not the overriding reaction governing the adsorption of 2,4-D in the systems studied.

TABLE 2
The Correlation of Soil Properties and Their Adsorption of 2,4-D

Soil components	R^2	Significance (%)
Organic matter	0.95	0.002
Pyrophosphate-extractable Al oxides	0.87	0.038
Pyrophosphate-extractable Fe oxides	0.82	0.100
Clay	0.12	20.259
Oxalate-extractable Al oxides	0.27	9.506
Oxalate-extractable Fe oxides	0.64	0.829
DCB-extractable Al oxides[a]	0.29	8.305
DCB-extractable Fe oxides	0.58	1.383

[a] Dithionite-citrate-bicarbonate.

TABLE 3
Rate Constants of the Adsorption of 2,4-D by Soils

Soil	Rate constants $\times 10^3$ (h^{-1}) Fast reaction	Slow reaction
Asquith	19.3	0
Balgonie	74.0	2.3
Melfort	122.7	1.5
Nobleford	58.8	0.43
Regina	57.6	0.65
Success	4.8	1.12

TABLE 4
Influence of Acidification of Soils on Their Adsorption of 2,4-D

Treatment	Suspension pH	2,4-D adsorbed per gram of soil (ng)
	Battrum Soil	
No organic acid	7.65	77
10^{-2} *M* citric acid	3.59	307
10^{-2} *M* malic acid	3.88	322
	Nobleford Soil	
No organic acid	8.02	92
10^{-2} *M* citric acid	4.38	185
10^{-2} *M* malic acid	5.00	158

Low-molecular-weight organic acids are being constantly introduced to soil environments through microbial metabolism, root exudation, canopy drop, and decay of plant and animal residues.[8] Their impact on the acidification of soils (Figure 1) deserves close attention. Furthermore, soils can be acidified through other mechanisms which include application of nitrogen,[2] phosphorus,[12] and sulfur[3] fertilizers, oxidation of pyrite,[13] deposition of acid rain,[3] and root respiration. Therefore, the development of soil acidity is related to soil parent materials, the advancement of the stage of soil formation, intensification of agriculture, and industrialization. Consequently, if there are no proper management practices, the extent of acidification of soils is deemed to increase as time goes by.

Acidification of soils results in the formation of sesquioxidic components which are rich in positive charge (Figure 1). Therefore, the accumulation of the dissociated 2,4-D in soils can be enhanced by acidification of soils (Table 4) as depicted in Figure 2. Some of the adsorbed 2,4-D on soil particles may be released to soil solution when land management practices are modified. The accumulation of 2,4-D in soils upon acidification and its potential dispersion to contaminate the biosphere should not be overlooked.

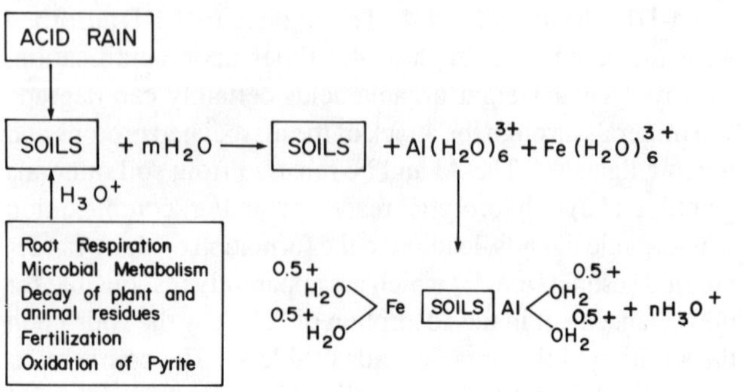

FIGURE 1. Mechanisms of acidification of soils.

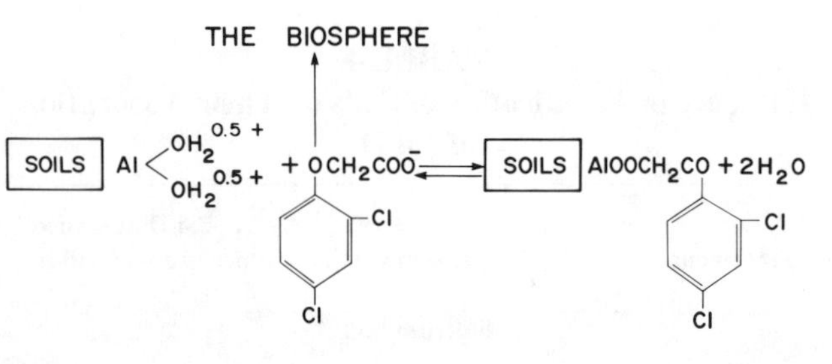

FIGURE 2. Accumulation of 2,4-D in soils upon acidification and its potential dispersion to the biosphere.

REFERENCES

1. **Bascomb, C.L.,** Distribution of pyrophosphate extractable iron and organic carbon in soils of various groups, *J. Soil Sci.*, 19, 251, 1968.
2. **Blevins, R.L., Thomas, G.W., and Cornelius, P.L.,** Influence of no tillage and nitrogen fertilization on certain soil properties after 5 years of continuous corn, *Agron. J.*, 69, 383, 1977.
3. **Coote, D.R., Dumanski, J., and Ramsey, J.F.,** An assessment of the degradation of agricultural lands in Canada, Research Branch, Agriculture Canada, L.R.R.I. No. 118, Ottawa, 1981.
4. **Day, P.R.,** Particle fractionation and particle-size analysis, in: *Methods of Soil Analysis, Part 1,* Black, C.A., Ed., American Society of Agronomy, Madison, WI, 1965, 545.
5. FAO/WHO, 1970 Evaluations of Some Pesticide Residues in Food, Food and Agriculture Organization/World Health Organization, Geneva, 1971, 59.
6. FAO/WHO, 1971 Evaluations of Some Pesticide Residues in Food, Food and Agriculture Organization/World Health Organization, Geneva, 1972.
7. **Greenland, D.J.,** Interaction between clays and organic compounds in soils. II. Adsorption of soil organic compounds and its effects on soil properties, *Soil Fert. (Commonw. Bur. Soil Sci.)*, 28, 521, 1965.
8. **Huang, P.M. and Violante, A.,** Influence of organic acids on crystallization and surface properties of precipitation products of aluminum, in *Interactions of Soil Minerals with Natural Organics and Microbes*, Huang, P.M. and Schnitzer, M., Eds., SSSA Special Publ. No. 17, Soil Science Society of America, Madison, WI, 1986, 159.
9. **Jackson, M.L.,** Soil chemical analysis — advanced course, published by the author, University of Wisconsin, Madison, WI, 1956.
10. **Jackson, M.L.,** *Soil Chemical Analysis,* Prentice-Hall, Englewood Cliffs, NJ, 1958.
11. **Kaufman, D.D.,** Degradation of pesticides by soil microorganisms, in *Pesticides in Soil and Water,* Guenzi, W.D., Ed., Soil Science Society of America, Madison, WI, 1974, 133.
12. **Lindsay, W.L. and Stephenson, H.F.,** Nature of the reactions of monocalcium phosphate monohydrate in soils. I. The solution that reacts with the soil, *Soil Sci. Soc. Am. Proc.*, 23, 12, 1959.
13. **McFee, W.W., Byrnes, W.R., and Stockton, J.G.,** Characteristics of coal mine overburden important to plant growth, *J. Environ. Qual.*, 10, 300, 1981.
14. National Research Council of Canada, Phenoxy Herbicides — Their Effects on Environmental Quality, Associate Committee on Scientific Criteria for Environment Quality, NRCC No. 16075, Ottawa, 1978, 400.
15. National Research Council of Canada, 2,4-D: Some Current Issues, NRCC No. 20647, Ottawa, 1983, 99.
16. **Schnitzer, M. and Kodama, H.,** Reactions of minerals with soil humic substances, in *Minerals in Soil Environments,* Dixon, J.B. and Weed, S.B., Eds., Soil Science Society of America, Madison, WI, 1977.
17. **Schwertmann, U.,** The differentiation of iron oxide in soils by a photochemical extraction with acid ammonium oxalate, *Z. Pflanzenernaehr. Dueng. Bodenkd.*, 105, 194, 1964.
18. **Schwertmann, U., Kodama, H., and Fischer, W.R.,** Mutual interactions between organics and iron oxides, in *Interactions of Soil Minerals with Natural Organics and Microbes,* Huang, P.M. and Schnitzer, M., Eds., SSSA Special Publication No. 17, Soil Science Society of America, Madison, WI, 1986, 223.
19. **Tan, K.H.,** Degradation of soil minerals by organic acids, in *Interactions of Soil Minerals with Natural Organics and Microbes,* Huang, P.M. and Schnitzer, M., Eds., SSSA Special Publ. No. 17, Soil Science Society of America, Madison, WI, 1986, 1.
20. U.S. Environmental Protection Agency, Criteria Document for 2,4-Dichlorophenoxy-Acetic Acid, Syracuse Research Corporation for Environmental Criteria and Assessment Office, Syracuse, NY, 1982.

Mechanism of Respiratory Failure in Organophosphate Intoxication in Nonhuman Primates

A. Anzueto, G. G. Berdine, G. T. Moore, C. D. White, and W. G. Johanson, Jr.,

INTRODUCTION

Organophosphate compounds are widely used as insecticides. These compounds act by irreversibly inhibiting acetylcholinesterase.[1,2] Toxic manifestations are due to accumulation of acetylcholine at both muscarinic and nicotinic receptors, resulting in cholinergic crisis and neuromuscular blockade. The cause of death is asphyxia; the mechanism has been attributed to bronchospasm, upper airway obstruction, bronchorrhea, respiratory muscle paralysis, and central apnea.[1,2] Meeter and Wolthuis[3] demonstrated that a $2 \times LD_{50}$ dose of pinacolyl methylphosphonofluoridate (Soman) in rats resulted in apnea lasting up to 7 h with minimal depression of diaphragm contractile force during tetanic contraction induced by phrenic nerve stimulation. They postulated that the apnea was central in nature. Adams et al.[4] showed that Soman, 38 μg/kg, produced cessation of efferent phrenic nerve traffic in guinea pigs. Spontaneous respiration was promptly restored by atropine sulfate, 4 mg/kg. Further, neuromuscular blockade of the diaphragm did not occur unless Soman, 400 μg/kg, was administered. We studied the respiratory effects of Soman in baboons to determine by what mechanisms this organophosphate agent produces apnea in primates.

METHODS

Animal Model

Animal care was conducted under the guidelines of the National Research Council's Guide for the Care and Use of Laboratory Animals. All protocols were reviewed and approved by the Animal Research Committee of the Southwest Foundation for Biomedical Research.

This animal model has been reported in detail previously.[5] In summary, adult baboons (*Papio* sp.) (11 to 24 kg) were anesthetized with sodium pentobarbital (30 mg/kg, i.v.) and endotracheally intubated. Additional pentobarbital (2 mg/kg) was administered periodically to maintain anesthesia. Pulmonary artery and systemic arterial catheters were placed aseptically to monitor blood pressures. Thermodilution cardiac output and arterial and mixed venous blood gases were also monitored. A thin-walled latex balloon-tipped catheter was positioned in the right pleural space for the measurement of pleural pressure (Ppl). A similar catheter was positioned via a small laparotomy incision in the abdomen for the measurement of abdominal pressure (Pab). The difference between the Ppl and Pab signals represented the transdiaphragmatic pressure (Pdi). Gas was monitored by an ultrasonic flow meter, and tidal volume (Vt) was obtained by the integration of flow signal.

The right phrenic nerve was isolated in the neck. The nerve was placed inside a bipolar electrode, and the signal was amplified, bandpass filtered (300 Hz to 2 kHz), and rectified (100 ms moving time average). The phrenic nerve electrode was connected to a digital pulse-generating system via a stimulus isolation unit. Pgas, Ppl, and Pdi were recorded during 2-s stimulations at 10, 20, 50, and 100 Hz.

CO_2 rebreathing was done by connecting the endotracheal tube to a 1-l bag containing 100% CO_2. Tracings of Vt, Ppl, Pgas, and Pdi were recorded while breathing ambient air and during closed-circuit rebreathing.

Experimental Protocol

Soman Studies

Twelve animals were studied. Each received Soman, 13.1 μg/kg (2 LD_{50}), via peripheral vein infusion over a 10-min interval. This dose was selected based on our previous experience[5] that 2 LD_{50} in the baboon produced severe hemodynamic and respiratory effects. One group of six animals (group A) had hemodynamics, pulmonary function tests, phrenic nerve recording, and CO_2 rebreathing studies done twice before Soman exposure, hourly over 4 h, and at 12 and 24 h after Soman exposure. In addition, all studies except rebreathing were performed at the time of apnea. This group also received atropine sulfate, 3 mg i.m., at the time blood pressure fell to half its baseline value to ameliorate the hemodynamic effects of Soman. A second group of six animals (group B) had hemodynamics, pulmonary function tests, phrenic nerve recording, and indirect diaphragm stimulation done twice before Soman exposure, and hourly up to 4 h after Soman exposure.

Control Studies

Three animals were studied under pentobarbital anaesthesia (group C). No Soman was administered. Hemodynamics, pulmonary function tests, phrenic nerve recording, CO_2 rebreathing, and diaphragm stimulation were performed hourly over a 4-h period.

RESULTS

Table 1 summarizes the hemodynamic and lung volume data for the three groups of animals. All animals exposed to Soman (groups A and B) became hypotensive and apneic within 10 min of completion of the infusion. There was no significant change in FRC or total lung capacity (TLC) in any group. The control animals (group C) showed no change in any parameter.

The initial respiratory effect of Soman was usually tachypnea, with a progressive decrease in tidal volume. All Soman-treated animals became apneic within 10 min, and required assisted mechanical ventilation for 2 to 3 h. Atropine appeared to have no effect on the apnea following Soman exposure. Apnea was associated with a cessation of phrenic nerve traffic (EPH) and diaphragm electromyographic (DEMG) activity. Figure 1 shows a typical tidal volume, EPH, and DEMG tracing.

All Soman-treated animals resumed spontaneous ventilation by 4 h after exposure. Vt was significantly lower than baseline (p <0.01), respiratory rate (RR) was significantly higher than baseline (p <0.01), and minute volume (MV) was not significantly changed throughout the experiment. There was no change in Vt, RR, or MV in the control animals over a 4-h period. Figure 2 shows the effect of Soman on the pattern of breathing gas containing 9% CO_2. There was no change in tidal volume (Vt_9), respiratory rate (RR_9), minute volume (MV_9), or transdiaphragmatic pressure (Pdi_9) in the control animals. In the Soman-exposed animals, Vt_9, MV_9, and Pdi_9 were reduced from baseline at 4 h (p <0.01); RR_9 did not significantly change throughout the experiment. By 12 h, all measured parameters were not significantly different from baseline values.

Figure 3 illustrates the effect of Soman on the Pdi generated during indirect diaphragm stimulation. Only the values obtained at baseline, 2- and 4-h post infusion, are shown for

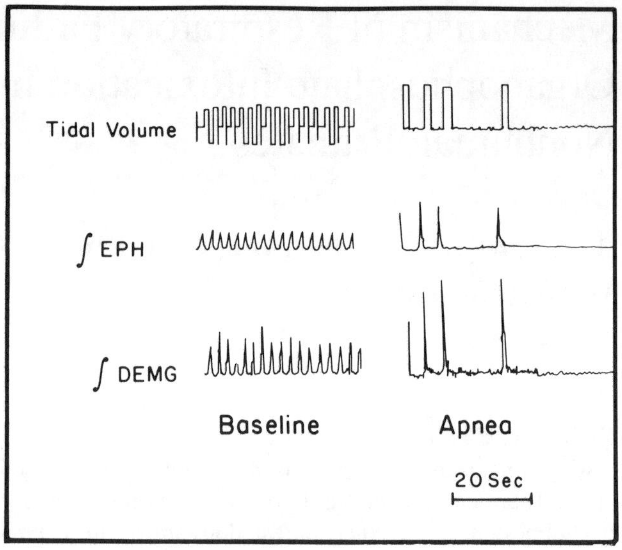

FIGURE 1. Vt, efferent phrenic nerve traffic (EPH), and diaphragm DEMG following Soman exposure. The panel shows tracing for an animal before Soman infusion and at the time of apnea following infusion.

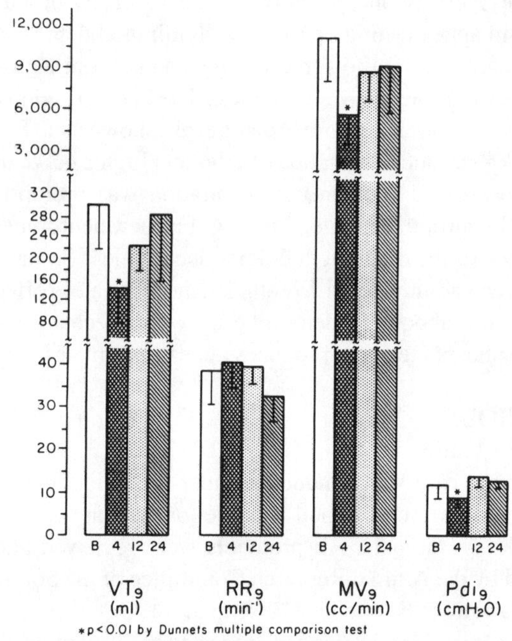

FIGURE 2. Ventilatory response to 9% CO_2 (see text for definition of symbols). Open bars are pre-Soman exposure. Shaded bars from left to right in each group of four are 4, 12, and 24 h postexposure. Bar heights are mean values and hashes represent one standard deviation. An * indicates that the measurement is significantly different from preexposure (p <0.05) by Dunnet's multiple comparison test. * = indicates p <0.01.

TABLE 1
Hemodynamics and Lung Volume

	MBP (mmHg)	CI (cc/s/kg)	FRC (ml)	TLC (ml)
Baseline				
A	127 ± 7	125 ± 8	613 ± 86	1929 ± 260
B	123 ± 6	130 ± 12	661 ± 90	1642 ± 200
C	118 ± 6	123 ± 11	645 ± 125	1978 ± 310
Apnea				
A	78 ± 16	100 ± 24	—	—
B	52 ± 12[a]	75 ± 15[a]	—	—
C	—	—	—	—
4 h				
A	106 ± 6	140 ± 10	594 ± 78	1885 ± 276
B	78 ± 8	99 ± 12	744 ± 148	1711 ± 241
C	119 ± 4	139 ± 19	692 ± 112	1978 ± 305

Note: Values are mean ± SEM.

[a] p <0.01 compared with baseline.

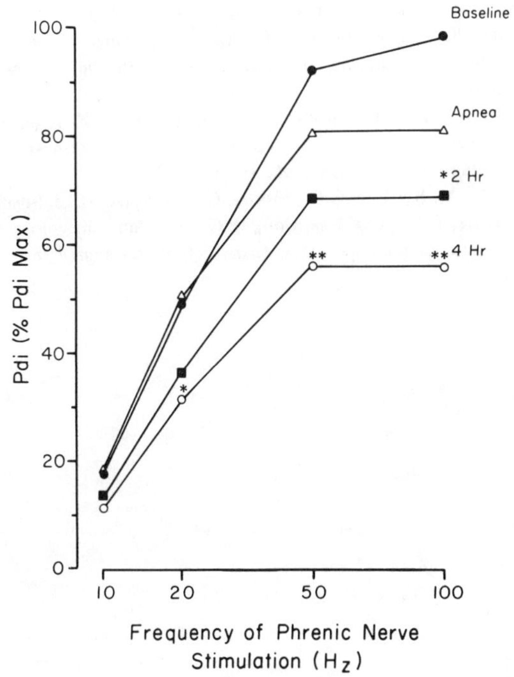

FIGURE 3. Indirect diaphragm stimulation curves. See text for details. Open circles are preexposure values, closed circles are at onset of apnea, open squares are 2 h post exposure, and closed squares are 4 h post exposure. An * indicates that value is significantly different from pre-exposure (p <0.05) by Dunnet's multiple comparison test. * = p <0.01.

clarity. Although there was a progressive decline in values at all frequencies over time, only the declines at 4 h for 20, 50, and 100 Hz, and the declines at 2 h for 100 Hz were statistically significant. No significant change in Pdi developed during indirect diaphragm stimulation in the control group over a 4-h period.

DISCUSSION

Critique of Experimental Design

Pentobarbital was chosen as the anesthetic for this experiment to minimize the severe mycolonic activity following Soman intoxication. Although pentobarbital could have resulted in depressed respiratory drive or reduced diaphragm function, the control animal group demonstrated that measurements of the pattern of breathing, Pdi during indirect diaphragm stimulation, and the ventilatory response to 9% CO_2 did not change during 4 h of pentobarbital anesthesia. Although additional anesthetic effects from 4 to 24 h were not tested in a control population, the fact that the ventilatory response to 9% CO_2 and the ventilatory pattern while breathing ambient air were progressively closer to baseline values at 12 and 24 h postexposure (see Figure 2) suggests that deterioration of ventilatory function due to prolonged barbiturate anesthesia was not likely.

Central Respiratory Effects of Soman

Apnea occurs within 10 min of Soman infusion and lasts for 3 to 4 h. This apnea corresponds to a cessation of efferent phrenic nerve traffic as seen in Figure 1. This central pattern of an absent efferent phrenic signal is markedly different from that seen following neuromuscular blockade. Further evidence of the central nature of Soman-induced apnea is that indirect diaphragm stimulation is not significantly depressed at this time (Figure 3). These data are consistent with previous work in the rat and rabbit.[3,4] Atropine had no effect on this central apnea, unlike the prompt resumption of respiration seen in the rabbit.[4]

Coincident with the central apnea following Soman exposure is significant hemodynamic shock. Although one cannot exclude hypotension as the etiology of this central apnea, the data suggest that this is not likely. The apnea persists for 3 to 4 h, whereas the blood pressure and cardiac output returned to near-normal levels within 2 h in the atropine-treated rebreathing group of animals. Respiratory control appears to be intact when spontaneous respiration resumes at 4 h as evidenced by a normal minute volume while breathing ambient air due to a tachypnea which compensates for small Vt. It seems unlikely that a hypotensive brainstem insult sufficient to abolish all respiratory effort for 4 h would immediately revert to normal respiratory control without a period of depressed respiratory drive reflected by low respiratory rates. A hypotensive brainstem insult lasting 3 to 4 h would not likely have such a consistent duration as seen in all 12 Soman-treated animals, despite a markedly different time course of blood pressure improvement between the rebreathing and diaphragm stimulation groups. Last, EEG recordings (data not shown) during this central apnea period are active, suggesting that any brainstem dysfunction is not due to global brain ischemia.

Peripheral Respiratory Effects of Soman

Soman appears to have a peripheral effect on respiration beginning approximately 2 h after exposure and lasting several hours. This effect is manifested as an abnormal ventilatory pattern upon recovery of spontaneous ventilation at 4 h, an abnormal ventilatory response to 9% CO_2, and a depressed indirect diaphragm stimulation curve. The spontaneous ventilatory pattern exhibited by the rebreathing group of animals at 4 h — low Vt, high RRs, and normal MV — could be due to respiratory muscle failure rather than abnormal ventilatory drive. Although the MV response to 9% $C0_2$ is depressed, it is probably also due to respiratory muscle failure rather than depressed central drive, since RRs are maximal both while breathing ambient air and 9% $C0_2$. It appears that the depressed response to 9% CO_2 is completely due to an inability to increase Vt. Further, the Pdi generated by indirect diaphragm stimulation is significantly depressed at this time. This peripheral respiratory failure has resolved by 12 h.

CONCLUSIONS

The immediate respiratory effect of Soman is suppression of respiratory drive, presumably through action on cholinergic receptors in the central nervous system. This effect lasts several hours. There is a delayed peripheral depression of respiratory muscle function by Soman or possibly a metabolite which lasts from 2 to 12 h. These effects are not related to changes in pulmonary mechanics.

REFERENCES

1. **Gliman, A.G., Goodman, L.S., and Gilman, A., Eds.,** *The Pharmacologic Basis of Therapeutics,* Macmillan, New York, 1980, 110.
2. **Namba, T., Nolte, C.T., Jackrell, J., and Grob, D.,** Poisoning due to organophosphate insecticides, *Am. J. Med.,* 50, 475, 1971.
3. **Meeter, E. and Wolthuis, O.L.,** The spontaneous recovery of respiration and neuromuscular transmission in the rat after anticholinesterase poisoning, *Eur. J. Pharmacol.,* 2, 377, 1968.
4. **Adams, G.K., III, Yamamura, H.I., and O'Leary, J.F.,** Recovery of central respiratory function following anticholinesterase intoxication, *Eur. J. Pharmacol.,* 38, 101, 1976.
5. **Anzueto, A., Berdine, G.G., Moore, G.T., Glesier, C., Johnson, D., White, C.D., and Johanson, W.G., Jr.,** Pathophysiology of Soman intoxication in primates, *Toxicol. Appl. Pharmacol.,* 86, 56, 1986.

The Fenitrothion Controversy: What is the Evidence?

D. J. Ecobichon

Currently, we have over 200 insecticides, 50 herbicides, and at least 20 fungicides to choose from in our search for selective chemicals to control insect, weed, and fungal infestations. Many of these agents have been registered for a long time and the toxicity data are inadequate by current standards. Public concern over seemingly unwarranted agricultural practices involving pesticides has resulted in demands for additional costly and extensive toxicity testing to provide assurances of safety for chemicals used for 20 to 30 years without evident adverse health effects.

The chemical that I would like to talk about today is not an agricultural insecticide in Canada, though it is used worldwide on various food crops and in malaria control programs. Fenitrothion (*0,0*-dimethyl *0*-[4-nitro-*m*-tolyl] phosphorothioate) is used in efforts to control an indigenous, lepidopterous, forest pest in eastern Canada, the spruce budworm (*Choristoneura fumiferana,* Clemens), that has been ravaging the forests from Manitoba eastward, destroying the mature stands of balsam fir and spruce required for softwood in paper manufacture.[1,2] I shall use this chemical to illustrate what is, unfortunately, becoming a recurring scenario with pesticides in Canada.

The use of any pesticide should be accompanied by an in-depth risk-benefit analysis. All too frequently, the benefits are considered only in terms of the availability of a suitable chemical on the market, the ease of application, cost, and the economic returns for an undamaged product. At the other extreme, there are groups who do not want to see any chemicals sprayed anywhere for any purpose, who refuse to believe any statements made by responsible government or agrochemical industry personnel, but who, at the same time, readily accept poor scientific evidence as gospel and thrive on the inflammatory and sometimes untruthful misinformation provided by the media. It was into this highly polarized atmosphere that I stepped in 1978 as consulting toxicologist for the New Brunswick Department of Health, some 10 years after fenitrothion had replaced DDT as the insecticide of choice in budworm control.[3,4]

In Figure 1 is shown the annual aerial forest spraying program in terms of the millions of hectares treated per year and in the amount of fenitrothion applied at a usual concentration of 140 to 280 g of active ingredient per hectare (210 g/ha being the most efficacious application rate). Agriculturally, fenitrothion is applied at five- to tenfold higher levels than that used in Canadian forests. In one particular year, 1979, fenitrothion was not used, partly as a consequence of public outcry about the perceived toxicity of this chemical ("a nerve gas"), real or imagined illness of all sorts, and genuine concern on the part of provincial health personnel about the quality of the toxicity data supporting the registration of this agent. Use of the agent was banned until the question of data quality was addressed by the manufacturer (the Sumitomo Chemical Company, Osaka, Japan) and additional toxicological studies could be done to provide information in line with current regulatory requirements. In Table 1 is shown a capsule summary of the data available at the time of the ban.[5] The data were "soft" at best and useless at worst for any complete and meaningful assessment of the safety of this chemical. The deficiencies, as the panel conducting the toxicity data assessment saw them, are shown in Table 2.[5]

The manufacturer was responsible enough to accede to our demands and initiated some of the studies requested. The

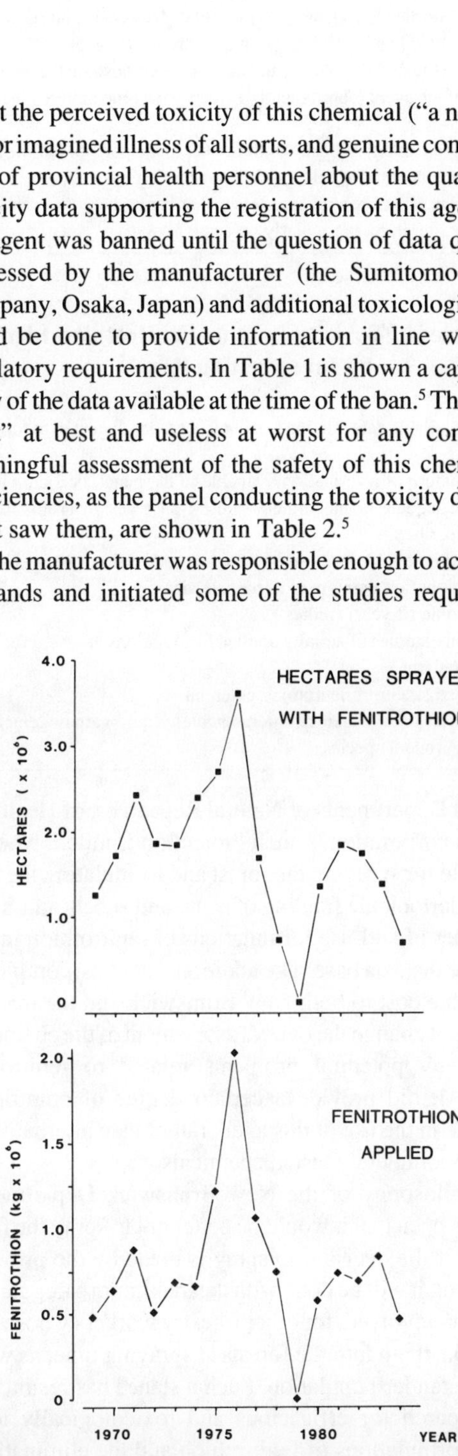

FIGURE 1. The extent of the aerial spraying program in New Brunswick as measured by the hectares of forest treated annually with fenitrothion and the total amount of active ingredient used annually at an average application rate of 210 g/ha.

TABLE 1
Available Data on Fenitrothion Toxicity — 1978

Acute oral toxicity studies (LD_{50}) for "pure" (99.9%) and technical (97%) grades of chemical in rats, mice, guinea pigs, cats, hens, etc., plus some information on formulations
Acute dermal studies on "pure" and technical material in a number of species
Chronic oral studies in rodents (rats) for 90 d, 16 weeks, and 63 weeks
Limited (one) studies involving the inhalation of the agent
Inconclusive studies in hens, rats, and rabbits to demonstrate the neuropathic potential of the agent (contradictory evidence was encountered in the open literature)
A limited *but* clean bill of health with respect to the potential of the agent to elicit mutagenic, teratogenic, or carcinogenic effects
Limited human exposure studies, equivocal or contradictory evidence being seen in the literature, particularly on acute toxicity.

TABLE 2
Deficiencies in Toxicity Data on Fenitrothion Identified by Investigating Panel[5]

Little or no available data relating to technical grade of agent or to formulated material
Superficial nature of information available to the panel, i.e., summaries of studies, inadequate or incomplete studies, poor design protocols inadequately described
Lack of good data on inhalation of technical material in two species, with particular attention being paid to particle size
Suitable chronic (2-year) studies
Accountability studies of aerially applied formulations to assess the drift of active ingredient
Adequate assessment of neurotoxic potential
Reconfirmation of mutagenicity, teratogenicity, and carcinogenicity in a second, nonrodent species

provincial Departments of Natural Resources, of Health, and the crown corporation, Forest Protection Limited, who were responsible for spraying the forest and formulating the insecticide, undertook the funding of acute and subchronic studies of a number of end-use formulations of fenitrothion in order to improve the data base. The additional studies, completed at considerable cost to both New Brunswick and the manufacturer, did not change the overall assessment of the chemical or pinpoint any potential problems related to fenitrothion. However, it did provide a certain degree of comfort and confidence in the use of this agent rather than alternative and perhaps incompletely tested chemicals.

The philosophy of the New Brunswick Department of Health has been that it would be better not to spray the forests at all, but if the decision to spray is made by the provincial government, it will be done with the maximum safeguards for the workers involved, for other forestry workers whose occupations take them into the forests at spraying time, as well as for the bystander population. Such a stance has resulted in a diligent search for efficacious and toxicologically tested, aqueous formulations of fenitrothion and the elimination of oil-based formulations in the province. It has involved the testing of several emulsifiers, initiating such studies in the laboratory, taking candidate chemicals to the field for testing in formulations, and frequently bringing them back into the laboratory again to be tested in combination with fenitrothion. It has involved the development of a system of reporting spills, oversprayings, etc., and their investigation, with environmental and health impact assessments being carried out.

A major task has been the initiation of aerial drift studies of different fenitrothion formulations with various types of aircraft under different climatic conditions. These have been essential to ascertain the conditions for the most efficacious application with minimal off-target loss of insecticide. The Department of Health has used these data to develop safety or setback zones from human habitation. Our current regulations for formulations used in the province are shown diagrammatically in Figure 2. Depending upon the type of aircraft used, ultralow volume (ULV) spraying of aqueous formulations of fenitrothion by agricultural-type aircraft is allowed up to a distance of 300 m (1000 ft) from homes. Within that 300-m distance, further protection may be afforded by spraying the biological control agent, *Bacillus thuringiensis,* application up to 150 m (500 ft) from homes by slow-flying agricultural aircraft of the Pawnee or Cessna type. Larger tracts of New Brunswick forest are sprayed by modified Grumman Avenger aircraft, but, because of their greater speed and reduced maneuverability, they are restricted to 1 mi from habitation for aqueous formulations and to 2 mi if oil-based formulations are used. No oil-based formulations of fenitrothion have been used in the past 5 years.

Despite these extensive toxicological assessments and innovative investigations, the public viewpoint was not mollified and the annual aerial forest spraying was deemed responsible for causing Reye's syndrome (a serious and frequently fatal encephalopathy in children), cancer, and birth anomalies in the province. The Department of Health commissioned a series of epidemiological task forces composed of renowned North American experts to study the interrelationships between aerial spraying with this rare neurological disease, oncogenicity, and human reproductive outcomes.[6-9] Epidemiological studies are exhaustive, expensive, and fraught with all sorts of methodological difficulties, particularly so if one is dealing with low incidence rates and a small population. None of these studies was conclusive due to the low statistical power of the data to detect rare events.

The cost of "proving" that fenitrothion is an efficacious and relatively safe insecticide, the "best of a bad lot" as it were, has been enormous. It has cost well in excess of $2,000,000 of New Brunswick tax dollars for the toxicological, epidemiological, and serial drift studies. In addition, one must add the costs of the various studies that were conducted by the manufacturer, the field research in monitoring the efficacy of different formulations in terms of pest mortality, egg mass counts, defoliation, tree mortality, environmental impact on wildlife, and the testing of different spray delivery systems and formulations. The benefits derived from this research can be translated into improved application techniques, improved and "safer" formulations, as well as policies developed by provincial government departments to make the spraying as safe as it is humanly possible to control. The bottom line is that

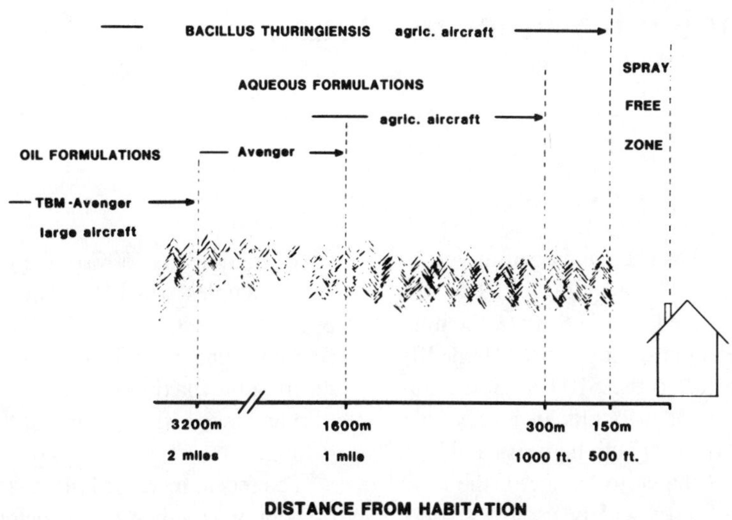

FIGURE 2. A schematic diagram depicting the current restrictions applied in New Brunswick to minimize the occurrence of overspraying of human habitation with fenitrothion. Slow-flying agricultural aircraft may spray an aqueous formulation of fenitrothion up to a distance of 300 m (1000 ft) from homes. Grumman Avenger (TBM) aircraft spraying aqueous fenitrothion formulations are restricted to 1.6 km (1.0 mi) from habitation, while all oil formulations sprayed by any aircraft are restricted to 3.2 km (2.0 mi) from homes. The biological control agent *(B. thuringiensis)* can be applied by agricultural aircraft up to a distance of 150 m (500 ft) from habitation, leaving a spray-free zone of 150 m that can be treated from the ground if necessary.

New Brunswick personnel had made the right choice in selecting fenitrothion as the best possible candidate insecticide. I submit that the province of New Brunswick would not have undertaken these extensive studies had the forest industry not been such an important factor in the provincial economy.

We need chemical pesticides, but their use exacts a price. The sort of activity that I have described cannot be carried out for every pesticide registered on a toxicity data base deemed inadequate in terms of current regulatory standards. One frequently hears the cry that "the chemical does not meet today's standards of toxicity testing". The costs of such studies, borne initially by the pesticide manufacturer, by representative provincial and federal governments, and the agricultural or forest industry, will be transferred to the public in the form of higher product costs and/or higher taxes. It is in the best interests of both industry and government to curb the excessive use (misuse) of pesticides, to introduce technological advances, to minimize application rates and off-target losses, and to develop rational guidelines for the protection of the population as well as the consumer.

REFERENCES

1. **Ecobichon, D.J.,** Aerial spraying of fenitrothion in forest programs: some problems and some solutions, *Can. J. Physiol. Pharmacol.*, 60, 1046, 1982.
2. **Symons, P.E.K.,** Dispersal and toxicology of the insecticide fenitrothion; predicting hazards of forest spraying, *Residue Rev.*, 68, 1, 1977.
3. **Anon.,** Fenitrothion: The Effects of Its Use on Environmental Quality and Its Chemistry, Publ. No. 14104, National Research Council of Canada, Ottawa, 1975.
4. **Anon.,** Fenitrothion: The Long-Term Effects on Its Use in Forest Ecosystems, Publ. No. 16073, National Research Council of Canada, Ottawa, 1977.
5. **Ecobichon, D.J., Chairman,** Fenitrothion: Assessment of the Toxicological Data, a report of the panel convened by the government of New Brunswick, 1978.
6. **Schneider, W.G., Chairman,** Forest Spray Program and Reye's Syndrome, report of the panel convened by the government of New Brunswick, 1976.
7. **Spitzer, W.O.,** Report of the New Brunswick task force on the environment and Reye's syndrome, *Clin. Invest. Med.*, 5, 203, 1982.
8. **Hatcher, J.D., Chairman,** Task Force on Chemicals in the Environment and Human Reproductive Problems in New Brunswick, report to the New Brunswick Department of Health, 1984.
9. **Spitzer, W.O.,** New Brunswick Task Force on the Environment and Cancer, report to the New Brunswick Department of Health, 1984.

Ethylene Dibromide Residues in Flour

G. McKay

Ethylene dibromide (EDB) has been and continues to be the subject of much controversy in the press. In fact, it has been at the center of a regulatory battle in the U.S. since the mid-1970s. It gained special attention during the 1981 medfly crisis in California which affected the $100 million citrus industry in that state. The major difficulty in justifying a complete ban of EDB is that it offers both high risk and high benefits. As such, it has been shown to be one of the most effective and, up until now, most widely used pesticidal fumigants for fruits and vegetables. Beyond these uses it is also employed as an antiknock additive for gasoline, as a fumigant for grain, and as a spot fumigant for milling machines.

In terms of risk assessment of EDB, most of the negative evidence concerning its use has come from animal studies. For a relatively comprehensive listing of EDB toxicity, the reader is referred to the literature search program of the National Library of Medicine which gives, free of charge, a citation listing "LS83-28" of 94 current citations on the toxicology of EDB. EDB has been shown to be carcinogenic, mutagenic, and to be a causative agent in reproductive disturbances in a number of animal studies. The available evidence indicates that the mechanism for its toxicity is mediated through covalent binding of dehalogenated metabolite(s) to cellular macromolecules, including DNA and subsequent damage.[1]

In 1980, the U.S. Environmental Protection Agency (EPA) proposed a suspension of the use of EDB as a fumigant for grain immediately, and for citrus fruits in mid-1983. However, it was not until February 3, 1984 that an actual ban on the sale, distribution, and use of EDB as a fumigant on stored grain and as a spot treatment of grain milling machinery occurred.

In Canada, EDB has been primarily used as a spot fumigant in the grain industry. In early 1984 EDB was deregistered by Agriculture Canada. However, most flour mills had stopped using EDB in late 1983. At that time residual testing began, and a report was released by the Health Protection Branch on March 9, 1984 relating the results of EDB levels in ready-to-eat foods and products normally baked before consumption.[2]

The results of the Health and Welfare Canada study are summarized in Table 1. In this report, acceptable levels of EDB in ready-to-eat products are given as those below 100 ppb. It is also stated that from 70 to 90% of EDB is lost during the baking process.[3] A summary of their data is shown in Table 2, wherein EDB levels in various flours are compared before baking and postbaking. As can be seen, although there is little correlation between EDB levels in the various flours and the resultant baked product, there is a reduction in EDB levels of >70% in more than 60% of the samples. More importantly, in those samples which contained high levels of EDB (>100 ppb), the reduction in the final baked product was >94%. Such findings tend to indicate that products made from flours meeting the maximum acceptable residue limits should contribute "insignificant" amounts of EDB to the diet after baking.

The recent increased interest in EDB prompted the development of a method for the determination of EDB in various flour products. The objectives were to (1) develop a rapid and reliable procedure with sufficient sensitivity to routinely monitor EDB levels in flour and/or flour products, and (2) to apply this procedure to flour samples obtained locally in Saskatoon and compare these to the national analytical results obtained to date.

The analytical gas chromatographic procedure based on electron capture detection has been previously published.[4] This procedure relies on the organic extraction of flour/water mixtures and uses an internal standard, 1-bromo-3-chloropropane, and is outlined in Figure 1. Recoveries of EDB at 10 and 100 ppb were 80.1 ± 2.8% (SD) and 84.4 ± 4.3%, respectively; recovery of the internal standard at the working concentration 500 ppb was 98.3 ± 6.7%. Calibration curves

TABLE 1
Summary of Health and Welfare Canada's EDB Analytical Results[2]

Food class	Analytical results
Ready-to-eat foods	
Citrus juices, vegetables, dried fruit, infant cereal	All samples contained no detectable levels
Baked products, breakfast cereal	97% contained <10 ppb with 80% no detectable levels (13 ppb highest level recorded)
Citrus fruit, fruits	No detectable levels in American citrus; >100 ppb found in shipment of Mexican oranges
Products normally baked	
Flour	No samples of national brands were found to contain more than 100 ppb (range 0—90 ppb); one regional brand in B.C. contained >100 ppb, removed voluntarily
Grain-based mixes	8% of 162 samples were found to contain legally unacceptable levels; these lots were removed voluntarily from store shelves by the manufacturer

TABLE 2
The Influence of Baking on EDB Residues in Flour and Baked Product (Biscuits)[3]

EDB conc (ppb)		
Flour	Biscuit	Reduction in EDB level (%)
6	0.5	92
9	1	89
11	3	73
11	14	—
13	3	77
14	ND[a]	100
14	15	—
18	6	67
21	1	95
27	9	67
28	12	57
30	16	47
34	27	21
53	69	—
67	3	95
98	2	98
1700	29	98
3400	143	96
4000	260	94
4000	66	98
4200	106	99

[a] ND = not detectable.

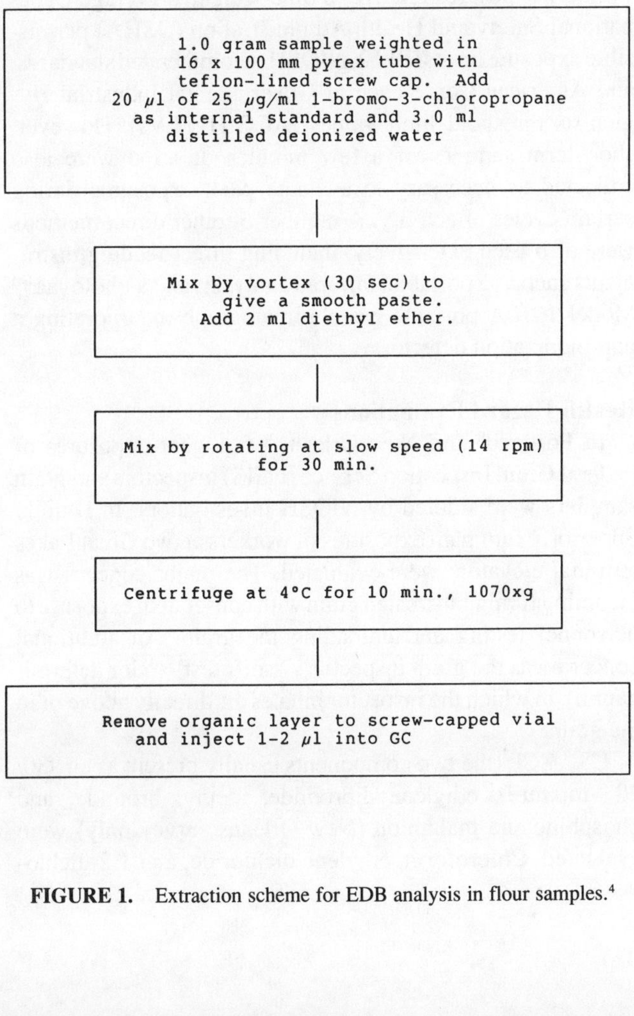

FIGURE 1. Extraction scheme for EDB analysis in flour samples.[4]

covering the range of 5 to 400 ppb were constructed by spiking blank flour with 100 µl of appropriate serial dilutions of EDB in *n*-pentane and 500 ppb internal standard. The chromatographic peaks were measured and the ratio of EDB/internal standard plotted against the concentration of EDB. The resultant curve was linear with a correlation of 0.9997 and overall mean coefficient of variation of 4.5%. The reliability of the procedure was assessed by using gas chromatography combined with mass spectrometry.

The results of the analysis of various flour products with this method are shown in Table 3. The results indicate the utility of this method for routine monitoring of EDB residues in cereal products. The levels found should not contribute "significant" levels of EDB to the diet when considererd in light of the losses of EDB during the baking process, as previously documented.[3] In addition, the reported levels of EDB in these products fall within the range reported for products analyzed by Health and Welfare Canada.[2]

TABLE 3
Analysis of Flour Products for EDB Residues in Saskatoon

Sample[a]	Mean EDB conc (ppb)	Standard deviation	C.V. (%)
F1	274.2	8.57	3.1
F2	405.3	10.19	2.5
F3	5.9	0.81	13.8
F4	50.1	0.49	1.0
F5	24.1	0.74	3.1
F6	22.6	1.25	5.5
F7	5.3	0.32	6.1
F8	6.1	0.53	8.7
F9	12.9	0.46	3.5
F10	4.1	0.42	10.3
F11	1.6	0.09	5.7
CM1	5.92	0.17	2.8
CM2	22.81	1.69	7.4
CM3	146.30	8.05	5.5
B11	ND[b]	—	—

[a] F = flour; CM = commercial white cake mix; B = baked buns.
[b] ND = not detectable.

REFERENCES

1. **Wiersma, D.A., Schnellman, R.G., and Sipes, I.G.,** in *Foreign Compound Metabolism,* Caldwell, J. and Paulson, G.D., Eds., Taylor & Francis, London, 1984, 53.
2. EDB Analytical Results Released, News Release, Health and Welfare Canada, March 9, 1984.
3. **Rains, D.M. and Holder, J.W.,** Ethylene dibromide residues in biscuits and commercial flour, *J. Assoc. Off. Anal. Chem.,* 64, 1252, 1981.
4. **McKay, G.,** Gas chromatographic determination of ethylene dibromide in flour products, *J. Assoc. Off. Anal. Chem.,* 68, 203, 1985.

Summary of Recent Environmental Assessments of Exposure to Grain Fumigants at Export, Inland, and Country Elevators

D. Zaebst, P. Morelli-Schroth, and L. Blade

INTRODUCTION

For many years, various chemicals have been used to protect harvested grain during storage in elevators, warehouses, and mills, and during transport in ships, trucks, and rail cars. Despite several early studies of exposures to a number of these chemicals, and indications of health effects, reports of illnesses due to occupational exposure (e.g., neuropathy and other neuropsychiatric changes) continue to occur.[1,2] In 1957 Paulus et al. reported high exposures (sometimes exceeding the capacity of respiratory canisters to remove the contaminant from inspired air) to a mixture of carbon disulfide (CS_2) and carbon tetrachloride (CCl_4) used to treat shelled corn.[3] A series of liver function tests on these employees, however, did not indicate liver damage.

Jones et al., in 1964, reported breathing zone concentrations of up to 11 ppm of phosphine during the addition of aluminum phosphide tablets to wheat prior to long-term storage or loading out on ships. Physiological responses exhibited by phosphine-exposed workers included chest tightness, vomiting, diarrhea, and central nervous system effects.[4]

More recently, concern about grain fumigant exposures was heightened by findings of potential human carcinogenicity of chemicals such as ethylene dibromide,[5] now banned for use as a grain fumigant by the U.S. Environmental Protection Agency.[6] Of more immediate concern, however, was the potential for illness due to acute exposures of grain samplers, inspectors, and other elevator workers to fumigants such as CS_2, CCl_4, ethylene dichloride, and a number of less-used grain fumigants.[2,7]

As a result, National Institute for Occupational Safety and Health (NIOSH) investigators conducted a number of environmental assessments in 1984 and 1985 aimed at determining the nature and extent of occupational exposures to grain fumigants. So far, all studies have been conducted at grain terminals and elevators; future plans include studies of other aspects of the grain industry (e.g., flour and feed mills). These studies also were aimed at determining whether the changing pattern of pesticide use, due to evolving regulations, might result in hazardous exposures different in quality, as well as quantity, from those in the past.

STUDIES OF PERSONAL FUMIGANT EXPOSURE

Personal exposures to airborne fumigants were typically evaluated using battery-operated, portable sampling pumps in conjunction with the suitable sorbent media specified by the respective NIOSH analytical method,[7] and manifolds which allowed the collection of several simultaneous samples. For example, standard 150-mg dual-section activated carbon tubes were used to sample at flow rates of 100 to 200 cc/min for CS_2, CCl_4, ethylene dibromide, ethylene dichloride, chloroform, and 1,2-dichloroethylene. Air concentrations of malathion and methoxychlor were evaluated at a flow rate of 200 cc/min using 13-mm preextracted, glass fiber filters followed in-line by Supelco Orbo 42™ glass tubes containing Chromosorb 102.[1] Airborne exposures to phospine were determined at a flow rate of 50 cc/min using glass tubes containing mercuric cyanide-impregnated silica gel, as specified by NIOSH Method S-332.[7]

Generally, a full-shift sampling methodology was used to evaluate exposures relative to time-weighted average Occupational Safety and Health Administration (OSHA) permissible exposure limits (PEL), NIOSH recommended standards, and American Conference of Governmental Industrial Hygienists threshold limit values (ACGIH TLVs). However, short-term samples of a few minutes duration were also collected as necessary to evaluate peak exposures during certain cycles of activity. A number of other direct methods were also used extensively, including direct-reading instruments such as a portable, infrared analyzer, and a Photovac™ Model 1010A portable gas chromatograph incorporating a photoionization detector.

Health Hazard Evaluations

In Portland[8] and New Orleans,[9] fumigant exposures of Federal Grain Inspection Service (FGIS) inspectors and grain samplers were studied by NIOSH investigators. In Duluth/Superior,[10] fumigant exposures of workers at two Great Lakes terminal elevators were evaluated. The major concern was incoming fumigant-treated grain with consequent exposure to personnel testing and unloading the grain. An additional concern was the grain inspector's "sniff test" (odor determination), in which the inspector inhales air directly above or in the grain.

CS_2, CCl_4 (the two components usually present as an "80/20" mixture), ethylene dibromide, methyl bromide, and phosphine and malathion (New Orleans survey only) were evaluated. Chloroform, ethylene dichloride, and 1,2-dichloroethane were not detected in any of the samples obtained in Portland and, thus, were not evaluated during the follow-up survey.

All fumigant levels were very low or not detected in the samples obtained (Table 1). This may reflect the low level of grain shipping or, more likely, sporadic treatment of grain. In some cases, these results also reflect a cautious approach by the terminal elevators involved. Some elevator personnel actively test grain, and do not immediately allow the grain to be unloaded into the elevator, or reject the shipment altogether. The testing is typically done using detector tubes or a direct-reading instrument such as a photoionization detector.

However, many grain shipments do not carry proper warning of the fact that the grain has been fumigated, the type of fumigant used, or the length of time elapsed since fumigation. As a result, episodes of reported acute illnesses in two instances led to requests in 1983 and 1984 to NIOSH for health hazard evaluations (HHE).[2,7] In addition, the potential for skin absorption (e.g., in grain samplers) of pesticides remains undocumented.

Important issues were raised in the HHE reports including the controversy surrounding the best method of testing grain for fumigants. One may draw an air sample either from within the grain mass or from the air space directly above the grain. The results of such short-term tests (e.g., in grain-mass sample) are sometimes being inappropriately compared to current occupational standards for personal exposures. Further research is required to determine the best ways of evaluating the extent of hazard of grain shipments on arrival, and the best means of decontaminating grain economically and quickly when necessary.

Survey Conducted by Contract to NIOSH-Liquid Fumigant Exposure

A study of occupational exposure to a liquid grain fumigant known as 80/20 (80% CCl/20% CS_2) was conducted through a contract with the University of Utah. A small country elevator and a large export terminal, both located in Texas, were studied.

At the export terminal, 37 bins of milo (grain sorghum) were fumigated with approximately 600 to 700 gal of 80/20 liquid fumigant. The fumigant was stored in a tank and pumped through a hose to the top of the elevator, where the nozzle was placed into the opening of the bin. The proper amount of fumigant was allowed to flow into the bin, at which time the hose was moved to the next bin. The task required 2 d, and involved up to four different job categories, three of which were sampled using charcoal tubes.

At the country elevator, four truck loads of milo were manually fumigated with 80/20 liquid fumigant by sprinkling can from a raised platform. CCl_4 and CS_2 concentrations were determined using both charcoal tubes and a direct-reading infrared analyzer.

Most exposures were within current OSHA standards, but were frequently in excess of NIOSH recommendations for both CCl_4 and CS_2 both for short-term and full-shift exposures (Table 2). The majority of the exposure in the country elevator occurred during the four 10-min periods of active fumigation; for example, CCl_4 concentrations ranged up to 378 mg/m^3, in excess of all short- term standards with the exception of peak OSHA standard (1260 mg/m^3, not to be exceeded for a maximum of 5 min).

Recommendations included installation of a pump and

TABLE 1
Personal Exposures to Fumigants during Three HHEs — 1984[7,9]

Compound	Geometric mean (μg/m^3) (N[a])			Standards/ guidelines	
	Portland	New Orleans	Duluth	NIOSH	OSHA PEL
CCl_4	480 (10)	1200[b] (26)	ND (37)	12,000	62,000
CS_2	ND (54)	ND (26)	ND (28)	3,000	62,000
Ethylene dibromide	0.48 (12)	—	ND (37)	340	3,800
Methyl bromide	ND (9)	—	—	60,000	80,000
Phosphine	—	ND (6)	—	—	400
Malathion	—	ND (5)	—	15,000	15,000

[a] Geometric mean of samples for which the result was above the limit of detection. If ND (not detected), then all samples were below the limit of detection. In parentheses are the numbers of samples quantifiable, or where ND is given, the total number of samples obtained.

[b] Not detected in samples from five of six field offices. The number represents a mean of four samples from one office.

TABLE 2
Summary of Personal Exposures to CS_2 and CCl_4 (NIOSH Contract No. 210-83-6002)[11]

Location	Concentration (mg/m^3)			
	CCl_4		CS_2	
	Range	GM[a]	Range	GM[a]
Export terminal				
Fumigator[b]	8.8—28.8	15.8	1.8—22.8	9.9
Fumigator helper	7.6—10.7	8.8	2.7—3.6	3.0
Manager/timer	5.0—8.8	6.9	1.8—2.7	2.1
Foreman (unloading trucks)	ND	ND	ND	ND
Grain inspector	ND	ND	ND	ND
Country elevator				
Fumigator[b]	38	—	15.6	—
Real time RbF measurements (IR)	265—378	302	60—132	99

[a] Geometric mean.

[b] Measurements made with a Miran™ Infrared Analyzer during 10-min periods of fumigation activity.

Standards/guidelines

CS_2	OSHA	8-h	TWA PEL:	62 mg/m^3
	NIOSH	8-h	TWA:	3 mg/m^3
CCl_4	OSHA	8-h	TWA PEL:	62 mg/m^3
	NIOSH	8-h	TWA:	12 (ceiling)

metering system at the country elevator, and establishment and training of a "fumigant safety coordinator", to be responsible for more consistent and safer work practices during active fumigation at the export terminal.

Industrywide Studies Program Surveys - 1985

In 1985 NIOSH's Industrywide Studies Program began a series of surveys to characterize current representative exposures to fumigants. To date, data on two of three separate surveys are available. The first survey studied the use of a liquid malathion/methoxychlor mixture used to fumigate the workhouse annex at the top of a small, largely empty, inland elevator prior to the receipt of major wheat shipments. The second, conducted later in the summer at the same elevator, evaluated exposures to phosphine gas during the addition of aluminum phosphide pellets to incoming wheat prior to storage.

During June to August 1985, wheat was received by truck from various other country and inland elevators in the area (approximately 30 to 50 trucks per day). The grain was lifted to the top level of the elevator (bin floor) and discharged onto a conveyor belt running the length of the floor. Aluminum phosphide tablets (approximately 57% aluminum phosphide, 43% inert ingredients) were dispensed through a plastic hose into the grain stream at a rate of 7/min at the point of discharge onto the belt. The treated grain was then conveyed to a tripper positioned to intercept and direct the grain into the designated storage bin.

Phosphine gas exposures were obtained over a 3-week period and represent 8-h, time-weighted average (TWA) exposures, or concentrations over a full shift in the areas measured (Table 3). Personal exposures of the bin floor operator ranged from 0.64 to 2.1 mg/m^3, while those of the supervisor (who spent ≤50% of his time inside the elevator) ranged from <0.02 to 0.78 mg/m^3. Geometric mean concentrations on the bin floor, as measured by area monitors (configured identically to the personal monitors), ranged from <0.02 to 1.92 mg/m^3.

TABLE 3
Summary of Phosphine Sampling Results During Use of Aluminum Phosphide Tablets (Cincinnati, OH, July 3—17, 1985)

	8-H TWA exposures (mg/m^3)[a]		
Job/area	**July 3**	**July 9**	**July 17**
Bin floor operator	2.1	2.2	0.64
Supervisor	0.78	0.43	<0.02
Area — bin floor	1.92[b]	1.25[b]	<0.02[b]
Basement — aerating	—	0.83	1.27

[a] Results of single samples.
[b] Geometric mean of concentrations measured in four areas of the bin floor.

Current standards and recommendations for phosphine gas
NIOSH: none
ACGIH TLV: 0.4 mg/m^3 8-h TWA; 1 mg/m^3, STEL
OSHA 8-h TWA PEL: 0.4 mg/m^3

Most of the phosphine concentrations were in excess of the current ACGIH TLV, and OSHA PEL of 0.4 mg/m^3 (NIOSH does not currently recommend an exposure limit for phosphine in air). The highest full-shift personal exposures measured were five times in excess of these recommendations and standards. Although no short-term measurements of phosphine concentrations were obtained to compare with the ACGIH short-term exposure limit of 1 mg/m^3, many of the 8-h, TWA exposures also exceeded this recommendation. At the time of the survey, only single-use, disposable dust respirators were being worn by exposed personnel.

The most noticeable trend in the data is the substantial drop in concentrations (both personal and area) during the 3-week period of the survey. Prior to the survey, it was expected that concentrations (and exposures) might rise gradually over time due to the increasing volume of treated grain stored in the bins, and to the general lack of room ventilation at the bin floor level. This expected trend did not occur; it is thus more likely that the exposures measured reflected more immediate factors such as the level of activity (volume of grain received, quantity of phosphide tablets used), moisture level of the grain, and temperature of the grain and air.

Ventilation measurements indicated generally no air movement in most areas of the elevator during the survey. As a result, recommendations were given to improve general ventilation in the bin floor area, and to install or extend ducts leading from bin aeration fans through the basement wall and outside the elevator.

Prior to the malathion/methoxychlor fogging at the same elevator (begun on Friday) the top (bin floor) was sealed (windows and doors closed). The fumigation process was conducted by filling three 5-gal cans (fitted with atomizing spray nozzles) with a premixed malathion/methoxychlor solution (approximately 13% in solvent). Compressed air lines were then attached to the cans. The cans were then placed in the area to be fogged and the compressed air turned on. The fumigator then left the area, allowed the fogging to proceed for about 2 h, and finally turned off the compressed air remotely (from outside the elevator). On Monday, the area was reentered, doors and windows opened, and normal elevator operations resumed.

On Friday, during the fumigation process, the fumigator's exposure was low (malathion: <0.5 μg/m^3; methoxychlor: 0.72 μg/m^3) over the 1-h period it took to prepare the solutions and conduct the operation. However, area samples (using pumps preprogrammed to start and stop automatically) obtained on Friday, Saturday, and Monday show a pattern of extremely high aerosol peak on Friday (9500 μg/m^3 malathion, 4400 μg/m^3 methoxychlor), followed by a sharp decline within 24 h (12.6 and 1.1 μg/m^3, respectively). Area concentrations on Monday were comparable to those on Saturday, and the personal samples obtained on Monday indicated the bin floor operator's exposures were comparable to the area results (Table 4). The exposures on Monday may

TABLE 4
Summary of Malathion and Methoxychlor Measurements during Elevator Bin Floor Fumigation by Fogging (Cincinnati, OH, June 7—10, 1985)

	Concentration (μg/m³)		
Job/area	During fumigation	22 h postfumigation	3 d postfumigation
Bin floor operator			
Malathion	—	—	9.3
Methoxychlor	—	—	1.9
Fumigator			
Malathion	<0.50	—	—
Methoxychlor	0.72	—	—
Area — bin floor[a]			
Malathion	9500	12.6	10.2
Methoxychlor	4400	1.1	1.3

[a] Geometric mean of two to three area samples.

Current standards and recommendations
Malathion: 15 mg/m³ (NIOSH, OSHA)
10 mg/m³ (ACGIH TLV)
Methoxychlor: 15 mg/m³ (OSHA PEL)
10 mg/m³ (ACGIH TLV)

have been due to dust suspended during normal operations, which had undoubtedly absorbed high levels of both pesticides. However, no respirable dust measurements were made to prove or discount this theory.

Although none of the airborne exposures approach any current recommended guidelines or standards for these compounds, continued exposure to the pesticides via contact with grain dust (skin absorption and possibly ingestion), as well as dust inhalation, is likely. The environmental persistence of methoxychlor especially may cause continued low-level exposure for months after the fumigation operation. The health significance of these exposures cannot at present be predicted.

CONCLUSIONS AND RECOMMENDATIONS

The HHE surveys conducted thus far suggest that exposures to grain fumigants resulting from the receipt of treated grain at elevators are generally low. However, as indicated in the original reports, these exposures are difficult to measure. This is due in part to the unpredictability of grain fumigation, e.g., how heavily the grain was treated, with which fumigant the grain was treated, and the elapsed time since treatment. Also, the screening methods used to determine whether grain is contaminated and whether it constitutes an exposure hazard were found to be crude and unreliable. It was recommended that these methods constitute an active area of further research.

Exposures of elevator workers to fumigants such as 80/20 (CCl_4/CS_2) and phospine during active treatment of grain at elevators can at times be in excess of current standards and recommended guidelines, posing significant acute, short-term hazards, and thus the potential for acute and chronic health effects. Although the sporadic nature of grain treatment may to some extent reduce the potential for chronic health effects, further environmental assessments should be conducted in a greater variety of workplaces (large and small elevators, flour and feed mills, etc.), and under a greater variety of environmental conditions. Such studies may shed light on the extent of exposures, as well as the need for additional research in the areas of control technology and health effects.

Finally, solid phosphine gas-generating products are increasing in use in many areas of the country, in part due to increasing regulation of some alternative fumigants, such as the liquid 80/20 formulations, ethylene dibromide, etc. Although all of these compounds are associated with their own peculiar hazard, it should not be assumed that the compound (aluminum phosphide) currently replacing these is necessarily safe under all conditions. Proper engineering controls (e.g., ventilation), appropriate work practices, and personal protective equipment (respirators, gloves, and clothing) are necessary when using any hazardous pesticide, including aluminum phosphide.

REFERENCES

1. **Peters, H.A., Levine, R., Matthews, C., Sauter, S., and Rankin, J.,** Carbon disulfide-induced neuropsychiatric changes in grain storage workers, *Am. J. Ind. Med.*, 3, 373, 1982.
2. **Ahrenholz, S.H.,** Health Hazard Evaluation — Superior, Wisconsin: Rep. No. 84-194-1549, National Institute for Occupational Safety and Health, Cincinnati, OH, 1985.
3. **Paulus, H.J., Lippmann, M., and Cohen, A.,** Fumigation of shelled corn with a mixture of carbon disulfide and carbon tetrachloride, *Ind. Hyg. Q.*, 18, 345, 1957.
4. **Jones, A.T., Jones, R.C., and Longley, E.O.,** Environmental and clinical aspects of bulk wheat fumigation with aluminum phosphide, *Am. Ind. Hyg. Assoc. J.*, 25, 376, 1964.
5. Current Intelligence Bull. 37, Ethylene Dibromide (revised), National Institute for Occupational Safety and Health, Cincinnati, Ohio, 1981.
6. Environmental Protection Agency, Decision and Emergency Order Suspending Registrations of Pesticide Products Containing Ethylene Dibromide, 49 FR 4452, February 6, 1984.
7. **Ahrenholz, S.H.,** Health Hazard Evaluation — Federal grain inspection service, Portland, Oregon: Rep. No. 83-375-1521, National Institute for Occupational Safety and Health, 1984.
8. National Institute for Occupational Safety and Health Manual of Analytical Methods, DHHS (NIOSH) publ. No. 84-100, 3rd ed., National Institute for Occupational Safety and Health, Cincinnati, OH, 1984.
9. **Ruhe, R.,** Health Hazard Evaluation — Federal Grain Inspection Service-USDA, New Orleans, Louisiana: Rep. No. 84-281-1607, National Institute for Occupational Safety and Health, 1985.
10. **Ahrenholz, S.H.,** Health Hazard Evaluation — grain elevators, Superior, Wisconsin, Duluth, Minnesota: Rep. No. 84-311-1575, National Institute for Occupational Safety and Health, 1985.
11. **Marano, D.E., White, K.L., Deer, H., and Alexander, G.,** Carbon tetrachloride/carbon disulfide exposures in grain fumigation, GDC-National Institute for Occupational Safety and Health Contract No. 210-83-6002, Final Report, unpublished, 1985.

Potential Consequences of a Fire in an Insecticide Storage Facility

C. Jeffries

INTRODUCTION

The first purpose of this presentation is to use a standard smoke- dispersion model to calculate the influence of wind and atmospheric stability conditions on ground level smoke concentrations of a hypothetical fire in an insecticide storage facility. The second purpose is to estimate the resulting prompt health impacts of such a fire.

Since a detailed account of our study has been published elsewhere,[1] only a summary is given here.

METHODS

The Gaussian plume model for continuous sources was used to model the distribution of the plume arising from the hypothetical fire. The storage facility was assumed to consist of two barns situated 300 m from village A and 15 km from city B. One barn was assumed to contain 11,300 kg of endrin and 48,100 kg of vehicle, cyclohexanone. The other barn was assumed to contain 222,500 kg of dimethoate and 308,500 kg of cyclohexanone.

A major simplifying assumption made in the study was that the insecticides were not altered in the fire, but were propelled intact into the atmosphere by combustion of the vehicle. Data from the toxicology literature were used to estimate the prompt health consequences of exposure to various levels of insecticide concentrations in smoke.

RESULTS

The mathematical model indicated that persons residing a few hundred meters from the storage facility could suffer a variety of adverse effects, including possible death from contact with smoke from a prolonged, low- temperature fire. Such a fire would justify evacuation of village A.

Knowledge of special hazards of cool toxic smoke as well as current atmospheric stability conditions could be useful to fire fighters. That is, fire fighters might wisely choose to promote a fast, hot fire which would propel toxicants high into the atmosphere, rather than risk local fumigation.

REFERENCES

1. **Jeffries, C. and Schiefer, H.B.,** Potential consequences of a fire in an insecticide storage facility, *Regul. Toxicol. Pharmacol.,* 5, 197, 1985.

Agricultural Chemical Hazards in Commercial Mushroom-Growing Operations

E. B. Reczek

INTRODUCTION

Medical and industrial hygiene reports regarding the health hazards faced by mushroom growers have focused on cases of exposure to inhaled organic antigens which elicit a response termed "mushroom worker's lung",[1-5] which in France has been recognized as a compensatable disease. It has been demonstrated that protection from environmental dusts which may provoke allergic alveolitis can be provided by inexpensive industrial dust respirators.[2,6,7] However, other aspects of occupational health at these agricultural worksites are less well characterized.

The successful commercial cultivation of edible mushrooms requires a controlled environment with fastidiously maintained levels of temperature, humidity, carbon dioxide, and ventilation, as well as constant vigilance for diseases and insect infestations. Throughout the production cycle, a mushroom crop may be subjected to a variety of pathogenic organisms such as fungi, bacteria, viruses, nematodes, and insect larvae. At the present time, 73 commercial pesticide preparations are registered by Agriculture Canada for use on edible mushrooms, the majority of which have active ingredients which are organophosphate (OP) pesticides. Other chemicals such as formaldehyde are routinely used in a regimen of general plant hygiene designed to prevent the spread of biological activity from one area of the plant to another. The hazards to farm workers from OP pesticides have been well documented in the literature,[8,9] and three commercial mushroom farms have been the subject of past studies by the National Institute for Occupational Safety and Health (NIOSH).[10-12] A recent paper has presented the results of industrial hygiene surveys for atmospheric ammonia and formaldeyde in an insect-rearing facility.[13] In this paper, preliminary industrial hygiene surveys for pesticides and for formaldehyde conducted in three commercial mushroom-growing operations, typifying the uses of pesticides in the industry in Alberta, are described.

PROCESS

The mushroom *Agaricus bisporus* is the most common commercial crop grown in seven plants in Alberta. The plants which were the subject of this study employed a standard five-step growing process using movable trays. Each firm employed between 25 and 150 workers, usually with more than one shift. A short description of the production steps and of the worker's tasks in each step will demonstrate the opportunities for chemical exposure.

Step 1 — Heavy equipment operators mix the raw material, usually outdoors, and add water to begin the composting process. The operators were exposed to high heat and humidity, particularly during the summer season, as specially designed composting machines, called "turners", mechanically invert the compost stacks and add water every 2 to 3 d. In some areas of the country, pesticides such as diazinon, malathion, or dichlorvos may be added to the compost as a guard against mushroom fly infestation,[14] although this was not done in Alberta plants.

Step 2 — The compost, which in some cases may be pesticide treated, is brought into the plant by a front-end loader, mixed with other ingredients in a separate soil shed area, and mechanically applied to trays or beds, usually made of wood. Before each application of soil, the mechanical tray line is manually washed and sanitized with a 2% formaldehyde solution, preceded in some plants by treatment of the tray- line room with pesticides such as diazinon. The soil shed is treated with formaldehyde solution before formulating the soil mixtures, usually by workers responsible for this part of the plant. Tray-line operators are employed during the filling of the beds, controlling the filling process, manually removing the excess soil from the sides of the beds, and stacking the beds before placement by forklift into cook-out rooms. Dermal exposure to treated soils is a possibility. A pasteurization of "cook-out" with steam eliminates parasitic organisms. Diazinon or dichlorvos may be sprayed in order to control flies during cool down of the compost. Methoxychlor, malathion, and/or pyrethrin may also be added to the compost immediately prior to the next step, for further fly control.

Step 3 — After pasteurization, the nutrient bed is seeded with commercial mycelium/grain preparation called "spawn". Spawn is worked into the bed surface mechanically, on a freshly sanitized tray line, or by hand. Care is taken to maintain as hygienic conditions as possible by following a regimen of disinfection or pesticide treatment of the area. Diazinon and pyrethrin were sprayed by growing-area workers during spawning in two of the plants under study and dichlorvos has also been used to control pests at this stage.[10] Workers have ample opportunity for exposure to treated soils, as well as exposure during chemical treatments and during the cleaning/disinfection of equipment and work areas as part of the general hygiene procedures.

Step 4 — A layer of pasteurized or chemically treated loam called the casing layer is applied to the beds in order to begin the change from the vegetative stage to the reproductive stage in which the fruiting or spore-bearing bodies of the mushroom are allowed to develop. Although the application of toxic chemicals such as methyl bromide to casing material has been discontinued, dichlorvos or pyrethrins are now used to treat casing soils. Recently, methoprene or malathion has also been recommended for application to the casing layer.[15] The treated mixtures are formulated by soil-shed workers, applied to the trays, and the trays returned to the growing rooms. Growing-area personnel may periodically apply benlate and/or zineb, especially where wooden beds are used, in order to prevent the growth of unwanted fungi. When mycelia reach the surface of the bed, ventilation rates are increased and temperature and carbon dioxide levels are dropped in the growing rooms, which causes the bundling of hyphae into pins, the reproductive body of the fungi. Provincial agricultural agencies recommend that diazinon, dimethoate, and/or malathion be used outside the growing rooms at this time, in hallways and corridors, to combat mushroom flies.

Step 5 — The fruiting bodies of the mushrooms develop into crops which are termed flushes. Growing-area workers, called watermen, carefully soak the mushroom beds with water containing chlorine and, often, benlate. The mushroom crop is very carefully monitored at this stage for infestations of pests, and any control measures are taken by growers as needed (Table 1). The crops are normally harvested by hand, by picking crews, when the mushroom cap is at its maximum size. Any diseased or fly-damaged mushrooms are discarded at this point and the affected part of the bed is treated with pesticide (one of diazinon, methoxychlor, dimethoate, or zineb). Growers in one plant used malathion for powder dusting and liquid fogging in the growing rooms. To prevent the development of mushroom disease, most growers allow the production of only three to four flushes before the houses are emptied. Again, cross-contamination of the various growing rooms is prevented by the pesticide spraying of central hallways and doorways and by the use of foot trays containing salt in the growing room doorways. Workers' hands and footwear are frequently disinfected by rinsing in chlorinated water. Once the decision has been made to discard a crop, the growing house is steamed to terminate any residual biological activity in the beds. Waste materials are disposed of in an area remote from the plant. Growing rooms are cleaned and sprayed with pesticide and/or formaldehyde prior to the introduction of another crop.

In all plants examined, formaldehyde was dispensed into buckets, diluted to 1.5 to 2%, and used to disinfect the soil-shed areas and the tray lines following each mechanical application of compost, spawn, or casing mix. Spawn rooms, growing rooms, and delivery trucks or other mechanical equipment were steam cleaned or washed and then disinfected with formaldehyde solutions as part of the plant's general hygiene procedures.

Initial walk-through surveys indicated that such commercial growing operations generally relied on respiratory protection and some form of personal protective equipment to prevent overexposure of workers to pesticides and to formaldehyde. Applicators usually wore boots, aprons, or rubber jackets and full face-piece cannister masks approved for pesticides, or for organic vapors in the case of formaldehyde applicators. One plant, however, did use half face-piece, air-purifying respirators for formaldehyde spraying. Outer protective pants or coveralls, gloves, and hats were often not worn during spraying and there was no mandatory washup following the operation in order to minimize dermal exposures. Programs of respiratory protection, respiratory fit testing, and biological monitoring of high risk groups of workers (compost barn and soil-room blenders and tray-line sprayers, spawn-room and growing-room sprayers) did not exist. Areas recently treated with chemical were often not posted to restrict entry.

TABLE 1
Diseases of Mushrooms, Their Causative Agents, and Treatments

Disease	Causative agent	Chemical treatment
Mycelial damage	Larvae of scarid, phorid, cecid flies	Diamethoate, diazinon Dichlorvos, malathion, pyrethrin, methoprene
Mite contamination	*Tyrophagus* mite	Malathion
Casing soil mildew:		
soft mildew and green mildew	*Dactylium dendroides* *Trichoderma koningi*	Formaldehyde, zineb, calcium hypochlorite
Bacterial blotch, pit, and mummy	*Pseudomonas* sp.	Chlorinated water
Dry bubble, brown spots	*Verticillium fungicola*	Zineb, benomyl, calcium hypochlorite, formaldehyde
Wet bubble	*Mycogene perniciosa*	Benomyl at casing, calcium hypochlorite, formaldehyde

CHEMICAL SAMPLING METHODS

Personal samples for malathion, diazinon, benomyl, zineb, and formaldehyde were taken in 1985 during spraying operations, at the breathing zones of soil -room and tray-line sprayers and of spawn-room and growing-room personnel (lapel samples), including the watermen. Since the greatest chemical exposures would initially be expected in the soil-shed and growing-area applicators, during fly control programs, sampling was concentrated on these workers.

In the case of both full shift and short-term pesticide samples, the previously calibrated sampling pumps drew a known volume of air through a 37-mm Teflon filter which was followed in series with a glass tube containing 75/37 Chromosorb 102. These were then analyzed by gas chromatography for malathion/diazinon or by high-performance liquid chromatography for benomyl. Zineb was estimated by the analysis for total zinc using atomic absorption spectrophotometry.

All formaldehyde samples were taken by impinger sampling using a 1% sodium bisulfite absorbing solution, reaction with chromotropic acid, and colorimetric measurement in the Alberta Occupational Health and Safety Laboratory, as per the NIOSH method (P + CAM 125).

RESULTS AND DISCUSSION

Personal samples in one plant were obtained from growers during malathion dusting and liquid fogging operations in the growing rooms and during benomyl fogging and watering operations executed by the watermen. At no time did the applicator's exposure levels, considering both individual pesticides and combined effects, approach the Alberta occupational exposure limit (OEL) of 10 mg/m^3 — time-weighted average for malathion and for benomyl. These workers' exposure levels were, in all cases, at least an order of magnitude lower than occupational health standards in effect at that time. Zineb levels were also extremely low (less than 0.3 mg/m^3 for STEL samples).

Data on the atmospheric concentrations of diazinon and formaldehyde in two commercial mushroom farms are presented in Tables 2 and 3. Short-term samples of growing-room applicators show a potential overexposure of the applicator to diazinon levels in excess of the 15-min OEL with one 15- to 20-min spraying of a growing room per applicator per day. At a frequency of application greater than this, for example, due to an infestation, daily chemical use at multiple sites would provide for a greatly increased applicator exposure, possibly resulting in depressed serum cholinesterase and neurological effects.

In Alberta, the hygienic limits for many OP pesticides carry the "skin" notation, as an indication that absorption through skin may provide a significant route of entry for that toxicant. Dermal exposure to pesticides would be significant for applicators wearing respirators, but not wearing protective clothing, as was seen during initial walk-through surveys.

TABLE 2
Commercial Mushroom Farms — Diazinon Exposure Levels, Ranges, and Means by Operation

Operation		Application time 8 h TWA[a] exposure (mg/m^3)	Application time 15 min TWA[a] exposure (mg/m^3)
Personal samples			
Growing room — fogging/dusting	Range	0.004—0.031	0.23—1.79
	Mean	0.020	0.94
	n	5	13
Area samples			
Spawn rooms/ growing rooms — fogging/dusting	Range	0.004—0.027	0.042—2.0
	Mean	0.012	1.07
	n	4	8
Alberta OELs for diazinon (mg/m^3)		0.1	0.3

[a] Time-weighted average.

TABLE 3
Commercial Mushroom Farms — Ceiling Formaldehyde Exposure Levels, Ranges, and Means by Operation (All Results Expressed as ppm)

Operation		Farm A	Farm B
Personal samples			
Tray-line washing	Range	0.87—4.81	4.4—11.9
	Mean	2.7	7.4
	n	7	7
Soil-shed washing	Range	6.5—10.1	—
	Mean	8.4	—
	n	8	—
Area samples			
Tray-line washing	Range	0.46—69	0.45—15.4
	Mean	12.4	4.5
	n	7	10
Soil-shed washing	Range	5.5—15.3	—
	Mean	9.3	—
	n	8	—
Alberta ceiling OEL (ppm)		2	

Complete suits of protective clothing and regular cleaning would prevent this. One plant's chlorinated rinse water used by growers showed detectable levels of malathion and diazinon. This practice of rinsing hands has been documented in others plants[10,11] and could provide additional exposure where significant pesticide contamination of the water is found. Rinsing under a flow of fresh chlorinated water which drains into the plant's waste system would prevent this exposure. It has been shown that excessively chlorinated water, some OP pesticides, and benomyl and zineb may be skin irritants and possible sources of sensitization.[10] Such exposures are easily prevented by the wearing of rubber gloves (unlined to facilitate removal of residues inside the glove), for example, by the watermen to prevent exposure to benomyl.

From the exposure of chemical applicators in mushroom-growing areas and given the amount of pesticide utilized, it is clear that programs of respiratory protection, including fit testing and biological monitoring of high-risk workers, are necessary. Use of protective clothing covering exposed skin, especially hands, must be mandatory and enforced, if necessary. Treated areas should be posted to prevent unauthorized entry, and accurate records should be kept of chemical use. Ample guidance from Departments of Agriculture and from Health and Safety Agencies is available concerning the use of agricultural chemicals and these should be strictly adhered to.

During the harvesting stage, the carbon dioxide levels in the growing rooms recommended by grower's guides commonly range between 800 and 1000 ppm, as opposed to 20,000 to 30,000 ppm during pinning, but these levels may become considerably higher in older plants built with roof vents or with meager ventilation in the houses. The growing rooms should be ventilated before growers or picking crews enter, to avoid headaches or other problems associated with exposure of workers to atmospheres containing elevated levels of carbon dioxide. It should be noted, however, that a number of such symptoms (headaches, weakness) may also be seen with chronic OP pesticide exposures.

As mentioned previously, the spray application of formaldehyde plays a major role in the routine hygiene of all commercial mushroom operations examined. On the basis of the measurements presented, it is not possible to conclude whether the workers responsible for plant hygiene in commercial mushroom farms as a job category are overexposed to formaldehyde. Interplant variability with respect to room sizes, etc. makes comparisons difficult. However, it can be seen that workers carrying out these general hygiene procedures have a high potential risk for overexposure to formaldehyde. Care must be taken in the mixing of dilutions for application and proper respiratory protective equipment and protective clothing must be worn by all workers involved in such operations. Air-purifying respirators are currently approved by NIOSH for use in atmospheres containing less than 30 ppm formaldehyde and a supplied air device will provide greater protection in those plant areas where high concentrations of formaldehyde are found. The potential solutions to worker overexposure include ventilation, personal protective equipment, and the substitution of less hazardous materials. The efficacy of other chemicals (quaternary ammonium compounds or chlorine or iodine-containing compounds) to replace formaldehyde as a topical disinfectant should be investigated. Departments of Agriculture have also warned of a possible reaction between hypochlorite and formaldehyde which may produce *bis*-chloromethyl ether, a potent carcinogen. This may be averted by administrative controls designed to preclude the mixing of these chemicals.

Last, some farms have turned to the use of metal trays for compost beds. Such trays are easily sanitized and would not be a source of larval or fungal contaminant as is often found in older commercial operations. In one such plant examined in Alberta, high efficiency filtered air and metal trays have eliminated most chemical use, except for the application of formaldehyde in general plant hygiene. However, the biological hazards, if any, which will arise from these operations should be investigated. Care must be taken that personal protective equipment and work practices eliminate worker exposure to biologically active materials during filter changes, maintenance, and similar operations.

CONCLUSIONS

With some exceptions, the levels of agricultural chemical exposure in commercial mushroom operations did not exceed current occupational health standards. Measurements were made during the spraying of spawn rooms and growing rooms with OP pesticides such as diazinon and during the spraying of formaldehyde solutions in soil sheds and tray-line areas as part of general in-plant hygiene. Recognizing that samples were limited in number, sampling showed that presently accepted exposure limits were exceeded during spraying in these areas. In large commercial operations with insufficient personal protective equipment for applicators, these activities would give the workers multiple exposures via respiratory and skin routes. Attention should be paid to the quantitation of exposures to these agents on the basis of occupational hygiene data. Formal programs of respiratory protection, personal protective equipment, and biological monitoring must be instituted to ensure the health of workers. Consideration should be given to newly emerging technology and to recently registered agricultural chemical products, as these have the potential to adversely affect worker health.

REFERENCES

1. **Sakula, A.,** Mushroom worker's lung, *Br. Med. J.,* 3, 708, 1967.
2. **Schulz, K.H., Felten, G., and Hausen, B.M.,** Allergy to the spores of *Pleurotus Florida* (letter), *Lancet,* 1, 29, 1974.
3. **Murphy, D.M.F., Atkinson, B.F., Nedwich, A., and Galgon, J.P.,** Mushroom worker's lung, *Am. Rev. Respir. Dis.,* 123, 97, 1981.
4. **Calderone, R.A., Russo, D., and Deahl, K.L.,** Serological studies of mushroom worker's lung disease, *Annu. Meet. Am. Soc. Microbiol.,* 82 (Abstr.), F32, 1982.
5. **Marland, P., Tabart, J., Bersay, Cl., Peninou-Gall, G., Bory, J., Labarre, C., and Elbaze, P.,** Le poumon du champignonniste, *Poumon Coeur,* 38, 371, 1982.
6. **Hendrick, D.J., Marshall, R., Fauxad, M.A., and Krall, J.M.,** Protective value of dust respirators in extrinsic allergic alveolitis. Clinical assessment using inhalation provocation tests, *Thorax,* 36, 917, 1981.
7. **Lacey, J., Nabb, S., and Webster, B.T.,** Retention of actinomycete spores by respirator filters, *Ann. Occup. Hyg.,* 25, 351, 1982.
8. **Popendorf, W.J. and Leffingwell, K.,** Regulating OP pesticide residues for farmworker protection, *Res. Rev.,* 82, 125, 1982.
9. **Griffith, J. and Duncan, R.C.,** Grower reported pesticide poisonings among Florida citrus fieldworkers, *J. Environ. Sci. Health,* 20, 61, 1985.
10. National Institute for Occupational Safety and Health, West Foods, Ventura, California, Health Hazard Evaluation Rep. No. HETA 81-366-1248, Cincinnati, OH, 1985.
11. National Institute for Occupational Safety and Health, Fillmore Dole Mushrooms, Castle and Cooke Foods, Fillmore, Utah, Health Hazard Evaluation Rep. No. HETA 81-138-1563, Cincinnati, OH, 1985.
12. National Institute for Occupational Safety and Health, Modern Mushroom Farms, Kennett Square, Pennsylvania, Health Hazard Evaluation Rep. No. HETA 84-255, Cincinnati, OH, 1985.
13. **Merkle, S.E.,** Ambient concentrations of formaldehyde and ammonia in a screwworm fly rearing facility, *Am. Ind. Hyg. Assoc. J.,* 46, 336, 1985.
14. **Cantello, W.W., Hendersen, D., and Argauer, R.J.,** Variation in sensitivity of mushroom strains to diazinon on compost treatment, *J. Econ. Entomol.,* 75, 123, 1982.
15. **Cantello, W.,** Control of a mushroom-infesting fly (Diptera: Sciaridae) with insecticides applied to the casing layer, *J. Econ. Entomol.,* 76, 1433, 1983.

Characterization of Causes of Reproductive Failure Using Rat Embryo Cultures

N. W. Klein

INTRODUCTION

The use of *in vitro* cultures of whole rat and mouse embryos for studies of teratology has increased considerably over the last several years. We have made use of midhead-fold-stage rat embryos (9.5 d of gestation with the morning after mating considered 0.5 d) and cultured them for 48 h following the technique originally developed by New et al. at Cambridge University.[1] The selection of this particular stage of rat embryo development to use for experiments was in part dictated by the technical ease and success that can be achieved with this stage, but this period also encompasses such critical steps in development as neural tube closure, formation of sense organs, the establishiment of body curvatures, and the initiation of a circulatory system which also makes this stage particularly useful.

It appears that four general approaches have been used with rat embryo cultures to study teratology. First, the simplest and most frequently used approach has involved the addition of a chemical directly to the culture medium. Second, in a slight modification of the direct addition approach a metabolic activating system such as a liver S9 fraction or hepatocyte cultures have been added along with the chemical to the culture medium. In a third approach, embryos have been exposed to a chemical within the intact animal for various periods of time and, subsequently, the embryos were removed from the uterus and placed into *in vitro* culture. Finally, we have made use of the fourth approach which has involved the culture of rat embryos on blood serum taken from an animal (rat, monkey, cow, or human subject) exposed to known or unknown teratogens.

Our use of rat embryos as a general bioassay to monitor serum for teratogens was based on two concepts. First, the use of a whole embryo in a teratogenic test rather than an organ or cell culture would provide the entire developmental repertoire to a test. Second, the use of serum from an exposed animal as the material to be tested might reveal as teratogens not only the administered chemical, but also metabolites of the chemical as well as long-term physiological changes caused by the chemical. Certainly, one could anticipate problems such as false positives, when, for example, a substance present in the serum would not pass through the placenta, or false negatives such as when the teratogen was a short-lived metabolite.

In our first experiments[2] rats were injected with various dosages of cadmium chloride or cyclophosphamide and after various periods they were sacrificed and bled. The blood serum was then used as a culture medium for head-fold-stage rat embryos taken from untreated pregnant rats. Depending on dosage level and time interval following dosing, embryos after 48 h of culture on these sera developed reproducible abnormalities and showed growth inhibition (protein and DNA accumulation). This suggested that rat embryo cultures could be used to monitor blood serum for teratogens. In addition, we found that it was quite critical to select embryos for experimentation within a narrow range of development or the test would not be reproducible. Thus, for all subsequent studies we have selected only midhead-fold-stage embryos for culture (we average two to three useful embryos per rat).

Soon after these initial studies, we found that rat embryos could be cultured on serum from some human subjects if the blood glucose levels were adjusted to 300 mg%.[3] Furthermore, rat embryo abnormalities were observed within serum samples taken from patients receiving chemotherapy for cancer or anticonvulsants for epilepsy. These studies suggested that rat embryos could be used to directly test human subjects. Hopefully the test would provide insights into the basis for human differences in teratogenic sensitivity, and provide a rational basis for the selection of animal models appropriate to the human condition, but more important it was hoped that this test would serve to identify unknown and previously unsuspected human teratogens. It now appears that these expectations will be realized.

We have studied two areas of high risk for reproductive failure. These include epilepsy and recurrent or habitual spontaneous abortion. Both of these conditions have been associated with long lists of possible etiologies, but successful diagnosis for individual subjects appears to be infrequent.

EPILEPSY

Nutrient Deficiencies

Rat embryos were cultured for 48 h on sera from 128 different epileptics receiving monotherapy with either carbamazepine, valproic acid, phenytoin, phenobarbital, or no drug.[4] The lowest frequency of teratogenic serum samples (samples that cause embryo abnormalities) was observed from subjects receiving phenobarbital, and this was comparable to the frequency observed with sera from epileptics receiving no drug therapy. However, regardless of the drug being used, it was not possible to relate the parent drug level in a serum sample to the teratogenicity of that sample (i.e., a dose-response relationship was not apparent).

Our first step in the analysis of this discrepancy was to dialyze the sera and retest the samples with the embryo cultures. To achieve this seemingly simple experiment it was necessary to first develop a nutrient mixture of essential small molecules that would be lost from the sera on dialysis. A mixture of water-soluble vitamins, essential amino acids, and glucose was formulated. As dialysis in conjunction with the use of this supplement overcame the teratogenicity of several serum samples, as a control the essential small nutrient mixture was added to sera also without prior dialysis. Much to our surprise, the teratogenicity of many serum samples was completely reversed. Indeed, of 53 teratogenic serum samples, 32 or 60% were completely corrected by the addition of the nutrient mixture without the need for dialysis. Furthermore, by sequentially eliminating nutrients from the mixture, it was possible to identify serum samples with specific vitamins, amino acids, or combinations of both that were essential to support normal rat embryo development. This suggested for the first time that rat embryo cultures might be useful to identify nutrient deficiencies. However, it is a large jump from rat embryos growing on a serum sample to a nutrient deficiency in the subject who had donated the serum. We will return to this critical question.

Phenytoin-Serum Protein Complex

From these first studies it appeared that many epileptics could have nutrient deficiencies which could be related to the anticonvulsant medication or to diet. We have been working on the possible causes for the apparent nutrient deficiencies. More recently we also became concerned with those teratogenic serum samples that remained teratogenic even after the addition of the vitamin and amino acid supplement. For this work we have concentrated on phenytoin because of its wide use. When serum samples were taken from rats receiving high dosages of phenytoin (up to 1 g/kg), serum teratogenicity was not observed. However, when an inhibitor of epoxide hydrase (TCPO or 1,2,epoxy 3,3,3,trichloropropane) was combined with the phenytoin gavage, serum teratogenicity was observed. This suggested, as others have noted,[5] that an arene oxide metabolite of phenytoin was the proximal teratogen. To avoid the complexities involved with the use of an inhibitor we turned to the use of rhesus monkeys and found a more appropriate human model. When monkeys were dosed by gavage with 275 mg/kg body weight phenytoin and bled after 10 h, their sera were teratogenic to cultured rat embryos. Abnormalities included failure of the neural tube to close, abnormal body curvature, anophthalmia, and inhibition of growth.

Next, we attempted to determine the basis for the teratogenicity of sera from these phenytoin-dosed monkeys. Previously, we found that phenytoin itself was not teratogenic when added to serum within therapeutic levels.[4] Considering first the possibility that the teratogenicity was due to a phenytoin metabolite, sera from dosed monkeys were extracted with ethyl acetate and the extract (less the ethyl acetate) was added to control monkey serum. The extract lacked activity so the possibility of protein-associated teratogenicity was next considered. Ammonium sulfate (70% v/v) precipitates of teratogenic sera were added to control monkey serum, and the precipitates from the phenytoin-dosed monkeys, but not control precipitates, were found to be teratogenic. Next, DEAE Sephacel ion exchange column chromatography was used with a stepwise elution, Tris-phosphate buffer (pH 7.4). Fractions at 0.1, 0.2, and 0.5 *M* were dialyzed, concentrated, and added at constant protein levels to control serum for embryo cultures. Only the 0.2 *M* fraction showed teratogenic activity.

To verify this finding and extend the observation, a polyclonal antibody to allantoin (the ring structure of phenytoin containing a free amine group) linked to hemocyanin was raised in rabbits. For Western blot analysis samples were run on SDS polyacrylamide gels, transferred to nitrocellulose membranes, and the membranes were incubated in the allantoin IgG and ^{125}I goat anti- rabbit IgG. Binding was visualized by autoradiography. By this technique a reactive protein was observed between transferrin and albumin at about 80,000 Da in the 0.2 *M* fraction obtained with the DEAE Sephacel chromatography. This protein was not observed in nonteratogenic monkey serum.[6]

The allantoin antibody has also been used to compare the movement of free phenytoin and the apparent phenytoin-protein complex within the cultured rat embryos. In these studies rat embryos were cultured for 24 h on sera taken from monkeys given phenytoin at 275 mg/kg, or phenytoin was added directly to control monkey sera at 50 μg/ml. Embryos were frozen and sectioned at 7 μm. The sections were treated with the allantoin antibody and then fluorescence conjugated goat anti-rabbit antibody. With directly added phenytoin, all fluorescence was limited to the extraembryonic yolk sac, while with serum from dosed monkeys, fluorescence was observed in the yolk sac as well as in the tissues of the embryo proper. These observations suggest that the teratogenicity of the phenytoin-protein complex may be related to the ability of the complex in comparison to free drug to reach the cells of the embryo proper.

Recently, the first phenytoin human serum sample was analyzed and the 80,000 Da protein was observed following Western blot analysis. This human teratogenic sample was selected for analysis because it was not corrected by the nutrient supplement. Thus, by using this approach we hope that it will soon be possible to distinguish those epileptics at risk for reproductive failure, because of a particular nutrient deficiency from those at risk, because of a high level of a teratogenic phenytoin-protein complex.

SPONTANEOUS ABORTION

Monkeys

Although the studies with phenytoin outlined above indicate the value of using monkeys in the study of teratogenicity, our initial studies with monkeys provided some insights that were less predictable. The first tests with monkey sera were conducted in collaboration with the primate facility at the University of California, Davis. Sera from 18 rhesus monkeys at three stages of their menstrual cycle were tested with our standard 48-h cultures of midhead-fold-stage rat embryos.[7] The stages of menstrual cycle did not influence rat embryo growth and development, but 2 of the 18 serum samples caused rat embryo abnormalities regardless of menstrual cycle stage. Subsequently, it was learned that these two serum samples that caused the abnormalities had been taken from monkeys which had not experienced successful pregnancy outcomes after years of mating trials. The other 16 of the Davis group had all experienced excellent reproductive histories. This observation was subsequently extended to a larger group of monkeys in conjunction with the primate facility of the University of Washington, and, again, the rat embryo culture test distinguishes sera from monkeys with poor reproductive histories (high-risk breeders) from those with excellent reproductive histories (low- risk breeders).

To determine the cause for the teratogenicity of these monkey serum samples, the amino acid and vitamin supplements were added first to the sera and the sera was retested by embryo culture. This failed to overcome the teratogenicity. Next, sera from the monkeys were dialyzed against control rat sera, and several were reversed by this treatment (five of eight tested). However, rather than continue the analysis of these dialyzable samples, we became most intrigued by those that failed to be reversed by dialysis and have concentrated our attention on this group. Turning to ammonium sulfate precipitation, fractions obtained at 0 to 35, 35 to 55, and 55 to 75% of saturation were tested by adding each fraction at constant protein to control serum. Embryo toxicity was essentially limited to the 0 to 35% fraction which, on SDS-polyacrylamide analysis, was found to contain the heavy and light chains of IgG.

To identify the rat embryo antigen reacting with the IgG from the monkeys, indirect immunofluorescence was used and the antibody was found to react with basement membranes and particularly Reichert's membrane. This membrane was made soluble, run on SDS polyacrylamide gel electrophoresis, transferred to nitrocellulose, treated with monkey sera, ^{125}I anti-monkey antibody exposed, and finally autoradiographed for visualization. The major reactive protein appeared to be laminin, an extracellular matrix protein. More recently, purified laminin obtained from Dr. H. Kleinman of the National Institute of Dental Research was injected into female monkeys with excellent reproductive histories and whose sera supported normal rat embryo development. Subsequent to the injection their sera became highly toxic to the cultured rat embryos and the sera were found to contain antibodies to laminin. These monkeys who had been injected with laminin appear to have a normal reproductive cycle, but have failed to have a successful pregnancy or become pregnant following the injection of laminin.

Humans

Naturally, an attempt was made to determine if sera from women suffering from recurrent abortions also might prove to be teratogenic in the rat embryo test. Serum samples were obtained from 52 women who had experienced 2 or more spontaneous abortions (the average was 4 spontaneous abortions with a range of 2 to 18).[8] When tested with rat embryo cultures, sera from 31 (60%) were found to be teratogenic. In contrast, sera from a control group of 15 pregnant women who had subsequent normal pregnancy outcomes gave a frequency of four (27%) teratogenic samples. The difference between the two groups (60 and 27%) was significant. Two additional findings from this study should be clearly noted. First, the frequency of teratogenic serum samples was directly related to the number of previous spontaneous abortions experienced by the subject. This relationship did not vary significantly from a straight line ranging from the 27% frequency observed for sera from zero abortions to 100% frequency for sera from those subjects who had experienced six or more spontaneous abortions. The second important finding regards nutrient supplementation. Of 26 teratogenic serum samples that were retested following supplementation with vitamins and amino acids, a striking 20 or 77% were completely reversed. Approximately half were corrected by vitamins alone and the other half by amino acids alone.

Nutrient responses had been observed with human sera from epileptics and now from spontaneous abortions; however, the relevance of these findings to the nutritional status of the serum donor has not been apparent. We have now completed an analysis with regard to this question for one subject. A patient of Dr. Pauline Olsen of Hartford, CT was 41 years old and had experienced two spontaneous abortions at approximately 50 d of pregnancy during the previous year. When her serum was tested with rat embryo cultures, embryos showed failure of neural tube closure. However, through extensive testing it was found that if three amino acids were added to the cultures normal neural tube closure occurred. The three amino acids were methionine, tryptophan, and arginine. Through dietary supplements (500 mg/d for each of the three amino acids) serum from the subject became capable of supporting normal rat embryo development without additional supplements. Subsequently, the patient became pregnant and in April of 1985 she gave birth to a normal child. However, analysis of her diet did not implicate deficiencies of protein or amino acids. Dr. R. Hillman of Washington University School of Medicine, St. Louis suggested the possibility that the patient had an amino acid absorption problem and offered to run fecal amino acid analysis. The subject showed extremely high levels of amino acids in her feces which clearly demonstrated a severe amino acid absorption defect.

Cows

Spontaneous abortions in dairy herds present a severe economic problem for the owner of the herd and could represent a potential for human food chain contamination. We recently found that rat embryos could be cultured on cow serum if the serum was supplemented with methionine. Recently, we tested sera from herds without reproductive problems and two herds where spontaneous abortions were a severe problem. The results have been striking, as extensive rat embryo abnormalities were observed with aborting herds and essentially no abnormalities were detected with non-aborting herds. We have started serum analysis with one herd. Preliminary findings suggested the presence of a dialyzable toxic substance in these sera samples. Chemical analysis of these sera samples has now been initiated.

REFERENCES

1. **New, D.A.T.,** Whole-embryo culture and the study of mammalian embryos during organogenesis, *Biol. Rev. Cambridge Philos. Soc.*, 53, 81, 1978.
2. **Klein, N.W., Vogler, M.A., Chatot, C.L., and Pierro, L.J.,** The use of cultured rat embryos to evaluate the teratogenicity of serum: cadmium and cyclophosphamide, *Teratology,* 21, 199, 1980.
3. **Chatot, C.K., Klein, N.W., Piatek, J., and Pierro, L.J.,** Successful culture of rat embryos on human serum: use in the detection of teratogens, *Science,* 207, 1471, 1980.
4. **Chatot, C.L., Klein, N.W., Clapper, M.L., Resor, S.R., Winger, W.D., Russman, B.S., Holmes, G.L., Mattson, R.H., and Cramer, J.A.,** *Epilepsia,* 25, 205, 1984.
5. **Martz, F., Failinger, C., and Blake, D.A.,** Phenytoin teratogenesis: correlation between embryopathic effect and covalent binding of putative arene oxide metabolite in gestational tissue, *J. Pharmacol. Exp. Ther.*, 203, 231, 1977.
6. **Clapper, M.L., Klein, N.W., and Silvester, S.A.,** Characterization of the teratogenic form of phenytoin in serum, *Fed. Proc.,* 44 (Abstr.), 1123, 1985.
7. **Klein, N.W., Plenefisch, J.D., Carey, S.W., Fredrickson, W.T., Sackett, G.P., Burbacher, T.M., and Parker, R.M.,** Serum from monkeys with histories of fetal wastage cause abnormalities in culture of whole rat embryos, *Science,* 215, 66, 1982.
8. **Gilles, P.A., Rhinehardt, A.E., Carey, S.W., Klein, N.W., Muechler, E., Miller, R.K., Mariona, F.G., and Greenstein, R.M.,** Sera from women with histories of recurrent spontaneous abortion induce abnormalities in whole rat embryo cultures, *Teratology,* 29 (Abstr.), 23A, 1984.

Cancer Risks in Agriculture

Ecogenetics and Agriculture

G. J. Brewer

INTRODUCTION

It is rewarding to see the topic of ecogenetics considered at an international symposium on health and safety in agriculture, because it symbolizes an awareness and recognition by the agricultural community of the special susceptibilities to environmental agents of certain of our population due to their genes. First, some definitions: what is "ecogenetics"? Let us start with the concept and definition of pharmacogenetics, because it is in pharmacogenetics that ecogenetics had its origin. Pharmacogenetics is the study of the role of inheritance in producing variation in the response to certain drugs. This is quite a broad definition, because it includes both discrete and continuous variation attributable to genes. For example, if one examines any biological parameter after drug administration, say blood level of the parent drug, a distribution of values will be seen in the "normal" population, perhaps in a given case a normally distributed or bell-shaped distribution. In a typical case the values in this distribution result from the interaction of many factors: rate of absorption, rates of clearance by the liver, rates of several metabolic routes, rate of loss in urine, etc. Each of these factors is likely to have some genetic influence, perhaps each factor being influenced by several genes as well as environmental influences. Thus, the person in the upper tail of the distribution is there in large part because of a constellation of genetic factors influencing the drug level to be higher. If high levels of the parent drug tend to be toxic, the people in the upper tail of the distribution are more likely to suffer toxicity. This, then, is one aspect of pharmacogenetics — that of the role of inheritance in producing continuous variation in response to drugs. Likewise, genes are important here: Vessell[1] has estimated the heritability of metabolism and/or disposition of several common drugs to be quite high — over 0.9.

In contrast, the more classical and dramatic concept of pharmacogenetics is that of discrete differences in response to drugs due to the effect of a single gene.

Proceeding from the example of pharmacogenetics, in 1971 I introduced the term "ecogenetics".[1] Broadly defined, it is simply the role of inheritance in producing variation in response to environmental agents. As I envisioned it, however, it was more narrowly the study of inherited variation in the response of humans to agents introduced into the environment by human activities. I had a specific example in mind when I introduced the term, and it might be useful to briefly review this and another example to set the stage. The first has to do with serum α_1-antitrypsin deficiency and pulmonary emphysema. Homozygotes for this deficiency are at a much increased risk for emphysema. There is some evidence to suggest that heterozygotes may be at an increased risk if they smoke cigarettes. Since pulmonary emphysema is an "urbanized" disease, we hypothesized years ago that urban air pollution might be inducing the disease in carriers of this gene.[2] The carrier frequency may be as high as 2 to 5% in many populations, and we postulated that since cigarette smoke and urban air pollution share many of the same kinds of pollutants, a genetic subgroup of our population might be at increased risk for emphysema due to what we are doing to our urban air.

The second example is aryl hydrocarbon hydroxylase (AHH) inducibility and lung cancer. The concept is that AHH, one of the cytochrome P-450 family of enzymes, can convert certain relatively harmless hydrocarbons to carcinogens. A paper several years ago indicated that the inducibility of AHH was under genetic control, and that high inducers had a much higher risk of lung cancer.[3] This area has subsequently been somewhat controversial, particularly the genetics of the system. However, several studies have tended to confirm that high inducibility is associated with increased rates of lung cancer. The implications are clear. When we pollute the environment with certain hydrocarbons, certain of us, those with genetically determined high inducibility of AHH, are at a greater risk of lung cancer.

With respect to drugs, if a drug is biologically active (as most presumably are), because there is always genetic variation in every biological system, that drug, if used enough, will somewhere, sometime, find a target population of people who are uniquely susceptible to its side effects because of their genotype. The extension to ecogenetics of this concept is straightforward. If an environmental agent is active, if there is human exposure as there almost invariably is, and if the biological system with which the agent interacts is an important one in the human, then that agent, be it pollutant, pesticide, or herbicide, will find a uniquely susceptible target population.

Now, as I indicated in the beginning, the definitions of ecogenetics are broader than inferred above by using the example of chemical agents. The sun causing skin cancer in people who are susceptible to the damaging effects of the sun because of their genes is a legitimate part of ecogenetics. However, here I wish to focus mostly on the extension of the concept of drug toxicity to chemical toxicity.

ECOGENETICS AND AGRICULTURE

The Paraoxonase System

Paraoxon is the active metabolite of a very commonly used organophosphorus insecticide, parathion. Due to the exposure of large numbers of agricultural workers, hundreds of deaths each year are attributed to this toxic agent.[4] Toxicity seems to be related to inhibition of essential acetyl cholinesterase, particularly in the central nervous system. Detoxification has been thought to be due to either or both of two mechanisms. One is covalent bonding to nonessential cholinesterases, such as those of the serum or the red cell. The other is hydrolysis by a serum enzyme called paraoxonase. Recently, Butler et al.[5] have shown in the rabbit that paraoxonase hydrolysis is a much more important detoxifying mechanism than covalent bonding. If the same is true in the human, it means that the level of serum paraoxonase may be a critical determinant of rate of detoxification of paraoxon. If humans vary in paraoxonase activity, it could be an important determinant of susceptibility to toxicity.

In this connection, exciting work in several laboratories[6-8] has revealed the presence of a genetic polymorphism in paraoxonase activity. The same enzyme has esterase activity toward aryl ester substrates, such as phenylacetate, as well as hydrolytic activity towards organic phosphates such as paraoxonase.[8] However, as La Du and Eckerson[9] have shown, arylesterase activity is unimodal in the population, while paraoxonase activity is trimodal. Careful human genetic studies of these three modes have revealed a single locus system, called ESA, with two alleles, A (low activity) and B (high activity). The frequencies of the three activities in the U.S. white population is estimated to be 50% low (AA), 43% intermediate (AB), and 7% high (BB).

In another paper, this same group[10] has used incubation studies with paraoxon to show that the degree of inhibition of serum cholinesterase is dependent on paraoxonase activity. Serum from people of A phenotype lose more cholinesterase activity than those of B phenotype, with AB type intermediate but closer to the B phenotype.

The hypothesis seems to me to be, then, as suggested by La Du and Eckerson,[10] that the serious toxicities including deaths from exposure to organic phosphate pesticides are occurring primarily in that portion of the population with genetically determined low activity of paraoxonase. It seems critical to me to test this hypothesis. If it is correct, then pretesting of agricultural workers prior to exposure would seem to be strongly warranted. In this connection at a symposium on ecogenetics and pharmacogenetics held in 1984, King said that at that point in time there was insufficient data from any of the pharmacogenetic and ecogenetic systems to indicate genetic screening related to the workplace.[11] I maintain that if we find that deaths and severe toxicity from paraoxon are concentrated in low paraoxonase individuals, we have our first example where screening and counseling should occur.

Other Agricultural Agents

As one leaves the above example of paraoxonase variation and paraoxon toxicity, one is immediately immersed in a sea of unknowns and speculation regarding other agricultural agents and their potential ecogenetic-type toxicity. As the examples in Table 1 illustrate, agents are often developed on the basis of their specific interactions with target and non-target enzyme systems — systems shared by the human and other mammalian organisms. Their potential toxicity involves many factors — degree of exposure and degree of intake (such as absorption) with a given exposure are two — but if part of the protective link is human metabolism of the agent, then variation in that metabolism will cause variation in susceptibility to toxicity.

TABLE 1
Mechanisms of Certain Agricultural Xenobiotics

Agent	Use	Potential toxicity
Organophosphates Carbamates and pyrethroid	Insecticides (insects have much less aryl esterase, paraoxonase, and carboxyl esterase activity than nontargets)	Low paraoxonase ? Mutant cholinesterase
Mexcarbamate	Against spruce budworm which activates by deacetylation, while higher organisms do not	? (The human has deacetylation systems)
Atrazine Propachlor	Herbicides — target plants unable to conjugate with GSH, while nontargets can	? Defective GSH transferase or other GSH abnormalities
Diclofop-methyl Clofop-isobutyl	Herbicides — target plants activate by aryl hydroxylation, while nontargets do not	? People with high AHH activity or inducibility

RECOMMENDATIONS

It might be useful for me to list some recommendations for consideration by the agricultural community for minimizing the potential toxicity of new and even old xenobiotics to be used for agricultural purposes. In doing this I must confess I am not burdened with very much knowledge about present practices and regulations as they refer to agriculture in either the U.S. or Canada. This may be advantageous in that a fresh, unbiased look at what should be happening may be desirable. On the other hand, you are free to label me naive if you choose.

One of the defects in toxicity studies, be they for drug or environmental agent evaluation, is the usually complete lack of any attention to genetic variation. Thus, in the case of a drug, toxicity may be evaluated in several species, but many of these may be highly inbred, leaving little room for identifying genetic-drug interactions. In the case of drugs, pharmacogenetic-type toxicity is left to be discovered after human exposure, usually through case reports. The same appears to be now occurring in agricultural ecogenetics. We have cases of people dying from parathion exposure. We will now investigate (I hope) and if we find that they are low paraoxonase types, action may ensue. Part of the sense of the recommendations listed below is to become more predictive of potential toxicity, thereby hoping to avoid a good deal of it.

First, the producer of a potential new agricultural agent presumably must prepare an environmental impact statement. If it does not already, that statement should include a detailed analysis of potential human toxicity, including ecogenetic-type toxicity.

Second, to evaluate ecogenetic-type toxicity, much attention should be paid to the mechanism of action and of detoxification of the agent and what is known about those mechanisms, and their genetic variation in the human (and other mammals). For example, if an agent is detoxified by acetylation or by paraoxonase, it is clearly predictable, without any human exposure, that there will be ecogenetic-type variation in toxicity.

Third, test systems should be set up to evaluate ecogenetic toxicity of every new agent. The first level of these should be *in vitro* to the extent possible. However, *in vivo* tests with volunteers are also a viable approach, in my opinion, with proper attention to prior animal experiments, dose, and safety. For example, we believe that we can now identify hemolytic potential to glucose 6-phosphate dehydrogenase (G6PD)-deficient patients from new drugs using an *in vitro* incubation system with mouse liver microsomes to provide hydroxylation. However, it is also possible to safely evaluate hemolysis in healthy G6PD-deficient volunteers.

Fourth, once ecogenetic-type toxicity is identified, if there is to be human exposure, there should be prescreening of those to be exposed, and counseling about the results. Exposure of those found to be susceptible should be minimized by job selection etc.

I recognize that these recommendations have much to be complained about — they are expensive, bureaucratically burdensome, and remove some of the agricultural expertise and creativity away from what appears to be more productive lines. However, consider the alternatives. The toxicity of an agent such as parathion is evaluated by exposing mice, rats, rabbits, dogs, etc. to the agent, and a level of exposure which is probably safe to humans established. However, because of inbreeding, small sample size or chance, discrete, genetic variation present in the human affecting the metabolic detoxification of the drug is not present in the animals studied. It is discovered only after several thousand humans die of parathion poisoning. This outcome generates a black eye for the company producing it, a black eye for the industry, and increased governmental intervention and regulation, to say nothing of lawsuits and of the human misery involved.

REFERENCES

1. **Vessell, E.,** Pharmacogenetics: multiple interactions between genes and environment as determinants of drug response, *Am. J. Med.,* 66, 183, 1979.
2. **Brewer, D.J.,** Human ecology, an expanding role for the human geneticist, *Am. J. Hum. Genet.,* 23, 92, 1971.
3. **Kellerman, G., Shaw, C.R., and Luyten-Kellerman, M.,** Aryl hydrocarbon hydroxylase inducibility and bronchogenic carcinoma, *N. Engl. J. Med.,* 289, 936, 1973.
4. **Hayes, W.J.,** *Pesticides Studied in Man,* Williams & Wilkins, Baltimore, 1982.
5. **Butler, E.G., Eckerson, H.W., and La Du, B.N.,** Paraoxon hydrolysis vs. covalent binding in the elimination of paraoxon in the rabbit, *Drug Metab. Dispos.,* 13, 640, 1985.
6. **Playfer, J.R., Eze, L.C., Bullen, M.E., and Evans, D.A.P.,** Genetic polymorphism and interethnic variability of plasma paraoxonase activity, *J. Med. Genet.,* 13, 337, 1976.
7. **Geldmacher, V. and Mallinckrodt, M.,** Polymorphism of human serum paraoxonase, *Hum. Genet.,* 1 (Suppl.), 65, 1978.
8. **Eckerson, H.W., Wyte, C.M., and La Du, B.N.,** The human serum paraoxonase/arylesterase polymorphism, *Am. J. Hum. Genet.,* 35, 1126, 1983.
9. **La Du, B.N. and Eckerson, H.W.,** The polymorphic paraoxonase/arylesterase isozymes of human serum, *Fed. Proc.,* 43, 2338, 1984.
10. **La Du, D.N. and Eckerson, H.W.,** Could the human paraoxonase polymorphism account for different responses to certain environmental chemicals?, in *Banbury Report 16: Genetic Variability in Responses to Clinical Exposure,* Cold Spring Harbor Laboratory, Cold Spring Harbor, NY, 1984, 167.
11. **Omenn, C.S. and Gelboin, H.V., Eds.,** *Banbury Report 16,* Cold Spring Harbor Laboratory, Cold Spring Harbor, NY, 1984.

Cancer, Genes, and Agriculture

H. H. McDuffie, J. A. Dosman, and D. J. Klaassen

Although neoplastic diseases seem to be ubiquitous, several studies have demonstrated that most healthy adult subjects[1-3] do not report having a first- degree relative (parent, sibling, child) with cancer. Families with sporadic cases are more frequent than families with two or more cases. At least four studies have demonstrated higher than expected numbers of cancers in relatives of lung cancer patients compared to relatives of their spouses, to relatives of age-, sex-, race-, and residence-matched controls, or to relatives of patients with cancer at other sites.[4-7] We have evaluated the importance of a positive family history of malignancy in relation to the risks associated with cigarette smoking and occupational exposure to hazardous substances.

We used a population-based registry, the Saskatchewan Cancer Foundation, to identify all patients with a diagnosis of primary lung cancer made between January 1979 and November 1983. The Saskatchewan Hospital Services Plan data bank was used to randomly select controls matched for age (±3 years), sex, and area of residence. Each individual was contacted by mail with a letter explaining the nature of the study, a consent form, and a short questionnaire requesting information on personal health, occupation, and smoking habits. They were also asked to identify first-degree relatives who had a diagnosis of neoplastic disease at any site or any pulmonary diseases. Reported neoplasms were documented where sufficient information was provided.

Responses from 931 patients or their next-of-kin and 1031 controls have been analyzed. The male/female ratio of patient responders was 3.98 and of controls 3.51. Eighty-seven (9.3%) patient families had at least one first- degree relative with primary lung cancer with a total of 103 cases. Fifty (4.8%) controls had one or more first-degree relatives with primary lung cancer, yielding a total of 56 cases ($p <0.005$) (Table 1). Each family was counted only once. The mean ages at diagnosis of male (64.6 ± 9.6, n = 67) and female patients (63.1 ± 11, n = 17) with a positive family history were not significantly different from those patients with a negative family history (66.7 ± 9.6, n = 610 and 62.8 ± 11, n = 149, respectively).

The smoking pattern of familial cases of primary lung cancer was similar to that of the total patient population. There were very few lifetime nonsmokers (4.8%) among them compared to 30% of controls with a positive family history. Current smokers among the patients had mean pack years of 46 ± 2.8, n = 32 and ex-smokers had mean pack years of 45.5 ± 5.2, n = 36. Estimated relative risks due to smoking (RR = 10.8) were much larger than those attributable to the familial effect (RR = 2.02).

Among males there were no significant occupational differences between patient and control groups (farmers 40.8 vs. 47.1%). No occupation other than farming was reported by more than four individuals in each group. The majority of females in both groups were homemakers.

There were no differences in the frequencies of exposures to herbicides, insecticides, fumigants, uranium, radiation, grains, arsenic, nickel, chromium, diesel fumes, or mining in male patients with and without a positive family history of lung cancer. Higher frequencies of exposure to potash (RR = 1.9) and welding (RR = 1.4) in the familial group were not significant. There were estimated relative risks of 2.5 for exposure to fungicides (95% CL = 1.3 to 4.9) and asbestos (95% CL = 1.2 to 4.2) among the familial group, although the absolute numbers exposed to each were small.

There were no differences in the distribution of lung diseases in families of individuals with a positive family history of lung cancer. One hundred and six patient families (11.4%) reported at least 1 case (total 134 cases) vs. 96 for control families (total 110 cases)($p >0.05$).

However, there were significant differences in the number of families (patients n = 227, vs. controls n = 208, Table 2) who reported cancer at sites other than the lung ($p = 0.025$). Neoplastic disease of the breast, colon/rectum, uterus, and prostate was the most frequently reported. There were many more nonlung cancers reported in sisters of patients than brothers, parents, or offspring, a reversal of the finding for

TABLE 1
Distribution of First-Degree Relatives of Patients and Community Control Subjects with Reported Primary Lung Cancer

	Patient		Control	
	n	%	n	%
Primary lung cancer				
One other case	77	8.30	45	4.4
Two other cases	6	0.64	4	0.39
Three other cases	2	0.02	1	0.01
Four other cases	2	0.02	0	
No other cases	844	90.6	981	95.2
Total families	931		1031	
Family history of lung cancer				
Positive	87	9.3	50	4.8
Negative	844	90.6	981	95.2
Total	931		1031	

Note: Chi square = 15.2, df = 1, $p <0.005$.

TABLE 2
Distribution of Cases of Cancer in Families of Patients with Lung Cancer and Community Controls

	Patients		Controls	
	n[a]	%	n[a]	%
Cancer site				
Breast	43	4.6	33	3.2
Colon/rectum	33	3.5	21	2.0
Uterus	12	1.3	4	.4
Prostate	9	1.0	17	1.6
Skin	11	1.2	12	1.2
Other known site	119	12.8	121	11.7
Affected				
Yes	227		208	
No	704		823	

Note: Chi square = 5.02, df = 1, p =0.025.

[a] n = number of families.

lung cancer. A male relative was more than twice as likely to be affected with primary lung cancer as a female relative regardless of the sex or patient/control status of the proband (Table 3). Table 3 also shows that the frequency of affected first- degree relatives is not affected by the sex of the proband. Among nonsporadic cases of primary lung cancer, we have observed more cases of sibling site concordance (64) than cases of parent/offspring concordance (36) (Table 4).

Male patients with a positive family history of lung cancer were less likely to report a chest injury (RR = 0.48, NS) and more likely to have arthritis (RR = 2.3, p <0.025) than the sporadic male patient group. There were no differences in the frequencies of reports of physician-diagnosed bronchitis, asthma, emphysema, pleurisy, tuberculosis, sinus trouble, hay fever, or heart disease. Overall, respiratory-related illness was no more frequent than among male patients with a negative family history.

In recent years, evidence that cancer has important genetic components has been accumulating from many scientific disciplines.[6-12] At the cellular level, a malignancy is an alteration in cell structure and characteristics which is passed to all the descendents of the transformed cell. Most theories of carcinogenesis relate alterations in DNA, or modifications of regulatory mechanisms governing DNA replication to the earliest stages of carcinogenesis. While neoplastic diseases appear to be ubiquitous, several studies[1-3] have demonstrated that most healthy adults do not have a first-degree relative with cancer. Albert et al., in a study of parents of mothers of newborns, found 66% reported no first-degree relatives with a malignancy at any site, 26% had one affected relative, and 8% had two or more affected relatives.[2] The respective values in self-selected volunteers were 47, 33, and 20%.[3] Thus, a minority of families may account for a disproportionate number of malignancies.

Tokuhata and Lilienfeld,[5] in their classic study of primary lung cancer patients and age-, sex-, race-, and residence-matched controls, found that 2.2% of patients' relatives had primary lung cancer and 7.8% had malignancies at other sites compared to 0.6 and 5.6% respectively, of the controls' relatives. In a high risk area of Louisiana,[4] 20% of patients' families vs. 8.3% of their spouses' families had a first-degree relative with primary lung cancer. In two studies comparing relatives of probands with lung cancer to relatives of probands with cancer at other sites,[6,7] Lynch et al. found a trend toward higher frequencies of lung cancer specifically in relatives of

TABLE 3
Gender of Lung Cancer-Affected Relatives of Patients with Primary Lung Cancer and Random Community Control Subjects with Positive Family Histories

	Patients		Controls	
	n	%	n	%
Male relatives	69	67.0	40	71.4
Female relatives	34	33.0	16	28.6
Male/female	2.03		2.5	

Note: Male relatives are twice as likely to be affected than female relatives of patients or community control subjects.

Relationship to Affected Relatives by Gender of the Proband Patient

	First-degree relatives of male patients (n)	First-degree relatives of female patients (n)
Father	19	6
Mother	7	4
Brother	34	9
Sister	18	3
Son	1	
Daughter	2	
	81 (78.6)	22 (21.4)

Note: The average number of first-degree relatives affected is not influenced by the sex of the proband, as males represent 79.2% of respondents. Affected male relatives of male probands represent 66.7% and affected male relatives of female probands represent 68.2% of the respective totals. Male relatives are twice as likely to be affected as female relatives of male or female probands.

TABLE 4
Relationship to Proband

Affected relative with primary lung cancer	Patients		Controls	
	n	%	n	%
Mother/father	36	35.0	16	28.6
Sister/brother	64	62.1	38	67.9
Offspring	3	2.9	2	3.6
Ratio parents/siblings		0.56		0.42

Note: Among the nonsporadic cases, there is more sibling concordance than parent-offspring concordance.

lung cancer probands. This trend reached significance in relatives of probands with adenocarcinoma of the lung. The Saskatchewan incidence of lung cancer is low. Cancer is a reportable disease. We found that 9.3% of patients had at least one first-degree relative with primary lung cancer.

If there are interactions between inherited susceptibility factors and environmental ones, the sibling concordance could reflect a large environmental component acting on one generation or an inherited propensity to develop malignancy which requires only a small environmental trigger. The parent/offspring pattern may reflect a greater hereditary and a lesser environmental influence. We were unable to evaluate the effect of relatives' smoking habits.

Cohen has postulated a common familial component as responsible for the development of primary lung cancer and chronic obstructive airways disease.[13,14] Tokuhata and Lilienfeld found excess mortality due to pulmonary disorders in relatives of the patients with lung cancer compared to age-, sex-, race-, and residence-matched controls.[5] Our data, based on a general question concerning lung diseases, do not confirm these observations.

Previous observations that female patients with primary lung cancer are more likely than males to show a familial etiological influence are not supported by our finding that the frequency of affected first-degree relatives is not influenced by the sex of the proband. One of the major studies was conducted in 1963 when the effect of changing female smoking habits was not as apparent. In our study, the estimated relative risk attributed to smoking is much larger than that due to familial factors for both sexes.

Cigarette smoke, which contains a variety of carcinogens including initiators, promotors, and activators, produces lung cancer in an estimated one in eight long-term heavy smokers.[15] Of course, many may die of other smoking-related disorders before the development of lung cancer due to its long latent period. However, the fact that many heavy smokers do not succumb to lung cancer even after prolonged exposure, while a small but significant proportion of passive smokers do, indicates that there are differences in host susceptibility. In this study, cigarette smoking habits did not differ between those patients with positive or negative family histories of this disease. The mean age at diagnosis was not significantly lower for patients with a positive family history, indicating that synergism between smoking habits and a positive family history was not a significant factor. The estimated relative risk due to cigarette smoking (RR = 10.8) is much larger than that due to a positive family history (RR = 2.02).

Farmers make up a significant proportion of our provincial population and the majority are current or ex-smokers.[16,17] The use of agricultural pesticides and fertilizers is extensive.[18] Several previously used chemicals have been classified by the International Agency for Research on Carcinogenesis as "probably carcinogenic in humans based on insufficient human evidence but sufficient evidence of carcinogenicity in animal studies".[19,20] Our studies and others have demonstrated that only a small percentage of farmers use masks when handling these substances. It is therefore probable that chemical inhalation occurs. Occupational exposures, with the exception of fungicides and asbestos, were not significantly different between sporadic male patients and those with a positive family history.

In conclusion, familial clustering of primary lung cancer occurs; the estimated relative risk of a positive family history is 2.02. The familial factor is independent of sex of the proband, and of occupation and occupational exposures. Cigarette smoking is a greater hazard than either occupation or family background. There is no apparent synergism between positive smoking and family histories, although there is a nonsignificant trend to a lower age at diagnosis among male patients with both factors. More male than female relatives are affected regardless of the patient/control status or the sex of the proband, and more siblings than parents are affected. These two findings point to a strong environmental influence in the etiology of this disease.

REFERENCES

1. **Lynch, H.R., Guirgis, H., Albert, S., and Brennan, M.,** Familial breast cancer in a normal population, *Cancer,* 34, 2080, 1974.
2. **Albert, S. and Child, M.,** Familial cancer in the general population, *Cancer,* 40, 1674, 1981.
3. **Albano, W.A., Lynch, H.T., Recabaren, J.A., et al.,** Familial cancer in an oncology clinic, *Cancer,* 47, 2113, 1981.
4. **Ooi, W.L., Mulvey, J.J., Chen, V.W., Wilson, A.F., Corcoran, K.A. and Rothchild, H.,** Familial aggregation of lung cancer, *Clin. Res.,* 29, 315A, 1981.
5. **Tokuhata, G.K. and Lilienfeld, A.M.,** Familial aggregation of lung cancer in humans, *J. Natl. Cancer Inst.,* 30(2), 289, 1963.
6. **Lynch, H.T., Fain, P.R., Albano, W.A., Ruma, T., Black, L., Lynch, J., and Shonka, M.,** Genetic/epidemiological findings in a study of smoking associated tumours, *Cancer Genet. Cytogenet.,* 6, 163, 1982.
7. **Lynch, H.T., Kimberling, W.J., Markvicka, S.E., Biscone, K.A., Lynch, J.A., Whorton, E., and Mailliard, J.,** Genetics and smoking associated cancers, a study of 485 families, *Cancer,* 57, 1640, 1986.
8. **Gilbert, F.,** Chromosomes, genes and cancer: a classification of chromosome abnormalities in cancer, *J. Natl. Cancer Inst.,* 71(6), 1107, 1983.
9. **Hamlyn, P. and Sikora, K.,** Oncogenes, *Lancet,* p. 326, 1983.
10. **Nebert, D.W.,** Pharmacogenetics: an approach to understanding chemical and biologic aspects of cancer, *J. Natl. Cancer Inst.,* 64(6), 1279, 1980.
11. **Bishop, J.M.,** Oncogenes and proto-oncogenes, *Hosp. Pract.,* August, 67, 1983.
12. **Strong, R.,** Genetic etiology of cancer, *Cancer,* 40, 438, 1977.
13. **Cohen, B.H.,** A common familial component in lung cancer and COPD, *Lancet,* 2, 523, 1977.
14. **Cohen, B.H.,** Chronic obstructive pulmonary disease: a challenge in genetic epidemiology, *Am. J. Epidemiol.,* 112(2), 274, 1980.
15. **Mulvihill, J.J.,** Host factors in human lung tumours: an example of ecogenetics in oncology, *J. Natl. Cancer Inst.,* 57, 3, 1976.
16. **Warren, C.P.W. and Manfreda, J.,** Respiratory symptoms in grain farmers: a brief report, in *Occupational Pulmonary Disease: Focus on Grain Dust and Health,* Dosman, J.A. and Cotton, D.J. Eds., Academic Press, New York, 1980, 383.
17. **Dosman, J.A., Graham, B.L., Cotton, D.J., et al.,** unpublished data, Humboldt area farmers.
18. Agriculture: Saskatchewan 1981 Census of Canada Catalogue, 96-

909, 1982.

19. International Agency for Research on Carcinogenesis, Monographs on the Evaluation of the Carcinogenic Risk of Chemicals to Man: Some Organochlorine Pesticides, Lyon, 1974, 5.

20. IARC Monographs on the Evaluation of the Carcinogenic Risk of Chemicals to Man: Some Fumigants, the Herbicides 2,4-D and 2-4,5-T, Chlorinated Dibenzodioxins and Miscellaneous Industrial Chemicals, Lyon, 1977, 15.

Lung Cancer in Saskatchewan Males

H. H. McDuffie, D. J. Klaassen, and J. A. Dosman

Blot and Fraumeni, in their study of the geographical distribution of primary lung cancer in the U.S. have concluded that the age-standardized rates increase with increasing urbanization.[1] Also, the rates in farming counties, defined as those with more that 25% rural farm residents, are substantially lower than in rural nonfarming counties. One might, therefore, infer that farmers are at low risk of developing this disease. This inference has not previously been tested in a population-based study with data on individual smoking habits. This led us to develop a study with the following objectives: (1) to test the hypothesis that a higher percentage of lung cancer patients in this province has a past history of farming than do randomly selected community control subjects matched for age, sex and residence; and (2) to evaluate exposure to environmental agents, including dusts, chemicals, and cigarette smoke.

The records of the Saskatchewan Cancer Foundation, a population-based agency, were used to provide a profile of each patient with a diagnosis of primary lung cancer made between January 1979 and November 1983. One of us extracted the following from each chart: names and addresses of patients and next-of-kin, ICDA code number, cell type and site of the neoplasm within the respiratory system, age at diagnosis, sex, previous personal health including documented cancer at any site and surgical operations, familial medical history, occupation, alcohol consumption, and smoking history.

A short questionnaire (available on request), a letter explaining the study, and a consent form were mailed to each patient or, if the patient was known to be deceased, to the next-of-kin. The questionnaire requested information on the personal health, smoking history, and occupation of the patient.

Control cases matched for age (±3 years), sex, and area of residence were generated at random from the Saskatchewan Hospital Services Plan register. The same questionnaire was sent to these subjects.

The province of Saskatchewan, located in western Canada, has a population of slightly more than 1 million. The two largest cities are Regina and Saskatoon with populations of 162,984 and 175,000, respectively. Industrial air pollution is minimal. Heavy industry, potash, and uranium mines have only recently commenced operation.

Saskatchewan offered a unique opportunity to conduct this study for several reasons. The Saskatchewan Cancer Clinics, in operation since 1932, register almost 100% of patients with lung cancer. The population is large enough to produce sufficient patients for statistical analyses, relatively stable in number, and small enough to attempt 100% patient inclusion. A significant proportion of our population are farmers and previous studies have indicated that a majority of them are current or ex-smokers. A majority of farmers in this province are grain, especially wheat, producers (63.9%).[2] Only 2.5% are dairy producers and 11% cattle producers. The use of agricultural pesticides and fertilizer is extensive. Several of the chemicals used have been classified by the IARC as "probably carcinogenic in humans based on insufficient human evidence but sufficient evidence of carcinogenicity from animal studies".[3-5] Of the residents, 95% are of European and British ancestry, 4% are Native Indian. The relative lack of industrial pollution both in ambient air and within buildings and factories removes a complicating factor present in most studies. The Saskatchewan Hospital Services Plan registers almost 100% of the province's residents and provides a pool from which age-, sex-, and residence-matched controls may be drawn.

Of 927 patients with lung cancer and 987 controls, there were 730 (78.7%) and 768 (77.8%) males, respectively. There were no significant differences in the frequencies of farmers between the two groups (Table 1). A possible bias toward respondents with a farm background may have existed in the control group due to interest generated among them by the explanation of the study contained in the correspondence. We attempted to counter this possible effect by including an addendum in the second mailing to nonresponders. We checked this possibility in the patient subset by telephoning 100 nonresponders to ascertain their occupation and found that an even higher percent were farmers.

Of respondents, 23 to 48% reported occupational exposure to grains, diesel fumes, herbicides, insecticides, or welding (Figure 1). Between 5 and 11% reported exposure to fungicides, fumigants, asbestos, mining or potash. Less than 5% were exposed to arsenic, nickel, chromium, uranium, or radiation. Significantly more patients than controls were exposed to grains ($p < 0.05$), fungicides ($p < 0.05$), and mining ($p < 0.05$). Estimates of relative risks were also calculated. While several of these were elevated sufficiently to produce statistical significance individually, they were small in comparison to the risk associated with a positive history of smoking.

Multiple exposure (muluse) is a variable which counts the number of hazardous substances or processes to which an individual is occupationally exposed. Among patients, farmers were exposed to significantly more classes of substances than other patients ($p < 0.001$) (Table 1). Comparing the distribution of this variable between patients and controls demonstrated no significant differences. Among farmers the

TABLE 1
A Comparison between Male Patients and Male Randomly Selected Community Controls of Broad Occupational Groups and Occupational Exposures

	Patients		Community control subjects	
	n	%	n	%
Farmers	343	47.3	370	48.2
Other known occupation	349	48.1	337	43.9
Unknown	33	4.5	61	7.9

$x^2 = 1.07$, df = 1, NS, unknowns were omitted

Muluse[a] (n)	Patients		Community controls	
	n	%	n	%
1—3	294	46.1	266	40.9
4—6	169	26.5	175	26.9
≥7	39	6.1	43	6.6
Unexposed	136	21.3	166	25.5
	638		650	

$x^2 = 4.57$, df = 3, NS

	Patients			
	Farmers		Other known occupations	
Muluse[a]	n	%	n	%
≤3	132	41.0	161	88.0
4—6	150	46.6	21	11.5
≥7	40	12.4	1	0.5
	322		183	

$x^2 = 107.14$, df = 2, $p < 0.001$

	Farmers only			
	Patients		Community controls	
Muluse[a]	n	%	n	%
1—3	129	41.0	134	40.4
4—6	148	47.0	158	47.6
≥7	38	12.1	40	12.0
Total farmers users	315		332	

$x^2 = 0.03$, df = 2, NS

[a] Muluse — a variable which counts the number of hazardous substances or processes selected because of known or suspected carcinogenicity or capacity to impair lung function.

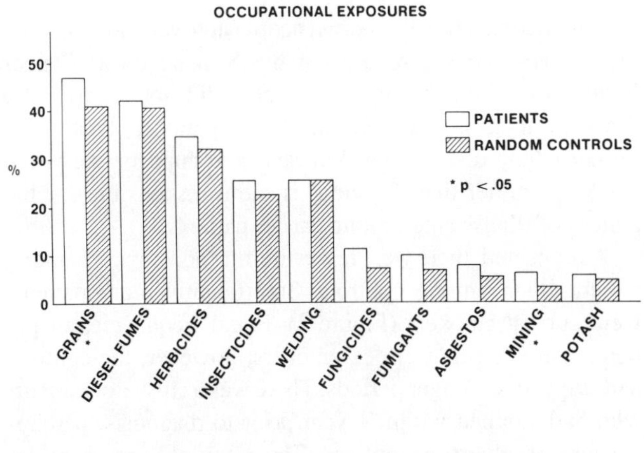

FIGURE 1. Random control subject questionnaire vs. patient questionnaire responses — occupational exposure, males.

multiple-use index between the two groups was equivalent. The interaction between exposures to hazardous substances and smoking pattern of these subjects was not significant. We did not have data on the specific chemicals used, the extent of use, the duration of exposure, or the frequency of accidental inhalations or spills on hands or clothing.

Similar frequencies of asthma, pleurisy, arthritis, hay fever, sinus trouble, and tuberculosis were reported by patients and controls (Figure 2). Patients reported significantly higher frequencies of bronchitis ($p < 0.001$), emphysema ($p < 0.001$), pneumonia ($p < 0.001$), chest injury ($p < 0.001$), farmer's lung disease ($p < 0.01$), and heart disease ($p < 0.01$). Several of these are smoking associated: bronchitis, emphysema, and heart disease; while chest injury and farmer's lung disease are not. There was also a higher frequency of previous or concurrent neoplasms at other sites among the patients. The most common second neoplasms were basal cell or squamous cell carcinoma of the skin (36), adenocarcinoma of the prostate (25), colon/rectum (16), lip (11), and bladder (11). In addition to these second neoplasms, 27 individuals have had a third

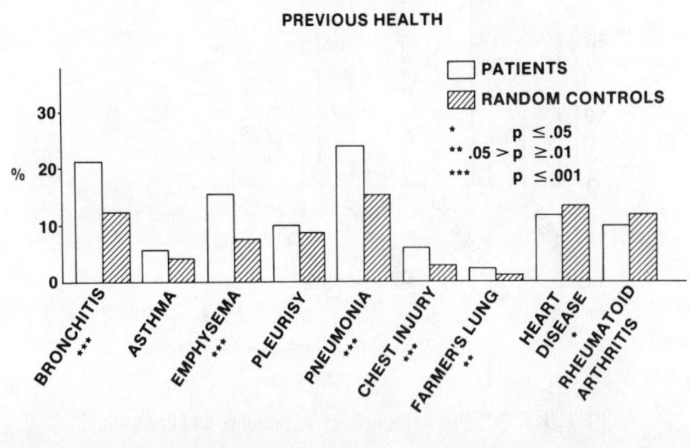

FIGURE 2. Random control subject questionnaire vs. patient questionnaire responses — previous health.

primary neoplasm. All reported neoplasms were documented by reference to the records of the Saskatchewan Cancer Foundation. Among patients, farmers did not report more illnesses than persons in other occupations, except for farmer's lung disease and skin cancer. It appears likely that smoking (rather than farming) is more responsible for the pattern of illness among lung cancer patients.

As expected, there were fewer lifetime nonsmokers among patients than among controls. Significantly more patients were current smokers (Figure 3). Equal frequencies of patients and controls were ex-smokers; however, the controls had stopped for longer periods. There was a cluster of patients who had stopped within 1 year prior to diagnosis, perhaps because of adverse symptoms. The mean pack years of patients was significantly higher than that of controls, although current smokers among male controls had a mean of 41.2 ± 1.8 (n = 129) and ex-smokers, 28.3 ± 1.1 (n = 318) pack years.

Contrary to several other studies, which concluded that farmers are more likely than persons in other occupations to be lifetime nonsmokers, we found no such dichotomy in our patient or community control data. The pack years of farmers were equivalent to those in other occupations. Among male patients, however, there were more lifetime nonsmoking farmers than expected, suggesting that there may be a risk factor associated with farming which is independent of smoking.

We also evaluated the possibility that male lifetime nonsmoking patients with lung cancer might be representatives of extreme environmental insults or especially susceptible due to familial factors as evidenced by a positive family history of

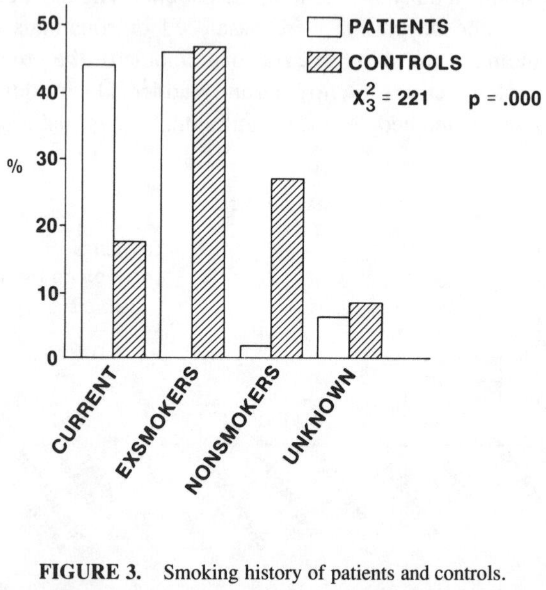

FIGURE 3. Smoking history of patients and controls.

neoplasms. The mean age at diagnosis was similar to that of the total male patient group. There were no differences in the distribution of multiple exposures in nonsmoking patients and nonsmoking controls. The numbers of nonsmoking male patients was too small to evaluate exposures to specific substances, although 73% were farmers. There were no first-degree relatives with lung cancer in the nonsmoking patient families and six cases in random control families. Thirty-eight percent of patient families had at least one first-degree relative with a neoplasm compared to 23% of control families (NS), both of which are in the same range as the total population data.

We have demonstrated that the association between farming and primary lung cancer in males is not a simple phenomenon. In this province, for this age group, a large proportion of men have a significant history of farming. The pesticides used and the pattern of use have been shifting dramatically. A minority of farmers use gloves, masks, and protective clothing when dealing with these potentially hazardous substances. This study did not collect data on the specific chemicals used, the extent of use, or the duration of use. Although some pesticides are suspected carcinogens and the use of agricultural chemicals is extensive, we found that the estimated relative risk due to occupational exposures to these substances is small in relation to the smoking hazard. This relative risk is apparently not synergistic with a positive history of smoking, as there is no diminution of the mean age at diagnosis among those who reported these exposures. The data support the theory that lung cancer develops more often in damaged tissues as evidenced by the significantly higher frequencies in patients with bronchitis,[6,7] emphysema, pneumonia, chest injury, and farmer's lung disease.

REFERENCES

1. **Blot, W.J. and Fraumeni, J.F.,** Geographic patterns of lung cancer: industrial correlations, *Am. J. Epidemiol.,* 103(6), 539, 1976.
2. Agriculture, 1981 Census of Canada, Saskatchewan Catalogue 96-909, 1982.
3. International Agency for Research on Carcinogenesis, Monogr. Evaluating the Risk of Chemicals to Humans: Some Halogenated Hydrocarbons, Vol. 20, Lyon, 1979.
4. International Agency for Research on Carcinogenesis, Monogr. on the Evaluation of the Carcinogenic Risk of Chemicals to Man, Some Fumigants, the Herbicides 2,4-D and 2,4,5-T, Chlorinated Dibenzodioxins and Miscellaneous Industrial Chemicals, Vol. 15, 1977, Lyon.
5. International Agency for Research on Carcinogenesis, Monogr. on the Evaluation of Carcinogenetic Risk of Chemicals to Man: Some Organochlorine Pesticides, Vol. 5, Lyon, 1974.
6. **Cohen, B.H.,** Is pulmonary dysfunction the common denominator for the multiple effects of cigarette smoking?, *Lancet,* November 11, 1024, 1978.
7. **Davis, A.L.,** Bronchogenic carcinoma in chronic obstructive pulmonary disease, *J.A.M.A.,* 235(6), 621, 1976.

Cancer Among Farmers and Farm Laborers

R. P. Gallagher, W. J. Threlfall, P. R. Band, and J. J. Spinelli

INTRODUCTION AND THE STUDY

Several reports from the U.S. and other parts of the world have indicated that individuals engaged in farming may be at elevated risks of certain types of cancer.[1-5] We have analyzed patterns of mortality for farmers and farm laborers using death registrations, and present here our data on cancer as a cause of death in these two occupational groups.

Data on 457,083 deaths occurring in British Columbia from 1950—1978 were obtained from the provincial Division of Vital Statistics (DVS). Age 20 years was selected as the lower limit for analysis, and after eliminating a small group of records with invalid cause of death or occupational codes, a total of 254,901 male and 165,913 female deaths were available for analysis. Only 87 female decedents were recorded as being farmers or farm laborers, and these were excluded from the analysis. A total of 28,032 deaths in male farmers and 2328 deaths in male farm laborers were recorded.

Cause of death information on death records was reconciled to the International Classification of Diseases (ICD-7) format for the study.[6]

Age-standardized proportional mortality ratios (PMR) were calculated for all male deaths for two age groups, age 20 and up and age 20 to 65 for farmers and farm laborers separately by the method specified in Monson.[7] Significance of elevated or reduced PMRs was assessed by comparison with the Poisson distribution.[8]

Because significantly elevated or lowered PMRs for major causes of death such as accidents and degenerative heart disease tend to affect individual cancer site PMR values, we also calculated age-standardized proportional cancer mortality ratios (PCMR) for each individual cancer site, and these are presented in the current report.

RESULTS

Proportional cancer mortality analysis showed that farmers dying at age 20 and up had significantly elevated risks for several cancers, including lip, stomach, pancreas, nasal sinus, prostate, and leukemia (Table 1). This group also showed significantly reduced mortality for other cancer sites including esophagus, colon, larynx, and lung. Several of the reduced risks are present in numerically important cancer sites, thereby generating a lowered overall PMR (PMR = 94, $p < 0.01$) for cancer in farmers. Analysis of the farmers who died at age 20 to 65 showed a similar cancer mortality pattern with the exception that liver cancer was elevated as was eye cancer; however, both findings were based on relatively few deaths (Table 2).

Farm laborers showed few elevated or lowered cancer sites, although those age 20 and up had a significantly lower value for lung cancer (Table 3). Farm laborers age 20 to 65 demonstrated an elevated risk for stomach cancer, similar to that seen in farmers (Table 4).

DISCUSSION

Studies of occupational mortality based on death registrations are subject to a number of well-known shortcomings, including imprecision in cause of death and occupational information.[9,10] Furthermore, proportional mortality analyses have both strong and weak points which have been well detailed in the literature.[4,7,11,12]

Despite the possible problems inherent in the type of analysis, several interesting findings concerning B.C. farmers and farm laborers have emerged. The elevated risks of death for stomach, lip, and prostate cancer and leukemia confirm, for Canada, findings which have been shown in U.S. farm-

TABLE 1
Cancer Deaths for Selected Sites[a]

ICD	Site	No. of deaths	PCMR	95% CI
140	Lip	17	203[b]	118, 326
141—144	Mouth	38	75	54, 103
150	Esophagus	68	63[c]	49, 80
151	Stomach	672	126[c]	116, 137
153	Colon	389	89[b]	80, 99
155.0	Liver	55	113	86, 147
157	Pancreas	334	112[b]	100, 125
160	Nasal sinus	14	186[b]	101, 313
161	Larynx	29	66[b]	44, 96
162.0, 162.1 162.8, 162	Lung	742	70[c]	65, 76
177	Prostate	764	120[c]	112, 130
192	Eye	10	154	73, 284
200, 202	Non-Hodgkins lymphoma	78	98	78, 123
201	Hodgkin's disease	27	90	59, 132
203	Multiple myeloma	68	109	86, 139
204	Leukemia	219	129[c]	112, 147

[a] Male farmers age 20 and up, B.C. 1950 to 1978.
[b] $p < 0.05$.
[c] $p < 0.001$.

TABLE 2
Cancer Deaths for Selected Sites[a]

ICD	Site	No. of deaths	PCMR	95% CI
140	Lip	2	159	19, 576
141—144	Mouth	6	62	22, 135
150	Esophagus	13	69	36, 118
151	Stomach	124	136[b]	114, 163
153	Colon	60	95	73, 123
155.0	Liver	22	173[c]	108, 262
157	Pancreas	68	121	95, 154
160	Nasal sinus	1	57	1, 319
161	Larynx	6	65	23, 143
162.0, 162.1, 162.8, 162	Lung	202	76[b]	58, 98
177	Prostate	40	126	92, 172
192	Eye	5	313[c]	101, 731
200, 202	Non-Hodgkins lymphoma	25	93	59, 137
201	Hodgkin's disease	11	79	39, 142
203	Multiple myeloma	17	124	72, 199
204	Leukemia	43	131	97, 177

[a] Male farmers age 20—65, B.C. 1950 to 1978.
[b] $p < 0.001$.
[c] $p < 0.05$.

TABLE 3
Cancer Deaths for Selected Sites[a]

ICD	Site	No. of deaths	PCMR	95% CI
140	Lip	—	—	—
141—144	Mouth	2	59	7, 210
150	Esophagus	7	95	37, 195
151	Stomach	46	126	94, 168
153	Colon	24	82	52, 123
155.0	Liver	4	107	29, 273
157	Pancreas	26	124	81, 183
160	Nasal sinus	1	176	4, 982
161	Larynx	3	95	19, 278
162.0, 162.1, 162.8, 162	Lung	60	76[b]	58, 98
177	Prostate	40	109	80, 149
192	Eye	2	399	48, 1441
200, 202	Non-Hodgkins lymphoma	9	111	50, 211
201	Hodgkin's disease	6	162	59, 354
203	Multiple myeloma	4	91	24, 233
204	Leukemia	13	100	53, 172

[a] Male farm laborers, age 20 and up, B.C. 1950 to 1978.
[b] $p < 0.05$.

ers.[12-14] The elevated PCMR findings for B.C. farmers age 20 and up for pancreatic and nasal sinus cancer are new and will require confirmation in other geographic areas. In younger farmers the elevated risks for primary liver cancer and eye cancers must be treated with some caution. Analysis of patterns of liver cancer in young farmers showed that the elevated risk occurred in the period 1950—1959. The PCMR dropped to 100 in 1960—1969, and was below unity in 1970—1978. While this may be due to changes in exposure to carcinogens over the 29-year period, it might equally be due to better death certification over the period, with fever metastatic liver deaths being incorrectly certified as due to primary liver cancer in the later periods. Likewise, the finding of an elevated risk for eye cancer in young farmers is based on only five deaths and must be treated with caution.

TABLE 4
Cancer Deaths for Selected Sites[a]

ICD	Site	No. of deaths	PCMR	95% CI
151	Stomach	19	203[b]	122, 318
153	Colon	6	89	32, 195
155.0	Liver	1	72	1, 402
157	Pancreas	7	122	48, 251
160	Nasal sinus	1	475	12, 2646
161	Larynx	1	105	2, 585
162.0, 162.1, 162.8, 162	Lung	20	75	45, 116
177	Prostate	5	171	55, 399
200, 202	Non-Hodgkins lymphoma	2	55	6, 201
201	Hodgkin's disease	3	112	23, 329
203	Multiple myeloma	3	217	44, 636
204	Leukemia	6	128	47, 280

[a] Male farm laborers, age 20—65, B.C. 1950 to 1978.
[b] $p < 0.01$.

The lowered risk of death from cancers of the larynx and lung as well as from esophageal cancer may be explained by lower levels of smoking and alcohol consumption among farmers. Several studies have, in fact, demonstrated that farmers' consumption of cigarettes and alcohol is lower than that of other occupational groups.[15,16]

The only tumors showing significantly elevated or lowered risks in farm laborers were stomach and lung cancer. It should be noted, however, that the number of deaths among farm laborers is very much lower than that of farmers. In addition, the higher PCMR for stomach cancer among young farm laborers may be at least in part due to the well-recognized inverse relationship between socioeconomic status and stomach cancer risk.[17,18]

Findings for cancer in farm workers must be interpreted cautiously, however, because most move out of this occupation and into more highly paid jobs as quickly as they are able; thus it is likely that many men who may have worked as farm laborers are not listed as such on death registrations. Thus, chronic disease with long latency periods such as cancer may be underrepresented in death registration studies.

REFERENCES

1. **Burmeister, L.F., Van Lier, S.F., and Isacson, P.,** Leukemia and farm practises in Iowa, *Am. J. Epidemiol.,* 115, 720, 1982.
2. **McMichael, A.J. and Hartshorne, J.M.,** Mortality risks in Australian men by occupational groups, 1968—1978, *Med. J. Aust.,* 1, 253, 1982.

3. **Milham, S., Jr.**, Occupational mortality in Washington State 1950—1979, DHHS Publ. No. (NIOSH) 83-116, Department of Health and Human Services, Cincinnati, OH, 1983.
4. **Gallagher, R.P., Threlfall, W.J., Jeffries, E., Band, P.R., Spinelli, J.J., and Coldman, A.J.**, Cancer and aplastic anemia in British Columbia farmers, *J. Natl. Cancer Inst.*, 72, 1311, 1984.
5. **Pearce, N., Smith, A., and Fisher, D.**, Malignant lymphoma and multiple myeloma linked with agricultural occupations in New Zealand — cancer-registry based study, *Am. J. Epidemiol.*, 121, 225, 1985.
6. World Health Organization, International classification of diseases, 7th revision, World Health Organization, Geneva, 1957, 75.
7. **Monson, R.**, *Occupational Epidemiology*, CRC Press, Boca Raton, FL, 1980, 86.
8. **Armitage, P.**, *Statistical Methods in Medical Research*, Blackwell Scientific, Oxford, 1971.
9. **Moriyama, I.M., Baun, W.S., Haenszel, W.M., and Mattison, B.F.**, Inquiry into diagnostic evidence supporting medical certification of death, *Am. J. Public Health*, 48, 1376, 1958.
10. **Engel, L.W., Strauchen, J.A., Chiazze, L., Jr., and Heid, M.**, Accuracy of death certification on an autopsied population with specific attention to malignant neoplasms and vascular diseases, *Am. J. Epidemiol.*, 111, 99, 1980.
11. **Kupper, L.L., McMichael, A.J., Symons, M.J., and Most, B.M.**, On the utility of the proportional mortality analysis, *J. Chronic Dis.*, 31, 15, 1978.
12. **Milham, S., Jr.**, Occupational mortality in Washington State 1950-1971, Vol. 1 to 3, DHEW Publ. No. (NIOSH) 76-175, U.S. Government Printing Office, Washington, D.C., 1976.
13. **Burmeister, L.F.**, Cancer mortality in Iowa farmers 1971—1978, *J. Natl. Cancer Inst.*, 66, 461, 1981.
14. **Blair, A. and White, D.W.**, Death certificate study of leukemia among farmers from Wisconsin, *J. Natl. Cancer Inst.*, 66, 1027, 1981.
15. **Sterling, T.D. and Weinkam, J.J.**, Smoking characteristics by type of employment, *Occup. Med.*, 18, 743, 1976.
16. **Pomrehn, P.R., Wallace, R.B., and Burmeister, L.F.**, Ischaemic heart disease mortality in Iowa farmers, *J.A.M.A.*, 248, 1073, 1982.
17. **Buell, P., Dunn, J.E., and Breslow, L.**, The occupational-social class risks of cancer mortality in men, *J. Chronic Dis.*, 12, 600, 1960.
18. **Cohart, E.M.**, Socio-economic distribution of stomach cancer in New Haven, *Cancer*, 7, 455, 1954.

Cancer Mortality in Iowa Farmers

L. F. Burmeister

CANCER PATTERNS IN IOWA FARMERS

Several mortality studies[1-7] have indicated that farmers may have elevated risks for some cancers. Most of those studies, however, were not completed in the midwestern states, an area where farmers constitute a relatively large occupational group (e.g., 18.2% in Iowa in 1970). Therefore, occupational analyses of Iowa deaths from 1971—1978 were completed to further evaluate the cancer risks in farmers.

The "usual occupation" as indicated on the death certificate was coded according to the job classifications established by the Department of Labor[8] and added to other information previously abstracted from the certificates onto computer tape. Although the occupational information was included for males and females for all causes of deaths, the analyses in this study were limited to the cancer deaths of white males, since there are few nonwhite farmers in Iowa and because detailed occupational information on females is lacking. Proportionate mortality ratios (PMRs) were computed and tested for significance, as suggested by Mantel and Haenszel,[9] for white males under the age 65, for those age 65 or older, and for all ages combined.

A total of 121,101 male death certificates from 1971 to 1978 were examined. The number of males with no entry regarding occupation was 3300 (2.7%). In addition, 385 males (0.32%) were coded as disabled, 72 (0.06%) were listed as retired, 1217 (1.0%) as children, and 2059 (1.7%) as students. The PMRs for males over age 65 were generally lower than for males under 65, except for those males who had genital-urinary cancers (Table 1). The PMR for all cancers combined was significantly less than 1.0 for farmers age 65 and older and for farmers for all ages. The number of farmers under age 65 who died of cancer during the 8-year period was 1282; for farmers over age 65 the number was 5120.

The combined PMRs that were significantly lower than 1.0 were for cancers generally considered to be smoking related. The most important was lung cancer, which, because of the numerous deaths it caused, was the primary reason that the overall PMR was significantly smaller than 1.0.

Lip cancer, which presumably is due to sun exposure, had the most highly elevated PMR. Skin cancer, also associated with sun exposure, had an elevated, but not statistically significant, PMR. However, the total number of deaths in farmers during the 8-year period from skin and lip cancers was only 125.

More important in terms of total deaths (1138) was the significantly elevated PMR for prostate cancer. Significantly high combined PMRs for Iowa farmers during 1971—1978 were also noted for stomach cancer, leukemia, lymphatic cancers, and multiple myeloma. The fact that the mortality rate for smoking-related cancers was significantly lower increased the possibility that cancers in some other sites would be significantly higher.

OTHER MIDWEST CANCER STUDIES

A study of Nebraska death certificates by Blair and Thomas[7] also indicated an increased mortality from leukemia in farmers. The Nebraska study is important since it was one of the first attempts to identify farming practices associated with leukemia. The farming practice in Nebraska most strongly associated with leukemia mortality was corn production.

In contrast to the suggestion that farming practices peculiar to corn production may be associated with leukemia, the possibility that dairy farming may be associated with leuke-

TABLE 1
Proportionate Mortality Ratios for White Male Farmers in Iowa, 1971—1978

Cancer type	PMR, <65	PMR, ≥65	Combined PMR[a]
Lip	4.65	1.51[b]	1.62[c]
Other buccal	0.91	0.79[b]	0.82[b]
Esophagus	0.91	0.67[c]	0.74[c]
Stomach	1.39[c]	1.09	1.14[c]
Large intestine	1.20[c]	1.00	1.03
Liver	1.01	0.87	0.90
Pancreas	1.32[c]	1.01	1.07
Other digestive	0.92	1.05	1.03
Lung	0.83[c]	0.76[c]	0.78[c]
Other respiratory	0.82	0.59[c]	0.65[c]
Bone and connective tissue	1.13	1.07	1.08
Skin	1.43[b]	1.02	1.13
Brain	1.20	0.98	1.10
Other[d]	1.50	0.97	1.14
Leukemia	1.11	1.10[b]	1.10[b]
Lymphoma	1.18	1.13[b]	1.14[b]
Multiple myeloma	1.71[c]	1.20[b]	1.27[c]
Other blood	1.42	0.99	1.06
Hodgkin's disease	1.32	1.16	1.22
Prostate	1.06	1.10[c]	1.10[c]
Bladder	0.83	0.93	0.92
Kidney	1.03	1.10	1.08
Other urinary and genital	0.65	1.07	0.86
Overall	1.02	0.96[c]	0.97[c]

[a] For males <65 and ≥65 years old.
[b] Statistically significant at $p < 0.05$.
[c] Statistically significant at $p < 0.01$.
[d] Includes eye, other nervous system, thyroid gland, and other endocrine gland cancers.

mia has also been suggested.[10-16] Not all epidemiologic studies evaluating the association of dairy cattle and leukemia have been positive, however.[17,18] Therefore, further research to identify suspected farming practices is needed.

Because of its high corn and dairy production, Iowa provided an excellent opportunity to compare the association of suspected farming practices with leukemia. Iowa ranked first in corn production by state in 1950 and 1960, compared to fifth in 1950 and third in 1960 for Nebraska.[19,20] Although Iowa's dairy production (number of cows and heifers 2 years old and over, kept for milk) did not rank as high nationally (fifth in both 1950 and 1960) as did corn production, it is much higher than that of Nebraska (19th and 18th in 1950 and in 1960, respectively). It may be that Nebraska's dairy production was not sufficiently large to allow any association with leukemia mortality to be detected; therefore, an Iowa death certificate study was completed to determine which farming practices are associated with leukemia mortality rates. The study was later expanded to include non-Hodgkin's lymphoma, multiple myeloma, stomach cancer, and prostate cancer.

The methods utilized in the Nebraska study were duplicated whenever possible to assure comparability of the two studies. The Nebraska study was based on deaths from 1957—1974, whereas the Iowa study was based on deaths from 1964 to 1978. The following steps, however, are common in the two studies: (1) the records of all white males, aged 30 years or more at time of death from leukemia as classified by the Seventh (code 204) and Eighth (codes 204 to 207) Revisions of the International Classification of Diseases (ICD),[21,22] were abstracted from computer listings; (2) for each leukemia death (1675 in Iowa compared to 1084 in Nebraska), 2 nonleukemic deaths were selected as controls. The controls were matched by county of usual residence, age at death (±2 years), and calendar year of death. When more than two eligible controls were identified for a case, the two controls included in the study were selected randomly. Only white males were included in both studies; (3) The usual occupation indicated on the Iowa death certificate was coded according to the 1960 Census[23] for the years 1964—1970, and according to the U.S. Bureau of Labor[8] for the years 1971—1978. The entire Nebraska study was based on occupations defined by the Bureau of the Census. However, both studies combined all "nonfarmer" occupations; (4) farm owners, tenants, and laborers were all considered "farmers".

Iowa counties were ranked according to 1949 crop and livestock production provided by the Iowa Crop and Livestock Reporting Service, and by pesticide usage provided by the 1964 agricultural census,[24] which is the earliest year for which pesticide usage data are available. Crop and livestock production figures from 1949 were used to provide a 15-year lag period from time of production until the first year included in the death certificate analyses. However, agricultural production in Iowa counties has changed very little during the past three decades. Spearman rank correlation coefficients for Iowa counties ordered for each of the major agricultural production figures in 1949 and 1974 were highly significant ($p < 0.001$).

Estimated relative risks and associated confidence intervals were computed as suggested by Miettinen.[25] All computations utilize the matching of two controls to each case.

The odds ratios for mortality in farmers were significantly elevated for each of the five cancers (Table 2). Table 2 also indicates if any patterns exist in the odds ratios according to year of birth, age at death, or year of death. Leukemia mortality and multiple myeloma mortality were elevated only in those born after 1890. Prostate cancer and non-Hodgkin's lymphoma are known to occur primarily in those over age 65;

TABLE 2
Odds Ratios for Mortality from Selected Cancer Types in White Male Iowa Farmers, 1964—1978

	Cancer type									
	Leukemia		Multiple myeloma		Non-Hodgkin's lymphoma		Prostate		Stomach	
Stratum	Odds ratio	No. of cases	Odds ratio	No. of cases	Odds ratio	No. of cases	Odds ratio	No. of cases	Odds ratio	No. of cases
Overall	1.24[a]	661	1.48[a]	550	1.26[a]	1101	1.19[a]	4827	1.32[a]	1812
Year of birth										
Before 1890	0.89	166	1.31	78	2.85[a]	154	1.52[a]	1539	1.33[a]	458
1890—1900	1.36[a]	290	1.56[a]	210	1.57[a]	336	1.27[a]	2081	1.33[a]	639
After 1900	1.39[a]	205	1.46[a]	262	0.85	611	0.76	1207	1.28[a]	715
Year of death										
1964—1970	1.15	344	1.51[a]	249	1.18	535	1.24[a]	2258	1.42[a]	973
1971—1978	1.36[a]	317	1.46[a]	301	1.34[a]	566	1.14[a]	2569	1.19	839
Age at death										
≤65 years	1.36[a]	116	1.46	150	0.72	447	0.60	492	1.40[a]	481
>65 years	1.21[a]	545	1.49[a]	400	1.75[a]	654	1.28[a]	4335	1.29[a]	1331

[a] Statistically significant at $p < 0.05$.

therefore, it was not unexpected that increased odds ratios were noted only in those born before 1901 and in those dying after age 65. There appear to be no discernible patterns among the stomach cancer odds ratios; however, the stomach cancer odds ratio was not significantly elevated in those dying after 1970.

AGRICULTURAL PRACTICES

To determine if associations exist between farmer mortality from the selected cancers, and specific agricultural practices, odds ratios were computed separately for those whose place of residence was in the highest 33 counties and for those whose residence was in the lowest 66 counties according to 1949 production figures for selected commodities. Odds ratios were computed according to birth cohort and high- and low-producing counties for each of the following agricultural commodities: egg-laying chickens, milk products sold, cattle, hogs, soybeans per acre, corn per acre, herbicide usage, insecticide usage, and fertilizer applied. Corn and soybeans per acre were used rather than total bushels since they are indicators of crop intensity. Table 3 includes only those commodities with significantly elevated odds ratios in one or more of the three birth cohorts in the high-producing counties with no simultaneous significant increase in the low-producing counties.

As might be expected from the low odds ratio for leukemia mortality in those born before 1890 (Table 2), there were no significantly elevated risks in this cohort. Soybeans per acre and corn per acre were associated with increased leukemia mortality risks among farmers for those born from 1890 to 1900. However, neither association was statistically significant for those born after 1900.

It is for those born after 1900 that the Iowa and Nebraska studies have the greatest inconsistencies. The odds ratios for farmers in Nebraska were significantly greater than 1.0 in those counties with the highest production of corn, cattle, and hogs, and highest number of acres treated with insecticides. None of these farming practices were associated with leukemia mortality for the youngest cohort in Iowa. Acres treated with herbicides and egg-laying chickens per county were statistically significant only in the high usage and production counties.

Multiple myeloma mortality exhibited the most consistent pattern across birth cohorts. There were significantly elevated odds ratios for the high-producing 33 counties and no significant increase in the odds ratios for the low 66 counties. Egg-laying chicken production, herbicide use, and insecticide use were consistently associated with elevated multiple myeloma odds ratios for the latter two birth cohorts (1890—1900, after 1900), which exhibited significantly elevated odds ratios.

Only farmers born before 1901 had elevated odds ratios for mortality from non-Hodgkin's lymphoma (Table 2). The odds ratios for those born before 1890 were significantly greater than 1.0 in both the high and low counties for each commodity production or usage. For the 1890—1900 birth cohort, egg-laying chicken production and herbicide usage were associated with non-Hodgkin's lymphoma mortality. In addition, milk products sold and hog production showed significant odds ratios only in the counties with the greatest production.

TABLE 3
Odds Ratios for Mortality from Selected Cancer Types in White Male Iowa Farmers According to Birth Cohort and Selected County Crop and Livestock Production, 1964—1978

Cancer type	Production categories by counties	Year of birth: Before 1890	1890—1900	After 1900
Leukemia	Soybeans/acre			
	33 highest	0.67	1.59[a]	1.40
	Others	1.03	1.23	1.38[a]
	Corn/acre			
	33 highest	0.83	1.63[a]	1.38
	Others	0.95	1.20	1.39[a]
	Egg-laying chickens			
	33 highest	1.10	0.95	1.53[a]
	Others	0.79	1.63[a]	1.31
	Insecticide usage			
	33 highest	0.73	1.32	1.13
	Others	0.95	1.38[a]	1.53[a]
	Herbicide usage			
	33 highest	0.77	1.10	1.60[a]
	Others	0.95	1.53[a]	1.28
Multiple myeloma	Egg-laying chickens			
	33 highest	17.00[a]	2.08[a]	2.14[a]
	Others	0.75	1.32	1.18
	Hog production			
	33 highest	1.90	1.83[a]	1.53
	Others	1.00	1.41	1.43
	Herbicide usage			
	33 highest	1.78	2.65[a]	2.43[a]
	Others	1.10	1.12	1.20
	Insecticide usage			
	33 highest	2.11	1.96[a]	2.00[a]
	Others	0.95	1.36	1.24
Non-Hodgkin's lymphoma	Egg-laying chickens			
	33 highest	3.54[a]	2.55[a]	0.91
	Others	2.54[a]	1.24	0.82
	Milk products sold			
	33 highest	4.10[a]	1.95[a]	0.65
	Others	2.45[a]	1.39	1.03
	Hog production			
	33 highest	2.71[a]	1.77[a]	0.69
	Others	2.93[a]	1.43	0.99
	Herbicide usage			
	33 highest	3.42[a]	2.19[a]	1.29
	Others	2.62[a]	1.31	0.68
Stomach	Milk products sold			
	33 highest	1.30	1.55[a]	1.25
	Others	1.38	1.11	1.33
	Cattle production			
	33 highest	1.74[a]	1.58[a]	1.27
	Others	1.07	1.21	1.29
	Corn per acre			
	33 highest	1.46	1.56[a]	1.29
	Others	1.26	1.22	1.28

[a] Statistically significant at $p < 0.05$.

Stomach cancer mortality had consistently elevated odds ratios among farmers across all three birth cohorts (Table 2). However, milk products sold, cattle production, and corn per acre were the only commodities with elevated odds ratios for one or more of the birth cohorts in the high-production counties and there was no evidence of elevated odds ratios in the low-production counties.

Prostate cancer is known to have a relatively late onset. This may explain the finding that only in those born before 1901 is there any evidence of increased mortality in Iowa farmers (Table 2). No separation according to any of the selected commodities in the first and third birth cohorts was noted. In the 1890 to 1900 cohort, the only commodities resulting in higher odds ratios in the high-production counties were egg-laying chickens, herbicide use, and insecticide use; however, the odds ratios were significantly elevated in both high- and low-producing counties for these commodities.

LEUKEMIA MORTALITY

The odds ratios for Iowa farmers for specific leukemia cell types are summarized in Table 4. Only those dying in 1968—1978 are included in Table 4, since the ICD Seventh Revision (used prior to 1968) may not separate acute lymphatic and acute myeloid leukemias. Both chronic lymphatic and unspecified lymphatic leukemias caused increased mortality in Iowa farmers.

Odds ratios for chronic and unspecified leukemias according to county agricultural production figures associated with combined leukemia types are summarized in Table 5. Chronic lymphatic leukemia was elevated in both the high- and low-producing counties for the five county characteristics included in Table 5. The strongest associations, both in magnitude of the odds ratios and differences in the high- and low-producing counties, were with soybeans per acre and acres treated with herbicides. Unspecified lymphatic leukemia mortality in farmers is associated with corn per acre, number of milk cows, and egg-laying chickens.

Of the six cancers with elevated mortalities in Iowa farmers, three (leukemia, non-Hodgkin's lymphoma, and multiple myeloma) were reticuloendothelial and hematological malignancies. Each of these reticuloendothelial cancers is relatively uncommon; consequently, their simultaneous increase may signify a common etiologic link. Although no single farming practice was associated with the reticuloendothelial neoplasms for each birth cohort, egg-laying chicken production, herbicide usage, and insecticide usage require further investigation.

It is possible that herbicides and/or insecticides may be carcinogenic to farmers. A recent case-control study from Sweden[26] has linked human soft tissue sarcomas with the phenoxyacetic acid pesticides (2-4D, 2-4-5T). Experimental animal evidence[17] also suggests that many pesticides are carcinogens or mutagens or may form carcinogens in contact with the soil.

It is not easy to rationalize the association of the reticuloendothelial neoplasms with poultry farming. There is precedent for this finding in previous work;[1,5,6] however, the studies have been contradictory.[6,7] Marek's disease, a herpes virus-induced malignant lymphoma, is widespread and highly contagious among chickens.[28] Although there is no current evidence of transmission of Marek's disease to humans or other primates, Marek's disease at least lends biological credibility to the

TABLE 4
Odds Ratios for Leukemia Mortality in White Male Iowa Farmers over 30 Years of Age for Specific Leukemia Types, 1968—1978

		Cases		Controls			
Leukemia type	ICD no.	No. of farmers	Farmers (%)	No. of farmers	Farmers (%)	Odds ratio	95% CL[a]
Acute lymphatic	204.0	28	28.9	69	35.6	0.68	0.37—1.23
Chronic lymphatic	204.1	132	45.8	207	35.9	1.70	1.22—2.38
Unspecified lymphatic	204.9	64	51.2	102	40.8	1.66	1.03—2.68
Acute myeloid	205.0	86	30.7	168	30.0	1.04	0.74—1.48
Chronic myeloid	205.1	46	32.6	90	31.9	1.04	0.65—1.65
Unspecified myeloid	205.9	36	36.7	80	40.8	0.80	0.45—1.41
Acute monocytic	206.0	10	32.3	19	30.1	1.07	0.44—2.59
Leukemia unspecified	208.0	31	41.3	60	40.0	1.07	0.58—1.97

[a] CL, confidence limits

TABLE 5
Odds Ratios for Leukemia Mortality in White Male Iowa Farmers over 30 Years of Age, according to Selected Leukemia Types and Selected County Crop and Livestock Production, 1968—1978

		Highest 33 counties		Other counties	
County characteristics	Leukemia type	Odds ratio	95% CL[a]	Odds ratio	95% CL
Soybeans/acre	Chronic lymphatic	2.21	1.33—3.66	1.39	0.97—2.00
	Unspecified lymphatic	0.62	0.25—1.52	2.49	1.38—4.48
Corn/acre	Chronic lymphatic	1.34	0.87—2.06	1.94	1.29—2.92
	Unspecified lymphatic	4.63	1.82—11.74	1.00	0.56—1.80
Milk cows	Chronic lymphatic	1.19	0.76—1.85	2.10	1.41—3.14
	Unspecified lymphatic	4.94	1.96—12.47	0.94	0.52—1.70
Egg-laying chickens	Chronic lymphatic	1.36	0.82—2.25	1.80	1.25—2.59
	Unspecified lymphatic	2.47	1.03—5.92	1.37	0.77—2.45
Herbicide usage	Chronic lymphatic	1.89	1.16—3.08	1.50	1.03—2.18
	Unspecified lymphatic	1.21	0.51—2.85	1.90	1.07—3.37

[a] CL, confidence limits.

association of chicken farming and reticuloendothelial neoplasms.

The nature of poultry farming has greatly changed during the past two to three decades in Iowa. Twenty to 30 years ago and before, nearly all farmers kept a laying flock, principally for personal use. Recently, a trend toward large confinement operations has been noted, with a resulting sharp decrease in the number of farmers with poultry operations. If poultry farming is a reason for elevated cancer mortality in farmers, the trend over the next several decades should be toward a relative decrease in farmer mortality from reticuloendothelial neoplasms.

The possible role of insecticides and herbicides in development of and subsequent mortality from selected cancers in farmers is one that has generated a high level of interest. Consequently, it seems prudent to evaluate mechanisms that may cause farmers to be at higher risk for some cancers due to possible pesticide exposure. One such mechanism may be the infusion of agricultural runoff (including pesticides) into shallow farm wells. It is hoped that future research will suggest other mechanisms that may play a role in the development of the suspected cancers in farmers. Only then is it likely that the appreciable risks currently facing farmers for mortality from selected cancers will be reduced.

REFERENCES

1. **Milham, S., Jr.**, Occupational mortality in Washington State, 1950—1971, Vol. 1, Rockville, MD, DHEW Publ. No. (NIH) 76-175-A, National Institute for Occupational Safety and Health, Cincinnati, OH, 1976.
2. The Registrar General's Decennial Supplement, England and Wales, 1961, Occupational Mortality Tables, Her Majesty's Stationery Office, London, 1971.
3. **Fasal, E., Jackson, E.W., and Klauber, M.R.**, Leukemia and lymphoma mortality and farm residence, *Am. J. Epidemiol.*, 87, 267, 1968.
4. Cancer Registry in Norway, The Incidence of Cancer in Norway, 1959—1961, Norwegian Cancer Society, Oslo, 1964.
5. **Milham, S., Jr.**, Leukemia and multiple myeloma in farmers, *Am. J. Epidemiol.*, 94, 507, 1971.
6. **Priester, W.A. and Mason, T.J.**, Human cancer mortality in relation to poultry population, by county, in 10 south-eastern states, *J. Natl. Cancer Inst.*, 53, 45, 1974.
7. **Blair, A. and Thomas, T.L.**, Leukemia among Nebraska farmers, a death certificate study, *Am. J. Epidemiol.*, 110, 264, 1979.
8. U.S. Department of Labor, Dictionary of Occupational Titles, 3rd ed., U.S. Government Printing Office, Washington, D.C., 1965.
9. **Mantel, N. and Haenszel, W.**, Statistical aspects of the analysis of data from retrospective studies of disease, *J. Natl. Cancer Inst.*, 22, 719, 1959.
10. **Khokhlova, M.P. and Rakhmamim, P.P.**, Comparative study on geographical distribution of human cattle leukosis, *Bibl. Haematol.*, 36, 654, 1970.
11. **Henricson, B. and Ringertz, N.**, Comparative studies on the geographical distribution of human and bovine leukemia, *Bibl. Haematol.*, 31, 331, 1968.
12. **Heath, C.W., Jr.**, Human leukemia, genetic and environmental clusters, *Bibl. Haematol.*, 36, 649, 1970.
13. **Lemon, H.M., Twiehaus, M.J., and Wilson, R.B.**, Symbiotic human and bovine lymphoma, *Proc. Am. Assoc. Cancer Res.*, 7, 41, 1966.
14. **Donham, K.J., Berg, J.W., and Sawin, R.S.**, Epidemiologic relationships of the bovine population and human leukemia in Iowa, *Am. J. Epidemiol.*, 112, 80, 1980.
15. **Wisinewski, D. and Weinreich, J.**, Lymphatische leukemie bei vater und sohn, *Blut*, 12, 241, 1966.
16. **Bartsch, D.C., Springer, F., and Fulk, H.**, Acute non-lymphocytic leukemia, an adult cluster, *J.A.M.A.*, 22, 1333, 1975.
17. **Kvarnfors, E., Henricson, B., and Hugosan, G.**, A statistical study on farm and village level on the possible relations between human leukemia and bovine leukosis, *Acta Vet. Scand.*, 16, 163, 1975.
18. **Preister, W.A., Oleinick, A., and Conner, G.H.**, Bovine leukosis and human cancer, *Lancet*, 1, 367, 1970.
19. U.S. Department of Agriculture, Agricultural Statistics 1950, U.S. GPO, Washington, D.C., 1950.
20. U.S. Department of Agriculture, Agricultural Statistics 1960, U.S. GPO, Washington, D.C., 1960.
21. World Health Organization, Manual of the International Statistical Classification of Diseases, Injuries and Causes of Death, 7th Revision, World Health Organization, Geneva, 1955.
22. World Health Organization, Manual of the International Statistical

Classification of Diseases, Injuries and Causes of Death, 8th Revision, World Health Organization, Geneva, 1967.

23. U.S. Bureau of the Census, 1960 Census of Population Classified Index of Occupations and Industries, U.S. GPO, Washington, D.C., 1960.
24. U.S. Bureau of the Census, 1964 Census of Agriculture, Vol. 1, Area Reports, U.S. GPO, Washington, D.C., 1966.
25. **Miettinen, O.S.,** Estimation of relative risk from individually matched series, *Biometrics,* 26, 75, 1970.
26. **Hardell, L. and Sandstrom, A.,** Case-control study: soft-tissue sarcomas and exposure to phenoxyacetic acids or chlorophenols, *Br. J. Cancer,* 39, 711, 1979.
27. **Eisenbrand, G., Ungerer, O., and Preussman, R.,** Formation of N-nitroso compounds from agricultural chemicals and nitrite, *IARC Sci. Publ.,* 9, 71, 1979.
28. **Sharma, J.M., Witter, R.L., Burmeister, B.R., and Landon, J.C.,** Public health implications of Marek's disease virus and herpes virus of turkeys. Studies on human and subhuman primates, *J. Natl. Cancer Inst.,* 51, 1123, 1973.

Agricultural Exposures, Allergies, and Risk of Multiple Myeloma

R. P. Gallagher, J. J. Spinelli, J. M. Elwood, and J. Moody

STUDIES OF MYELOMA

Several studies both from the U.S. and other parts of the world have reported farming to be associated with increased risk of multiple myeloma.[1-5] In British Columbia we conducted a small case-control study of multiple myeloma in order to investigate the influence of medical history, smoking, alcohol consumption, occupation, and a number of other factors on risk of the disease. In this brief article we summarize our findings concerning medical history and occupation.

THE BRITISH COLUMBIA STUDY

A total of 84 myeloma patients were interviewed for the study, including 49 males and 35 females. In order to maximize numbers of patients available for study we elected to interview all surviving myeloma patients (prevalent cases) seen at the A. Maxwell Evans Clinic (AMEC) in Vancouver during the years 1972—1978 (total 32 patients), as well as the 52 British Columbia cases referred to the AMEC from 1979—1981 (incident cases).

Virtually all patients not included in the study had died before contact by our interviewer. No patients refused interview.

Two groups of controls were selected. The first consisted of 84 AMEC patients with head and neck tumors, and the second were patients treated at AMEC for a variety of other tumors, including digestive system (26 patients), basal cell skin (10), breast and female genital (27), male genital (7), hematopoietic (13), and brain (1). All the controls had been treated from 1977—1980. Cases were matched to a single control from each of the two groups by sex, age within 5-year age groups, and by year of diagnosis within 5 years.

Cases and controls were interviewed by a trained interviewer using a standard questionnaire. A total of 119 different variables were examined in the interview, but special emphasis was placed on occupation and presence of prior autoimmune diseases. Because the study was conducted as an initial overview of possible etiologic factors, relatively little detail was collected on individual variables.

Data were analyzed using a conditional logistic regression model for matched case-control data sets as specified by Breslow and Day.[6]

RESULTS

Comparisons revealed that the marital status and educational levels of myeloma cases were similar to those of the nonhead and neck control group. However, the head and neck group appeared to have a lower educational level, perhaps indicating reduced socioeconomic status in comparison to the other groups.

For most analyses the two control groups were combined giving two controls per patient, but for several variables, notably alcohol and tobacco use, only the nonhead and neck controls were used, as these factors are known to be associated with head and neck tumors.

Myeloma patients reported more prior allergies than controls as shown in Table 1. As well, a greater number of cases reported prior treatment for allergies including special diets, drugs, and injections. The character of the symptoms associated with the reported allergies also differed between the myeloma cases and controls. Nearly one half of the symptoms reported by controls were described as breathing difficulties, while less than 20% of myeloma patients with allergies reported this symptom. Myeloma patients described primarily skin rashes, swelling, and hives, but specific information on allergies was not available. No difference was seen between the prevalent and incident groups of patients for allergy associations. A higher proportion of myeloma patients reported prior primary myxedema than did controls. No association was seen with diseases of autoimmune origin including systemic lupus erythematosis, rheumatoid arthritis, or ankylosing spondylitis.

Occupational information was recorded for each job held by study subjects for 1 year or more and, as well, each respondent was questioned using a checklist of industries and occupations common to the province of British Columbia. Examination of the detailed occupational data revealed an elevated risk of myeloma in individuals who worked in farming occupations as indicated in Table 2. Nevertheless, no further association was seen between length of time spent in

TABLE 1
Medical Factors in Myeloma

Factor	Myeloma cases (%)	Controls (%)	Odds ratio	*p* value	95% CI
Prior allergies	28.6	11.3	3.1	<0.001	1.6, 6.3
Prior allergy treatment	13.1	4.2	3.5	0.02	1.1, 10.6
Prior myxedema	7.1	1.2	5.0	0.04	1.0, 25.7

From Gallagher, R.P., Spinelli, J.J., Elwood, J.M., and Skippen, D.H., *Br. J. Cancer,* 48, 853, 1983. With permission.

TABLE 2
Occupational Factors in Myeloma

Factor	Myeloma cases (%)	Controls (%)	Odds ratio	*p* value	95% CI
Farm workers	36.9	22.6	2.2	0.01	1.2, 4.0
Exposure to processed grain	10.7	4.8	2.4	0.09	0.9, 6.6
Work in dusty conditions	45.2	32.1	2.1	0.02	1.1, 4.0

From Gallagher, R.P., Spinelli, J.J., Elwood, J.M., and Skippen, D.H., *Br. J. Cancer,* 48, 853, 1983. With permission.

agricultural work and risk of myeloma. The associations, as well, appeared to be primarily in the incident myeloma cases, with little association in the prevalent myeloma case group. This may be simply due to the small number of prevalent cases (32), but might also indicate a selection bias in the prevalent caseload.

Unfortunately, we did not collect data on the type of farming respondents were engaged in. Exposure to processed grain as recorded on the occupational checklist showed a nonsignificantly increased odds ratio. Since work with processed grain tends to be dusty, we created an index of dust exposure for each job and this was applied to the detailed occupational data. When exposures were totaled, myeloma patients appeared to have greater exposure to dusty workplace conditions than controls. However, the association disappeared when farming as an occupation was removed from the analysis. No significant associations were found between myeloma and other dusts including asbestos, textile, metal, wood, coal, or silica. A number of other factors were reviewed for association with myeloma in our study but with negative results.[7]

DISCUSSION

The fact that only 20% of the cases seen at AMEC from 1972—1981 were interviewed for our study raises the possibility that our results are due to a sampling or selection bias. The fact that the 32 prevalent cases from 1972—1978 had survived until our study began in 1979 indicates that their disease process was less severe or that they had a better response to treatment than cases who had died. Although the odds ratio for allergies was similar in the prevalent and incident groups, the association with farming was found primarily in the incident group. It is worth bearing in mind that our small study also examined a large number of factors (119) and it is possible that some of our significant findings may be due to chance.

In spite of possible drawbacks to our study, the elevated risk for myeloma in farmers appears to confirm prior findings in several U.S. studies.[1-3]

The U.S. studies showed that farmers residing in counties with heavy insecticide and herbicide use, or with large chicken populations, were at highest risk.[2,3] No detail was available in our study as to type of farming engaged in.

Our finding of an increased risk of multiple myeloma in individuals with prior allergies is of interest because of case reports which have appeared in the literature on myeloma in allergy suffers.[8]

It should be emphasized that our study is small and our findings have not been confirmed by other groups to date. Future etiologic studies of myeloma and farming should collect detailed data of exposures to pesticides including both insecticides and herbicides.

REFERENCES

1. **Milham, S., Jr.,** Leukemia and multiple myeloma in farmers, *Am. J. Epidemiol.,* 94, 307, 1971.
2. **Cantor, K.P. and Blair, A.,** Farming and mortality from multiple myeloma: a case-control study with the use of death certificates, *J. Natl. Cancer Inst.,* 72, 251, 1984.
3. **Burmeister, L.F., Everett, G.D., Van Lier, S.F., and Isacson, P.,** Selected cancer mortality and farm practices in Iowa, *Am. J. Epidemiol.,* 118, 72, 1983.
4. **Agu, V.U., Christensen, B.L., and Buffler, P.A.,** Geographic patterns of multiple myeloma, racial and industrial correlates, state of Texas, 1969—71, *J. Natl. Cancer Inst.,* 65, 735, 1980.
5. **Pearce, N., Smith, A., and Fisher, D.,** Malignant lymphoma and multiple myeloma linked with agricultural occupations in a New Zealand cancer registry-based study, *Am. J. Epidemiol.,* 121, 225, 1985.
6. **Breslow, N.E. and Day, N.E.,** *Statistical Methods in Cancer Research Vol. I, The Analysis of Case-Control Studies,* IARC Scientific Publ. No. 32, IARC, Lyon, 1980, 192.
7. **Gallagher, R.P., Spinelli, J.J., Elwood, J.M., and Skippen, D.H.,** Allergies and agricultural exposure as risk factors for multiple myeloma, *Br. J. Cancer,* 48, 853, 1983.
8. **Woodroffe, A.J.,** Multiple myeloma associated with long history of hyposensitisation with allergen vaccines, *Lancet,* 1, 99, 1972.

Lymphatic and Hematopoietic Cancer Among Farmers

A. Blair, K. Cantor, R. Gibson, G. Everett, L. Schuman, L. Burmeister, S. Van Lier, and W. Blattner

INTRODUCTION

Farming is a physically demanding occupation requiring many skills usually associated with other occupations, including machinery repairing, welding, applying pesticides, painting, and operating heavy equipment. During the performance of these tasks, farmers may come in contact with pesticides, solvents, fuels and oils, dusts, zoonotic viruses, microbes, and fungi. Included among these agents are a number of known or suspected carcinogens. Contact with these potential carcinogenic agents has raised concern that farmers may be at higher risk of cancer than other occupational groups.

Although several cancers have been reported to be excessive among farmers, lymphatic and hematopoietic cancers have received more attention than other tumors. The National Cancer Institute has conducted a series of studies to evaluate the role of agricultural exposures in the origin of these tumors. In this paper we review results from geographic correlations, occupational surveys, and death certificate case-control studies, and we present some preliminary findings from an interview case-control study of leukemia and non-Hodgkin's lymphoma (NHL) conducted in Iowa and Minnesota.

GEOGRAPHIC CORRELATIONS

Geographic correlations were used as the first step to investigate the relationship between mortality from lymphatic and hematopoietic cancers and agricultural exposures. In these investigations the mortality rates for individual counties were correlated with agricultural and other characteristics of the counties. Geographic patterns of mortality from lymphatic and hematopoietic cancers show high rate areas in rural regions in the central U.S.[1-4] For leukemia, a striking band of high-rate counties runs through the central states from North Dakota to Texas.[1] For NHL and multiple myeloma, high-rate counties tend to be located in the far west and north central states. Mortality rates for leukemia were positively associated with production of certain agricutural commodities including corn, cotton, and wheat, and with pesticide use.[2] No significant associations were seen between these agricultural factors and multiple myeloma or NHL.[3,4] A positive association with socioeconomic status was seen for all three tumors. Although these geographic mortality studies suggest that agricultural factors may be involved in the origin of leukemia, they lack the specificity to identify etiologic agents. Geographic correlations are also susceptible to the ecologic fallacy because the correlations are based on characteristics of geographic units and not of individuals. The individuals in the unit developing the disease may not possess the characteristic attributed to the unit. Studies based on individuals are required for hypothesis testing.

BROAD OCCUPATIONAL SURVEYS

Surveys evaluating mortality or morbidity risks for large numbers of occupations, including farmers, are available from several countries.[5-16] These surveys support the hypothesis that farmers experience elevated risks for leukemia, NHL, and multiple myeloma. The surveys typically report disease risks for many occupations and were not designed to specifically evaluate the cancer-farming hypothesis; thus, they provide little detail on farm practices. They do, however, indicate the consistency of the farming and cancer association around the world.

To summarize results from multiple studies, we have tabulated the proportion of studies that show relative risks greater than 1.00. Seven of 11 studies have elevated relative risks for leukemia ranging from 1.01 to 1.22, with two studies showing statistically significant excesses. Six of the 10 studies with information on NHL had estimates of relative risks greater than 1.00, with risks ranging from 1.01 to 2.15. With one exception, the elevated estimates of relative risk of NHL among farmers were all less than 1.20. Seven of 9 studies with information on multiple myeloma showed excesses among farmers, with relative risks ranging from 1.03 to 3.09, and three were statistically significant.

Although these broad occupational surveys lack detailed information on agricultural exposures, they are valuable. They show that the association between farming and lymphatic and hematopoietic cancer is not restricted to a single geographic area, but, in fact, occurs in several countries.

DEATH CERTIFICATE CASE-CONTROL STUDIES

Findings from the geographic correlation studies and broad occupational surveys encouraged us to further evaluate the risks of lymphatic and hematopoietic cancers among farmers through a series of death certificate case-control studies. This study design was chosen because it provided some occupational information on each subject which was not available in the geographic correlation studies, and because such studies are inexpensive and can be completed quickly. Studies of leukemia were conducted in Nebraska[17] and Wisconsin[18] and NHL and multiple myeloma in Wisconsin.[19,20]

These midwestern states have large farm populations and their vital records offices had coded occupation and industry information from death certificates which was available on computer tapes. Specific agricultural exposures were inferred from the level of agricultural production for the county of residence of the subject. For example, farmers from major corn-producing counties were compared to farmers from other counties. In Nebraska the risk of leukemia was greatest among farmers born after 1900 and/or dying before age 65. No striking risk was associated with farmers residing in major dairy cattle-producing counties. The strongest correlation with an agricultural factor was with corn production, suggesting that exposure to pesticides may be involved. Although the data are thin, further analysis of the Nebraska data by leukemia cell type suggested that the myeloid leukemias may be more strongly associated with corn production than lymphatic leukemias.[21]

The pattern of leukemia risk among farmers in Wisconsin paralleled that seen among Nebraska farmers, although relative risks were generally lower. Farmers residing in major dairy counties in Wisconsin were at higher risk than those from other counties. The risk was elevated, however, among dairy and nondairy farmers alike, suggesting that dairy activities themselves may not be the crucial factor.

In Wisconsin, NHL was elevated among farmers, particularly those residing in counties where pesticides were heavily used. As with leukemia, the association with farming was stronger among farmers born more recently and dying at younger ages than in other groups. Risks were greater among farmers for reticulum cell sarcoma than for other histologic types of NHL. Classification of NHL by cell type from death certificates, however, is fraught with difficulty. For multiple myeloma, unlike the pattern for leukemia and NHL, farmers born earlier and dying after age 65 were at higher risk than other ages and years of birth groups. A particularly striking finding was the elevated risk among farmers from heavy insecticide-using counties who were born after 1906. This was the only situation where the relative risk among farmers born more recently was greater than among farmers born earlier. These younger farmers may be more likely to have engaged in modern, chemical farming than farmers born earlier.

These death certificate case-control studies further support the hypothesis that agricultural factors are associated with the origin of lymphatic and hematopoietic cancer.

CASE-CONTROL INTERVIEW STUDIES: PRELIMINARY RESULTS

To identify specific agricultural exposures that may be involved in the origin of leukemia and NHL, we initiated a population-based case-control interview study in Iowa and Minnesota and another in Kansas. Both studies were designed to provide detailed information on specific farm activities and exposures among a series of incident cases and population controls. The study in Kansas was specifically designed to evaluate the role of herbicides in the origin of Hodgkin's disease, NHL, and soft-tissue sarcoma, while the study of leukemia and NHL in Iowa and Minnesota focused on other exposures in addition to herbicides. Only preliminary results from this latter study will be presented here. The states of Minnesota and Iowa were chosen because their mortality and incidence rates for leukemia and NHL exceed the national average, because they have large agricultural populations, and because they were able to rapidly ascertain incident cases for interview.

Diagnosed were 728 cases of leukemia and 780 cases of NHL among white men 30 years or older between 1981 and 1983. Controls (n = 1543) were frequency matched to cases by date of birth, vital status, state of residence, and date of death, if the case was deceased. For living cases under age 65 at diagnosis, controls were selected by a two-stage random-digit telephone dialing process. For living cases over age 65, controls were selected from the medicare files of the Health Care Finance Administration. State vital records were used to select deceased controls for deceased cases. Of the cases ascertained, 87% were interviewed, and approximately 80% of the controls. All cases received a pathology review by a panel of pathologists.

Study subjects or their next-of-kin were interviewed to obtain information on residential history, drinking water sources, nonfarm occupational history, smoking and alcohol use, use of pasteurized dairy products, hobbies, medical conditions including allergies, family history of cancer, and farm activities. The detailed agricultural portion of the questionnaire covered crops and animals raised, and use of specific insecticides, herbicides, and fungicides. Relative risks for various agricultural exposures were estimated by the odds ratio (OR), considering nonfarmers as nonexposed.

Preliminary results indicate that working or living on a farm after the age of 18 was not associated with NHL or leukemia (OR of 1.1 and 1.0, respectively). ORs for living on a farm before age 18 were also approximately 1. It is not clear why these results differ from those from death certificate case-control studies where rather consistent elevated risks among farmers of about 1.25 have been observed. Analysis of living and deceased cases gave identical results. In this interview study, complete work histories were available to define farming, while in death certificate studies farmers were identified strictly from the notation on the death certificate. In the interview study, subjects who left the farm early in their working career would be classified as farmers, but from death certificate studies they may not. No clear pattern of risk by recency of farming is evident. Those farming for less than 20 years were at lower risk if they ceased farming before 1960 than those stopping after 1960. For those who farmed more than 20 years, the risk was greater among those who stopped before 1960. These patterns do not seem to explain the discordant results from death certificates and interview studies.

Death certificate studies of leukemia in Nebraska[17] and Wisconsin[18] indicate that farmers dying in earlier years were at higher risk than those dying more recently and may indicate that the risk associated with agriculture is lessening. This could explain the lack of an association in the interview study, since cases were diagnosed after 1980, which is a few years after the period covered by the death certificate studies. However, contrary to the Nebraska study,[17] death certificate studies of leukemia in Iowa[22] and multiple myeloma[20] and NHL[19] in Wisconsin show higher risks among farmers dying more recently than those dying in earlier years.

ORs by duration of farming show few statistically significant elevations and no clear pattern of rising risk of leukemia or NHL with duration of farming. In fact, short-duration farmers appear to have a slightly higher risk of leukemia than long-duration farmers. Considering relative risks for farmers by type of crops grown, only chronic lymphatic leukemia (CLL) showed a significant association with a crop (OR = 1.4 for corn), which was contrary to the death certificate studies where myeloid leukemia seemed more closely associated with corn production. No significant associations were seen between crops and NHL, although the OR among corn-producing farmers was 1.4 for small cell lymphoma.

No significant excess risks for leukemia overall or NHL were associated with type of livestock raised. Significantly elevated ORs for CLL were seen among farmers who raised chickens (OR = 1.4; 95% CI = 1.0 to 1.9), dairy cattle (OR = 1.4; 95% CI = 1.1 to 2.0), and swine (OR = 1.4; 95% CI = 1.1 to 1.9). Death certificate studies in Nebraska[17,21] and Iowa[22] have also shown association between lymphatic leukemia and chicken and dairy farming.

Analyses of use of pesticides have just begun. Leukemia overall was unrelated to pesticide exposures. However, CLL and NHL (especially small cell lymphoma) showed significantly elevated ORs among farmers first exposed to pesticides 20 or more years before interview. Relative risks appear to be elevated among those exposed to DDT, chlordane, dieldrin, dichlorvos, and 2,4,5-T. Additional analyses are needed to confirm these preliminary findings.

SUMMARY

Despite consistent evidence from geographic correlations, broad occupational surveys, and death certificate studies suggesting an association between farming and lymphatic and hematopoietic cancer, we see little evidence of an overall increased risk of leukemia and NHL among farmers from an interview case-control study. Chronic lymphatic leukemia did occur more frequently among farmers raising certain types of livestock and engaged in corn production. Analysis of these data by pesticide use is underway, but is not yet complete. Preliminary results suggest that farmers who used specific pesticides 20 or more years before interview may be at elevated risk for CLL and NHL.

EDITOR'S NOTE

Since the preparation of the foregoing manuscript, additional evidence has been published by Hoar et al.[23] on agricultural herbicide use and risk of lymphoma and soft tissue sarcoma from a population-based controlled study carried out in Kansas. They found that herbicide farm use was associated with non-Hodgkin's lymphoma (OR 1.6) and that the relative risk increased significantly with number of days of herbicide exposure per year (men exposed to herbicides more than 20 d/year had a sixfold increase of non-Hodgkin's lymphoma (OR 6.0) relative to nonfarmers. Those who frequently mixed or applied herbicides had an OR of 8.0. Excesses were associated with the use of phenoxyacetic acid herbicides. This study appears to confirm reports from Sweden and the U.S. that non-Hodgkin's lymphoma is associated with farm herbicide use, particularly phenoxyacetic acids.

REFERENCES

1. **Mason, T.J., McKay, F.W., Hoover, R., Blot, W.J., and Fraumeni, J.F., Jr.**, *Atlas of Cancer Mortality for U.S. Counties: 1950—1969*, Publ. No. (NIH) 75-780, U.S. Department of Health, Education, and Welfare, Washington, D.C., 1975.
2. **Blair, A., Fraumeni, J.F., Jr., and Mason, T.J.**, Geographic patterns of leukemia in the United States, *J. Chron. Dis.*, 33, 251, 1980.
3. **Cantor, K.P. and Fraumeni, J.F., Jr.**, Distribution of non-Hodgkin's lymphoma in the United States between 1950 and 1975, *Cancer Res.*, 40, 2645, 1980.
4. **Blattner, W.A., Blair, A., and Mason, T.J.**, Multiple myeloma in the United States, 1950—1975, *Cancer,* 48, 2547, 1981.
5. **Blair, A., Malker, H., Cantor, K.P., Burmeister, L., and Wiklund, K.**, Cancer among farmers: a review, *Scand. J. Work. Environ. Health,* 11, 397, 1985.
6. **Milham, S., Jr.**, Occupational Mortality in Washington State, 1950—1979, DHHS Publ. No. (NIOSH) 83-116, Department of Health and Human Services, Cincinnati, OH, 1983.
7. **Petersen, G.R., and Milham, S., Jr.**, Occupational Mortality in the State of California, 1959—1961, DHEW Publ. No. (NIOSH) 80-104, Department of Health, Education and Welfare, Rockville, MD, 1980.
8. **Burmeister, L.F.**, Cancer mortality in Iowa farmers, 1971—1978, *J. Natl. Cancer Inst.*, 66, 461, 1981.
9. **Gallagher, R.P., Threlfall, W.J., Jeffries, E., Band, P.R., Pinelli, J., and Coldman, A.J.**, Cancer and aplastic anemia in British Columbia farmers, *J. Natl. Cancer Inst.*, 72, 1311, 1984.
10. **Guralnick, L.**, Mortality by Occupation and Cause of Death, Vital Statistics Special Rep. 53(3), Public Health Service, Department of Health, Education and Welfare, Washington, D.C., 1963.
11. **Walrath, J., Rogot, E., Murray, J., and Blair, A.**, Mortality Patterns Among U.S. Veterans By Occupation and Smoking Status, NIH Publ. No. 85-2756, U.S. Government Printing Office, Washington, D.C., 1985.
12. **Fox, A.J. and Goldblatt, P.O.**, Longitudinal Study of Sociodemographic Mortality Differentials, 1971—1975, Office of Population Censuses and Surveys, Her Majesty's Stationery Office, London, 1980.
13. **Howe, G.R. and Lindsay, J.P.**, A follow-up study of a ten-percent sample of the Canadian labor force. I. Cancer mortality in males, 1965—73, *J. Natl. Cancer Inst.*, 70, 37, 1983.
14. **Decoufle, P., Stanislawizyk, K., Houten, L., Bross, I.D.J., and Viadana, E.**, A Retrospective Survey of Cancer in Relation to Occupation, DHEW (NIOSH) Publ. No. 77-178, Department of Health, Education and Welfare, Cincinnati, OH, 1977.
15. **Williams, R.R., Stegens, N.L., and Goldsmith, J.R.**, Association of cancer site and type with occupation and industry from the Third National Cancer Survey Interview, *J. Natl. Cancer Inst.*, 59, 1147, 1977.
16. **Dubrow, R. and Wegman, D.**, Occupational Characteristics of Cancer Victims in Massachusetts 1971—1973, National Institute for

Occupational Safety and Health, Public Health Service, U.S. Department of Health Human Services, Cincinnati, OH, 1982.

17. **Blair, A. and Thomas, T.L.**, Leukemia among Nebraska farmers: a death certificate study, *Am. J. Epidemiol.*, 110, 264, 1979.
18. **Blair, A. and White, D.**, Death certificate study of leukemia among farmers from Wisconsin, *J. Natl. Cancer Inst.*, 66, 1027, 1981.
19. **Cantor, K.P.**, Farming and mortality from non-Hodgkin's lymphoma: a case-control study, *Int. J. Cancer*, 29, 239, 1982.
20. **Cantor, K.P. and Blair, A.**, Farming and mortality from multiple myeloma: a case-control study with the use of death certificates, *J. Natl. Cancer Inst.*, 72, 251, 1984.
21. **Blair, A. and White, D.**, Leukemia cell types and agricultural practices in Nebraska, *Arch. Environ. Health*, 40, 211, 1985.
22. **Burmeister, L.F., Van Lier, S.F., and Isacson, P.**, Leukemia and farm practices in Iowa, *Am. J. Epidemiol.*, 115, 720, 1985.
23. **Hoar, S.K., Blair, A., Holmes, F.F., Boysen, C.D., Robel, R.J., Hoover, R. and Fraumeni, J.F.**, Agricultural herbicide use and risk of lymphoma and soft-tissue sarcoma, *J.A.M.A.*, 256, 1141, 1986.

A Census-Based Record Linkage Study of Canadian Farmers

H. I. Morrison, Y. Mao, L. Ritter, and D. T. Wigle

INTRODUCTION

There is almost no information on the possible human health effects of chronic "low-level" exposure to pesticides, but there is an increasing demand for human epidemiological studies on the possible health risks to individuals exposed to agricultural chemicals. Canada's farming industry utilizes highly sophisticated chemical techniques. Farm operators are exposed to a wide variety of chemicals, most of which have not been tested for long-term human health effects. Some, such as Alachlor and Captan, are known mutagens and animal carcinogens.[1,2]

Farmers comprise the largest population subgroup with direct exposure to pesticides. Farmers have been found to be at an increased risk from certain types of cancer, such as lip, skin, stomach, prostate, brain, leukemia, and multiple myeloma.[3-8] These associations, with the exception of cancers of the lip and skin which are related to sun exposure, remain unexplained.

Two censuses are conducted every 5 years in Canada: a Census of Population and an Agricultural Census. In 1971 one third of all Canadians filled out a "2B" population census form which was considerably more detailed than the "2A" form filled out by the remaining two thirds of Canadians. In 1971 any Canadian who sold at least $50 worth of livestock or produce and farmed at least 1 acre of land was required by law to complete an Agricultural Census form. In that year, 366,128 farms were identified in the Census of Agriculture. The vast majority of these were privately owned and operated. Most questions on the two censuses related to estimating population and economic variables. They were certainly not designed with future epidemiologic studies in mind. However, many questions were asked which could be of benefit in better understanding the long-term health risks, if any, associated with the use of agricultural chemicals. Table 1 contains an abbreviated list of variables contained on either the agricultural and/or the long (2B) population census form.

In 1972 the Census of Agriculture group of Statistics Canada instituted a program to link the 1971 Census of Agriculture to the 1971 Census of Population. The purpose of this computerized record linkage was to produce a data base which would enable the cross classification of the agricultural characteristics of farm operations with socioeconomic characteristics of farm operators and households.[9] However, neither of these files contain the names and addresses of farm operators. Because farms are viewed as businesses, names and addresses of farmers are captured from census records and maintained in a central farm registry. Farmers are the only group of Canadians whose names are data captured from census records.

PILOT STUDY

In 1983 the Health Protection Branch of the Department of National Health and Welfare, in conjunction with Statistics Canada, initiated a pilot study to test the feasibility of a record linkage study of Canadian farmers using census data. Information on 1000 farmers aged 70 years or more, identified from the 1971 linked Agriculture-Population Census file, was linked to the 1971 Central Farm Registry in order to capture names and addresses. This manual linkage was performed through matching on a series of identifiers including a unique household/farm number. With farmers properly identified, this data set was then manually linked to the Mortality Database operated by Statistics Canada. Proportionate mortality ratios (age 70+) for the cohort and for Canada were calculated. As a way of assessing the quality of the record linkage, an alive follow-up was performed by linking farmers' records to the 1981 Central Farm Registry. Based on the findings of the pilot study, it was believed to be most appropriate to relink the 1971 Census of Population with the 1971 Census of Agriculture, and then add the names and addresses from the 1971 Central Farm Registry. This data set will then be linked to both mortality and cancer morbidity records.

Of the 1000 farmers age 70 years and older in 1971, 349 were found in the Mortality Database. This is 63.5% of the

TABLE 1
Variables of Interest from the 1971 Linked Agriculture-Population Records

Census of population		Census of agriculture
Demographic	**Socioeconomic**	
Age	Birth place	Location of farm
Sex	Citizenship	Size of farm
Marital status	Date of immigration	Types of crops grown by acreage
Number of children	Ethnic origin	Acreage sprayed by insecticides
	Religion	Acreage sprayed by herbicides
	Language	Number of farm animals by type
	Education	
	Economic status	
	Housing characteristics	
	Source of drinking water	

expected number (565) of deaths based on a life table for Canadian males. Proportionate mortality ratios for this cohort are found in Table 2. The proportions of deaths due to lung cancer and chronic obstructive lung disease were smaller than for Canada as a whole, while those for stomach cancer, prostatic cancer, multiple myeloma, aplastic anemia, and motor vehicle traffic accidents were higher.

Smoking status for farmers by region is presented in Table 3. Levels were highest in Quebec and lowest in British Columbia. Smoking status by crop type (Table 4) revealed that tobacco farmers had the highest proportion of current smokers.

MAIN STUDY

The main study will involve the assemblage of a cohort of roughly 360,000 Canadian farmers identified by the 1971 censuses. Record linkage to the Mortality Database of Statistics Canada will use the Generalized Iterative Record Linkage System (GIRLS).[10]

The first step will be to link the 1971 and 1981 central farm registries. Those farm operators who link to the 1981 registry will then be linked to the 1981 Census of Population to retrieve month of birth. The 1971 Census of Agriculture will be relinked to the 1971 Census of Population. Name, address, and month of birth (when available) from the 1971 farm registry and the 1981 census of population will then be added to the linked 1971 agriculture-population file. This cohort will be linked to the Mortality Database (1971—1984). Those links which are "questionable" will be separated out, and month of birth will then be abstracted from the microfilmed 1971 Census of Population. With month of birth added, these "questionables" will be relinked to the mortality data base.

Cigarette-smoking status will be assessed on a subgroup through an annual survey of Canadian farmers conducted by the Agriculture Division of Statistics Canada. Questions relating to current smoking status, including number of cigarettes smoked daily, were added to the 1984 National Farm Survey. This survey derives its sampling frame from the central farm registry and is annually administered to 20,000 farmers. Although this survey will not provide individual smoking levels for farmers in the 1971 linkage study, it will allow estimates of smoking level according to crop type and pesticide usage and will be used in a future study of smoking and mortality among farmers.

TABLE 2
Proportionate Mortality Ratios of Farmers' Cohort and Canada (Age 70+)

	Farmers' cohort (1971—1981)	Canada (1977)
Stomach cancer	2.9	1.6
Large intestine cancer	2.6	2.0
Lung cancer	2.9	5.4
Prostatic cancer	5.4	3.1
Multiple myeloma	0.9	0.3
Aplastic anemia	0.6	0.05
Ischemic heart disease	41.3	35.9
Cerebrovascular disease	8.9	11.0
Chronic obstructive lung disease	1.7	2.7
Motor vehicle traffic accidents	1.1	0.5
Accidental falls	1.1	1.1

TABLE 3
Canadian Farmers by Smoking Habits and Region (%)

	Atlantic	Quebec	Ontario	Prairies	B.C.	Canada
Never smoked	42.4	41.1	51.0	45.5	51.0	46.3
Ex-smoker	28.0	26.2	21.7	23.7	30.3	24.2
Pipe or cigar	3.7	2.6	7.0	2.1	2.6	3.6
Cigarette	24.9	27.7	18.4	26.6	14.7	23.9
Combination	1.1	2.5	1.9	2.0	1.4	2.0
All current smokers	29.6	32.7	27.3	30.8	18.7	29.5

From 1984 National Farm Survey.

TABLE 4
Canadian Farmers' Smoking Status by Crop Type (%)

	Crop type				
Smoking status	Spring wheat	Potatoes	Tobacco	Fruit trees	All
Never smoked	46.9	39.5	39.0	49.7	46.3
Ex-smoker	22.8	25.6	15.1	29.6	24.2
Pipe or cigar	2.2	3.6	1.6	2.5	3.6
Cigarette	26.1	27.8	32.3	16.9	23.9
Combination	2.1	3.5	11.9	1.4	2.0
Current smokers	30.3	34.9	45.8	20.0	29.5

From 1984 National Farm Survey.

DISCUSSION

A pilot study has demonstrated the feasibility of linking the 1971 Census of Population, Census of Agriculture, and Mortality Databases. One finding of the pilot study was a lower-than-expected number of deaths among farmers. There are three possible explanations for this. One is that farmers as a whole have a more favorable mortality experience than that of the general population.[3] A second possible explanation is the "healthy worker effect". Individuals still employed at age 70+ are bound to be healthier than the general public of the same age and to have lower risk of mortality. A third possible explanation would be the data base. A failure to make a correct link tends to occur when identifying information for an individual is either scant or incorrect. This is most likely to occur when the individual is old as was the situation in the pilot study. The addition of month of birth for "questionable" links and their relinkage should improve the linkage rate for the main study over the pilot study. In addition, the linkage to the 1981 Central Farm Registry will provide an alive follow-up. However, a farmer will only appear in the 1981 registry if he or she is still farming in that year. It will be useful to see how many farmers are reported as alive on the 1981 registry, but for whom death certificates are found.

Proportionate mortality ratios from the pilot study demonstrate a pattern of mortality which has been reported previously[3-8] for farmers, i.e., low lung cancer mortality and higher than normal mortality from stomach and prostatic cancer as well as from multiple myeloma and aplastic anemia.

Given the importance of cigarette smoking in the etiology of many cancers in man, it is highly desirable to have smoking status on individuals in analyzing any cohort. As is usually the case in retrospectively assembling a cohort from existing data, this information is lacking in the present study. However, results from the 1984 National Farm Survey will provide cigarette- smoking information on various classes of farmers. Of course, what would be of greater benefit than current data would be a knowledge of smoking patterns among farmers over the past 30 to 40 years. Because this is not available current data are used as a proxy, with the assumption that basic patterns have remained over time. This information should be of considerable benefit in analyzing cigarette smoking-related causes of death. The results of the 1984 survey indicate that farmers as a whole smoke less than the general Canadian public. Evidence of this can be seen in the findings of the pilot study, with smoking-related causes of death such as lung cancer and chronic obstructive lung disease accounting for a smaller proportion of deaths than among the general population. Regional differences among farmers follow a pattern similar to that noted for Canadians as a whole.

The main study should help to identify health problems which may characterize Canadian farmers. The large size of the cohort will allow multivariate analysis of the large number of farm and socioeconomic variables available from census records. It is anticipated that health problems identified will be explored in subsequent in-depth case-control studies.

REFERENCES

1. **U.S. EPA,** U.S. Position Document #1, Environmental Protection Agency, Alachlor, December 1984.
2. U.S. DHEW, Bioassay for Captan for Possible Carcinogenicity, National Cancer Institute Carcinogensis Technical Report Series No. 15; Public Health Service, U.S. Department of Health, Education and Welfare, 1977, 38p.
3. **Burmeister, L.,** Cancer mortality in Iowa farmers, 1971—78, *J. Natl. Cancer Inst.,* 66, 461, 1981.
4. **Milham, S.,** Leukemia and multiple myeloma in farmers, *Am. J. Epidemiol.,* 94, 207, 1971.
5. **Gallagher, R.P., Threlfall, W.J., Jeffries, E., Band, P.R., Spinelli, J., and Coldman, A.J.,** Cancer and aplastic anemia in British Columbia farmers, *J. Natl. Cancer Inst.,* 72, 1311, 1984.
6. **Wiklund, K.,** Swedish agricultural workers, a group with a decreased risk of cancer, *Cancer,* 51, 566, 1983.
7. **Delzell, E. and Gufferman, S.,** Mortality among white and nonwhite farmers in North Carolina, 1976—1978, *Am. J. Epidemiol.,* 121, 391, 1985.
8. **Choi, N.W., Schuman, L.M., and Gullen, W.H.,** Epidemiology of primary central nervous system neoplasms. I. Mortality from primary central nervous system, neoplasms in Minnesota, *Am. J. Epidemiol.,* 91, 238, 1970.
9. **Freeman, W.G.,** An introduction to the agriculture-population linkage program, *Can. Farm Econ.,* 11, 9, 1976.
10. **Howe, G.R. and Lindsay, J.,** A generalized iterative record linkage computer system for use in medical follow-up studies, *Comput. Biomed. Res.,* 14, 327, 1981.

Dermatoses and Zoonoses

Epidemiology of Occupational Skin Disease in Agriculture

C. G. T. Mathias

Descriptive epidemiology is a useful tool for identifying occupational risks and establishing priorities for occupational health research. In 1978 a report prepared by the Occupational Safety and Health Administration Advisory Committee on Cutaneous Hazards, based on results of annual surveys performed by the Bureau of Labor Statistics (BLS), established that agricultural and manufacturing industries had the highest relative risks for occurrences of occupational skin disease (4.1 for each of these industrial classifications). Other than this high relative risk, little else is known about the epidemiology of occupational skin disease in agriculture. This review will focus on the descriptive epidemiology of occupational skin disease in three areas, with particular reference to agriculture: (1) rates and risks by industry classification, (2) casual agents by industry classification, and (3) disability by industry classification. This analysis is based on 1974 and 1979 data generated by the Doctor's First Report of Injury System, compiled annually by the California State Department of Industrial Relations, Division of Labor Statistics and Research.

Table 1 illustrates the numbers and proportions of occupational disease cases by major industry classification. In 1974 and 1979 skin disease in agriculture accounted for 14.4% (2052 of 14,283) and 13.0% (2065 of 15,894) of all occupational skin disease cases, respectively. Although skin disease accounted for 45.3 and 38.7% of all occupational illnesses for all industry classifications for these years, the number of skin disease cases as a percentage of all illnesses within agriculture was disproportionately high, accounting for 70.0 and 64.9%, respectively. These proportions may be contrasted to manufacturing, where skin disease accounted for only 43.0 and 36.9% of all occupational illness occurrences.

Case rates for occupational skin disease may be calculated for industry classifications from employment estimates for the respective classifications, provided by the California Employment Development Department. These rates are depicted in Table 2. Approximately 3 to 4% of the California work force is employed in the agricultural industry. Case rates per 1000 employees in agriculture for 1974 and 1979 were 6.7 and 6.1, respectively. These rates compare with 3.5 and 3.4 for manufacturing, and 1.8 and 1.6 for all industries combined. Data for 1979 indicate that 81.5% of the agricultural industry cases occurred in job categories subclassified within agricultural crop production (Table 3), which employed approximately 75% of all agricultural workers. Conversely, agricultural activities not related to crop production (including animal and horticultural services, forestry, and fishing) accounted for only 18.5% of all cases, but employed only about 5% of the total agricultural workforce.

The relative risks of occupational skin disease by major industry classification (Table 4) for 1974 and 1979 may be derived from the calculated rates in Table 3. The relative risk for agricultural employees was almost double that for manufacturing employees, and greater than four times the risk for all nonagricultural employees considered together. In contrast, the relative risk for manufacturing employees was

TABLE 1
Numbers and Proportions[a] of Occupational Disease Cases by Industry Classification, California, 1974 Vs. 1979

Occupational disease	Agriculture		Nonagriculture		Manufacturing		Nonmanufacturing		All industries	
	1974	1979	1974	1979	1974	1979	1974	1979	1974	1979
Number										
Skin	2,052	2,065	12,231	13,829	5,876	6,723	8,407	9,171	14,283	15,894
Nonskin	878	1,118	16,507	24,009	7,795	11,479	9,590	13,648	17,385	25,127
All	2,930	3,183	28,738	37,838	13,671	18,202	17,997	22,819	31,668	41,021
Proportion (%)										
Skin	70.0	64.9	42.6	36.5	43.0	36.9	46.7	40.2	45.3	38.7
Nonskin	30.0	35.1	57.4	63.5	57.0	63.1	53.3	59.8	54.7	61.3
All	100	100	100	100	100	100	100	100	100	100

[a] Expressed as % of all occupational disease within each industry classification.

From California Department of Labor Statistics Research.

TABLE 2
Rates[a] of Occupational Disease by Industry Classification, California, 1974 Vs. 1979

	Agriculture		Nonagriculture		Manufacturing		Nonmanufacturing		All industries	
Occupational disease	1974	1979	1974	1979	1974	1979	1974	1979	1974	1979
Rates										
Skin	6.66	6.10	1.56	1.43	3.47	3.36	1.30	1.15	1.75	1.59
Nonskin	2.85	3.30	2.11	2.48	4.60	5.73	1.49	1.71	2.14	2.51
All	9.51	9.40	3.67	3.91	8.07	9.09	2.79	2.85	3.89	4.10

[a] Per 1000 employees. Numerators derived from Table 1. Denominator estimates (amounts in thousands) supplied by the California Employment Development Department for 1974/1979 as follows: agriculture (308.1/338.6); manufacturing (1694.1/2002.7); all industries (8142.4/10,003.2).

TABLE 3
Rates of Occupational Skin Disease in Agriculture by Industry Subclassification, California, 1979

	Number of cases	Rates[a]
Agricultural production	1683	6.64
Agricultural services	382	4.49
Pest control, other agricultural chemicals services	52	
Other crop and soil services	85	
Veterinary and other animal services	12	
Landscape and horticultural services	227	
Forestry and fishing	6	

[a] Per 1000 employees. Denominator estimates (amounts in thousands) obtained from California Employment Development Department as follows: agricultural production (253.6); agricultural services (85.0).

TABLE 4
Relative Risks[a] of Occupational Disease by Industry Classification, California, 1974 Vs. 1979

	Agriculture vs. manufacturing		Agriculture vs. nonagriculture		Manufacturing vs. nonmanufacturing	
Occupational disease	1974	1979	1974	1979	1974	1979
Risk						
Skin	1.8	1.8	4.3	4.3	2.7	2.9
Nonskin	0.58	0.58	1.4	1.3	3.1	3.4
All	1.1	1.0	2.6	2.4	2.9	3.2

[a] Derived from calculated rates in Table 2.

TABLE 5
Occupational Skin Disease in Agriculture by Causal Agent, California, 1979

Causal agent	Number of cases	Total cases (%)
Plant/animal products (inedible)[a]	1028	63.2
Agricultural chemicals	331	20.4
Food products (edible)	192	11.8
Soaps, detergents, cleaners	13	0.8
Dusts	11	0.7
Nonsolvent petroleum products	11	0.7
All others	39	2.4
Total[b]	1625	100

[a] Poison oak (789 cases) comprises 76.8% of total cases within this causal agent category.

[b] Distribution of causal agents for cases where no causal agent was specified (total of 440) has been assumed to be similar to the distribution for specified cases. Cases where no causal agent was specified were subsequently excluded, leaving a total of 1625 remaining cases for analysis.

roughly three times greater compared to non-manufacturing employees considered as a whole.

Table 5 lists the numbers of occupational skin disease cases in 1979 by causal agent for agriculture. Assuming that causal agents for unspecified cases (total of 440) were distributed similarly to causal agents for the specified cases, these unspecified cases may be dropped from analysis; the proportions illustrated in Table 5 (based on 1625 analyzed cases) give a reasonable estimate of the rank order of importance of these causal agents. Inedible plant and animal products, agricultural chemicals, and edible food products accounted for 95.4% of the total cases. Poison oak accounted for 76.8% of the total cases within the inedible plant and animal product causal agent category.

Causal agent subclassifications for these three major causal agent categories are listed in Table 6. Pesticides caused the overwhelming number of cases due to agricultural chemicals. Fruits and nuts, followed by vegetables, produced most of the cases secondary to edible food products.

Disability as measured by the proportion of occupational disease cases which lost time from work is presented in Table 7 by major industry classification. In 1974 and 1979, 24.5 and 26.6% of occupational skin disease cases in agriculture lost time from work, compared to 15.4 and 20.7% of cases in manufacturing, respectively. The average proportions of lost workday cases in 1974 and 1979 for all industries were 19.8 and 23.7%. Thus, it appears that agricultural workers are more likely to lose time from work, secondary to occupational skin disease, than workers in other industries.

The descriptive epidemiology which has been presented suggests that future research should identify the variables responsible for the high rates and high relative risks of

TABLE 6
Occupational Skin Disease in Agriculture by Three Leading Causal Agents, California, 1979

Causal agent	Number of cases	Total cases[a] (%)
Plant/animal products (inedible)	1028 (total)	63.2
Poison oak	789	48.5
Shrubs, plants, trees	183	11.3
Weeds	21	1.3
Flowers	19	1.2
Leather, hides, feathers, fur	3	0.18
Lumber, wood products	1	0.06
Other	12	0.74
Agricultural chemicals	331 (total)	20.4
Fungicides, herbicides, insecticides, rodenticides	183	11.3
Sulfur	37	2.3
Halogenated hydrocarbons	28	1.7
Fertilizers	13	0.8
Organophosphates	8	0.49
Carbamates	4	0.25
Phenolics	2	0.12
Pyrethrums	1	0.06
Other	55	3.4
Food products (edible)	192 (total)	11.8
Fruits and nuts	113	7.0
Vegetables	69	4.2
Grain and feed	4	0.25
Food preservatives	4	0.25
Fish, seafood	1	0.06
Spices, food additives	1	0.06

[a] See legend from Table 5. Cases where no causal agent was specified have been excluded from final analysis.

TABLE 7
Disability from Occupational Disease by Industry Classification, California, 1974 Vs. 1979

Occupational disease	Number of LWC[a]		Total cases[b]		Total cases (%)	
	1974	1979	1974	1979	1974	1979
Agriculture						
Skin	409	473	1,669	1,775	24.5	26.6
Nonskin	389	519	682	921	57.0	56.4
All	798	992	2,351	2,696	33.9	36.8
Manufacturing						
Skin	782	1,244	5,062	6,009	15.4	20.7
Nonskin	1,910	3,781	6,186	9,732	30.9	38.9
All	2,692	5,025	11,248	15,741	23.9	31.9
All Industries						
Skin	2,365	3,232	11,936	13,622	19.8	23.7
Nonskin	5,154	9,166	13,354	20,311	38.6	45.1
All	7,519	12,398	25,290	33,933	29.7	36.5

[a] LWC = lost workday cases.
[b] Cases where lost workdays not specified have been excluded from analysis.

agricultural workers for development of occupational skin disease. Pesticides have been identified as major causal agents; the variables responsible for pesticides-associated skin disease require further elucidation.

The Role of Farm Animals in the Control of Zoonotic Skin Diseases in Man

K. R. Armstrong and K. Post

INTRODUCTION

Physicians and veterinarians have a common interest in zoonotic skin disease, and if they join forces a marked reduction is possible in the spread of these diseases from farm animals to man.

Nonhuman hosts and their environments can be a very important reservoir of the agents of zoonotic skin diseases. Zoophilic dermatophytes are emphasized in this article. The viral, bacterial, and arthropod agents are discussed to a lesser extent. The procedures which will reduce the reservoir of infection and decrease the spread to humans will be stressed.

DERMATOPHYTOSIS (TINEA, RINGWORM)

Dermatophytosis is an important disease to physicians and veterinarians. The dermatophyte agents are often classified into three ecological groups; anthropophilic, geophilic, and zoophilic. Only those agents in the zoophilic groups are discussed in this paper.

Zoophilic dermatophytes which are normally animal parasites are usually the most important etiologies of dermatophytosis (tinea) in man in agricultural areas in temperate and colder climates, and their incidences are related to animal husbandry practices.

Studies in New Zealand and Poland involving agricultural workers revealed that 70% of the dermatophytes recovered from man were zoophilic fungi.[1]

The common dermatophytes isolated at the Western College of Veterinary Medicine in Saskatoon that cause ringworm (tinea) in farm animals and pets are various species of *Trichophyton* and *Microsporum: T. mentagrophytes, T. verrocosum, T. equinum* and *M. canis. T. mentagrophytes* and *T. verroscosum* are common causes of skin infection in the rural population in this area.[2]

The common nonhuman hosts are cattle, dogs, and horses. The occasional hosts are sheep, pigs, and wildlife — especially deer and moose. The people at most risk are farm families (especially the children), stockyard and abattoir workers, veterinarians, animal attendants, and also hunters.

Clinical Findings in Animals

1. Cattle — The incidence of infection is very high in calves and yearlings that are housed or are in feedlots during the winter (Figure 1). The disease is commonly ignored by livestock owners unless the animals are for sale or used for show purposes. This correlates with a higher incidence in man during the winter. The typical cattle lesion found about the head is a raised annular gray-white crust. Lesions may coalesce and occasionally may be distributed over the entire body of the calf.[3]
2. Dogs and cats — The most common fungi, *M. canis,* produces a thick, circumscribed, tightly adherent gray crust. *Trichophyton* species cause a similar lesion, but may also be more generalized with dry, scaly, diffuse lesions.[4]
3. Horses — The lesions are of two types, superficial and deep. The superficial lesions have a diffuse moth-eaten appearance with desquamation and alopecia. The deep lesions are small foci of inflammation and suppuration.[3]
4. Pigs — The lesions are a progressing ring of inflammation surrounding a scabby alopecic center.[3]
5. Sheep — The lesions are round and alopecic with a grayish crust.[3]

Clinical Findings in Man

T. verrucosum frequently causes the most severe reaction in man known as kerion. As a general rule zoophilic dermatophyte infections are more severe than the infections with the anthropophilic species. The lesions are typically a severe pustular folliculitis with loss of hair and are usually found in exposed areas of the body, head, or arms.[2,5]

Diagnosis

The method of diagnosis is direct microscopy and culture. For control purposes, it is important to establish if the dermatophyte is anthropophilic, zoophilic, or geophilic.

Transmission of Zoophilic Dermatophytes

The direct exposure to an infected animal is one important mode of transmission. The existence of asymptomatic carrier animals occurs especially in cats. Inanimate objects play a major role in transmission; animal brushes and combs, saddles, saddle blankets, and harnesses are frequently contaminated. The immediate animal environment of the kennel, barn, shed, or corral and its contents, especially bedding, may also be highly contaminated. The arthrospores and hyphae can remain viable in the environment for years provided they are kept dry and cool. Therefore, premises and inanimate objects may be infective for periods of up to 6 or 7 years.

Control of Zoophilic Dermatophytes

The elimination of zoophilic dermatophyte infections is

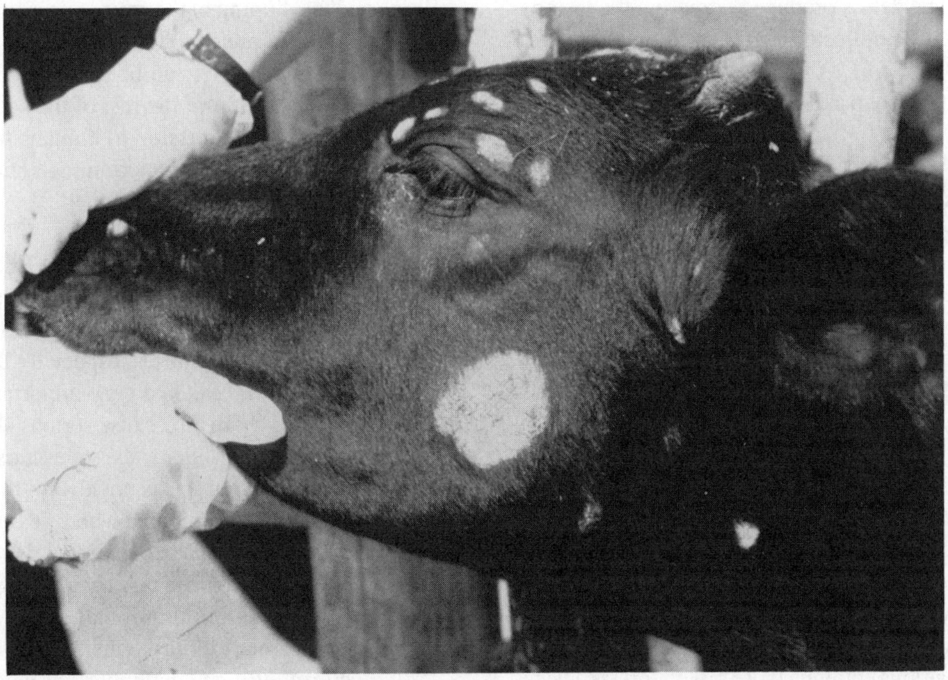

FIGURE 1. *T. verrucosum* infection in a young cow. *T. verrucosum* may produce severe skin infections in man.

impossible, but a dramatic reduction in the spread of infection may occur if control is focused on reducing the reservoir of infection both in the animals and the environment. Control programs usually fail in veterinary cases unless the environment is included as an essential part of the program. The literature indicates that the same situation exists in man. A major effort to control tinea infections in a South Pacific island with mass treatment of the population with griseofulvin failed because the population was reinfected from the environment.[5]

The basic principles of control are the same in veterinary medicine as in human medicine — the isolation and treatment of infected and in-contact animals. The treatment of individual lesions in large farm animals consists of removing the crust with a stiff brush and the application of a fungicidal or fungistatic ointment. The ointments commonly prescribed in human and small animal medicine are not economically practical for large animal use. Many products with variable efficacy are used. Solutions of iodine or hexetidine have been used successfully and copper naphthematae 37.5% (Kopertox®) is commonly used in cattle, but is not recommended for horses. The ointment the first author prescribes is 4% thiabendazole in paraffin applied at 3-d intervals for three to four treatments. This ointment, which is made up at the Veterinary College pharmacy (W.C.V.M.), is not commercially available. If the lesions are widely spread, then the animal should be bathed with an iodine scrub (Betadine®) at weekly intervals. A 0.01% solution of Natamycin as a spray has been reported to give excellent results using 0.75 to 1.5 l per spraying and applied twice at 4-d intervals.[7] However, our experience with this drug in two cattle trials has been disappointing. It has not been approved in Canada. Systemic treatment for large animals is only approved for horses using griseofulvin orally at 7.5 mg/kg body weight daily for 10 d.[3] This is an effective but expensive treatment. Small animal cases are treated locally and systemically. Many topical ointments are used after the lesion area has been clipped and cleaned. A 2% cream or solution of miconazole is very effective. Griseofulvin is often used orally at 30 to 150 mg/kg body weight, daily for a minimum of 4 to 6 weeks, depending upon the form of the drug. Ketoconozole, also very effective but expensive, is used at a dose of 10 to 30 mg/kg orally per day.[4]

The environment of the infected animals is an essential aspect of control, as untreated premises remain sources of infection for years. The stables, sheds, corrals, or kennels should be cleaned out and all bedding burned or manure spread on fields used for grain crops. The surfaces should then be sprayed with a new vapor-phase formaldehyde product (Lysofume®). It may also be used on halters, ropes, forks, shovels, etc. The products previously used, like Captan and regular formalin, are hazardous. Do not use Lysofume® on blankets or saddles. These should be cleaned with a detergent wash, followed by sodium hypochlorite (0.25%) on the blanket and saddle soaps on the saddle.

If possible, separate marked grooming utensils and blankets should be available for use on infected animals and these should not be used on noninfected animals.

To minimize the spread of infection to workers, the use of long-sleeve gloves and a high standard of personal hygiene are encouraged if contact with infected animals cannot be avoided. Adults, not children, should be performing the necessary treatments. The debris produced preparing the lesions for treatment should be collected and burned or discarded. Immediately after the work is completed, workers should wash with a fungistatic soap and change their outer clothing.

VIRAL ZOONOTIC SKIN INFECTIONS

Viral zoonotic skin infections occur occasionally in farm families, veterinary students, and clinicians. The most common etiologies are the pox viruses. The nonhuman hosts are cattle which get the disease pseudocowpox, and sheep which commonly get contagious ecthyma (sore mouth). The latter disease has also been seen in goats, reindeer, and muskox. These viruses produce the diseases of "milker's node" and "orf" in man.

Clinical Findings in Animals

1. Cattle — Pseudocowpox is seen occasionally in dairy cattle here. The lesions, up to ten per teat, are characterized by a heavy scab which falls off in about 10 d leaving a circular or horseshoe-shaped ring of small scabs.[3]
2. Sheep — Contagious ecthyma (sore mouth, orf) is a very common disease of sheep, especially lambs. The lesions develop on the mucocutaneous junctions at the oral commissures and then spread to the muzzle. They are characterized by thick, discrete scabs packed very closely together forming a plaque with fissures. They crumble easily but are difficult to remove from the underlying area of granulation and inflammation.[3]

Clinical Findings in Man

The lesions of orf begin as a small, red or reddish blue papule which enlarges to form a flat-topped hemorrhagic pustule, often crusted in its umbilicated center.[8] The clinical differential diagnosis of orf from milker's node is impossible.[8]

Diagnosis

These diseases may be diagnosed on the clinical findings and confirmation by complement-fixation test, tissue culture, and more often by electron microscopy.

Transmission of Zoonotic Viruses

Pseudocowpox virus is commonly spread by direct contact to milker's hands, by the milking machine teat cups, and the washcloths used to prepare the cow's udder for milking.

Contagious ecthyma (orf) virus is spread by direct contact to an animal attendant hand-feeding affected lambs and by contaminated inanimate objects. The virus withstands drying and is capable of surviving at room temperature for at least 15 years.[3]

Control of Zoonotic Viruses

The isolation of affected animals is important in both diseases if it is recognized in the early stage. In the case of pseudocowpox, the affected cows should be milked last, and the use of separate paper towels for washing the cows prior to milking and the dipping of the entire teat in virucidal disinfectant after milking are recommended. Contagious ecthyma (orf), if recognized early, can be controlled effectively with autogenous vaccination of the rest of the sheep flock.

To prevent the transmission to humans the use of rubber gloves when handling the affected animals and a high standard of personal hygiene are recommended.

BACTERIAL ZOONOTIC SKIN INFECTIONS

Bacterial zoonotic skin infections are rare unless there is very heavy contamination of the source of infection, or they may be associated with injuries, especially bites (Table 1).

Staphylococcus aureus is a very important cause of skin infections in man. It can also cause serious skin infections in animals, especially dogs. Schwabe states most surveys reveal it to be host-specific, but this is not always the case, as Phage type 80/81 has been incriminated where heavy environmental contamination produced an occupational risk of human infection. Transmission involving poultry meat-processing plants, swine farms where antibiotic-resistant staphylococcus was a problem, and veterinary clinic kennels are given as examples.[9]

Salmonella dublin and *typhimurium* have been isolated from skin lesions of veterinarians handling infected cattle.[8] The Western College of Veterinary Medicine had an interesting case of skin infection due to *S. saint-paul*. It was isolated from the skin of two students and one clinician suffering folliculitis after working on an obstetrical case involving a dead fetus which also yielded the organism.[10]

Erysipelothrix insidiosa, Francisella tularensis, and *Bacillus anthracis* are causes of skin diseases in man in which the infection may have originated from animals or animal products.[8] These organisms often cause acute or peracute septicemia in animals. Bites and injuries often play a major role in transmission.

Diagnosis of the etiology is based on bacteriological culture and then the appropriate antibacterial agent chosen for therapy.

Control of Bacterial Zoonotic Skin Disease

The prompt cleaning and disinfection of any wounds and a strong emphasis on personal hygiene are highly recommended to reduce bacterial skin diseases. Environmental disinfection would be very important in *Staphylococcus aureus* and *Salmonella* infections which have originated from dogs or pigs. Vaccination programs will greatly reduce the

TABLE 1
Bacterial Zoonotic Skin Infections

Etiology	Main nonhuman host
Staphylococcus aureus	Cattle, dogs
Salmonella saint-paul	Cattle
Bacillus anthracis (Anthrax)	Cattle, sheep
Erysipelothrix insidiosa (Erysipelas)	Swine, turkey
Francisella tularensis (Tularemia)	Wild mammals and birds primarily

incidence of erysipelas in pigs and turkeys or anthrax in all farm animals. The use of protective gloves is important when handling rodents to prevent tularemia and by those with a high risk of contracting other zoonotic bacterial skin diseases.

Arthropod Zoonotic Skin Infections

Arthropod skin infections are common in animals and most species of mites are very host-specific. However, *Sarcoptes scabiei* is not as host-specific as other species and may be a problem. The nonhuman hosts are swine, where it is very common; also cattle, sheep, and dogs are infected occasionally.

The lesions produced in animals by this mite are red papules and erythema with a loss of hair initially, followed by thick brown scabs over thickened, wrinkled skin. Severe pruritis causes scratching and biting which results in excoriation of the area.[3]

The diagnosis is confirmed by finding the mite in skin scrapings viewed under the microscope.

The mites are spread by direct contact and also by inanimate objects such as grooming brushes and bedding.

The infected animals should be sprayed with an organophosphate insecticide and the treatment repeated in 2 weeks. Human transmission may be controlled by good personal hygiene after handling the infected animals. One may have some temporary skin irritation if infected.

The important procedures necessary to reduce the spread of zoonotic skin diseases from farm animals and pets to man are (1) reduction in the size of the reservoir of infection by treating the affected and in-contact animals and by the use of vaccination in those diseases where applicable — orf, erysipelas, and anthrax; (2) the decontamination of infected fomites and the environment of affected animals by thorough cleaning and disinfecting; (3) the use by animal attendants of rubber or protective gloves to reduce direct contact when it is essential to handle infected animals; and (4) the use of a high standard of personal hygiene immediately after handling infected animals and the prompt cleaning and disinfection of any wounds which may have occurred.

The collaboration of physicians and veterinarians to implement these disease control procedures would markedly reduce the spread of infection from farm animals and pets to man.

REFERENCES

1. **Prochacki, H.,** Mycological flora isolated from people in Poland, *Mycopathologia,* 40, 65, 1970.
2. **Hogan, D.,** personal communication, 1985.
3. **Blood, D.C., Radostits, O.M., and Henderson, J.A.,** *Veterinary Medicine,* 6th ed., Balliere Tindall, London, 1983, 837.
4. **Muller, G.H., Kirk, R.W., and Scot, D.W.,** *Small Animal Dermatology,* 3rd ed., W.B. Saunders, Philadelphia, 1983, 243.
5. **Hubbert, W.J., McCullough, W.F., and Schnurrenberger, P.R.,** *Diseases Transmitted from Animals to Man,* 6th ed., Charles C Thomas, Springfield, IL, 1975, 469.
6. **Oldenkamp, E.P.,** Natamycin treatment of ringworm in cattle in the United Kingdom, *Vet. Rec.,* p. 105, 1979.
7. **Oldenkamp, E.P.,** Treatment of ringworm in horses with Natamycin, *Equine Vet. J.,* p. 11, 1979.
8. **Rook, A., Wilkindon, D.S., and Eblinh, F.J.G.,** *Textbook of Dermatology,* Vol. 1, 3rd ed., Blackwell Scientific, Oxford, 1979, 581.
9. **Schwabe, C.W.,** *Veterinary Medicine and Human Health,* 3rd ed., Williams & Wilkins, Baltimore, 1984, 194.
10. **Pantekoek, J.F.C.A., Rhodes, C.S., and Saunders, J.R.,** Salmonella folliculitis in veterinarians infected during obstetrical manipulation of a cow, *Can. Vet. J.,* 15 (4), 123, 1974.

Dermatophyte Infections in Farmers and Other Rural Residents in Saskatchewan

D. J. Hogan and W. T. Martin

Occupational skin diseases are the most common occupational illnesses. Skin infections are second in importance to contact dermatitis as a cause of occupational skin disease.[1] The most common occupational skin infections are dermatophyte infections (ringworm).[1] Dermatophyte infections are defined as superficial skin infections due to the genera *Microsporum, Trichophyton*, and *Epidermophyton* of the Fungi Imperfecti. Zoophilic species primarily infect animals. Most zoophilic species are also capable of infecting man. Characteristically, zoophilic dermatophytes produce highly inflammatory reactions in man. These reactions may heal with permanent scarring. The most severe pattern of reaction is known as a kerion[2] (Figure 1). Anthropophilic dermatophytes primarily infect man where they tend to produce relatively mild chronic infections.[2] Certain zoophilic dermatophytes have been reported to infect predominantly rural populations and agricultural workers. Since there are few recent studies on these dermatophyte infections in rural areas in North America, we decided to analyze 205 cases of dermatophyte infections, proven by culture, seen at University Hospital, Saskatoon, Canada between 1972 and 1985.

T. mentagrophytes and *T. verrucosum* infections were significantly associated with rural residence, whereas relatively few rural patients with *T. rubrum* infections were seen in our hospital (Table 1).

Urban patients were significantly more likely to present with dermatophyte infections of the feet and nails. Rural patients were significantly more likely to present with dermatophyte infections of the face and of the trunk.

Farmers had a significantly higher incidence of dermatophyte infections with *T. verrucosum* and a lower incidence of *T. rubrum* infections than other occupational groups.

The peak number of positive dermatophyte isolates per 3-month period between January 1973 and December 1984 was in the July-through-September period. *T. verrucosum* isolates, however, were most common in the 3-month period between January and March.

Ten dermatophyte infections were diagnosed as kerions in

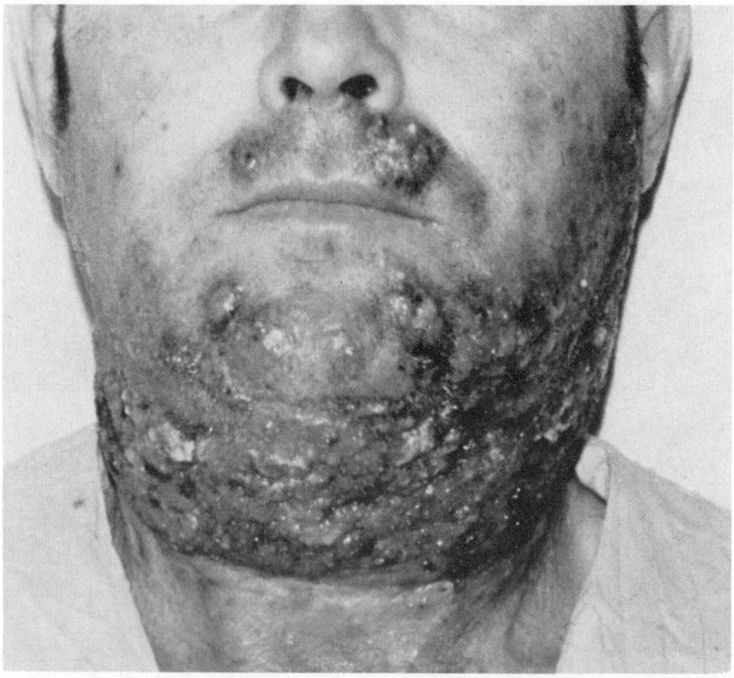

FIGURE 1. Extensive inflammatory skin lesions on the face and neck due to skin infection with *T. verrucosum*.

TABLE 1
Distribution of Dermatophyte Isolates According to Residence of Patient

Dermatophyte species	Residence of patient: Rural	Urban
T. mentagrophytes[a]	49.4	30.8
T. rubrum[a]	17.3	44.2
T. verrucosum[a]	25.9	2.5
E. floccosum	7.4	9.2
M. canis[a]	0	13.3
Total	100%	100%

[a] Statistically significant.

our series. Six kerions were associated with *T. mentagrophytes,* three with *T. verrucosum,* and one from *T. rubrum.* Six kerions were on the face.

A study similar in design to ours was performed 30 years ago in northeastern Michigan by Georg et al.[3] They classified cases of dermatophytosis due to *T. verrucosum* and *T. mentagrophytes* as rural ringworm. These infections predominantly involved exposed areas of the body and were often highly inflammatory. A similar trend was noted in our study.

Chatterjee et al., in a recent study in India, isolated dermatophytes in 5.3% of rural and 3.4% of urban inhabitants.[4] Zoophilic dermatophytes accounted for 55.9% of rural and 13.5% of urban dermatophyte infections. This difference was due primarily to *T. verrucosum* infections.

T. mentagrophytes is one of the most common dermatophyte species to infect man and animals.[5,6] This dermatophyte is commonly isolated from small rodents.[5,6] These rodents may contaminate grains and straw with *T. mentagrophytes.* Farmers working with wheat contaminated with *T. mentagrophytes* have been reported to develop skin infections.[7] *T. mentagrophytes* was the most common dermatophyte isolated amongst farmers and rural patients in our study. It was the dermatophyte most frequently isolated from infections of the face and trunk in our series. A similar pattern was noted in Czechoslovakia where *T. mentagrophytes* is the most common dermatophyte to infect agricultural workers.[5] In North America *T. mentagrophytes* is more common in rural than urban areas.[7] It is the most frequently isolated dermatophyte from man in Saskatchewan and Alberta.[8,9]

T. rubrum is an anthropophilic fungus. It has never been isolated from soil and only rarely from animals.[7] *T. rubrum* is the predominant dermatophyte in urban areas of North America.[7] *T. rubrum* was the most common dermatophyte in our urban patients. It was the most common cause of tinea pedis, tinea manuum, and onychomycosis in our series and other studies in North America.[7]

T. verrucosum is the main cause of ringworm in cattle.[10] *T. verrucosum* infections are much more common in cattle and in man during the winter months as documented in our and other studies.[3,6,10] *T. verrucosum* may remain viable in the environment (the ground, woodwork of cattle stalls, scratching posts, etc.) for over a year.[7] This affords the opportunity for indirect as well as direct transmission of this dermatophyte to the skin of man. *T. verrucosum* continues to be a major cause of tinea facei and tinea corporis in Saskatchewan. Suppurative folliculitis of the face and scalp is a frequent feature of ringworm due to *T. verrucosum.* Fortunately, immunity to reinfection develops in most individuals following an initial infection with *T. verrucosum.*[2]

M. canis is a zoophilic fungus. Cats and dogs are the main reservoir of this fungus.[6] Most infections are acquired from direct contact with an infected animal.[6] After childhood, infections are more common in women. In our study *M. canis* infections predominantly occurred in women residing in Saskatoon.

Our study indicates that farmers and rural residents are at higher risk than other groups of developing severe dermatophyte infections of exposed body areas. *T. verrucosum* and *T. mentagrophytes* are the main pathogenic dermatophytes in farmers and other rural residents of Saskatchewan. Infections with these dermatophytes remain common in farmers and rural populations because of the ubiquitous nature of *T. mantagrophytes* and the continued presence of *T. verrucosum* infections in cattle.

REFERENCES

1. **Meneghini, C.L.,** Occupational microbial dermatoses, in *Occupational and Industrial Dermatology,* Maibach, H.I. and Gellin, G.A., Eds., Year Book Medical Publishers, Chicago, 1982, 95.
2. **Roberts, S.O.B. and MacKenzie, D.W.R.,** Mycology, in *Textbook of Dermatology,* Rook, A., Wilkinson, D.S., and Ebling, F.J.G., Eds., Blackwell Scientific, Oxford, 1986, 885.
3. **Georg, L.K., Hand, E.A., and Menges, R.A.,** Observations on rural and urban ringworm, *J. Invest. Dermatol.,* 27, 335, 1956.
4. **Chatterjee, A., Chattopadhyay, D., Bhattacharya, D., et al.,** Some epidemiologic aspects of zoophilic dermatophytosis, *Int. J. Zoon.,* 7, 19, 1980.
5. **Chmel, L., Buchvald, J., and Valentova, M.,** Spread of *Trichophyton mentagrophytes var. gran.* infections to man, *Int. J. Dermatol.,* 14, 269, 1975.
6. **Kaplan, W.,** Epidemiology and public health significance of ringworm in animals, *Arch. Dermatol.,* 96, 404, 1967.
7. **Kane, J. and Krajden, S.,** Dermatophytosis, in *Occupational Mycoses,* Salvo, D., Ed., Lee & Febiger, Philadelphia, 1983, 143.
8. **Congly, H.,** Epidemiology of dermatophyte infections in Saskatchewan, chap. 77, this volume.
9. **Serhon, A.S. and Garg, A.K.,** A 13-year study of dermatophytic infections in Alberta, Canada, *Mykosen,* 29, 255, 1986.
10. **Rook, A.J., Frain, and Bell, W.,** Cattle ringworm, *Br. Med. J.,* p. 1198, 1954.

Epidemiology of Dermatophyte Infections in Saskatchewan

H. Congly

As there has been no previous published report on the occurrence of dermatophyte infections in Saskatchewan, this study was undertaken to determine the relative frequency of the etiologic agents and their distribution in relation to age, sex, and seasonal variation.

From 1979 to 1984, the numbers of dermatologic specimens and dermatophytes showed a steady increase (Figure 1). The annual percentage of positive findings varied from 11.6 to 17.3%, with an average of 13.4%. Nine species of dermatophytes were identified from 1275 isolates (Table 1).

Trichophyton mentagrophytes was most frequently isolated (44.3%), followed by *T. rubrum* (25.9%), *T. verrucosum* (16.1%), *Epidermophyton floccosum* (6.1%), and *Microsporum canis* (5.9%). Other dermatophytes, including *M. persicolor*, *M. gypseum*, *M. equinum*, and *T. violaceum*, were isolated very infrequently.

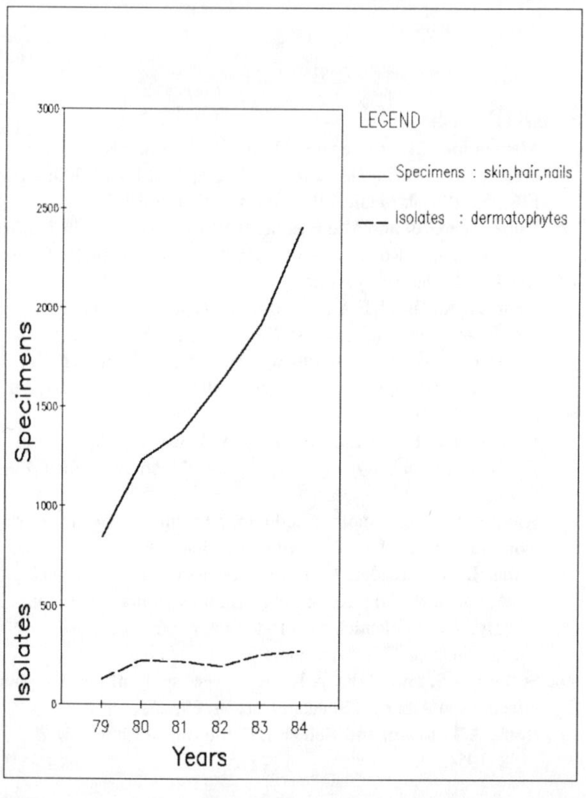

FIGURE 1. Annual frequencies of dermatologic specimens and dermatophyte isolates, 1979—1984.

The distribution of dermatophytes according to age, sex, and clinical types of infection for Figures 2 to 4 was based upon complete information only available between 1982 and 1984.

Tinea corporis was the most common clinical form of ringworm infection followed by tinea pedis, tinea faciale, tinea cruris, onychomycosis, tinea manuum, and tinea capitis.

Figure 2 shows the distribution of dermatophytes according to different clinical forms of ringworm infection. *T. mentagrophytes* was the most common cause of tinea corporis, followed by *T. verrucosum* and *M. canis*. These three dermatophytes were also responsible for tinea faciale and capitis, but with lower frequencies than those observed in tinea corporis. *T. rubrum* was the principal cause of tinea pedis, onychomycosis (tinea unguium), and tinea cruris. *E. floccosum* was a causative agent of tinea cruris secondary only to *T. rubrum*.

The distribution of dermatophytes in relation to age and sex is shown in Figure 3. Infections occurred in patients of all age groups, but were more prevalent in those between the ages of 21 and 30 years. Children under 10 years of age were also very susceptible. There was a general predominance of males (454) over females (235).

Figure 4 shows tinea capitis in relation to age of patients between 1979 and 1984. This condition occurred most frequently in children under 10 years of age, but rarely in adults over 31 years. Children of either sex were equally susceptible to this infection.

Data on the occurrence of dermatophytes according to seasonal variations during 1979—1984 showed that the numbers of dermatophytes were generally high in summer and fall, and declined gradually during winter and spring.

The results of this study demonstrate the importance of *T. mentagrophytes* as a major etiologic agent of dermatophyte

TABLE 1
Dermatophyte Species Isolated from Human Patients in Saskatchewan, 1979—1984

Anthropophilic	Zoophilic	Geophilic
T. rubrum	*T. mentagrophytes*	*M. gypseum*
T. violaceum	*T. verrucosum*	
E. floccosum	*M. canis*	
	M. persicolor	
	M. equinum	

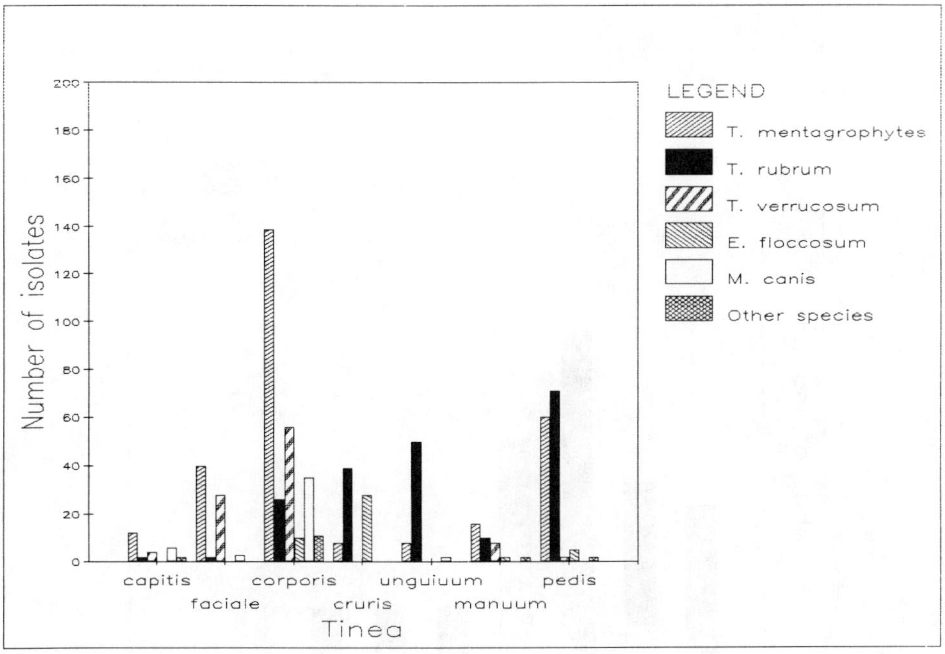

FIGURE 2. Distribution of dermatophytes in relation to clinical forms of infection, 1982—1984.

infection in Saskatchewan. It accounted for 45% of total isolates from skin lesions and was the main cause of tinea corporis, faciale, and capitis. Combined with *T. verrucosum* and *M. canis*, the three dermatophytes constituted 62% of total isolations, whereas *T. rubrum* and *E. floccosum* accounted for 32%. The preponderance of zoophilic fungi over anthropophilic species suggests that indigenous domestic and wild animals are most likely the major sources of dermatophytes causing human infections in Saskatchewan. Opportunity for contact with pet and farm animals suggests that patients of either sex and of all age groups were equally affected by zoophilic dermatophytes as observed in this study.

Available data suggest there is regional variation in the prevalence of etiologic agents of dermatophyte infections in Canada. *M. audouinii* was the predominant fungus in Ontario in 1952.[2] *T. rubrum* was the most frequently isolated dermatophyte in Montreal and British Columbia.[3,4] *M. canis* was a major cause of infection among children in Montreal,[5] and in Newfoundland it also was the most prevalent dermatophyte.[6] In Alberta, *T. mentagrophytes* was the predominant organism.[7] In this study, *T. rubrum* was the second most commonly isolated fungus and frequently responsible for tinea pedis, cruris, and unguium. This finding agrees with that reported from Alberta.[7] *M. audouinii* was not encountered in the present study. This organism has been replaced by *T. tonsurans* and *M. canis* as the most frequent cause of tinea capitis in many North American cities.[8] The isolation rate of *M. canis* in this survey was considerably lower (5.9% of total isolations) than that reported from Montreal (76.5%).[5] The high incidence of *M. canis* infections among the urban population combined with evidence of interhuman transmission of this pathogen suggests a growing tendency for this zoophilic dermatophyte to gradually become anthropophilic.

The findings of this study indicate the effect of age and sex on the type and frequency of dermatophyte infections. The high percentage of scalp infection found mainly in children may be due to a lack of sebum production in these patients. Tinea pedis, cruris, and unguium occurred most frequently in adults aged 21 to 60 years and showed a male preponderance. High exposure rates, type of clothing, and hygenic practices are factors contributory to higher infections among males. In children and young adult females the frequency of *M. canis* infections was high. More frequent and closer contact with dogs and cats and the possible transmission of this pathogen from mothers to young children may account for the high incidence of *M. canis* infection in these children.

This study also demonstrates that there was no marked correlation between seasonal variations and the total distribution of dermatophytes. The overall numbers of dermatophytes isolated were generally high in the summer and fall, and slightly lower in the winter and spring. Warm and humid conditions in the summer and early fall would be favorable for dermatophyte infections. Among the five common dermatophytes encountered in the present study, *T. verrucosum* was the only species showing a correlation between its distribution and seasonal variations. The numbers of *T. verrucosum* isolated from human lesions were highest during the winter and early spring months, but declined in summer and fall. Similarly, the high incidence of cattle ringworm infections generally occurred during winter and spring; infections were usually low during summer and early fall.[9]

The importance of zoophilic fungi as the major cause of dermatophyte infections in Saskatchewan has been demonstrated in this study. Domestic animals, particularly pet and farm animals, are the main sources of human infections in both home and farm environments. Wild animals as a reservoir of

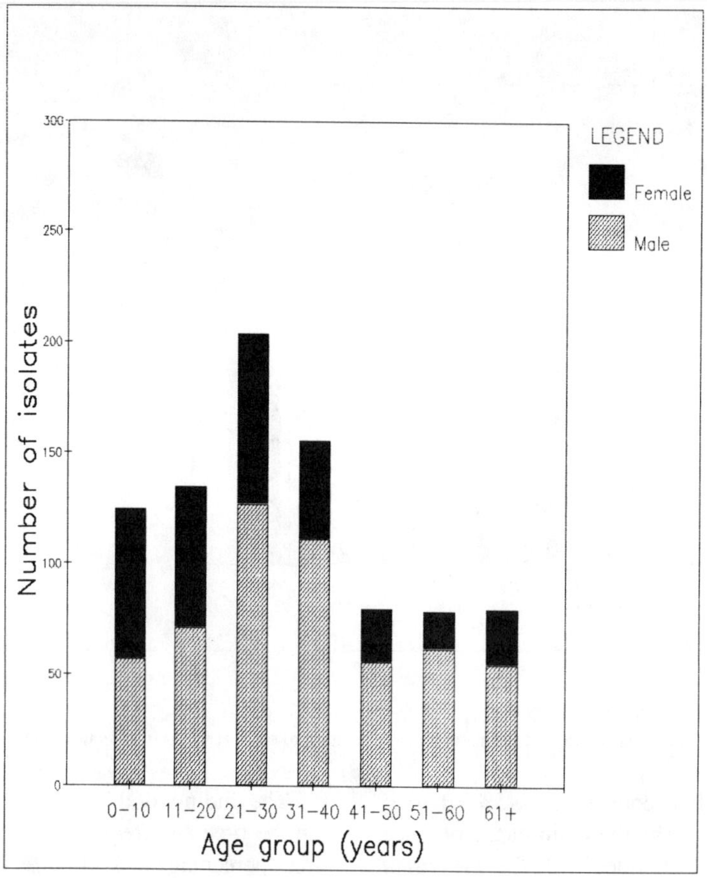

FIGURE 3. Distribution of dermatophytes according to age and sex, 1982—1984.

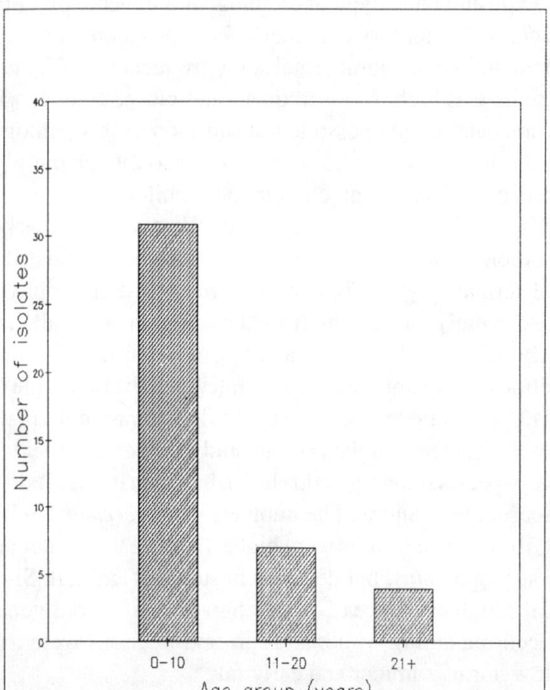

FIGURE 4. Fungal isolates from the scalp in relation to age, 1979—1984.

animal ringworm would also contribute to such infections. The public health significance, as well as economic consequence of dermatophyte infections of man and farm animals in this province, deserves further investigation.

REFERENCES

1. **Rebell, G. and Taplin, D.,** *Dermatophytes. Their Recognition and Identification,* 2nd ed., University of Miami Press, Coral Gables, 1970.
2. **Fisher, J.B. and Wrong, N.M.,** Fungous infections of the skin, hair and nails, *Can. Med. Assoc. J.,* 67, 398, 1952.
3. **Dion, W.M. and Kapica, L.,** Isolation of dermatophytes, Candida species and systemic fungi from dermatologic specimens in Montreal, 1963 to 1973, *Can. Med. Assoc. J.,* 112, 712, 1975.
4. **Bowmer, E.J.,** Dermatomycoses in British Columbia, *Can. J. Public Health,* 57, 32, 1966.
5. **Poisson, M., Delage, G., Martineau, B., and Ricard, P.,** Dermatophytes isoles à un hûpital pediatrique à Montreal de 1954 à 1979, *Can. Med. Assoc. J.,* 127, 134, 1982.
6. **Ross, J.B., Butler, R.W., Cross, R.J., and Fardy, P.W.,** A retrospective study of dermatophyte infection in Newfoundland for the period 1962—1968, *Can. Med. Assoc. J.,* 104, 492, 1971.
7. **Padhye, A.A. and Sekhon, A.S.,** Dermatophytoses in Alberta (1959—1971), *Can. J. Public Health,* 64, 180, 1973.
8. **Rippon, J.W.,** *Medical Mycology: The Pathogenic Fungi and the Pathogenic Actinomycetes,* 2nd ed., W.B. Saunders, Philadelphia, 1982, 154.
9. **Georg, L.K., Hand, E.A., and Menges, R.A.,** Observations on rural and urban ringworm, *J. Invest. Dermatol.,* 27, 335, 1956.

Cutaneous Exposure of Pesticide Spray Applicators

K. Yoshida

INTRODUCTION

The technology of pest control has made remarkable progress during the past decade. Contributing factors were introduction of advanced pesticide formulations with high efficacy and safety and development of high-performance application equipment.

The worker exposure to pesticide may be assessed directly or indirectly. The direct methods involve measurement of inhalation and dermal exposure by employing personal samplers in the worker's breathing zone, and by using a patch method. The indirect methods involve quantification of the active ingredient or its metabolites in body tissue (or fluids) or excreta (usually urine) of the exposed subjects. The literature indicates that direct methods tend to result in poor quality data; thus, the results obtained may not be as realistic as actual exposure, though the major routes of bodily entry may be identified. On the other hand, the indirect method may produce results more accurately insofar as the total bodily intake is concerned.[1]

Cutaneous penetration of pesticides occurs from direct contact with the spray liquid, contact with contaminated clothing or gloves, and diffusional contact with the skin, while a certain amount of once-deposited compound is removed before penetration. After the absorption and metabolic processes, the compound is excreted through renal or biliary mechanisms and residual pesticide retained in the human body may be accumulated.

Throughout these processes, there exists a significant discrepancy of information among commonly used pesticides. So much has been studied about insecticides; however, so little is known about herbicides. This paper is intended to present a discussion on these aspects, together with the presently used method of dermal exposure measurement and a recommendation for worker exposure control.

NATURE OF SPRAY APPLICATOR EXPOSURE

Routes of Bodily Entry

In pesticide spray application, the chemicals may enter the body via three routes: (1) inhalation of respirable spray droplets or vapors produced from volatile spray liquid; (2) dermal entry by direct contact with the liquid or splash, contact with contaminated surfaces, or diffused skin deposition of airborne droplets; and (3) accidental oral entry of pesticides by eating, drinking, or smoking of cross-contaminated foods and beverages or smoking materials.

The nature of actual field exposure takes the form of mixed exposure through a combination of the three routes of entry. A primary concern of pesticide exposure researchers is which route is predominant in spray- applicator exposure so that appropriate measures of worker exposure control may be established.[2] It is known that an extraordinary deposition of pesticide occurs persistently in workers' hands whether protective gloves are used or not. A practice of hand washing after exposure, therefore, is very important in exposure control.[3]

Multiple Exposure

In the reality of field spray application, workers are exposed not only to a single chemical or active ingredient of the compound, but also to multiple compounds in the technical formulations used as emulsifiers, adjuvants, carrier liquids, surfactants, or evaporation suppressants. Therefore, actual exposure takes place as mixed exposure to active ingredients and additives. In investigating problems associated with exposure to impurities contained in contaminated formulation, it was found that some of the contaminants in the technical formulation may be more acutely toxic than the parent pesticides.[4] Very little research has been conducted with respect to workers exposed to multiple pesticides.

In multiple exposure, the effect of a toxic substance is enhanced by exposure of the affected organism to another substance. Under such circumstance, synergism could occur between the two substances. Where the effect of a toxic substance is reduced by exposure of the affected organism to another substance, antagonism may occur. Sometimes, varying the time interval between exposure to the toxic substance and to the secondary substance can produce results varying from synergism to antagonism or vice versa.

MECHANISMS OF CUTANEOUS PENETRATION AND ABSORPTION

Routes of Penetration

The skin is considered as a passive membrane with a primary function of restricting the diffusion of substances into and out of the body. The skin, in simplest terms, can be described as three basic layers: epidermis, dermis, and subcutaneous fat. The outermost layer of the epidermis is the stratum corneum, the major barrier to percutaneous absorption. Below this exists a viable epidermal layer, the most metabolically active layer in skin.

Penetration may be defined as a process of movement of chemical compounds from the outside ambient air (absorption) or direct contact, into the skin, but not beyond this stage.

Skin Permeability

Permeability of chemical compounds through the skin may

be affected by the following internal factors:[5] (1) diffusion through the membranes, (2) morphological pathways including intra-appendageal or intercellular route, (3) molecular diffusion, (4) apportionment of diffusional resistance to skin layers,[6] (5) degree of circulation, (6) sebum as a barrier, (7) transient diffusion or role of skin appendages, (8) regional variations in permeability,[7] and (9) the skin color. Methods of determining permeability for various chemical compounds have been introduced. There exists very little information on the skin permeability of pesticides, in particular, of phenoxy herbicides.[8] Effects of surfactants and solvents on the permeability of the epidermis were studied by using lipid solvents, hydrogen-bonding solvents, and anionic surfactants.[9] The penetration of pesticides without solvent is dependent on the water solubility rather than on the oil solubility. The penetration characteristics of water-soluble pesticides were demonstrated on the hydrophobic and hydrophilic regions of the body.[10] In the practical aspect of pesticide application, the more serious factor to be considered may be the idiosyncratic properties of physically damaged or disease-affected stratum corneum.

Anatomic Variation

The effect of anatomic region on the absorption of malathion was studied *in vivo*. There were enormous differences in penetration between sites. The forearm allowed the least penetration and the scrotum the greatest. The scrotum region would appear to have almost no barrier to penetration of the chemical tested. By having the forearm value of density at unity (reference region), the penetration values on the palm and ball of the foot ranged from one to two, while the scalp, angle of the jaw, and forehead resulted in three, and the value on axilla was found to be seven.

Absorption Mechanisms

The role of the skin in the absorption of industrial chemicals is commonly recognized. For some compounds of an organic nature, the contribution of cutaneous absorption to the total absorption cannot be neglected or excluded. In cutaneous absorption, either transepidermal or transfollicular passage may occur; for organic compounds the transepidermal route seems to be of paramount importance.[11]

Factors affecting percutaneous absorption are: (1) temperature and moisture of body surface, (2) state of ionization, (3) acidity of solution, (4) water/lipid solubility, (5) chemical structure of the compound, (6) molecular size, and (7) species variation.[12]

The cutaneous absorption of organic compounds increases both with increasing temperature and with increasing moisture of the skin, and therefore becomes facilitated in warm seasons.[13] The absorption of organic compounds may follow the surface contamination of the skin (or work clothing). For some compounds, it may occur directly from the gaseous phase.

Under industrial conditions, absorption of a substance from a contaminated skin surface may result in two basic ways of contact absorption: (1) absorption of even relatively volatile substances in continuing repetition of dermal contact, and (2) slow absorption of nonvolatile compounds that occasionally contaminate the skin and remain on the skin for prolonged periods. However, information on the cutaneous absorbability of chemical substances after direct contact with the skin does not provide a common yardstick for qualitative comparisons.

Absorption of some compounds can occur directly through the skin from the gaseous phase. The mechanism of cutaneous absorption of organic vapors is not well understood. The rate of absorption may be roughly proportional to the ambient concentration of the vapors.[14] There is no evidence to suggest that even the lightest gases pass through the stratum corneum via skin appendages.[15]

Absorption Measurement

The technique most widely applicable is the measurement by analytical or isotope techniques of the rate of disappearance of a test substance from the site of its topical application. *In vitro* measurements are often useful as a guide to the planning of *in vivo* tests. A method for determining the percutaneous LD_{50} of a pesticide was proposed by using laboratory rats. However, the relevance of *in vitro* data with only limited numbers of compounds in percutaneous absorption study was discussed.[16]

Metabolism and Excretion

The ability to metabolize toxic substances is affected by individuals and species. It is this variability which makes possible the development of selective pesticides. Once a substance is absorbed, excretion occurs mainly in the urine or bile; biliary excretion is more effective than renal. However, it may also occur in expired air, sweat, breast milk, saliva, gastrointestinal secretions, genital secretions, and normal turnover of hair, nails, and skin.

Urine is one of the most frequently analyzed biological specimens, due to the ease of collection and the presence of nearly all exogenous chemicals or their metabolites in amounts which are often proportional to the absorbed dose. Theoretically, blood is the body fluid which should show the best correlation with the amount of substance absorbed regardless of the route of entry, the degree of retention, and the severity of the effect. Due to practical considerations, however, such as the difficulty of obtaining frequent specimens, the low concentrations produced, and its general unsuitability as an analytical matrix, blood is often avoided in favor of urine or breath.

Implications

Among various formulations of pesticides available on the market, formulations with high lipid solubility are more readily absorbed. In general, emulsions are more easily absorbed than the wettable powders. Pesticide formulations with pH values under 5 and exceeding 8 are skin irritants. Since the alkalies are keratin solvents and the acids are protein precipitants, formulations at these pH values tend to destroy the stratum corneum of the epidermis.

Formulations containing petroleum products, emulsifiers, surfactants, or organic solvents are considered to be more hazardous than the products in granular form. Some of these formulations, such as organophosphate insecticides, may persist for long periods in the skin and accumulation may lead to dermal toxicity and the development of delayed skin irritation after repeated exposure.

Environmental factors and personal habits also play an important role in predisposition to pesticide toxicity. Those who have poor cleaning habits, disregard protective clothing, or wear soiled clothes for prolonged periods are frequently the victims of pesticide poisoning by dermal absorption. Percutaneous absorption of organophosphate insecticides was studied in detail earlier; however, few studies have been made with phenoxy or carbamate herbicides.[17]

The pharmacokinetics of phenoxy compounds are not fully understood; however, it is known that some salt formations inhibit oxidative phosphorylation and oxygen uptake in liver mitochondria, and adversely affect peripheral nerve function. Phenoxy compounds are absorbed in nearly all organs; however, they show higher concentrations in the kidney. They were not significantly stored in fat and are excreted in urine, usually within hours or, at the most, days of exposure.[18]

METHOD OF DERMAL EXPOSURE MEASUREMENT

Body Surface Area

The surface area of the human body is calculated by using Du Bois' formula: (surface area in m^2) = (body weight in $kg^{0.425}$) × (height in $m^{0.725}$) × 0.001718. A table of total surface area for age, sex, and racial origin has been produced for cancer chemotherapy and burn treatment.[19] The measuring method of the human body has been established.

In the practical application of Du Bois' formula for pesticide exposure assessment, regional surface areas used, in percent of surface area to the total for a man 175 cm tall, 78 kg weight at total surface area of 1.92 m^2, are head (5.6%), neck (1.2%), shoulders (6.8%), chest (8.0%), back (8.0%), hips (9.1%), thighs (19.4%), calves (13.5%), feet (6.4%), upper arms (9.7%), forearms (6.7%), and hands (5.6%).[2] Values of deposition density of active ingredients on an individual patch are multiplied by the regional surface area to obtain the regional deposition of pesticide.

Patch Method

The degree of dermal exposure is determined directly by employing a patch method.[2] Dermal patches are located over critical body regions prior to the exposure and the density of the active ingredients or dye tracer in the spray liquid deposited on the patches may be determined through solvent extraction and trace element analysis. In typical exposure studies, 14 patches are used both inside and outside of protective clothing. Filter papers of 45mm diameter of glass fiber or nitrocellulose material are mounted as a sandwich in 75 × 125-mm adhesive dressing with a 35-mm circular window. The filter surface is impregnated with ethylene glycol by micropipetting in order to enhance the deposition and retention of droplets. Petri dishes commonly used for colony counts are used as film badges by mounting the same filter paper and locating them on the outside of the chest region.

After the exposure period, all patches are collected and stored in a cooler for transport. At the same time, exposure subjects are asked to wash their hands in isopropyl alcohol. The hand wash water is later analyzed for the active ingredient. A particular region commonly used as the reference in pesticide exposure study is the right forearm.[7]

Skin Scrub Method

In typical field measurement of spray-applicator exposure, one of the disadvantages of the patch method is accidental destruction or total loss of the patches due to the severeness of body motion by study subjects. The patch method may be supplemented by employing a skin-scrubbing method.[21] The recovery rate of fluorescent dye mixed in a water emulsion of herbicide was 72% by using 1.5 ml of ethanol for 10 cm^2 of skin scrubbing.

Fluorescent Tracer Dye Method

The assessment of dermal exposure was made by computer-based video imaging coupled with fluorescent tracers. Nontoxic fluorescent dyes were introduced into the spray liquid, and the skin deposition density was visualized under long-wave UV light. The video imaging technique measures the deposition by means of a video camera, digitizes the light signal, displays it on a CRT screen, and stores data for analysis. The system can detect a nanogram of deposition density and can be used to evaluate protective clothing efficiency.[22]

Permeable Membrane Method

Studies on the use of such a membrane are still in progress and no compatible method has been established.

WORKER EXPOSURE CONTROL

Protective Clothing

The studies on penetration of pesticide through work clothing and gloves have been reported as an extension of permeability of textile or other materials with hazardous substances such as solvents.[23] The solubility factors of pesticide formulation are another factor to affect penetration.[10] For gloves used in pesticide spray application, the penetration or challenge liquid (acetone) concentration inside was highest at regions where the latex neoprene film was thinnest. These regions include the back, palm, and interstices between the fingers. The study concludes that palm backs are appropriate spots to use for glove permeation tests.[24]

The residual amount of pesticides in contaminated work clothing is becoming one of the serious concerns, because this creates secondary exposure by dermal contact. The effect of laundry practice on the residual amount of pesticide with various textures of work clothing has been investigated.[25] The emission mechanism of pesticide residues in the form of particles or vapor from contaminated work clothing was also investigated.[26]

The protection factors of work clothing may be defined as a ratio of deposition density on the surface of clothing to the skin. In a series of aerial spray trials with low volatile phenoxy herbicides, values of regional protection factors of coveralls of cotton-polyester blend were 5 to 20 at the chest and 10 at the forearm (reference) region.[27] Results of laboratory tests with organophosphorus insecticide applied on various textiles confirmed the above figures.[28]

Barrier Creams

Any material that would decrease the skin penetration would be of value; however, limited use of such creams has been seen in pesticide application.

Closed Application System

The practice of isolation may be seen in the introduction of direct injecting-type equipment with a closed system. The concentrate is introduced into the dispersal system without manual loading or mixing, thus avoiding the chance of direct contact by the operator.[29]

Emergency Treatment

Emergency situations in pesticide exposure can be handled by health care professionals.

CONCLUSIONS

1. In ground-spray application of phenoxy herbicides, which are most commonly used in grain farming on the Prairies, it is known that the major route of entry is through the skin.
2. In the field, actual applicator exposure is not single exposure to active ingredients, but also exposure to additives and carrier liquids.
3. Cutaneous penetration of insecticides was already studied extensively; however, little is known about the phenoxy herbicides.
4. Results from the direct exposure assessment method must be supplemented by results from the indirect method for realistic assessment of applicator exposure.
5. Further studies are needed to determine the cutaneous penetration rate of newly introduced pesticides and phenoxy herbicides.
6. The development of advanced methods of measuring cutaneous exposure to pesticides is urgently required.

REFERENCES

1. **Bristol, D.W. and Lewis, R.G.,** Direct and indirect methods for exposure assessment, in *Determination and Assessment of Pesticide Exposure, Studies in Environmental Science,* Vol. 24, Elsevier, Amsterdam, 1984, 79.
2. **Franklin, C.A. and Spear, R.C.,** Correlation of urinary pesticide metabolite excretion with estimated dermal contact in the course of occupational exposure to guthion, *J. Toxicol. Environ. Health,* 7, 715, 1981.
3. **Kazen, C., Bloomer, A., Welch, R., Oudbier, A., and Price, H.,** Persistence of pesticides on the hands of some occupationally exposed people, *Arch. Environ. Health,* 29, 315, 1974.
4. **Rosival, L. and Batora, V.,** Consequences for field exposure of impurities and pesticide formulations, in *Field Worker Exposure During Pesticide Application, Studies in Environmental Science,* Elsevier, Amsterdam, 1980, 163.
5. **Scheuplein, R.J. and Blank, I.H.,** Permeability of the skin, *Physiol. Rev.,* 51, 702, 1971.
6. **Marzulli, F.N.,** Barriers to skin penetration, *J. Invest. Dermatol.,* 26, 284, 1965.
7. **Maibach, H.I., Felman, R.J., Milby, T.H., and Serat, W.F.,** Regional variation in percutaneous penetration in man, *Arch. Environ. Health,* 23, 208, 1971.
8. **Felman, R.J. and Maibach, H.I.,** Percutaneous penetration of some pesticides in man, *Toxicol. Appl. Pharmacol.,* 28, 126, 1974.
9. **Scheuplein, R. and Ross, L.,** Effects of surfactants and solvents on the permeability of epidermis, *J. Soc. Cosmet. Chem.,* 21, 853, 1970.
10. **Dedek, W.,** Solubility factors affecting pesticide penetration through skin and protective clothing, in *Field Worker Exposure During Pesticide Application, Studies in Environmental Science,* Elsevier, Amsterdam, 1980, 47.
11. **Scheuplein, R.J.,** Mechanism of percutaneous absorption. I. Routes of penetration and the influence of solubility, *J. Invest. Dermatol.,* 45, 334, 1965.
12. **Grasso, P. and Lansdown, A.G.B.,** Methods of measuring, and factors affecting, percutaneous absorption, *J. Soc. Cosmet. Chem.,* 24, 481, 1972.
13. **Stoughton, R.B.,** Percutaneous absorption — influence of temperature hydration, *Arch. Environ. Health,* 2, 551, 1965.
14. **Scheuplein, R.J.,** Mechanism of percutaneous absorption, II. Transient diffusion and the relative importance of various routes of skin penetration, *J. Invest. Dermatol.,* 48, 79, 1967.
15. **Riihimaki, V. and Pfaffli, P.,** Percutaneous absorption of solvent vapours in man, *Scand. J. Work Environ. Health,* 4, 73, 1978.
16. **Franz, T.J.,** Percutaneous absorption — on the relevance of in-vitro data, *J. Invest. Dermatol.,* 64, 190, 1975.
17. **Nabb, D.P., Stein, W.J., and Hayes, W.J.,** Rate of skin absorption of parathion and paraoxon, *Arch. Environ. Health,* 12, 502, 1966.
18. **Nash, R.G. and Kearney, P.C.,** Agricultural applicators exposure to 2,4-dichlorophenoxyacetic acid, in *Pesticide Residue and Exposure, ACS Symp. Ser. 1982,* American Chemical Society, Washington, D.C., 1982, 119.
19. **Gehane, E.A. and George, S.L.,** Estimation of human body surface area from height and weight, *Cancer Chemother. Rep.,* 54, 225, 1970.
20. **Harris, A.,** *Human Measurement,* Heineman Educational, London, 1978.
21. **Yoshida, K. and Li, K.Y.R.,** Skin scrub method for determining dermal exposure of herbicide spray applicators, *Am. Ind. Hyg. Assoc.,* Paper No. 157, 1983.
22. **Fenske, R.A. and Spear, R.C.,** Evaluation of dermal exposure to chemical hazards, *Am. Ind. Hyg. Assoc.,* Paper No. 272, 1984.
23. **Williams, J.R.,** Permeation of glove materials by physiologically harmful chemicals, *Am. Ind. Hyg. Assoc. J.,* 40, 877, 1979.
24. **Bernardinelli, S.P. and Nall, R.H.,** Site-specific whole glove chemical permeation, *Am. Ind. Hyg. Assoc. J.,* 46, 60, 1985.
25. **Rigakis, K.B.,** Residual amount of pesticide in work clothing after various laundry processes, private communication, College of Home Economics, University of Alberta, 1985.
26. **Yoshida, K. and Atiemo, M.,** Crystalline particle emission from work clothing contaminated with 2,4-D dimethyl amine spray, *Am. Ind. Hyg. Assoc.,* Paper No. 45, 1985.
27. **Franklin, C.A. and Yoshida, K.,** Effect of various factors on exposure of workers involved in the aerial application of herbicides, *Trans. Am. Conf. Govt. Ind. Hyg.,* p. 97, 1982.
28. **Orlando, J. and Henry, M.,** The development of evaluation of protective apparel for pesticide applicators, in *Determination and Assessment of Pesticide Exposure, Studies in Environmental Science,* Elsevier, Amsterdam, 1984, 53.
29. **Knaak, J.B. and Maddy, K.T.,** Safety effectiveness of closed transfer, mixing — loading, and application equipment in preventing exposure to pesticides, *Arch. Environ. Contam. Toxicol.,* 9, 231, 1980.

Epidemiology of Pesticide-Related Skin Disease in California Agriculture: 1978—1983

M. A. O'Malley, C. G. T. Mathias, and M. J. Coye

Since 1972 the U.S. Bureau of Labor Statistics (BLS) has consistently identified agriculture as the industry division at the highest risk of occupational skin disease.[1] We evaluated (1) the distribution of skin disease associated with chemical exposure in the major subdivisions of agriculture, and (2) the pesticides most frequently associated with skin disease in agricultural workers.

The first approach focuses attention chiefly on identifying work settings where improvements in pesticide handling practices may be necessary. The second approach focuses attention on specific pesticides which deserve (1) careful clinical and toxicologic study to evaluate their properties of skin irritation and skin sensitization, and (2) a critical assessment of the work practices involved in handling them. The possible role of confounding by nonchemical exposures was examined by comparing the distribution of chemically induced skin disease with both the distribution of systemic chemical poisonings and the distribution of skin disease due to naturally occurring vegetation.

Of lost work time skin disease associated with agriculture chemicals, 556 cases were reported to the BLS Supplementary Data System[2] by the California Department of Industrial Relations[3] during the 5 years of the study period (1978—1983, excluding 1982). This compared to the 1288 cases of pesticide-associated skin disease reported to the Pesticide Illness Surveillance Program (PISP) operated by the California Department of Food and Agriculture (CDFA) during the same period.[4] Based on the data reported to the Supplementary Data System (SDS) the mean rate of pesticide associated skin disease was 2.4 cases per 10,000 California agricultural workers. Based on the PISP data the rate was 5.5 cases per 10,000 workers. The difference in rates can be accounted for by differences in reporting criteria, since at least 1 d of lost work time was required for reporting to the SDS and lost work time was not required for the reporting of cases to PISP.

The subdivisions of California agriculture at highest risk of skin disease associated with agricultural chemicals are shown in Table 1. Ten of 24 Standard Industrial Classification (SIC) groups,[5] representing 3.6% of the total agricultural workforce during the study period, had no reported cases of disabling skin disease associated with pesticide exposure. Comparison of data on pesticide-related skin disease with data on pesticide-related systemic poisoning showed a significant agreement, as measured by rank correlation, between the distribution of the outcomes ($R = 0.83, p < 0.01$). However, there was no significant agreement between the SIC groups at high risk for plant- related skin disease and skin disease associated with agricultural chemicals ($R = 0.20, p > 0.05$).

Although the BLS workers' compensation data file (SDS) gave very little information on specific chemicals associated with occupational skin disease in the agricultural workforce, detailed information was available from the CDFA's PISP program. It is probable that most of the chemicals responsible for the BLS cases were, in fact, pesticides, since work by Baginsky[6] has shown that less than 8% of skin disease due to agricultural chemicals in California is associated with non-pesticide agents. More than 140 individual active pesticide ingredients were associated with one or more episodes of skin disease reported to PISP. However, of 1288 cases of skin disease in agricultural workers reported to the PISP system during the study period, eight active ingredients or mixtures accounted for 51.5% of the cases (Table 2).

Table 3 shows the frequency of reported pesticide-associated dermatitis among subgroups of the agricultural workforce, listed according to the PISP occupational categories. The PISP data show that occupation appears to be significantly associated with dermatitis due to certain pesticides. For example, 28.0% (146 of 522 total) of cases among fieldworkers were associated with exposure to sulfur, compared to 6.4% (49 of 766) of cases among nonfieldworkers ($p < 0.001$ by chi square). Propargite, in contrast, appeared to account for a larger proportion of dermatitis among crop service workers (24.2%—168 of 622 cases) than among other agricultural workers (11%—73 of 666 cases — $p < 0.001$ by chi square). Finally, the occupation of nursery workers, identified from the BLS data as being the subgroup of the agricultural workforce at highest risk of disabling pesticide-associated skin disease, is associated with a disproportionate number of cases due to several pesticides including benomyl and dienochlor ($p < 0.001$ by chi square for each).

Results of our study illustrate the uses of surveillance data in setting priorities for control of occupational skin disease due to pesticides and other agricultural chemicals. We have identified several SIC groups which appear to be at elevated risk of skin disease due to agricultural chemicals, with the single group at highest risk being horticultural specialties (018). PISP data show, moreover, that skin disease in these workers is associated with a relatively unique group of pesticides, including several well- recognized contact allergens.[7] Further work will be necessary to ascertain whether the skin disease associated with the use of benomyl or other active ingredients associated with disabling skin disease is due to (1)

TABLE 1
Distribution of Pesticide-Related Skin Disease (Dermatitis)[a] in California Agriculture: 1978—1983

SIC group[b]	Total employed[c]	Pesticide-related dermatitis		Systemic pesticide illness		Plant-related dermatitis	
		Cases	Rate[d]	Cases	Rate[d]	Cases	Rate[d]
018 Horticulture	125,826	64	5.1	150	11.9	200	15.9
072 Crop services	177,246	84	4.7	252	14.2	34	1.9
025 Poultry/eggs	41,670	16	3.8	38	9.1	6	1.4
016 Vegetables/melons	258,492	72	2.8	308	11.9	96	3.7
019 General crop farms	290,856	74	2.5	172	5.9	98	3.4
071 Soil prep services	8,046	2	2.5	16	19.9	0	0.0
017 Fruits/trees/nuts	603,615	138	2.3	226	3.7	354	5.9
076 Farm labor services	354,669	60	1.7	82	2.3	144	4.1
013 Nongrain cash crop	135,401	22	1.6	80	5.9	22	1.6
024 Dairy farms	61,669	8	1.3	38	6.2	6	1.0
021 Nondairy livestock	34,877	4	1.1	24	6.9	14	4.0
011 Cash grains	26,538	2	0.8	40	15.1	8	3.0
075 Nonvet animal services	24,309	2	0.8	12	4.9	2	0.8
078 Landscaping services	127,655	8	0.6	60	4.7	394	30.9
082 Forest nurseries	37	0	0.0	0	0.0	2	541.0
091 Commercial fishings	11,862	0	0.0	0	0.0	0	0.0
029 General livestock	10,375	0	0.0	2	1.9	10	9.6
074 Veterinary services	48,931	0	0.0	4	0.8	0	0.0
084 Gathering forest products	76	0	0.0	0	0.0	0	0.0
085 Forestry services	2,026	0	0.0	0	0.0	12	59.2
081 Timber tracts	478	0	0.0	0	0.0	0	0.0
027 Animal specialties	10,571	0	0.0	6	5.7	8	7.6
092 Fish hatcheries	481	0	0.0	0	0.0	0	0.0
097 Hunting and trapping	66	0	0.0	0	0.0	0	0.0
Total	2,355,802	556	2.4	1,510	6.4	1,418	6.0

[a] Based on cases reported to the U.S. BLS's (for 1978, 1979, 1980, 1981, and 1983) by the California Department of Industrial Relations from tabulations of Employers' First Reports of Occupational Injury or Illness. All cases involved one or more lost work days.
[b] SICs as used in SIC Manual, 1972; with 1977 supplement; the Statistical Policy Division, Office of Management and Budget, Executive Office of the President: U.S. Government Printing Office.
[c] Employment figures are derived from unemployment insurance records by the California Department of Employment Development. Reported figures are the sum of mid-third quarter employment (peak agricultural employment) for the study period (1978, 1979, 1980, 1981, and 1983).
[d] Rates are reported as number of cases/10,000 workers employed.

the inherent capacity of these compounds to cause skin irritation or sensitization, (2) inadequate pesticide-handling practices, or (3) some factor associated with nursery crops per se, such as a large surface area which might increase effective worker exposures to foliar residues.

It is important to note here some of the limitations of the PISP and BLS data in terms of sensitivity and specificity. The specificity of both data sources is limited by the quality of the case documentation, which customarily amounts only to a brief description of the circumstances of exposure. This type of limited information is most problematic in interpreting reports of skin disease in fieldworkers, for whom it may be difficult to accurately assess when or how the exposure which precipitated the attack of skin disease took place. It may be impossible under many circumstances to separate skin disease possibly due to a crop per se from skin disease due to a pesticide residue on the crop. This dilemma is epitomized by the long-standing difficulty in assessing the nature of skin rashes among grape harvesters which have frequently been attributed to sulfur exposure.[8]

Our results provide some reassurance regarding the specificity of the BLS reporting system with regard to one potentially significant confounding factor — exposure to naturally occurring plants and vegetation such as poison oak. Data in Table 1 show that there is an extremely low correlation between the distribution of skin disease associated with plants and skin disease associated with pesticide exposure. Thus, although cases of poison oak may occasionally be confused with cases of skin disease due to pesticides, exposure to poison oak does not appear to account for the distribution of pesticide associated skin disease in the agricultural workforce.

A second limitation common to the PISP and BLS data is possible underreporting of pesticide-associated illnesses. For systemic pesticide illnesses, the exact degree of underre-

TABLE 2
Pesticides Associated with Skin Disease in California Agriculture: PISP[a] Data 1978—1983[b]

Pesticide	Cases	Percent of total	Cumulative percent
Propargite	241	18.7	18.7
Sulfur	195	15.1	33.9
Glyphosate	53	4.1	38.0
Propargite/sulfur	51	4.0	41.9
Methyl bromide	43	3.3	45.3
Benomyl	30	2.3	47.6
Captan	26	2.0	49.6
Petroleum products	24	1.9	51.5
Cyhexatin	21	1.6	53.1
Captan/sulfur	18	1.4	54.5
Dinitrophenol	18	1.4	55.9
Ethylene dibromide (EDB)	15	1.2	57.1
Paraquat	15	1.2	58.2
Methomyl	13	1.0	59.2
Diazinon	12	0.9	60.2
Ziram	12	0.9	61.1
Captafol	11	0.9	62.0
DD mixture	11	0.9	62.8
Chlorothalonil	10	0.8	63.6
Dicofol	10	0.8	64.4
Captan/DCNA/sulfur	9	0.7	65.1
Acephate	8	0.6	65.7
Carbaryl	8	0.6	66.3
Dichloronitroaniline (DCNA)	8	0.6	66.9
Malathion	8	0.6	67.5
Naled	8	0.6	68.2
Dienochlor	8	0.6	68.8
Triadimefon	8	0.6	69.4
Dimethoate	6	0.5	69.9
Fenbutatin oxide	6	0.5	70.3
Anilazine	5	0.4	70.7
Benomyl/captan	5	0.4	71.1
Maneb	5	0.4	71.5
Simazine	5	0.4	71.9
Metam-sodium	5	0.4	72.3
Zineb	5	0.4	72.7
Other	168	13.0	85.7
Unknown	184	14.3	100.0
Total	1,288	100.0	100.0

[a] PISP operated by the CDFA; all reported cases are included in published PISP reports, regardless of the amount of time lost from work.
[b] Reporting period excludes cases from 1982.

porting has long been a matter of controversy,[9,10] while for skin disease, no estimates of the degree of underreporting have yet been made. In assessing the distribution of pesticide-associated skin disease, however, underreporting is a less significant problem than reporting biases which may affect some SIC groups more than others. For instance, the elevated rate associated with crop service workers (072) could be attributed to the routine medical surveillance to which these workers are subject.[11] Conversely, factors which may adversely affect reporting, such as illegal immigrant status, may be more common among some portions of the agricultural workforce than others. Unfortunately, this type of reporting bias cannot readily be evaluated from existing data sources.

TABLE 3
Frequency of Pesticide-Related Skin Disease by Agricultural Occupation: PISP[a] Data 1978—1983[b]

Occupation	Cases	Percent of total	Cumulative percent
Crop service workers	693	53.8	53.8
Ground applicator	314	24.4	
Mixer/loader	126	9.8	
Fumigator	71	5.5	
Unspecified applicator	58	4.5	
Tractor driver/irrigator	54	4.2	
Cleaner/repairer	30	2.3	
Flagger	23	1.8	
Packer/processor	17	1.3	
Fieldworker	522	40.5	94.3
Nursery worker	73	5.7	100.0
Total	1288	100.0	100.0

[a] PISP operated by the CDFA; all reported cases are included in published PISP reports, regardless of the amount of time lost from work.
[b] Reporting period excludes cases from 1982.

To conclude, it should be emphasized that the limitations of existing surveillance data sources need not preclude effective disease-control efforts. This has been demonstrated for infectious diseases[12] and there is little reason to conclude it will not also prove true for occupational diseases which are relatively acute. While efforts to improve surveillance data for pesticide-associated illnesses have obvious merit, they should be accompanied by field work aimed at control of illnesses in high risk subgroups of the agricultural workforce which have already been identified.

REFERENCES

1. Division of Periodic Surveys, U.S. Bureau of Labor Statistics (1972—1984), unpublished data.
2. Bureau of Labor Statistics, U.S. Department of Labor, Division of Periodic Surveys, Chapter 17, Occupational Health Statistics, reprint from Handbook of Methods, Bull. 2134-1, 1983.
3. Division of Labor Statistics and Research, California Department of Industrial Relations, personal communication, 1985.
4. Worker Health and Safety Unit, California Department of Food and Agriculture, Pesticide Illness Reports, 1983 HS-1186, 1981 HS-985, 1980 HS-849, 1979 HS-712, 1978 HS-65.
5. Statistical Policy Division, Office of Management and Budget, Executive Office of the President, Standard Industrial Classification Manual, with 1977 suppl., U.S. Government Printing Office, Washington, D.C., 1972.
6. **Baginsky, E.,** Occupational skin disease in California (with special reference to 1977), California Department of Industrial Relations, January 1982.
7. **Cronin, E.,** *Contact Dermatitis,* Churchill Livingstone, London, 391.
8. **Winter, C.K. and Kurtz, P.H.,** An investigation into factors influencing grape worker susceptibility to skin rashes, California Department of Food and Agriculture Worker Health and Safety Unit HS-1093, September 1983.

9. California Department of Food and Agriculture, Pesticide Illness Reports, HS-985, 1981.
10. **Kahn, E.,** Pesticide related illness in California farmworkers, *J. Occup. Med.*, 18, 693, 1976.
11. California Department of Food and Agriculture, Extracts from California Agricultural Code, 1985.
12. **Thacker, S.B., Keewhan, C., and Brachman, P.S.,** The surveillance of infectious disease, *J.A.M.A.*, 249, 1181, 1983.

Risk Factors for Skin Cancer in Saskatchewan Farmers

D. J. Hogan

Although skin cancer is the most common form of cancer in North America, it is the form of cancer for which there is the least precise epidemiologic information.[1] Saskatchewan is one of the few areas of the world where accurate registration of all cases of melanoma and nonmelanoma skin cancer (NMSC) is performed. NMSCs (squamous cell carcinoma [SCC] and basal cell carcinoma [BCC]) are the most common malignancies in Saskatchewan. The incidence of NMSC in Saskatchewan is 170.6 per 100,000 men and 120.4 per 100,000 women.[2] This is similar to the estimated incidence of skin cancer in Caucasians in the U.S..[3] The rate of NMSC has doubled in Saskatchewan in the last 15 years.

NMSC, in particular SCC of the skin, is believed to reflect DNA damage to epidermal cells accumulated over relatively long periods of time.[1] The large majority of Saskatchewan farmers are chronically exposed to the carcinogenic effects of ultraviolet light. Studies from England,[4,5] but not from Sweden or Finland,[6,7] have identified the occupation of farming as a significant risk factor for NMSC. A recent questionnaire survey, conducted by our section of dermatology, of 881 cases of skin cancer and 1250 age-, sex-, and location-matched controls in Saskatchewan revealed that farming was a significant risk factor for NMSC, even though any association between farming and NMSC would be minimized by our matching controls to the residence of the cases. Multivariate analysis of significant risk factors for skin cancer in our study revealed that farming was the chief exogenous risk factor for NMSC and was the second strongest risk factor for NMSC overall. In our study there was a statistically significant association between 2,4-D use and the development of skin cancer, but this association was not significant after multivariate analysis of all significant risk factors for skin cancer was performed. Multivariate analysis also indicated that the risk factors of outdoor work and working outdoors more than 6 h/d in summer were secondary in importance to the occupation of farming itself.

Saskatchewan has the second highest incidence in the world of SCC of the lip (Figure 1).[2,8] This cancer is more prone

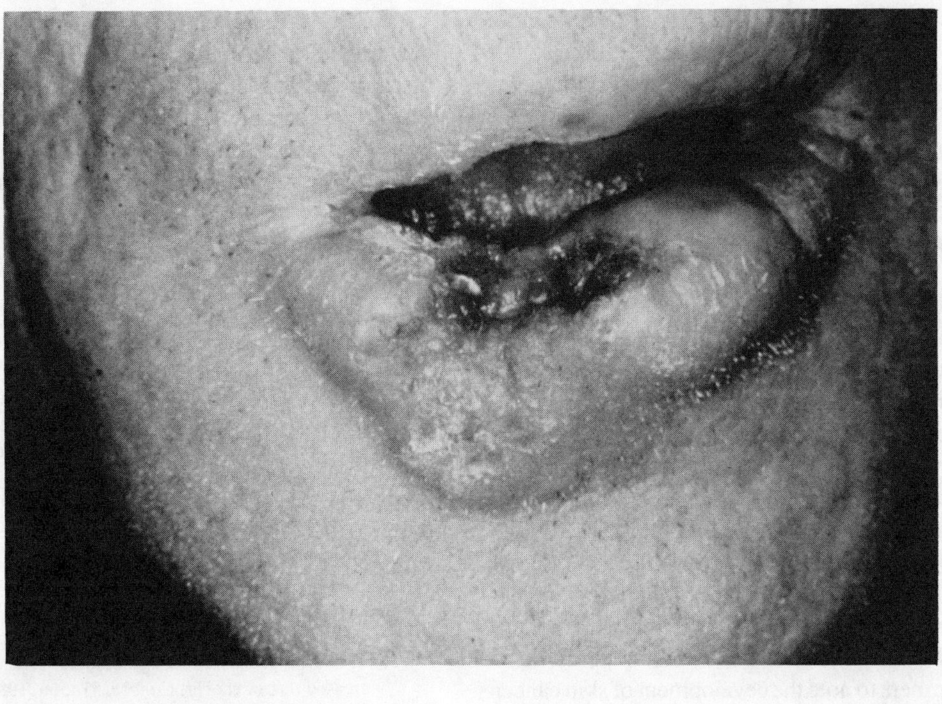

FIGURE 1. Advanced squamous cell carcinoma of the lower lip in a male farmer. This carcinoma is more prone to metastasize than other forms of nonmelanoma skin cancer.

to metastasize than other forms of NMSC. Previous reports have emphasized that SCC of the lip is associated with rural residence.[8] Saskatchewan farmers are unlike most other groups identified to date, for they are at an elevated risk of not only SCC of the lip, but also NMSC of other sites. Our study and the majority of published studies do not identify farmers as being an occupational group at elevated risk of cutaneous malignant melanoma.[9]

Farming was the strongest risk factor identified in our study for SCC of the skin in women. This finding probably reflects the active role that women play in farming on the typical family-run farm in Saskatchewan. A previous study in England noted an excess number of SCC of the arms and especially of the ears in male farmers.[4] There was no statistical difference in skin cancer sites between farmers and nonfarmers with skin cancer in our study.

In our study there were no differences between farmers with skin cancers and farmers without skin cancer regarding 2,4-D use, number of years of 2,4-D use, use of protective clothing when using pesticides, use of arsenical insecticides, number of years of use of arsenical insecticides, hours spent outdoors in summer and winter for both occupational and recreational purposes, size of farm, and number of years of farming. Farmers with skin cancer compared to farmers without skin cancer, however, were more likely to have many of the phenotypic risk factors for skin cancer identified in previous studies. The main phenotypic characteristics identified as risk factors for NMSC in our study were a family history of skin cancer, prominent freckles in childhood, light skin color, history of severe sunburn, and skin that sunburned easily and suntanned poorly. Family history of skin cancer was the strongest risk factor identified for NMSC in our study. Our study is the first to identify family history of skin cancer as such a strong risk factor for NMSC.

Ultraviolet light exposure is believed to account for most cases of SCC and a majority of cases of BCC.[10] It has been argued that acute trauma to the skin may provoke the development of skin cancer.[11] Trauma was identified by cases as a preceding event in a minority of cases of NMSC in our study. The importance of trauma was difficult to evaluate because of the varying types of trauma and the varying periods of time between the occurrence of the trauma and the apparent development of the non-melanoma skin cancer. A minority of respondents noted that their skin cancer developed in an area of frostbite injury. Usually frostbite injuries occurred 20 to 30 years before the development of skin cancer. The ear was significantly more likely to be the site of BCC or SCC in cases with a prior history of frostbite at the site of skin cancer. Cases of SCC of the skin were significantly more likely to state that their skin cancer occurred in an area of frostbite than were cases of BCC of the skin. Farmers were significantly more likely than nonfarmers to note the development of skin cancer at previous sites of trauma or frostbite in our study.

It is unfortunate that large numbers of individuals with the phenotypic risk factors for NMSC engage in farming where they are exposed heavily to sunlight. Since it is unlikely that these individuals will stop farming solely because of the development of NMSC, which usually can be cured with appropriate treatment, work practices which decrease exposure of susceptible skin to the carcinogenic effects of sunlight are advocated for these workers. Simple protective clothing such as long-sleeved shirts, trousers, and hats is of assistance. It must be emphasized that it requires fairly heavy clothing, such as denim, to adequately protect the skin from the carcinogenic effects of sunlight. The damaging wavelengths of ultraviolet light are of maximum intensity in the middle of the day, so work should be scheduled before 11:00 a.m. and after 3:00 p.m. to minimize exposure of the skin to carcinogenic ultraviolet light. Workers must be reminded that appropriate skin protection is also required on cloudy days, since all clouds except very dense storm clouds transmit significant amounts of carcinogenic ultraviolet light. Considerable exposure to carcinogenic ultraviolet light also occurs in the shade.

The use of sunblocks should be recommended to individuals with phenotypic risk factors for skin cancer who engage in farming. The higher the sun protection factor (SPF) the better the protection from ultraviolet light. Susceptible individuals should use a sunblock with a SPF of at least 15. The regular use of these sunblocks should be commenced at as early an age as possible. The regular use of sunblocks with a SPF of 15 from childhood on has been predicted to decrease the incidence of NMSC by 98%.[12] If susceptible individuals carried out the recommendations suggested, NMSC would no longer be the most common form of cancer in Caucasians in many areas of the world. Regular periodic medical examination of farmers and agricultural workers at risk for the development of skin cancer assists the detection and treatment of premalignant and malignant skin lesions. Early diagnosis and treatment reduces the morbidity and mortality of NMSC. Medical examinations should be combined with advice regarding appropriate skin protection from sunlight. Since only a minority of farmers presently use sunblocks, such advice may have to be reinforced frequently by health professionals and through public education.

REFERENCES

1. **Emmett, E.A.,** Occupational skin cancer: a review, *J. Occup. Med.,* 17, 44, 1975.
2. The Saskatchewan Cancer Foundation, Annual Report 1984—1985.
3. **Scotto, J., Fears, T.R., and Fraumeni, J.F.,** Incidence of Non-Melanoma Skin Cancer in the United States, U.S. Department of Health and Human Services, April, 1983.
4. **Whitaker, C.J., Lee, W.R., and Downes, J.E.,** Squamous cell skin cancer in North-West of England, 1967—69, and its relation to occupation, *Br. J. Ind. Med.,* 36, 43, 1979.
5. **Atkin, M., Fenning, J., Heady, J.A., et al.,** The mortality from cancer of the skin and lip in certain occupations, *Br. J. Cancer,* 3, 1, 1949.
6. **Wiklund, K.,** Swedish agricultural workers, a group with a decreased risk of cancer, *Cancer,* 51, 566, 1983.
7. **Lindquist, C., Teppo, L., and Pulkala, E.,** Occupations with low risk of lip cancer show a high risk of skin cancer of the head, *Community Dent. Oral Epidemiol.,* 9, 247, 1981.
8. **Douglass, C.W. and Gammon, M.D.,** Reassessing the epidemiology of lip cancer, *Oral Surg.,* 57, 631, 1984.

9. **Austin, D.F. and Reynolds, P.,** Occupations and malignant melanoma of the skin, in *Epidemiology of Malignant Melanoma,* Gallagher, R.P., Ed., Springer-Verlag, Berlin, 1986, 38.
10. **Urbach, F., Rose, D.B., and Bonnem, M.,** Genetic and environmental interactions in skin carcinogenesis, in *Environment and Cancer,* Williams & Wilkins Co., Baltimore, 1972, 355.
11. **Stoll, H.L. and Crissey, J.T.,** Epithemioma from single trauma, in *Cancer Dermatology,* Helm, F., Ed., Lea & Febiger, Philadelphia, 1979, 25.
12. **Stern, R.S., Weinstein, M.C., and Baker, S.G.,** Risk reduction for non-melanoma skin cancer with childhood sunscreen use, *Arch. Dermatol.,* 122, 537, 1986.

Principles of Safety in Agriculture and Related Industries

Statistical Description of Agricultural Injuries in the U.S.

M. L. Cohen, M. B. Moll, P. W. Maley, and H. I. Linn

Agriculture has been perhaps the single most important economic activity of mankind throughout the history of civilization. The systematic, efficient production of food and related products has, in fact, allowed the development of modern civilization, with its urban and industrial centers. Farms in the U.S. have been getting larger and more productive, producing abundant food for U.S. citizens and for much of the rest of the world as well.[1] Increased agricultural production has its negative aspects as well. Improvements in farming techniques and methods, while making some aspects of agricultural production safer, have been attended by a host of hazards related to use of large and complex equipment and a diverse array of pesticides and other chemicals.

Agriculture is as much a way of life as it is an occupation, a fact that creates unique safety risks for workers. Farm owners and agricultural workers typically perform a wide range of tasks during highly irregular hours that become long work weeks during busy periods of the year.[2] They often work alone, in areas isolated from medical facilities. Many agricultural workers are part-time on the farm while also holding nonagricultural jobs. Farm laborers are often poorly trained in safe procedures for performing tasks that are intrinsically extremely hazardous, and many farm owners cannot afford modern injury prevention control technologies.

Small family farms often are exempt from federal and state regulations. The Occupational Safety and Health Act does not cover self-employed farmers or farms with fewer than 11 employees.[3] Workers' compensation laws vary widely from state to state in their coverage and reporting of agricultural injuries.[1] Typically, farm owners with small payrolls or few or no hired employees are exempt from workers' compensation laws. Consequently, routine collection of data on agricultural injuries at the national level is restricted to farms with payrolls or labor forces above minimum levels set by state workers' compensation legislation.

The National Safety Council published data in 1982 from a survey of 32 states, showing that only 22.3% of agricultural injuries were incurred by hired help, while 73.4% involved members of the farm family.[4] Unfortunately, our agricultural injury data primarily reflect the injury experience of hired workers employed in relatively large-scale agricultural operations, admittedly only a fraction of the total agricultural injury picture. However, it is probable that hired workers and farm family members engaged in agricultural production are exposed to similar hazards and working conditions.

In 1982 there were nearly 870,000 farms employing hired labor (down about 9% from 1978) and nearly 4.9 million hired agricultural workers (down about 12% from 1978) in the U.S.[5] (Table 1). In both 1978 and 1982 nearly 60% of all hired agricultural workers were employed on farms with ten or more workers, while about one fourth worked on farms with four or less workers.[3]

Injury incidence rates for hired agricultural workers are higher than for the general private sector (Figure 1). Although employment in agriculture is holding fairly stable with perhaps a slight decrease, injury rates and trends are increasing.[3] As Figure 2 indicates, these trends are projected to continue at least through 1990. The data represented for 1986 through 1990 are projections based upon a least-squares linear regression of employment on injury rates using the available 1977—1985 data from the U.S. Bureau of Labor Statistics.

U.S. workers' compensation data for 1976 to 1983 were analyzed to determine the natures and sources of agricultural injuries.[6] Selected data for crops (including tree fruits and vegetables), livestock, and agricultural services sectors of the industry are presented in Table 2. Agricultural services include establishments primarily engaged in performing soil preparation services, crop services, veterinary or other animal-related services, farm labor and management services, and landscape and horticultural services for farm owners on a fee or contract basis. Comparisons of injuries on general farms (crops) vs. dairy farms are presented in Table 3.

In agricultural industries, whether crop production, livestock production, or services, it may be inferred that the hired

TABLE 1
Hired Agricultural Labor in the U.S.: U.S. Farm and Labor Populations, 1978 and 1982

	Farms with hired labor		Hired workers	
	1982	**1978**	**1982**	**1978**
1 worker	279,119	289,327	279,119	289,327
2 workers	168,855	181,485	337,710	362,970
3—4 workers	187,824	205,150	638,302	698,219
5—9 workers	138,161	162,235	867,596	1,024,501
10 or more workers	95,878	113,708	2,733,130	3,124,177
Total	869,837	951,905	4,855,857	5,499,194

From U.S. Department of Census, Census of Agriculture, 1978—1982, 1982.[5]

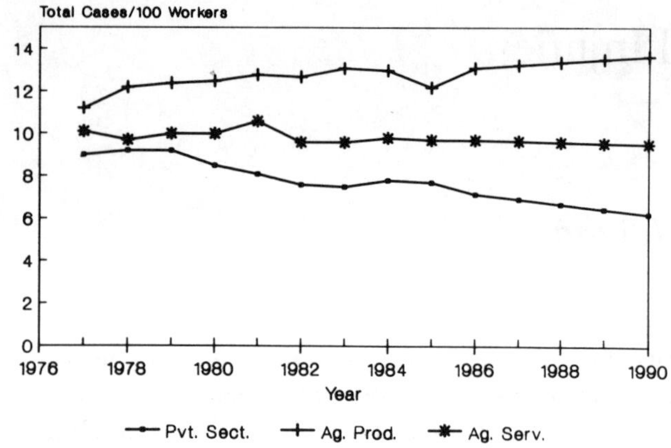

FIGURE 1. Incidence rates and trends — agriculture vs. private sector injuries.

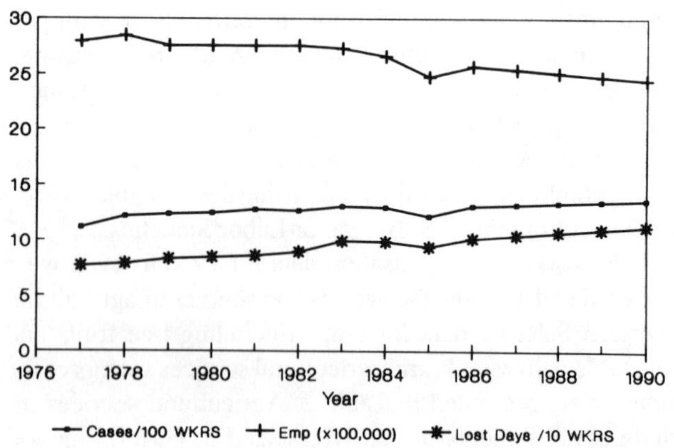

FIGURE 2. Agricultural production — trends in employment, cases and lost days.

agricultural workers most often injured are 15- to 30-year-old male laborers who have been on the job less than 6 months at the time of injury. Typically, these workers suffer from sprains/strains or cuts/lacerations involving the extremities, with the source of the injury most likely to be the work surface, livestock, or hand tools. In addition, the percentage of injured workers involved in livestock production is fairly evenly distributed during the year, while there is a greater percentage of workers being injured during the summer and fall in agricultural services industries and in crop production industries.

These same patterns of injuries described above appear to hold for "general crop" farms and for dairy farms, where the major difference appears to be the source of the injuries. In general crop production, the working surface appears as the main source of injuries, while in dairy farms it is, not surprisingly, the livestock.

In the U.S. agricultural industry there are similar percentages of injury type, source, and body part in different types of farming because the equipment and work activity hazards are similar. As might be expected, laborers suffer 60 to 70% of the injuries, because there are so many more laborers than other job categories among hired agricultural workers, and because the work they do appears to be intrinsically more hazardous. Work surfaces, hand tools, and vehicles are the most common sources of injury because these are the hazards most agricultural workers are exposed to on a daily basis. Hands, arms, legs, and feet account for nearly half of all injuries, and the back accounts for another 15 to 20%, because the majority of tasks performed by agricultural workers stress these parts of the body to their limits. More than half of all injuries occur in the spring and summer, because these are the seasons of peak agricultural production activity for crop or service industries (Figure 3 — percentage of total injuries by month).

Our analysis leads us to conclude that the types and sources of injuries to hired agricultural workers are quite predictable. Predictable injuries are also preventable. The injury toll among hired agricultural workers is simply far too high, and is amenable to intervention.

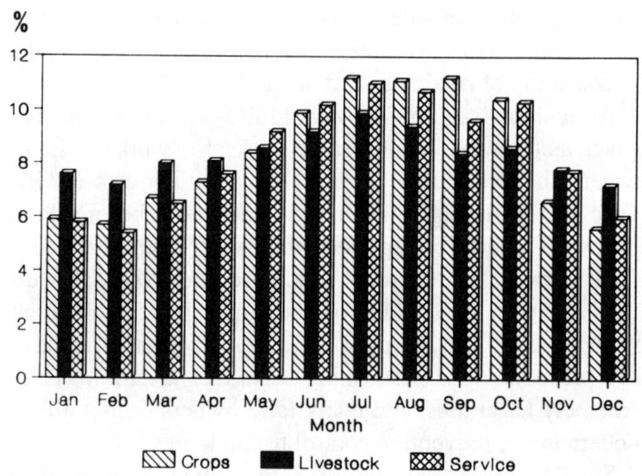

FIGURE 3. Injuries to agricultural workers by month of year, 1976—1983.

TABLE 2
Characteristics of Injuries to Hired Agricultural Workers in Crops, Livestock, and Services Sectors

	Crops (%) (n = 175,413)		Livestock (%) (n = 68,814)		Service (%) (n = 94,434)
Sex					
Male	86.0		89.3		87.3
Female	14.0		10.7		12.7
Age					
<15	0.5		0.4		0.2
15—30	51.2		54.4		62.4
31—45	25.9		25.7		23.4
46—60	17.4		15.0		11.2
61>	5.0		4.5		2.8
Length of service					
<6 months	43.6		46.1		52.1
6—11 months	15.2		19.3		16.7
1—2 years	13.6		12.5		11.1
2—3 years	6.3		6.2		5.6
3—4 years	4.7		3.9		3.3
Totals	83.4		88.0		88.8
Body part injured (top five for each sector)					
Upper extremity	25.7		26.9		27.9
Wrist/hand	20.2		20.8		22.0
Arms	4.8		5.2		5.0
Lower extremity	20.7		23.3		21.4
Ankle/foot	10.4		11.3		10.4
Legs	9.4		11.0		10.4
Back	18.0		16.0		16.6
Head	13.1		11.7		11.6
Eyes	8.2		5.6		6.3
Multiple parts	7.0		5.6		7.3
Totals	84.5		83.5		84.8
Occupation (top five in each sector)					
Laborers	73.6	Laborers	71.6	Laborers	62.4
Nonclassified	6.9	Foremen	6.2	Nonclassified	8.0
Foremen	5.6	Nonclassified	7.8	Foremen	7.1
Truck drivers	2.8	Truck drivers	2.6	Machine operators	5.9
Farm implement mechanics	2.0	Farm implement mechanics	0.8	Truck drivers	1.7
Totals	90.9		89.0		85.1
Type of accident					
Struck by	22.7		28.1		25.9
Overexertion	19.5		15.5		19.1
Fall from elevation	11.4		11.1		10.2
Struck against	8.9		10.0		9.4
Caught in/under	7.2		9.1		7.7
Totals	69.7		73.8		72.3
Source of injury (top five in each sector)					
Work surface	15.8	Animal live	16.4	Work surface	14.4
Hand tools	10.4	Work surface	16.4	Hand tools	13.1
Vehicles	9.6	Metal items	9.0	Plant/tree	8.9
Tractors	2.3				
Metal items	9.1	Containers	8.2	Vehicles	8.4
				Tractors	1.1
				Planes	0.3
		Vehicles	7.3	Machines	7.8
Containers	8.6	Tractors	1.1	Agricultural	2.3
Totals	53.5		58.1		52.6
Nature of injury (top five in each sector)					
Sprain/strain	31.2	Sprain/strain	27.6	Sprain/strain	28.5
Cut/laceration	17.7	Cut/laceration	17.2	Cut/laceration	21.7
Contusion	10.7	Contusion	14.3	Nonclassified	10.9
Nonclassified	10.2	Fracture	12.4	Contusion	9.6
Fracture	8.4	Nonclassified	9.0	Fracture	9.2
Totals	78.2		80.5		79.9

TABLE 3
Characteristics of Injuries to Hired Agricultural Workers: General Farms (Crops) Compared to Dairy Farms, 1976—1983

	General farms (%) (n = 45,584)		Dairy farms (%) (n = 19,479)
Sex			
Male	92.4		94.6
Female	7.6		5.4
Age			
<15	0.8		0.5
15—30	51.2		56.2
31—45	25.1		25.1
46—60	16.5		14.8
61>	6.4		3.4
Occupation			
Laborers	72.4	Laborers	79.6
Nonclassified	10.6	Nonclassified	7.5
Foremen	5.6	Foremen	4.9
Truck drivers	1.9	Truck drivers	1.5
Farm implement mechanics	1.6	Delivery men	1.1
Totals	92.1		94.6
Type of Accident			
Struck by	22.1	Struck by	32.7
Overexertion	17.8	Overexertion	14.1
Fall from elevation	11.8	Struck against	10.1
Caught in/under	10.4	Caught in/under	9.5
Struck against	8.1	Fall to lower level	9.1
Totals	70.2		75.5
Source of injury			
Work surface	15.9	Animal, live	23.6
Vehicles	11.1	Work surface	15.2
Tractors	3.5		
Metal items	11.6	Metal items	8.3
Machines	9.6	Containers	8.0
Agricultural	6.7		
Hand tools	8.2	Machines	6.6
		Agricultural	3.6
Totals	56.4		61.7
Nature of injury			
Sprain/strain	28.7	Sprain/strain	25.2
Cut/laceration	16.1	Cut/laceration	16.9
Fracture	13.1	Contusion	15.2
Nonclassified	11.5	Fracture	12.5
Contusion	9.7	Nonclassified	11.5
Totals	79.1		81.3
Body part			
Upper extremity	24.1		29.1
Wrist/hand	18.9		22.2
Arms	4.6		5.8
Lower extremity	23.7		23.3
Ankle/foot	11.9		11.6
Legs	10.8		11.0
Back	18.2		14.8
Head	10.5		10.2
Eyes	5.6		4.9
Multiple parts	6.1		6.2
Totals	82.9		83.6

REFERENCES

1. OSHA, *U.S. Agricultural Safety and Health Statistics,* Coordination and Analysis, Division of Statistical Studies, Occupational Safety and Health Administration, Washington, D.C., December 1978.
2. National Safety Council, Accident Facts, Chicago, 1981.
3. Occupational Injuries and Illnesses in the United States by Industry, Bureau of Labor Statistics, U.S. Department of Labor, Washington, D.C., 1975—1983.
4. National Safety Council, Farm Accident Survey Report, Chicago, 1982.
5. U.S. Department of Census, Census of Agriculture, Washington, D.C., 1978—1982, 1982.
6. Supplementary Data System, Bureau of Labor Statistics, U.S. Department of Labor, Washington, D.C., 1976—1983.
7. **Field, W.D. and Tormoehlen, R.L.,** Analysis of fatal and non-fatal farm accidents involving children, presented at the American Soc. of Agricultural Engineers 1982 Winter Meet. in Chicago, December 14 to 17, 1928.

Mortality from Accidents in Farmers and Farm Laborers in British Columbia

J. J. Spinelli, R. P. Gallagher, P. R. Band, and W. J. Threlfall

INTRODUCTION

In recent years, several studies have demonstrated a high accident mortality rate for farmers compared to all other occupations.[1-3] As part of a study to delineate risks of death for major occupational groups, including farmers, in British Columbia, risks of death due to accidents for farmers and farm laborers were analyzed.

METHODS

Information on 457,083 deaths that occurred in British Columbia from 1950 through 1978 was obtained from the Division of Vital Statistics. After eliminating deaths before age 20, and deaths with invalid cause of death or occupational codes, a total of 254,901 male and 165,913 female deaths were available for analysis. Of the male decedents, 28,032 were farmers or farm managers and 2328 were farm laborers. Of the female decedents only 62 were farmers or farm managers and 25 were recorded as farm laborers, and thus females are excluded from this analysis.

It was found that farm workers have very different mortality patterns from farmers and farm managers.[4] Elevated proportions of deaths due to tuberculosis and pneumonia indicate that laborers are generally of a lower socioeconomic status than farmers. Furthermore, farm laborers are generally a transient population who are often incorrectly classified on death certificates.[5] For these reasons farm laborers were analyzed separately from farmers and farm managers.

Cause-of-death information for the 29-year period was coded according to the sixth, seventh, and eighth revisions of the International Classification of Diseases (ICD).[6-8] All causes of death codes were reconciled to the ICD-7 format for this study. The usual occupation was coded according to the 1951 and 1961 Canadian Occupation Manual and reconciled to the 1961 format.

Age-standardized proportional mortality ratios (PMRs) for 160 causes of death were calculated for male farmers and farm laborers dying at ages 20 years and over and also for ages 20 to 65 by the method of Monson.[10] PMRs were calculated for the entire 29-year period and also separately for deaths in the years 1950—1959, 1960—1969, and 1970—1978. The significance of PMR values was assessed by comparison with the Poisson distribution.[11] Results for deaths at ages 20 to 65 and deaths at ages 20 and over were similar, and hence, this paper will concentrate on the results for the 20-and-over age group.

RESULTS

British Columbia farmers demonstrated PMRs as expected for the most common cause of death, arteriosclerosis and degenerative heart disease (PMR = 101) (Table 1). Mortality from all cancers combined was slightly lower than expected (PMR = 94), mortality from cerebrovascular accidents was slightly higher than expected (PMR = 109), and mortality from all accidents was as expected (PMR = 101).

Farm workers in British Columbia had significantly lower than expected PMRs for arteriosclerosis and degenerative heart disease (PMR = 91) and all cancers (PMR = 78) (Table 1). Mortality from cerebrovascular accidents was as expected (PMR = 102) and mortality from all accidents was higher than expected (PMR = 129).

For both farmers and farm laborers, statistically significant elevated and reduced PMRs were found for different accident types. For farmers, significantly elevated PMRs were found for motor vehicle accidents (PMR = 122); environmental accidents which include deaths caused by animals, food, and

TABLE 1
Major Causes of Death in Male Farmers and Farm Laborers in B.C., 1950—1978

Cause of death	ICD-7 no.	No. of deaths	PMR	95% confidence interval
		Farmers		
Arteriosclerosis and degenerative heart disease	420—422	11,128	101	99, 103
All cancers	140—205	4,602	94[a]	91, 98
Cerebrovascular accident	330—334	3,493	109[b]	105, 113
All accidents	E800—936	1,203	101	95, 107
All deaths		28,104		
		Farm Laborers		
Arteriosclerosis and degenerative heart disease	420—422	763	91[a]	84, 98
All cancers	140—205	326	78[b]	69, 87
Cerebrovascular accident	330—334	220	102	89, 117
All accidents	E800—936	276	129[a]	114, 146
Pneumonia	490—493, 525	124	146[a]	122, 175
Tuberculosis	002—019	28	220[b]	146, 319
All deaths		2,328		

[a] $p < 0.01$.
[b] $p < 0.05$.

water deprivation, and exposure (PMR = 234); machinery accidents (PMR = 300); and gun- related accidents (PMR = 226) (Table 2). Significantly lowered PMRs were found for water transport accidents (PMR = 50), aircraft accidents (PMR = 55), drug or alcohol poisoning (PMR = 25), and poisoning due to chemicals, metals, or solvents (PMR = 35). PMRs for other types of accidents were within expected limits.

For farm workers, significantly elevated PMRs were found for motor vehicle accidents (PMR = 129), railway accidents (PMR = 268), burns (PMR = 270), environmental accidents (PMR = 652), and accidental drowning (PMR = 202) (Table 3). Significantly lowered PMRs were found for water transport accidents (PMR = 24), aircraft accidents (PMR = 12), and being struck by a falling object (PMR = 41).

In order to examine trends in the risk of accidental death, PMRs were calculated for each accident type over three time periods, 1950—1959, 1960—1969, and 1970—1978. For farmers, significant increasing trends in PMR occurred in all accidents combined, water transport accidents, poisoning from gases, and machinery accidents (Table 4). No significant decreasing trend in any accident category was found.

For farm workers, significantly increasing trends in PMR occurred in poisoning from gases and machinery accidents (Table 4). There was also a significant decreasing trend in motor vehicle accidents.

Detailed information on the type of accident was examined using the fourth digit of the ICD code which, prior to 1969, gave the location where the accident occurred. For farm workers the small number of deaths due to any individual accident type precludes detailed analysis, while for farmers several interesting points emerge. Of the machinery accidents where the place of accident was recorded, 36 out of 44, or

TABLE 2
Mortality from Accidents for Male Farmers in B.C., 1950—1978

Cause of death	ICD-7 no.	No. of deaths	PMR	95% confidence interval
Railway accidents	E800—802	10	56	27,104
Motor vehicle or road	E810—845	424	122[a]	110, 135
Water transport	E850—858	23	50[a]	31, 75
Aircraft	E860—866	12	55[b]	28, 96
Poison: drugs or alcohol	E870—880	13	25[a]	13, 44
Poison: chemicals	E881—888	8	34[a]	14, 67
Poison: gases	E890—895	25	99	64, 147
Falls	E900—904	342	91	82, 102
Burns	E916	71	104	82, 132
Environmental factors	E926—928, 930—935	48	234[a]	176, 311
Drowning	E929	50	78	59, 104
Struck by falling object	E910	32	78	55, 110
Machinery	E911—912	66	300[a]	235, 382
Firearms or explosion	E915, 919	19	226[a]	136, 354

[a] $p < 0.01$.
[b] $p < 0.05$.

TABLE 3
Mortality from Accidents for Male Farm Laborers in B.C., 1950—1978

Cause of death	ICD-7 no.	No. of deaths	PMR	95% confidence interval
Railway accidents	E800—802	11	268[a]	133, 479
Motor vehicle or road	E810—845	127	164[a]	137, 195
Water transport	E850—858	3	24[a]	4, 71
Aircraft	E860—866	1	12[a]	0.67
Poison: drugs or alcohol	E870—880	6	54	19, 119
Poison: chemicals	E881—888	2	36	4, 130
Poison: gases	E890—895	5	115	37, 268
Falls	E900—904	24	81	51, 121
Burns	E916	26	270[a]	176, 396
Environmental factors	E926—928, 930—935	16	652[a]	327, 1059
Drowning	E929	28	202[a]	134, 292
Struck by falling object	E910	5	41[b]	13, 96
Machinery	E911—912	8	127	54, 251
Firearms or explosion	E915, 919	1	39	0.216

[a] $p < 0.01$.
[b] $p < 0.05$.

TABLE 4
Trends in Mortality from Accidents for Male Farmers and Farm Laborers in B.C., 1950—1978

Cause of death	PMR (no. of deaths) Overall	1950—59	1960—69	1970—78	Test for trend
Farmers					
All accidents	101	93 (440)	100 (390)	113 (373)	a
Motor vehicle or road	122	122 (147)	124 (147)	133 (130)	
Water transport	50	29 (6)	43 (7)	99 (10)	a
Poison: gases	99	54 (13)	145 (5)	263 (7)	a
Environmental factors	234	147 (13)	311 (19)	289 (16)	
Machinery	300	178 (26)	280 (20)	611 (20)	a
Farm laborers					
All accidents	129	109 (85)	151 (105)	127 (86)	
Motor vehicle or road	164	211 (45)	184 (50)	113 (32)	a
Poison: gases	115	28 (1)	113 (1)	359 (3)	a
Environmental factors	652	346 (4)	563 (4)	1104 (8)	
Machinery	127	32 (1)	88 (2)	445 (5)	a

[a] $p < 0.01$.

89%, occurred on the farm (Table 5). In addition, the proportion of machinery accidents occurring on the farm increased in the 1960s over the 1950s.

Accidents due to environmental factors could be separated into three main categories: animal-caused, exposure to cold, and others including exposure to excessive heat, starvation, and lightning (Table 6). The proportion of the environmental

TABLE 5
Number and Percentage (in Parentheses) of Machinery Accidents by Location for Each Decade for Male Farmers in B.C.

Location	1950—1959	1960—1969	1970—1978
Farm	18 (72)	18 (95)	0 (0)
Other (including home)	7 (28)	1 (5)	0 (0)
Unspecified	1	1	19

TABLE 6
Number and Percentage (in Parentheses) of Accidents Caused by Environmental Factors by Type for Each Decade for Male Farmers in B.C.

Type	1950—1959	1960—1969	1970—1978
Animal	8 (62)	3 (16)	3 (19)
Exposure to cold	2 (15)	9 (47)	13 (81)
Other	3 (23)	7 (37)	0 (0)

deaths that was due to animals decreased over time, whereas the proportion due to exposure to cold increased.

Accidental poisonings were also examined in detail to determine if pesticide exposure was a significant cause of accidental death in farmers. Prior to 1969, pesticide exposure could not be determined directly from the cause of death code; however, out of 11 poisoning deaths in farmers, none were recorded as occurring on the farm and in only one death was the type of poison unspecified and the location not recorded. After 1969, pesticide exposure was coded as a separate category, and out of ten recorded accidental poisoning deaths, only one was due to pesticides.

For farm laborers, only eight poisoning deaths were recorded in the 29-year period. Prior to 1969, there was one poisoning death recorded as due to alcohol poisoning. After 1969, seven poisoning deaths were recorded with one death due to pesticides.

DISCUSSION

Mortality studies using death registrations have a number of well-known shortcomings. Cause of death information available through death registrations is much less precise than that found in clinical records. Moriyama et al. analyzed available clinical data on 2122 deceased individuals and found that the death certificates recorded the most probable cause of death in only 80% of cases.[12] Accidental deaths were not examined; however, it seems reasonable to assume that accidental deaths are better recorded than other causes, as all accidental deaths in British Columbia are reviewed by the coroner. Diagnostic and coding errors arise in any large data file, but these should be fairly random, and there is no reason to believe that errors are specific to any single occupational group and would bias our analysis.

The quality of occupational information on death registrations has been assessed by Peterson and Milham, who have shown that the occupation recorded on the certificate is correct in 75% of deaths and gives a closely related occupation for another 10% when compared with data from next-of-kin interviews. Thus, the recorded occupation on the death certificate should be a good indicator of the usual occupation of the deceased.[13]

Use of the PMR statistic in studies of occupational mortality has positive and negative aspects. The major disadvantage is that the PMR gives no information on the overall force of mortality within an occupational group.[1] Since the overall PMR for each group must equal 100, a markedly high or low PMR for a common cause of death, such as heart disease, will have the effect of lowering or raising the PMR for uncommon causes of death. Examination of the major causes of death indicates that this is likely not a major problem in this study, since farmers have a PMR close to expected for the most common cause of death, arteriosclerosis and degenerative heart disease. Farmers do have a slightly lowered PMR for all cancers, but a corresponding slightly elevated PMR for cerebrovascular accident.

Farm workers do have lowered PMRs for major causes of death; however, this is likely due to the high accident and pneumonia mortality rather than the high accident rate being due to lowered heart disease and cancer mortality.

PMR avoids some of the disadvantages of the other common comparative mortality measure, the standardized mortality ratio (SMR). It does not require externally generated information on either the population at risk or the number of years of follow-up for the group at risk, and it avoids the "healthy worker effect" evident in comparisons of mortality in working groups to that of the general population.[10]

The major interpretational difficulty arising in a death certificate study is that none of the life-style factors that might confound the associations discovered can be controlled. However, this may be of less importance for findings concerning accidental death, because with the exception of alcohol use and to a lesser extent smoking (fires), accidents are not thought to be strongly related to life-style factors.

Finally, in a study in which a summary measure is used to compare mortality from many causes of death in large numbers of occupational groups, some statistically significant elevated risks can be expected to appear by chance alone. The consistency of most of our findings with those of several previous studies implies that our findings are likely true indications of risk in farmers and farm workers.[1,2]

The finding for farm laborers of an elevated PMR for all accidents is consistent with other mortality studies. Milham reported an elevated risk of accidental death for Washington State farm workers (PMR = 132).[1] Also, in two independent studies of California farm workers, Stubbs et al. found a significantly higher-than-expected all-accidents PMR of 130 and Peterson and Milham found a significantly elevated all-accidents PMR of 142.[2,3]

The finding of an overall PMR for accidents in farmers as expected has also been reported by Milham for Washington farmers (PMR = 103).[1] However, Stubbs et al. (PMR = 122) and Peterson and Milham (PMR = 111) found significantly elevated all-accidents PMRs for California farmers.[2,3]

The increasing trend in accidental death in farmers was not found by Milham, who demonstrated no trend in all-accidents mortality for either farmers or farm workers.[1]

The major accident risk for farmers according to number of deaths was motor vehicle accidents. Unfortunately, all motor vehicle accidents occurring on public roads are classified in this category without regard to type of vehicle. Thus, we are unable to ascertain whether the particular accident occurred in a passenger car or a farm vehicle such as a tractor.

The finding of a significantly elevated PMR for motor vehicle accidents for farmers is consistent with other mortality studies. Stubbs et al. and Peterson and Milham found PMRs of 163 and 137 in two separate studies of California farmers, and Milham found a PMR of 113 for Washington State farmers.[1-3] Similarly, the elevated motor vehicle accident PMR for farm laborers has been demonstrated in other mortality studies.[1-3]

The decreasing trend in motor vehicle accidents in farm laborers was not demonstrated by Milham, who found no trend in motor vehicle accidents in either farmers or farm laborers.[1]

The finding of an increased risk of machinery accidents and an increasing trend in PMR in both farmers and farm workers is consistent with Milham's findings.[1] This trend is most likely due to the increased mechanization in farming over the 29-year interval.

For all other individual accident types, with the exception of aircraft accidents, the B.C. and Washington State studies demonstrate similar PMRs. We demonstrated a significant deficit of deaths due to aircraft accidents in farmers, whereas Milham found a nonsignificant PMR of 110.[1]

A major cause of accidental death in both farmers and farm workers (which was not analyzed in any of the three previously mentioned mortality studies) is death from environmental factors. The high PMR for farmers may be a reflection of the outdoor nature of the occupation, and the extremely high risk of death (PMR = 652) for farm laborers may additionally be due to the generally poor living conditions of farm workers. Other types of accidental death present only for farm workers which may be further indications of poorer living standards are accidental death due to burns and drowning. At this time no information is available to further examine this issue.

It is clear that farmers and farm workers are certainly exposed to pesticides more than the general population; however, this increased exposure does not seem to translate into an increased accident risk. In order to assess the long-term effect of such exposures, other causes of death, in particular cancer and aplastic anemia, need to be considered. An examination of cancer risks in farmers and farm laborers from our mortality study has been examined elsewhere.[4,14]

Due to the inherent problems with proportional mortality studies already discussed, findings from this study must be interpreted with caution. However, these results do indicate several areas of concern regarding the health of farmers and farm workers.

REFERENCES

1. **Milham, S., Jr.,** Occupational Mortality in Washington State 1950—1979, Publ. 83-116, U.S. Department of Health and Human Services, Washington, D.C., 1983.
2. **Stubbs, H.A., Harris, J., and Spear, R.C.,** A proportionate mortality analysis of California agricultural workers, 1978—1979, *Am. J. Ind. Med.,* 6, 305, 1984.
3. **Peterson, G. and Milham, S., Jr.,** Occupational Mortality in the State of California 1959—61, U.S. Department of Health, Education and Welfare, Washington, D.C., 1980.
4. **Gallagher, R.P., Threlfall, W.J., Spinelli, J.J., and Band, P.R.,** Occupational mortality patterns among British Columbia farm workers, *J. Occup. Med.,* 26, 906, 1984.
5. **Carlson, M.L. and Peterson, G.R.,** Mortality of California agricultural workers, *J. Occup. Med.,* 20, 30, 1978.
6. World Health Organization, Manual of the International Statistical Classification of Diseases, Injuries and Causes of Death, 6th Revision, World Health Organization, Geneva, 1948.
7. World Health Organization, International Classification of Diseases, 7th Revision, World Health Organization, Geneva, 1957.
8. National Center for Health Statistics, International Classification of Diseases, 8th Revision, U.S. Government Printing Office, Washington, D.C., 1968.
9. Occupational Classification Manual, 1961 Census of Canada, Statistics Canada, Ottawa, 1961.
10. **Monson, R.,** *Occupational Epidemiology,* CRC Press, Boca Raton, FL, 1980.
11. **Armitage, P.,** *Statistical Methods in Medical Research,* Blackwell Scientific, Oxford, 1971.
12. **Moriyama, I.M., Baun, W.S., Haenszel, W.M., and Matison, B.F.,** Inquiry into diagnostic evidence supporting medical certification of death, *Am. J. Public Health,* 10, 1376, 1958.
13. **Peterson, G.R. and Milham, S., Jr.,** Hodgkin's disease mortality and occupational exposure to wood (brief communication), *J. Natl. Cancer Inst.,* 53, 957, 1974.
14. **Gallagher, R.P., Threlfall, W.J., Jeffries, E., Band, P.R., Spinelli, J.J., and Coldman, A.J.,** Cancer and aplastic anemia in British Columbia farmers, *J. Natl. Cancer Inst.,* 72, 1311, 1984.

Safety Risks of Ontario Farmers

J. R. Kinley

These comments attempt to respond to the assigned topic, but go beyond it in some respects to report on the work of the Ontario Task Force on Health and Safety in Agriculture.

The Task Force was established in October 1983 in response to a specific request made by the farming industry to the Ministers of Agriculture and Food and of Labour. To a considerable degree the origins of the request were in the fact that "farming operations" were excluded from the coverage of Ontario's *Occupational Health and Safety Act* when it was approved by the legislature in 1978. This exclusion gave rise to continuing concern about the health and safety protection of farm workers.

In brief, the assignment of the Task Force was to assess (1) whether or not there is a need for greater occupational health and safety protection than presently exists for people who work on Ontario farms, and (2) if needed, what measures are most likely to provide that protection.

The other important dimension of the assignment was to specify the need as experienced by farmers, unpaid farm family members, and hired farm workers.

The Task Force discovered quickly that existing data sources would not permit effective description of the health and safety experiences of the three categories of farm workers specified. Also, establishing clear relationships between the experience information available and major variables that appear to shape farm work health and safety performances proved a difficult task. A survey which might have overcome some of these problems was designed but proved too costly to gain funding approval. As a consequence the information sources used were (1) the existing health and safety experience data; (2) exploration of approaches to farm work health and safety in other jurisdictions; (3) arrangement for seven special review reports or memoranda on topics of priority concern, such as machinery risks and chemical-related health problems; (4) public hearings for the receipt of both written and oral briefs in 11 farm community centers; (5) discussion with subject area specialists at Task Force meetings; and (6) staff study of relevant literature, legislation, and opinions.

There is potential for extensive discussion of each of the information sources listed, but only two or three matters of particular concern are noted. The available lost time injury data are derived from claims approved by the Workers' Compensation Board. It shows the injury experience of hired farm workers reasonably well, but there are serious questions about the extent to which it reflects that of farmers and unpaid workers. The result is that conclusions based on the data are limited and vulnerable to serious challenge. The fatality data available are reasonably satisfactory in terms of coverage, but need to be collected against much more rigorous criteria of what constitutes a farm work-related fatality. In its use of these data, the Task Force applied considerably more rigorous standards than are used in the information-gathering process and reflected in the use of the data in most press reports. Despite this purging of the data, the fatality experience in farming is comparable to that in mining (Figure 1).

The third point has to do with occupational health. Although the Task Force sought expert advice and spent a considerable proportion of its funds in the area, it had to work with almost no work-related health experience information on Ontario farm workers. It obtained only the most general and cautiously given opinions from health professionals as to what the priority health problems of farm work are.

By drawing on the various information bases mentioned, the Task Force reached an overall conclusion that there is a need to improve the occupational health and safety experience of people working on Ontario farms. The fatality experience is clearly excessive and is described in comparison with other industries in the figure on the next page. Several sources show that the risk-taking associated with farm fatalities approaches the unbelievable and that other jurisdictions have found means of radically reducing the incidence of them. The lost-time injury experience indicates a gradual worsening of performance on farms relative to that in other industries. However, there are grounds for thinking that this change is mainly a reflection of better reporting. The levels of concern expressed by practically all segments of the industry in the public hearings strongly reinforced the indication of need suggested by the data. Finally, the review studies identified specific areas of urgent concern. Against this general position the Task Force identified and made recommendations pertaining to specific hazards or risk.

Perhaps it is important to note that the work done revealed little if any new health and safety issues relating to farm work. It did, however, make it possible to set clear priorities on known hazards and placed emphasis on others, which, while known, have received little attention.

Machinery-related risks unquestionably have the highest priority. Figure 2 shows that they account for more than three quarters of all farm work fatalities. Those associated with tractors contribute 50% of the total. In comparison, approximately a quarter of lost-time injuries are attributed to machinery, but this statistic is suspect because of the limited coverage of farmers as opposed to hired workers in the lost-time injury data.

While tractors are a factor in 50% of all farm work fatalities, rollovers account for 60% of the tractor-related

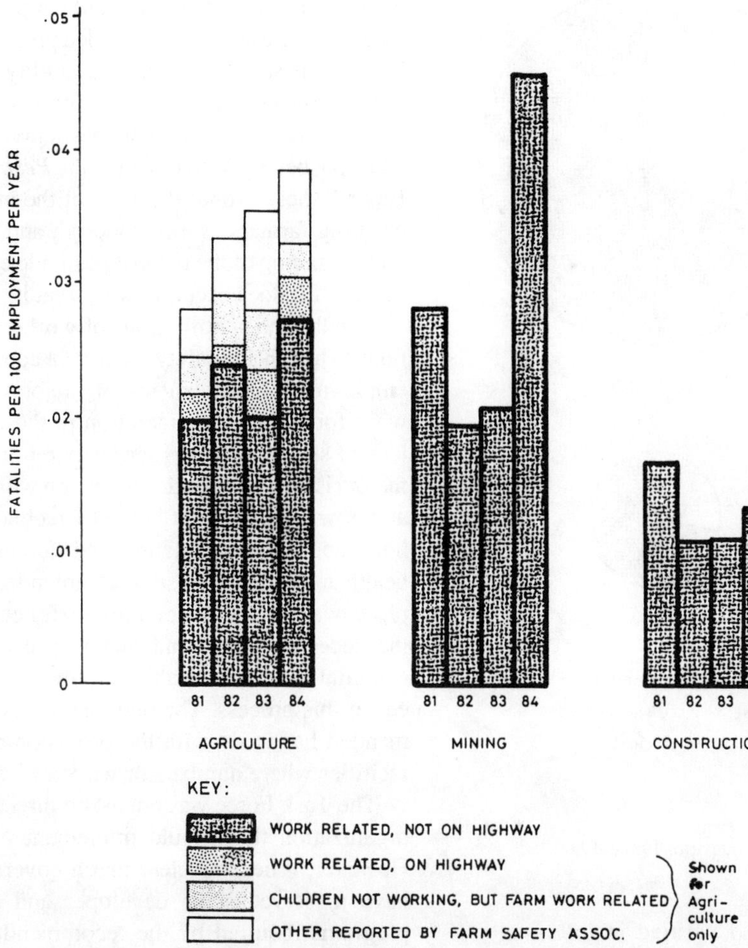

FIGURE 1. Comparison of fatality rates.

share — 134 out of 226 in the 10 years, 1975—1984. The second major area of tractor accidents is in runover of operators, passengers, or bystanders. Children are involved in many of these. Other major sources of machinery-related fatalities and injuries are workers becoming caught in operating machines, workers being struck by a reversing vehicle, and inadequate precautions when performing maintenance. An important factor in many machinery accidents is failure to use the shields or guards provided with most farm machines and which are intended to be in place when the machine is operating.

In response to the clear evidence of safety risks associated with machines, the Task Force has recommended strong, corrective measures. These include mandatory requirements for (1) rollover protection and seat belts on all but specially excluded farm tractors; (2) automatic power cutoff switches activated when the operator leaves the normal control station on all self-propelled farm machines sold in Ontario; (3) automatic backup beepers on all new self-propelled farm machines; (4) shields and guards on new farm machinery that conform to standards to be specified in Ontario; and (5) operating farm machines only when the shields and guards supplied with the machines are in place.

If acted on, these recommendations will save lives and prevent serious injuries. Judged on these criteria the rollover protective structure recommendation is unquestionably the most important one in the report. It is followed closely in this report by the recommendations on shielding and guarding. In addition to the mandatory recommendations, seven others relate to influencing machinery manufacturers to improve the safety features of machinery design, encouraging training in operation and maintenance, developing voluntary auditing programs, and vigorously promoting awareness of safety hazards.

A second major area of health and safety risks has to do with structures, fields, and yards, that is, the fixed facilities of the farms. As compared to machinery, these contribute to relatively few fatalities, but, insofar as the Task Force could determine, to a high proportion of lost-time injuries. The principal risks associated with these features of farms arise from heights, gases, noise, dust, contagious diseases, animal behavior, and farm use of public roadways. While a number of major risks are easily identified, the evidence of their impact on worker health and safety is not as fully demonstrated as in the case of machinery. Nevertheless, the risks associated with silo gas, exposed silo ladders, and open

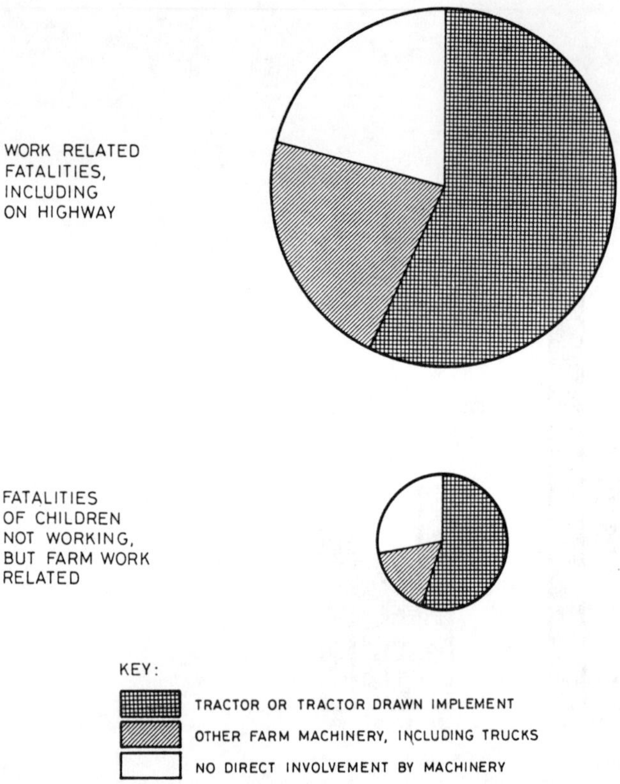

FIGURE 2. Sources of fatalities (Ontario, 1981—1984).

manure lagoons are apparent and verified in information available to the Task Force. However, regulation of when and how farmers enter silos did not seem useful to the Task Force and it recommended major emphasis on education and awareness instead. In contrast, fences are clearly the practical means of protecting people, especially children, from falling into and drowning in manure pits, and a mandatory requirement that they be in place is proposed. A similar suggestion is made with respect to cages on exposed silo ladders. Many of the risks arise from fixed facilities related to health rather than safety, and long-term impact rather than causing acute responses. In these circumstances, information and protective equipment are the appropriate means of protection. Recommendations are made to ensure that these are available to the people at risk — health professionals, police, and others who need to be informed or protected to do work relating to farms.

Farm chemicals are a major source of concern and, if they are not properly handled, of health and safety risk. Because of the uncertainty and controversy that surrounds the use of pesticides on farms, particular care was taken to recognize the potential hazards they present and to propose reasonable and practical protection against them. The recommendations are directed toward knowledgeable and careful use. However, 3 of the 12 recommendations would impose mandatory requirements on the use of recommended protective equipment when handling pesticides, on the availability of such equipment in pesticide dealerships, and on the observance of recommended reentry times on fields where pesticides have been applied. In addition, better enforcement of the *Pesticides Act* is endorsed. Beyond these proposals, most of the others have to do with ensuring farmers', farm workers', and health professionals' understanding of the risks of pesticide exposure, how to limit it, and, if it does occur, how to treat it.

Finally, the contributions of workers and working conditions to health and safety risks were considered. The different employment arrangements of major segments of the farm work force are a consideration in these concerns. Also, the lack of survey data was seriously felt in dealing with human factor risks. The ones discussed are work or skill proficiency and how it has been affected by technological change, long hours and fatigue, sanitation, and worker participation in farm health and safety. The six recommendations on these matters place heavy emphasis on farmers for ensuring that they have the necessary skills and that they establish the conditions essential to safe and healthful farm work and involve employees in this process. The one mandatory requirement recommended has to do with the provision of suitable sanitation facilities where numbers of workers are employed.

The Task Force was not asked directly to recommend the organization that should implement its various proposals. However, it became clear that a government agency would have to serve as the developer and administrator of the programs required by the recommendations. Although the workers covered fall within the responsibilities of the two ministries mentioned, the Task Force believes that health and safety in the farming industry should be treated as a comprehensive single program. Consequently, it suggested that an agency established jointly by the Ministries of Labour and of Agriculture and Food be given the task.

The assignment required discussion of the appropriate means of health and safety protection for farmers. Most farmers and farm organizations stated a strong preference for reliance on education. People representing seasonal workers opted for comprehensive health and safety legislation. As is already clear, the Task Force believes that different means of protection are needed to deal with different risks. It has suggested a substantial amount of regulation, but in recognition of the major element of self-employment in farming, it has placed the main thrust of protection on educational and awareness measures. At the same time, it has stated that a much higher level of farmer participation in health and safety than has occurred in the past is needed. If after 5 years there is little or no evidence of improvement in the farm work health and safety experience in Ontario, the Task Force suggests that stronger measures be put in place.

Use of Job Design to Reduce Mechanical Stress on the Back

D. W. Badger

In the U.S., disability (lost work or restricted duty) attributed to low back pain costs an estimated 91 million lost work days per year and more than $14 billion in lost productivity, medical costs, disability payments, and lawsuits. Workers' Compensation payments alone cost employees more than $250 million/year. The cause of back pain is obscure and multifactoral,[1] making preventive actions at the workplace difficult to plan and implement. Many such programs focus upon workers who routinely perform manual materials- handling tasks (e.g., warehouse workers, nursing aids, etc.), since such workers often show unusually high disability rates associated with back pain.[2] A common denominator among such high risk groups appears to be mechanical stress imposed on the back during the application of force (lifting, carrying, etc.).

Traditionally, attempts to reduce the incidence and severity of low back pain have centered upon either the selection of workers for heavy manual tasks based upon medical or other evaluation, or by training workers to use "correct" lifting procedures. The use of medical criteria (histories and/or examinations) to select workers most fit to perform heavy manual tasks and eliminate those predisposed to back injury and thus unfit for such jobs does not appear to influence back disability rates. Snook[3] examined compensatable back injuries among policyholders at Liberty Mutual Insurance Co. and concluded that there were no significant differences in back injury rates between those policyholders who used medical evaluation of workers as selection criteria and those who did not. Montgomery[4] reviewed the literature on the use of prehire back X-rays and concluded that such examinations have little predictive value for identifying potential back injury claimants. Rowe[5] stated that less than 10% of men who become low back disabiity problems can be identified at the time of hiring by present diagnostic procedures. Thus, while medical examinations are useful to identify gross inadequacies in health status of prospective workers, their use in identifying individuals likely to develop back problems seems problematical.

Other selection techniques have been investigated. While most studies indicate that anthropometric factors are not good predictors of those at high back injury risk, tests designed to measure worker performance capacity may have better predictive value. In particular, measurements of static strength[6,7] and dynamic strength[8-10] may have potential value for selecting workers with reduced risk of low back disability, particularly if the employee's work capacity is matched with job demands. Controlled trials are required to evaluate the effectiveness of such selection procedures.

Since the 1930s, use of correct lifting techniques to minimize back stress has been widely advocated, and most training programs have emphasized this concept.[11] The accepted method has been variously described but can be simply stated as lifting with a straight back and bent knees. With few exceptions, training in the use of such lifting techniques has had little influence on reducing back injury.[3,11-13]

In the past decade, however, renewed interest in training has occurred which appears to go beyond the "straight back, bent knee", postural approach. Many back schools and training programs have sprung up in the U.S. Different lifting postures and/or techniques are advocated by the various schools; most emphasize the importance of keeping the load close to the body with the upper trunk as erect as possible, and lifting smoothly without twisting. Such factors are sound from the standpoint of reducing mechanical stress to the spine, and workers should be encouraged to follow such work practices. However, the wide variety of manual materials-handling tasks workers perform certainly precludes a single posture suitable for all such tasks. In addition, two other factors impact upon the effectiveness of such training: how to train workers to use such postures consistently, and, of primary importance, workplace designs which impose stressful postural demands on the employee and preclude the use of good lifting practices. It is this latter factor, improving job design, that offers to be an effective method for reducing low back disability.[1,3,5,13-15] Improved design of manual materials- handling tasks can provide marked reduction in mechanical stress to the musculoskeletal system. For at least a portion of these tasks, recommendations are available for the identification and control of such hazardous postures.

The National Institute for Occupational Safety and Health (NIOSH) has published a Work Practices Guide for Manual Lifting,[14] that contains recommendations for lifting tasks which minimize risk of spinal column overloading and the development of excessive fatigue. The document provides methods for identifying jobs with high-risk potential and suggestions for modifying the job to reduce the risk. Based upon epidemiological (injury rates), biomechanical (spinal forces), physiological (metabolic), and psychophysical (maximum acceptable lift, muscle strength) data, the document develops composite equations for defining acceptable load limits. The equations take into account not only the weight of the load, but also its horizontal and vertical location at the start and end of the lift, the height of lift, and the frequency of lift. Values of these task variables are obtained from an on-site analysis of the job.

Three zones of lifting conditions are defined: (1) acceptable lifting conditions, (2) administrative controls required, and (3) hazardous lifting conditions. The action limit (AL) separates the first two lifting conditions, and the maximum permissible limit (MPL) separates the second from the third condition.

$$AL\ (Kg) = 40\ Kg\ (15/H)\ (1.0\text{-}0.004\ V\text{-}75)\ (0.7 + 7.5/D)\ (1 - F/F_{max})$$

$$MPL\ (Kg) = 3\ AL$$

where H = horizontal location (cm) forward of midpoint of ankles at the start of lift — $15 > H < 80$; V = vertical location (cm) at starting of lift — $0 > V < 175$; D = vertical travel distance (cm) (start to end) — $25 > D < (200 - V)$ (if $D < 25$, set $D = 25$); F = observed frequency of lift (lifts per min); and F_{max} = maximum frequency which can be sustained (depends upon lift posture and duration of lift).

AL values are based upon the criteria that; (1) over 99% of male and 75% of females have muscle strengths to perform such lifts; (2) loads of this magnitude will result in compressive forces of 350 kg (3450 N) or less on the L5/S1 intervertebral disk; and (3) metabolic rates for repetitive lifting of these loads will not exceed 3.5 kcal/min (23% of aerobic capacity in most fit young males and 33% of aerobic capacity in most fit young females).

Lifts that exceed the MPL will (1) require muscle strengths that 25% of males and less than 1% of females possess; (2) produce compressive forces of more than 650 kg (6370 N) on the L5/S1 disk; and (3) require metabolic rates in excess of 5.0 kcal/min (33% of aerobic capacity in most fit young males and 50% of aerobic capacity in most fit young females).

The above guidelines are intended to apply only to two-handed lifts of compact objects located in front of the body. However, a significant percentage of lifting in industry is of this type. Moreover, such tasks are common to most occupations. For example, estimates of the Wisconsin work force engaged in manual materials-handling tasks include 60% of 230,000 craft workers, 60% of 200,000 service workers, 90% of 72,000 laborers, and 100% of farm owners and laborers.[16]

Examples of the applicability of NIOSH guidelines to typical industrial tasks include the following.

Case 1. Loading or Unloading of Stock Bins (warehousing, etc.)

Figure 1 illustrates a palletizing task carried out by an order selector in a grocery warehouse. The restricted height of the stock shelves requires the worker to bend and reach for cases stored on the bottom shelf. Analysis of one of the most hazardous tasks yielded the following information:

H = 60 cm, V = 10 cm, D = 160 cm, F = 2/min
F_{max} = 15/min (stooped posture, 1 h duration)
Loads actually lifted, 10.8 kg, 16.2 kg

Inserting these values into the equation yields the following:

FIGURE 1. Order selector making a long, low reach for stock. The horizontal reach distance (H) exceeds 75 cm.

$$AL = 40\ (15/60)\ (1\text{-}.004 \times 65)\ (.7 + 7.5/16)\ (1 - 2/15) = 4.8\ Kg$$

$$MPL = 14.4\ Kg$$

The data indicate that for the lifting configuration described above, loads weighing less than 4.8 kg are considered acceptable, while loads that exceed 14.4 kg are regarded as hazardous, yet the worker is required to routinely lift loads which approach or even exceed the MPL. To modify the job and reduce the risk of injury associated with this task, several options can be considered, with varying levels of complexity (and cost). The extent of the problem in this physically demanding job[17] may dictate the appropriate solution. Injuries and lost time may not be the only factors: often productivity is an important consideration.

Reducing the weight of the load in this case must be discarded as unrealistic. Since cartons are handled rapidly as single units, mechanical lifting aids also cannot be regarded as a practical solution to the problems. However, inspection of the data indicate that both the AL and MPL can be raised significantly if the horizontal distance of the load from the worker can be reduced. In fact, if the load can be brought closer by sliding the carton to within 30 cm with a hook and then lifted, the AL is raised to 9.6 kg, the MPL to 28.8 kg. Lifting the required cases now becomes an acceptable task. Such a simple solution has, in fact, been used by several large grocery warehouses as a means to reduce back strain. (It should be noted, however, that close supervision is required, since lifting the box rather than sliding it appears to save time. In addition, hooks often get misplaced, and workers seldom take the time to find a replacement.)

Other, more costly solutions are also available. Since the most demanding physical task required of the order selector involves lifting from the bottom shelves, placing casters on the bottom row of pallets enables the worker to eliminate long, low reaches by moving the pallet out of the storage area before lifting. Better organization of storage (heavier objects at waist height etc.) or reduction in pallet size may be an additional option. For other examples of job design for such tasks, see Konz[18] or Rodgers.[19]

Case 2. Loading, Unloading of Stock Baskets

Figure 2 shows a task seen throughout most manufacturing operations. Typically, loaded stock baskets are brought to a machine by forklift and placed on the floor next to the worker. The operator bends, selects a part, moves it to the bench, performs an operation, and places the part in another basket. As the worker empties the basket, the reach distance for parts increases, since the basket is too heavy to reposition manually. Several factors contribute to the potential for fatigue and likelihood of back strain: (1) as the worker empties the basket, usually from front to back, the horizontal reach distance increases, since the stock basket is usually too heavy to reposition manually; (2) each lift is from floor to bench height; (3) the sides of the basket prevent getting close to the load. Often the worker must lift awkwardly, placing unusual strain on back muscles; (4) as the basket is emptied, parts shift and become wedged. Often significant effort is required to free the part so that it may be lifted from the basket.

Job analysis of a "worst case" lift (the most distant part in the basket) gave the following data:

$H = 65$ cm, $V = 10$ cm, $D = 75$ cm, $F = 1/\text{min}$, $F_{max} = 12/\text{min}$
Actual weight lifted = 14 kg

Calculation of AL and MPL values give the following:

$$AL = 40\ (15/65)\ (1-.004 \times 65)\ (.7 + 7.5/75)\ (1-1/12) = 4.9 \text{ kg}$$

$$MPL = 14.7 \text{ kg}$$

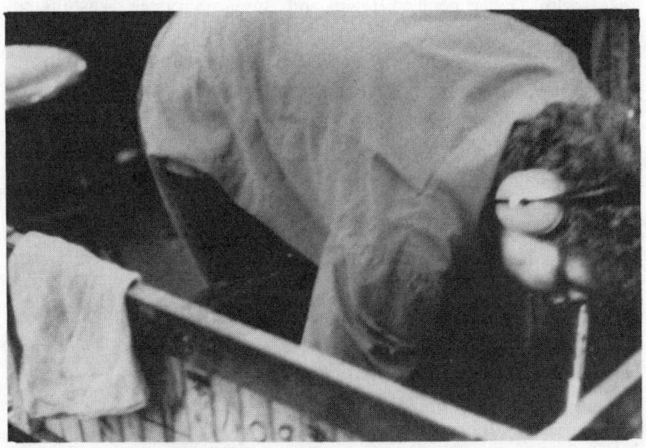

FIGURE 2. Excessive back strain resulting from bending to obtain parts from a stock basket. The sides of the basket restrict foot placement, requiring the worker to make a back lift.

For this "worst case" lift, the worker lifts a load close to the MPL. In addition, other factors given above (restricted motion of legs, extra exertion required to lift edged parts) place unusual, fatiguing demands upon the worker. Clearly, this task required modification. Again, different options are available. Figures 3 and 4 present two relatively inexpensive solutions which modify H, V, and D simultaneously. Figure 3 illustrates a simple castered stock dolly which positions the stock basket at waist height, inclined at about a 30° angle. Use of this inexpensive device raises the load from the floor and allows the operator to position it to reduce the horizontal reach distance.

An alternate approach, shown in Figure 4, employs a load leveler to reposition the load at waist level closer to the operator. (Also note the use of a roller conveyor to move the

FIGURE 3. An inclined stock dolly that aids in positioning parts closer to the operator.

FIGURE 4. A load leveler can reposition the stock at waist height, reducing the vertical lift distance (V). Also note the use of a roller-conveyor which eliminates the need for lifting the filled stock basket.

load away from the machines. Both devices eliminate the problems of an awkward, fatiguing back lift and extra effort required to extract wedged parts from the stock basket.)

The two cases given above illustrate the use of NIOSH recommendations for the identification and control of lifting tasks which impose excessive strain on the back. Two other factors, postural fatigue and workplace clutter, also increase the risk of injury, and should be taken into consideration when evaluating the job. Figure 5 illustrates an example of an awkward static posture likely to result in excessive back strain.[11] Other examples include working with excessive shoulder abduction, improper workbench height (either high or low), or excessive reach distances to perform a bench task. Often such postures result from poor work station design, and may be modified by repositioning equipment and/or stock.

Finally, slips and falls account for an estimated 4 to 10% disability.[5] Cluttered or crowded aisles (Figure 6) contribute to such accidents; often simple good housekeeping practices can eliminate this factor as a source of risk.

FIGURE 5. Faulty workplace design that requires an awkward static working posture.

FIGURE 6. Unnecessary crowding in a warehouse. Such cluttered workspace increases the risk of slips and falls, often resulting in injury to the back.

REFERENCES

1. **Troup, J.D.G.,** Causes, prediction and prevention of back pain at work, *Scand. J. Work Environ. Health,* 10, 419, 1984.
2. **Klein, B.D., Jensen, R.C., and Sanderson, L.M.,** Assessment of Workers' Compensation claims for back strains/sprains, *J. Occup. Med.,* 26, 443, 1984.
3. **Snook, S.H.,** A study of three preventive approaches to low back injury, *J. Occup. Med.,* 20(7), 478, 1978.
4. **Montgomery, C.H.,** Preemployment back X-rays, *J. Occup. Med.,* 16, 527, 1976.
5. **Rowe, M.L.,** *Backache at Work,* Perinton Press, Rochester, NY, 1983.
6. **Chaffin, D.B., Herrin, G.D., and Keyserling, W.M.,** Preemployment strength testing — an updated position, *J. Occup. Med.,* 20, 403, 1978.
7. **Keyserling, W.M., Herrin, G.D., Chaffin, D.B., Armstrong T.J., and Foss, M.L.,** Establishing an industrial strength testing program, *Am. Ind. Hyg. Assoc. J.,* 41(10), 730, 1980.
8. **Pytel, J.L. and Kamon, E.,** Dynamic strength test as a predictor for maximal and acceptable lifting, *Ergonomics,* 24(9), 663, 1981.
9. **Kamon, E., Kiser, D., and Pytel J.L.,** Dynamic and static lifting capacity and muscular strength of steelmill workers, *Am. Ind. Hyg. Assoc. J.,* 43(11), 853, 1982.
10. **Kroemer, K.H.E.,** Testing individual capability to lift material: repeatability of a dynamic test compared with static testing, *J. Saf. Res.,* 16, 1, 1985.
11. **Brown, J.R.,** Manual Lifting and Related Fields: An Annotated Bibliography, Labour Safety Council of Ontario, Ontario Ministry of Labour, Ontario, Canada, 1972.
12. **Dehlin, O., Hendenrud, B., and Horal, J.,** Back symptoms in nursing aids in a geriatric hospital, *Scand. J. Rehabil. Med.,* 8, 47, 1976.
13. **Troup, D.,** Back injuries can be prevented, *Health Saf. Work,* 15, 1982.
14. DHHS/NIOSH, Work Practices Guide for Manual Lifting, Technical Rep. 81-122, National Institute for Occupational Safety and Health, Department of Health and Human Services, Washington, D.C., 1981.
15. A proposed national strategy for the prevention of musculoskeletal injuries, in Proposed National Strategies for the Prevention of Leading Work-Related Diseases and Injuries, Association of Schools of Public Health/National Institute for Occupational Safety and Health, Washington, D.C., 1986.
16. Wisconsin Bureau of Research and Statistics, 1974.
17. **Garg, A., Hagglund, G., and Mericle, K.,** Physical fatigue and stresses in warehouse operations, Final Report, CDC/NIOSH Contract 210-81-6008, July 1983.
18. **Konz, S.,** *Work Design: Industrial Ergonomics,* 2nd ed., John Wiley & Sons, New York, 1983.
19. **Rodgers, S.H.,** *Working With Backache,* Perinton Press, Rochester, NY, 1984.

Prevention of Hearing Loss Among Farmers

V. S. Ejercito, D. A. Hansen, and W. E. Pierce

It has been suspected that farmers suffer a greater degree of hearing loss than do individuals engaged in other occupations. In a population of patients referred to the audiology department of the Marshfield Clinic for hearing tests, both male and female farmers showed a greater percent hearing loss than did the nonfarm population (Figure 1). At the frequency of 4000 H, the hearing loss for male and female farmers is 54 and 32 dB respectively. For nonfarm males and females, the hearing loss at 4000 H is 30 and 22 dB, respectively. Moreover, these hearing losses are even more striking among younger men (Figure 2). Among farmers, the hearing loss is not confined to the older men. At 4000 H there is essentially no difference in the hearing loss between the various age groups of farmers, while hearing loss increases with age in the other populations.

It cannot be determined from these data if this difference in hearing loss is the result of a real difference, or the reflection of a delay in seeking medical attention on the part of farmers so that only those with severe hearing loss seek medical attention.

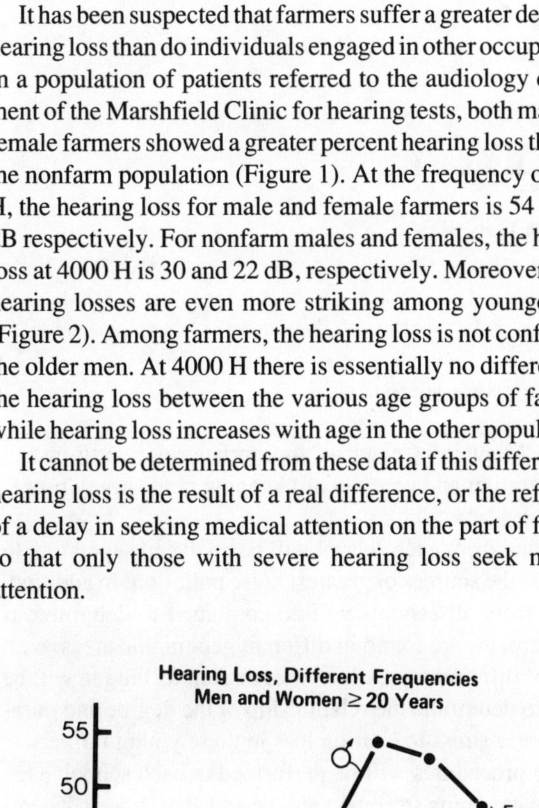

FIGURE 1. Hearing loss frequencies, men and women ≥20 years.

There may be many sources of noise which contribute to farmers' hearing loss. Because of the rapid changes in farm equipment and the diversity of utilization of such equipment, all sources may not be known. Further, there may be marked differences in noise associated with different types of farm operations as well as with the various types, makes, uses, and vintage of equipment. Thus, to properly evaluate noise-induced hearing in farmers, it is important to include farms and farmers from a wide variety of operations, and degree of sophistication of their operation.

We are fortunate in that the glaciers that traversed Wisconsin have produced a natural laboratory to study a diversity of discrete farming practices. Within a 50 mi radius of Marshfield, there are farms that are limited to dairying, hog, or poultry production, or those that produce only cranberries, potatoes, peas, beans, or cereal grains. There is pulpwood and lumber production as well as the world's largest ginseng gardens.

We considered it important to evaluate young farmers to determine when the onset of hearing loss occurs and to identify the causative factors. We chose as our study population vocational agricultural students from the high schools in central Wisconsin. These students are organized in a national organization of the Future Farmers of America (FFA). The FFA's accomplishments are legendary as is their leadership.

The vocational agriculture teachers of this area were approached to enlist their cooperation in conducting this study and all were willing to participate. The studies will be of a long-term longitudinal nature.

The first step will be the identification of individuals with noise- induced hearing loss and the source of their noise exposure. This will be followed by the development of an educational program to make young farmers more aware of the dangers of noise. The final phase of this program will be to evaluate preventive equipment and practices.

Under the leadership of the vocational agriculture teachers and with the guidance of the staff of the Farm Medicine Center and members of the Audiology Department of the Marshfield Clinic, the following will be accomplished.

Hearing test screens will be conducted annually on each student. A van equipped with audiological equipment is used to conduct these screens at each school. If a hearing problem is suspected in a student, more definitive tests are conducted to determine the extent and etiology of the problem. Other members of the student's family are encouraged to participate in hearing tests.

A 25% sample of the farms of the students in each school will be surveyed each year for noise. The farms selected to be

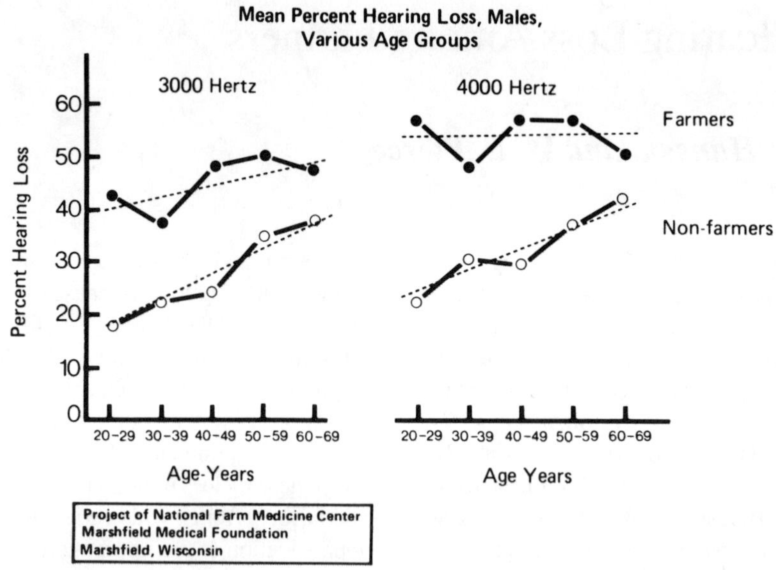

FIGURE 2. Mean % hearing loss, males, various age groups.

surveyed will include five farms of the students with the greatest hearing loss. The other farms will be randomly selected from the remaining. Each year a different set of farms will be evaluated.

The students of each school will develop lists of noise sources on farms to be surveyed. The survey will encompass all sources of noise — all mechanical equipment in all phases of operation; animals; ventilation equipment; as well as stereos, radios, and televisions. Noise dosimeters which collect, store, and process sound intensity and duration are used to obtain these data. Other data will be recorded as appropriate. Farms will be surveyed by teams of students made up of the student from the farm and three others whose farms are not surveyed. Teams shall be assigned by the vocational agriculture teacher to ensure proper balance of maturity and competence. Each team will be indoctrinated and supervised by the vocational agriculture teacher, a member of the Audiology Department of the Marshfield Clinic, and a member of the National Medicine Center as appropriate relative to instrument operation and care, record keeping, and survey procedures.

The data from each school will be analyzed separately to determine the sources of greatest noise pollution. In addition, the data from all schools will be combined to determine if similar stresses are found in different geographic areas, with similar or different operations. Regression technique will be utilized to determine the relationship of the degree and duration of noise stress to hearing loss in these young farmers.

These procedures will be performed in each school, each year. The incoming freshmen will be added each year. Therefore, as the study progresses we shall have continuous records of annual hearing tests on all vocational agriculture students in this area. If a student's hearing deteriorates, further evaluation will be made.

Over 500 students from 14 schools are currently being evaluated. These numbers will ultimately be expanded to over 1000 students from 30 schools.

Health Considerations in the Manufacture of Granular Fertilizers

D. Jasper

INTRODUCTION

Western Co-operative Fertilizers in Calgary (WCFL) manufactures granular monoammonium phosphate and prilled ammonium nitrate for sale in western Canada. These two materials are sold separately or blended together in any proportion required. If a "complete" fertilizer is required, potash (potassium chloride) is also blended in as necessary.

The manufacturing complex consists of an ammonia plant, a nitric acid plant, an ammonium nitrate plant, a carbon dioxide recovery plant, two sulfuric acid plants, two phosphoric acid plants, and an ammonium phosphate plant, along with the associated storage and shipping facilities. The complex described is fairly typical of a granular fertilizer manufacturing unit. Note: the term "granular" in this paper is used to describe solid fertilizer, in general, whether actually granulated or prilled.

Various potential health hazards may be encountered by operators and maintenance workers in such a chemical plant. This paper briefly discusses some of these hazards.

HAZARD LIST

Possible hazards include corrosive materials (e.g., sulfuric acid, nitric acid, phosphoric acid), noxious gases (e.g., ammonia, nitrogen dioxide, carbon monoxide, sulfur dioxide, sulfur trioxide, hydrogen sulfide, hydrogen fluoride), melt (e.g., molten ammonium nitrate), radioactivity (e.g., α, β, and γ radiation from the ^{238}U decay series), dusts (e.g., catalyst dusts [vanadium pentoxide], asbestos, phosphate rock dust, ammonium phosphate dust, ammonium nitrate dust, nucleating agent [bentonite], coating agent for ammonium nitrate [proprietary]), welding fumes, and physical agents (e.g., noise, heat stress, heavy lifting).

LOW POTENTIAL EXPOSURE HAZARDS

Normal plant operation ensures that risk of significant exposure to anhydrous ammonia, nitric acid, ammonium nitrate melts or solutions, sulfuric acid, phosphoric acid, nitrogen dioxide, carbon monoxide, sulfur dioxide, sulfur trioxide, and the various catalyst dusts is minimal. However, plant leaks do occur from time to time and protective acid-resistant suits, glasses/goggles, and respiratory protection may be required. Various kinds of protective clothing and equipment may be necessary on an ongoing basis in some plants (e.g., sulfuric acid).

Various catalysts are used in the processes. These materials do not find their way into the final product, but can constitute a hazard to workers removing or installing the charges. Such catalysts as vanadium pentoxide (V_2O_5), cobalt oxide (CoO), molybdenum oxide (MoO_3), iron oxide (Fe_2O_3), chromium oxide (Cr_2O_3), zinc oxide (ZnO), copper oxide (CuO), and nickel oxide (NiO) often break down to produce a dusty version of the original charge. Generally, dust masks or other respiratory protection along with disposable coveralls and gloves are necessary for this type of operation.

RADIOACTIVITY

The phosphate rock used for making phosphoric acid contains about 100 ppm of natural uranium, of which about 99.3% is the ^{238}U isotope. Its radioactive decay series results in a mixture of isotopes in the rock in radioactive equilibrium. During the phosphoric acid process, ^{238}U and ^{226}Ra are separated, the uranium going with the phosphoric acid and the radium going out to the gypsum disposal ponds along with the gypsum ($CaSO_4$ and $RaSO_4$ are chemically similar). The phosphoric acid is "loaned" to a uranium extraction company adjacent to the WCFL plant and is returned almost uranium-free. Thus, the amount of radioactivity associated with uranium in the final granular phosphate fertilizer is very low. However, α, β, and γ radiation related to the radium in the gypsum is of concern and has to be monitored throughout the phosphoric acid plants. The ^{226}Ra activity in the gypsum is approxiately 925 mBq/g (25 pCi/g). With respect to the amount of gypsum in the storage ponds (about 1200 t/d for 20 years), even this low concentration has given cause for concern with regard to housing development in the plant vicinity.[1] Workers constructing gypsum dykes on ponds are also subject to exposure to low radiation. However, the prime areas of concern in the plant proper occur wherever there are large amounts of gypsum indoors and where there is a potential for it building up inside pipes and vessels. The general workroom air must therefore be monitored for alpha radiation (resulting from the ^{226}Ra, ^{222}Rn gas and the particular "radon daughters": ^{218}Po, ^{214}Pb, ^{214}B, and ^{214}Po). Normal industrial ventilation has been found to be quite adequate for keeping radon and radon daughter concentrations at a satisfactory level. Atomic Energy Control Act Regulations allow a maximum exposure to α emitters (radon and radon daughters) for nonradiation workers of 0.4 WLM ("working level months") per year. Measurements at WCFL have shown that, in regions of the plant with highest potential for exposure to these isotopes, the greatest dose rate would be 0.12 WLM.[2] Build-up of scale on vessels and in pipes can give rise to β and γ radiation if the scale contains significant amounts of gypsum and hence, ^{226}Ra. Pipe and vessel walls effectively shield operators from the β radiation, except for maintenance workers cleaning scale or dismantling piping. In these cases, the area is checked out prior to commencement of the job. The Alberta government has set guideline limits for disposal of scale material with γ readings above 2.5 mR/h, and warning notices must be displayed at levels of one tenth of the above. γ-Radiation surveys in WCFL's plant have shown levels of 0.018 to 0.040 mR/h. Continuous occupancy in these areas for a 40-h week would result in an additional dose above background of about 300 μSv/year (30 mrem/year), which is about 1/17th of the allowed level for nonradiation workers. Thus, it can be seen that, although the presence of uranium in the rock constitutes a potential health hazard, in practice the levels of measured radioactivity in phosphate production do not pose any significant problems. The removal of uranium from the phosphoric acid by WCFL's neighbor also results in WCFL's monoammonium phosphate being virtually radiation-free, with no resultant handling problems.

DUST HAZARDS

Asbestos

Much of the original insulation on pipes and vessels at WCFL was asbestos. As maintenance on such plant parts is an ongoing operation, awareness of the hazards of asbestos insulation and how to deal with it safely are very important to the health of workers. WCFL has written policy and procedures on this, detailing how in-plant maintenance involving asbestos is to be handled. Replacement with nonasbestos-containing insulation is now part of maintenance policy.

Phosphate

Phosphate rock as shipped to WCFL has a wide range of particle sizes. After unloading the railcars at Calgary, the rock goes into storage until used, when it is ground further in a ball mill. Phosphate rock occurs in many places in the world. In North America, the principal sources are Florida and North Carolina (87%), Tennessee (3%), and the states of Utah, Wyoming, and Idaho (10%). Rock from the latter three locations is generally referred to as "western rock" and differs from Florida rock in a number of ways, e.g., particle-size distribution and organic carbon content. Chemically, the rock is fluorapatite, $Ca_5(PO_4)_3$ (OH,F,C1). Health concerns from the western rock used by WCFL include the 12% silica content, the amount of rock finer than 325 mesh (45 um) (11 to 50%), and the hydrogen sulfide equivalent (this latter aspect will be covered later).

Because of the potential exposure of workers to significant amounts of silica-containing dust and asbestos fibers, a medical surveillance program, better known as the Fibrosis Programme, has been set up at WCFL, in accordance with the Asbestos and Silica Regulations of Alberta's Occupational Health and Safety Act.[3-5] Of the 310 employees at WCFL Calgary, 194 are on the Fibrosis Programme (including operations supervisors, all phosphate division operators, some nitrogen division operators, all shipping personnel, all laboratory personnel, and all maintenance tradesmen). In summary, the medical surveillance program consists of:

1. Pulmonary function test (PFT), which must be done by a government-approved pulmonary function technician
2. P/A chest X-rays, to be interpreted by a radiologist
3. Work and respiratory history on each worker

Under the program, testing must be done every 2 years, and after exposure for 10 years, must be done every year. WCFL has its own Occupational Health Service and therefore is able to do more than the minimum required by the regulations. For example, if a problem is identified with a worker, he will be referred to his family doctor for assessment. Then the results are sent, as per the regulations, to the Director of Medical Services (Alberta Worker Health, Safety and Compensation, Occupational Health and Safety Division). When a problem is detected, Medical Services will then contact the worker directly and obtain permission to release the information to the family doctor, who is then responsible for providing follow-up care for the worker.

The free silica in the rock dust has been identified as quartz, although on one occasion a small amount of cristobalite was present, also. The phosphate rock-handling areas in the plant are monitored on an ongoing basis for respirable dust levels (both work area and personal monitoring being carried out). Because of the variable silica content of the dust, permissible exposure limits are more restrictive than for "nuisance" dusts. For example, the occupational exposure limit (also commonly referred to as threshold limit value, TLV) for respirable nuisance dust is 10 mg/m^3 (8 h time-weighted average). If the dust contains 10% quartz, this is reduced to 0.8 mg/m^3.[6] As measured values in these work areas are usually significantly higher than these limits, employees must always wear government-approved dust masks when working there.

Monoammonium Phosphate

The manufacture of granulated monoammonium phosphate also creates a dust hazard for some workers. In the plant proper, however, the material is often moist and therefore dust loading of the work environment is low. The crushing and screening process does produce some dust, but monitoring in this area of the plant has indicated that dust masks, in general, need not be worn. In the shipping area, where repeated handling can cause breakdown or abrasion, workers occasionally find themselves exposed to levels above the permissible, requiring the wearing of dust masks. Although manufacturers make every attempt at producing a dust-free or abrasion-resistant product, some dust will always be produced upon rehandling. Therefore, suppliers frequently treat their granulated product with a dust-control agent. Distributors and farmers should not encounter significant dust problems from a product so treated. For the most part, dusts from phosphate fertilizers generally contain less than 2% silica and therefore can be basically considered as nuisance dust. A modifier to this, however, is the fact that some products may be slightly acidic thereby increasing the tendency of this dust to irritate the eye, mucous membranes or the respiratory system.

Ammonium Nitrate

Ammonium nitrate can be either granulated or prilled. At WCFL, the material is prilled. In either case, modern production methods have resulted in ammonium nitrate generally being a much harder product than previously, with substantial resistance to breakdown on handling. Any dust which is produced through breakdown of the nitrate tends to be quite irritating, as it is generally slightly acid (with a pH of about 6). Because of the hygroscopic nature of ammonium nitrate and, hence, its strong caking tendencies, manufacturers coat the product with some form of anticaking agent. Typically, these are kaolin clays or diatomaceous earth applied at a rate of about 1%, with the net result that these types of nitrate fertilizers tend to be quite dusty. These dusts can contain fairly substantial amounts of free silica and can therefore be considered hazardous to some extent. Other manufacturers (WCFL included) have moved away from this type of coating, resulting in either a no-dust or at least a low-dust product. Generally the newer anticaking agents have low toxicity and are applied at lower rates (as low as 0.07%). Thus, the health and nuisance hazards are considerably reduced. The harder ammonium nitrate is often produced via the addition of a hardening agent (e.g., bentonite clay) and the workers in the ammonium nitrate-producing plant are often required to wear dust masks when working with this additive.

NOXIOUS GASES

Of the various gases which are produced during the course of granular fertilizer manufacture, only gaseous fluorides (mostly hydrogen fluoride, HF, but also lesser amounts of fluosilicic acid, H_2SiF_6) and hydroden sulfide pose any real health threat. Because of the nature of the processes, sulfur dioxide, sulfur trioxide, ammonia, and nitrogen dioxide have been categorized low risk in this paper, even though they are quite hazardous.

Hydrogen Fluoride

This gas (along with varying amount of fluosilicic acid) is evolved in the phosphoric acid process (phosphate rock mixed with sulfuric acid) because the rock contains about 3% fluoride. As a considerable amount of this process is open to atmosphere, primarily at the gypsum filtration stage, the workroom air will contain some of these gases. It is important to keep all pollution- scrubbing equipment, suction fans, and ventilation equipment in good working order so that concentrations of fluorides can be kept below the allowable maximum level of 2.5 mg/m^3 (8 h time-weighted average).[7]

Because of the potential for overexposure to these gases, a comprehensive environmental and medical surveillance program is carried out at many phosphate manufacturing facilities (including WCFL). This includes regular urine fluoride-level monitoring, personal air monitoring, and area air monitoring. By carefully assessing and correlating the results from these tests, problems can be readily spotted and the appropriate action taken to eliminate the situation. Such action might include improving ventilation, use of approved masks, or process improvements to reduce the emissions to the work environment.

Hydrogen Sulfide

This gas can be evolved in the phosporic acid process, along with the previously discussed hydrogen fluoride. The majority occurs in the very early rock digestion stages, so that it does not tend to be a problem with normal rock and providing the pollution control scrubber is kept in good operating condition. However, the level of hydrogen sulfide equivalent in the rock is variable; 0.08% is the average value, and does not cause serious problems. Levels of 0.10% and upwards put a severe strain on the equipment and exposure to dangerous amounts of hydrogen sulfide is possible. The 8-h time-weighted average limit for H_2S is 14 mg/m^3.[7] The highly toxic nature of this gas requires that the areas with greatest potential for exposure be fitted with audible alarms and emergency gas mask equipment for rescue and escape as necessary. Regular air monitoring is also carried out and spot checks can be rapidly carried out by workers using detector tubes. Careful maintenance of ventilation equipment is also necessitated. It should be mentioned that this tendency to produce hydrogen sulfide is not present in all phosphate rock. "Western rock" appears to be potentially the worst in this regard, primarily because it is required to go through a high-temperature calcination stage after mining. The evolution of hydrogen sulfide proceeds very readily when present. Therefore, very little, if any, remains in the final granulated ammonium phosphate fertilizer.

Because of the hazards posed by the various gases and dusts described above, it is imperative that breathing equipment, in general, be capable of being correctly fitted and worn. Therefore, the whole plant complex at WCFL, including the shipping and storage division, has been designated a "no beard" area.

CONCLUSION

This paper has discussed some of the physical and chemical agents which constitute potential health hazards to workers in a typical solid chemical fertilizer manufacturing plant. If proper attention is not given on a continuous basis toward protecting workers from these agents, then serious health effects could result. Such problems could include silicosis, asbestosis, fluorosis, and various cancers of the lungs and gastrointestinal tract. In practice, the most significant ailments encountered are bronchitis, asthma, occupational dermatitis, and noise-induced hearing loss. It should be noted that many trades personnel working in a chemical plant such as described in this paper are also subject to other hazards such as welding fumes, high heat, etc. These are not specific to the fertilizer industry but are often encountered in many other heavy industries. Responsible ongoing monitoring of the work environment, coupled with good occupational health and safety practices, ensures that workers in this industry are effectively at no greater risk than workers in many other heavy industries.

REFERENCES

1. **Ferry, R.A., and Jasper, D.,** Radon emission from phosphogypsum an its potential impact on a proposed residental development, in Canadian Phosphate Technology Conf., Western Co-operative Fertilizers Ltd., Pocatello, ID, unpublished, 1982.
2. **Jasper, D.,** Radioactivity resulting from the presence of uranium in phosphate rock, in Canadian Phosphate Technology Conf., Western Co-operative Fertilizers Ltd., Calgary, Alberta, unpublished, 1978.
3. Occupational Health and Safety Act, Government of the Province of Alberta, office consolidation, Queen's Printer, Edmonton, Alberta.
4. Asbestos Regulation, Alberta Regulation 7/82, Queen's Printer, Edmonton, Alberta.
5. Silica Regulation, Alberta Regulation 9/82, Queen's Printer, Edmonton, Alberta.
6. Threshold Limit Values for Chemical Substances and Physical Agents in the Work Environment and Biological Exposure Indices with Intended Changes for 1984—85, ACGIH, Cincinnati, OH, 1985, 34.
7. Chemical Hazards Regulation, Alberta Regulation 8/82, Queen's Printer, Edmonton, Alberta, 9.

Principles of Safety in the Formulation of Agricultural Chemicals

W. M. Haag, Jr.

SUMMARY

The National Institute for Occupational Safety and Health (NIOSH), as part of its Control Technology Assessment Program, initiated a study in 1977 for pesticide manufacturing and formulating processes. This study provided the major input for guidelines being prepared by a United Nations interagency effort chaired by the World Health Organization. This text is a distillation of those guidelines which are in final review for publication.

Safety and health principles recognize that an effective control system considers mechanisms for control at the source of emissions, in the workplace, and at the worker. It is important to monitor the normal operation to detect any malfunctions as they occur in order that corrective actions can be initiated. The proper education and training of both supervisory management and production employees are essential for safe handling of agricultural chemicals.

INTRODUCTION

While the hazards of pesticide application are well known and solutions to prevent adverse effects have been documented, the formulation of agricultural chemicals has received much less attention.[1-4] Formulators generally receive concentrated active ingredients and dilute them with "inert" extenders to yield various product forms, such as dusts, powders, granules, pellets, emulsifiable concentrates, or aerosols. Formulations may be of the dry or wet type.

In the NIOSH control technology assessment study, the most frequently reported exposures occur during the transfer of materials (e.g., receipt, drum, and barrel filling) and during packaging and shipping. A second activity of even higher exposure involves maintenance and repair of equipment, process monitoring, and product sampling.

Despite the diversity of plants in terms of type and volume of production, technology, or size of the workforce, almost all formulation facilities share the following features: (1) equipment is generally used for more than one product, therefore cleaning is required; (2) a wide range of toxic and flammable hazards is associated with the chemicals used; and (3) most reaction, blending, and mixing processes are batch operations, which may contribute to additional exposure.

In general, successful health and safety programs include a system of control described in the following text.

CONTROL THROUGH MANAGEMENT AND TRAINING

Commitment from top management is probably the most important factor in the implementation of an effective control system. Through education and training, both management and employees must be made aware of the hazards they face and of the control measures which are to be followed both in routine and in emergency situations.

The education and training of all personnel must consist of three parts to be a complete program. First, the prejob assignment should involve a basic review of the different types of materials to be handled, the associated hazards, signs, and symptoms of exposures. Second, upon job assignment, training should focus on proper safe operation of process equipment, procedures for materials handling, use of personal protective equipment, and methods for detecting the presence or release of a toxic material in the workplace. Finally, ongoing training for reinforcement and to understand new procedures required by process or product changes is necessary.

CONTROL THROUGH TECHNOLOGY

Most occupational hazards require controls — first at the point of origin of the hazard, second in the general workplace environment, and third at the individual worker. These controls must function as an integrated coordinated system in order to provide worker protection under normal operating conditions as well as under conditions of maintenance or process upset. Controls supplied at, or near, the point of origin of the hazard are preferred to minimize the spread of exposure.

Control at the Source

Replacement of hazardous materials with less hazardous materials should be a high priority. Quite frequently process modification is an important approach to emission control. Raw materials specifications or process chemistry may have to be changed so that toxic contaminants or by-products are not present or formed. Processes should be designed to operate as close as possible to (and preferably somewhat below) ambient temperatures and pressures.

Equipment should be selected to contain as much as possible all hazardous materials, to afford ease of cleaning, and to require low maintenance. A process component known to be prone to breakdowns should be discarded in favor of a more reliable component.

Quick coupling connectors on transfer pipes or solid conveyors will permit easier cleaning. Low speed rotation is likely to produce less wear and to require less maintenance; the noise output is also less. All filling machines must automatically close when deenergized.

Isolation involves the use of a barrier between a hazard source and those who might be affected by the hazard. The barrier may be provided by a physical structure, by distance, or by a time interval.

Local exhaust ventilation is probably the most commonly used and certainly one of the most efficient and versatile controls available provided it is properly designed, installed, and used. The objective of a local exhaust ventilation system is to capture air contaminants, such as dusts or solvent vapors, at the source, and to avoid their dispersion and contamination of the work environment and their being breathed by workers.

Control at the Workplace

Workplace control measures are important adjuncts to control measures which are applied at the source. The importance of maintenance and housekeeping cannot be overemphasized. It is absolutely crucial for a number of reasons: (1) a dirty or untidy plant creates an attitude of carelessness in the workforce and demonstrates the low importance given by management to careful work practices; (2) surface contamination, whether or not it is visible, can be one of the major sources of skin contamination; and (3) equipment leaks or spills, if not attended to promptly, may constitute important sources of both air and surface contamination.

Vacuum cleaning is widely used in formulation plants. If vacuum cleaners are not working, brooms and shovels may have to be used, but with utmost care so as not to disperse a cloud of dust. Dry sweeping is not recommended. Damp sweeping is often proposed to minimize raising dust and subsequent air contamination.

Liquid spills should be cleaned up by absorbent material and placed into containers for disposal. In some cases, decontamination with decontaminant solution may be necessary.

The operating areas should be well lighted and ventilated. No waste or extraneous material of any kind should be allowed to accumulate in the work area except in containers designed for that purpose. A product-labeled bag or container should never be used as a receptacle for floor sweepings or waste material.

Maintenance activities are inherently hazardous, with substantial opportunity for exposures to toxic chemicals. Maintenance programs should include: (1) inspection, by trained personnel and on a regular basis, of all equipment in the plant; (2) recording of equipment performance in logs that are regularly reviewed to detect any deterioration in performance; (3) regular and routine service and adjustment of equipment; and (4) repair of breakdowns, as soon as possible.

General ventilation refers to the removal (or supply) of air for the purpose of either thermal comfort control or dilution of contaminated air with uncontaminated air, or both. General ventilation used as means of diluting air contaminants is not to be compared in efficiency to local exhaust ventilation, since it does not remove the contaminants but only dilutes them to currently acceptable levels.

Control at the Worker

Controls may also be applied near or worn by the individual worker. Isolation of the worker to prevent exposure is generally achieved through remote control of the process, such as in the control room of an automated process. Alternatively, the worker can wear full protective clothing with a self-contained breathing apparatus or an air-supplied device.

Carelessness and operator error can circumvent the best of control systems. Training and experience are necessary to ensure proper equipment use, adherence to safety operating procedures, and good occupational hygiene practices. Instructions should include specific methods for reducing exposures, vessel entry requirements, and designation of assembly points in disasters.

Whenever personal protective devices or special working clothes are required, management should be responsible for their provision. Important aspects of any personal protective equipment program are the fitting of the equipment and education of workers on its use and importance, as well as its cleaning and maintenance. Cleaning is extremely important, since contaminated protective gear may become a source of exposure.

Respiratory protection must be available within any pesticide plant in case of fire, toxic chemical release, and/or closed space entry. Disposable dust masks are widely used; however, these may only be acceptable when less toxic dusts (and no vapor or gases) are involved. Considering the strain represented by wearing most types of respiratory protection on a continuous basis, particularly in hot climates, all efforts should be made to reduce hazardous exposures through engineering controls.

MEASUREMENT

Effective monitoring/warning programs are important to the proper functioning and maintenance of the total control system. Ambient air sampling around the plant perimeter should be an integral part of the monitoring program. Although the concentration of an airborne contaminant is a fundamental factor for the assessment of occupational exposures, it is not the only one, since the uptake of any harmful agent is affected by other factors such as the time of exposure, routes of entry, and physical activity of workers.

Preplacement health examinations identify health conditions which may make the worker specifically susceptible to the hazardous substances or conditions of work, or which may prevent the worker from using the required personal protective devices. Periodic health examinations identify new health conditions which might have developed since preplacement examination or the previous periodic examination. The frequency in which these examinations occur depends upon the level of exposure, health conditions, and results of preceding examinations. Usually, 1-year intervals are considered as being satisfactory.

STORAGE

Storage areas should have a firm, impermeable base such as concrete surrounded by a containment sill. Asphalt is not recommended because it softens in hot climates and under the influence of certain solvents.

In general, all pesticides should be stored in such a way that there is always sufficient space for fire-fighting access. Storage on pallets is recommended. For a variety of reasons, it is desirable to segregate products and ingredients within the warehouse. Storage tanks must be located in an impermeable, walled area to contain a volume of liquid at least as great as that contained by the largest of the tanks.

DISPOSAL

In the disposal of waste products, the plant must comply with national and local legislation. Frequently, controlled incineration in an incinerator built for the specific purpose is used. However, the fumes and ash resulting from this incineration should not create additional hazards. If an effective incinerator is not available, waste can be disposed of in a special landfill site approved for this purpose. Disposal by deep burial can only be done where surface or groundwater cannot be contaminated. Records must be kept of disposals and the location of burial sites by those responsible for the formulation plant. A copy should be provided to the appropriate governmental office.

SUMMARY

It is possible to prevent safety and health problems in the formulation of agricultural chemicals. The first consideration is to design the production facility and to locate it based upon consideration of known safety and health principles. These principles recognize that an effective control system considers mechanisms for control at the source of emissions, in the workplace, and at the worker. It is important to monitor the normal operation to detect any malfunctions as they occur in order that corrective actions can be initiated. The proper education and training of both supervisory management and production employees are essential for the safe handling and disposal of agricultural chemicals.

REFERENCES

1. U.S. EPA, Recognition and Management of Pesticide Poisonings, 3rd ed., U.S. Environmental Protection Agency, Washington, D.C., January 1982.
2. International Labour Office, *Guide to Health and Hygiene in Agricultural Work,* 1979.
3. Safe Use of Pesticides, WHO Technical Rep. Ser., No. 634, World Health Organization, Geneva, 1979.
4. Health Hazard Evaluation Control in the Pesticide Formulating and Manufacturing Industry, Symp. Proc., National Institute for Occupational Safety and Health, Washington, D.C., April 1981.

Protection Standards for Farmworkers in Texas — Protecting Farmworkers from Pesticides

R. D. White, V. Kozak, and R. Mulder

The Texas Department of Agriculture (TDA) has had the authority to adopt regulations to protect farmworkers and all Texans from being unnecessarily exposed to pesticides. Not until Jim Hightower took office in January 1983, however, did TDA finally decide to do something.

TDA recognizes that in order to maintain our system of relying on the few to feed the many, we must have the considerable aid of chemicals to get the job done. While these chemicals provide many benefits when properly used, they are also capable of producing disastrous results when used improperly. Our goal is to gain the benefits of the chemicals, without exposing human life and wildlife to potentially serious consequences.

After a lengthy public hearing schedule involving input from thousands of Texas citizens all across the state, TDA adopted new Pesticide Health and Safety Standards in January 1985. Opposition groups, including chemical company lobbyists and the lobbyists for the Texas Farm Bureau, immediately criticized the new standards, claiming they were unworkable and unenforceable.

TDA embarked on a public awareness campaign in an attempt to clear up the considerable amount of misinformation that was being spread about the new regulations. We sent TDA officials to meet with every group or organization in the state wanting to get more information about the new standards. We also vowed to continue listening to valid concerns about how the standards were being implemented and agreed to amend the standards where any improvements could be made.

The new standards made several significant improvements. First, they allowed neighbors living immediately next to a field being sprayed by an airplane to be notified before the spraying takes place. Neighbors do *not* have the right to stop the spraying, but they do have the opportunity to take whatever precautionary measures they choose.

The new standards also establish reentry intervals for those workers who are likely to have substantial contact with the pesticide residues while engaging in labor-intensive activities. In addition, the standards give uniform procedures to follow concerning the warnings which must be given to farmworkers and specific information on medical symptoms to watch for and what actions to take if those medical symptoms appear.

The "prior notification" section gives adjoining neighbors and some others the right to know in advance when a spraying is scheduled to take place, and pertains to aerial applications only. While that may appear to be discriminatory against aerial applicators, it has been TDA's experience that most of the problems with pesticides landing off-target involve aerial applications.

The ultimate responsibiilty for giving prior notice belongs to the farm operator, that is, the person in charge of the crop in the field being sprayed. The requesting neighbor must give to the farm operator his or her name, address, home and business phone numbers, and hours available, and the location of the field being sprayed. The request can be limited to certain pesticides, for instance, only defoliants or only organophosphates. The request for prior notification must be put in writing and preferably sent by registered mail.

The neighbor's property must be immediately adjacent to the field being sprayed, and, in addition, the neighbor must live or work in a structure that is no more than 1/4 mi from the field to be sprayed. Certain institutions can also request prior notification. Nursing homes, schools when children are present, hospitals, in-patient clinics, and day care centers if located within 1/4 mi of the field to be sprayed by an airplane can ask to be notified before the spraying starts. Note that while these institutions do have to be within 1/4 mi of the field being sprayed, they do not have to be immediately adjacent to the field. Farm labor camps licensed by the State Health Department or located on property owned by the farm operator are to be notified automatically.

Some people have medical conditions which can be made worse by exposure to pesticides. These medically affected persons can also request prior notification. However, their medical condition must be verified by a licensed physician and then certified by TDA. If for some reason their request is denied, TDA must state in writing why the individual in question does not meet the requirements. The medically affected must live within 1/4 mi of the field to be sprayed by an airplane, but, again, they do not have to live immediately adjacent to the field.

There are a number of ways in which prior notification can be accomplished. The easiest is by telephone. A personal visit, a post card or warning flags can also be used. However, warning flags are not to be used in notifying the medically affected.

Prior notification is to be given at least the day before the planned spraying. Emergency applications can also be made, but the attempt to notify the medically affected must be made at least 1 h before the spraying is scheduled to begin. Warning flags, generally designed to notify a large number of neighbors requesting prior notification, should also be used during emergency applications.

The warning flags must be 2 ft above the crop or 5 ft above the ground, whichever is higher, and they must be placed within 650 yd of each neighbor requesting prior notification. In the event of unusally tall crops, such as corn or sugar cane, the flags can be located at a specific predetermined position, which can be more than 650 yd away provided they are still visible. The flags should be placed in the fields within 24 h of the spraying and must be removed within 48 h.

The telephone number of the farm operator must be on or near the flag. The farm operator, if asked, must also be able to provide the name of the pesticide and the intended date and approximate time of the scheduled application. The flag must be at least 18 in. wide and 24 in. high. In addition, the universal symbol is at least 18 × 24 in. and consists of a 12-in.-diameter red circle with a red diagonal bar 1-in. in width placed from the 2 o'clock position to the 8 o'clock position across the circle. The words "DANGER-PESTICIDES" shall appear in black letters at the top of the flag or sign and the words "PELIGRO-PESTICIDAS" shall be at the bottom in lettering at least 1 1/2 in. high (Figure 1).

If the farm operator has difficulty reaching an individual who requests prior notification, he or she is encouraged to contact the nearest TDA district office or call the toll-free number 1-800-TDA-REGS. TDA employees will then attempt to contact the medically affected persons and record the fact that an effort to comply with the standards was made.

There are some changes in this section that went into effect in August 1985. Under the original proposal, schools, nursing homes, day care centres, schools when children are present, hospitals, and farm labor camps were to be notified automatically, without requesting it, if they were located within 1/4 mi

FIGURE 1. Universal symbol.

of the field. Farm labor camps are still notified automatically, but the other institutions are now in the same category as all other neighbors. However, if a school administrator refuses to seek prior notice, TDA can require it if a parent has, with good cause, made that request.

One final note on prior notification. Each request automatically expires at the end of each year. Those individuals who wish to keep receiving notification must renew their requests after January 1 of each year.

The "reentry intervals and labor-intensive activities" section of the new regulations is aimed directly at protecting the farmworkers in the field as well anyone else who is likely to have substantial contact with the pesticide residues. In addition, this section applies to all pesticides, no matter whether they are applied by airplane or with ground rigs. The key is whether or not the individual is likely to have substantial contact with the pesticide residues while engaging in labor-intensive activities, regardless of how those residues arrived.

TDA requires the farm operator to find out the reentry interval for each pesticide. Some of the information is conflicting. Therefore, TDA has decided that in those instances in which the Environmental Protection Agency (EPA) and TDA reentry intervals are not consistent, whichever reentry interval is longer shall prevail.

Both TDA and EPA have, for a number of years, set appropriate reentry intervals for specific pesticides on a crop-by-crop basis. A national minimum reentry interval was established in 1974 to prohibit all workers from entering a field until the pesticide spray has dried or the pesticide dust has settled, unless the workers wear appropriate protective clothing. Other reentry intervals have also been established, and TDA has the responsibility to establish certain reentry intervals as a part of the new standards.

Under the revised standards which went into effect in August 1985, TDA established a 24-h reentry interval for those chemicals which have not met certain minimum testing requirements established by EPA (Table 1). In July 1988, in a dramatic departure from their long standing one-product-at-a-time approach to regulation, EPA proposed regulations based on this concept. In general, the test data include the toxicity of the chemicals and their chronic effects when encountered by farmworkers in labor-intensive activities. This procedure is aimed at encouraging the chemical companies to get their test data in and at making sure the individuals working with those chemicals which have not been adequately tested are protected.

A 24-h reentry period is now in effect for those pesticides with the signal word "DANGER" printed on the label. TDA has assigned only one pesticide a 7-d reentry interval, and that is ethyl parathion. A few others have been assigned 48-h reentry intervals.

The new standards also give TDA the authority to set Texas-specific reentry intervals. This is necessary since the standards established by EPA may or may not be appropriate for use in the Texas climate.

Under this program, TDA will first survey every chemical

TABLE 1
Reentry Intervals

24-h reentry	48-h reentry Chemical (trade name)	7-d reentry
Whenever "danger" is printed on pesticide label	Azinphos-methyl (Guthion)	Ethyl parathion
Applies to toxicity category 1 formulations only	Carbophenothion (Trithion)	
	Demeton (Systox)	
	Dicrotophos (Bidrin)	
	Disulfoton (Di-Syston 8)	
Also includes EPN and Phosalone under federal EPA regulations	Endosulfan (Thiodan)	
	Endrin (Endrex)	
	Ethion (Ethodan)	
	Methidathion (Supracide)	
All pesticides with registered agricultural uses for labor-intensive crops for which testing has not been completed (effective 11/1/85)	Mevinphos (Phosdrin)	
	Monocrotophos (Azodrin)	
	Oxydemeton-methyl (Metasystox-R)	
	Phorate (Thimet)	
	Phosphamidon (Dimecron)	

used in labor-intensive crops. The EPA-established level of toxicity of each compound in the pesticide will be determined, and then those data will be compared with figures representative of Texas. Because the pesticide residues decompose at different rates depending on the chemical makeup of the pesticide and various weather factors, including the amount of sunlight, temperature, rainfall, humidity, and dust in the air, new reentry intervals may have to be established.

Whether or not the residues endanger workers depends on how much residue gets on the worker, how quickly the worker's body absorbs it through the skin or lungs, the toxicity of the pesticide, and the individual's susceptibility. The goal is to come up with a reentry interval that will keep the workers out of the fields until the residues no longer pose a danger, or, if workers must reenter a "hot" field, to make them wear the appropriate protective clothing.

The new standards encourage farm operators to keep the farmworkers out of the field until the reentry interval has expired. If that is done, the farm operator has complied with all of the requirements and does not need to post any warning signs unless the reentry interval exceeds 7 d.

If the farmworkers are sent back into the field before the reentry interval has expired, a number of steps must be taken first. The farmworkers can go back in only if reentry signs have been posted and only if the farmworkers have received a clear warning directing them not to reenter without wearing the appropriate protective clothing. In addition, the farmworkers must be given the trade and chemical names of the pesticide, the date when it is safe to reenter without protective clothing, and specific health information including instructions to wash themselves and their clothing, what medical symptoms to watch for, and what to do if those symptoms appear. This information can be given either in person or in writing, including the use of warning signs to go alongside the reentry signs.

The reentry signs should be in English and Spanish and must be posted at the normal entry and exit points in the field. They must also include the universal symbol and at least two phone numbers where the farm operator and/or applicator can be reached.

TDA is committed to making sure farmworkers are protected. We are especially interested in making sure the farmworkers are not ordered to reenter a treated field before the reentry interval has expired. Any farmworker who believes he or she has been instructed to engage in some activity which violates the new standard is urged to file a complaint with TDA.

Although the farm operator is not required to provide the farmworkers with the apprropiate protective clothing, he or she may choose to, since that is the only way the farmworkers can be sent back into the field before the reentry interval has expired (Table 2). The definition of "appropriate protecting clothing" varies according to the circumstances. The first thing to do is check the label on the pesticide container. In most cases, protective clothing means a minimum of clean, loose-fitting one- or two-piece garment of closely woven or impermeable material, plus shoes, socks, hat, and impermeable gloves.

Examples of impermeable gloves are rubber, vinyl and plastic-coated gloves. Cloth, leather, and paper gloves are not acceptable.

TABLE 2
Appropriate Protective Clothing If Reentry Is To Occur

After reentry interval expires: normal work clothes
More than 24 h after spraying but before reentry period expires: a minimum of clean, loose-fitting one- or two-piece garment of closely woven or impermeable material plus shoes, socks, hat, and impermeable gloves
After the pesticide spray has dried or after the pesticide dust has settled, but less than 24 h after spraying: whatever is called for on the label for the applicator, but pesticide respirator not required
Before the pesticide spray has dried or pesticide dust has settled: whatever is called for on the label for the applicator including a respirator, if specified

TDA continues to live up to its promise to listen to every reasonable proposal and make changes wherever they are needed to improve the new standards. Meanwhile, implementation of the standards has progressed smoothly. A lawsuit was filed in March 1985 to prevent the regulations from taking effect, but the plaintiffs repeatedly requested delays, and in the meantime the new standards were in effect for one crop season with no problems. Subsequently, the regulations were successfully defended in court by the Attorney General of Texas.

Despite claims that the new standards would be unenforceable and unworkable, as well as a costly burden on the producer, those predictions proved to be untrue. TDA maintained from the beginning that we were simply putting in writing common sense practices which good neighbors already follow. Farmworkers and neighbors are being protected, farm operators are continuing to produce with little inconvenience, and all Texans are benefiting from a program that makes Texas a better and safer place to live and work.

The Practitioner's Role in Safety on the Farm

R. Marchant

Farm accident fatalities in Saskatchewan have been on the increase over recent years. In both 1980 and 1981 a total of 11 people were killed as a result of farm accidents. The figure in 1982 was 13; in 1983, 17; and in 1984, 21. The most susceptible age group for all these years was the over-50 group. From a total of 73 people killed in farm accidents over the years 1980—1984, 35 were in this age group. Using age groups with a 10-year span, the age group with the least number of fatalities was the 31—40 age group which had four fatalities. A far higher percentage of men than women are the victims of farm fatalities.

An accident ratio has been calculated by Bird in *Management Guide to Loss Control.*[1] It shows that for every serious or disabling accident, there are 10 minor injuries, 30 incidents of property damage, and 600 near miss accidents.

The recorded hospitalized accidents occurring on the farm in the years 1981 and 1982 were 101 and 104, respectively, for age groups 0 to 14 and 15 to 24 years. These groups were composed of males and females, the males being more susceptible to various types of accidents — poisoning, falls, fire — than females. Farming is generally a family business, so serious accidents occur in all age groups.

Pesticides are responsible for a large number of farm accidents each year and the numbers have been rising steadily since 1978, with a slight decline in 1980 and 1981. In 1978 there were 36 pesticide poisonings in Saskatchewan; in 1983 the total was 148. Only one death occurred (in 1978) in the time frame 1978—1983. The most susceptible age group is the under-4-year-olds, suggesting that better storage and safety procedures could reduce the number of poisonings in this age group.

Chemicals are sprayed in spring, summer, and fall with the majority being used in the summer.

In spring, preseeding spraying takes place, followed by insecticide and fungicide treatments at seeding time. Then a preemergent chemical is applied. This is all done in a very short time period, usually it takes a couple of weeks.

In summer, herbicide as well as more insecticide and fungicide is applied. A desiccant is sprayed on at harvest time.

In the fall, preplant chemicals are applied plus fertilizers.

A survey of Battleford farmers was undertaken to find out what protection they used when applying chemicals on their farmland. Ten farmers were questioned; the average number of herbicides and insecticides used was 9 with the maximum being 14. The majority wore protective suits, hats, and rubber gloves. Less than half of the farmers wore a mask, respirator, or rubber boots. Seven farmers ate on the job. They all washed their hands, took a shower, and changed their clothes after work. Four wore clean clothes daily. Five of the ten farmers had a university education. This level of protection, although not adequate, is considered to be better than the average farmer when spraying and handling pesticides.

Besides chemical agents, farmers are exposed to various physical hazards in the course of their work:

1. Ultraviolet radiation from the sun can cause sunburn and skin cancer. This can be prevented by the use of sunscreen or covering the exposed skin.
2. Noisy farm equipment is a cause of noise-induced hearing loss. Noise can be lessened by the use of the earmuffs, a cab on a tractor, and by keeping equipment in good repair.
3. Vibration from farm equipment can cause back and gastrointestinal problems. A well-sprung seat can help with this problem.
4. Extremes of temperature can cause either frost-bite or heat-stroke and cramps. Appropriate clothing is essential in the Saskatchewan climate, and plenty of water and rest stops are necessary in hot weather. An air-conditioned tractor is helpful in this regard.
5. Accidents with such items as hydraulic air radiators or super cold anhydrous ammonia can be prevented by using the necessary safety rules and equipment, or hiring a contractor.

Accidents caused by biological hazards can be decreased by providing farmers with information on zoonoses, stressing the importance of cleanliness after working with animals, e.g., washing hands and not eating, drinking, or smoking in animal-confinement areas, and wearing a face mask where dust is generated. Animal waste should be disposed of well away from dwellings.

Accidents caused by ergonomic factors may involve improperly designed tools or work area, improper lifting or reaching, poor visual conditions, or repeated motions in an awkward position. Some equipment has easily removable safety guards or no provision for the engine to switch off when the equipment is not in use.

Most farms have large amounts of mechanical equipment, sometimes as many as 25 pieces of heavy equipment, e.g., tractors, combines, grain augers, balers. Most maintenance is done by the farmer once the warranty has run out. A major problem is that many farmers remove safety guards as soon as they purchase equipment.

TABLE 1
Farm Fatality Statistics by Sex

Sex	1979	1980	1981	1982	1983	1984
Male	24	7	14	13	17	18
Female	1	4	0	0	0	3

The practitioner's role in farm safety is one of counseling to work towards lessening the number of accidents and fatalities on the farm. The five basic principles of industrial hygiene may be used as a framework to encourage prevention rather than treatment.

In order of priority, the five principles are

1. Elimination, e.g., getting rid of dangerous equipment
2. Substitution, e.g., substituting equipment or practices which have caused accidents in the past
3. Isolation, e.g., wearing appropriate clothing for weather conditions
4. Ventilation, eg., in silos and pig barns
5. Personal protective equipment

By working together, the practitioner and the farmer can minimize accidents on the farm and make it a safer place for all age groups.

REFERENCES

1. **Bird, F.E.,** *Management Guide to Loss Control,* Institute Press, Atlanta, 1974.
2. Poison Control Center Statistics, Statistics Canada, 1985.

Health and Safety in Agriculture — The Farmer's Role in Safety on the Farm

S. M. Leier

Before I could begin to address the topic of the farmer's role in safety, I went to *Webster's Dictionary*[1] to see who he defined as a farmer. He says a farmer is "a person who pays a fixed sum for some privilege or source of income", "a person who cultivates land, crops or raises livestock", or "a yokel or a bumpkin". Well, for my purposes I will disregard Webster's attempt at humor and deal with the person who cultivates the land or raises livestock.

In Saskatchewan we have approximately 64,000 farmers cultivating approximately 51 million acres of land.

The environment in which the farmer works, often up to 20 h a day, abounds with machinery, livestock, tools, and chemical and electrical hazards. In order to live and work safely it is essential that he become aware of the dangers of his environment.

The farmer is what he is because his farm is his land, his home, his workplace. He is the manager, the technician, worker, and person who, along with his family, makes it all work.

He has no overseer. The government or a board of directors does not tell him how to manage and operate his farm. He is his own boss, but along with this privilege is the fact that he often works long hours alone.

The majority of farmers are regarded as conservative-minded, but the nature of their business compels them to take high risks. They are gamblers!

I have learned from discussing safety with farmers that there are two main roles in which they perceive themselves. The first is to be a safe practitioner, one who puts into practice the safety rules and guidelines set out by the educators, implement manufacturers, and chemical dealers. The second is that of an educator. It should be his responsibility to teach his family and any others in his employ the safe practice principles.

Most farmers spend so much time on their equipment that they feel confident to deal with almost any situation. Why, then, did we have 24 farm fatalities in Saskatchewan in 1984?

As in other walks of life, we see the total range of personalities among farmers. At one end of the spectrum is the patient, meticulous type and at the other is the "gotta go Joe", impatient, have-to-have-it-done-yesterday type. In combination with these personality types, we have the main factor that jeopardizes rural safety. It is the critical time factor. Certain elements of farming must be completed within a very tight time frame. This is again complicated by other influencing factors. For example, spraying time depends upon the maturity of the crop, the maturity of the pest, and the right weather and wind conditions. All of these factors must occur simultaneously to ensure an effective and efficient operation. In the fall, if we are fortunate enough to get crop maturity and good weather conditions occurring at the same time, then harvest is full speed ahead. This time "squeeze" multiplies the safety risk greatly.

If a union worker does not finish his quota or assigned task at the end of his shift, then another worker on the following shift will complete the job or he can pick up where he left off on the next assigned day. The worker still gets paid. If a farmer doesn't get his crop sprayed or harvested or his cattle to market at the desired time, he suffers the monetary loss. If this happens with regularity, then the security of his job, home, and ultimately his family is threatened.

The farmer must finance his own way, thus the quality of his equipment will depend on his past monetary success. Any personal protective or safety equipment will depend on his commitment to safety. If he has just come through a couple of tough years, as is the case with southern Saskatchewan farmers right now, he may be looking to where he can cut corners. This unfortunately may result in not purchasing those safety goggles, chemical gloves, or a respirator.

A second obstacle to the farmer fulfilling his role in safety is the lack of safety education opportunities. Again, there is a comparison to industry. In most task-oriented jobs the worker is initially walked through the procedures by a skilled supervisor. Hazards of the job and safe operational procedures are outlined. Emphasis is placed on safety equipment which is provided by the employer and often its use is a condition of employment. The farmer has the onus directly on himself for studying the operator's manuals and using the provided safeguards, for devices such as shields, cages, etc. Whether or not he uses safety equipment for a certain task will depend entirely on him. He is his own Occupational Health Committee, supervisor, and employee. Very few people will know or care if he operates outside of the safety guidelines, unless he becomes an accident statistic or, more drastically, a fatality.

Prior to 1970 not a great deal of thought or too much action was given to agricultural health and safety in Canada. As the number of tragic accidents increased and were reported by the news media, farmers themselves began giving more thought to their health and safety. In 1973 a group of Ontario farmers formed the Ontario Farm Safety Organization so that they and their employees could have accident coverage by the Workers Compensation Board. They also wished to actively promote safety to the agricultural industry and reduce the rural accident rate.

Today, several provinces have active farm-safety organizations. Alberta, Ontario, and P.E.I. are three such provinces with full-time staff members providing a variety of educational programs.

Saskatchewan does not have such an organization. Despite a 5-year effort of the Saskatchewan Safety Council to obtain funds from various departments of the provincial government to develop a farm-safety program, no such funds or programs have been forthcoming. As an interim measure the Safety Council has produced a nine-segment farm-safety video program. Copies of this video have been distributed to all of the Saskatchewan Agricultural Representatives and can be borrowed for family viewing, free of charge.

In 1983 the Saskatchewan Safety Council conducted a questionnaire survey at the Canadian Western Agribution in Regina. This survey was to establish if the rural community saw a need for a farm-safety program. All of the completed forms indicated a desire by these people to have such a program in their communities. The biggest response came from the women.

It is not surprising that farm women would show more concern about safety than their husbands. The farm woman is, for the most part, the support person, the homemaker. She is generally not part of the actual machine operation or maintenance and therefore has little control over accidents. She is often the first on the accident scene. She may have to shut off equipment, administer emergency first aid, and make decisions related to a life-and-death situation. It is therefore vital that we develop programs to train farm women to deal with emergency situations.

The Ontario Farm Safety Organization has designed and offers such a program for farm women. "What to do" in case of an accident is the central theme of the safety workshops. They offer safety films, slides, discussions, and hands-on training sessions. As a result of the publicity about the first workshop, the association has received requests from all over North America.

Our large geographical area and scattered population in Saskatchewan makes it difficult for the farmer to access safety programs. The Wascana Institute, Agricultural Division in Regina is offering 1-d farm-safety and family-safety programs through its Outreach program. This winter I will be traveling for several weeks conducting workshops on various topics. The topics will be determined by the requests of the participants. Suggestions have been farm-implement safety, safe handling of chemicals, stress, and emergency first aid. To date we have 20 requests for these workshops.

If our children are to grow up learning how to live and work safely, it is necessary that they become aware of the danger of their environment. Nowhere is that more evident than on the farm. Everyday the health and safety of rural children are threatened by machinery, livestock, tools, chemicals, and electricity. Most accidents happen while children are playing in the farmyard, particularly in spring and summer. Over one half of these accidents have severe or permanent consequences. The majority of the children are injured by farm machinery and nearly all others are injured by livestock. However, over 80% of fatalities involve machinery.

During 1984 in Saskatchewan four children between 1 and 10 years of age died in farm accidents. Two were run over by tractors driven by relatives. A 2-year-old fell out of a truck box and was run over by a truck or trailer following it. A 7-year-old boy was playing alone on a rock picker and fell off into the sprockets.

Each year numbers change a little, but basically the stories are the same. All of these accidents are preventable, but the success lies in each adult and child knowing the hazards on the farm and doing his best to prevent these accidents.

The statistics illustrate dramatically the need to educate rural children in farm safety. As a supplement to the teaching done by parents, Alberta, Ontario, and P.E.I. have farm-safety programs in the rural schools for grades 1 to 6. These programs include safety films, coloring books, games, and poster contests. These are designed to aid students in identifying hazards on their farm and in discussing farm safety. These provinces are to be commended for their efforts and hopefully at the next gathering such as this, more provinces will have followed their lead.

Despite the majority of farmers being experienced, knowledgeable of their equipment, and concerned about the safety of their family, even the most experienced operator may become an accident statistic.

According to Tyler, professor of psychology at Brandon University, "one of the problems in safe behavior is that of developing the individual's understanding of the probabilities for unpleasant and undesirable outcomes in a situation, before they move into that situation. Knowing this, an individual's skill can then be developed or adjusted to the realities of the situation and the situations' potential for undesirable results could be eliminated."[4]

These abilities, namely, the ability to identify the dangerous aspects of particular situations and the ability to develop skills appropriate for dealing with these, are acquired or learned and are basic to safe behavior.

The role of repetition in learning is clear. The more a behavior is repeated in the presence of specific cues, the more readily that behavior will occur in conjunction with these cues, and the less an individual will be aware of his response. In short, that behavior becomes unconscious and the individual functions consistently in a specific situation without being significantly aware of it. A practical example of this would be seat belts. A student taking driver training will learn during class that part of the safe-driving procedure is to buckle up upon getting into the vehicle. When doing the practical part of the training, a warning buzzer will sound or the instructor will remind the student to buckle up. After driving for a while, the driver who has learned under these conditions will automatically buckle up upon entering a vehicle even if that vehicle does not have a prompting buzzer.

How do we get farmers to take training programs that

would develop safe behavior patterns in the operation of farm machines? Possibly one way would be to have farm accidents dramatized by the media. TV and poster campaigns would be especially useful, followed by information programs available throughout the province. It may be necessary to use tragedies to make positive gains. This approach has been very successful with the Safety Grad program, Seat Belt program, and Drinking and Driving program.

The human factor can be implicated in approximately 90% of all machine-related fatalities. One of the major factors is indifference. This happens when individuals possess the information and skills basic to safe behavior, but choose to behave dangerously. Factors such as anxiety, stress, worry, fatigue, hunger, medication, and alcohol all constitute a state of arousal, reducing an individual's capacity to recognize dangerous situations; for example, a quarrel with your wife will reduce your alertness.

There is very little effort made to communicate to the farmer what is known about the effects of arousal and his safe behavior. This lack of effort results from a lack of safety promotion with the individual farmer.

Public awareness regarding farm stress was initiated this year by a group of high school drama students. Based on a survey of farmers conducted by the Mental Health Association of North Battleford, Saskatchewan, these students wrote and produced a play which deals with the stress of financial and marital problems of a young farm couple. It ends with the death of the husband in a suicide cum farm accident.

The play stresses the importance of recognizing signs of stress and not being too proud to seek outside help.

The old phrase that "farming is not just a job, it is a life-style" is true. It is not always a safe and healthy life-style, but there is a growing desire to make it safer. Desire is not enough, though; there must be people willing to organize, promote, and lobby government and organizations for funding. Training in farm safety must be made available and accessible to farm people. As the theme from several Alberta Farm Safety brochures states, "Cultivate Safety Harvest Life".

REFERENCES

1. *Webster's New Collegiate Dictionary,* Thomas Allen and Son, Toronto, Ontario.
2. *Farmsafe,* 10, 4, 1985.
3. Coroner's Records, Farm Fatalities, Government of Saskatchewan, Saskatchewan, 1984.
4. **Tyler, E.,** Speech Canada Safety 10th Annu. Conf., Winnipeg, Manitoba, October 1978.
5. Alberta Agriculture, *Teacher's Guide to Farm Safety.*
6. Youth learn about stress first hand, *West. Prod.,* July 11, 1985.
7. Alberta Agriculture, Farm Safety Branch, *Farm Employee Safety Series.*

Compensation Programs for Farmers

G. F. Elliott

Workers' Compensation in Canada is a no-fault work injury insurance scheme financed completely by assessments from eligible employers. It is a provincial responsibility, so that ten provinces and two territories in Canada have separate Workers' Compensation Acts administered by independent boards. The 12 acts are similar but not identical. Inasmuch as Workers' Compensation legislation was originally designed to compensate for accidental injury in heavy industry, many occupations, including agriculture, were originally excluded. With the passage of time, more and more occupations have come under the umbrella of Workers' Compensaton legislation; however, agriculture has been one of the last.

In the province of Saskatchewan, agricultural employees have been covered on application of the employer since 1971, and all farmers have been covered on application since 1973; however, even now only 1000 out of 65,000 to 70,000 farmers are covered.

Table 1 shows statistics for the whole of Canada.

When I tried to acquire equivalent statistics from each of the Workers' Compensation Boards (WCBs), I ran into several problems. First, each jurisdiction places occupations in groups with similar accident experience so that different agricultural enterprises are placed in occupation groups that contain other industries, and it is not possible to separate the statistics. There is no uniformity in the types of statistics that are developed by the several boards. Variations in legislation affect the statistics; for example, Alberta does not require no-time-loss claims to be reported. For these reasons, the information that I have shown on Table 1 is not completely accurate and, as you can see, much is missing.

The reason for the delay in bringing the agricultural industry under Workers' Compensation is found in the nature of the industry. The typical Saskatchewan farm is owned and operated by a single family which employs no full-time help and only rarely employs part-time help. All members of the family, both sexes, contribute to the farm work from childhood on.

For compensation purposes, employers are assessed on a percentage of payroll. Farmers often wait until the end of the year in order to learn whether or not they can pay themselves and their families any wages. They are understandably reluctant to commit themselves to pay a cash assessment in advance of knowing the bottom line of the balance sheet. In this regard, considerable thought and effort is being expended in an attempt to create a unique method of assessment that will ensure coverage for farmers, and at the same time be actuarily sound.

The topic that I was assigned is "Compensation Programs

TABLE 1
Summary of WCB Coverage by Province[a]

	B.C.	Alta.	Sask.	Man.	Ont.	Que.	N.B.	N.S.	P.E.I.	Nfld.	Y.T.
Voluntary coverage	Farmer	Yes	Yes	Yes	Farmer		Farmer	Yes		Farmer	Farmer
Compulsory coverage	Worker	No	No	No	Worker		Worker	No		Worker	Worker
No. of farms	?	56,700	67,000	?	?		?	?			?
No. covered	?	500	1,092	94	23,000		?	8%		80%	?
Lower age limit	10 years	No	No	No			16	No		No	No
Assess. rates (%)	3	7.15	3	3.31	3.65		2.2	2		1.9	3
				1.39	5.47						
					5.07						
Annual claims	623	449	163	176	5,696		?	166		?	0
No time loss	367	0	47	66	2,168		?	86		?	0
Time loss	599	449	111	110	3,409		?	76		?	0
Fatalities	1	0	1	0	12		?	0		?	0
Permanent impairment	17	11	4	5	183		?	0		?	0
Average rate PI	?	?	10%	?	?		?	0		?	0
Average lost days	40	?	36.7	37.9	44		?	0		?	0
Occupational disease	28	1	0	3	2.9%		?	?		?	0

[a]Direct communication

Note: Abbreviations: B.C. — British Columbia, Alta. — Alberta, Sask. —Saskatchewan, Man. — Manitoba, Ont. — Ontario, Que. — Quebec, N.B. — New Brunswick, N.S. — Nova Scotia, P.E.I. — Prince Edward Island, Nfld. — Newfoundland, and Y.T. — Yukon Territory.

for Farmers". I must, therefore, outline what compensation is available to farmers. In this I will restrict myself to outlining what is available under Saskatchewan legislation which, as I have mentioned, is similar to, but not identical to legislation in other parts of Canada.

At present, farmers in Saskatchewan pay an assessment of 3% of payroll for employees or 3% of the amount of coverage that they elect for themselves. In return for this, they are covered for any injury arising out of and in the course of their employment. Injury in this context includes any disease arising out of the occupation.

Coverage includes medical treatment, hospital costs, drugs and appliances, as well as travel costs in certain circumstances when they must travel away from the home community for diagnosis or treatment. In addition, while totally disabled from work, they receive wage replacement at the rate of 90% of net income. Net income is gross income minus unemployment insurance and Canada pension premiums and estimated income tax. Maximum gross income coverage is $48,000/year at present.

In order to ensure maximum recovery in the optimum time, a physical rehabilitation program is available for those who require it.

In the eventuality that an injured worker is anticipated to have an employment problem due to work restrictions from work injury, he is entitled to vocational rehabilitation assistance. This usually involves a period of receiving job search allowance and might extend to retraining.

If, because of the work injury, the worker suffers an ongoing wage loss, he is entitled to wage-loss compensation at the rate of 90% of the difference between preaccident earnings and current earnings, until age 65. Anyone who receives benefits for 24 months has a sum, equal to 10% of the benefits paid, placed into an annuity fund which will be available at age 65. This is meant to compensate for any decrease of pension plan contribution due to the work injury.

Anyone who is left, after a reasonable recovery period, with measurable, permanent, functional impairment as a result of work injury will be assessed and will receive a single cash payment representing a percentage of $20,000 but not less than $1,000.

For the employers, there is an additional benefit in that if they provide Workers' Compensation coverage, injured employees may not sue them for the effects of work injury.

Table 2 shows the areas of injury which were reported by the Saskatchewan farmers in 1984.[2]

You will note that the areas subjected to injury are hands, arms, legs, feet, back, and head, which are subject to trauma with few or no claims due to chemical or other environmental damage. There were no claims for hearing loss.

We know that acoustic trauma is a major hazard on mechanized farms. We know that farmers handle large volumes of toxic chemicals. We know that their respiratory tracts are exposed to dusts of many kinds such as pollens, molds, etc. Why were there no claims for hearing loss, chemical toxicity, cardiac or respiratory disease, or dermatitis?

TABLE 2
Saskatchewan Agricultural Workers' Claims Settled in 1984 by Area of Injury[a]

Area of injury	Claims	Area of injury	Claims
Head	9	Pelvis	1
Eyes	9	Upper leg	1
Neck	6	Knee, lower leg, and ankle	31
Shoulder	8	Foot and toes	13
Upper arm	1	Cardiovascular and respiratory	—
Elbow, lower arm, and wrist	10		
Hand and fingers	38	Dermatitis and other industrial diseases	—
Dorsal spine	4		
Thorax	3	Hearing	—
Lumbar spine	24	Multiple injuries	1
Abdomen including herniae	4		
		Total	163

[a] Saskatchewan Workers' Compensation Board Annual Report 1984.

The answer is, undoubtedly, multifaceted. It is certainly possible that farmers and their attending physicians are not recognizing all cases of chemical toxicity. It is possible that attending physicians do not expect farmers to have Workers' Compensation coverage and, therefore, do not advise them to claim for compensation. It is certainly true that farmers are reluctant to stop working for any reason and, therefore, it is possible that they do not claim for nondisabling injuries.

Regardless of why farmers have not been claiming for environmental disease, the main reason that farmers are not compensated for work injury is due to their failure to avail themselves of Workers' Compensation coverage by applying for it in advance.

Farmers are perpetual optimists. They always expect next year to be better and they don't expect work injury to happen to them. In spite of this, those who have studied the incidence of work injury and the cost of insurance are convinced that Workers' Compensation is the least expensive insurance possible and is a good investment for farmers. When you consider that Workers' Compensation claim costs can run to hundreds of thousands of dollars per injury and lawsuits by unprotected workers could cost a farmer hundreds of thousands of dollars, can farmers afford not to seek Workers' Compensation coverage?

In order to correct the present lack of farm coverage, two things are required. First, there is need for ongoing educational programs to make farmers aware of the availability and value of Workers' Compensation coverage. Second, efforts must continue in an attempt to work out a unique, actuarily sound, alternate method of assessment of farmers so that farm coverage can become compulsory, as it is in most industries. Prestigious organizations such as the Saskatchewan Federation of Agriculture and the Saskatchewan Association of Rural Municipalities have come out in favor of universal coverage for farmers.

Workers' Compensation is not a social welfare scheme that dispenses money from general taxation. Whatever method of finance is arrived at must be actuarily sound and must be acceptable to farmers.

While it is not my purpose to burden you with WCB administrative problems, you will appreciate the difficulties that would be encountered if farm coverage were compulsory. The problem of predicting income has been mentioned. In addition, who is a farm worker? A 5-year-old who gathers eggs? Can you realistically and fairly apply a lower age limit for farm workers? When is a farmer at work and when is he not at work? He goes to town for groceries and purchases repairs on the same trip. If he drops into the bar to see a neighbor about sharing work, is he still in employment? Even if a grain farmer has been injured during his work season, is he suffering a wage loss while he is toasting his toes in a warm climate during the winter?

In order to arrive at actuarily sound assessment methods, there must be valid statistics. Inasmuch as WCB does not have these, some method must be worked out for recording accurately the incidence of trauma and environmental disease in farmers. We hope that the Occupational Health and Safety Branch of the Department of Labour, with the aid and cooperation of the Department of Agriculture and Department of Health, will be able to fill this gap.

Considering the vagaries of farm income, which is affected by weather, pestilence, and world markets, will there ever be an actuarily sound assessment method at rates that farmers can, and will, pay unless there is a potential for subsidy? If there is subsidy, where should it come from? Should it come from industrial employers through WCB, through grants from municipal, provincial or federal governments, or from agriculture itself through checkoffs on agricultural product sales? I do not perceive an early resolution of all of the problems involved in extending Workers' Compensation coverage to all agricultural workers, but I am convinced that efforts must continue.

REFERENCES

1. Saskatchewan Workers' Compensation Acts, 1979, and amendments, Available Queen's Printer, Regina, Saskatchewan, 1979.
2. Saskatchewan Workers' Compensation Board Annual Report 1984, Available Workers' Compensation Board, Regina, Saskatchewan, 1984.

Injuries in Agriculture and Related Industries

Injuries on Dairy Farms in Otsego County, NY: Results of Active Surveillance for 1 Year in an Established Cohort

D. S. Pratt, L. Stallones, D. Darrow, and J. J. May

INTRODUCTION

Agriculture enjoys the dubious distinction of being one of the most dangerous major occupations. The U.S. National Safety Council estimates that in 1980 agriculture was the most dangerous work, followed by mining/quarrying and construction.[1] The evaluation of injuries sustained in farming is often collected from death certificates, emergency room and hospital retrospective records, or from passive reporting of mishaps.[2-4] This tends to bias the data toward more severe events. Our clinical experience with dairy farmers in Otsego County, NY, over the last 5 years suggested that many injuries do not go to hospitals or emergency rooms, that few are fatal, and most are ignored by these stoic people.

Because passive surveillance systems seldom produce an accurate picture of health phenomena, we chose to actively pursue the occurrence of injuries in dairy farms in our area.[5,6] We hoped that by following a single group of farms for a prolonged period we could get a more accurate picture of the incidence and severity of dairy farm related injuries.

METHODS

Otsego County is located in rural, upstate New York. It has a population of 59,000; the major industry is dairy farming with over 540 farms listed by the New York State Department of Agriculture and Markets. Our hospital is located in Otsego County and affords us an excellent opportunity to work with these farmers.

In 1982, with the cooperation of the New York State Department of Agriculture and Markets, we contacted by letter every farm in Otsego County. We requested that farmers reply to our letter if they were interested in helping us learn more about the health and safety of dairy farming. Ultimately, 475 farmers representing 220 farms agreed to a telephone surveillance study of injuries and hazards on the farm. We chose this method because results of telephone surveys compared to mailings appear quite similar and are much less cumbersome.[7] Each farm family gave consent by phone and in writing. The 220 self-selected farms were compared to the entire county and found to have a similar distribution of livestock (Table 1). The herd similarly suggested that our sample was likely to be representative of the entire county.

Beginning in the late spring of 1984, each of the 220 farms was called every other month. The same person made all the calls using a rehearsed format and a battery of follow-up questions when and if an injury had occurred. A total of 1314 phone contacts were made with the cohort. Calls continued through 1 year. Data were stored on computer for analysis later. All injuries were self-defined. No attempt was made to prompt reporting. Quite often the calls were anticipated and the information ready.

RESULTS

Of the 220 farms which began in the study, only 1 left during the year. Complete data were available on 219 farms for 12 months. There were 140 injuries on 93 farms. The injuries represented an attack rate of 42 injuries per 100 farms per year or 42%. There were no injuries on 126 farms (58%), one injury on 61 farms (28%), two on 23 farms (11%), three in 5 locations (1.4%), and four and five on 2 farms each (1.0%). There was no difference in size of farms where accidents occurred and where they did not.

Injuries occurred every month of the year. The spring had both the lowest number (April, 3) and the highest (May, 20). Nearly 50% of all injuries occurred in the middle 1/3 of the year (May to August). Figure 1 shows this distribution and draws attention to the perils in May.

Injuries occurred every day of the week, but weekends had the most injuries. The day of the week of the recorded injury was definite in only 32% of cases. Thus, our sample is quite small and subject to question. Nonetheless, almost four times as many injuries were reported for Saturday (11) as for Wednesday (3). This is shown in Figure 2. The recall for the time of day (morning vs. afternoon) was good (greater than 95%). Of the injuries we recorded, 60% occurred in the afternoon and 40% in the morning. The injuries we uncovered were related to mechanical equipment 60% of the time, to

TABLE 1
Validating the Sample Surveyed

Number of Cows milked	Cohort 1982 (%)	AG and market survey 1978 (%)
10—19	1	5
20—49	56	58
50—99	39	33
100—199	4	4
200+	0	0
Total	100	100

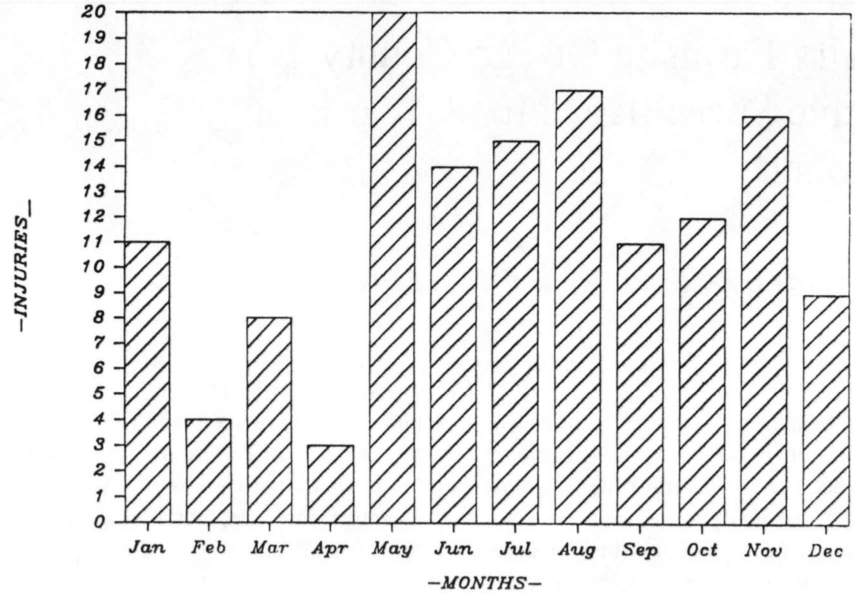

FIGURE 1. Distribution of injuries by month. Otsego County, NY, 1984 to 1985.

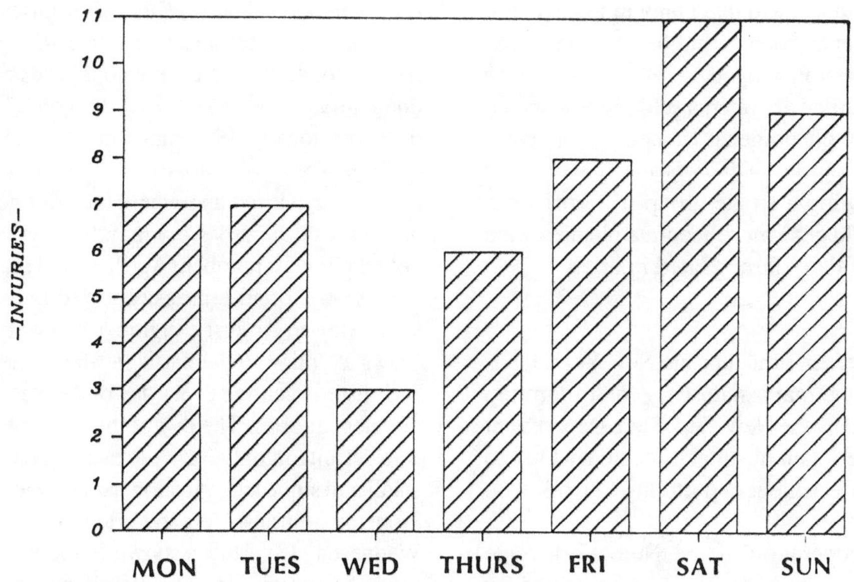

FIGURE 2. Distribution of injuries by day. Otsego County, NY, 1984 to 1985.

livestock 39% of the time, and to chemicals in 1%. Our single fatal injury was sustained on a backhoe when it overturned and killed its operator. The injury with the longest disability was related to a bull attacking a farmer, inflicting multiple trauma.

During the year of surveillance, 117 individuals were injured. These persons represented nearly 25% of all the farm workers or an annual attack rate for injuries of 25%! Men were far more numerous in the work force and were injured proportionately; 85% of the injuries were sustained by men. The vast majority of men had a single accident (84%), 12 individuals had two injuries, 3 men had three, and 1 unfortunate 60 year old had four mishaps. Of the women, 17 were hurt and only 1 woman had two injuries.

The difficulties encountered by men required medical care 62% of the time, while women went to a hospital or clinic in 37% of instances. Mechanical injuries were more serious than animal injuries for both groups, with 70% of mechanical mishaps requiring care and only 40% of those related to animal trauma. Mechanical injuries were also twice as likely to cause a lost-time problem. A total of 407 d was lost during the year secondary to 35 different injuries. The average lost-time event lasted 11 d. Ten of the injured persons filed claims for disability.

DISCUSSION

The data reported above covering 219 farm-years experience strongly suggest that dairy farming in upstate New York is very dangerous indeed. About 1/4 of the entire work force we studied sustained an injury during the year. The figures are even worse for the individual farm in that 40% of farms had almost one injury per year.

Our impression had been that many injuries were trivial. In the survey almost 2/3 of all mishaps required medical attention, strongly refuting our contention. Although the severity of reported injuries also reflects reporter bias, we are impressed that 60% were serious enough to get professional help. The single death that occurred does not allow a fair estimate of the frequency of deaths on dairy farms. If extrapolated for a population of 100,000 farms, it would approach 200 deaths per year — an astronomical figure. Goodman et al. found an annual death rate of 54/100,000 for their highest risk group.[2] Our sample is simply too small to allow valid annual fatality rates to be estimated.

We found that late spring and summer were the most dangerous times. This observation agrees with that of Goodman and others.[2,8,9] Our peak in May was not clearly attributable to the initiation of machine use for the growing season. The May injuries were nearly evenly distributed between mechanical- and animal-related mishaps.

The frequency of injuries by day suggests that weekends are dangerous — especially Saturdays. Again, our observations about weekend agricultural injuries corroborate those of Goodman et al. and Smith.[2,9] We speculate that the weekend peak is caused by either the rush to finish the work to get needed time off or perhaps because children or part-time workers are involved. Our data do not support the latter postulate.

One additional observation we made about our injuries is that nearly 2/3 occurred in the afternoon. This supports previous published data on deaths.[2,9] Here the likely explanation is fatigue, a common problem in occupational health.

The study also revealed that although most of our farmers take care of at least 50 head of cattle, often enclosed for months, mechanical injuries are more common than animal mishaps. Powers, in the pre-World War II period, at our hospital made the same observation for over 310 hospitalized injuries.[10] The mechanical injuries sustained by farm workers were also more serious than animal injuries, with 70% of the former and 40% of the latter requiring medical attention. Mechanical injuries were twice as likely to cause lost time from work. The average time out for a lost-time accident was 11.6 d. This is an impressive finding given that most farmers can ill afford to be out even a single day.

Our data confirm that dairy farming, like all other agriculture, is dangerous. Even when we tried to correct for biased reporting, we found very high attack rates of serious mishaps. The work on agricultural injuries has far to go to detect real causes and to allow rational interventional strategies to be mapped out. Much larger, detailed epidemiologic studies must be undertaken to establish reliably the determinants of the very real dangers.

REFERENCES

1. National Safety Council Accident Facts, 1981 ed., National Safety Council Publications, Chicago, 1981.
2. **Goodman, R.A., Smith, J.D., Sikes, R.K., et al.,** Fatalities associated with farm tractor injuries: an epidemiologic study, *Public Health Rep.*, 100, 329, 1985.
3. **Helgeson, S.D. and Milham, S., Jr.,** Farm workers electrocuted when irrigation pipes contact powerlines, *Public Health Rep.*, 100, 325, 1985.
4. **Saari, K.M. and Aine, E.,** Eye injuries in agriculture, *Acta. Ophthalmol. Suppl. (Copenhagen)*, 161, 42, 1984.
5. **Simball, A.M., Thacker, S.B., and Levy, M.E.,** Shigella surveillance in a large metropolitan area: assessment of a passive reporting system, *Am. J. Public Health*, 70, 164, 1980.
6. **Marier, R.,** The reporting of communical diseases, *Am. J. Epidemiol.*, 105, 587, 1977.
7. **Siemiatycki, J., Campbell, S., Richardson, L., and Aubert, D.,** Quality of response in different population groups in mail and telephone surveys, *Am. J. Epidemiol.*, 120, 302, 1984.
8. **Pollock, J.,** Accident and injury analysis — 1983, Onondaga County and New York State Accident Survey, personal communication.
9. **Smith, T.L.,** Deaths associated with agriculture, New York Vital Statistics Review April to June 1980, New York State Health Department, Albany, 1980.
10. **Powers, J.H.,** The hazards of farming, *J.A.M.A.*, 113, 1375, 1939.

Agricultural Injuries to the Upper Extremity

V. Bowen

INTRODUCTION

Agriculture is a hazardous industry.[1,2] Several published reports describe the epidemiology, morbidity, and mortality of farm-related injuries.[3-5] Additional studies present data from Canadian farming regions.[6-9]

Saskatchewan farmers are highly mechanized and use large, motorized machines. At certain times of year there are outbreaks of farm-related accidents. This study aimed to collect data on the circumstances of accidents amongst those patients that presented to our hospital with farm-related upper extremity injuries. The method was to retrospectively review the charts of all patients admitted to the University Hospital in Saskatoon with agriculture-related injuries to the hand and upper extremity between April 1, 1983 and March 31, 1985.

RESULTS

There were 68 patients; 60 were male, 8 were female, and their ages ranged from 4 to 77 years. Their occupations are listed in Table 1. Hand dominance was only recorded in the records of 44 patients; 41 were right-hand dominant, and 3 were left-hand dominant. Of these 44, 23 injured their dominant hand, 18 their nondominant hand, and 3 suffered bilateral injuries. In addition to these there were an additional 3 patients who injured both upper extremities, making a total of 6 bilateral injuries.

The time of accident was recorded in 52 charts. Injuries occurred at almost every hour of the day and night, but there were peak periods between 1000 and 1100 and 1400 and 1500. No single day of the week stood out as being different from any other, although injuries seemed less likely on Tuesdays and Sundays. Agricultural trauma occurred every month of the year, but there was a peak incidence during the harvest season from August to October.

The causes of injury are listed in Table 2. Although grain augers head the list, the farm tractor is really the most dangerous machine, as all the power take-off injuries were on tractors. Machines with powered moving components such as grain augers, tractors, swathers, combine harvesters, and hay balers caused most injuries. Nonpowered machines such as cultivators, harrows, and seed drills were only involved in two accidents — one where a tractor pulling a harrow ran over a farmer and another where a rope attached to a seed drill caused a ring avulsion injury.

Ten patients were younger than 18 years old. Two, both under 10, reached in to touch moving parts. Two fell from a height — a haystack and a horse. One other suffered a ring avulsion when her horse bolted. Only four were injured by powered machines that they were operating in an unsupervised situation.

The 68 patients suffered every possible kind of injury, but certain patterns of injury were common to specific machines.

TABLE 1
Occupation of Injured Patients

Farmer	37
Farm wife/female farmer	5
Farm child	
Boy	7
Girl	3
Part-time farmer	1
Electrician	1
Mechanic	1
Plumber	1
Carpenter	1
Painter/decorator	1
Machine shop foreman	1
Agricultural inspector	1
Fire protection laborer	1
Commercial fisherman	1
Driving instructor	1
Farm service job	
Farm machinery salesman	2
Fertilizer salesman	2
Stockyard laborer	1
Grain elevator worker	1

TABLE 2
Cause of Injury

Grain auger	13
Power take-off	9
Swather	7
Combine harvester	6
Hay baler	5
Tractor (not PTO)	5
Fall from height	5
Rope avulsion injury	3
Rock picker	2
Chicken-singeing burn	2
V-belt (not on above machines)	2
Grain-cleaning machine	1
Forage harvester	1
Grain mill	1
Chop making machine	1
Grinding wheel	1
Sheetmetal	1
Butcher knife	1
Crushed by pig's head	1
Electrical burn (not grain auger)	1

Grain augers caused injuries in three ways. Hands caught in the rotating screw (8/13) suffered multilevel dorsal lacerations to the fingers. Severity of injury varied from skin only to complete amputation. Persons struck by the crank handle (3/13) suffered fractures and dislocations in the hand. Contact with overhead power lines (2/13) resulted in widespread electrical burns. Auger injuries showed no preference to dominant or nondominant limbs and tended to occur during the summer, fall, and early winter. There were no children involved in grain auger accidents. The youngest person to be injured by one of these machines was 19.

Power take-offs caused injury when an entangled item of clothing pulled the victim into the rotating shaft (9/9). The usual injury was an avulsion amputation of the dominant thumb (4/9 complete, 2/9 incomplete, and 1/9 open dislocation of an MCPJ) when a glove was caught. There were, however, 2/9 major limb avulsions when a sleeve became entangled. Power take-off injuries almost always occurred in the winter months when farmers were wearing layers of bulky clothing.

Swathers caused injury by two different methods. Contact with the sickle blade (5/7) caused sharp lacerations which were of variable severity and usually confined to a single finger. Extremities caught in a V-belt drive (2/7), not only in swathers, but also in other machines, suffered from crush injuries to the distal part of multiple fingers. Usually these involved damage to the nail, friction burns, untidy lacerations to volar, and dorsal skin, and bone and joint disruption. Swather injuries in general, and V-belt injuries, in particular, tended to involve dominant extremities and six out of seven occurred in August.

Combine harvester accidents usually involved V-belt drives (4/5) and were no different to those of swathers. One patient lacerated the dorsum of his hand on the sheet metal casing. All combine harvester injuries were seen during the harvest season of August, September, and October. Although some combine harvesters are self-propelled, many utilize tractor power take-offs, but there were no combine harvester power take-off injuries, most likely because gloves and bulky long-sleeved clothing were not worn at this time of year.

Hay balers caused injury in three ways. The V-belt drive caused typical injuries (1/5). Patients caught in the pickup or roller mechanism (2/5) suffered very severe major limb amputations which, in one case, was bilateral. The victim's second arm was taken into the machine as he attempted to extract the first. (This also happened to the patient injured in the grain-cleaning machine.) In 2/5 accidents, fingers were injured by being struck by the chassis. Once a hay baler fell from a jack, and once a trailer hitch was involved. The V-belt injury affected the dominant hand, but the crushing injuries affected the nondominant hand. (This was also noted in the rock picker accidents.) Hay baler injuries were only seen in September and October.

Tractors, in addition to causing power take-off injuries, were involved in five other accidents. Three farmers received severe crush injuries after being run over. One of these subsequently died in University Hospital — the only death in the series. A swinging door crushed one hand and in another case a crank handle caused a true chauffeur's fracture to the distal radius. Tractor injuries were not related to hand dominance. All occurred during August, September, and October.

Five patients injured themselves falling from a height. Two fell from a haystack and the others from a fence, a horse, and a barn roof. Apart from one who received a ligamentous injury to the thumb, they all suffered closed wrist and forearm fractures, and in two cases these were bilateral. The other four bilateral injuries in the study were two major limb avulsions previously mentioned (one in a hay baler and one in a grain-cleaning machine), one man who crushed all eight fingers in a swather V-belt drive, and one with electrical burns sustained while moving a grain auger. Falls from a height occurred both in summer and winter.

Management of the injuries by the 68 patients was as varied as the injuries themselves. In general, standard modern hand surgery procedures were used. Twelve patients underwent microneurovascular operations. Three other potential candidates did not undergo replantation because of damage to the amputated parts. Seven of the patients who underwent replantation had suffered avulsion amputations from power take-offs. Three patients from the series are candidates for future thumb reconstruction by microneurovascular toe-to-thumb transfer.

DISCUSSION

Farm accidents in Saskatchewan have previously been described by Huston and Smith[7] and Simpson.[9]

In this study many of the data concerning injured patients are similar to those of other series.[3-6,9] The majority of patients were male and specific accident types varied with the season. The bimodal time curve was again noted with similar peaks at late morning and early to midafternoon. It was again noted that machines with powered moving parts were involved in most accidents, whereas nonpowered machines were relatively safe.

There were some differences from the previous series. Only 13 out of 68 patients were outside the 18- to 65-year age group, and there was not an accident peak on any particular day of the week. In addition, some types of farm injury, reported in the literature, were not seen because the equipment responsible for them is rarely used in Saskatchewan (corn-picker injuries,[10] irrigation pipe electrocution[11]).

This review has shown grain auger injuries to be common, but in contradiction to previous studies[12-14] none of the injuries involved children. There were also very few electrocution burns from contact between augers and power lines. These findings suggest that the recent publicity given to grain auger injuries may have been effective.

In our power take-off injuries the prevalence of thumb amputations was greater than in earlier reports,[15,16] but hay baler injuries were similar to those described previously.[17]

Apart from power take-off injuries, tractor-related accidents[18] did not bring a large number of patients to the hospital.

In particular, there were no casualties from tractor rollover accidents — a commonly reported mechanism of accident.

Since Simpson's 1980 paper,[9] a Hand Surgery Service with microsurgical specialists has been started at University Hospital. This has not only increased the number of hand injuries referred to the center, but has also changed the types of injury seen. For instance, many avulsion amputations of the thumb that were managed by simple wound closure in small community hospitals are now referred for replantation. Similarly, the sharp lacerations from swather sickle blades are highly suitable for microneurovascular repair and these also now tend to be referred to University Hospital. These facts demonstrate the inaccuracy of this type of hospital-based retrospective study by showing how results can change simply by altering the kind of treatment offered. It is important to realize that these figures may only be the tip of the iceberg, as they do not include people who do not seek medical attention, those managed in local community hospitals, those referred elsewhere, or those that die of their injuries and never get to our hospital. Nevertheless, until data collection for all agricultural injuries can be improved, these data serve to bring our attention to the kinds of injuries that are occurring on our farms.

REFERENCES

1. **Cliff, K.S.,** Agriculture — the occupational hazards, *Public Health,* 95, 15, 1981.
2. **Jackson, F.C.,** Farm and ranch injuries in West Texas, *Tex. Med.,* 79, 51, 1983.
3. **Calandruccio, R.A. and Powers, J.H.,** Farm accidents, *Am. J. Surg.,* 78, 652, 1949.
4. **Muckala, K.A.,** Farm accidents and their prevention, *Minn. Med.,* 50, 1477, 1967.
5. **Young, H.H. and Ghormley, R.K.,** Accidents on the farm, *J.A.M.A.,* 132, 768, 1946.
6. **Howell, J.M. and Smith, E.S.O.,** An agricultural accident survey in Alberta, 1970, *Can. J. Public Health,* 64, 36, 1973.
7. **Huston, A.F. and Smith, C.,** Farm accidents in Saskatchewan, *Can. Med. Assoc. J.,* 100, 764, 1969.
8. Ontario farm accident survey, March 1, 1959 to February 29, 1960, Ontario Department of Agriculture, Toronto.
9. **Simpson, S.G.,** Farm machinery injuries, *J. Trauma,* 24, 150, 1984.
10. **Campbell, D.C., Bryan, R.S., Cooney, W.P., et. al.,** Mechanical cornpicker hand injuries, *J. Trauma,* 19, 678, 1979.
11. Irrigation-pipe associated electrocution deaths — Washington, *Morbid Mort. Wkly. Rep.,* 32, 169, 1983.
12. **Beatty, M.D., Zook, E.G., Russel, R.C., and Kinkead, L.R.,** Grain auger injuries: the replacement of the corn-picker injury, *Plast. Reconstr. Surg.,* 69, 96, 1982.
13. **Grogono, B.J.S.,** Auger injuries, *Injury,* 4, 247, 1972.
14. **Letts, R.M. and Gammon, W.,** Auger injuries in children, *Can. Med. Assoc. J.,* 118, 519, 1978.
15. **Kalenak, A., Gordon, S.L., Miller, S.H., et al.,** Power take-off injuries, *J. Trauma,* 13, 775, 1973.
16. **McElfresh, E.C. and Bryan, R.S.,** Power take-off injuries, *J. Trauma,* 13, 775, 1973.
17. **Gainor, B.J.,** Hay baler trauma to the upper extremity: a roller injury, *J. Trauma,* 23, 1069, 1983.
18. Farm-tractor associated deaths - Georgia, *Morbid. Mort. Wkly. Rep.,* 32, 481, 1983.

Amputations in Farming

J. P. Kostuik

INTRODUCTION

In industry, which includes farming, the etiology of amputations is primarily traumatic. They are frequently caused by loose clothing which can easily be entrapped in machinery such as power take-off mechanisms, augers, and other motorized implements. The injuries are frequently crushing in nature and often contaminated with dirt, oil, and other foreign materials.

UPPER EXTREMITY AMPUTATIONS

Vascular disease, the most common cause for lower extremity amputations, is rarely a cause for upper extremity amputations. Traumatic upper limb amputations may be clean-cut if caused by sharp objects, but are more frequently crushing in nature. As a result, microvascular surgery for reattachment of limbs is often not possible.

Preservation of the distal end of the radius and ulna or the distal end of the humerus or more proximal humerus is rarely of functional value. In addition, preservation of the most proximal inch of the radius and ulna is of little value. Ideally, surgical technique should involve the use of incisions through healthy skin with intact sensation. Sensation is of paramount importance in the upper extremity. There must be adequate soft tissue padding over bony prominences and an amputation should preferably be as distal as possible.

It is important to remember hand function, as this is the most important functional unit in the upper limb being used for power grip and various kinds of pinch. The upper extremity has the ability to work in all degrees of freedom simultaneously, whereas a user of a prosthesis can rarely do this, performing generally one, and at the most two, degrees of freedom.

Amputations involving complete excision of the scapula and upper extremity generally result from tumor surgery. Traumatic amputations sometimes occur through the shoulder joint with preservation of the scapula. Ideally, amputations at this level require the use of a harness for maintenance of the prosthesis or retention of the prosthesis on the body. The patient in this situation is best fitted with an electric elbow and hand. The shoulder is usually passively manipulated.

The 1950s saw the advent of the external control prosthesis rather than the usual cable control prosthesis. These were initially gas controlled but more recently can be electrically or myoelectrically controlled. Major advances are constantly being undertaken in this field.

Fitting of prostheses for above-elbow amputations depends upon the occupation. Quite frequently, cable-controlled elbow and a hook- or cable-controlled hand are satisfactory. However, many younger amputees are capable of learning myoelectric or electric controls for both elbow and hand.

Retention of a small amount of proximal humerus is of no functional value nor is an amputation through the elbow. Similarly, preservation of only a few centimeters of the proximal radius and ulna is of little functional value. Ideally, an amputation at the elbow should be performed a few inches above the joint.

The distal end of the radius and ulna is of little functional use because of the large bony prominences which may cause pain if preserved. Nowadays, amputations done below the elbow are ideally fitted with myoelectric or electric controls with an interchangeable hand and sometimes an electric hook. However, many people originally felt that these did not have sufficient power, so that cable-controlled prostheses with hook- or cable-controlled hand are still frequently used. Many of these terminal devices have removable interchangeable components so that specific functions can be carried out. Hand cosmesis remains a major problem and it is difficult to match the prosthetic hand glove coloring to that of a human skin, although recent improvements have shown promise. Frequently, upper extremity amputees wish to pronounce themselves as amputees and, consequently, may wear a glove which is usually made of leather.

Retention of the carpus can be of value in hand amputations. These patients are ideally fitted with a spatula-like prosthesis, although a review in children showed that these are rarely retained. The remaining carpus must be mobile for use.

Amputations in the fingers and hand are a very specialized subject. Today, with the advent of microvascular surgery, preservation of fingers with good sensation is possible. The amputation of a single thumb, especially if of a guillotine type, is best repaired microsurgically. A single digit amputation is a less clear indication for microvascular surgery. Amputation of multiple fingers usually results in some form of microvascular surgery being performed.

Prosthetic devices for hand and partial hand amputations are generally of a cosmetic nature, but may be of some functional value if opposition between the digits and thumb can be achieved.

LOWER EXTREMITY AMPUTATIONS

The most common cause for lower extremity amputation is a dysvascular limb. This is usually secondary to major vessel arteriosclerosis combined with diabetes in 50% of cases.

Other cause for amputations are trauma, tumor and infection.

The major surgical problem in a dysvascular limb is the selection of amputation level. If the level of amputation is too high, decreased function is noted. If too low, there may be problems with wound healing. Vascular disease accounts for 80% of lower limb amputations; 10% of patients with intermittent claudication undergo amputation. The dysvascular amputee is not always elderly. If a dysvascular amputee survives 5 years, he has almost a 50% chance of losing the other leg.

Decisions concerning the surgical level of amputation should take into consideration pathological, anatomical, surgical, prosthetic, and personal factors. Amputation is rarely just another stage in the treatment of arterial occlusive disease in a dysvascular limb. The surgeon must always secure a mandate from the patient for the lowest feasible level, recognizing that this may fail and subsequently be revised to a higher level. Preservation of the knee is extremely important in order to allow for reasonable function.

In the 1960s, it was said that if you had gangrene of your large toe, this would subsequently require an above knee amputation and this is what was practiced. In 1965, at the Toronto General Hospital, 65% of all amputations in dysvascular patients were performed above the knee. This has reversed such that between 75 and 80% are now below knee.

Many techniques have been used to help decide the level of amputation. Arteriography and oscillography have been found to be of little value. The assessment of skin temperature, aided by thermocup or thermography, has not yet shown value. The use of radioactive isotopes, however, despite being time consuming, has demonstrated great accuracy.

Experimental intracellular oxygen tension measurements have demonstrated that they will almost certainly become the ultimate test, but at the present time the assessment of skin blood pressures using Doppler methods is the most preferred technique, provided it is used in combination with good clinical assessment. In practice, a Doppler blood pressure of 40 Hg or more will allow for adequate skin healing.

Hind quarter amputations or hip disarticulations are usually carried out for tumor but may, on occasion, be done for vascular disease and rarely for trauma. Above knee amputations are generally best performed in the mid or distal end of the femur. These are usually for trauma or for failure of healing in a below knee amputation. Healing problems may be encountered in below knee amputations because of neuropathy, muscle atrophy, and impaired sensation in the skin. In the foot, these same factors may cause ulceration and deformity — complications that eventually lead to amputation.

Reconstructive or conservative foot surgery is very satisfactory in diabetics where soft tissue problems are secondary to small vessel disease. Generally speaking, however, diabetics with ischemic feet should not be treated any differently from nondiabetics.

The rehabilitation of bilateral lower limb amputees is difficult. As a rule, if a single limb amputee has to have a second amputation, the newest amputation should be performed as distal as possible.

SURGICAL CONSIDERATIONS

Traumatic amputations are uncommon in the farming community. In those that do occur, there are some differences between children and adults. The selection of level in children should aim to preserve all possible length and attempt to include the epiphyses or growing ends of the long bone. In children skin grafts on amputation stumps will survive and can bear weight. One must always consider the proximal joint in a child, as this not only will grow, but may have been damaged at the time of the injury as well. In adults with traumatic amputations, the level is not always as important as it is in children. In adults, end-bearing amputations are ideal if possible; skin flaps will survive but skin grafts do not, and, in addition, they do not tolerate weight bearing; proximal joints should be conserved even in the presence of proximal fractures. It should be remembered that large bones will, with time, shrink and become small ones. In the presence of a dirty wound, if an amputation is to be performed, flaps should be formed and the wound left open. Secondary closure is carried out later.

Indications for early amputation are dysvascular limbs, traumatically dysvascular or devitalized tissue, neurogenic loss, and loss of support. Indications for late amputation are deformity, pain, neurogenic, ulceration, infection, and lack of function. It is important to remember that amputation need not be performed through the most proximal fracture in a multitraumatized lower extremity. If the more proximal fracture can be fixed, the amputation should only be carried out at the lowest possible level where skin may survive. Vascular repair may preserve length and allow for more distal amputation and must be considered in any traumatically dysvascular limb. Microvascular surgery with the use of muscle pedicles may, together with improved nerve transplants, allow for a more distal amputation.

If there is any question of vascularity or viability in a dysvascular limb, always wait for demarcation before deciding the level of amputation.

The world has come a long way with rehabilitation of the amputee from the day of Bacchus or Long John Silver. Sir W. Ferguson said in 1865, amputation is "one of the meanest yet one of the greatest operations in surgery: mean, when described to what better may be done; great (when it) is the only step to give comfort and prolonged life ...", and this remains true today. Dr. Gingras, a well-known Canadian, has stated that prescription of the lower extremity prosthetic appliance is far from being the answer to the rehabilitation problem of the amputee, and a team concept with the patient being the center of the team requiring support from family, family practioner, medical staff, nursing staff, therapist, public health nurse, vocational rehabilitation people, and social workers must be considered. Despite this, however, we will frequently see either the amputee who functions extremely well on his own without aid or, conversely, the person who does not respond to anything. The idea of a team concept today has resulted in a marked improvement in rehabilitation of the amputee in both the upper and lower extremity.

Farm Machinery Accidents in Children

R. M. Letts

INTRODUCTION

Unfortunately, farm machinery accidents in children are all too common. In most prairie treatment centers, farm machinery accidents are the most common cause of traumatic amputation, exceeded only by congenital causes of limb deficiency. With increasing farm mechanization, it does not appear that this type of trauma will decrease in the foreseeable future. Indeed, the farm family as a whole is extremely vulnerable to this type of trauma because of the proximity of family life to the work place and the extreme familiarity with complicated equipment that frequently leads to a lack of respect towards this type of potentially dangerous farm machinery. In North America, grain auger and power take-off injuries account for 50% of all fatalities from machinery accidents, although farmers account for only 7% of the population. Thus, the farm family is vulnerable to severe injury, not only in the fall when the crop is being harvested, but also throughout the entire year, since farm work is ongoing throughout the year.

AUGER AND POWER TAKE-OFF INJURIES

Power take-off and auger injuries are the most common types of trauma encountered in children in the farm environment in Manitoba. There is a unique, hypnotic fascination of the grain disappearing into the auger for the inquisitive child, and unfortunately the result is frequently severe extremity injury. There are few "minor injuries" sustained when a limb comes into contact with this type of machinery. The end result is usually devastating soft tissue and bone trauma or even complete ablation of the limb itself (Figure 1).

There is an inexplicable lack of publications related to this type of common prairie injury. A review of the Manitoba experience revealed a total of 32 power take-off or auger injuries over a 10-year period. In the Juvenile Amputee Clinic in Winnipeg, 17 children are currently being followed for amputation sustained in farm machinery accidents. Because of the inherent nature of the auger and power take-off, once an extremity becomes entangled, injury is sustained from multiple laceration and avulsion of all or a portion of the extrem-

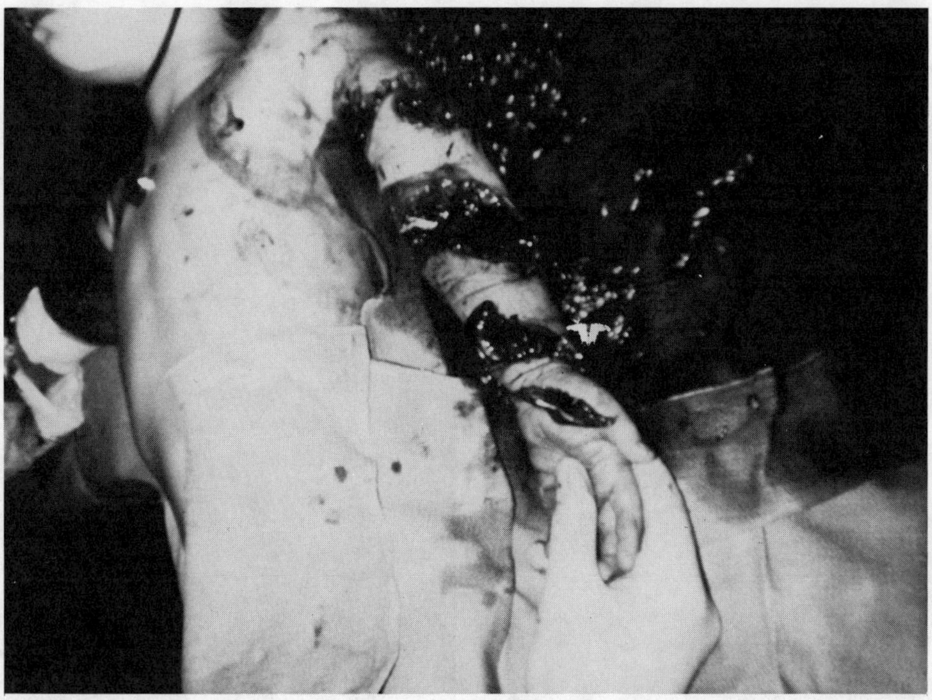

FIGURE 1. A 4-year-old boy who, while playing around an auger, became entangled in the machinery. The limb had to be amputated above the elbow. Note the lacerations of the screw, one more revolution of which would have severed his shoulder and neck. (From Letts, R.M. and Gammon, W., *Can. Med. Assoc. J.*, 118, 519, 1978. With permission.)

ity. The end result is very severe injury to the limb, often requiring high-level amputation; of the 17 patients, 10 have either above knee or above elbow amputations. Those children who manage to escape amputation are frequently left with a severe degloving injury, where the skin has been completely detached, frequently requiring multiple skin grafts with associated bone and muscle damage. Those patients who have sustained amputation of the limb frequently have associated soft tissue trauma in the region of the stump, resulting in poor or inadequate skin coverage, also necessitating grafting and frequently causing future prosthetic fitting problems. (Figures 2A and B).

ENVIRONMENTAL CHARACTERISTICS OF AUGER INJURIES

To determine some of the environmental characteristics surrounding children injured in farm machinery accidents, a special questionnaire was sent out to all farm families to whom such an accident had occurred to one of the children. Since most of these children were attending treatment, either in the Juvenile Amputee Clinic or as a result of their major trauma, their response was virtually 100%. There is frequently a misconception that children are injured on the farm while helping their fathers with the running of the farm. Although this is true in some instances, in the majority of cases these children were injured simply while playing in the environs of the farm yard within 100 or 200 yd of their home. In only four children was the injury sustained while actually performing some type of farm chore. The remainder of the 32 were either observing, playing in the farm yard, or simply "along for the ride". Over 60% of the injuries occurred in the fall, related to increased use of farm machinery at harvest time. Summer was the next most common season of injury, with winter being the least. The question was asked of all families whether a safety device was in use at the time of the accident and in over 50% of instances, there was no safety device available. In the other 25%, the safety device was not in use and in the remainder, there was some sort of safety device which may or may not have been in place at the time of the accident. Many comments to the effect that the safety devices were removed to increase the flow of grain were received. Frequently, the safety device is not really encompassing enough to keep a child's arm or leg out of the auger and really only prevents large extremities or the body from being entangled within the machinery itself. Of the children injured, 65% were found to be under 7 years of age, again emphasizing the finding that injury to the children in the farm environment is seldom associated with the child being used to assist in the farm operation.

DEGLOVING INJURIES

Although amputations are a frequent result of auger and power take-off injuries, there may be associated soft tissue trauma resulting in avulsion of skin or so-called degloving injuries, or these injuries may present unassociated with any amputation. The most common cause of the degloving injury is a shear-type injury or avulsion injury. This is frequently encountered in children when a wheeled vehicle runs over an extremity resulting in the skin and soft tissue being pulled over

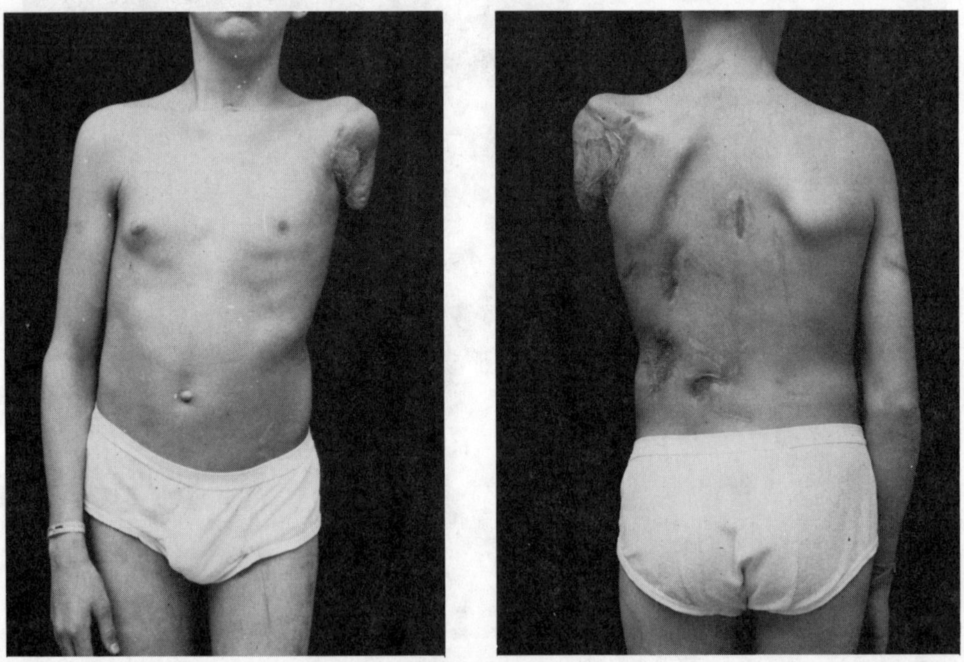

A B

FIGURE 2. The marked scarring often encountered in healed auger injuries resulting in difficult prosthetic fitting.

the bone and disruption of the underlying subcutaneous tissue and vascular interconnections. Since skin derives its main nutrition from perforating vessels from the subcutaneous tissues, this skin, when completely degloved, almost always becomes gangrenous.

There are three types of degloving injuries that may occur as the result of farm machinery accidents. These are (1) complete anatomical degloving in which the skin is actually removed from the limb, (2) concealed degloving in which the underlying attachment of the skin is avulsed from the soft tissue, but there is no disruption in the skin itself; and (3) a combination of the anatomical and concealed varieties.

It is essential that the treatment team recognize these injuries, especially the concealed type, since immediate defatting and reapplication of such skin as a full thickness skin graft is the most efficacious way to deal with this particular problem. In the concealed variety, examining the skin for excessive mobility will frequently make the diagnosis. One should also be suspicious in those extremities with tire marks over them, indicating an extremity has been run over.

Frequently, these major soft tissue degloved extremities will have associated fractures, and in such instances the fractures are best treated with skeletal traction or an external fixator which allows the skin to be carefully monitored even after it has been replaced as a full thickness graft. Should further split thickness grafting be required, this can be more easily accomplished without the encumbrance of plaster (Figure 3).

Since degloving injuries are most commonly associated with tire trauma, the provision of backup audible signals on farm equipment and vehicles should be standard for those pieces of mobile equipment used frequently on the farm.

TRACTOR TRAUMA

Tractor accidents are the most common cause of child mortality in the farm environment. There is little that one can say about this type of trauma except that it is frequently related to the inexperience of the operator. Indeed, such trauma is a common cause of mortality even amongst experienced farmers, and it can only be emphasized that adequate safety instruction must be given to any child old enough to participate in the operation of farm vehicles. The most notorious cause of tractor turnover, that of operating a vehicle on a slope, needs to receive extra emphasis. Regular continuing education of proper safety techniques for the entire family must be implemented. The family farm is still, in most areas, a family operation, and as such it is not sufficient just to include the farmer himself in the education program. It must include his wife and those older children who will actually be a contributing part of the farm operation.

ALL TERRAIN VEHICLES

The all terrain vehicle (ATV) burst on the scene several years ago and although not an intricate part of the farm operation, it is becoming a very popular recreational vehicle in those environments with adequate space and opportunity to operate them — namely, the farm environment. In some ways the ATV is analogous to a tractor on a smaller scale. It has the same inherent instability on rough ground but even more so. The three-wheeled vehicles are very unstable and require

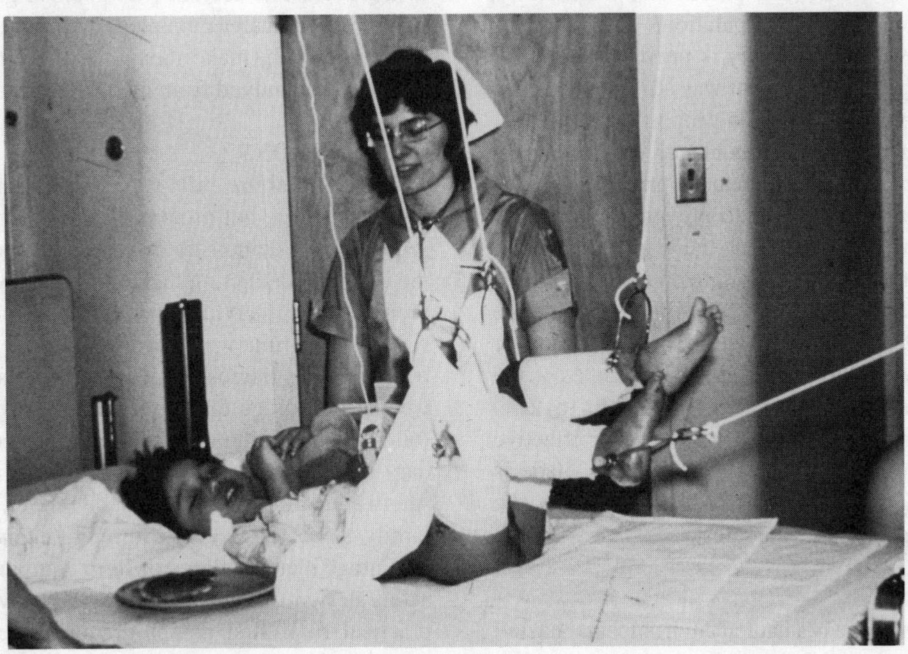

FIGURE 3. Degloved extremities are best managed in skeletal traction, enabling the limb to be closely monitored. (From *J. Paed. Orth.*, 6, 193, 1986. With permission.)

considerable skill in their operation. Children of a very young age are being allowed and, indeed, encouraged to operate this type of vehicle by safety-conscious parents who would under no circumstances allow their child to operate comparable farm machinery. Unfortunately, the injury capability of the ATV is equal to or exceeds that of most farm equipment. The increased numbers of ATV accidents that are being seen across the country in hospital emergencies are somewhat analogous to high trauma rates noted with the introduction of the snowmobile. Hopefully, with proper legislation, safety equipment, and education programs, trauma from the ATV will decrease in a manner similar to snowmobile trauma, although at the present time this does not seem to be occurring.

ROTARY BLADE INJURIES

A number of farm machinery implements possess rotary blades, with the most common, of course, being the rotary lawn mower. Farm children are usually more susceptible to rotary blade injuries simply because the farm environment is conducive to large lawns and usually the larger types of power mower are purchased by the farm family. The riding mower is a particular hazard, especially when children are riding with an operator on the mower.

In a review of 31 children sustaining rotary blade injuries, we found the most common cause of injury occurred in children who were simply playing near a mower and in trying to get out of the path of the mower had a limb come into contact with the blade. Other types of rotary mower injuries are sustained by missile fragments shot out from the mower or by the child being actually struck by the mower without coming into contact with the blade itself. Since most rotary mowers rotate at 3000 rpm, it is virtually impossible to extricate a limb before the blade has had several passes at the extremity, resulting in severe laceration or amputation.

Almost every lawn mower injury is preventable. In our series, responsibility for the accident was attributed to man, machine, or environmental factors in proportions of 29:1:1. Improvements in machine design could be enforced in an attempt to minimize errors in human judgment. We have suggested the following guideline to prevent lawn mower trauma: (1) educate children about the dangers of lawn mowers, (2) prevent children from riding on tractor mowers, (3) use the grass catcher whenever possible to reduce the number of projectiles, (4) avoid unnecessary use of power wheel-driven mowers, (5) have hand controls mounted on the handles so that the mower can be turned off quickly, (6) keep the blade sharp so that it can be operated at a lower effective speed; because the kinetic energy of the rotating blade is directly proportional to the square of the blade's velocity, the lowest operating speed will be the safest.

DISCUSSION

Farm machinery trauma is a natural environmental hazard to everyone working in the farm environment. Although it may never be completely eradicated, it can certainly be decreased from its current level of carnage.

In most industrial situations, great strides have been made in eliminating or reducing occupational hazards, largely through regulations that are enforced by employers, union, and government agencies. Such regulatory mechanisms, however, are frequently not operative in the case of the farmer, whose safety precautions are self-imposed and who is free to operate with or without them as he chooses, usually with meager instruction. Safety shielding for power take-off shafts, cutting blades, and augers is usually an optional accessory in Canada and, because of its cost and the real or imagined decrease in efficiency of the auger, is usually not obtained by farmers. Not so in the U.S.A., Britain, and Australia, where farm equipment manufacturers are under more stringent regulations to provide safe equipment for the user.

There is a need for more research in safety engineering and design for farm equipment. Protective devices currently available are inadequate for preventing an extremity from becoming entangled.

Proper control of auger injuries in children lies in preventative rather than therapeutic measures. However, once prevention has failed, the initial management of such major injuries rests not in the hands of the emergency team of specialists at a modern hospital, but with one of the child's relatives out in the field, at some distance from the nearest medical aid. For this reason, as well as the overall high rate of farm accidents, it is essential that all farmers and their families take courses in first aid. At present, such courses are few and in most areas nonexistent. The typical farm in Canada is still a family operation; hence, all members of the family, including children, should receive adequate safety and first aid instructions. The declining numbers of general practitioners who live in rural areas, coupled with the modern area-centered clinic approach, make todays farmer, unlike the industrial worker, far removed from medical assistance.

CONCLUSIONS

We can and are influencing the magnitude of the farm injury problem, but progress is slow. The military medical services have demonstrated well the importance of locating the injured, conducting the appropriate evacuation, and providing backup services to reduce the mortality for major wounds. It is time we provided a similar service to the farm battlefield. As interested individuals in the farm health field, we can encourage farm groups and governments to do the following: (1) determine the true incidence of trauma, secondary to accidents with farm equipment; (2) provide better safety education, not only for the farmer, but also the entire family since the average farm is a family operation; (3) encourage manufacturers of farm equipment to improve the safety design of their machinery; (4) develop effective and practical rural first aid programs; (5) encourage realistic rehabilitation programs for the injured farm member so that the farm operation may continue.

REFERENCES

1. **Grogono, B.J.,** Auger injuries, *Injury,* 4, 247, 1972.
2. **Howell, J.M. and Smith, E.S.O.,** An agricultural accident survey in Alberta, 1970, *Can. J. Public Health,* 64, 36, 1973.
3. **Houston, A.F. and Smith, C.,** Farm accidents in Saskatchewan, *Can. Med. Assoc. J.,* 100, 764, 1969.
4. **Knapp, L.W., Jr.,** Occupational and rural accidents, *Arch. Environ. Health,* 13, 501, 1966.
5. **Letts, R.M. and Gammon, W.,** Auger injuries in children, *Can. Med. Assoc. J.,* 118, 519, 1978.
6. **Letts, R.M. and Coeary, J.,** The child and the snowmobile, *Can. Med. Assoc. J.,* 113, 1061, 1975.
7. **Letts, R.M. and Mardirosian, A.,** Lawn mower injuries in children, *Can. Med. Assoc. J.,* 116, 1151, 1977.
8. **Lucas, G.L. and Wirka, H.W.,** Orthopaedic problems in agricultural trauma, *Wis. Med. J.,* 64, 471, 1965.
9. **Muckala, K.A.,** Farm accidents and their prevention; an epidemiological approach, *Minn. Med.,* 50, 477, 1967.
10. **Power, J.H.,** Hazards to health; the hazards of farming, *N. Engl. J. Med.,* 270, 839, 1964.

Child-Operated Motorcycle Accidents: The Pattern, Severity, and Sequelae of Injuries

K. Yong-Hing, E. Howlett, and J. S. Owen

INTRODUCTION

In recent years, a broad range of new recreational vehicles has appeared on the market. These small-frame motorized vehicles, two- and three-wheeled, are now being promoted for use by children as young as 5 years old.

While there have been many studies of injuries in adult motorcycle accidents (MCAs), there have been few published studies of child-operated MCAs.[1-9] In a report on MCAs in teenage males, Avery discussed only 15- to 19-year-old victims.[6] King reported MCAs in a country area in Australia and included 14 children under 16 years old.[2] Speca and Cowell reported the injuries in 21 adolescents, ranging in age from 9 to 16 years, sustained while operating minibikes, motorcycles, and go-carts in Delaware.[7] Sherman and MacKinnon reported 24 children. Half were injured while scrambling at official club meetings and the other half while riding off the road.[8]

We reviewed the charts of children admitted to the University Hospital, Saskatoon for injuries sustained in child-operated MCAs in order to determine the pattern, severity, and sequelae of the injuries. We also hoped to obtain an indication of the number of children involved, the circumstances of the accidents and perhaps arrive at conclusions regarding measures that might prevent these injuries.

MATERIALS AND METHODS

We carried out a retrospective review of the children under 16 years of age admitted to the University Hospital, Saskatoon in the 6 years 1978—1983 for injuries sustained in MCAs. This group was selected because 16 years is the age at which a Canadian becomes eligible for a licence to operate a motorcycle. Each child was either the operator of the motorcycle or was a pillion rider (passenger) behind another child.

The children were riding two- or three-wheeled motorized vehicles which were referred to as motorcycles, motorbikes, trail bikes, competition bikes, minibikes, dirt bikes, or all-terrain vehicles. No children were on go-carts.

In listing the injuries, we omitted several minor injuries such as broken teeth, lacerations, abrasions, and burns. We excluded children whose injuries were treated on an outpatient basis and who were not admitted. We also excluded two children who were admitted for injuries sustained while riding pillion behind an adult operator.

We obtained from the hospital medical record statistics the number of children under 16 years admitted for injuries sustained in accidents as passengers in closed vehicles (cars, vans, and trucks) and the number of adults admitted following MCAs in the same 6 years.

RESULTS

There were 51 admissions following 51 MCAs involving 50 children — one boy was admitted twice following two accidents 12 months apart. The age of the children ranged from 6 to 15 (Table 1). There were 45 boys and 6 girls. Seventy-six percent of the admissions were in the months May to August.

In the same period of 6 years, 162 children were admitted following accidents in closed vehicles (cars, vans, and trucks) and 146 adults were admitted following MCAs.

The majority (87%) of the injuries were to the head and extremities with few injuries to the trunk (13%) (Table 2).

TABLE 1
Age Distribution of Patients Involved in 51 Child-Operated Motorcycle Accidents

Age (years)	Number of children
6	1
7	0
8	1
9	0
10	3
11	5
12	5
13	13
14	6
15	17

TABLE 2
Anatomical Region of Injuries in 51 Child-Operated Motorcycle Accidents

	Number of children
Cranio-facies	22
Upper extremity	17
Lower extremity	26
Chest	4
Abdomen	2
Pelvis	2
Spine	2

Many injuries were serious (Table 3) and they had serious sequelae in 12 children (Table 4). Average hospital stay was 14.4 d (range 1 to 73). There was an average of 1.4 surgical procedures per admission.

Twelve of the 51 MCAs occurred because of a collision with a car, truck, or van and the pattern of injury in these victims was different from the other 39 accidents which were the result of falls. The injuries in these 12 children were more often to the head and trunk. There were eight head, two chest, two abdomen and two pelvic injuries. All three deaths were in this group.

Thirty-two of the 51 accidents occurred off the road on private land in fields, trails, dirt tracks, or backyards; 13 occurred on the road; and in the remaining 6 the site of the accident was unknown.

DISCUSSION

University Hospital, Saskatoon is a secondary and tertiary referral center which attracts more seriously injured children. However, a comparison of the 51 child-operated MCA admissions with 162 child admissions for injuries sustained in closed vehicle accidents, and 156 adult admissions for MCAs in the same period supports the conclusion that child-operated MCAs occur in worrying numbers. University Hospital is only one of five similar-sized referral hospitals in Saskatchewan (population 1 million). Furthermore, the coroner of Saskatchewan statistics shows that 14 children under 16 years died in accidents related to all-terrain vehicles and motorcycles in the 3 years 1981—1983.[10] This experience in Sakatchewan is unlikely to be unique.

TABLE 3
List of Injuries in 51 Motorcycle Accidents

Skull fracture	6
Intracranial hemorrhage	4
Mandible fracture	5
Orbit fracture	1
Clavicle fracture	4
Forearm fracture	6
Humerus fracture	3
Finger amputation	1
Femur fracture	11 (1)
Tibia fracture	17 (6)
Floating knee	6
Compartment syndrome	2
Femoral artery transection	2

TABLE 4
Permanent Sequelae Occurred in 12 of the 51 Children

Death	3
Amputation	2
Impaired intellect	3
Partial deafness	1
Diplopia	1
Knee instability	1
Ulnar nerve palsy	1

Stilwell, in a survey of the incidence of accidents in motorcycle scrambling in Great Britain, concluded that the sport is no more dangerous than other childhood activities.[9] However, in that survey, the accidents occurred in races that were organized, controlled, and supervised by adult officials. Sherman and MacKinnon found that the severity of injuries was the same in children injured while scrambling and those injured in off-the-road accidents.[8] The results of our study support Speca and Cowell's conclusion that child-operated MCAs lead in many instances to serious injuries with lifelong disabilities, and not infrequently, death.[7]

The pattern of injury in the group of 39 children whose accidents followed a fall and who had few injuries to the chest, abdomen, pelvis, and spine was different from the usual pattern of injury in adult urban MCAs. However, the 12 children whose MCAs followed a collision had an injury pattern and severity that resembled the adult urban MCAs.

Accidents are the leading cause of death among children in industrialized countries[11] and the rate of death from accidents among children 1 to 14 years is higher in Canada than in 19 other developed countries.[12] The results of our study constitute the five Ws of this type of research: who, what, when, where, and why? Armed with these data, we should consider what preventive measures are possible and practicable.

The nature of the majority of the injuries that occurred, fractures and crush injuries, suggest that protective clothing such as elbow pads and knee pads, even if their use could be enforced, would not affect the injury pattern, severity, or sequelae. The frequency of use of helmets in our patients could not be studied because of incomplete data.

Thirty-two of the 51 accidents in our study occurred off the road and would, therefore, not have been prevented by existing legislation which only affects the operation of motorcycles on highways. Whereas the laws restrict the operation of motorcycles on highways to licenced individuals, the only legislative control of the operation of motorcycles by children (or adults) off the public road is concerned with the protection of landowner's property and creating a nuisance.

Laws to control the operation of motorcycles by children or adults off the public road would be difficult to introduce, as they would encroach on the highly sensitive area of restriction of personal freedom. In addition, the policing of such laws would be difficult. Educational programs which warn the public of the health hazards of motorcycles have an effect that is short-lived and are probably no match against the glamour of advertisements in the media. Safety programs, especially if they were sponsored by government departments, would lend tacit approval of these vehicles for underaged children.

The authors have come to the conclusion that in the same manner and for the same reasons that the importation, manufacture, and distribution of unsafe children's toys are banned by federal legislation, laws should be introduced to stop the importation, manufacture, or sale of small-frame two- and three-wheeled motorized vehicles.

REFERENCES

1. **Trinca, G.W. and Dooley, B.J.,** The pattern of motorcycle injuries sustained by motorcyclists in Victoria in 1974 and 1975, *Aust. N.Z. J. Surg.*, 49, 203, 1979.
2. **King, P.,** The pattern of injury in motorcycle injuries in a country area, *Aust. N.Z. J. Surg.*, 49, 207, 1979.
3. **Zetta, J.P., Zetta, P., and Thanasophon, B.,** Injury patterns in motorcycle accidents, *J. Trauma,* 19, 833, 1979.
4. **Hodad, J.P., Echave, V., Brown, R.A., et al.,** Motorcycle accidents: a review of 77 patients treated in a three-month period, *J. Trauma,* 16, 550, 1976.
5. **Balcerak, J.C., Pacione, K., and States, K.,** Moped, minibikes and motorcycle accidents: associated injury problems, *N.Y. State J. Med.*, 78, 628, 1978.
6. **Avery, J.G.,** Motorcycle accidents in teenage males: a modern epidemic, *Practitioner,* 222, 369, 1979.
7. **Speca, J.M. and Cowell, H.R.,** Minibike and motorcycle accidents in adolescents: a new epidemic, *J.A.M.A.*, 232, 55, 1975.
8. **Sherman, K. and MacKinnon, J.,** Motorcycling injuries in children, *Br. Med. J.*, 289, 877, 1984.
9. **Stilwell, J.H.,** Motorcycle scrambling injuries in boys, *Br. Med. J.*, 1(6115), 1758, 1978.
10. **Garner, J.W.,** White paper: all terrain (off road) vehicles act questionnaire, Highways and Transportation, Saskatchewan, 1984.
11. Health of the worlds's children, *W.H.O. Chron.*, 33, 119, 1979.
12. **Millar, W.J.,** Accidental mortality of Canadian children, *Chronic Dis. Can.*, 2, 33, 1981.

Injuries from Farm Animals

P. Fretz

INTRODUCTION

There is considerable information available in the veterinary literature regarding veterinary zoonoses and public health hazards associated with farm animals. However, little, if any, information is available regarding injuries sustained by veterinarians from farm animals.[1] The purpose of this paper is to discuss the prevalence, cause, and nature of injuries sustained by veterinarians from farm animals.

ETHOLOGY OF FARM ANIMALS

"Ethology is the scientific study of the behaviour of animals in their natural or usual environments."[2] The customary environment of the farm animal is, of course, domestication. This study is, therefore, highly relevant to veterinary medicine. Veterinary ethology is the study or behavior of domesticated species when used particularly as a means of appraising health of function. Behavior of the mature animal is a mixture of inherited, or innate, and acquired components. Innate components consist of simple reflexes, compound responses, and complex behavior patterns. Acquired components consist of conditioned reflexes, learned responses, and general habits. These various components interact, generating a wide range of behaviors.

The domesticated large farm animals (cattle and horses) possess a relatively sophisticated herd-oriented society. The herd represents a complex social fabric with highly organized elements of communication, territorialism, dominance, and symbiotic relationships. In essence, within the herd both complex herd-oriented behavioral systems and individual interreactions exist. In their natural setting horses and cattle derive comfort and tranquility from this structured herd society. Ethologists now recognize that many individuals within the herd demonstrate motivated behavior. These animals will express pain, hunger, fear, rage, excitement, and anxiety in response to the appropriate stimulus.[2]

The domestication process imposes limits on the herd size and influences this social fabric. Overcrowding and isolation are areas that have been addressed because of their implication in the field of animal science and animal production. For the veterinarian, isolation of the individual animal is important in terms of restraint, examination, and treatment. These situations destroy the sanctity of the herd, and obstruct the communication process and the social order. This can lead to anxiety and a well-demonstrated pathologic behavioral pattern.[2]

Much of the behavior of the individual domesticated large animal is concerned with self-maintenance. Highly successful self-maintenance is the basis of animal productivity. Activities involved in self-maintenance can be broken into eight primary categories: reflex responses, feeding, kinesis, exploration, association, body care, territorialism, and rest. Reflex responses and territorialism are the two areas within this field that are of most concern to veterinarians.

Reflex behavior is one of the first lines of defense for the individual animal. Reflex responses typically occur promptly, are carried out rapidly, and are reasonably predictable. The suddenness of reflex responses is extremely important to veterinarians and their predictability makes preventative measures possible.

In general, territorialism is "bourgeois" in the natural setting. It simply entails proprietary behavior in respect to defense of all parts of the animal's home range and it is directed primarily against members of the same species. However, territorialism is responsible for much of the agonistic behaviour elicited in the "fight or flight" response. Obviously, if one reduces the animal's territory or imposes isolation, any comfort derived from the herd and home range will be obliterated, generating major behavioral changes. These behavioral changes must be anticipated by the veterinarian, who will find that the responses from the bombardment of stimuli may no longer be as predictable as before. A change in the nature of the beast from "fright and flight" to "fright and fight" may occur.

There is one unconditioned reflex which accentuates the "fright and fight" response that veterinarians should be aware of. The "reflex escape struggle" is readily seen when animals are placed in close confinement. This reflex is not a conditioned or learned phenomenon. It can be expected from most animals when abruptly removed from their domicile and placed in small quarters (chutes or stocks) and exposed to multiple stimuli. For example, animals that are loaded, shipped (individually), and restrained in a chute will possess an extremely active (agitated) reflex escape response. This response may even take precedence over their own well-being.

SURVEY

All members (167) listed in the Saskatchewan Veterinary Medical Association Register were surveyed by letter to determine the prevalent cause and nature of the injuries that they had received from farm animals. Of the 68 that responded (41%), 64 (38% of those questioned or 94% of those who answered the questionnaire) responded positively, indicating they had had one or more notable injuries. Four responded negatively (3% of those questioned or 6% of those who answered the questionnaire), indicating that they had not had any injuries from farm animals. This questionnaire did not consider the type of employment (large animal practice, small animal practice, government service, academic employment, etc.).

All those who responded positively recorded multiple injuries. Although there were no deaths to veterinarians (two reports recorded ranchers being killed), several veterinarians were forced to leave large animal practice due to chronic degenerative joint disease as a result of being kicked in the knee.

The injuries noted in the 64 positive responses were categorized:

1. There were 32 lacerations which occurred while performing elective minor surgical procedures such as castrating and dehorning. These lacerations were usually of little significance and involved the veterinarians hands. However, one individual was kicked while performing a cesarean section on a cow. As a result he stabbed his opposite elbow joint and severed the radial nerve.
2. There were 18 injuries sustained when using improper restraint, as seen when hands, arms, and legs become entangled in ropes and halters or when tying cattle and horses to fences and rails that were inadequate for the situation. These injuries varied from minor bruises, four accounts of broken arms and fingers, and two accounts of loss of fingers.
3. There were 56 injuries sustained when working in and around poorly constructed or inadequately designed facilities. In general, a poor chute complex provided inadequate rear restraint or a poorly designed head gate. These injuries were difficult to categorize since they were so diverse in nature. The vast majority of injuries involved being kicked in the lower extremities.
4. There were 26 accidents that occurred when loading and off-loading cattle and horses into a chute complex. These accidents usually involved crushing-type injuries with the individuals inside the chute complex. The vast majority of these accidents involved bruised or broken ribs.

The injuries varied from minor cuts and bruises, repeated kicks to the knees that led to chronic degenerative problems, to life-threatening situations which usually involved rope entanglements, crushing injuries within a chute complex, or severe kicks to the head and chest. Cattle accounted for the majority of the injuries; however, when horses were mentioned the injuries were severe.

VETERINARIANS' SELF-PERCEIVED ROLE

A subjective assessment of the survey would reveal that veterinarians are their own worst enemy. Veterinarians involved in large animal practice perceive that those who use their services expect them to be animal handlers, "cowboys — ranch hands", and this leads to a "macho profile". Veterinarians are expected to restrain, manipulate, and handle the individual large animal. This is usually performed in chute complexes; however, many times these chute complexes and associated facilities are inadequately designed and poorly constructed, making the job of the veterinarian extremely dangerous.

PRECAUTIONS

It is a fact that animals are examined and treated individually, and therefore veterinarians must be aware of animal behavioral patterns and should possibly take formal courses in ethology. Veterinarians should be aware of the predicted response of animals when placed in unusual situations and bombarded with stimuli. They should not develop a false sense of security or overfamiliarity with large animal patients. They should take time to think through situations to survey the possible responses of the animal. Most importantly, veterinarians should refuse to work in poorly constructed and inadequately designed facilities which may endanger their health and/or life.

CONCLUSIONS

The injuries recorded in this survey categorically reflect either working in conditions known to be inadequate or failure of the veterinarian to appreciate and respond to the predicted (anticipated) behavioral reflexes of the farm animal. One must remain constantly cognizant of the possible and predictable response that a stimulus or bombardment of stimuli may invoke. Failure to do so when working with a beast ranging from 1000 to 2000 lbs could prove fatal.

REFERENCES

1. **Schwabe, C.W.,** *Veterinary Medicine & Human Health,* 3rd ed., Williams & Wilkins, Baltimore, 1984.
2. **Fraser, A.F.,** *An Introduction to Veterinary Ethology,* 1st ed., University of Saskatchewan Press, Saskatoon, 1979.

Degenerative Knee Disease among Dairy Farmers

C. Anderson, P. S. Treuhaft, W. E. Pierce and E. P. Horvath

INTRODUCTION

One of us (P.S.T.) noted an apparently disproportionate number of dairy farmers among patients with degenerative knee disease at the Marshfield Clinic, a large rural group practice in central Wisconsin. We studied medical records to determine if dairy farmers did have a higher prevalence of clinical degenerative knee disease than workers in other occupations.

METHODS

Degenerative knee disease was defined as osteoarthritis of the tibiofemoral joint or meniscal damage, which was not attributed either to a single trauma, the effects of condylar fracture, or inflammatory joint disease. Meniscal damage was diagnosed by arthrogram or surgery. The study population was comprised of males aged 20 to 60 who had been continuously enrolled in the Greater Marshfield Community Health Plan (GMCHP) for the last 5 years. This prepaid health plan has about 64,000 members, of which 6200 met the criteria for inclusion in the study population. Most medical care and all orthopedic care for GMCHP are delivered at the Marshfield Clinic. Joint medical records of the clinic and the adjoining St. Joseph's Hospital document all inpatient and outpatient care in a single folder.

A computer search identified 741 patient records that included a diagnosis or history of joint pain, internal derangement of the knee, knee sprain, or degenerative arthritis of the knee. Each of these records was reviewed to determine if it met our definition of degenerative knee disease. The date of onset was calculated from the date of the initial physician's note of knee symptoms and the duration of symptoms recorded in that note.

For each case, a male control subject, born within 1 year of the case, was randomly selected from the study population. Patients were excluded from the control group if no notation of their occupation or employer was recorded, or if they received all of their care at one of the small GMCHP clinics and, thus, had no record at the Marshfield Clinic.

We recorded the primary occupation of each case and matched control at the date of onset in the case. This date was selected in order to determine the occupation before any job changes in response to knee symptoms. The occupation was found on a social history form or in the physician's notes. The subject was coded as a farmer if farming was noted as full-time or part-time occupation. Since the type of farming was not consistently noted, and dairy farming was sometimes mentioned only in reference to symptoms during milking, all farmers were considered together in the analysis. However, dairy farming is the predominant form of agriculture in this region. If the subject was not a farmer, the occupation listed as full-time or listed first was recorded. Nonfarm occupations were grouped into two groups: blue collar workers, including service workers, craftsmen, operatives, and laborers, and white collar workers, including professionals, managers, sales people, and clerks.

For both cases and controls, the height and weight measurements closest to the date of onset of symptoms in the case were used. These were summarized by the body mass index,[1] which is equal to the mass in kilograms divided by the square of the height in meters.

The data were analyzed using conditional logistic regression. This technique estimates the simultaneous effect of several risk factors on disease occurrence, while accounting for the matching of cases and controls. Risk factors are assumed to have a multiplicative effect when they occur in the same subject. The statistic provided by this analysis is the prevalence odds ratio, which estimates the ratio of the incidence rate among subjects with the risk factor to the incidence rate in a reference group without the risk factor, assuming the risk factor is unrelated to recovery or death and is not influenced by the disease. In this study, the reference group for occupational comparisons was white collar workers, so that the odds ratio for white collar workers was always 1.00.

RESULTS

We identified 176 cases of degenerative knee disease. One hundred nineteen had clinical evidence of osteoarthritis, 26 had meniscal damage, and 31 had both. The mean age of cases and controls was 44.4 (±8.8 SD) years at the time of onset in the case, and 51.3 (±7.3 SD) years at the time they were selected for this study. Primary occupation was known for 170 pairs and is shown in Table 1.

TABLE 1
Primary Occupations of Cases of Degenerative Knee Disease and Controls, Marshfield, WI, 1984

	Cases	Controls
Farmers	90	55
Dairy	67	27
Unspecified	23	28
Blue collar	59	62
White collar	25	53
Total	170	170

The relationship of occupation to degenerative knee disease is shown in Figure 1. In each age group, the odds for farmers and blue collar workers are compared to the odds for white collar workers. The odds ratios for farmers in the three age groups were similar, ranging from 3.60 for farmers aged 50 to 60 to 5.20 for farmers aged 40 to 49. For ages 20 to 60, the odds ratio for farmers was 4.30 ($p = 0.0001$). At age 20 to 39, the prevalence in blue collar workers was approximately that in white collar workers (odds ratio = 0.66), but by age 50 to 60, the prevalence in blue collar workers rose to approximately that in farmers (odds ratio = 3.87).

Height and weight for both the case and control subject were known for 135 pairs. As shown in Table 2, the cases and controls were the same height, but the cases were 9.7 lbs heavier and, thus, had a significantly greater body mass index ($p = 0.02$).

In 133 pairs, data on occupation, height, and weight were available for both the case and control. We examined the simultaneous effect of occupation, body mass index, the interaction of body mass index with occupation, and the interaction of age with occupation and body mass index in these subjects. Of the interactions, only blue collar work by age was significant. Removing the other interactions did not alter the odds ratios for farming or blue collar work. Removing body mass index, leaving only occupation and the interaction of blue collar work and age, slightly increased the odds ratio for farming (from 3.28 to 3.60), somewhat decreased the odds ratio for blue collar work (1.60 to 1.42), and did not alter the odds ratio for the interaction of blue collar work by age in years (1.11). This analysis confirmed the results of the stratified analysis shown in Figure 1, which was based on 170 case-control pairs.

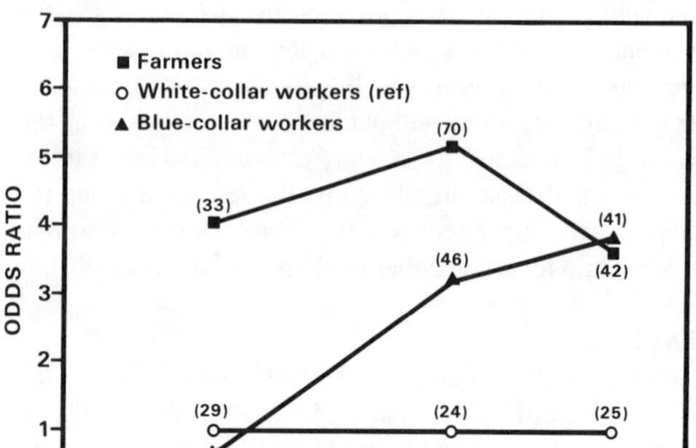

FIGURE 1. Prevalence odds ratio for degenerative knee disease, by occupation and age, Marshfield, WI, 1984. The number of subjects represented by each data point is shown in parentheses.

TABLE 2
Body Build of Degenerative Knee Disease Cases and Controls, Marshfield, WI, 1984

	n	Height (in.)	Weight (lb)	Body mass index (kg/m²)
Cases	135	69.5	200.9	29.2
Controls	135	69.3	191.2	27.9
p (two-tailed t test)		0.58	0.03	0.02

DISCUSSION

The finding of an interaction between blue collar work and age was not expected. Blue collar work includes a variety of activities, some of which may have a delayed effect compared to farming. However, in this community many older blue collar workers may have been farmers in the past. We did not abstract data on past occupations.

This study included only subjects who sought medical care. If farming influenced the symptoms of degenerative knee disease, or if farmers were more or less likely than other workers to seek medical care, farmers may have had a different probability of becoming cases in this study. We selected subjects who had been members of GMCHP for 5 years or more in order to assure that subjects had ample opportunity to seek care. In a follow-up study we plan to compare the severity of radiographic changes in farmers and nonfarmers.

No previous study has specifically addressed the occurrence of degenerative knee disease in farmers. When all joints are considered together, farmers have been shown to have higher rates of hospitalization and disability due to osteoarthritis than are found among other workers.[2,3]

Mechanical loading, in particular impulsive loading, has been shown experimentally to cause changes in articular cartilage that are characteristic of early osteoarthritis.[4,5] High rates of osteoarthritis have been reported in specific joints of workers subjected to mechanical loading, including elbows and knees of British coal miners;[6,7] elbows of workers using vibrating tools;[6,8-10] fingers, elbows, and knees of dock workers;[11] and hands of cotton mill workers.[12,13] Only in cotton mill workers has an association been demonstrated between patterns of usage in specific joints and the occurrence of osteoarthritis.[13]

We do not know what aspect of farming is responsible for this increased prevalence of degenerative knee disease among farmers in our study population. The majority of them are dairy farmers, but they also grow their own hay and silage and are exposed to a variety of hazards. Two possible explanations for the increased prevalence are acute trauma and chronic biomechanical stress. Most dairy farmers have been kicked by cows, and repeated acute trauma may have a cumulative effect on degenerative knee disease. However, we attempted to exclude cases in which the symptoms originated with a single trauma.

Frequent squatting during machine milking may be a source of biomechanical stress, and should be investigated as a possible cause of degenerative knee disease among dairy farmers. In the flat barns that are still used on almost all Wisconsin dairy farms, farmers squat to wash the udders, to apply and remove the vacuum milking unit, to dip the udders in a protective solution, as well as to adjust the vacuum milking unit or massage the cow's bag. Since cows are usually milked twice daily, farmers squat at least eight times daily for each cow. The average Wisconsin dairy farm has 42 cows.[14]

While the effect of specific farming practices on degenerative knee disease is not known, measures do exist to decrease the stress on the knee during machine milking. The ideal workstation for milking is a milking parlor, in which cows stand in an elevated stall during milking, and the farmer can reach the udders comfortably while he stands. However, the addition of a milking parlor is probably not economically feasible for most small farms. Some farmers sit on a one-legged strap-on stool while milking. The stool hangs loosely behind the farmer when he stands. Kneeling on one knee, which is protected by an industrial knee pad, is another alternative to squatting. While kneeling has its own hazards, such as prepatellar bursitis,[15,16] the forces are distributed over a larger area by an appropriate pad.

CONCLUSIONS

Farmers in our study population have an increased prevalence of clinical degenerative knee disease, which may be specifically attributable to dairy farming. Further biomechanical and epidemiologic research is needed to determine the effects of specific farming practices, such as squatting during machine milking, on this disease. Currently, simple prudent means, such as strap-on stools and knee pads, can be used to decrease the stress on the knees of dairy farmers.

REFERENCES

1. **Keys, A., Fidanza, F., Karvonen, J., Kimura, N., and Taylor, H.L.,** Indices of relative weight and obesity, *J. Chronic Dis.*, 25, 329, 1972.
2. **Burkhart, J.A., Egleston, C.F., and Voss, R.J.,** The Rural Health Study: A Comparison of Hospital Experience between Farmers and Nonfarmers in a Rural Area of Minnesota, Publ. No. 78-184, National Institute for Occupational Safety and Health, Cincinnati, 1978.
3. **Kennedy, J. and Fischbach, T.J.,** Occupational Characteristics of Disabled Workers, Publ. No. 80-145, National Institute for Occupational Safety and Health, Cincinnati, 1980.
4. **Radin, E.L.,** Mechanical aspects of osteoarthritis, *Bull. Rheum. Dis.*, 26, 7, 1976.
5. **Radin, E.L., Swann, D.A., Paul, I.L., and McGrath, P.J.,** Factors influencing articular cartilage wear *in vitro, Arthritis Rheum.*, 25, 974, 1982.
6. **Lawrence, J.S.,** Rheumatism in coal miners. III. Occupational factors, *Br. J. Ind. Med.*, 12, 249, 1955.
7. **Kellgren, J.H. and Lawrence, J.S.,** Osteo-arthritis and disc degeneration in an urban population, *Ann. Rheum. Dis.*, 17, 388, 1958.
8. **Hunter, D., McLaughlin, A.I.G., and Perry, K.M.A.,** Clinical effects of the use of pneumatic tools, *Br. J. Ind. Med.*, 2, 10, 1945.
9. **Bovenzi, M., Petronio, L., and DiMarino, F.,** Epidemiological survey of shipyard workers exposed to hand-arm vibration, *Int. Arch. Occup. Environ. Health*, 46, 251, 1980.
10. **Wright, V.,** Biomechanical factors in the development of osteoarthritis, epidemiological studies, in *Epidemiology of Osteoarthritis*, Peyron, J.G., Ed., Ciba-Geigy, Rueil-Malmaison, France, 1981.
11. **Partridge, R.E.H. and Duthie, J.J.R.,** Rheumatism in dockers and civil servants: a comparison of heavy manual and sedentary workers, *Ann. Rheum. Dis.*, 27, 559, 1968.
12. **Lawrence, J.S.,** Rheumatism in cotton operatives, *Br. J. Ind. Med.*, 18, 270, 1961.
13. **Hadler, N.M., Gillings, D.B., Imbus, H.R., Levitin, P.M., Makuc, D., Utsinger, P.D., Yount, W.J., Slusser, D., and Moskovitz, N.,** Hand structure and function in an industrial setting: influence of three patterns of stereotyped, repetitive usage, *Arthritis Rheum.*, 21, 210, 1978.
14. **Spencer, C.D., Davies, J., Lippert, E., Row, J.S., Kriesel, R.F., Heiser, M., and Thompson, W.,** *1984 Wisconsin Agricultural Statistics*, Wisconsin Agriculture Reporting Service, 1984.
15. **Tanaka, S., Smith, A.B., Halperin, W., and Jensen, R.,** Carperlayere's knee (letter), *N. Engl. J. Med.*, 307, 1276, 1982.
16. **Bracken, T.H.,** Genu genuflectorum (letter), *N. Engl. J. Med.*, 308, 1107, 1983.

Low Back Pain in the Farmer — An Outline

P. R. Dupuis

INTRODUCTION

The Problem of Back Pain in the General Population

The number of patients seeking treatment for low back pain (LBP) is steadily increasing.[1-3] The prevalence of back pain has been reported to be anywhere from 5 to 25% and the incidence from 30 to 90%.[4-9] It has been shown that 85 to 90% of patients with attacks of LBP will have recovered within 8 to 10 weeks. Those 10 to 15% of back pain patients off work for more than 8 weeks account for about 80% of the cost[10,11] and get 30% of the total of compensation dollars. Pheasant reviewed the admission data from California for the year 1974.[12] He then extrapolated his findings to the national level and estimated that about 721,000 of the hospital discharges for that year in the U.S. were due to back pain-related problems. The average length of stay was 9 d for a projected estimate of 6,482,340 patient days and 200,000 operations. Even if most pathologies causing LBP and sciatica are minor in nature and self-limiting in time, the cost of treatment, the impact on health care delivery and the effects on industry are becoming a terrible burden for society.[2,9,13]

Back Pain: The Farmer vs. Other Occupations

Very little is known about back pain in the farmer and most of what has been published has been anecdotal. However, a series of papers by Magora[14-16] on the relation between LBP and occupation, which was published in the mid-1970s, gives us a good insight into the problem. He evaluated both subjectively and objectively 429 patients with LBP from a total of 3316 subjects belonging to eight preselected occupations: bank clerk, post office clerk, bus driver, policeman, nurse, farmer, light industry worker, and heavy industry worker.

Age at Onset of Back Pain Symptoms

In 48% of farmers age at onset of symptoms was 30 or less. This is comparable to heavy industry workers (50%), nurses (46%), and bank clerks (63%). The rest of the selected occupations showed fairly even distribution of percentage of onset for each decade.

Relation Between Period Since Onset of LBP and Occupation

There was no statistical difference between groups. Seventy to eighty percent of persons will see their back pain begin within 1 to 7 years of starting their job. The peak distribution interval for farmers seemed a little later than other occupations at the 4- to 7-year interval.

Relation Between Type of Onset of LBP and Occupation

Except for bus drivers, policemen, and heavy industry workers where in the majority the onset was sudden, all other occupations were relatively similar. The bank clerk was strikingly different in that in only 10.5% of cases was the onset sudden. It is interesting to note that in almost 48% of farmers the type of onset was sudden and that in 25% the type of onset could not be defined.

Relation Between Subjective Assessment of the Cause of LBP and Occupation

Weightlifting and bending were the most commonly incriminated causes of LBP. In farmers, lifting and sudden back motion were incriminated equally. It is very interesting to note that a high percentage of persons in each occupation could not relate their back pain to any specific cause.

Relation Between Aggravating Factors and Occupation

There was no significant difference between groups and most noted standing, bending, and coughing as definite aggravating factors. It is interesting to note that farmers were the only occupational group that did not think of their work as an aggravating factor.

Relation Between Occupation Neurological and/or Root Tension Signs

Farmers, policemen, and bus drivers were found to have the least neurological findings. Most neurological signs were found in light and heavy industry workers, and the most extensive neurological findings in light and heavy industry workers and nurses.

ACTIVITIES RELATED TO BACK PAIN IN THE FARMER AND OTHER HEAVY WORKERS

Farm Equipment Operation and Prolonged Sitting

Most farmers operate heavy equipment for prolonged periods of time from late April to late October, more so during seeding and harvesting. Prolonged sitting and driving have been correlated with an increase in the risk of developing disk disease.[18-20] The exact mechanism is unknown. The loading of a disk is a complex mechanism. There are two ways by which a disk or other viscoelastic structures absorb energy. The first is by the phenomenon of 'creep'. As the load is applied, the disk space will slowly get thinner with the passage of time and serial measurements will yield a time/displacement curve. One can verify this by measuring oneself upon getting out of bed in the morning and remeasuring onself just before going to bed at night. What happens is that in the upright posture, the constant weight of the body acts upon the disk during the hours of the day, deforming it slowly. During the hours of the night, the disk is unloaded and relaxes back to original height. This is probably related to reimbibition of water expressed out of the disk during the day. The second mechanism is the one that is involved in absorbing the constant and repeated axial pounding produced by driving machinery over a bumpy field for hours at a time. It is called "hysteresis". This is the phenomenon by which the disk or other viscoelastic structures absorb energy when subjected to temporary or repeated axial loading. The energy is dissipated by the deformation of the structure. Less energy is released by the structure in recovering from the effects of an applied force than the energy needed to deform it. One can see that by repeatedly loading a disk, hysteresis will slowly diminish thereby decreasing the shock-absorbing qualities of the disk. Unfortunately, as hysteresis decreases the likelihood of permanent injury to the disk structure increases.

Lifting, Bending, Twisting

Farmers often have to work in awkward positions around very dangerous equipment and must lift heavy weights not held in ergonomical positions. This produces abnormal stresses on the spine. Abnormal mechanical stresses seem to be closely related to the development of LBP and certainly to its recurrence and maintenance of symptoms. Clinically, every one of us experiencing back pain has more pain when the back is mechanically stressed. One way of measuring stresses applied to the spine is to measure the pressures generated within the disk while performing different tasks or assuming different postures. Nachemson[21,22] and his group in Sweden obtained such *in vivo* measurements from normal volunteers. Measurements were obtained from the L3 disk space. His findings are reported in Table 1. It is interesting to note that lifting, especially with rotation or arm extended, even with a small weight produces almost five times the standing-at-ease disk pressure. Repeatedly lifting or bending in an abnormal position can decrease the hysteretic properties of the disk more rapidly than the mechanically correct posture. Also of interest is the fact that patients with back pain due to disk disease tend to report worsening with the sitting posture, which seems to correlate with Nachemson's findings that sitting pressures are 40% higher than standing-at-ease pressures.

TABLE 1
Pressure Generated Within the Disk

Position	N[a]
Supine	
Awake	250
Semi-Fowler	100
Traction 500 N	0
Tilt table 50°	400
Arm exercises	500
Upright sitting, no support	700
Sitting, lumbar support, inclined 110	400
Standing at ease	500
Coughing	600
Straining	600
Forward bend	
20	600
40	1000
20 with 20 kg	1200
Bilateral leg lift	800
Sit-up exercises	1200
Lifting 10 kg back straight	1700
Lifting 10 kg back bent	1900
Holding 5 kg arm extended	1900
Forward flexed 20, rotated 20 with 10 kg	2100

[a] Newton = force necessary to give a mass of 1 kg an acceleration of 1 m/s^2. 1 N = 0.1 kp = 2.25 lbf.

Pushing, Pulling

Davis[23] and Davis and Troup[24] studied the effects of pushing, pulling, and lifting on intraabdominal pressures. It was observed that elevated pressures were present when pushing and low pressures were present when pulling. This was associated with tense abdominal muscles when pushing and tense paraspinal muscles when pulling. Pulling produces a forward bending moment which is resisted by the paraspinal muscles. Since these muscles are situated near the vertebral center of rotation, their lever arm is very short. They therefore have to develop very high tension to resist the forward bending moment. This produces high loads on the disk. In contrast, pushing produces an extension or backward bending moment. The abdominal muscles are far away from the vertebral center of rotation. They need to develop much lower tension to resist the extension moment, since their lever arm is so much longer. This produces much smaller loads on the disk. They have also shown that pacing the push, pull, or lift is very important because high accelerations tend to yield very high pressures.

CONCLUSIONS

All of the findings in Magora's study tend to underline the fact that farmers are most often stricken with mechanical LBP. This seems to be corroborated by our personal experience here in Saskatoon. Of our rural patients, 75 to 80% had been treated successfully over the years by their local chiropractor or physical therapist for recurrent bouts of intermittent low back and leg pain where back pain strikingly was dominant, before referral to the Back Clinic. This, in our experience, usually indicates that the origin of back pain is of mechanical origin. This is radiologically related to moderate to severe degenerative changes in the lumbar and lumbosacral motion segments. Interestingly, as noted before by Katevuo et al.,[25] the anterior part of the motion segment seems to be more at risk in these patients (spondylosis). However, this cannot be related to severity of presenting symptoms.[26]

REFERENCES

1. **Andersson, G.B.J.,** Epidemiologic aspects of low back pain in industry, *Spine,* 6, 53, 1981.
2. **Andersson, J.A.D.,** Back pain and occupation, in *The Lumbar Spine and Back Pain,* Jayson, M., Ed., Pitman Medical, London, 1980, 57.
3. **Magora, A.,** An investigation of the problem of sick leave in the patient suffering from low back pain. *Ind. Med. Surg.,* 38, 398, 1969.
4. **Andersson, J.A.D.,** Low back pain — cause and prevention of long term handicap (a critical review), *Int. J. Rehabil. Res.,* 3, 89, 1981.
5. **Hirsch, C., Jonsson, B., and Lewin, T.,** Low back symptoms in a Swedish female population, *Clin. Orthop. Relat. Res.,* 63, 171, 1969.
6. **Hult, L.,** The Munkford investigation, *Acta Orthop. Scand. Suppl.,* 16, 1954.
7. **Hult, L.,** Cervical, dorsal, and lumbar spinal syndromes, *Acta Orthop. Scand. Suppl.,* 17, 1954.
8. **Nachemson, A.L.,** Low back pain. Its etiology and treatment, *Clin. Med.,* 7, 18, 1971.
9. **Nachemson, A.L.,** The lumbar spine: an orthopaedic challenge, *Spine,* 1, 59, 1976.
10. **Benn, R.T. and Wood, P.H.N.,** Pain in the back; an attempt to estimate the size of the problem, *Rheumatol. Rehabil.,* 14, 121, 1975.
11. **Pheasant, H.C.,** The problem back, *Curr. Pract. Orthop. Surg.,* 7, 89, 1977.
12. **Pheasant, H.C.,** Backache — its nature, incidence and cost, *West. J. Med.,* 126, 330, 1977.
13. **Rowe, M.L.,** Low back pain in industry, a position paper, *J. Occup. Med.,* 11, 161, 1969.
14. **Magora, A.,** Investigation of the relation between low back pain and occupation. IV. Physical requirements: bending, rotation, reaching and sudden maximal effort, *Scand. J. Rehab. Med.,* 5, 186, 1973.
15. **Magora, A.,** Investigation of the relation between low back pain and occupation. V. Psychological aspects, *Scand. J. Rehab. Med.,* 5, 191, 1973.
16. **Magora, A.,** Investigation of the relation between low back pain and occupation. VI. Medical history and symptoms, *Scand. J. Rehab. Med.,* 6, 81, 1974.
17. **Magora, A.,** Investigation of the relation between low back pain and occupation. VII. Neurologic and orthopaedic conditions, *Scand. J. Rehab. Med.,* 7, 146, 1975.
18. **Kelsey, J.L. and Ostfeld, A.M.,** Demographic characteristics of persons with acute herniated lumbar intervertebral disc, *J. Chron. Dis.,* 28, 37, 1975.
19. **Kelsey, J.L. and Hardy, R.J.,** Driving motor vehicles as a risk factor for acute herniated lumbar intervertebral disc, *Am. J. Epidemiol.,* 102, 63, 1975.
20. **Kelsey, J.L.,** An epidemiological study of acute herniated lumbar intervertebral disc, *Rheumatol. Rehab.,* 14, 144, 1975.
21. **Nachemson, A.,** The load on lumbar disc in different positions of the body, *Clin. Orthop. Relat. Res.,* 45, 107, 1966.
22. **Nachemson, A.,** Disc pressure measurements, *Spine,* 6, 93, 1981.
23. **Davis, P.R.,** The use of intra-abdominal pressure in evaluating stresses in the lumbar spine, *Spine,* 6, 90, 1981.
24. **Davis, P.R. and Troup, J.D.G.,** Pressures in the trunk cavities when pulling, pushing and lifting, *Ergonomics,* 7, 465, 1964.
25. **Katevuo, K., Aitasalo, K., Lehtinen, R., and Pietilla, J.,** Skeletal changes in dentists and farmers in Finland, *Community Dent. Oral Epidemiol.,* 13, 23, 1985.
26. **Magora, A.,** Relation between the low back pain syndrome and X-ray findings. I. Degenerative osteoarthritis, *J. Rehab. Med.,* 8, 115, 1976.

Spinal Cord Injuries in Agricultural Workers

M. Khan

INTRODUCTION

Work is a common cause of spinal cord injury. Studies from various countries have shown that a significant number of cord injuries occurred in occupational pursuits ranging from 20 to almost 50% of cord injuries.[1] Sutton[2] and Chesire[3] have reported an incidence of about 20% in two different regions in Australia, and Gehrig[4] found an incidence of 35% in Switzerland. In England, Frankel et al.[5] found an incidence of 27% and Damanski[6] recorded that occupational cord injuries comprised almost 50% of the cases.

Not surprisingly, the type of occupational hazards causing cord injuries varies considerably from one location to another. This was noted by Peerless and Schweigel[7] who reported an incidence of 45% of cord injuries due to logging accidents in workers of British Columbia.

This study suggests a threefold increase in the incidence of spinal cord injuries due to farming accidents relative to other work-related accidents within the last few years. These have all been complete cord injuries with major fracture-dislocations and very limited potential for neurological recovery.

MATERIALS AND METHODS

The records of patients with acute spinal cord injuries admitted to the major hospitals in Saskatoon and Regina during the years 1980—1985 were reviewed. In addition, the records of all patients with spinal cord injuries admitted to the University Hospital in Saskatoon during the years 1955—1979 were analyzed. Regina hospitals are referral centers for the southern half of Saskatchewan, which is slightly more populated than the northern half which is serviced by the Saskatoon hospitals. The population of Saskatchewan remains fairly stable over the past 20 years at approximately 900,000 to 1 million.

RESULTS

It is estimated that within the last 5 years, about 140 patients suffered from acute and complete spinal cord injuries from various causes (Tables 1 and 2). The incidence of spinal cord injuries has progressively and significantly increased over the past 20 years in Saskatchewan. Road accidents remain the most common cause of acute spinal cord injury. However, spinal cord injuries from farming accidents have dramatically increased relative to other causes. For example, between the years 1955—1979, about 460 patients sustained severe cord injuries of whom 38 resulted from farming accidents. Within the last 5 years, 140 patients sustained cord injuries, 20 as a result of farming accidents. Of these, 11 cases occurred within the past three years. Of these 11 patients, 10 sustained cord injuries as a result of accidents with front-end loaders while lifting bales of hay weighing between 500 and 1000 lb. Other types of farming accidents causing spinal cord injuries include rolling tractors, falling from barns, and falling off horses.

About 70% of cord injuries were confined to the cervical region, while the remaining 30% were localized to the thoracic or thoracolumbar regions. A few patients had multiple level involvement. All patients were males ranging in age from 20 to 70 years with a preponderance of men over 50 years of age. The accidents occurred on farms. The mortality rate during the first 2 years from the time of the injury is approximately 25% for cord injuries due to farming accidents as compared to 15% for cord injuries secondary to other causes.

DISCUSSION

Major injuries to the cervical spinal cord, especially at an older age, carry a high mortality rate. The morbidity from all spinal cord injuries is great because of the devastating effects such injuries have on most body functions. Apart from the serious psychology and social sequelae on patients and relatives, such injuries invariably have profound economical impact on society. It is estimated that the cost to society of each new quadriplegic patient amounts to hundreds of thousands of dollars in lifelong medical costs and lost earnings.

This study suggests a marked preponderance of males over

TABLE 1
Causes of Spinal Cord Injury (1955—1979)

Causes	%
Traffic	44
Industry and lumber	26
Farm related	8
Others	22

TABLE 2
Causes of Spinal Cord Injury (1979—1984)

Causes	%
Traffic	36
Industry and lumber	28
Farm related	15
Others	21

females with spinal cord injury from all causes. With respect to occupational and farming accidents causing spinal cord injuries, all patients were male. The vast majority of patients with cord injuries resulting from farming accidents were middle-aged or older men. This may be a factor in explaining the higher mortality rate in this group of cord-injured patients compared to other cord-injured groups.

Within the past 5 years there has been a preponderance of front-end loader accidents causing acute and complete spinal cord injuries among the farming accidents. These machines are used to lift large and heavy objects such as bales of hay weighing about 500 to 1000 lb. The large bucket carrying such objects tends to topple after reaching a certain critical angle. About 75% of patients who sustained cord injuries from this type of accident were quadriplegic. The remaining types of farming accidents causing spinal cord injuries include rolling tractors or other farm machinery, falling from barns, or falling off horses and are listed in Table 3.

It is not the intent of this paper to define the pathophysiology of acute spinal cord injury. Nevertheless, it would seem appropriate to note that severe trauma to the spinal cord rarely causes complete anatomic transection of the cord initially,[8] although it usually causes immediate and complete cessation of cord function, or spinal shock. The initial physical forces of trauma initiate a series of pathological changes that result in almost complete anatomic transection of the cord at the injury site. Experimental spinal cord injury studies indicate early hemorrhage in the spinal cord in every form of acute cord injury.[9,10] This is followed by severe posttraumatic ischemia leading to infarction of the cord.[11] Regeneration of neural tissue sufficient to restore cord function has not been possible. However, it may be possible to prevent or reduce cord ischemia by steroids, mannitol, diuretics, local hypothermic or normothermic irrigation, and hyperbaric oxygen.[12]

Patients suffering from acute and complete spinal cord injuries from all causes were managed according to established principles in most neurosurgical centers. These included immediate hemodynamic and respiratory support in an attempt to prevent or reduce posttraumatic ischemia and infarction of the spinal cord. Steroids, mannitol, and diuretics have been used, although not consistently. Spinal immobilization was necessary for spinal instability to prevent further cord or root injury and allowed for rapid mobilization of patients. This was achieved by closed methods including skull tongs and traction to reduce fracture- dislocation, and halo devices or operative fusion anteriorly or posteriorly, with or without internal fixation devices. Persistent cord compression required surgical decompression either anteriorly or posteriorly. No patient was treated with hypothermic irrigation or hyperbaric oxygen. Apart from some return of root function, no patient admitted with the clinical diagnosis of complete cord injury showed any signficant return of cord function. Mortality from complete spinal cord injuries due to farming accidents remains extremely high, rating between l5 and 25% within the first 5 years after injury. Deaths were often attributed to renal and pulmonary complications.

Prevention of all spinal column and spinal cord injuries is not possible. However, increased public awareness of the hazards associated with use of various farm machinery may well reduce or prevent most spinal cord injuries from farming accidents. The medical profession should take a more active role in communicating and educating farmers as well as manufacturers of farm machinery regarding the dangers of using such machines as well as possible improvement in their mechanics and design.

TABLE 3
Types of Arming Accidents Causing Spinal Cord Injuries (1979—1984)

Front-end loaders	12
Rolling tractors	4
Falling from barns	2
Falling off horses	2
Total	20

REFERENCES

1. **Kurtrzke, J.F.,** Epidemiology of spinal cord injury, *Exp. Neurol.,* 48, 163, 1975.
2. **Sutton, N.G.,** Injuries of the spinal cord, in *The Management of Paraplegia and Tetraplegia,* 1st ed., Butterworth, London, 1973.
3. **Chesire, D.J.E.,** The complete and centralized treatment of paraplegia. A report on the spinal injuries centre for Victoria, Australia, in Proc. 16th Annu. Spinal Cord Injury Conf., 1967, 39.
4. **Gehrig, R. and Michaelis, L.S.,** Statistics of acute paraplegia and tetraplegia on a national scale in Switzerland, *Paraplegia,* 12, 145, 1968.
5. **Frankel, H.L. et al.,** The value of postural reduction in the initial management of closed injuries of the spine with paraplegia and tetraplegia, *Paraplegia,* 7, 179, 1969.
6. **Damanski, M.,** The paraplegic patient as a social problem, *Paraplegia,* 2, 169, 1964.
7. **Peerless, S.J. and Schweigel, J.F.,** The medical and economic fate of twenty-nine industrial spinal cord injured patients, *Paraplegia,* 12, 145, 1974.
8. **Sandler, A.N. and Tator, C.H.,** Review of the effect of spinal cord trauma on the vessels and blood flow in the spinal cord, *J. Neurosurg.,* 45, 638, 1976.
9. **Dohrmann, G.J., Wagner, F.C., and Bucy, P.C.,** The microvasculature in transitory traumatic paraplegia: an electron microscopic study in the monkey, *J. Neurosurg.,* 35, 263, 1971.
10. **Khan, M. et al.,** Hemorrhagic changes in experimental spinal cord injury models, *Can. J. Neurol. Sci.,* 10, 161, 1983.
11. **Tator, C.H. and Rowed, D.W.,** Current concepts in the immediate management of acute spinal cord injuries, *Can. Med. Assoc. J.,* 121, 1453, 1979.
12. **Tator, C.H.,** Acute spinal cord injury: a review of recent studies of treatment and pathophysiology, *Can. Med. Assoc. J.,* 107, 143, 1972.

High-Pressure Injection Injuries in Farmers

G. H. F. Johnston

INTRODUCTION

Traditionally recognized as injuries of the automotive industry, high-pressure injection injuries of the hand are rarely encountered outside the industrial heartlands of the U.S. and Canada. However, when they do occur outside this sphere, the results of neglected or inadequate treatment are so devastating to ultimate hand function that any person who uses high-pressure gun equipment should be alerted to the inherent danger of that equipment, as well as how to respond should an accidental injection occur. Equally cognizant should be the family practitioners and surgeons to whom these injuries present. Like the calm before the storm, these injections frequently appear innocuous and inconsequential, but they progress to become excruciatingly painful lesions, with amputation and significant long-term disability all too often the dissatisfying outcome.

Farmers who maintain their own farm equipment are a population at risk for high-pressure injection injuries. Schoo et al.[1] described a typical history and injury course in a farmer's wife.

Experience at University Hospital in Saskatoon reflects the rarity of this entity. Of eight high-pressure injection injuries of the hand treated in the hospital over a 10-year period between 1975 and 1985, only two were in farmers. As expected, the others were generally painters or labourers by trade.

Case 1

A 30-year-old male farmer grasped a hydraulic hose that contained oil under pressure while working on a tree grasper. He sustained an injection to the tip of his nondominant left thumb. Examination within hours of injury revealed a thumb with extensive swelling and erythema. Thumb motion was restricted. There was a punctate entry wound on the radial aspect of the pulp. At surgery, marked edema of the palmar fat to the level of the IP joint was noted. Intravenous antibiotics were maintained. The part was salvaged, but, at the present time, the patient is too early for final disability assessment.

Case 2

A 66-year-old male farmer suffered a high-pressure injection of aluminum paint and varsol into the pulp of his nondominant left index finger. He subsequently developed severe pain that progressively extended proximally into the palm. At surgery, paint was seen to have extended from the site of injection to the level of the metacarpophalangeal joint. The digit was initially salvaged but the ultimate viability of the fingertip was in doubt.

DISCUSSION

Schoo et al.,[1] in 1980, published an extensive review of 127 cases culled from the North American literature and described the causes, natural history, and recommended treatment.

Paint and grease guns emit a fine spray at pressures ranging from 600 to 10,000 psi.[2] With partial nozzle blockage, pressures may increase to over 12,000 psi.[3] The most commonly injected agents are grease and paint which have been reported in approximately 60% of cases. Hydraulic fluid, diesel fuel, and paint thinner are seen each in about 5% of injections.

The injured are typically young working males, in their mid-30s. Generally injuries occur early in the person's work experience, or as a result of carelessness in the more seasoned worker. Classically, the mode of injury is one of inadvertent injection of the nondominant index finger while the operator is cleaning the nozzle. Of injections in Schoo's review, 45% were into the index finger, another 40% divided equally between the palm and the middle finger.

Rees,[4] in 1937, described the clinical course of a high-pressure injection injury of diesel fuel. An innocuous pinpoint entry site, so typical of these injuries, is often the only early sign. No clues to the actual severity of the injection are initially evident. Over the subsequent 24 h, rapid swelling and intense pain develop. Lymphadenitis, leukocytosis, and ischemia follow. Acute inflammation coupled with tissue distension may rapidly embarrass the circulation. Waters et al.[5] documented injection injuries in experimental animals describing pathophysiology of the acute inflammatory response. If the part is salvaged, subacute and chronic inflammatory changes may also compromise ultimate hand function.

Once injected, the material disperses along the paths of least resistance. Bone, tendon, and pulleys are dense and so relatively resistant to injection. They generally deflect the injectate into the surrounding soft tissues. Nerves and vessels, as well as the membranous portions of the fibro-osseous sheath of the flexor tendons, provide conduits for passage of the material.

Treatment should be aimed at both the injected chemical and the inflammation. Surgical debridement of the injected material is mandatory. Awareness of the tracking characteristics and thorough inspection of the neurovascular bundles and tendon sheaths proximal to the areas of soft tissue necrosis are necessary to avoid missing "skip" areas of deposition. Decompression should be wide, and the wounds left open or loosely closed over drains.

Parenteral antibiotics are administered before and after operation. Systemic corticosteroids in high doses are also given,[6,7] but they should be regarded as an adjunct to surgery and never used as an alternative.

Details of the patient's tetanus immunization status should be sought and prophylaxis given as necessary.

Although all agree that surgery is the cornerstone of treatment, it must be realized that surgery neither guarantees a good outcome nor salvage of the part. Stark[8] noted that salvage was significantly higher if surgical intervention occurred within 10 h of the injection. X-rays may be useful in determining proximal extension.[9]

Amputation of the injured part can be expected in about 50% of high-pressure injections.[1] Unless pressures of injection exceed 7000 psi, the amount of pressure has not been found to correlate with the incidence of amputation. The material injected, however, significantly affects the outcome. Paint thinner, diesel fuel, and paint are particularly likely to cause sufficient inflammatory reaction to necessitate amputation. Grease and hydraulic fluid are much less harmful.

Salvage of the part in itself does not constitute victory. Chronic inflammatory changes with granuloma, oleoma, sinus formation, and associated fibrosis may leave the hand and fingers seriously disabled. Late surgery in the form of oleoma excision, scar revision, tenolysis, and late amputation are common necessities. Gelberman and Posch[10] concluded that the disability time appeared to be directly related to the time elapsed until late amputation of the involved digit. In their series, the average disability time was 7 months. Digital injections carry a worse prognosis than injections in other sites.

Thus, in conclusion, high-pressure injection injuries of the hand are rare. Farmers who service their own equipment utilizing high-pressure guns are at risk. Both the worker and the family physician should be well versed as to the severity of the injury despite its very innocuous beginning. Immediate, aggressive surgical decompression, debridement and drainage, antibiotics, and corticosteroids are the principles of treatment. Amputation rates of the order of 50% are to be expected, and late problems together with serious disability plague many of the remainder.

REFERENCES

1. **Schoo, M.J., Scott, F.A., and Boswick, J.A.,** High pressure injection injuries of the hand, *J. Trauma,* 20, 229, 1980.
2. **Dial, D.E.,** Hand injuries due to oil at high pressure, *J.A.M.A.*, 110, 1747, 1938.
3. **Ramos, H., Posch, J.L., and Lie, K.K.,** High pressure injection injuries of the hand, *P.R.S.*, 45, 221, 1970.
4. **Rees, C.E.,** Penetration of tissue by fuel oil under high pressure from diesel engine, *J.A.M.A.*, 109, 866, 1937.
5. **Waters, W.R., Penn, I., and Ross, H.M.,** Airless paint gun injuries of the hand: a clinical and experimental study, *P.R.S.*, 39, 613, 1967.
6. **Kaufman, H.D.,** High pressure injection injuries, the problems, pathogenesis and management, *Hand,* 2, 63, 1970.
7. **Phelps, D.B., Hastings, L.H., and Boswick, J.A., Jr.,** Systemic corticosteroid therapy for high-pressure injection injuries of the hand, *J. Trauma,* 17, 206, 1976.
8. **Stark, H.H., Ashworth, C.R., and Boyes, J.H.,** Paint-gun injuries of the hand, *J.B.J.S.*, 49, 637, 1967.
9. **O'Reilly, R.J. and Blatt, G.,** High pressure injection injury, *J.A.M.A.*, 233, 533, 1975.
10. **Gelberman, R.H., Posch, J.L., and Jurist, J.M,** High-pressure injection injuries of the hand, *J.B.J.S.*, 57A, 935, 1975.

Electrical Burns in Farmers

L. R. Chasmar

The first reported electrical injury was in 1746 when two Dutch physicists accidentally discharged a Leyden jar between themselves. The first recorded death occurred in 1879 when a stagehand in Lyons, France died while installing some lighting in a theater.[1] The record room of the University Hospital in Saskatoon produced 24 charts of patients admitted with electrical burns in the 5-year period January 1, 1979 to December 31, 1984. Of these, only 5 could be directly linked to agriculture. The numbers are not large, but the injuries are 100% preventable and the cost in human disability and death is enormous.

The severity of electrical injury is determined by: voltage of current, amperage of current, resistance of tissues, type of circuit, duration of contact, and pathway of current.[2]

Electrical burns are divided into high-voltage injuries, greater than 1000 V, and low-voltage injuries, lower than 1000 V. In the event of lower skin resistance, household current of 110 V can be fatal. Electrical burns are classified by voltage because it is the only factor that is constant in electrical injury. Ohm's law (voltage = amperage x resistance) relates voltage to amperage and resistance. There is a very narrow range of safety between the amperage of household current (1 to 10 mA) and that which causes respiratory arrest (20 to 50 mA).

The nature of the local lesion produced in an electrical burn is due more to the resistance at the point of contact than to any other single item.[3] When the skin is dry and contact with the electrode is unstable, an electric arc is set up between the electrode and the tissue. The temperature of such an arc is sufficient to melt steel. Tissues are affected in proportion to their resistance to the current — bone, fat, tendon, dry skin, muscle, blood vessels, and nerve, in that order. The local lesion in electrical burns may be anything from the most minute electric current mark on the skin to aseptic necrosis of the whole limb.

Alternating current is more dangerous to life than direct current. The reversing electrical impulses produced by the alternating current are thought to throw the muscles into tetanic spasm causing the victim to close his grip on the electrodes from which he cannot voluntarily let go. The longer the victim is in contact with the current, the more severe will be the injury. The pathway that the electrical current takes will determine the type of injury, as well as the complications one can expect.

Except in bathtub accidents in which there may be no marks on the patient, one usually finds one or more entry wounds and one or more exit wounds. The most common entry points are the feet. Direct contact, particularly with high voltage, almost always results in severe burns. Flash burns from an electrical explosion tend to be more superficial. Arc burns, where the current arcs to patient from a high-voltage line, are usually very deep burns. Secondary thermal burns can occur if the clothing becomes ignited. Occasionally, a metal ring, wristband, or belt buckle will become part of the circuit, be heated up, and cause a thermal burn secondarily.

Cardiopulmonary arrest may occur and either ventricular fibrillation or asystole may be seen. After resuscitation, one could be faced with an arrhythmia or bundle branch block. There may be EKG evidence of coronary ischemia. Cardiac enzymes may be elevated. Respiratory arrest is common with slow recovery. It is, therefore, advisable to continue artificial respiration for an hour or more.

The destruction of muscles and blood results in myoglobinuria and hemoglobinuria. Prompt fluid resuscitation maintaining a good urinary output is the key to preventing myoglobinuric renal failure.

The skull is a frequent entry point for electrical current. The high resistance of the skin, fat, and bone results in destruction of these tissues, but once the current reaches the brain and spinal fluid it is dissipated rapidly through the body. This results in reduced coagulation beneath the dura in most electrical accidents. The current passing through the brain results in ataxia and delayed cataract development. There can be direct brain or spinal cord damage with paralysis. Peripheral nerves are also damaged by passage of electric current. Median and ulnar nerve paralysis occurs with electrical burns of the flexor aspects of the wrists, a common site for damage. These may be lesions in continuity, in which case recovery is the rule. Deeper burns at the wrists result in segmental nerve loss with paralysis and distal sensation loss.

Curling's ulcers are more common with electrical burns than with thermal burns. Mesenteric thrombosis, gallbladder necrosis, and small and large bowel injuries can occur particularly if abdominal burns are present. Prolonged ileus is a clue to their diagnosis.

Lightning injury, while uncommon, kills 150 to 300 persons annually in the U.S. Many survive, disproving the popular belief that being struck by lightning is fatal. Many victims will respond to vigorous initial resuscitation. If the patient survives the initial shock, extensive muscle damage may be complicated by myoglobinuria. This should be suspected and early treatment of the lightning victim should include administration of osmotic diuretics, large volumes of fluid, and alkalinization of the urine.[4] The agricultural workers must avoid high ground during electrical storms.

The typical farm accident causing electrical injury or death involves a metal object, usually the unwieldy grain auger, contacting an overhead power line. This is more likely to occur in the farmer's yard than where the power line crosses an open field. Other objects which have contacted power lines are aluminum ladders, metal grain bins, television aerials, machinery, and eavestroughing. These accidents are 100% preventable, and underground hydro lines on the farm would virtually eliminate this type of disaster.

Extrication of the victim is only safe when the power has been shut off. It may be better to leave the victim alone than to have the rescuers become victims as well. Gloves, ropes, and wooden poles all have their pitfalls.

As in any other emergency, one pays attention first to the *A*irway with cervical spine control, the *B*reathing, and the *C*irculation. Muscle contractions intense enough to cause vertebral fractures are a possibility. Large volumes of Ringers lactate will be required to overcome hypovolemia and to dilute the myoglobin and avoid myoglobinuric renal failure. Heat generated by the passage of electrical current is dissipated by the high flow rate of the arteries. The intima can be damaged resulting in edema of vessel walls in the postburn period. Sludging and thrombosis can occur as a delayed phenomenon giving the impression of progressive necrosis. Veins often become thrombosed during or shortly after the injury. This causes massive edema of the undrained extremity. Escharotomy may be necessary to salvage distal tissue.

An EKG is important in the initial management. Burn ileus mandates a nasogastric tube. Fractures and dislocations are common and must be ruled out by appropriate X-rays. A careful neurological examination is essential. Tetanus prophylaxis is administered according to standard protocols. Appropriate and extensive debridement of necrotic tissue is the best preventative measure against tetanus and other infections. Grafting is undertaken once one has a viable base for the graft.

Electrical burns are classified as major burns. Transfer to a Burn Center familiar with their care is recommended. These injuries often result in multiple amputations of entire extremities, loss of occupation, changes in body image, as well as marital and family problems. Well coordinated therapy requires a burn surgeon, experienced nursing care, physical and occupational therapy, social and psychiatric counseling for the patient and family, and special prosthesis training.

In conclusion, electrical burns in agricultural workers are not common, but when they occur, death or severe multiple injuries often result. Overhead power lines combined with an assortment of unwieldy farm equipment have been implicated. It is clear that electrical disasters in agriculture can be almost totally prevented. The single most effective preventative measure would be to place power lines within the farm perimeter underground.

REFERENCES

1. **Bernstein T.,** Theories of the cause of death from electricity in the late nineteenth century, *Med. Instrum.*, 9, 267, 1975.
2. **Cooper, M.,** *Emergency Med. Clin. North Am.*, 2 (3), 489, 1984.
3. **Chasmar, L.R.,** Electrical burns, *Can. Med. Assoc. J.*, 97, 453, 1967.
4. **Artz, C.P., Moncrief, J.A., and Pruitt, B.A.,** *Burns: A Team Approach*, W.B. Saunders, Philadelphia, 1979, 353.

Stress and Psychiatric Problems in Rural Areas

Farm Stress: Research Considerations

L. M. Haverstock

INTRODUCTION

The 20th century has been called the Age of Anxiety — the Century of Stress. In recent years, the interaction of psychosocial and biological factors in health and disease has become an important area of scientific research.[1] The approach of merely correlating stressful life events and subsequent disease, however, has yielded unimpressive results. Investigators interested in stress and human health are faced with a variety of situational, interpersonal, and sociocultural contexts.[2]

Such challenges are evident in farm stress research. Studies in this area have focused primarily upon reporting that rural people have been subject to increasingly rapid and significant changes that have seriously affected their levels of stress. Some researchers offer suggestions for coping and/or delivery of services. Few attempts have been made to examine rural people and the uniqueness of their situations, basic interpersonal skills, as well as their sociocultural backgrounds. Such data are necessary for health and agricultural service providers to gain insights into those in the farming community who are at greatest risk for mental and/or physical illness as a result of stressful events. The first step for health professionals and researchers is to conceptualize the individuality of farmers as a group. Farming is more than an occupation. Farm stress and its manifestation in psychiatric symptomology are different from mental disorders as outlined by the *DSM-III* (revised). The changing face of agriculture has resulted in thousands of North America's rural population struggling to adjust to the simultaneous losses of heritage, future, community, income, pride, and control. A major task for health professionals is to fully understand the health needs of rural individuals and communities in order to facilitate appropriate interventions and preventive services. The ultimate goal is to reduce the risk of rural people becoming a charge rather than an asset to society. In the process, the likelihood of long-term physical and mental illness is lessened.

AN HISTORICAL PERSPECTIVE

Crises in rural North America have been evident since homesteaders broke the land. At the turn of the century, the federal government of Canada sought to populate the Western prairies by affording settlers the opportunity to own land following a 3-year tenure of tilling a portion of it. Free land, however, did not ensure success. Between 1911 and 1931, nearly six out of ten farmers left the land before securing titleship. Fowke[3] points out that possessing the deed often led the settler to use it as collateral for credit "which, fully reasonable in a period of inflation, was wholly unrealistic in the succeeding periods of deflation and agricultural collapse".[4] A 1938 study illustrated that the farming community faced and recovered from the challenges of dwindling markets, heavy monetary losses with an accompanying exhaustion of reserves, mounting burden of private and public debt, abondoned farms, depreciation of machinery, buildings, and equipment, as well as a sharp decline in the standard of living in the entire agricultural population.[5]

Rural living has never been free of fluctuating land prices, and unpredictable stressors such as variable weather conditions, crop and animal diseases, machinery breakdowns, and time pressures. What is it then about the current agricultural situation that has mental health professionals, medical specialists, politicians, journalists, directoral boards of trade, economists, bankers, and farmers talking about "the farm crisis"? Where do those interested in understanding the problems of rural people begin?

STRESS IN FARMING: GROUPS TO BE INVESTIGATED

Williams[6] described the typical profile of farmers who have recently reached retirement age. These people began their farming careers "with the horse-drawn mower; (they) hand-milked cows, skimmed the milk, and ground grain, slopped hogs, and raised dual purpose hens that laid eggs for one season before they were popped into the pot." Having lived through the Great Depression, many of them learned about frugality, cautiousness, and adopted a skeptical view of good fortune.[7] However, they also participated in the prosperity created by the World War II with the advent of more secure markets, and were drawn into a new era of technological changes, fertilizers, and chemical sprays unknown to their parents. Those who adjusted to these developments made decisions along the way that resulted in the expansion of their operation. Farms got larger and machinery more powerful. Neighbors needed each other less and isolation grew. Success was often contingent upon the efforts of the entire family, within which there was no separation of economic and social roles.[8] This interdependence amongst farm family members is unique and has been a crucial component in the survival of a rural way of life.[9] Many present-day retirees worked for their fathers and, in some cases, their grandfathers. Until recently, there has been an expectation of the continuance of this family tradition.

Williams[6] pointed out, however, that changes in one generation of farming have been greater than since its inception. In the 1970s, some well-established farmers, and others just starting, purchased land not unlike their predecessors. Given the interest rates at the time, the high price of grain, and the advice of lending institutions who encouraged many to buy twice what they were requesting, farmers bought inflated land or borrowed purchasing power against their hollow land value. When land prices dropped, farmers were left thousands of dollars in debt at high interest rates. In 10 years, Canadian farm debt increased 450%.[11] At the present time, a "farmer may have the farm enterprise vertically integrated with a service industry, may be running a pork factory, driving a $100,000 tractor, and consulting with a tax lawyer and production specialist".[7] To further complicate the situation, Canadian farmers of the 1980s are prisoners of a food trade war between the U.S. and the European Economic Committee with the knowledge that there is little likelihood of a truce in the near future. A barrage of information from governments, policymakers and breakers, farm organizations, and agricultural experts is ever present. Advocates for consumers, animal rights activists, conservationists, and media presentations of the world's starving speak to rural people more frequently and with greater clarity because of "split second" communication systems. As well, improved transportation has taken farmers from their local communities and introduced them even more directly to the increased selection in services, activities, and items that their urban counterparts enjoy.

All of this has come with some cost to rural people. Several writers in the area of farm stress have generated lists of stressors inherent in farming. As reported earlier, many stressors have remained unchanged. The most significant differences that warrant study appear to lie in the individual farm family member's perceived lack of control and the intrapsychic of interpersonal difficulties that accompany this. Younger farmers have been found to experience higher levels of stress as manifested in greater alcoholism and substance abuse of prescribed medications, marital breakdowns, and suicide rates. This has been attributed to the sense of failure associated with the potential or real loss of their rural way of life. Farmers have stated that they feel as though they have betrayed the hard efforts of their ancestors who contributed to the farm in the past, as well as forfeiting the future of upcoming generations. As more farmers experience economic hardship, greater numbers seek ways in which to keep the operation afloat. The added income is described by many as "disappearing into a bottomless pit of debt owed to lending institutions". Walker and Walker[11] indicated that more stress is evident amongst those who have off-farm employment and younger farmers generally comprise this group.

Farm women also report significantly higher levels of stress as they struggle to balance their domestic duties, keep farm records, care for the family, keep the yard and garden, and act as a workhand when hired help is no longer affordable. A significant proportion of them have off-farm employment as well. In a study conducted by farm women,[12] younger farm women (aged 25 to 34 years) reported the greatest degree of stress. Mental fatigue was listed as the most common symptom of stress and was defined as "an emotional state of indifference". Frustration, the result of helplessness, was stated as the second most common reaction, with irritability being the third. The latter was reported as seriously affecting family relationships and marital conflict was identified as a stressor in itself. It was found that indecisiveness, insomnia, and numerous physical complaints were also present. All of these symptoms could have considerable impact on safe operation of machinery and vehicles, and particularly on interpersonal relationships.

Traditionally, the role of farm women has been undermined in the male-dominated socioculture. In many cases family farms are passed in patriarchal fashion from father to son. According to Haley,[13] rural women get entangled in a complex intergenerational system of bonds and loyalties that makes them significantly different from their urban counterparts. The increased need for day care because of changing demands cannot be addressed through urban solutions and the problem is compounded by long-held beliefs that every farm woman "should" be able to care for her own house and family without "outside" help. An added area of concern involves violence, in which the farm family's traditional response to "keeping their troubles private" results in many farm women not seeking assistance. Isolation further complicates the situation.

Rural children are an integral part of the family system and often are ignored in times of crisis because of their parents' inability to provide attention while under stress. Research that examines the physical and mental health needs of farm children in high-risk families is woefully lacking. In the present writer's clinical practice, there are increasing accounts of young children hiding toys following farm foreclosures because of the fear of having their "trucks and tractors taken away like daddy's". Higher incidences of behavioral problems are suspected among children whose farming parents are known to be in financial trouble, but there have been no studies from which to cite data. School personnel recount situations in which uninformed youngsters are told by knowing classmates about a pending seizure of their family's farm equipment, cattle, or grain. These reports have been supported by parents who admit that they have never discussed the seriousness of their situation with each other, let alone their children. At the other end of the spectrum, some children have stated that all they have ever heard at home were complaints about financial institutions or the severe money problems of the family. Many parents have expressed that their greatest fear is that their adolescent/young adult children will harm themselves because of despondency and feelings of hopelessness related to their being no future in farming for them. Rural communities in the province of Saskatchewan are making efforts to curb alcohol abuse amongst their young people because they perceive it as a serious problem.

One might speculate the older, more financially secure farmers would be unaffected by the farm crisis. This indeed, is not the case. In practice, most younger farmers rely on the financial assistance of their parents in order to begin farming. Retirement, in itself, is stressful. For "lifetime" farmers, there are complex considerations. There are a growing number of senior farmers who are faced with the dilemma of whether to retire with some equity or to help an adult child get established. Some who have already sold their farms may find themselves in financial difficulty if the farm proves unsuccessful. Many retired farmers who cannot separate their identities from the farm invest their savings into the farm that they have already sold. It should be noted that some older farmers with substantial financial security report high levels of stress and hopelessness because of what they view as the demise of family farming.

Because of the distinctiveness of rural families and the intensity of the problem, it appears that no one escapes unscathed in the farm crisis. There are no winners — neighbors, local communities, even the largest centers that inherit the "farm family-gone-under" or the remnants of a farm family that are now headed by a single parent pay a price. The realities are that many rural people are becoming lost to society as a resource. The economic costs can be estimated hypothetically; the human costs cannot. To more fully understand stress reactions and what is needed to alleviate stress and/or improve adaptability, each one of the groups discussed requires investigation through scientific inquiry.

FURTHER CONSIDERATIONS

In the book *Stress and Human Health,* Hamburg[14] discussed the paradoxes of investigating stress. How do researchers examine susceptibility to virtually identical life stressors? Are there ways to assess vulnerability, to link emotional experiences and tissue damage, to use stress to promote health, or to identify biological mechanisms or therapeutic interventions to protect people from its harm? Clinical and epidemiological research has demonstrated unequivocally that highly stressed individuals are at greater risk for physical and mental illness. However, simple cause-effect concepts of stress and disease ignore the significance of interactions among stressors, reactions, mediators, and consequences.

A decade ago, Smith et al.[15] found that farming ranked as the 12th most stressful occupation in a field of 136 different careers. Most would agree that agricultural conditions have deteriorated in recent years. The result has been an overwhelming increase in the numbers of "hurting" rural people and a startling rise in suicides amongst farmers. To illustrate, the suicide rate for Kentucky farmers rose 29% from 1983 to 1984 as the rate among nonfarm populations decreased. Missouri rates tripled in the early 1980s among farm families.[16] Similar findings have been recorded in Canada. An Alberta suicidiologist reported that there were 40.6 suicides per 100,000 in 1984 compared with 15.5 per 100,000 in 1981. In Alberta, the suicide rate for the general population was 18.1 per 100,000 in 1984. Comparisons with unemployment data have shown that the Canadian province with the highest unemployment has the lowest suicide rate. This information raises questions regarding farmers and how they conceptualize what it is that they do. It supports the earlier contention that farming is not an occupation.

Many who scorned their farming neighbors in trouble 2 years ago now find themselves in the same situation. There were and continue to be those, both within and outside agriculture, who believe that the answer to farm stress is $7.00 per bushel grain. Greater economic security could benefit most people, but it will not heal the emotional damage of those who have suffered severe stress. "Studies of the health effects of natural or man-made disasters suggest that about half of the physically unharmed survivors experience an acute emotional, physical, or psychosocial consequence; many of these victims suffer long-term changes in health."[14] It appears that once the process of severe stress has begun, no single intervention will directly counteract the effects. The contention is discussed by Pearlin et al.,[17] who purported that life events, chronic life strains, self concepts, coping, and social supports fuse to form a process of stress. If the model is applied to farmers in crisis, what began as an economic problem escalates to situations in which farmers question their own abilities. Criticisms from others, especially from fellow farmers, could erode self-esteem. It is this diminished sense of self that results in heightened vulnerability to stress and its subsequent symptoms (e.g., anxiety, depression, physical complaints). According to Pearlin et al., the individual's ability to cope was reduced because of the changes in self-esteem and mastery.

The work of Kobasa et al.[2] raises questions about the initial "resistance resources" or "hardiness" of individuals prior to stressful life events. They list resources such as constitutional strengths (i.e., good family history with regard to illness), social supports (i.e., social contacts and status), health practices, and personality dispositions such as commitment, control, and challenge. The theories of both Pearlin et al. and Kobasa et al. lend themselves to research with farmers. The former is more in keeping with programs that have been created to serve rural people in crisis in the U.S. To date, Canadian services have been sporadic because of the delay in the crisis reaching various regions. Although a prompt response to farmers in need is warranted, invaluable information could be gathered by utilizing the Kobasa et al. theory, in the hope that preventative efforts could be made in the future. It is important for health professionals to recognize and stabilize immediate/potential crises and to assist rural people in learning to accept help. As health service providers, we must be responsive to the special needs of the farming population. As those who wish to develop new knowledge in this area, answers lie in replicable, reliable research of behavior.

"Over the course of centuries, science has wrenched area after area from the provinces of philosophy, religion, and literature — first the motions of the heavenly bodies, then the nature of matter, then the life forms on earth and their evolution; and finally the working of the human body. Human behavior is only the latest area of encroachment."[18]

REFERENCES

1. **Lipowski, Z.J.,** Psychosomatic medicine: past and present. I. Historical background, *Can. J. Psychol.,* 31, 2, 1986.
2. **Kobasa, S.C., Maddi, S.R., and Kahn, S.,** Hardiness and health: a

prospective study, *J. Pers. Soc. Psychol.*, 42 (1), 168, 1982.

3. **Fowke, V.C.,** *The National Policy and the Wheat Economy,* University of Toronto Press, Toronto, 1978.
4. **Fowke, V.C.,** *The National Policy and the Wheat Economy,* University of Toronto Press, Toronto, 1978, 285.
5. **Wilson, C.F.,** *A Century of Canadian Grain Government Policy to 1951,* Western Producer Prairie Books, Saskatoon, 1978, 248.
6. **Williams, C.M.,** The Rural Environment in Transition, a paper delivered at the Canada Conf. Ser., Halifax, Nova Scotia, Canada, December 1987.
7. *Am. Psychol. Assoc. Monitor,* 16 (4), 1985.
8. **Rosenblatt, P.C. and Anderson, R.M.,** Interaction in farm families: tension and stress, in *The Family in Rural Society,* Coward, R.T. and Smith, W.M., Jr., Eds., Westview Press, Boulder, CO, 1981.
9. **Gordon, E.,** Stress in the family farm: implications for the rural human service worker, *Can. Home Econ. J.,* 37 (1), 1987.
10. **Mitchell, D.,** Banking on the family farm, *Briarpatch,* 13 (10), 7, 1984.
11. **Walker, J.L. and Walker, L.J.,** Self-reported stress symptoms in farmers, *J. Clin. Psychol.,* in press.
12. **Ireland, G.,** *The Farmer Takes a Wife: A Study by Concerned Farm Women,* Concerned Farm Women, Chesley, Ontario, 1983.
13. **Haley, E.,** Hard Times for Farm Women, unpublished paper, 1987.
14. **Hamburg, D.A.,** An outlook on stress research and health, in *Stress and Human Health: Analysis and Implications of Research,* Elliott, G.R. and Eisdorfer, C.E., Eds., Springer Verlag, New York, 1982.
15. **Smith, M., Culligan, M., and Nurell, J.,** A review of the National Institute for Occupational Safety Health and Stress Research, paper presented at a conference on job stress, University of Southern California, Los Angeles, 1977.
16. **Cecil, H., and Hannon, R.,** Stress: country style — a program to serve farmers and their families, *Hum. Serv. Rural Environ.,* 11 (2), 36, 1987.
17. **Pearlin, L.I., Leiberman, M.A., Menaghan, E.G., and Mullan, J.T.,** The stress process, *J. Health Soc. Behav.,* 22, 337, 1981.
18. **Stenovich, K.E.,** *How to Think Straight About Psychology,* Scotts, Foresman, Glenview, IL, 1986.

Rural Psychiatry: A Definition of the Field

C. N. Larson, S. Kuperman, and R. E. Smith

HISTORICAL AND EPIDEMIOLOGICAL PERSPECTIVES ON RURAL AND URBAN DIFFERENCES IN PSYCHIATRIC DISORDERS

Over the past two centuries, the trend toward increasing urbanization and the population shift from rural to urban areas has been a topic of concern for many political and philosophical writers. With growing rates of industrialization and technological change, some have argued that the increasing stresses of urban life may lead to higher rates of psychiatric illness than in rural areas.

A considerable number of studies have attempted to look at the differences between rural and urban population in terms of psychiatric disorder. Dohrenwend and Dohrenwend identified nine studies conducted in different cultures and geographical areas including Japan, Taiwan, Nigeria, Iceland, and Finland.[1] Their conclusion was that the rate of psychiatric disorder was generally higher in urban than rural areas, although when specific diagnostic categories were considered, functional psychoses and manic-depressive psychoses were more common in rural areas. On the other hand, rates for personality disorder and neuroses were higher in urban areas.

The reports of Mueller[2] and Helgason[3] are consistent with the findings of the Dohrenwends' with higher rates of manic-depressive psychoses in rural areas and supported the findings of Sethi et al.,[4] Brownet al.,[5] and Rutteret al.[6] which found depression to be higher in urban areas.

Further support for the notion that affective disorder rates may be lower in a rural population comes from a preliminary analysis of the NIMH — Clinical Research Branch Collaborative Program on the Psychobiology of Depression. From this Coryell et al. found that the lifetime rate of affective illness among 1090 interviewed relatives of depressed and manic probands was considerably lower in Iowa, a rural state, than in the other four urban research centers.[7] Among the various affective disorder diagnoses, only primary unipolar depression was significantly less in Iowa.

The differences are largely explained on the basis of selective patterns of migration as opposed to environmental causation, although objective data are scant. On the other hand, Mueller[2] notes that recent studies have found a relationship between depression and the level of social support in the community and notes that certain groups in urban areas are more susceptible to isolation, including housewives, the elderly, and the disabled. As well, suicide rates appear to be higher in densely populated areas, dropping in more rural areas, except in sparsely populated regions where the rates again rose.[2]

EPIDEMIOLOGY AND RURAL PSYCHIATRY

Several examples of the importance of epidemiology to rural psychiatry exist, but the discussion will be limited to suicide rates, the developing field of behavioral toxicology, and studies of rural and urban attitudes toward psychiatric illness.

Suicide

Suicide is frequently listed among the leading causes of death in this country, and thus poses a major public health problem.[8] It is extremely difficult to make generalizations without taking into consideration socioeconomic status and multiple regional and demographic factors. However, suicides in largely rural, possibly isolated populations such as Nevada and Wyoming are twice the national average of 30 per 100,000 population in males aged 35 to 74, while in more urban states suicides are approximately half the national average. Although generalization from these types of data is of limited value, the data do raise questions pertinent to the practice of psychiatry in rural areas. Higher rates of suicide in mountainous and some rural regions have been explained in terms of social isolation or patterns of migration, but low rates of recognition because of low care-giver population, low willingness to accept care, and low rates of intervention must be considered as well. However, environmental, interpersonal, or even biological factors unique to the rural or mountainous regions could be of importance. This must be kept in mind by psychiatrists who practice in rural areas.

Behavioral Toxicology - Implications for Psychiatry

In an age of increased biochemical and biological sophistication, we are fast approaching an era where it may be possible to more clearly elucidate and understand mechanisms of interaction between environmental toxicants, the brain, and behavior. Although the field of environmental and behavioral toxicology is still in its infancy, we do have some well-studied examples of this type. The three best examples are chronic lead exposure leading to neurochemical effects of lead toxicity, chronic manganese poisoning which may produce a Parkinson syndrome in humans, and the neurologic effects produced by several other heavy metals.[9-11] Although there are many farming and agricultural chemicals with known peripheral neurological effects, little is known about central brain effects of chronic low-level exposure on the mental status of farm families and other rural residents.

It has been observed, for example, that some grain storage workers have neuropsychiatric changes including nervousness, mood swings, extrapyramidal symptoms of Parkinsonian type, peripheral neuropathy, cerebellar signs, and sleep apnea along with cognitive impairment. These were felt most likely related to carbon disulfide and malathion exposure. In other studies, organophosphate pesticides have been implicated in causing anxiety and depression in exposed farmworkers.[12,13]

The importance of nutrition and its effect on susceptibility to the development of disease is becoming better understood, and poor nutritional status apparently can greatly increase chances for neurotoxicity. Certain dietary deficiencies are known to increase the toxicity of a number of pesticides, including carbamatecarbaryl, parathion, thalidomide, captan, heavy metals such as lead and cadmium, and atmospheric contaminants including ozone.[14] Although our society seems to benefit greatly from the vast number of pharmaceutical, industrial, and agricultural chemicals which have entered our environment in recent years, potential effects on the central nervous system are extremely difficult to assess. The study of environmental toxicants and behavior is deserving of systematic investigation and certainly a psychiatrist in rural areas needs to observe for and be able to identify the mental and behavioral changes which may accompany exposure.

Attitudes Toward Mental Health and Psychiatry

In the 1950s, studies by Star and others brought attention to the lack of understanding, anxiety, and rejection which characterized the public's conception of mental illness.[15] This negative attitude tended to continue through the 1960s with the average person feeling that psychiatry was not a true medical subspecialty.[16-21]

The 1970s saw the public beginning to accept the belief that mental illness was indeed a medical illness and not just the result of a "morally weak" individual. An interesting exception to this finding was the study of Lee et al., which found few of the investigated individuals with mental health problems in rural North Carolina considered the mental health center as a resource.[22] This appeared to be due to their fear of being labeled mentally ill, and they instead approached their local medical doctor for help. Although the general level of sophistication and understanding regarding mental illness and its treatment has improved greatly, there is still a need for continued education concerning mental illness in order to reach the vast number of individuals who suffer without treatment, and this may be even more important in the rural areas.

PROVISION OF CARE TO RURAL AREAS

The differences between rural and urban mental health delivery systems become noticeable immediately upon investigation. According to a survey by Gertz et al., major problems identified in providing mental health services to a rural area are geography, entry of the mentally ill into the mental health system, and lack of professional and other types of mental health resources.[23] However, it has been demonstrated that community leaders and nonmentally ill individuals in the community are able to recognize the benefits of the community mental health center. However, as mentioned above, Gertz's survey emphasizes that the "clients" of a rural mental health center felt inhibited about using its services.[23] The issue of the stigmatization of mental illness and fears of confidentiality and scarce resources requires that the skills of rural practitioners be somewhat more flexible and creative than their urban counterparts.[24] Lee et al.[22] and Daniels[25] all identify that urban mental health systems are more comprehensive and serve larger populations, while rural systems often provide more community education due to limited resources.

Unfortunately, there appear to be difficulties in attracting and maintaining psychiatrists in rural areas. Talbott emphasizes that rural psychiatrists are predominantly supported by public funds,[26] and that an important factor attracting them into the public services is an interest in public health and their wish to change a substandard system into something better. Unfortunately, many of these psychiatrists leave the system, disenchanted because of a wide variety of restrictions, lack of support systems, and constraints on performance, secondary to fiscal and legislative controls, and to the civil service procedures. In addition, more nonphysicians have been given a greater role in administrative responsibility, eroding the psychiatrists' involvement in this area.[26] Tucker et al. emphasizes that the majority of psychiatrists who leave rural New Hampshire to practice in urban areas do so because of low salaries, negative cultural factors, professional isolation without a chance for professional growth, and at times, difficulties maintaining boundaries between themselves and the community, where many nonmedical daily contacts are made with patients, their relatives, or potential patients.[27] He proposes the establishment of regional academic training initiatives for psychiatric residents to provide a supply of professionals to these areas. He further suggests providing continuing education and active professional stimulation for psychiatrists already established there. The factors which contribute to dissatisfaction and low autonomy need to be looked at in rural areas, so as to restore and improve the position of the psychiatrist to a more satisfying position with greater opportunity for professional growth than is currently available.

PSYCHIATRIC RESEARCH REQUIREMENTS WITHIN THE RURAL SETTING

The recent past has demonstrated that allocations of substantial funding and indiscriminate establishment of more traditional programs to rural settings are not totally effective nor economically feasible. Thus, new research is needed. Corrective measures for the 1990s must be based upon new knowledge. The higher prevalence of bipolar affective disorder and the high rate of completed suicide in sparsely populated areas, the effectiveness of public education in demystifying psychiatric illness in rural areas, and the assessment of behavioral toxicology seem to be four important areas where research thrusts could be given priority. However, before appropriate planning and implementation of programs can be instituted, the population of potentially ill individuals must be identified through epidemiological studies. Certainly, within this population there will be a portion under therapy, while another portion will not currently be receiving any assistance from the medical community. These groups need to be identified so that effective intervention can be proposed. For example, the elderly may represent the most underserved rural population because of their high need for numerous human services that often are scarce. Finally, although preventive psychiatry is in its infancy, successful efforts at secondary and tertiary prevention of relapse and/or subsequent episodes of illness have proven efficacy by utilizing maintenance therapy with various pharmacological and some psychotherapeutic interventions. It would be important to know whether the rural population currently is receiving maximum benefits from the knowledge that has been gained in this particular area. Since the burden of care for the psychiatrically ill in rural areas is largely on the local physician, it is mandatory to assess whether patients are receiving the optimal therapeutic and medical intervention currently available.

THE FUTURE PRACTICE OF RURAL PSYCHIATRY

For effective implementation of psychiatric care to a low density population setting, it is imperative that the care be both time and cost effective. These constraints mandate that a biological approach in psychiatry be encouraged for the rural population. The utilization of the medical model for psychiatric illness should enhance greater acceptance by all, including the local physician, community leaders, and the patients themselves. The biological approach would also maximize those aspects of preventive psychiatry that have been shown to produce a difference. Since a large proportion of the rural population identifies their family physician as the individual whom they approach when psychiatric problems/illness arise, it seems appropriate to identify this group of physicians as the primary deliverers of psychiatric care. Acceptance of psychiatric therapeutic principles based upon the same model appears to be an accomplishable goal. The family physician tends to have a positive attitude toward psychiatric education and principles and, thus, should be a willing student in this area. Thus, the major burden of psychiatric care in the rural setting could be met by the physicians currently in place, but augmented with psychiatric consultation and education to update the quality of care. As mentioned earlier, recruitment of psychiatrists into rural areas is fraught with problems, so it is important, as Tucker suggests, to support physicians already locally placed by facilitating interactions with a regional or university Department of Psychiatry. The university would supply education and training, and serve as a source for referral of complex or nonresponsive cases. The local physician could be assisted by paraprofessionals in the mental health field in his office or in a separate facility, as currently endorsed by the community mental health concept. Then the problem becomes not a shortage of psychiatrists in a rural setting, but a shortage of psychiatric intervention, which is correctable by utilizing the family physican and other therapists, who are often already in place within the physician's clinic. Fauman's study[28] of urban internists emphasizes the frequency of psychiatric problems, particularly depression, in nonpsychiatry practice and notes that approximately 20% of the physicians' time was spent in treating psychiatric problems. This supports the notion that primary care physicians are already treating psychiatric disorders but need the necessary psychiatric and mental health support.

SUMMARY

Psychiatry in a rural setting appears to be unique from several standpoints. Data currently suggest that there may be a difference in the prevalence and incidence of certain psychiatric illnesses between rural and urban settings. The data are fragmentary and further research is needed. There is currently a marked shortage of psychiatrists in the rural areas and this shortage appears to be growing. On the other hand, the shortage of the family physician is decreasing, so it would appear appropriate that these physicians be recognized and utilized as significant contributors to the delivery of psychiatric care. To make advances, rural psychiatry must be the focus of research efforts to strengthen the currently available data base. It is only from these research efforts that appropriate programs and interventions can be proposed.

REFERENCES

1. **Dohrenwend, B.P., and Dohrenwend, B.S.,** Psychiatric disorders in urban settings, in *American Handbook of Psychiatry,* Vol. 2, Caplan, G., Ed., Basic Books, New York, 1974, 426.
2. **Mueller, D.P.,** The current status of urban-rural differences in psychiatric disorder: an emerging trend for depression, *J. Nerv. Ment. Dis.,* 169, 18, 1981.
3. **Helgason, T.,** Epidemiology of mental disorders in Iceland, *Acta Psychiatr. Scand.,* 40, 1, 1964.
4. **Sethi, B.B., Kathawat, S.S., and Gupta, S.C.,** Depression in India, *J. Soc. Psychol.,* 91, 3, 1973.
5. **Brown, G.W., Davidson, S., Harris, T., et al.,** Psychiatric disorders in London and North Uist, *Soc. Sci. Med.,* 2, 367, 1977.
6. **Rutter, M.B., Yule, B., Quinton, D., et al.,** Attainment and adjustment in two geographic areas. III. Some factors accounting for area differences, *Br. J. Psychiatry,* 126, 520, 1975.
7. **Coryell, W., Winokur, G., and Andreasen, N.,** Effect of case definition on affective disorder rates, *Am. J. Psychiatry,* 138, 1106, 1981.
8. Metropolitan Life Insurance Company, Regional variations in mortality from suicide, *Stat. Bull. Metrop. Life Insur. Co.,* 54, 2, 1973.
9. **Hanin, I.,** Central neurotransmitter function and its behavioral correlates in man, *Environ. Health Perspect.,* 26, 135, 1978.
10. **Needleman, H.L.,** The neuropsychiatric implications of low level exposure to lead, *Psychol. Med.,* 12, 461, 1982.
11. **Weiss, B.,** The behavioral toxicology of metals, *Fed. Proc.,* 37, 22, 1978.
12. **Peters, H.A., Levine, R.L., et al.,** Carbon disulfide-induced neuro psychiatric changes in grain storage workers, *Am. J. Ind. Med.,* 3, 373, 1982.
13. **Levine, H.S.,** Behavioral effects of occupational exposure to organophosphate pesticides, in Behavioral Toxicology, Washington, D.C., U.S. DHEW, Publ. No. (NIOSH) 74-125, National Institute for Occupational Safety and Health, 154.
14. **Mahaffey, K.R. and Vanderveen, J.E.,** Nutrient toxicant interactions: susceptible populations, *Environ. Health Perspect.,* 29, 81, 1979.

15. **Star, S.,** *Psychiatry the Press and the Public: Problems in Communication,* Bloomberg, W., Lyons, L., Morse, R., et al., Eds., American Psychiatric Association, Washington, D.C., 1956.
16. **Cumming, E. and Cumming, J.,** *Closed Ranks: An Experiment in Mental Health Education,* Harvard, Harvard University Press, Cambridge, 1957, 51.
17. **Nunnally, J.,** *Popular Conceptions of Mental Health: Their Development and Changes,* Rinehart and Winston, New York, 1961, 232.
18. **Lemkaw, P. and Crocett, G.,** An urban population's opinion and knowledge about mental illness, *Am. J. Psychiatry,* 118, 692, 1962.
19. **Dohrenwend, B., Bernard, V., and Kolb, L.,** The orientations of leaders in an urban area toward problems of mental illness, *Am. J. Psychiatry,* 118, 683, 1962.
20. **Crocetti, G.M., Spiro, H.R., and Siassi, I.,** *Contemporary Attitudes Toward Mental Illness,* University of Pittsburgh Press, Pittsburgh, 1974, 128.
21. **Padilla, E., Elinson, J., and Perkins, M.,** The public image of mental health professionals and acceptance of community mental health services, *Am. J. Public Health,* 56, 1524, 1966.
22. **Lee S, Gianturco E, and Eisdorfer C**, Community mental health center accessibility: a survey of the rural poor, *Arch. Gen. Psychiatr.*, 31, 335, 1974.
23. **Gertz, B., Meider, J., and Pluckhan, M.,** A survey of rural community mental health needs and resources, *Hosp. Commun. Psychiatr.*, 26, 816, 1975.
24. **Edgerton, J. and Bentz, W.,** Attitudes and opinions of rural people about mental illness and program services, *Am. J. Public Health,* 59, 470, 1969.
25. **Daniels, D.,** The community mental health center in the rural areas: is the present model appropriate, *Am. J. Psychiatry,* 124, 32, 1967.
26. **Talbott, J.,** Why psychiatrists leave the public sector, *Hosp. Commun. Psychiatr.*, 32, 778, 1977.
27. **Tucker, G., Turner, J., and Chapman, R.,** Problems in attracting and retaining psychiatrists in rural areas, *Hosp. Commun. Psychiatr.*, 32, 118, 1981.
28. **Fauman, M.A.,** Psychiatric components of medical and surgical practice. II. Referral and treatment of psychiatric disorders, *Am. J. Psychiatry,* 140, 760, 1983.

The Epidemiology of Mental Illness in Rural Areas

D. Blazer and L. K. George

INTRODUCTION

Psychiatric epidemiology has traditionally searched for differences in the prevalence of emotional problems across urban/rural boundaries. The motivation behind these scientific inquiries may derive from an idealistic search for Shangri-la, the ideal supportive community described by Hilton in *Lost Horizon*.[1] In this community, individuals lived an easy, secure life "off the land" with ample leisure and universal well-being. This idealistic view has been translated into a number of theoretical models, most of which conceive the rural environment as more culturally homogeneous, supportive, and stable. In other words, persons in rural settings are more socially integrated. In contrast, popular speculations, such as those of Frank,[2] have suggested that the city naturally houses more individuals suffering from psychiatric disorders than do more peaceful and pastoral rural environments.

Over the past 50 years, these hypotheses have been tested in a variety of studies. For example, Faris and Dunham,[3] in a study of mental hospital admissions in Chicago, found that persons suffering from schizophrenia tended to cluster in the "tough" inner city described by James Clavell in his exposition of Chicago through the tragic character of *Studs Lonigan*.[4] The original conclusions derived from this study were that the pressures of low socioeconomic status and the stress of the urban environment contributed to the etiology of this most serious of psychiatric disorders.

The changing landscape of the rural environment, however, decreased the emphasis of psychiatrists and psychiatric epidemiologists upon the potential benefits of rural life. The homogeneous rural community, from the hillsides of New Hampshire through the plains of Saskatchewan to the San Joaquin Valley in California, over the past 30 years has undergone dramatic changes. Windmills have been replaced by the satellite dish as a symbol of the farming family. The rural community is no longer isolated from information, being instantaneously made aware through television and radio of events not only within state and province, but the world itself. Next, cross-migration has significantly made an impact on the character of the rural resident. Many children of farming families have moved to urban settings as the opportunities for economic well-being have diminished dramatically on the farm. In contrast, professionals, businessmen, and especially retirees are returning to rural areas to escape perceived pressures of the urban environment. Though not actual, this "return" is perceived as a return to one's "roots". Improved transportation, as exemplified by the interstate highway system in the U.S. and its counterpart in Canada, has made access to urban settings, such as Saskatoon, readily available to rural residents for shopping, health care, and even cultural and recreational activities. As sociologists and anthropologists have explored the rural environment, pockets of poverty, crime, and stress have been uncovered that rival any inner city. In conclusion, the rural life of 50 years ago, still idealized by the media, does not exist.

Investigators at Duke University Medical Center, working with the National Institute of Mental Health Epidemiologic Catchment Area (ECA) Program, believe this to be an ideal time to reevaluate the question of urban/rural differences in psychiatric disorders.[5]

First, the potential urban/rural differences can be assessed using diagnostic tools based on the more recent typology of psychiatric disorders outlined in the Diagnostic and Statistical Manual-III (DSM-III). Second, the differences can be explored in a geographic area where urban and rural communities are contiguous. Comparisons in the past have been made across hundreds of miles, and therefore regional and cultural differences could not be adequately controlled.

Specifically, the investigators present findings from the Piedmont Health Survey, a community survey of nearly 4000 individuals in Piedmont County of North Carolina. As noted above, this study was designed to compare contiguous counties to control for cultural and regional differences that have plagued previous urban/rural comparisons. We also have chosen to use the standard approach of defining "urban" as a county or group of contiguous counties which contains at least one city of 50,000 inhabitants or more. Durham County is part of a standard metropolitan statistical area (SMSA). Demographic characteristics of the five counties studied highlight the contrasts between Durham County and the four rural counties. For example, the population per square mile in Durham County is 503 vs. 81 for the rural counties. The unemployment rate is approximately 8% in the rural counties, double that found in Durham County, and persons below the poverty level in the rural counties reached 21%, based on 1979 estimates. Housing adequacy is decreased in the rural counties, with 14% having incomplete plumbing, 55% with no central heating, and 17% with no telephone service.

By utilizing the Diagnostic Interview Schedule, relevant symptomatic information for generating specific psychiatric diagnoses was collected in this large-scale urban/rural comparison. In addition, potential confounding factors, such as age, sex, race, marital status, socioeconomic status, and whether individuals had moved were controlled.

METHOD

This study is part of the National Institute of Mental Health

Epidemiologic Catchment Area Program.[5] Of the five ECA sites, the North Carolina sample contained the highest number of rural residents and was designed so that urban/rural comparisons could be made. In the Piedmont Health Survey, a sample was drawn to yield 3000 community respondents and an additional 900 interviews of elderly (60+ years of age) community residents. Though the survey included both community and institutional residents, this chapter reports results from community respondents only. An in-depth description of the area surveyed and the specifics of the sampling and interview procedures have been presented previously.[6]

There were 3798 interviews usable for the study in the first wave, a response rate of 79%. The analyses presented are based on data weighted to reflect the adult population estimates of prevalence for a series of psychiatric disorders. The weighting procedure took into account household probability selection, nonresponse, and the 1980 census demographic profile of adults in the five-county catchment area. For significance tests, the data were downweighted to the original number of subjects, but the weighted adjustments remained. Each participant in the survey underwent a 2-h interview, which included the Diagnostic Interview Schedule (DIS).[7] The DIS was designed to generate the diagnosis of a series of psychiatric disorders, mostly from Axis I of DSM-III. The method by which the DIS generates a DSM-III diagnosis is to elicit the elements of the diagnosis, including symptoms and their severity, frequency, and distribution over time. Though diagnoses can be generated for both current and lifetime prevalence, this paper reports only 6-month prevalence by utilizing a computer algorithm based on those criteria outlined in DSM-III for each of the relevant diagnoses.

RESULTS

The demographic profile of respondents by urban/rural residence in the Piedmont Health Survey is presented in Table 1.

Weighted figures are presented. As can be seen, the unweighted demographic characteristics are skewed, reflecting the elderly oversample.[8] Otherwise, the sample respondents do not differ from the general population of these five counties.

The unweighted and weighted crude prevalence rates of a series of psychiatric disorders by urban/rural residence are presented in Table 2.

As can be seen, major depressive episodes are more common in the urban area, whereas cognitive deficit and alcohol abuse and/or dependence are more common in the rural area. The increased prevalence of cognitive deficit in the rural area is dependent primarily upon the increased age and decreased education of rural respondents. Generalized anxiety disorder was not included in the first round of the Piedmont Health Survey interviews. At the second personal interview, the diagnosis of generalized anxiety disorder was included and therefore an estimate could be made of crude urban/rural differences. These analyses were based on 2925 respondents whose interviews were usable one year after the initial survey.

TABLE 1
Demographic Profile of Community Subjects Participating in the Piedmont Health Survey (N = 3798)

	Rural			Urban			Total		
	No.	%	Weighted[a] %	No.	%	Weighted %	No.	%	Weighted %
Total	1936	51	44	1862	49	56	3798	100	100
Age									
18—24	156	8	17	217	12	19	373	10	18
25—44	560	29	37	656	35	43	1216	32	40
45—64	553	29	29	496	27	26	1049	28	27
65+	667	34	17	493	26	13	1160	31	14
Sex									
Male	718	37	44	771	41	47	1489	39	46
Female	1218	63	56	1091	59	53	2309	61	54
Race									
White	1143	59	53	1256	68	69	2399	63	62
Nonwhite	791	41	47	604	32	31	1395	37	38
Marital Status									
Married with spouse	1001	52	59	899	48	56	1900	50	57
Separated/divorced	243	13	11	262	14	12	505	13	12
Widowed	466	24	11	350	19	8	816	21	9
Nonmarried	226	12	19	351	19	24	577	15	22
Education									
Less than high school	1095	57	49	696	38	29	1791	47	38
High school graduate	837	43	51	1160	62	71	1997	53	62

[a] The weighted percentage adjusts the actual response percentage to take into account household probability selection, response rate, and demographic characteristics, and the elderly oversample of the 196,790 persons in the five-county survey area who were 18+ years of age. (Total five-county population, based on 1980 census data, was 269,863.)

TABLE 2
6-Month Prevalence for Selected DIS/DSM-III Psychiatric Disorders Stratified by Residence and Weighted to Reflect the Demographic Composition of the Population Surveyed in the Piedmont Health Survey

Disorder (unweighted no. of cases)	Prevalence by residence (%)			
	Urban	Rural	Total	*p*
Major depressive episode (65)	2.70	0.93	1.90	<0.01
Dysthymia (65)	1.40	1.60	1.50	NS
Alcohol abuse and/or dependence (90)	2.60	4.20	3.30	<0.01
Cognitive deficit (60)	1.20	2.10	1.60	<0.05
Generalized anxiety disorder (259)	8.84	7.26	8.12	NS

As can be seen, generalized anxiety disorder is one of the more common diagnoses identified in the urban/rural study, yet there was no significant urban/rural difference.

Because many demographic characteristics may confound simple urban/rural differences in the prevalence of psychiatric disorder, a number of procedures, including stratification and logistic regression analysis, were used to determine the relative contribution of these potential confounders to urban/rural variations. Though these results are presented elsewhere,[6] they may be summarized as follows. The increased prevalence of major depressive episodes (but not dysthymia) in the urban area persists even when age, marital status, sex, race, education, and whether the person has moved within the last year are controlled, simultaneously. In contrast, alcohol abuse and/or dependence is more common in the rural counties when age, sex, race, marital status, and socioeconomic status are controlled.[9] Therefore, simple demographic characteristics do not explain the general findings of urban/rural differences in psychiatric disorders, and more detailed investigative efforts will be required to clarify these urban/rural differences further.

CONCLUSION

Urban/rural differences continue to exist for some, but not all, psychiatric disorders as defined by DSM-III and operationalized by the DIS. The findings are not unidirectional, in that some disorders, such as depression, are more common in urban settings, whereas other disorders, such as alcohol abuse/dependence, are more common in rural settings. The differences found are of the same magnitude as found in studies previously referred to from the 1950s and 1960s. These differences cannot be explained on the basis of demographic factors, such as age, sex, race, and socioeconomic status. Nevertheless, the differences are more pronounced for certain age, sex, and race groups. For example, the urban/rural differences in major depression are much more pronounced for white females under the age of 40 than for any other demographic group.

Both urban and rural environments are more complex than the traditional picture would suggest. Many have commented upon the changing characteristics of the farm, with farms becoming larger, more mechanized, and requiring fewer persons for their operation. In addition (at least in the southeastern U.S.), low technological industry is moving into the rural areas to capture the labor market that is available secondary to the changing personnel needs in agriculture. Therefore, the 1980s provide an ideal time to investigate not only the potential noxious effects of rural living, but the potentially advantageous effects as well. As many persons move into rural settings to "escape" the pace of urban living, these studies may be especially relevant. Rural Canada and the U.S. are composed of different types of individuals in complex relationships, and therefore rural Canada and America can no longer be captured by a simple picture of the farmer working his or her field.

Nevertheless, urban/rural differences in psychiatric disorders persist in this changing environment. One must be careful not to translate these initial research findings into policy decisions too quickly and thus follow the mistakes of the 60s. Nevertheless, recent epidemiologic surveys and ongoing investigative efforts will provide a rich opportunity to understand further the role of the social and cultural environment of the farmer in the development of a variety of psychiatric disorders, whether that role be protective or facilitative.

REFERENCES

1. **Hilton, J.,** *Lost Horizon,* W. Morrow, New York, 1934.
2. **Frank, L.K.,** Society is the patient, *Am. J. Sociol.,* 42, 335, 1936.
3. **Faris, R.E.L. and Dunham, H.W.,** *Mental Disorders in Urban Areas: An Ecological Study of Schizophrenia and Other Psychoses,* University of Chicago Press, Chicago, 1939.
4. **Farrell, J.T.,** *Studs Lonigan,* Vanguard Press, New York, 1979.
5. **Regier, D.A., Myers, J.K., Kramer, M., Robins, L.N., Blazer, D.G., Hough, R.L., Eaton, W.W., and Locke, B.Z.,** The NIMH Epidemiologic Catchment Area Program, *Arch. Gen. Psychiatry,* 41, 934, 1984.
6. **Blazer, D.G., George, L.K., Landerman, R., Pennybacker, M., Melville, M.L., Woodbury, M., Manton, K.G., Jordan, K., and Locke, B.,** Psychiatric disorders: a rural/urban comparison, *Arch. Gen. Psychiatry,* 42, 651, 1985.
7. **Robins, L.N., Helzer, J.E., Groughan, J., and Ratcliff, K.S.,** National Institute of Mental Health diagnostic interview schedule: its history, characteristics, and validity, *Arch. Gen. Psychiatry,* 38, 381, 1981.
8. **Fillenbaum, G.,** Piedmont Health Survey of the Elderly: Area and Population Characteristics, report submitted to the National Institute on Aging, 1985.
9. **Blazer, D., Crowell, B.A., Jr., and George, L.,** Increased alcohol-related problems in the rural South, paper submitted for publication.
10. **George, L.K., Hughes, D.C., and Blazer, D.G.,** Urban/rural differences in the prevalence of anxiety disorders, *Am. J. Soc. Psychiatry,* in press.

Urban/Rural Differences in Depressive Disorders: A Quebecois Comparison

V. Kovess

INTRODUCTION

Depressive disorders are the most common psychiatric disturbances encountered in surveying general populations and differences between rural and urban rates have been claimed, although they are controversial. If such differences are genuine, their comprehension will certainly bring some insight into mental health in rural areas in terms of protective or vulnerability factors. Mueller[1] found that rural communities have lower rates of depression than do urban centers, when studies used elaborate diagnostic techniques. Unfortunately, as Mueller,[1] Weissman and Klerman,[2] and Neff[3] point out, comparison among studies made in rural and urban areas is not easy because of the different methodologies used, including lack of uniform psychiatric diagnostic criteria and problems in defining what is an urban or rural population.[1-3] In addition, Neff[3] pointed out that demographic variables, other than residence, may, in fact, explain the reported mental health differences. The different rates of depression found between urban/rural areas may be explained in terms of social support, the absence of confiding relationships, or insufficient support from persons identified as important figures. It has been considered that in rural communities, a "sense of community" still exists and may enhance the above.[1,4,5]

METHOD

The goal of this study was to replicate the findings of urban/rural differences outlined in other studies, and to understand what additional psychosocial factors may contribute to these apparent differences. The study compared an urban population of 183,000 people, the catchment area of Verdun, Quebec, with the rural, 50,000 population of Rimouski. While both populations were predominantly Francophone, Verdun included 25% Anglophones, while the Rimouski population was predominantly French and contained only 5% Anglophones. The instrument used to assess depression in this population was a well-validated mental health symptom check list, which yields a depression factor. Although not generating a DSM III diagnosis of depression, as would the standard Diagnostic Interview Schedule (DIS), the validity of the "depressive factor" mentioned above was confirmed through a comparative reinterview of 167 subjects with the DIS. This demonstrated a satisfactory correlation between the two sets of "diagnoses". The study design used the U.S. "epidemiologic catchment area" research as a model, so the results can be discussed in such a way as to compare the findings with those reported by Blazer et al.[8] in the Piedmont Health Survey.

Population sampling was undertaken similarly, but not identically, in the two catchment areas. The electoral list was the most complete data base for Verdun, while the municipal taxation list was the best for the Rimouski region. In the Verdun region, 887 households were surveyed, with an 80.8% completion rate, while in the Rimouski region, 857 households were covered, representing 93.3% success. Following the model used by the Canadian Health Survey, one household member was interviewed regarding the general health of all household members, by a survey assistant, and then a self-administered questionnaire was left, to be completed individually by each person in the household over 15 years of age. Using reminders and recontact, 83.6% of the self-administered questionnaires in the Verdun region were returned and 93.3% in the Rimouski area.

The 2-h household interview covered typical health survey matters, including disability, medical consultations, accidents, and chronic diseases. Depression was included as a chronic disease. The demographic data included mobility, number of years in school, marital status, employment status, activity, index, income, religious attendance, marital satisfaction, presence of a close person, and life events during the past year. A sociability index evaluated how leisure time was utilized, including frequency of participation in parties, friendship, or family gatherings. Mental health was assessed by scales from the self-administered questionnaire which included two scales: (1) the Ilfeld version of the Hopkins Symptom Checklist,[9] which identified four factors — depression, anxiety, aggressivity, and cognitive problems; and (2) a revised version of a scale developed by Dupuy.[10]

RESULTS

The demographic profile shows that the urban population was older, had more women, was more mobile, and had a larger number of divorced/separated individuals, while unemployment was higher in the rural setting (Table 1). Religion appears to be more important in the rural areas. The pattern of sociability tends to be more complex for urban individuals, with the data towards a U-shaped curve where urban people have either a very high or low level of sociability compared to rural people. In terms of leisure activity, it appears that people in urban settings were more isolated and, similarly, participated less in parties and family gatherings than their rural

TABLE 1
Demographic Profile of Community Subjects Participating in the Quebec Pilot Health Survey

Characteristic	Rural	Urban	Chi square	Probability
Age	N = 1815	N = 1476		
18—24	436 (24.0%)	349 (23.6%)		
25—44	703 (38.7%)	505 (34.2%)		
45—64	464 (25.6%)	395 (26.8%)	13.373	0.005
»65	212 (11.7%)	227 (15.4%)		
Sex	N = 1815	N = 1476		
Male	917 (50.5%)	676 (45.8%)		
Female	898 (49.5%)	800 (54.2%)	7.086	0.009
Marital Status	N = 1680	N = 1414		
Married	1066 (63.5%)	833 (58.9%)		
Separated/divorced	45 (2.7%)	92 (6.5%)		
Widowed	87 (5.2%)	108 (7.6%)	36.193	0.000
Nonmarried	482 (28.7%)	381 (26.9%)		
Education	N = 1810	N = 1460		
«6 years of school	315 (17.4%)	227 (15.5%)		
7—11	878 (48.5%)	761 (52.1%)		
12—14	420 (23.2%)	314 (21.5%)	4.826	N.S.
»15	197 (10.9%)	158 (10.8%)		
Moved within past year	N = 1815	N = 1476		
No	1656 (92.1%)	1260 (85.4%)		
Yes	159 (8.8%)	216 (14.6%)	27.238	0.0000
Employment status	N = 1733	N = 1327		
Keeping house	398 (23.0%)	315 (23.7%)		
Student	180 (10.4%)	105 (7.9%)		
Not working	448 (25.9%)	207 (15.6%)	67.606	0.000
Working	531 (30.6%)	549 (41.4%)		
Retired	176 (10.2%)	151 (11.4%)		

counterparts and had fewer people close to them, although social satisfaction differences were not obvious (Table 2).

In contrast to other studies, the prevalence rates for diagnosed depression were quite comparable in the urban and rural settings. The rates were 4.5% for rural and 6.4% for the urban areas when major depression and dysthymia were grouped together. The lack of rural/urban differences in depressive symptomatology also held up utilizing the Ilfeld global score, the Dupuy scale global score, the reported number of consultations to general practitioners during the past year, and even utilizing a question which tapped the general perception of health.

When major depression and dysthymia were separated and the small city of Rimouski treated separately in analyses, differences in prevalence of major depression between urban/rural areas appeared. The rates were 3.7% for urban, 2.9% rural, and 1.1% in the small city. As expected, being a woman ($F = 37.68, p < 0.001$), being divorced or separated ($F = 22.23, p < 0.001$), being single ($F = 3.40, p < 0.001$), or having a lower education ($F = 11.12, p < 0.000$) was associated with a greater prevalence of the depression syndrome. The residence variable took on greater significance when other more powerful variables were added in a multiple regression model (Table 3).

When analysis was performed on subgroups, urban women aged 25 to 44 and 45 to 64 demonstrated substantially higher levels of depression than comparable women from rural settings. As well, urban single, separated, or divorced individuals were notably more depressed than were their rural counterparts. Depression in urban women appeared to be associated with less religious involvement, while in rural women, there was a greater possibility for depression to be associated with spending a great deal of leisure time alone. Interestingly, in a rural setting, women with the most and least church attendance (a U-shaped relationship) were more depressed; in the urban setting, women with less religious involvement were more depressed.

CONCLUSIONS

The previously reported differences in prevalence of depression between urban and rural settings were replicated, only when more sophisticated statistical models were applied and further distinctions made. As a matter of fact, the differences applied only for specific subgroups. This likely explains why they were not detected in the more global analyses. Interestingly, a subgroup of "isolated" women have been found to be more depressed living in rural areas. It is worth noting that when the rural/urban distinction was trichotomized into rural without the small town of Rimouski, then with Rimouski and urban Verdun, striking differences appear in sociodemographic, as well as in the psychological variables. However, questions could be raised as to which population is truly representative of the urban or rural populations.

TABLE 2
Religion and Sociability for Community Subjects in Participating in the Quebec Pilot Health Survey

	More than once a week	Once a week	At least once a month	Less than monthly but at least once a year	Never
Sociability					
Rural (N = 1683)	252 (15.0%)	470 (27.9%)	551 (32.7%)	345 (20.5%)	65 (3.9%)
Urban (N = 1424)	276 (19.4%)	364 (25.6%)	429 (30.1%)	260 (18.6%)	95 (6.7%)
				Chi square: 25.908 Probability 0.0000	
	Almost all of it by myself	**More than half of it by myself**	**About half of it by myself and half of it with others**	**More than half of it with others**	**Almost all of it with others**
Leisure type					
Rural (N = 1666)	112 (6.7%)	175 (10.5%)	388 (23.3%)	481 (28.9%)	510 (30.6%)
Urban (N = 1405)	111 (7.9%)	145 (10.3%)	397 (28.3%)	308 (21.9%)	444 (31.6%)
				Chi square: 23.406 Probability: 0.0000	
	More than once a week	**Once a week**	**More than once a month**	**More than once a year**	**Never**
Church attendance					
Rural (N = 1709)	167 (9.8%)	566 (33.1)	314 (18.4)	512 (30.0%)	150 (8.8%)
Urban (N = 1428)	77 (5.4%)	256 (17.9%)	146 (10.2%)	522 (36.6%)	427 (29.9%)
				Chi square: 321.952 Probability: 0.0000	
	Of great importance	**Fairly important**	**Of little importance**	**Of no importance**	
Religion importance					
Rural (N = 1697)	846 (49.9%)	615 (36.2%)	173 (10.2%)	63 (3.7%)	
Urban (N = 1420)	476 (33.5%)	557 (39.2%)	250 (17.6%)	137 (9.7%)	
				Chi square: 124.187 Probability: 0.0000	

TABLE 3
Depression Syndrome (Logistic Regression)

Term	F	*p*
Sex	37.68	0.0001
Divorced/separated	22.23	0.0001
Education	11.12	0.0001
Single	3.40	0.0001
Age	2.88	0.0346
Rural/urban	2.19	0.1394
Widowed	1.63	0.2018

In fact, it may be very difficult to generalize from our rural community to other rural communities and, certainly, still more difficult to do so about urban/rural differences. These preliminary results cannot be claimed to be relevant outside of Quebec, and, possibly, not even beyond the local regions studied. Certainly, there was a strong tendency for urban women to report more depression than rural women, and this applies not only to marital status, but also at any level of education. On the other hand, for males, the interaction with social variables and depression appears to be much more complex or not affected by the variables studied. Beyond the gender differences, some similarities to Prudo et al.'s[11] findings were uncovered. Lower income levels seem to be important for the presentation of depression in urban areas, while in rural areas lack of social integration and low church attendance appear, at least in women, to be related to susceptibility to depression. In terms of social integration, it is possible that isolated rural women are more vulnerable to depression, since isolated urban women may find social outlets more easily or have supports that are not available to rural women. However, isolation may not mean the same thing to rural and urban residents and certainly depression may result from an interaction with such other factors as the powerful effect of divorce in urban women.

Our study has attempted to identify some of the complexities of urban/rural differences and has emphasized that other sociodemographic variables must be kept in mind when identifying groups of people at risk. At-risk groups that must be researched further are urban women at certain ages, particularly if separated or divorced, and rural women, particularly with low religious and social affiliations, so that if vulnerability is replicated, protective interventions can be developed.

REFERENCES

1. **Mueller, D.,** The current status of urban/rural differences in psychological disorder: an emerging trend for depression, *J. Nerv. Ment. Dis.*, 169, 18, 1981.
2. **Weissman, M. and Klerman, G.L.,** Sex differences and the epidemiology of depression, *Arch. Gen. Psychiatry,* 34, 98, 1977.
3. **Neff, J.A.,** Urbanicity and depression reconsidered: the evidence regarding depressive symptomatology, *J. Nerv. Ment. Dis.*, 171, 546, 1983.
4. **Brown, G.W. and Harris, T.,** *Social Original Depression: A Study of Psychiatric Disorder in Women,* London Tavistock, 1978.
5. **Wirth, L.,** Urbanism as a way of life, *Am. J. Sociol.*, 44, 3, 1938.
6. Canada Health Survey, Health and Welfare Canada, 1981.
7. **Robins, L.N., Helzer, J.E., Croughan, J., and Rarcliff, K.S.,** National Institute of Mental Health Diagnostic Interview Schedule: its history, characteristics and validity, *Arch. Gen. Psychiatry,* 38, 381, 1981.
8. **Blazer, D., George, L., Landerman, R., et al.,** Psychiatric disorders, a rural/urban comparison, *Arch. Gen. Psychiatry,* 42, 651, 1985.
9. **Ilfeld, F.W.,** Psychological status of community residents along major demographic dimensions, *Arch. Gen. Psychiatry,* 35, 716, 1978.
10. **Dupuy, H.J.,** The Research Edition of the General Psychological Well-Being Schedule, Mimeo, 1980.
11. **Prudo, R., Brown, G.V., and Harris, T.,** Psychiatric disorder in a rural and an urban population, sensitivity to loss, *Psychol. Med.*, 11, 601, 1981.

Priorities in Mental Health Planning

A. H. Leighton

INTRODUCTION

One of the gravest problems facing those who are concerned with mental illnesses — whether as patients, patients' families, or professionals, and whether in country or in city — is an inability on the part of the society to create and maintain adequate systems of care for people with these disorders. The history of psychiatry in the 20th century is largely a history of effort and failure in this regard. In the pages to follow, I shall first point out some of the major barriers that appear to be responsible for these failures, and then go on to suggest that the establishment of priorities is a highly desirable step.

BARRIERS TO ADEQUATE CARE

High Rates of Prevalence

The overall rate of clinically definable mental illnesses in the population at large is much higher than has ever been reflected in policy-making. Studies conducted in the U.K., U.S., Norway, Sweden, and numerous other countries now show prevalence rates that range from about 15 to about 25%.[1,3,4] Our own studies in Nova Scotia indicate a rate of about 20%.[1]

The size of these numbers immediately raises questions of validity. During the last 30 years, and particularly during the last 15, there has been a growing effort in Britain and North America to render diagnoses in psychiatry more standardized and objective. One can, therefore, at the present time say that the illnesses reflected in the above rates are those recognized by practicing clinicians, and exclude people with the milder, reactive conditions that are sometimes referred to as "worried well" or as "problems of living". The criteria for inclusion are in the main those of the International Classification of Diseases-9 and the Diagnostic and Statistical Manual-III (DMS-III) of the American Psychiatric Association.

The Heterogeneity of Mental Illnesses

Not only do mental illnesses as a group have high prevalence rates, but they also constitute an exceedingly protean array of phenomena. The disorders seen in psychiatric hospitals, outpatient services, and private practice have many different symptomatic patterns and causes that require different forms of treatment and care. Consider, for example, the range from Alzheimer's through psychoneurosis to personality disorder. Consider, too, the fact that not only are these syndromes dissimilar, but so also are their etiologies (to the extent they are known), comprising many different combinations of genetic, traumatic, infectious, degenerative, psychological, societal, and cultural influences. The same applies to the variation of responses to treatment which extend from good to intractable. High frequency and heterogeneity together make for an extremely difficult set of problems.

The Emotional Reactions People Have to Mental Illnesses

Encounter with mental illnesses, especially with the psychoses, commonly provokes strong, fearful emotions in the nonill. Even mere discussion of the topic may have this effect. There appears to be something primordial and existential about this, similar to the fear of death. The very thought of madmen and madwomen has an element of the gothic and dire about it, like the gibbering witches on the blasted heath in *Macbeth.* Despite an element of fascination, people seem to react to these feelings as they do to other unpleasant emotions and display such well-known mechanisms as avoidance, denial and sublimation.

In addition to inspiring fear, some mental illnesses can appear attractive, even to the point of being simulated by people who want the status, self-image, and attention conferred. It was not so long ago, for instance, that among some upper class groups, having a psychoneurosis that required psychoanalysis was considered an indicator of how psychologically complex and interesting one was. It marked him or her as composed of something more sophisticated than ordinary human clay.

All told, it would seem that the topic of mental illness has a capacity for disturbing the rational faculties of the nonill.

The Lord Ronald Syndrome

This syndrome consists in trying to do everything and ending up with underachievement. The name was suggested some years ago by Grinker[5] after Stephen Leacock's hero, Lord Ronald, who "flung himself on his horse and rode madly off in all directions". Characteristically, too many goals are embraced in this syndrome, or the goals are out of reach of society's resources, or both. The whole sequence is, at least in part, traceable to the emotions mentioned above as evoked by the topic of mental illness.

What we see in history is that policies and plans have often been made without regard to the magnitude of the problem they were proposed to tackle. As a rule, no information has been sought or utilized regarding the numbers of people in the population who would be candidates for care in the proposed facilities — whether hospitals, mental hygiene clinics, or mental health centers. There has also been a lack of knowledge as to what proportion of the potential clients had severe, chronic disabilities and would require long-term care. As a consequence, large numbers of mentally ill people are now underserved, badly served, or not served at all. There is also reason to suspect that in many places the least disabled consume an inappropriate amount of whatever resources are available.

Rapid Social Change

There are two aspects of accelerating social change that add up to being a serious barrier to adequate care for mentally ill people. One is the possibility that those theories are correct which hold that this phenomenon, because of the stresses it produces, does generate higher prevalence rates, thus increasing the magnitude of the first barrier sketched. It is certain that the impact of abrupt change on such psychological processes as identity and orientation can be very disturbing.

Consider the following illustration. In his autobiographical book, Thomas, Chancellor of Memorial Sloan-Kettering Cancer Centre in New York, has this to say:

> The town I was born and raised in disappeared. The only trace left behind to mark the location of the old clapboard house we lived in is the Long Island Railroad which still penetrates and crosses the town through a deep ditch, and somewhere alongside the ditch, behind a cement wall, is the backyard of my family's house. All the rest is gone. The yard is now covered by an immense apartment house. The whole block, and the other blocks around where our neighbors' clapboard houses and backyard gardens were, are covered by apartment houses, all built fixed to each other as though they were a single syncytial structure. The trees, mostly maples and elms, are gone. The church my family went to, most Sunday mornings, is still there, looking old and beat up, with a sign in the front indicating that it is no longer the Dutch Reformed Church and now is Korean Protestant. I drove down this block, darkened now like a tunnel by the apartment buildings set close to the curb on each side, and saw nothing to remind me of any part of my life.[6]

The second and more direct aspect of rapid social change that constitutes a barrier is the way this process twists and disorganizes the care-giving services. What often happens is that the services are not appropriate to the needs they have to meet because the situation is no longer the same as it was at the time the services were designed. Efforts at revision suffer the same fate: by the time they come into effect they too are out of date.

THE NEED FOR PRIORITIES

The long-range requirement for overcoming the barriers to caring for people with mental illnesses is, first of all, for more knowledge than we now have regarding these disorders and regarding the social and psychological reactions they elicit; and second, for the spread of this knowledge to the public, writers for the media, policy makers, administrators, and all others who exert influences on what happens. A component of this is the need to eradicate many false notions that currently prevail, such as the notion that all mental illnesses are essentially the same things. As Josh Billings observed long ago, "It ain't so much ignorance that hurts folks as knowing so dern many things that ain't so."

The short-range, immediate requirement is to do the best possible with the resources available. This requirement has three roots: the high prevalence rates, the low level of resources, and the devastating effect of that heedless horseman, Lord Ronald.

From our studies of health services and illness rates in a rural county in Nova Scotia[1] and from examination of the relevant literature, it seems to me that the first and most essential step in the task of establishing priorities is to develop clear understanding and consensus with regard to the criteria on which priorities might be based.

Sartorius of the World Health Organization has proposed three categories of criteria that deserve close attention. They are treatability, prevalence, and disability.[7] They fit very well with the conclusions we have reached from our study of a mental health center[1] and from epidemiologic surveys, but I would include a fourth category, namely, seriousness.

It is fortunate that in seeking criteria we have some advantages today that did not formerly exist. Among these is the current surge of interest in diagnosis that has risen in North American psychiatry during the last 15 years and taken specific shape in DSM-III. There is an implicit hierarchy of seriousness among some of the resultant diagnostic categories, so that schizophrenia, for example, is more serious than anxiety due to the fact that the prognosis is by and large much worse. DSM-III categories, therefore, could be taken as one set of criteria in terms of which priorities could be defined and given weights.

Disability is an area in which scientific instrumentation — both clinical and epidemiologic — is as yet poorly developed and poorly utilized. The evolution of criteria for disability measurement therefore will require methodological attention. This is an area for clinical research.

With regard to treatability, there are now medications and forms of treatment (such as chlorpromazine, lithium, and behavior modification) that are effective with a significant proportion of cases in some of the diagnostic categories. Research, not only with regard to medications, but also with regard to psychotherapy, is proceeding at a lively rate, and one can expect that in the foreseeable future there will be results. Criteria in regard to treatability, therefore, need to be established with an eye to frequent revision and modification of the weights used in determining priorities.

The prevalence of disorders in the population provides, as we have noted, the basis for still another group of criteria. These could be based on the common-sense notion that, other criteria aside, the more widespread a disorder in the population, the more it should command attention. Prevalence rates become particularly relevant when they are combined with measures of disability. It is an area, however, in which much more thinking and active research are needed, and, again, the development of this could be an important contribution from research.

With these four criteria in mind, I would like to suggest that the time is now at hand for holding consensus-developing conferences and for the making of position papers among professionals and mental health associations with a view to ultimate crystallization of standard, codified priorities. Epidemiological researchers make it clear that general practitioners and family physicians should be participants in these activities, because more people with mental illness are seen by these physicians than are seen by specialists.[8-10] Thus, at a given time in most places, some 2 to 3% of the population are being seen by mental health services, while primary care givers are seeing two or three times that many.

In offering these suggestions for symposia and position papers on consensus about criteria for priorities, I do not underestimate the disagreements that will arise. There are, however, grounds for some optimism that in time, these can be overcome. Many participants in the mental health field are now much more practical in their aims than formerly; primary care givers are moving toward a more frank, less evasive, and less idiosyncratic way of dealing with mental illnesses; academic psychiatry is becoming less ideological and more objective and research oriented; and epidemiology has developed to a point where it is now capable of gathering both heuristic and practical information about the distribution of mental illnesses in populations, about the factors associated with them, and about the impact of mental health services on their courses and outcomes.

Cities and rural areas have many similarities in regard to these basic problems, but they differ in one in particular: almost universally, rural areas are much less adequately served. Furthermore, the more remote a settlement from an urban center, the worse the discrepancy is apt to be. This fact calls for particular attention in developing priorities for rural areas. It needs to loom large in the viewpoints of policy makers.

As things stand now, policies by and large are based more on the hopes and opinions of the planners than on scientific realities, but it is the realities, not the opinions, that determine the outcomes. A task for psychiatry is to bring the opinions and the realities closer together. The development of criteria for priorities that are based on data — clinical, experimental, epidemiological — could be a step in that direction.

REFERENCES

1. **Leighton, A.H.,** *Caring for Mentally Ill People, Psychological and Social Barriers in Historical Context,* Cambridge University Press, New York, 1982.
2. **Murphy, J.M.,** Continuities in community-based psychiatric epidemiology, *Arch. Gen. Psychiatry,* 37, 1215, 1980.
3. **Regier, D.A., Goldberg, I.D., and Taube, C.A.,** The De Facto U.S. Mental Health Service System, a public health perspective, *Arch. Gen. Psychiatry,* June, 35, 1978.
4. **Dohrenwend, B.P., Dohrenwend, B.S., Gould, M.S., et al.,** *Mental Illness in the United States,* Praeger Publishers, New York, 1980.
5. **Grinker, R.R., Sr.,** Psychiatry rides madly in all directions, *Arch. Gen. Psychiatry,* March, 10, 1964.
6. **Thomas, L.,** *The Youngest Science, Notes of a Medicine-Watcher,* Viking Press, New York, 1983, 1.
7. **Sartorius, N.,** Priorities for research likely to contribute to better provision of mental health care, *Soc. Psychiatry,* 12, 171, 1977.
8. **Hinkle, L.E., Jr. and Wolfe, H.G.,** Health and social environment: experimental investigations, in *Explorations in Social Psychiatry,* Leighton, A.H., Clausen, J.A., and Wilson, R.N., Eds., New York Basic Books, New York, 1957, 115.
9. **Barquero, J.L.V, Munoz, P.E., and Jaurequi, V.M.,** The interaction between physical illness and neurotic morbidity in the community, *Br. J. Psychiatry,* 139, 328, 1983.
10. **Kessler, L.G., Tessler, R.C., and Nycz, R.G.,** Co-occurrence of psychiatric and medical morbidity in primary care, *J. Fam. Pract.,* 16 (2), 319, 1983.

Innovations in Mental Health Service Delivery in Rural Areas

W. G. Hollister

For certain, agricultural life is no longer that mythical, simple, rustic, serene life fostered by stereotypes drawn from the past. Today agricultural living has become a pressure-laden complicated business, a fast- pace endeavor beset with economic pitfalls, tough decisions to make, multiple relationships to maintain, as well as long hours of tense work with high risks of failure.

Even though these increased pressures of daily living drive a few of our rural people into major mental breakdowns, even larger numbers of persons are becoming overstressed, nervous, tense, and/or depressed by these stressors. Some of these individuals are becoming what we call in military medicine the "walking wounded". These are persons on our farms, grasslands or woodlands, or in our small towns who are struggling just to keep going but who are not so sick or overwhelmed that they require hospitalization. Nevertheless, they are casualties of the stresses induced by modern rural life and deserve services to ease their kind of suffering.

Yet there seems to be a strange reluctance to provide help for these walking wounded. In the field of physical health, we have long accepted responsibility to provide care for the lacerations, the fractures, the minor disabilities, as well as for cases of cancer or heart disease. Unfortunately, we seem to be slower in accepting responsibility to care for the "deeply cut personalities", the fractured relationship, or the emotional spasms that also create human suffering.

Because of the increased stresses affecting rural families, some profound changes are taking place in the way that rural people are using and what they expect of mental health services. As elsewhere in the past, rural residents only thought that the words "mental health" meant serious mental illness. Mental health services were only to be used when someone went what they called "crazy" and could no longer be tolerated at home. This very limited and constricted view of mental health services is gradually giving way to a broader concept. Today an increasing proportion of the population is becoming interested in the health emphasis in mental health as well as in more adequate care for the mentally ill. More and more people are getting interested in staying mentally and emotionally healthy, in avoiding emotional breakdown, in coping with stress, and enriching the loving bonds in family life that protect us from illness.

Today's better-educated and more sophisticated rural people are more keenly aware that psychological problems and relationship breakdowns can profoundly disable daily life and peoples' capacity to work productively. As amusing as it may seem, the soap operas on TV, the articles in women's magazines, as well as the daily paper are educating farm women and men that emotional problems can disrupt a marriage, ruin one's physical health, and lead right on into depressions or nervous breakdowns. Our rural people are well into the so-called "psychological era" and as a result they need and they request a broader array of mental health services provided to them locally.

What is this change that is taking place? Rural families still want up-to-date care for those with severe mental illness, but also they want mental health services to include help for persons overwhelmed by stress, for persons caught up in overuse of alcohol and drugs, for families torn apart by conflict and for individuals whose emotional tension is making them physically sick. In a recent set of studies Bass and Ozarin[1] of the National Institute of Mental Health in the U.S. have documented a significant change in the kind of people who come into local community mental health centers seeking help. When these local facilities first opened, their case loads mainly consisted of very ill persons still in the community and living with their families. Over the last two decades the actual numbers of such cases coming into centers has increased, but note that the proportion of the total admissions of such severely ill persons has fallen to only 23.4% of all the people coming in for services. Increasingly, the majority of people (60.4%) presenting themselves for care are not psychotic or do not have neurotic problems, but they are persons suffering from stress reactions, relationship breakdowns, emotional and body tension problems, as well as behavior disorders or life crises, although 12.8% of the presentations were for alcoholism and 3.4% for mental retardation. Let's not dismiss their needs too lightly; these people are also suffering. They deserve and they want care. Although some psychiatric and political leaders would like to restrict mental health services to the psychotic or drug abusers, the people in our towns and farms are literally "voting with their feet" to have a broader range of mental health services provided to them. They are presenting themselves at the doors of our local service units requesting care for the suffering, anxieties, and fears they are experiencing from these less severe, but still disabling problems. To summarize, there is a progressing change in the kinds of problems the public is bringing to community mental health services and what kinds of services the public wants. It is no longer sufficient to provide just a local psychiatric illness clinic, we must also provide services for both illness and for health maintenance.

NEEDED INNOVATIONS IN MENTAL HEALTH SERVICES

In the very rural state of North Carolina where I work, our mental health team of Drs. W. G. Hollister, J. W. Edgerton, N. Mazade, F.T. Miller, J. Aponte, and W. Bentz became aware of this shift in service demands as we personally worked out in rural areas. We launched a nationwide study to see if others confirmed our experience. In 1977 Miller and Mazade[2] of our team polled 257 mental health centers and found most of them reported a shift away from providing only clinical services, and now were making increased use of nonclinical, educational psychosocial interventions. Follow-up interviews showed that this shift in the kind of services being provided was due to several factors: (1) increased numbers of less severe problems presenting, as described above; (2) a 200 to 250% increase in case loads; (3) decreases in budgets that sharply curtailed the number of clinicians they could hire; and (4) the need to use what clinician time they had left for the very severe cases.

Intrigued by the significance of this shift in services provided, we began visiting other rural programs and surveying the literature to find out "what are these alternative nonclinical services that are increasingly being used?" Soon we had an extensive list of such intervention ideas, a classification of which is presented below. We also discovered that (1) no one local service was using more than three or four of these various approaches; that (2) several of the approaches have been evaluated as helpful but (3) almost no one had combined all these approaches into a broad comprehensive program of services and then evaluated the outcome on the people served.

Thus, in 1979 we sought the aid of the U.S. National Institute of Mental Health to conduct a study that (1) would pull together a large series of these alternative kinds of help into a more comprehensive, organized, budgeted and staffed program; (2) would evaluate its impact on persons seeking help and compare its outcome with traditional clinical care; and (3) would field test such alternative services in rural areas as a supplement to the clinical care already being delivered there.

We have now completed our field study and our outcome evaluation and have reported our experience in a new book called *Alternative Services in Community Mental Health: Programs and Processes,* published by our own University of North Carolina Press in Chapel Hill.[3] Let me share some highlights of our experience.

Our first task was to define and classify what we came to call our Alternative Care Services. We ended up with programs in our rural counties that tried to provide the following kinds of services:

1. Tutorial mental health education tailored to an individual or couple, or family: individualized patient education focusing on the problem or growth objectives motivating the bid for help
2. Group mental health education experiences, both cognitive and affective: groups may contain both patients and general public attendees. Topics include general information on behavioral, interpersonal, and stress-related problems. The process may involve experience in emotional-support giving, relationship building, and building sensitivity to feelings without crossing over into psychotherapy; e.g., groups for singles, divorcees, parents, adolescents, abused individuals, etc.
3. Self-help: the use of membership in mutual aid groups without professional leadership to provide education, support, and resource mobilization
4. Formal social support: the use of an organized group or person-to-person support by a designated helper who is usually backed up by an agency; e.g., Big Brothers
5. Informal social support: planned mobilization of family, friends, or someone previously experienced in the client's problems to provide support; e.g., building up an extended family network
6. Rehabilitation: services to improve or restore a person's functioning in society, involving training in social, vocational, homemaking, and/or interpersonal and social skills; includes partial hospitalization, resocialization training, and use of sheltered workshops
7. Home care or care in an alternative residence: to help maintain, protect, or conserve the independence of a person
8. Personal services: specific assistance with money management, tax preparation, shopping aid, house repair or maintenance, legal aid, and nutritional or recreation resources
9. Crisis services: mobilizing or helping persons use community crisis resources
10. Case management: mobilizing and monitoring resources and care with and for the client

After assembling this more comprehensive set of interventions, we endeavored to tap the creativity of rural mental health staffs so they could contribute their own rural "know-how" on the best ways to adapt such a problem into rural and small-town culture and resources.

Third, we had to deal with the reality that most mental health staffs are trained to work as "independent agents" and were not used to reaching out and sharing responsibilities for care with other agencies, disciplines, and lay groups in town. We found they were often not fully aware of the other helping agencies in the community. Much effort was spent in staff training, building up relationships with other helping resources in the community, and in setting up citizen self-help and support groups, so a more complete range of care became available to meet the complex needs of some rural people presented to us for care.

Fourth, we found most rural mental health workers already doing some kinds of alternative care. We found local staffs receptive to doing alternative care, but often unsure what specifically to do for a patient that had come to them. As result, we took the 17 most common relationship, behavior or stress problems our patients brought in for care, and developed a set of 17 learning guides called *Learning Experiences for People with Problems*.[4] These guides can be used by local staff members to create learning experiences during the interview or at-home activities that focus directly on the problem the patient faces. We have these guides available for mental health workers who might wish to use them. Problems most frequently met by patients presenting with moderately severe problems were, in adults, marital conflicts, depressed mood, alcohol dependence, anxiety, child problems, passive dependence, inferiority feelings, problems with the opposite sex, impulsive acts, lack of friends, and loss of family. In children, the problems were parent conflicts, misbehavior, depressed mood, inability to work at school, anxiety, and impulsive acts.

What were the results? What was the impact on patients?

To evaluate such a more comprehensive program of alternative services, as a first step we took into the study only the patients coming in with moderately severe problems. Severely ill were placed immediately into clinical care. Of those in the study, half were placed into the usual clinical care programs they had always been given before, while the other half went into alternative care services. Over 3 years only 17 of 26 cases had to be later put into combined clinical and alternative care to meet emerging needs.

Patient follow-up studies and interviews revealed the following. First, patients liked the alternative care approach and did not consider it second-class care. Some felt it made more sense for their problems. Second, with the exception of a few "die hard" clinicians, most care givers found alternative services feasible to give and reasonably effective. Finally, outcome studies with a battery of psychiatric and psychological assessment tools showed that: (1) alternative care and clinical care approaches were equally effective; (2) the patient outcomes of alternative care delivered by clinically trained staff and nonclinically trained staff were equal in effectiveness; and (3) the amount of staff time to give clinical care or to give alternative care (including mobilizing community resources) was roughly equal. We anticipated that once a rural program has had 2 to 3 years to build up its alternative care-giving resources, the considerable time we had to devote to building up care-giving resources will lessen and alternative care will eventually require less time per patient.

Basically, we feel we have demonstrated the feasibility, effectiveness, and potential economic savings of adding a "second track" or set of alternative services to work alongside of, and be interlocked with, existing clinical services. We think that such an alternative service addition to local units offers a more complete repertoire of services that can be more completely adapted to the changing mental health needs emerging in our rural (and urban) populations.

We challenge you to test this out for yourselves, to see if the addition of alternatives services to clinical services makes your program more adapted to the realities and needs of a rural population.

REFERENCES

1. **Bass, R.D. and Ozarin L.D.,** Community Mental Health Center Programs: What Is Past Is Prologue, paper delivered at American Psychiatric Association, Toronto, Ontario, Canada, May 1977.
2. **Miller, F.T., Mazade, N., et al.,** Trends in mental health programming, *Am. J. Commun. Psychol.*, 61, 191, 1978.
3. **Hollister, W.G., Edgerton, J.W., and Hunter, R.H.,** *Alternative Services in Community Mental Health: Programs and Processes*, University of North Carolina Press, Chapel Hill, 1985.
4. *Learning Experiences for People with Problems,* University of North Carolina, UNC Press, Chapel Hill, NC.

Neurotoxicity Testing of Agricultural Workers Exposed to Pesticides

D. A. Otto

INTRODUCTION

The farmer has been one of the major beneficiaries of the chemical revolution that has occurred over the past half century. The development of new fertilizers, pesticides, herbicides, fungicides, and other agricultural chemicals has dramatically increased crop yields and reduced crop losses due to insects and disease. Agrochemicals, however, frequently have hidden costs that may not become apparent for months or even years after they have been introduced on the market. Hazards to human health in the form of acute and delayed neurotoxicity resulting from pesticide exposure are the hidden costs that will be addressed in this paper.

Anger[1] recently pointed out that direct nervous system effects have been used as one of the bases of exposure standards for 167 of the 588 industrial chemicals for which "safe" exposure standards have been recommended. Although the discipline of neurobehavioral toxicology has grown steadily during the past decade, it is still an infant discipline within the larger field of toxicology. The challenge, therefore, is to highlight the relevance of neurobehavioral toxicology in the health assessment of agricultural workers.

ACUTE NEUROTOXICITY OF ORGANOPHOSPHORUS PESTICIDES

Organophosphorus (OP) pesticides are widely used today in agriculture. Exposure to OP pesticides can produce acute cholinergic symptoms in humans as well as insects that the compounds are designed to eradicate. A bewildering variety of autonomic nervous system dysfunctions result. The resulting symptoms are directly attributable to the action of OP pesticides on cholinergic neural transmission, specifically, inhibition of acetylcholinesterase (AChE).[2] This enzyme, released postsynaptically, breaks down acetylcholine, and its inhibition leads to the accumulation of acetylcholine, which results in disruption of peripheral muscarinic, nicotinic and central AChE receptors through paralysis of cholinergic transmission. A wide range of symptoms can thus be produced by the inhibition of AChE at cholinergic synapses throughout the nervous system (Table 1).

TABLE 1
Signs and Symptoms of OP Pesticide Poisoning

Site affected	Manifestations
Parasympathetic Autonomic Nervous System (Muscarinic Receptors)	
Exocrine glands	Increased salivation, lacrimation, perspiration
Eyes	Miosis (pinpoint and nonreactive), ptosis, blurring of vision, conjunctival injection, "bloody tears"
Gastrointestinal tract	Nausea, vomiting, abdominal tightness, swelling and cramps, diarrhea, tenesmus, fecal incontinence
Respiratory tract	Excessive bronchial secretions, rhinorrhea, wheezing, edema, tightness in chest, bronchospasm, broncho-constriction, cough, bradypnea, dyspnea
Cardiovascular system	Bradycardia, decrease in blood pressure
Bladder	Urinary frequency and incontinence
Somatic Motor Nerve Fibers (Nicotinic Receptors)	
Skeletal muscles	Muscle fasciculations (eyelids, fine facial muscles), cramps, diminished tendon reflexes, generalized muscle weakness in peripheral and respiratory muscles, paralysis, flaccid or rigid tone
	Restlessness, generalized motor activity, hyperreactivity to acoustic stimuli, tremulousness, emotional liability, ataxia
Central Nervous System (Acetylcholine Receptors)	
Midbrain, cortex	Drowsiness, lethargy, fatigue, mental confusion, inability to concentrate, headache, pressure in head, generalized weakness
	Coma with absence of reflexes, tremors, Cheyne-Stokes respiration, dyspnea, convulsions, depression of respiratory centers, dyanosis

Adapted from Ecobichon, D.J. and Joy, R.M., *Pesticides and Neurological Disease,* CRC Press, Boca Raton, FL, 1982, 166.

ORGANOPHOSPHORUS-INDUCED DELAYED NEUROTOXICITY (OPIDN)

Some OP esters produce a different and devastating nervous system syndrome called "OP-induced delayed neurotoxicity" or OPIDN.[3,4] The classical OPIDN syndrome entails progressive weakness and ataxia leading in severe cases to flaccid paralysis. The symptoms begin distally in the lower limbs and may spread to the spinal cord and upper limbs.

* This manuscript has been reviewed by the Health Effects Research Laboratory, U.S. Environmental Protection Agency and approved for publication. Mention of trade names or commercial products does not constitute endorsement or recommendation for use.

Although peripheral nerve damage may recover slowly, spinal cord degeneration is likely to produce irreversible spasticity, ataxia, or quadriplegia. The extent of involvement and severity of symptoms depend on the specific compound as well as the length and intensity of exposure.

The cardinal features of OPIDN include: (1) a characteristic delay in the onset of clinical symptoms (6 to 21 d after exposure); (2) a marked species difference in sensitivity (cats, horses, sheep, birds, and humans are very susceptible, while rodents and rabbits are not); and (3) a stereotypic morphological pattern of action (Wallerian or dying-back axonopathy of distal nerves with progressive degeneration back toward the cell body).

There is little correlation between the acute cholinergic and delayed neurotoxic effects of OP pesticides. In fact, measurement of AChE inhibition is of little value in predicting the occurrence of OPIDN.[3] OP pesticides[2,4] including leptophos, fenthion, isofenphos, EPN, trichlorfon, and methamidiphos have been reported to induce delayed neurotoxicity. Workers in plants where leptophos was being manufactured experienced a variety of puzzling neurological symptoms that were initially diagnosed as multiple sclerosis. A National Institute for Occupational Safety and Health health hazard evaluation in 1976[5] discovered the OPIDN syndrome, and this led to withdrawal of the pesticide from the markets in the U.S. and Egypt.[6]

Fenthion, a fumigant, fungicide, and dermal pesticide in large animals, lead to a 2-year progression of polyneuropathies, ataxia, and paresthesias in veterinary workers.[3,7]

Both fenthion and leptophos were approved for production and application in agriculture before discovering the serious human health hazard of delayed neurotoxicity associated with these compounds. Is there a more pervasive hazard among the hundreds of OP pesticides currently used in agriculture? Neurobehavioral monitoring of agricultural workers chronically exposed to OP pesticides would be prudent to prevent further unexpected incidents of acute or delayed neurotoxicity.

NEUROTOXICITY TESTING METHODS

The neurotoxicity testing methods to be reported are psychomotor, psychophysical, and electrophysiological. The psychomotor category includes traditional behavioral tests of sensorimotor integration, motor coordination, and memory. The psychophysical group involves tests of auditory, visual, and tactile sensitivity, while the electrophysiological tests include sensory measures derived from scalp or dermal-recorded electrical activity from the brain or peripheral nerves. These are appropriate for the evaluation of the range of neurobehavioral symptoms associated with pesticide poisoning in humans. Many are recommended by the World Health Organization.[12]

Psychomotor Test Batteries

Table 2 lists standardized, validated, and reliable neurobehavioral tests recommended by the World Health Organization[8] for neurotoxicity testing in the workplace. The functional nature of each test, a minimal or "core" screening battery, and additional or alternative tests for in-depth assessments are specified. The digit symbol, digit span, and block design subscales of the Wechsler Adult Intelligence Scale (WAIS), and the Santa Ana Dexterity Test, in particular, have proven to be sensitive to a variety of chemicals.[9] This latter test is simple and rapid to administer, but is not available commercially. However, it may be obtained from the Division of Biomedical and Behavioral Science, National Institute for Occupational Safety and Health, 4676 Columbia Pkwy, Cincinnati, OH 45226 U.S.

Psychophysical Tests

Simple and inexpensive psychophysical tests of tactile sensitivity and thermal sensitivity have recently been developed specifically for occupational assessment including one from Sensortek Inc. (Clifton, NJ 01013 U.S.).[10]

The loss of tactile sensation is a good early measure of neurotoxicity (peripheral neuropathy) from OP pesticides, since large-diameter axons that conduct vibratory information are susceptible to damage from chemicals such as acrylamide, lead, and alcohol.[11,12] Traditional clinical and electrophysiological tests such as nerve conduction velocity are relatively insensitive to small, slower-conducting nerve fibers.[13] In addition, since small, slower, temperature-conducting fibers are often affected first, the quantitative assessment of thermal sensation offers a useful adjunct for clinical neurological and neurotoxicological evaluations.[13] A thermal- sensitivity tester is also available from Sensortek for neurological and occupational screening.

The visual system is also unusually vulnerable to chemical exposure.[14] Simple assessment of visual acuity with a Snellen eye chart is useful in neurotoxicity screening, while more complex, contrast-sensitivity testing[15] of the eye's ability to discriminate spatial frequencies is another potentially useful technique.[15] This testing has been demonstrated in the assessment of Minimata disease (methyl mercury poisoning).[16]

Electrophysiological Testing Procedures

Physiological recording of electrical activity in central nervous and peripheral nervous systems from skin or scalp is important. In the case of lead exposure which can cause demyelinating lesions and peripheral neuropathy in adults, motor and sensory nerve conduction velocity (NCV) measures provide a useful and sensitive method for monitoring and evaluating exposed workers.[17] Demyelinating lesions clearly slow nerve conduction. However, if toxicant exposure produces selective axonal degeneration, but more than 50% of nerve fibers are left intact, then NCV measures are unlikely to show any demonstrable early effect.[13,15] In the case of the distal axonopathy characteristic of OPIDN, NCV measures would be unlikely to show any effect until irreversible damage was done. Tactile or thermal-sensitivity testing should provide a more sensitive and useful early sign of OP-induced delayed neuropathy.

TABLE 2
Proposed Psychological Test Methods[a]

Functional domain	Screening battery minimum program	Additional and alternative methods for a standard battery	Tests for in-depth studies or tests to be developed further
Motor speed	Finger tapping		Two-plate tapping
Motor steadiness	Flanagan coordination or mira	Aiming Tremometer	Body steadiness
Reaction time		SRT (60—100 signals) CRT	Series of RT tasks with increasing complexity
Perceptual-motor speed	Digit symbol, WAIS	Bourdon-Wiersma Neisser Identical numbers	Combinations of different perceptual motor speed test
Motor coordination	Santa Ana or some other dexterity test	Michigan eye-hand coordination Purdue pegboard	Other psychomotor tests
Visual cognitive functions	Visual retention, WMS	Block design, WAIS Embedded figure Benton visual retention Symmetric drawing	Picture completion, WAIS Object assembly (WAIS) Figure classification Memory-for-design test
Verbal abilities	Digit span, WAIS, WMS Logical memory, WMS Similarities, WAIS	Associative learning, WMS Stroop color naming text	Various memory span tests Memory tests with delayed recall Tests for auditory and semantic processing Verbal comprehension tests
Attention/vigilance		PASAT (paced auditory serial attention test) Continuous performance test	Audiovisual vigilance test Other tests for attention and vigilance
Moods, emotions	PMS (profile of mood states)	Taylor manifest anxiety scale EPI (Eyesenck personality inventory)	Rorschach

[a] Presented at WHO International Meeting on the Prevention of Neurotoxic Illness in Working Population, meeting at NIOSH, Cincinnati, OH, May 1983.

From Bleecker, M.L., *Arch. Environ. Health,* 39, 213, 1984. With permission.

Sensory-evoked potential (SEP) tests may also be useful for neurotoxicity assessment, although little human data in exposure to OP pesticides are available.[12,18] The development and clinical application of SEP tests has proceeded rapidly in parallel with advances in microprocessor technology during the past decade. SEPs provide noninvasive, objective, and reliable measures of the functional receptors to cortical projection areas in localizing lesions from sensory pathways, such as in multiple sclerosis.[19] SEPs have been used extensively in animal models[20] to characterize the neurotoxicity of a wide range of chemicals, and comparability of SEPs in species as disparate as rat and man emphasizes its possible utility in human neurotoxicology.

The first SEP measure is the brainstem auditory-evoked potential (BAEP), one of the few SEP phenomena for which the neural origins have been relatively well established.[21] The correlation between anatomy and phenomenology is sufficiently good that the BAEP can be used to localize auditory pathway lesions.[19] Since neurotoxicants tend to produce symmetrical, diffuse effects on central nervous system function,[22] BAEPs should provide a sensitive index of neurotoxicity, particularly in the case of chemicals that produce demyelinating effects. The BAEP is particularly useful, since it is not affected by changes in attention; for example, some anesthetics do not alter BAEP latencies.[20] As well, it is immune to cultural bias since no verbal mediation is required, and BAEPs are simple to record and the latency measures are easy to make and interpret. Preliminary evidence from BAEP use in children at risk for lead poisoning[22] provides evidence for utility in future neurotoxicity testing (Figure 1).

Second, the pattern reversal visual-evoked potential (PREP), where dark and light squares of a checkerboard pattern are reversed repetitively on a video display, is recommended over the use of the flash visual-evoked potential (FEP) in the diagnosis of multiple sclerosis.[23] Although the FEP has been used widely for toxicity testing in animal models,[20] the PREP may be a better candidate for human neurotoxicity testing.

The third, and often considered ideal for neurotoxicity testing, is the somatosensory-evoked potential (SSEP). Anatomically, the long peripheral nerves are accessible for measurement at many points along the sensory pathway. By stimulating a distal point, both peripheral and central nerve conduction measures can be obtained simultaneously.

SEPs have not yet been used to assess the effects of pesticide exposure in agricultural workers. The clinical sensitivity and utility of these measures, however, suggest that SEP methods could provide a useful adjunct in agricultural health assessment. Evoked potential technology, however, is more expensive and requires more highly trained personnel to administer than the behavioral and psychophysical tests described above.

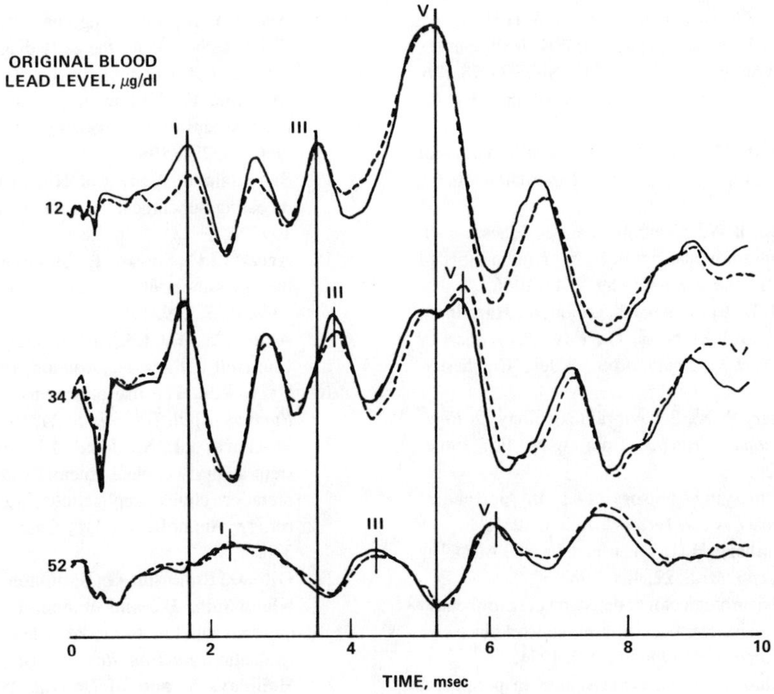

FIGURE 1. BAEPs from three children with varying lead-exposure history. Replicate averages of 2000 trials recorded binaurally using 60dBSL clicks are shown for each child. Latencies of waves III and V increased linearly with blood lead levels obtained 5 years earlier. (Adapted from Otto, D., Robinson, G., Baumann, S., Schroedoer, S., Mushak, P., Kleinbaum, D., and Boone, L., *Environ. Res.*, 168, 1985.)

CONCLUDING COMMENTS

A brief overview of neurobehavioral toxicology has been presented, beginning with the neurotoxicity of OP pesticides. The acute and delayed effects of OP pesticide exposure were used to illustrate the crucial importance of neurobehavioral symptoms and measures in agricultural health. Since pesticides are usually designed to attack the central and peripheral nervous systems of insects, it is not surprising that careless application or accidental exposure to these poisons can be severely debilitating or lethal to humans.

The insidious, delayed effect of exposure to OP pesticides, initially considered to pose no hazard to humans at low concentrations, is another question. Incidents in which exposure to leptophos and fenthion resulted in the unexpected development of delayed peripheral neuropathies were reviewed. The effects of chronic low-level exposure to pesticides have not been adequately studied in humans. The marked species specificity of OPIDN underscores the need to study the consequences of chronic exposure in agricultural workers.

Improved health monitoring programs are needed in rural areas for early detection of subclinical effects of pesticide exposure. Appropriate remedial steps can often be taken before irreversible damage has been done. By the time clear-cut clinical symptoms of OPIDN appear, the damage is often irreversible. The earliest complaints expressed by workers exposed to pesticides are frequently neurobehavioral signs such as headache, excessive fatigue, numbness of extremities, visual disturbances, or emotional signs such as increased irritability, anxiety, and depression. Awareness of these symptoms as early signs of pesticide poisoning is crucial among health care professionals in agricultural communities.

Several of the measures described above — vibratory, thermal, and contrast sensitivity testing, in particular — are well suited to worker screening and monitoring programs. The psychomotor and psychophysical testing procedures are simple to administer and score, require little time to perform, are of light weight and portable for use in field studies, and have proven to be sensitive to occupational and environmental toxicants. Wider application of neurobehavioral tests in agricultural health research, screening, and monitoring could significantly reduce the health risks of pesticide exposure in the agricultural community.

REFERENCES

1. **Anger, W.K.,** Neurobehavioral testing of chemicals: impact on recommended standards, *Neurobehav. Toxicol. Teratol.,* 6, 147, 1984.
2. **Ecobichon, D.J. and Joy, R.M.,** *Pesticides and Neurological Diseases,* CRC Press, Boca Raton, FL, 1982.
3. **Cherniack, M.G.,** Toxicological Screening for Organophosphorus Induced Delayed Neurotoxicity: Complications in Toxicity Testing, NIOSH Reports/NTIS No. PB 86-107703, National Institute for Occupational Safety and Health, Washington, D.C., 1985.
4. **Abou-Donia, M.B.,** Organophosphorus-induced delayed

neurotoxicity, *Annu. Rev. Pharmacol. Toxicol.*, 21, 511, 1981.

5. **Xintaras, C., Burg, J.B., Tanaka, S., et al.**, NIOSH Health Survey of Velsicol Pesticide Workers, DHEW Publ. (NIOSH) 78-136, Department of Health, Education and Welfare, Washington, D.C., 1978.
6. **Aou-Donia, M.B., Othman, M.A., Tantawy, G., Khalil, A.Z., and Shawer, M.F.**, Neurotoxic effects of leptophos, *Experientia*, 30, 63, 1974.
7. **Knowles, C.O. and Arlo, B.W.**, Metabolism and residues associated with dermal and intramuscular application of radio-labelled fenthion in dairy cows, *J. Econ. Entomol.*, 59, 1346, 1966.
8. **Johnson, B.L., Baker, E.L., El Batawi, M., Gilioli, R., Hanninen, H.,Seppalainen, A. M., and Xintaras, C., Eds.**, *Prevention of Neurotoxic Illness in Working Populations*, Wiley, Chichester, 1987.
9. **Johnson, B.L. and Anger, W.K.**, Behavioral toxicology, in *Environmental and Occupational Medicine*, Rom, W.N., Ed., Little, Brown, Boston, 1983, 329.
10. **Bleecker, M.L.**, Quantifying sensory loss in peripheral neuropathies, *Neurobehav. Toxicol. Teratol.*, 7, 305, 1985.
11. **Arezzo, J.C. and Schaumberg, H.H.**, The use of the OPTACON as a screening device, *J. Occup. Med.*, 22, 461, 1980.
12. **Bleecker, M.L.**, Clinical neurotoxicology: detection of neurobehavioral and neurological impairments occurring in the workplace and the environment, *Arch. Environ. Health*, 39, 213, 1984.
13. **Krarup, C. and Buchthal, F.**, Conduction studies in peripheral nerve, *Neurobehav. Toxicol. Teratol.*, 7, 319, 1985.
14. **Merigan, W.H. and Weiss, B., Eds.**, *Neurotoxicity of the Visual System*, Raven Press, New York, 1980.
15. **Arden, G.B.**, Measuring contrast sensitivity with gratings: a new simple technique for the early diagnosis of retinal and neurological disease, *J. Am. Optom. Assoc.*, 50, 35, 1979.
16. **Mukuno, K., Ishikawa, S., and Okamura, R.**, Grating test of contrast sensitivity in patients with Minimata disease, *Br. J. Opthalmol.*, 65, 284, 1981.
17. **Seppalainen, A.M. and Hernberg, S.**, Sensitivie technique for detecting subclinical lead neuropathy, *Br. J. Ind. Med.*, 29, 443, 1972.
18. **Arezzo, J.C., Simson, R., and Brennan, N.E.**, Evoked potentials in the assessment of neurotoxicity in humans, *Neurobehav. Toxicol. Teratol.*, 7, 299, 1985.
19. **Aminoff, M., Ed.**, *Electrodiagnosis in Clinical Neurology*, Churchill Livingstone, London, 1980.
20. **Dyer, R.S.**, The use of sensory evoked potentials in toxicology, *Fundam. Appl. Toxicol.*, 5, 24, 1985.
21. **Stockard, J.J., Stockard, J.E., and Scharbrough, F.W.**, Brainstem auditory evoked potentials in neurology: methodology, interpretation, clinical applications, in *Electrodiagnosis in Clinical Neurology*, Aminoff, M., Ed., Churchill Livingstone, London, 1980, 370.
22. **Otto, D., Robinson, G., Baumann, S., Schroedoer, S., Mushak, P., Kleinbaum, D., and Boone, L.**, Five-year follow-up study of children with low-to-moderate lead absorption: electrophysiological evaluation, *Environ. Res.*, 38, 168, 1985.
23. **Halliday, A. and McDonald, W.**, Visual evoked potentals, in *Clinical Nuerophysiology*, Stalberg, E. and Young, R., Eds., Buttersworths, London, 1981, 228.

Index

INDEX

A

B

E

F

N

O

Q

R

S

T

U

V

W

X

Y

Z